ENCYCLOPEDIA OF MATHEMATICS AND ITS APPLICATIONS

FOUNDED BY G.-C. ROTA

Editorial Board
P. Flajolet, M. Ismail, E. Lutwak

Volume 108

Combinatorial Matrix Classes

ENCYCLOPEDIA OF MATHEMATICS AND ITS APPLICATIONS

Combinatorial Matrix Classes

RICHARD A. BRUALDI
University of Wisconsin, Madison

CAMBRIDGE
UNIVERSITY PRESS

CAMBRIDGE UNIVERSITY PRESS
Cambridge, New York, Melbourne, Madrid, Cape Town, Singapore, São Paulo

Cambridge University Press
The Edinburgh Building, Cambridge CB2 2RU, UK

Published in the United States of America by Cambridge University Press,
New York

www.cambridge.org
Information on this title: www.cambridge.org/9780521865654

First published 2006

Printed in the United Kingdom at the University Press, Cambridge

A catalog record for this publication is available from the British Library

ISBN-13   978-0-521-86565-4 hardback
ISBN-10   0-521-86565-4 hardback

Contents

Preface

In the preface of the book *Combinatorial Matrix Theory*[1] (CMT) I discussed my plan to write a second volume entitled *Combinatorial Matrix Classes*. Here 15 years later (including 6, to my mind, wonderful years as Department of Mathematics Chair at UW-Madison), and to my great relief, is the finished product. What I proposed as topics to be covered in a second volume were, in retrospect, much too ambitious. Indeed, after some distance from the first volume, it now seems like a plan for a book series rather than for a second volume. I decided to concentrate on topics that I was most familiar with and that have been a source of much research inspiration for me. Having made this decision, there was more than enough basic material to be covered. Most of the material in the book has never appeared in book form, and as a result, I hope that it will be useful to both current researchers and aspirant researchers in the field. I have tried to be as complete as possible with those matrix classes that I have treated, and thus I also hope that the book will be a useful reference book.

I started the serious writing of this book in the summer of 2000 and continued, while on sabbatical, through the following semester. I made good progress during those six months. Thereafter, with my many teaching, research, editorial, and other professional and university responsibilities, I managed to work on the book only sporadically. But after 5 years, I was able to complete it or, if one considers the topics mentioned in the preface of CMT, one might say I simply stopped writing. But that is not the way I feel. I think, and I hope others will agree, that the collection of matrix classes developed in the book fit together nicely and indeed form a coherent whole with no glaring omissions. Except for a few reference to CMT, the book is self-contained.

My primary inspiration for combinatorial matrix classes has come from two important contributors, Herb Ryser and Ray Fulkerson. In a real sense, with their seminal and early research, they are the "fathers" of the subject. Herb Ryser was my thesis advisor and I first learned about the class $A(R, S)$, which occupies a very prominent place in this book, in the fall of 1962 when I was a graduate student at Syracuse University (New York).

---

[1] Authored by Richard A. Brualdi and Herbert J. Ryser and published by Cambridge University Press in 1991.

In addition, some very famous mathematicians have made seminal contributions that have directly or indirectly impacted the study of matrix classes. With the great risk of offending someone, let me mention only Claude Berge, Garrett Birkhoff, David Gale, Alan Hoffman, D. König, Victor Klee, Donald Knuth, H.G. Landau, Leon Mirsky, and Bill Tutte. To these people, and all others who have contributed, I bow my head and say a heartfelt thank-you for your inspiration.

As I write this preface in the summer of 2005, I have just finished my 40th year as a member of the Department of Mathematics of the University of Wisconsin in Madison. I have been fortunate in my career to be a member of a very congenial department that, by virtue of its faculty and staff, provides such a wonderful atmosphere in which to work, and that takes teaching, research, and service all very seriously. It has also been my good fortune to have collaborated with my graduate students, and postdoctoral fellows, over the years, many of whom have contributed to one or more of the matrix classes treated in this book. I am indebted to Geir Dahl who read a good portion of this book and provided me with valuable comments.

My biggest source of support these last 10 years has been my wife Mona. Her encouragement and love have been so important to me.

Richard A. Brualdi
Madison, Wisconsin

1

Introduction

In this chapter we introduce some concepts and theorems that are important for the rest of this book. Much, but not all, of this material can be found in the book [4]. In general, we have included proofs of theorems only when they do not appear in [4]. The proof of Theorem 1.7.1 is an exception, since we give here a much different proof. We have not included all the basic terminology that we make use of (e.g. graph-theoretic terminology), expecting the reader either to be familiar with such terminology or to consult [4] or other standard references.

1.1   Fundamental Concepts

Let
$$A = [a_{ij}] \quad (i = 1, 2, \ldots, m; j = 1, 2, \ldots, n)$$

be a matrix of m rows and n columns. We say that A is of *size* m by n, and we also refer to A as an m by n matrix. If $m = n$, then A is a *square* matrix of *order* n. The elements of the matrix A are always real numbers and usually are nonnegative real numbers. In fact, the elements are sometimes restricted to be nonnegative integers, and often they are restricted to be either 0 or 1. The matrix A is composed of m row vectors $\alpha_1, \alpha_2, \ldots, \alpha_m$ and n column vectors $\beta_1, \beta_2, \ldots, \beta_n$, and we write

$$A = \begin{bmatrix} \alpha_1 \\ \alpha_2 \\ \vdots \\ \alpha_m \end{bmatrix} = [\beta_1 \, \beta_2 \, \ldots \, \beta_n].$$

It is sometimes convenient to refer to either a row or column of the matrix A as a *line* of A. We use the notation A^{T} for the transpose of the matrix A. If $A = A^{\mathrm{T}}$, then A is a square matrix and is *symmetric*.

A zero matrix is always designated by O, a matrix with every entry equal to 1 by J, and an identity matrix by I. In order to emphasize the size of these matrices we sometimes include subscripts. Thus $J_{m,n}$ denotes the m by n matrix of all 1's, and this is shortened to J_n if $m = n$. The notations $O_{m,n}, O_n,$ and I_n have similar meanings.

A *submatrix* of A is specified by choosing a subset of the row index set of A and a subset of the column index set of A. Let $I \subseteq \{1, 2, \ldots, m\}$ and $J \subseteq \{1, 2, \ldots, n\}$. Let $\bar{I} = \{1, 2, \ldots, m\} \setminus I$ denote the *complement* of I in $\{1, 2, \ldots, m\}$, and let $\bar{J} = \{1, 2, \ldots, n\} \setminus J$ denote the complement of J in $\{1, 2, \ldots, n\}$. Then we use the following notations to denote submatrices of A:

$$A[I, J] = [a_{ij} : i \in I, j \in J],$$
$$A(I, J] = [a_{ij} : i \in \bar{I}, j \in J],$$
$$A[I, J) = [a_{ij} : i \in I, j \in \bar{J}],$$
$$A(I, J) = [a_{ij} : i \in \bar{I}, j \in \bar{J}],$$
$$A[I, \cdot] = A[I, \{1, 2, \ldots, n\}],$$
$$A[\cdot, J] = A[\{1, 2, \ldots, m\}, J],$$
$$A(I, \cdot] = A[\bar{I}, \{1, 2, \ldots, n\}], \text{ and}$$
$$A[\cdot, J) = A[\{1, 2, \ldots, m\}, \bar{J}].$$

These submatrices are allowed to be empty. If $I = \{i\}$ and $J = \{j\}$, then we abbreviate $A(I, J)$ by $A(i, j)$.

We have the following partitioned forms of A:

$$A = \left[\begin{array}{c|c} A[I, J] & A[I, J) \\ \hline A(I, J] & A(I, J) \end{array} \right], \quad A = \left[\begin{array}{c} A[I, \cdot] \\ \hline A(I, \cdot] \end{array} \right],$$

and

$$A = \left[\begin{array}{c|c} A[\cdot, J] & A[\cdot, J) \end{array} \right].$$

The $n!$ permutation matrices of order n are obtained from I_n by arbitrary permutations of its rows (or of its columns). Let $\pi = (\pi_1, \pi_2, \ldots, \pi_n)$ be a permutation of $\{1, 2, \ldots, n\}$. Then π corresponds to the permutation matrix $P_\pi = [p_{ij}]$ of order n in which $p_{i\pi_i} = 1$ $(i = 1, 2, \ldots, n)$ and all other $p_{ij} = 0$. The permutation matrix corresponding to the inverse π^{-1} of

π is P_π^{T}. It thus follows that $P_\pi^{-1} = P_\pi^{\mathrm{T}}$, and thus an arbitrary permutation matrix P of order n satisfies the matrix equation

$$PP^{\mathrm{T}} = P^{\mathrm{T}}P = I_n.$$

Let A be a square matrix of order n. Then the matrix PAP^{T} is similar to A. If we let Q be the permutation matrix P^{T}, then $PAP^{\mathrm{T}} = Q^{\mathrm{T}}AQ$. The row vectors of the matrix $P_\pi A$ are $\alpha_{\pi_1}, \alpha_{\pi_2}, \ldots, \alpha_{\pi_m}$. The column vectors of AP_π are $\beta_{\pi'_1}, \beta_{\pi'_2}, \ldots, \beta_{\pi'_n}$ where $\pi^{-1} = (\pi'_1, \pi'_2, \ldots, \pi'_n)$. The column vectors of AP^{T} are $\beta_{\pi_1}, \beta_{\pi_2}, \ldots, \beta_{\pi_n}$. Thus if P is a permutation matrix, the matrix PAP^{T} is obtained from A by *simultaneous permutations of its rows and columns*. More generally, if A is an m by n matrix and P and Q are permutation matrices of orders m and n, respectively, then the matrix PAQ is a matrix obtained from A by *arbitrary permutations of its rows and columns*.

Let $A = [a_{ij}]$ be a matrix of size m by n. The *pattern* (or *nonzero pattern*) of A is the set

$$\mathcal{P}(A) = \{(i,j): a_{ij} \neq 0, \quad i = 1, 2, \ldots, m, \quad j = 1, 2, \ldots, n\}$$

of positions of A containing a nonzero element.

With the m by n matrix $A = [a_{ij}]$ we associate a combinatorial configuration that depends only on the pattern of A. Let $X = \{x_1, x_2, \ldots, x_n\}$ be a nonempty set of n elements. We call X an *n-set*. Let

$$X_i = \{x_j: a_{ij} \neq 0, j = 1, 2, \ldots, n\} \quad (i = 1, 2, \ldots, m).$$

The collection of m not necessarily distinct subsets $X_1, X_2, \ldots, X_m$ of the n-set X is the *configuration* associated with A. If P and Q are permutation matrices of orders m and n, respectively, then the configuration associated with PAQ is obtained from the configuration associated with A by relabeling the elements of X and reordering the sets $X_1, X_2, \ldots, X_m$. Conversely, given a nonempty configuration $X_1, X_2, \ldots, X_m$ of m subsets of the nonempty n-set $X = \{x_1, x_2, \ldots, x_n\}$, we associate an m by n matrix $A = [a_{ij}]$ of 0's and 1's, where $a_{ij} = 1$ if and only if $x_j \in X_i$ $(i = 1, 2, \ldots, m; j = 1, 2, \ldots, n)$.

The configuration associated with the m by n matrix $A = [a_{ij}]$ furnishes a particular way to represent the structure of the nonzeros of A. We may view a configuration as a *hypergraph* [1] with vertex set X and *hyperedges* $X_1, X_2, \ldots, X_m$. This hypergraph may have repeated edges, that is, two or more hyperedges may be composed of the same set of vertices. The *edge–vertex incidence matrix* of a hypergraph H with vertex set $X = \{x_1, x_2, \ldots, x_n\}$ and edges $X_1, X_2, \ldots, X_m$ is the m by n matrix $A = [a_{ij}]$ of 0's and 1's in which $a_{ij} = 1$ if and only if x_j is a vertex of edge X_i $(i = 1, 2, \ldots, m; j = 1, 2, \ldots, n)$. Notice that the hypergraph (configuration) associated with A is the original hypergraph H. If A has exactly two 1's in each row, then A is the edge–vertex incidence matrix of a *multigraph* where a pair of distinct vertices may be joined by more than one edge. If no two rows of A are identical, then this multigraph is a *graph*.

Another way to represent the structure of the nonzeros of a matrix is by a bipartite graph. Let $U = \{u_1, u_2, \ldots, u_m\}$ and $W = \{w_1, w_2, \ldots, w_n\}$ be sets of cardinality m and n, respectively, such that $U \cap W = \varnothing$. The *bipartite graph* associated with A is the graph $\mathrm{BG}(A)$ with vertex set $V = U \cup W$ whose edges are all the pairs $\{u_i, w_j\}$ for which $a_{ij} \neq 0$. The pair $\{U, W\}$ is the *bipartition* of $\mathrm{BG}(A)$.

Now assume that A is a nonnegative integral matrix, that is, the elements of A are nonnegative integers. We may then associate with A a *bipartite multigraph* $\mathrm{BMG}(A)$ with the same vertex set V bipartitioned as above into U and W. In $\mathrm{BMG}(A)$ there are a_{ij} edges of the form $\{u_i, w_j\}$ $(i = 1, 2, \ldots, m; j = 1, 2, \ldots, n)$. Notice that if A is a (0,1)-matrix, that is, each entry is either a 0 or a 1, then the bipartite multigraph $\mathrm{BMG}(A)$ is a bipartite graph and coincides with the bipartite graph $\mathrm{BG}(A)$. Conversely, let BMG be a bipartite multigraph with bipartitioned vertex set $V = \{U, W\}$ where U and W are as above. The *bipartite adjacency matrix* of BMG, abbreviated *bi-adjacency matrix*, is the m by n matrix $A = [a_{ij}]$ where a_{ij} equals the number of edges of the form $\{u_i, w_j\}$ (the *multiplicity* of (u_i, v_j)) $(i = 1, 2, \ldots, m; j = 1, 2, \ldots, n)$. Notice that $\mathrm{BMG}(A)$ is the original bipartite multigraph BMG.

An m by n matrix A is called *decomposable* provided there exist nonnegative integers p and q with $0 < p + q < m + n$ and permutation matrices P and Q such that PAQ is a direct sum $A_1 \oplus A_2$ where A_1 is of size p by q. The conditions on p and q imply that the matrices A_1 and A_2 may be vacuous[1] but each of them contains either a row or a column. The matrix A is *indecomposable* provided it is not decomposable. *The bipartite graph* $\mathrm{BG}(A)$ *is connected if and only if A is is indecomposable.*

Assume that the matrix $A = [a_{ij}]$ is square of order n. We may represent its nonzero structure by a digraph $D(A)$. The vertex set of $D(A)$ is taken to be an n-set $V = \{v_1, v_2, \ldots, v_n\}$. There is an arc (v_i, v_j) from v_i to v_j if and only if $a_{ij} \neq 0$ $(i, j = 1, 2, \ldots, n)$. Notice that a nonzero diagonal entry of A determines an arc of $D(A)$ from a vertex to itself (a *directed loop* or *di-loop*). If A is, in addition, a nonnegative integral matrix, then we associate with A a *general digraph* $\mathrm{GD}(A)$ with vertex set V where there are a_{ij} arcs of the form (v_i, v_j) $(i, j = 1, 2, \ldots, n)$. If A is a (0,1)-matrix, then $\mathrm{GD}(A)$ is a digraph and coincides with $D(A)$. Conversely, let GD be a general digraph with vertex set V. The *adjacency matrix* of $\mathrm{GD}(A)$ is the nonnegative integral matrix $A = [a_{ij}]$ of order n where a_{ij} equals the number of arcs of the form (v_i, v_j) (the *multiplicity* of (v_i, v_j)) $(i, j = 1, 2, \ldots, n)$. Notice that $\mathrm{GD}(A)$ is the original general digraph GD.

Now assume that the matrix $A = [a_{ij}]$ not only is square but is also symmetric. Then we may represent its nonzero structure by the *graph*

---

[1] If $p + q = 1$, then A_1 has either a row but no columns or a column but no rows. A similar conclusion holds for A_2 if $p + q = m + n - 1$.

$G(A)$. The vertex set of $G(A)$ is an n-set $V = \{v_1, v_2, \ldots, v_n\}$. There is an edge joining v_i and v_j if and only if $a_{ij} \neq 0$ $(i, j = 1, 2, \ldots, n)$. A nonzero diagonal entry of A determines an edge joining a vertex to itself, that is, a *loop*. The graph $G(A)$ can be obtained from the bipartite graph $\mathrm{BG}(A)$ by identifying the vertices u_i and w_i and calling the resulting vertex v_i $(i = 1, 2, \ldots, n)$. If A is, in addition, a nonnegative integral matrix, then we associate with A a *general graph* $\mathrm{GG}(A)$ with vertex set V where there are a_{ij} edges of the form $\{v_i, v_j\}$ $(i, j = 1, 2, \ldots, n)$. If A is a (0,1)-matrix, then $\mathrm{GG}(A)$ is a graph and coincides with $G(A)$. Conversely, let GG be a general graph with vertex set V. The *adjacency matrix* of GG is the nonnegative integral symmetric matrix $A = [a_{ij}]$ of order n where a_{ij} equals the number of edges of the form (v_i, v_j) (the *multiplicity* of $\{v_i, v_j\}$) $(i, j = 1, 2, \ldots, n)$. Notice that $\mathrm{GG}(A)$ is the original general graph GG. A general graph with no loops is called a *multigraph*.

The symmetric matrix A of order n is *symmetrically decomposable* provided there exists a permutation matrix P such that $PAP^{\mathrm{T}} = A_1 \oplus A_2$ where A_1 and A_2 are both matrices of order at least 1; if A is not symmetrically decomposable, then A is *symmetrically indecomposable*. The matrix A is symmetrically indecomposable if and only if its graph $G(A)$ is connected.

Finally, we remark that if a multigraph MG is bipartite with vertex bipartition $\{U, W\}$ and A is the adjacency matrix of MG, then there are permutation matrices P and Q such that

$$PAQ = \begin{bmatrix} O & C \\ C^{\mathrm{T}} & O \end{bmatrix}$$

where C is the bi-adjacency matrix of MG (with respect to the bipartition $\{U, W\}$).[2]

We shall make use of elementary concepts and results from the theory of graphs and digraphs. We refer to [4], or books on graphs and digraphs, such as [17], [18], [2], [1], for more information.

1.2 Combinatorial Parameters

In this section we introduce several combinatorial parameters associated with matrices and review some of their basic properties. In general, by a *combinatorial property or parameter of a matrix* we mean a property or parameter which is invariant under arbitrary permutations of the rows and columns of the matrix. More information about some of these parameters can be found in [4].

Let $A = [a_{ij}]$ be an m by n matrix. The *term rank* of A is the maximal number $\rho = \rho(A)$ of nonzero elements of A with no two of these elements on a line. The *covering number* of A is the minimal number $\kappa = \kappa(A)$ of

---

[2]If G is connected, then the bipartition is unique.

lines of A that contain (that is, *cover*) all the nonzero elements of A. Both ρ and κ are combinatorial parameters. The fundamental minimax theorem of König (see [4]) asserts the equality of these two parameters.

Theorem 1.2.1

$$\rho(A) = \kappa(A).$$

A set of nonzero elements of A with no two on a line corresponds in the bipartite graph $\mathrm{BG}(A)$ to a set of edges no two of which have a common vertex, that is, pairwise vertex-disjoint edges or a *matching*. Thus Theorem 1.2.1 asserts that in a bipartite graph, the maximal number of edges in a matching equals the minimal number of vertices in a subset of the vertex set that meets all edges.

Assume that $m \leq n$. The *permanent* of A is defined by

$$\mathrm{per}(A) = \sum a_{1i_1} a_{2i_2} \ldots a_{mi_m}$$

where the summation extends over all sequences $i_1, i_2, \ldots, i_m$ with $1 \leq i_1 < i_2 < \cdots < i_m \leq n$. Thus $\mathrm{per}(A)$ equals the sum of all possible products of m elements of A with the property that the elements in each of the products occur on different lines. The permanent of A is invariant under arbitrary permutations of rows and columns of A, that is,

$$\mathrm{per}(PAQ) = \mathrm{per}(A), \text{if } P \text{ and } Q \text{ are permutation matrices.}$$

If A is a nonnegative matrix, then $\mathrm{per}(A) > 0$ if and only if $\rho(A) = m$. Thus by Theorem 1.2.1, $\mathrm{per}(A) = 0$ if and only if there are permutation matrices P and Q such that

$$PAQ = \begin{bmatrix} A_1 & O_{k,l} \\ A_{21} & A_2 \end{bmatrix}$$

for some positive integers k and l with $k+l = n+1$. In the case of a square matrix, the permanent function is the same as the determinant function apart from a factor ± 1 preceding each of the products in the defining summation. Unlike the determinant, the permanent is, in general, altered by the addition of a multiple of one row to another and the multiplicative law for the determinant, $\det(AB) = \det(A)\det(B)$, does not hold for the permanent. However, the *Laplace expansion* of the permanent by a row or column does hold:

$$\mathrm{per}(A) = \sum_{j=1}^{n} a_{ij}\mathrm{per}(A(i,j)) \quad (i = 1, 2, \ldots, m);$$

$$\mathrm{per}(A) = \sum_{i=1}^{m} a_{ij}\mathrm{per}(A(i,j)) \quad (j = 1, 2, \ldots, n).$$

We now define the widths and heights of a matrix. In order to simplify the language, we restrict ourselves to $(0,1)$-matrices. Let $A = [a_{ij}]$ be a $(0,1)$-matrix of size m by n with r_i 1's in row i ($i = 1, 2, \ldots, m$). We call $R = (r_1, r_2, \ldots, r_m)$ the row sum vector of A. Let α be an integer with $0 \le \alpha \le r_i$, $i = 1, 2, \ldots, m$. Consider a subset $J \subseteq \{1, 2, \ldots, n\}$ such that each row sum of the m by $|J|$ submatrix

$$E = A[\cdot, J]$$

of A is at least equal to α. Then the columns of E determine an α-*set of representatives* of A. This terminology comes from the fact that in the configuration of subsets $X_1, X_2, \ldots, X_m$ of $X = \{x_1, x_2, \ldots, x_n\}$ associated with A (see Section 1.1), the set $Z = \{x_j : j \in J\}$ satisfies

$$|Z \cap X_i| \ge \alpha \quad (i = 1, 2, \ldots, m).$$

The α-*width* of A equals the minimal number $\epsilon_\alpha = \epsilon_\alpha(A)$ of columns of A that form an α-set of representatives of A. Clearly, $\epsilon_\alpha \ge |\alpha|$, but we also have

$$0 = \epsilon_0 < \epsilon_1 < \cdots < \epsilon_r \tag{1.1}$$

where r is the minimal row sum of A. The widths of A are invariant under row and column permutations.

Let $E = A[\cdot, J]$ be a submatrix of A having at least α 1's in each row and suppose that $|J| = \epsilon(\alpha)$. Then E is a *minimal α-width submatrix* of A. Let F be the submatrix of E composed of all rows of E that contain exactly α 1's. Then F cannot be an empty matrix. Moreover, F cannot have a zero column, because otherwise we could delete the corresponding column of E and obtain an m by $\epsilon_\alpha - 1$ submatrix of A with at least α 1's in each row, contradicting the minimality of ϵ_α. The matrix F is called a *critical α-submatrix* of A. Each critical α-submatrix of A contains the same number ϵ_α of columns, but the number of rows need not be the same. The minimal number $\delta_\alpha = \delta_\alpha(A)$ of rows in a critical α-submatrix of A is called the α-*multiplicity* of A. We observe that $\delta_\alpha \ge 1$ and that multiplicities of A are invariant under row and column permutations. Since a critical α-submatrix cannot contain zero columns, we have $\delta_1 \ge \epsilon(1)$.

Let the matrix A have column sum vector $S = (s_1, s_2, \ldots, s_n)$, and let β be an integer with $0 \le \beta \le s_j$ ($1 \le j \le n$). By interchanging rows with columns in the above definition, we may define the β-*height* of A to be the minimal number t of rows of A such that the corresponding t by n submatrix of A has at least β 1's in each column. Since the β-height of A equals the β-width of A^{T}, one may restrict attention to widths.

We conclude this section by introducing a parameter that comes from the theory of hypergraphs [1]. Let A be a $(0,1)$-matrix of size m by n. A (*weak*) t-*coloring* of A is a partition of its set of column indices into t sets $I_1, I_2, \ldots, I_t$ in such a way that if row i contains more than one 1, then $\{j : a_{ij} = 1\}$ has a nonempty intersection with at least two of the sets $I_1, I_2, \ldots, I_t$ $(i = 1, 2, \ldots, m)$. The sets $I_1, I_2, \ldots, I_t$ are called the *color classes* of the *t*-coloring. The (*weak*) *chromatic number* $\gamma(A)$ of A is the smallest integer t for which A has a t-coloring [3]. In hypergraph terminology the chromatic number is the smallest number of colors in a coloring of the vertices with the property that no edge with more than one vertex is monochromatic. The *strong chromatic* number of a hypergraph is the smallest number of colors in a coloring of its vertices with the property that no edge contains two vertices of the same color. The *strong chromatic number* $\gamma_s(A)$ of A equals the strong chromatic number of the hypergraph associated with A. If A is the edge–vertex incidence matrix of a graph G, then both the weak and strong chromatic numbers of A equal the chromatic number of G.

1.3   Square Matrices

We first consider a canonical form of a square matrix under simultaneous permutations of its rows and columns.

Let $A = [a_{ij}]$ be a square matrix of order n. Then A is called *reducible* provided there exists a permutation matrix P such that PAP^{T} has the form

$$\begin{bmatrix} A_1 & A_{12} \\ O & A_2 \end{bmatrix}$$

where A_1 and A_2 are square matrices of order at least 1. The matrix A is *irreducible* provided that it is not reducible. A matrix of order 1 is always irreducible. Irreducibility of matrices has an equivalent formulation in terms of digraphs. Let D be a digraph. Then D is *strongly connected* (or *strong*) provided that for each ordered pair of distinct vertices x, y there is a directed path from x to y. A proof of the following theorem can be found in [4].

Theorem 1.3.1 *Let A be a square matrix of order n. Then A is irreducible if and only if the digraph $D(A)$ is strongly connected.*

If a digraph D is not strongly connected, its vertex set can be partitioned uniquely into nonempty sets each of which induces a maximal strong digraph, called a *strong component* of D. This leads to the following canonical form with respect to simultaneous row and column permutations [4].

Theorem 1.3.2 *Let A be a square matrix of order n. Then there exist a permutation matrix P of order n and an integer $t \geq 1$ such that*

$$PAP^{\mathrm{T}} = \begin{bmatrix} A_1 & A_{12} & \ldots & A_{1t} \\ O & A_2 & \ldots & A_{2t} \\ \vdots & \vdots & \ddots & \vdots \\ O & O & \ldots & A_t \end{bmatrix}. \tag{1.2}$$

where $A_1, A_2, \ldots, A_t$ are square, irreducible matrices. In (1.2), the matrices $A_1, A_2, \ldots, A_t$ which occur as diagonal blocks are uniquely determined to within simultaneous permutations of their rows and columns, but their ordering in (1.2) is not necessarily unique. □

Referring to Theorem 1.3.2, we see that the digraph $D(A)$ is composed of strongly connected graphs $D(A_i)$ ($i = 1, 2, \ldots, t$) and some arcs which go from a vertex in $D(A_i)$ to a vertex in $D(A_j)$ where $i < j$. The matrix A is irreducible if and only if $t = 1$. The matrices $A_1, A_2, \ldots, A_t$ in (1.2) are called the *irreducible components* of A. By Theorem 1.3.2 the irreducible components of A are uniquely determined to within simultaneous permutations of their rows and columns. The matrix A is irreducible if and only if it has exactly one irreducible component.

The canonical form (1.2) of A is given as a block upper triangular matrix, but by reordering the blocks so that the diagonal blocks occur in the order $A_t, \ldots, A_2, A_1$, we could equally well give it as a block lower triangular matrix. Thus *we may interchange the use of lower block triangular and upper block triangular.*

We now consider a canonical form under arbitrary permutations of rows and columns.

Again let A be a square matrix of order n. If $n \geq 2$, then A is called *partly decomposable* provided there exist permutation matrices P and Q such that PAQ has the form

$$\begin{bmatrix} A_1 & A_{12} \\ O & A_2 \end{bmatrix}$$

where A_1 and A_2 are square matrices of order at least 1. According to our definition of decomposable applied to square matrices, the matrix A would be decomposable provided in addition $A_{12} = O$. A square matrix of order 1 is partly decomposable provided it is a zero matrix. The matrix A is *fully indecomposable* provided that it is not partly decomposable. The matrix A is fully indecomposable if and only if it does not have a nonempty zero submatrix $O_{k,l}$ with $k + l \geq n$. Each line of a fully indecomposable matrix of order $n \geq 2$ contains at least two nonzero elements. The covering number $\kappa(A)$ of a fully indecomposable matrix of order n equals n. Moreover, if $n \geq 2$ and we delete a row and column of A, we obtain a matrix of order $n - 1$ which has covering number equal to $n - 1$. Hence from Theorem 1.2.1 we obtain the following result.

Theorem 1.3.3 *Let A be a fully indecomposable matrix of order n. Then the term rank $\rho(A)$ of A equals n. If $n \geq 2$, then each submatrix of A of order $n - 1$ has term rank equal to $n - 1$.* □

A collection of n elements (or the positions of those elements) of the square matrix A of order n is called a *diagonal* provided no two of the elements belong to the same row or column; the diagonal is a *nonzero diagonal* provided none of its elements equals 0. Thus Theorem 1.3.3 asserts that a fully indecomposable matrix has a nonzero diagonal and, if $n \geq 2$, each nonzero element belongs to a nonzero diagonal.

We have the following canonical form with respect to arbitrary row and column permutations [4].

Theorem 1.3.4 *Let A be a matrix of order n with term rank equal to n. Then there exist permutation matrices P and Q of order n and an integer $t \geq 1$ such that*

$$PAQ = \begin{bmatrix} B_1 & B_{12} & \ldots & B_{1t} \\ O & B_2 & \ldots & B_{2t} \\ \vdots & \vdots & \ddots & \vdots \\ O & O & \ldots & B_t \end{bmatrix}, \quad (1.3)$$

where $B_1, B_2, \ldots, B_t$ are square, fully indecomposable matrices. The matrices $B_1, B_2, \ldots, B_t$ which occur as diagonal blocks in (1.3) are uniquely determined to within arbitrary permutations of their rows and columns, but their ordering in (1.3) is not necessarily unique. □

The matrices $B_1, B_2, \ldots, B_t$ in (1.3) are called the *fully indecomposable components* of A. By Theorem 1.3.4 the fully indecomposable components of A are uniquely determined to within arbitrary permutations of their rows and columns. The matrix A is fully indecomposable if and only if it has exactly one fully indecomposable component.

As with the canonical form (1.2), the canonical form (1.3) of A is given as a block upper triangular matrix, but by reordering the blocks so that the diagonal blocks occur in the order $B_t, \ldots, B_2, B_1$, we could equally well give it as a block lower triangular matrix.

A fully indecomposable matrix of order $n \geq 2$ has an inductive structure which can be formulated in terms of $(0, 1)$-matrices (Theorem 4.2.8 in [4]).

Theorem 1.3.5 *Let A be a fully indecomposable $(0, 1)$-matrix of order $n \geq 2$. There exist permutation matrices P and Q and an integer $m \geq 2$ such that*

$$PAQ = \begin{bmatrix} A_1 & O & O & \ldots & O & E_1 \\ E_2 & A_2 & O & \ldots & O & O \\ O & E_3 & A_3 & \ldots & O & O \\ \vdots & \vdots & \vdots & \ddots & \vdots & \vdots \\ O & O & O & \ldots & A_{m-1} & O \\ O & O & O & \ldots & E_m & A_m \end{bmatrix} \quad (1.4)$$

where each of the matrices $A_1, A_2, \ldots, A_m$ is fully indecomposable and each of $E_1, E_2, \ldots, E_m$ contains at least one 1. □

Note that a matrix of the form (1.4) satisfying the conditions in the statement of the theorem is always fully indecomposable.

Let $A = [a_{ij}]$ be a fully indecomposable matrix of order n. Then A is called *nearly decomposable* provided whenever a nonzero element of A is replaced with a zero, the resulting matrix is not fully indecomposable. Nearly decomposable matrices have an inductive structure which we formulate in terms of $(0, 1)$-matrices [4].

Theorem 1.3.6 *Let A be a nearly decomposable $(0, 1)$-matrix of order $n \geq 2$. Then there exist permutation matrices P and Q and an integer m with $1 \leq m < n$ such that*

$$
PAQ = \left[
\begin{array}{cccccc|c}
1 & 0 & 0 & \ldots & 0 & 0 & \\
1 & 1 & 0 & \ldots & 0 & 0 & \\
0 & 1 & 1 & \ldots & 0 & 0 & \\
\vdots & \vdots & \vdots & \ddots & \vdots & \vdots & F_1 \\
0 & 0 & 0 & \ldots & 1 & 0 & \\
0 & 0 & 0 & \ldots & 1 & 1 & \\
\hline
& & F_2 & & & & A_1
\end{array}
\right]
$$

where

   (i) *A_1 is a nearly decomposable matrix of order m,*

  (ii) *the matrix F_1 contains exactly one 1 and this 1 is in its first row and last column,*

 (iii) *the matrix F_2 contains exactly one 1 and this 1 is in its last row and last column,*

 (iv) *if $m \neq 1$, then $m \geq 3$ and the element in the last row and last column of A_1 is a 0.*

A simple induction shows that for $n \geq 3$, a nearly decomposable $(0, 1)$-matrix of order n has at most $3(n - 1)$ 1's [4].

The matrix A of order n is said to have *total support* provided each of its nonzero elements belongs to a nonzero diagonal. A zero matrix has total support as it satisfies the definition vacuously. If $A \neq O$, then we have the following result [4].

Theorem 1.3.7 *Let A be a square, nonzero matrix of order n. Then A has total support if and only if there are permutation matrices P and Q such that PAQ is a direct sum of fully indecomposable matrices. If A has total support, then A is fully indecomposable if and only if the bipartite graph $\mathrm{BG}(A)$ is connected.*

In the canonical form (1.3) of a nonzero matrix of total support, the off-diagonal blocks B_{ij} $(i < j)$ are zero matrices.

1.4 An Existence Theorem

Consider nonnegative integral vectors

$$R = (r_1, r_2, \ldots, r_m) \text{ and } R' = (r'_1, r'_2, \ldots, r'_m),$$

and

$$S = (s_1, s_2, \ldots, s_n) \text{ and } S' = (s'_1, s'_2, \ldots, s'_n),$$

where

$$r'_i \leq r_i \ (i = 1, 2, \ldots, m) \text{ and } s'_j \leq s_j \ (j = 1, 2, \ldots, n).$$

Consider also an m by n nonnegative integral matrix

$$C = [c_{ij}] \quad (i = 1, 2, \ldots, m; j = 1, 2, \ldots, n).$$

With this notation we have the following matrix existence theorem of Mirsky [13, 14].

Theorem 1.4.1 *There exists an m by n integral matrix $A = [a_{ij}]$ whose elements and row and column sums satisfy*

$$0 \leq a_{ij} \leq c_{ij} \qquad (i = 1, 2, \ldots, m; j = 1, 2, \ldots, n),$$

$$r'_i \leq \sum_{j=1}^{n} a_{ij} \leq r_i \qquad (i = 1, 2, \ldots, m),$$

$$s'_j \leq \sum_{i=1}^{m} a_{ij} \leq s_j \qquad (j = 1, 2, \ldots, n)$$

if and only if for all $I \subseteq \{1, 2, \ldots, m\}$ and $J \subseteq \{1, 2, \ldots, n\}$,

$$\sum_{i \in I, j \in J} c_{ij} \geq \max \left\{ \sum_{i \in I} r'_i - \sum_{j \notin J} s_j, \sum_{j \in J} s'_j - \sum_{i \notin I} r_i \right\}. \tag{1.5}$$

This theorem is derived by Mirsky using transversal theory but it can also be regarded as a special case of more general supply–demand theorems in network flow theory [8, 4]. As a result, the theorem remains true if the integrality assumptions on R, R', S, and S' are dropped and the conclusion asserts the existence of a real nonnegative matrix. The integrality assumptions on other theorems in this and the next section can be dropped as well, and the resulting theorems remain valid for the existence of real matrices. The following corollary is an important special case of Theorem 1.4.1.

Corollary 1.4.2 *Let $R = (r_1, r_2, \ldots, r_m)$ and $S = (s_1, s_2, \ldots, s_n)$ be nonnegative integral vectors satisfying $r_1 + r_2 + \cdots + r_m = s_1 + s_2 + \cdots + s_n$. There* *exists an m by n nonnegative integral matrix $A = [a_{ij}]$ whose elements and row and column sums satisfy*

$$0 \le a_{ij} \le c_{ij} \qquad (i = 1, 2, \ldots, m; j = 1, 2, \ldots, n),$$

$$\sum_{j=1}^{n} a_{ij} = r_i \qquad (i = 1, 2, \ldots, m),$$

$$\sum_{i=1}^{m} a_{ij} = s_j \qquad (j = 1, 2, \ldots, n)$$

if and only if for all $I \subseteq \{1, 2, \ldots, m\}$ and $J \subseteq \{1, 2, \ldots, n\}$,

$$\sum_{i \in I} \sum_{j \in J} c_{ij} \ge \sum_{j \in J} s_j - \sum_{i \notin I} r_i, \qquad (1.6)$$

equivalently,

$$\sum_{i \in I} \sum_{j \in J} c_{ij} \ge \sum_{i \in I} r_i - \sum_{j \notin J} s_j, \qquad (1.7)$$

$\square$

The following corollary is an important special case.

Corollary 1.4.3 *Let $R = (r_1, r_2, \ldots, r_m)$ and $S = (s_1, s_2, \ldots, s_n)$ be nonnegative integral vectors satisfying $r_1 + r_2 + \cdots + r_m = s_1 + s_2 + \cdots + s_n$. Let p be a positive integer. There exists an m by n nonnegative integral matrix $A = [a_{ij}]$ whose elements and row and column sums satisfy*

$$0 \le a_{ij} \le p \quad (i = 1, 2, \ldots, m; j = 1, 2, \ldots, n),$$

$$\sum_{j=1}^{n} a_{ij} = r_i \quad (i = 1, 2, \ldots, m) \text{ and } \sum_{i=1}^{m} a_{ij} = s_j \quad (j = 1, 2, \ldots, n)$$

if and only if for all $I \subseteq \{1, 2, \ldots, m\}$ and $J \subseteq \{1, 2, \ldots, n\}$,

$$p|I||J| \ge \sum_{j \in J} s_j - \sum_{i \notin I} r_i, \qquad (1.8)$$

equivalently,

$$p|I||J| \ge \sum_{i \in I} r_i - \sum_{j \notin J} s_j. \qquad (1.9)$$

$\square$

If in this corollary we assume that $r_1 \ge r_2 \ge \cdots \ge r_m$ and $s_1 \ge s_2 \ge \cdots \ge s_n$, then (1.9) is equivalent to

$$pkl + \sum_{i=k+1}^{m} r_i - \sum_{j=1}^{l} s_j \ge 0 \quad (k = 0, 1, \ldots, m; l = 0, 1, \ldots, n). \qquad (1.10)$$

The special case of this corollary with $p = 1$ is due to Ford and Fulkerson [8] and will be given a short proof in Chapter 2.

1.5 An Existence Theorem for Symmetric Matrices

Theorem 1.4.1 can be used to obtain existence theorems for symmetric integral matrices. The key for doing this is the following lemma of Fulkerson, Hoffman, and McAndrew [9] for matrices with zero trace and its extension to matrices with no restriction on the trace due to Brualdi and Ryser [4]. We observe that if $A = [a_{ij}]$ is a symmetric integral matrix of order n each of whose main diagonal elements equals 0, then

$$\sum_{i=1}^{n}\sum_{j=1}^{n} a_{ij} = 2 \sum_{1 \leq i < j \leq n} a_{ij},$$

an even number.

Lemma 1.5.1 *Let $R = (r_1, r_2, \ldots, r_n)$ be a nonnegative integral vector and let p be a positive integer. Assume that there is a nonnegative integral matrix $B = [b_{ij}]$ of order n whose row and column sum vectors equal R and whose elements satisfy $b_{ij} \leq p$ $(i, j = 1, 2, \ldots, n)$. Then there exists a symmetric, nonnegative integral matrix $A = [a_{ij}]$ of order n whose row and column sum vectors equal R and whose elements satisfy $a_{ij} \leq p$ $(i, j = 1, 2, \ldots, n)$. If $r_1 + r_2 + \cdots + r_n$ is an even integer and B has zero trace, then A can also be chosen to have zero trace.*

Corollaries 1.4.2 and 1.4.3 and Lemma 1.5.1 immediately give the next theorem [4]. If A is a symmetric matrix of order n, then so is PAP^{T} for every permutation matrix P of order n. Thus the assumption that R is nonincreasing in the theorem is without loss of generality.

Theorem 1.5.2 *Let $R = (r_1, r_2, \ldots, r_n)$ be a nonincreasing, nonnegative integral vector and let p be a positive integer. There exists a symmetric, nonnegative integral matrix $A = [a_{ij}]$ whose row and column sum vectors equal R and whose elements satisfy $a_{ij} \leq p$ $(i, j = 1, 2, \ldots, n)$ if and only if*

$$pkl - \sum_{j=1}^{l} r_j + \sum_{i=k+1}^{n} r_i \geq 0 \quad (k, l = 0, 1, \ldots, n).$$

(Note that the domain of k and l in the preceding inequality may be replaced with $0 \leq l \leq k \leq n$.)

Corollary 1.5.3 *Let $R = (r_1, r_2, \ldots, r_n)$ be a nonincreasing, nonnegative integral vector.*

(i) *There exists a symmetric $(0,1)$-matrix whose row sum vector equals R if and only if*

$$kl - \sum_{j=1}^{l} r_j + \sum_{i=k+1}^{n} r_i \geq 0 \quad (k, l = 0, 1, \ldots, n).$$

(ii) *There exists a symmetric $(0,1)$-matrix with zero trace whose row sum vector equals R if and only if*

$$kl - \min\{k,l\} - \sum_{j=1}^{l} r_j + \sum_{i=k+1}^{n} r_i \geq 0 \quad (k,l = 0,1,\ldots,n).$$

$\square$

For symmetric matrices with zero trace we also have the following theorem of Chungphaisan [6] that follows from Theorem 1.4.1 and Lemma 1.5.1. The equivalence of the two conditions in Theorem 1.5.4 is shown in [4].

Theorem 1.5.4 *Let $R = (r_1, r_2, \ldots, r_n)$ be a nonincreasing, nonnegative integral vector such that $r_1 + r_2 + \cdots + r_n$ is an even integer, and let p be a positive integer. There exists a symmetric, nonnegative integral matrix $A = [a_{ij}]$ of order n with zero trace whose row sum vector equals R and whose elements satisfy $a_{ij} \leq p$ $(i,j = 1,2,\ldots,n)$ if and only if*

$$\sum_{i=1}^{k} r_i \leq \sum_{i=1}^{k} \min\{r_i, p(k-1)\} + \sum_{i=k+1}^{n} \min\{r_i, pk\} \quad (k = 1, 2, \ldots, n),$$

equivalently,

$$pk(k-1) \geq \sum_{i=1}^{k} r_i - \sum_{i=k+1}^{n} \min\{r_i, pk\} \quad (k = 1, 2, \ldots, n).$$

Theorem 1.5.4 contains the following theorem of Erdős and Gallai [7] as a special case.

Theorem 1.5.5 *Let $R = (r_1, r_2, \ldots, r_n)$ be a nonincreasing, nonnegative integral vector such that $r_1 + r_2 + \cdots + r_n$ is an even integer. There exists a symmetric $(0,1)$-matrix with zero trace whose row sum vector equals R if and only if*

$$\sum_{i=1}^{k} r_i \leq k(k-1) + \sum_{i=k+1}^{n} \min\{k, r_i\} \quad (k = 1, 2, \ldots, n).$$

The theorems of Chungphaisan and of Erdős and Gallai are usually stated in graph-theoretic language of degree sequences. One can consult [16] for a survey on degree sequences.

1.6 Majorization

Let $X = (x_1, x_2, \ldots, x_n)$ and $Y = (y_1, y_2, \ldots, y_n)$ be two n-vectors of real numbers which are assumed to be nonincreasing:

$$x_1 \geq x_2 \geq \cdots \geq x_n \text{ and } y_1 \geq y_2 \geq \cdots \geq y_n.$$

Then X is *majorized by* Y, denoted $X \preceq Y$, provided that their partial sums satisfy

$$\sum_{i=1}^{k} x_i \le \sum_{i=1}^{k} y_i \quad (1 \le k \le n)$$

with equality for $k = n$; equivalently

$$\sum_{i=k+1}^{n} x_i \ge \sum_{i=k+1}^{n} y_i \quad (0 \le k \le n-1)$$

with equality for $k = 0$.

Let $\tau = \sum_{i=1}^{n} x_i = \sum_{i=1}^{n} y_i$. If X and Y in addition have positive integral elements, then X and Y are *partitions* of τ. Majorization is a partial order on nonincreasing n-vectors, or nonincreasing vectors with the same sum τ. In the latter case, the n-vector $(\tau/n, \tau/n, \ldots, \tau/n)$ is the unique minimal element with respect to majorization, and the n-vector $(\tau, 0, \ldots, 0)$ is the unique maximal element.

If the vectors X and Y have different sizes, then we can still compare X and Y for majorization by appending 0's to one of the vectors so as to obtain vectors of the same size. Also the assumption that X and Y are nonincreasing vectors can be eliminated provided we first reorder the components of the vectors so as to be in nonincreasing order before we compute partial sums. Thus if $X = (x_1, x_2, \ldots, x_n)$ and $Y = (y_1, y_2, \ldots, y_n)$ are two n-vectors, and $X' = (x_{[1]}, x_{[2]}, \ldots, x_{[n]})$ and $Y' = (y_{[1]}, y_{[2]}, \ldots, y_{[n]})$ are obtained from X and Y, respectively, by rearranging components in nonincreasing order, then X is *majorized* by Y provided X' is majorized by Y'. Under these circumstances majorization is not a partial order since $X \preceq Y$ and $Y \preceq X$ imply only that X and Y are rearrangements of one another.

Now let $R = (r_1, r_2, \ldots, r_m)$ be a nonnegative integral vector of size m, and let n be an integer with $r_k \le n$ $(k = 1, 2, \ldots, m)$. Then the *conjugate* of R is the nonnegative integral vector $R^* = (r_1^*, r_2^*, \ldots, r_n^*)$ of size[3] n where

$$r_k^* = |\{i : r_i \ge k, i = 1, 2, \ldots, m\}| \quad (k = 1, 2, \ldots, n).$$

The conjugate of R has the same sum as R and is always nonincreasing. Moreover, it does not depend on the order of the elements of R. We also note that

$$\sum_{i=k+1}^{n} r_i^* = \sum_{i=1}^{n} (r_i - k)^+ \quad (k \ge 1),$$

where for a real number a, $a^+ = \max\{0, a\}$.

---

[3]There is a certain arbitrariness in the size of the conjugate R^* in that it can be any integer which is at least as large as the maximum element of R. This only affects the number of trailing 0's in R^*.

The conjugate of R can be pictured as follows. Consider an array of m rows which has r_i 1's in the initial positions of row i $(i = 1, 2, \ldots, m)$. The *conjugate* R^* of R is the vector of column sums of this array. For example, let $R = (5, 4, 2, 1, 1)$ and $n = 5$. Then using the array

$$
\begin{array}{ccccc}
1 & 1 & 1 & 1 & 1 \\
1 & 1 & 1 & 1 & \\
1 & 1 & & & \\
1 & & & & \\
1 & & & &
\end{array}
$$

we see that $R^* = (5, 3, 2, 2, 1)$.

Conjugation is related to majorization as given in the following elementary result.

Theorem 1.6.1 *Let $X = (x_1, x_2, \ldots, x_n)$ and $Y = (y_1, y_2, \ldots, y_n)$ be nonincreasing, nonnegative integral vectors with $X \preceq Y$. Then $Y^* \preceq X^*$.*

Proof. Because $X \preceq Y$ we have

$$
\sum_{i=p+1}^{n} x_i \geq \sum_{i=p+1}^{n} y_i \quad (p = 0, 1, \ldots, n - 1).
$$

Let k be a positive integer, and let p be the largest index for which $x_p > k$. Then

$$
\sum_{i=1}^{k} x_i^* = \sum_{i=1}^{n} \min\{x_i, k\} = pk + \sum_{i=p+1}^{n} x_i
$$

$$
\geq pk + \sum_{i=p+1}^{n} y_i
$$

$$
\geq \sum_{i=1}^{n} \min\{y_i, k\} = \sum_{i=1}^{k} y_i^*.
$$

It now follows that $Y^* \preceq X^*$. $\qquad\qquad\qquad\qquad\qquad\qquad\qquad\qquad\quad\square$

The following result shows how a majorization between a pair of vectors implies a majorization between pairs of vectors derived from them [10].

Theorem 1.6.2 *Let $X = (x_1, x_2, \ldots, x_n)$ and $Y = (y_1, y_2, \ldots, y_n)$ be nonincreasing, nonnegative integer vectors with $X \preceq Y$. Let X' be obtained from X by reducing p positive elements in positions $i_1, i_2, \ldots, i_p$ by 1. Also let Y' be obtained from Y by reducing p positive elements in positions $j_1, j_2, \ldots, j_p$ by 1. If $i_1 \leq j_1, i_2 \leq j_2, \ldots, i_p \leq j_p$, then $X' \preceq Y'$.*

Proof. We prove the result by induction on p. First assume that $p = 1$ and set $i_1 = i$ and $j_1 = j$. By hypothesis, $i \leq j$. Let $X' = (x_1', x_2', \ldots, x_n')$

be obtained from X by replacing x_i with $x_i - 1$, and let $Y' = (y'_1, y'_2, \ldots, y'_n)$ be obtained from Y by replacing y_j with $y_j - 1$. If X' and Y' are nonincreasing vectors, then $X' \preceq Y'$ follows from $X \preceq Y$ and $i \leq j$.

If X' and Y' are not both nonincreasing, we proceed as follows. First we arrange X' and Y' to be nonincreasing by assuming that component i' of X has been reduced by 1 to get X' and that component j' of Y has been reduced by 1 to get Y'. Here $i' \geq i$, $j' \geq j$, and

$$x_i = x_{i+1} = \cdots = x_{i'} > x_{i'+1}, \ y_j = y_{j+1} = \cdots = y_{j'} > y_{j'+1}, \quad (1.11)$$

where $x_{i'+1} = 0$ if $i' = n$ and $y_{j'+1} = 0$ if $j' = n$. If $i' \leq j'$, then as above $X' \preceq Y'$ follows from $X \preceq Y$. Suppose that $i' > j'$. Then we have

$$1 \leq i \leq j \leq j' < i' \leq n. \quad (1.12)$$

Assume to the contrary that $X' \preceq Y'$ does not hold. Then there is an integer k satisfying

$$j' \leq k < i' \text{ and } x_1 + x_2 + \cdots + x_k = y_1 + y_2 + \cdots + y_k. \quad (1.13)$$

From $X \preceq Y$ we obtain

$$\begin{aligned} x_1 + x_2 + \cdots + x_{k-1} &\leq y_1 + y_2 + \cdots + y_{k-1}, \text{ and} \\ x_1 + x_2 + \cdots + x_{k+1} &\leq y_1 + y_2 + \cdots + y_{k+1}. \end{aligned} \quad (1.14)$$

From (1.13) and (1.14) we get

$$x_k \geq y_k \text{ and } x_{k+1} \leq y_{k+1}, \quad (1.15)$$

and from (1.11), (1.12), and (1.13) we get

$$x_k = x_{k+1}. \quad (1.16)$$

Now (1.15), (1.16), and $y_k \geq y_{k+1}$ imply

$$y_k = x_k = y_{k+1}. \quad (1.17)$$

If $k = j'$, this contradicts (1.11). Otherwise $j' < k$ and we have from (1.17) and (1.13) that

$$x_1 + x_2 + \cdots + x_{k-1} = y_1 + y_2 + \cdots + y_{k-1}. \quad (1.18)$$

We may now repeat the argument with k replaced with $k-1$. After a finite number of steps, (1.11) is contradicted. This completes the argument for $p = 1$.

We now proceed by induction on p. Suppose that $i_1 < i_2 < \cdots < i_p$ and $j_1 < j_2 < \cdots < j_p$. Let X'' and Y'' be obtained from X and Y by reducing components $i_2, \ldots, i_p$ of X and components $j_2, \ldots, j_p$ of Y by 1. By the induction assumption we have $X'' \preceq Y''$. We may rearrange X'' and Y'' into nonincreasing order without disturbing the i_1 position of X''

or the j_1 position of Y''. Applying the argument used for $p = 1$ to these rearrangements shows that $X' \preceq Y'$. □

1.7 Doubly Stochastic Matrices and Majorization

A nonnegative matrix $A = [a_{ij}]$ of order n is called *doubly stochastic* provided each of its line sums equals 1:

$$\sum_{j=1}^{n} a_{ij} = 1 \quad (i = 1, 2, \ldots, n), \text{ and } \sum_{i=1}^{n} a_{ij} = 1 \quad (j = 1, 2, \ldots, n). \quad (1.19)$$

The equations (1.19) are equivalent to the matrix equation

$$AJ_n = J_nA = J_n.$$

From this it follows that the product of two doubly stochastic matrices of order n is also a doubly stochastic matrix of order n, and that a *convex combination*

$$cA + (1 - c)B \quad (1 \leq c \leq 1)$$

of two doubly stochastic matrices A and B is also doubly stochastic.

The set (convex polytope), in real n^2-dimensional space $\Re^{n^2}$, of doubly stochastic matrices of order n is denoted by Ω_n. The matrix $(1/n)J_n$ each of whose elements equals $1/n$ is a doubly stochastic matrix in Ω_n. The integral doubly stochastic matrices of order n are precisely the permutation matrices of order n. Recall that an *extreme point* of a convex polytope Φ is a point x of Φ that cannot be expressed as a nontrivial convex combination of points of Φ: $x \neq cy + (1 - c)z$ where $0 < c < 1$ and $y, z \in \Omega_n$ with $y \neq z$. The nonnegativity property of doubly stochastic matrices implies that if a doubly stochastic matrix A is expressed as a nontrivial convex combination of doubly stochastic matrices B and C, then the patterns of B and C are contained in the pattern of A. We also note that if A is a doubly stochastic matrix, and P and Q are permutation matrices, then PAQ is also a doubly stochastic matrix.

Each permutation matrix of order n is an extreme point of Ω_n. The theorem of Birkhoff [5] asserts that Ω_n has no other extreme points.

Theorem 1.7.1 *The extreme points of Ω_n are precisely the permutation matrices of order n. If A is a doubly stochastic matrix of order n, then there exist permutation matrices $P_1, P_2, \ldots, P_t$ and positive real numbers $c_1, c_2, \ldots, c_t$ with $c_1 + c_2 + \cdots + c_t = 1$ such that*

$$A = c_1P_1 + c_2P_2 + \cdots + c_tP_t.$$

Proof. Since the permutation matrices are extreme points of Ω_n the two assertions in the theorem are equivalent. Let $A = [a_{ij}]$ be a doubly stochastic matrix of order n and assume that A is an extreme point of Ω_n. Consider the bipartite graph $\mathrm{BG}(A)$ representing the nonzero pattern of A. Suppose this graph has a cycle. By alternating $+1$'s and -1's on this cycle, we can construct a matrix C of order n with elements equal to $0, \pm 1$ such that the nonzero pattern of C is contained in the nonzero pattern of A and each line sum of C equals 0. It follows that there is a small positive number ϵ such that $A \pm \epsilon C$ are doubly stochastic matrices. Since

$$A = \frac{1}{2}(A + \epsilon C) + \frac{1}{2}(A - \epsilon C),$$

A is not an extreme point of Ω_n. Thus $\mathrm{BG}(A)$ does not have any cycles and hence is a forest of trees. Since each line sum of A equals 1, it follows easily that each of these trees is a tree with exactly one edge, and therefore that A is a permutation matrix. $\qquad\square$

The following corollary is an immediate consequence of Theorem 1.7.1.

Corollary 1.7.2 *The term rank of a doubly stochastic matrix of order n equals n.* $\qquad\square$

This corollary can be proved directly by first showing that the covering number of a doubly stochastic matrix of order n equals n and then invoking Theorem 1.2.1 of König. This leads to a different proof of Theorem 1.7.1 (see [4]).

We now develop a connection between majorization of vectors and doubly stochastic matrices.

Lemma 1.7.3 *Let $Y = (y_1, y_2, \ldots, y_n)$ be a vector of size n, and let $A = [a_{ij}]$ be a doubly stochastic matrix of order n. Then the vector $X = YA = (x_1, x_2, \ldots, x_n)$ is majorized by Y.*

Proof. We may choose permutation matrices P and Q such that YP and XQ are in nonincreasing order. Since $X = YA$ implies that $(XQ) = (YP)(P^{\mathrm{T}}AQ)$ and since $P^{\mathrm{T}}AQ$ is also a doubly stochastic matrix, we may assume that $x_1 \geq x_2 \geq \cdots \geq x_n$ and $y_1 \geq y_2 \geq \cdots \geq y_n$.

Let k be an integer with $1 \leq k \leq n$. Then

$$\sum_{j=1}^{k} x_j = \sum_{j=1}^{k} \sum_{i=1}^{n} y_i a_{ij} = \sum_{i=1}^{n} y_i e_{ik}, \qquad (1.20)$$

where

$$0 \leq e_{ik} = \sum_{j=1}^{k} a_{ij} \leq 1 \text{ and } \sum_{i=1}^{n} e_{ik} = k. \qquad (1.21)$$

Since $e_{in} = 1$ $(i = 1, 2, \ldots, n)$, we see that $\sum_{j=1}^{n} x_j = \sum_{i=1}^{n} y_i$. Using (1.20), (1.21), and the nonincreasing property of Y, we get

$$
\sum_{j=1}^{k} x_j - \sum_{i=1}^{k} y_j = \sum_{i=1}^{n} y_i e_{ik} - \sum_{i=1}^{k} y_i
$$

$$
= \sum_{i=1}^{n} y_i e_{ik} - \sum_{i=1}^{k} y_i + y_k \left(k - \sum_{i=1}^{n} e_{ik} \right)
$$

$$
= \sum_{i=1}^{k} (y_i - y_k)(e_{ik} - 1) + \sum_{i=k+1}^{n} e_{ik}(y_i - y_k)
$$

$$
\leq 0.
$$

Hence $X \preceq Y$.   □

We now turn to the converse of Lemma 1.7.3. Let n be a fixed positive integer, and let i and j be integers with $1 \leq i < j \leq n$. Let $T_n^{(i,j)}$ be the permutation matrix corresponding to the permutation of $\{1, 2, \ldots, n\}$ which interchanges i and j and leaves every other integer fixed. A doubly stochastic matrix of the form

$$
cI_n + (1 - c)T_n^{(i,j)} \quad (0 \leq c \leq 1) \tag{1.22}
$$

is called an *elementary doubly stochastic matrix*. Note that if $c = 0$ then (1.22) is the permutation matrix $T_n^{(i,j)}$ which interchanges i and j; if $c = 1$, then $T_n^{(i,j)} = I_n$.

If $Y = (y_1, y_2, \ldots, y_n)$, then $Y(cI_n + (1 - c)T_n^{(i,j)})$ is the n-vector

$$
(y_1, \ldots, y_{i-1}, cy_i + (1 - c)y_j, y_{i+1}, \ldots, y_{j-1}, (1 - c)y_i + cy_j, y_{j+1}, \ldots, y_n)
$$

which differs from Y in at most two components.

We next prove a lemma due to Muirhead [15] and Hardy, Littlewood, and Pólya [11] which gives a strong converse of Lemma 1.7.3.

Lemma 1.7.4 *Let $X = (x_1, x_2, \ldots, x_n)$ and $Y = (y_1, y_2, \ldots, y_n)$ be nonincreasing vectors of size n such that $X \preceq Y$. If X and Y differ in t components, then there are $k \leq t - 1$ elementary doubly stochastic matrices $A_1, A_2, \ldots, A_k$ such that $X = Y A_1 A_2 \ldots A_k$.*

Proof. If $X = Y$, then $A_1 A_2 \ldots A_k$ is an empty product equal to I_n. Now assume that $X \neq Y$. Then $t \geq 2$, and there exists an index i such that $x_i < y_i$ and then an index $j > i$ such that $x_j > y_j$. We let i and j be the smallest and largest such indices, respectively. Also let

$$
\epsilon = \min\{y_i - x_i, x_j - y_j\} \text{ and } c = 1 - \frac{\epsilon}{y_i - y_j}.
$$

Then $0 < c < 1$ and the vector $Y' = Y(cI_n + (1-c)T_n^{i,j}) = (y_1', y_2', \ldots, y_n')$ satisfies

$$Y' = (y_1, \ldots, y_{i-1}, y_i - \epsilon, y_{i+1}, \ldots, y_{j-1}, y_j + \epsilon, y_{j+1}, \ldots, y_n).$$

By Lemma 1.7.3, $Y' \preceq Y$. We also observe that $y_i' = x_i$ if $\epsilon = y_i - x_i$ and $y_j' = x_j$ if $\epsilon = x_j - y_j$. Therefore Y' differs from X in at most $t - 1$ components. That $X \preceq Y'$ easily follows from the facts that $X \preceq Y$, that Y' and Y differ in only the two components i and j where $i < j$, that $y_i' + y_j' = y_i + y_j$, and that $y_i' \geq x_i$. We conclude that $X \preceq Y'$ and that X and Y' differ in at most $t - 1$ components. It follows inductively that $X = Y'A$ where A is a product of $k' \leq t - 2$ elementary doubly stochastic matrices. Hence $X = Y(cI_n + (1-c)T_n^{(i,j)})A$, from which the lemma now follows.   □

A consequence of the proof of Lemma 1.7.4 is the following corollary for nonnegative integral vectors.

Corollary 1.7.5 *Let $X = (x_1, x_2, \ldots, x_n)$ and $Y = (y_1, y_2, \ldots, y_n)$ be nonincreasing, nonnegative integral vectors of size n such that $X \preceq Y$. Then there exist nonincreasing, nonnegative integral vectors*

$$X_0 = X, X_1, \ldots, X_{p-1}, X_p = Y$$

of size n such that

$$X_0 \preceq X_1 \preceq \cdots \preceq X_{p-1} \preceq X_p$$

and X_i and X_{i+1} differ in exactly two coordinates.   □

Combining Lemmas 1.7.3 and 1.7.4 we get the following theorem of Hardy, Littlewood, and Pólya [11].

Theorem 1.7.6 *Let X and Y be two vectors of size n. Then $X \preceq Y$ if and only if there is a doubly stochastic matrix A such that $X = YA$.*

The theory of majorization is thoroughly developed in [12].

References

[1] C. Berge, *Graphs and Hypergraphs*, North-Holland, Amsterdam, 1973 (translation and revision of *Graphes et hypergraphes*, Dunod, Paris, 1970).

[2] B. Bollobás, *Modern Graph Theory*, Grad. Texts Math. **184**, Springer, 1998.

[3] R.A. Brualdi and R. Manber, Chromatic number of classes of matrices of zeros and ones, *Discrete Math.*, **50** (1984), 143–152.

[4] R.A. Brualdi and H.J. Ryser, *Combinatorial Matrix Theory*, Cambridge U. Press, Cambridge, 1991.

[5] G. Birkhoff, Tres observaciones sobre el algebra lineal, *Univ. Nac. Tucumán Rev. Ser. A*, (1946), 147–151.

[6] V. Chungphaisan, Conditions for sequences to be r-graphic, *Discrete Math.*, **7** (1974), 31–39.

[7] P. Erdős and T. Gallai, Graphs with prescribed degrees of vertices (in Hungarian), *Mat. Lapok*, **11** (1960), 264–274.

[8] L.R. Ford, Jr. and D.R. Fulkerson, *Flows in Networks*, Princeton U. Press, Princeton, 1962.

[9] D.R. Fulkerson, A.J. Hoffman, and M.H. McAndrew, Some properties of graphs with multiple edges, *Canad. J. Math.*, **17** (1965), 166–177.

[10] D.R. Fulkerson and H.J. Ryser, Multiplicities and minimal widths for (0,1)-matrices, *Canad. J. Math.*, **14** (1962), 498–508.

[11] G.H. Hardy, J.E. Littlewood, and G. Pólya, *Inequalities*, Cambridge U. Press, Cambridge, 1934 (1st ed.), 1952 (2nd ed.).

[12] A.W. Marshall and I. Olkin, *Inequalities: Theory of Majorization and Its Applications*, Academic Press, Orlando, 1979.

[13] L. Mirsky, Combinatorial theorems and integral matrices, *J. Combin. Theory*, **5** (1968), 30–44.

[14] L. Mirsky, *Transversal Theory*, Academic Press, New York, 1971.

[15] R.F. Muirhead, Some methods applicable to identities and inequalities of symmetric algebric functions of n letters, *Proc. Edinburgh Math. Soc.*, **21** (1903), 144–157.

[16] G. Sierksma and H. Hoogeven, Seven criteria for integer sequences being graphic, *J. Graph Theory*, **15** (1991), 223–231.

[17] D.B. West, *Introduction to Graph Theory*, 2nd ed., Prentice-Hall, Upper Saddle River, 2001.

[18] R.J. Wilson, *Introduction to Graph Theory*, Longman, London, 1985.

2

Basic Existence Theorems for Matrices with Prescribed Properties

In this chapter we prove some existence theorems for matrices with special properties. We do not attempt to give a detailed account of such theorems. Rather, we concentrate mainly on existence theorems that guarantee the nonemptiness of the classes of matrices that we study in depth in this book. Some existence theorems for special matrices are derived in [5] as consequences of fundamental network flow theorems. We do not invoke network flow theory, preferring in this chapter to derive our theorems using more elementary mathematical arguments.

2.1   The Gale–Ryser and Ford–Fulkerson Theorems

Let $A = [a_{ij}]$ be an m by n nonnegative matrix. Let

$$r_i = a_{i1} + a_{i2} + \cdots + a_{in} \quad (i = 1, 2, \ldots, m)$$

be the sum of the elements in row i of A, and let

$$s_j = a_{1j} + a_{2j} + \cdots + a_{mj} \quad (j = 1, 2, \ldots, n)$$

be the sum of the elements in column j of A. Then, as used in Chapter 1,

$$R = (r_1, r_2, \ldots, r_m)$$

is the *row sum vector* of A, and

$$S = (s_1, s_2, \ldots, s_n)$$

is the *column sum vector* of A. The vectors R and S satisfy the fundamental equation

$$r_1 + r_2 + \cdots + r_m = s_1 + s_2 + \cdots + s_n, \tag{2.1}$$

since both sides equal the sum $\tau = \tau(A)$ of all the elements of A.

Now let $R = (r_1, r_2, \ldots, r_m)$ and $S = (s_1, s_2, \ldots, s_n)$ be nonnegative vectors satisfying (2.1), and assume that τ, the common value in (2.1), is not zero. The m by n matrix $A = [a_{ij}]$ where

$$a_{ij} = \frac{r_i s_j}{\tau} \quad (i = 1, 2, \ldots, m; j = 1, 2, \ldots, n)$$

has row sum vector R and column sum vector S, and all of its elements are positive if R and S are positive vectors.

We can also inductively construct an m by n nonnegative matrix $A = [a_{ij}]$ with row sum vector R and column sum vector S. If $m = 1$, then

$$A = \begin{bmatrix} s_1 & s_2 & \ldots & s_n \end{bmatrix}$$

is the unique such matrix. If $n = 1$, then

$$A = \begin{bmatrix} r_1 & r_2 & \ldots & r_m \end{bmatrix}^{\mathrm{T}}$$

is the unique matrix. Now assume that $m > 1$ and $n > 1$. We let $a_{11} = \min\{r_1, s_1\}$. First suppose that $a_{11} = r_1$. Let $a_{12} = \ldots = a_{1n} = 0$, and define $R' = (r_2, \ldots, r_m)$ and $S' = (s_1 - r_1, s_2, \ldots, s_n)$. Then

$$r_2 + \cdots + r_m = (s_1 - r_1) + s_2 + \cdots + s_n.$$

Proceeding inductively, there exists an $m - 1$ by n matrix A' with row sum vector R' and column sum vector S'. The matrix

$$\begin{bmatrix} r_1 & 0 & \ldots & 0 \\ & & A' & \end{bmatrix}$$

is a nonnegative matrix with row sum vector R and column sum vector S. If $a_{11} = s_1$, a similar construction works. The matrix inductively constructed in this way has at most $m + n - 1$ positive elements.

We denote by

$$\mathcal{N}(R, S)$$

the *set of all nonnegative matrices with row sum vector R and column sum vector S*. Partially summarizing, we have the following theorem.

Theorem 2.1.1 *Let $R = (r_1, r_2, \ldots, r_m)$ and $S = (s_1, s_2, \ldots, s_n)$ be nonnegative vectors. Then $\mathcal{N}(R, S)$ is nonempty if and only if the equation (2.1) holds.* $\qquad\square$

We observe that if R and S are nonnegative integral vectors satisfying the fundamental equation (2.1), then the algorithm preceding Theorem 2.1.1 produces a nonnegative integral matrix A with row sum vector R and column sum vector S. Let

$$\mathcal{Z}^+(R, S)$$

denote the *set of all nonnegative integral matrices with row sum vector R and column sum vector S.*

Theorem 2.1.2 *Let $R = (r_1, r_2, \ldots, r_m)$ and $S = (s_1, s_2, \ldots, s_n)$ be non-negative integral vectors. Then $\mathcal{Z}^+(R, S)$ is nonempty if and only if (2.1) holds.* ☐

If we impose other requirements on the elements of A, then the fundamental equation (2.1) need not suffice for the existence of a matrix A with the required row and column sum vectors. For instance, there is no 4 by 3 matrix with row sum vector $R = (3, 3, 2, 1)$ and column sum vector $S = (4, 4, 1)$ each of whose elements equals 0 or 1. We denote by

$$\mathcal{A}(R, S)$$

the *class*[1] *of all $(0, 1)$-matrices with row sum vector R and column sum vector S.*

Gale [11] and Ryser [20] independently proved the basic theorem for the existence of a (0,1)-matrix with row sum vector R and column sum vector S. Ryser used induction and direct combinatorial reasoning; Gale used the theory of network flows. This theorem is also derived in [5] using network flows and is also derived in [3]. See also [22]. We give the simple inductive proof in [4]. There is no loss in generality in assuming that R and S are nonincreasing vectors.

Theorem 2.1.3 *Let $R = (r_1, r_2, \ldots, r_m)$ and $S = (s_1, s_2, \ldots, s_n)$ be non-increasing vectors of nonnegative integers. Then $\mathcal{A}(R, S)$ is nonempty if and only if S is majorized by the conjugate R^* of R, that is,*

$$S \preceq R^*. \tag{2.2}$$

Proof. Assume there is an m by n $(0,1)$-matrix A with row sum vector R and column sum vector S. Then $s_1 + s_2 + \cdots + s_n = r_1 + r_2 + \cdots + r_m$. If A is partitioned as

$$[\, B_1 \,|\, B_2 \,] \qquad (B_1 \text{ is } m \text{ by } l),$$

then the number of 1's in B_1 does not exceed the number of 1's in the first l columns of the matrix obtained from A by sliding the 1's in each row as

---

[1] By tradition, beginning with Ryser, $\mathcal{A}(R, S)$ is referred to as a class, rather than a set, of matrices.

far to the left as possible; that is, $\sum_{i=1}^{l} s_i \leq \sum_{i=1}^{l} r_i^*$ $(l = 1, 2, \ldots, n-1)$. Hence (2.2) holds.

We prove the converse by contradiction. Choose a counterexample with $m+n$ minimal, and for that $m+n$, with $\tau = r_1 + r_2 + \cdots + r_m$ minimal.

Case 1: $\sum_{i=1}^{l} s_i = \sum_{i=1}^{l} r_i^*$ for some l with $0 < l < n$. It is an immediate consequence that

$$R_1 = (r_1', \ldots, r_m') = (\min\{l, r_1\}, \ldots, \min\{l, r_m\}) \text{ and } S_1 = (s_1, \ldots, s_l),$$

we have $S_1 \preceq R_1^*$, and for the vectors

$$R_2 = (r_1 - r_1', \ldots r_m - r_m'), \text{ and } S_2 = (s_{l+1}, \ldots, s_n),$$

we have $S_2 \preceq R_2^*$. By minimality there exist a $(0, 1)$-matrix A_1 with row sum vector R_1 and column sum vector S_1, and a $(0,1)$-matrix A_2 with row sum vector R_2 and column sum vector S_2. Hence the matrix

$$\left[\, A_1 \,\middle|\, A_2 \,\right]$$

has row sum vector R and column sum vector S, a contradiction.

Case 2: $\sum_{i=1}^{l} s_i < \sum_{i=1}^{l} r_i^*$ for all l with $0 < l < n$. By minimality, $r_m \geq 1$ and $s_n \geq 1$. Let $R' = (r_1, \ldots, r_{m-1}, r_m - 1)$ and $S' = (s_1, \ldots, s_{n-1}, s_n - 1)$. Then $S' \preceq R'^*$. By minimality again there is an m by n matrix $A' = [a_{ij}']$ with row sum vector R' and column sum vector S'. If $a_{mn}' = 0$, then changing this 0 to a 1 we obtain a matrix with row sum vector R and column sum vector S, a contradiction. Suppose that $a_{mn}' = 1$. Since $r_m - 1 \leq n - 1$, there is a q such that $a_{mq}' = 0$. Since $s_q \geq s_n$, there is a p such that $a_{pq}' = 1$ and $a_{pn}' = 0$. Interchanging 0's with 1's in this 2 by 2 matrix gives a matrix A'' with row sum vector R' and column sum vector S', and we get a contradiction as before. $\square$

If we assume that $r_i \leq n$ $(i = 1, 2, \ldots, m)$, then $S \preceq R^*$ if and only if

$$\sum_{i=1}^{m} (r_i - k)^+ \geq \sum_{j=k+1}^{n} s_j \quad (0 \leq k \leq n-1), \tag{2.3}$$

with equality if $k = 0$.

Consider the class $\mathcal{A}(R, S)$ of all $(0,1)$-matrices with row sum vector $R = (r_1, r_2, \ldots, r_m)$ and column sum vector $S = (s_1, s_2, \ldots, s_n)$. Theorem 2.1.3 gives a simple necessary and sufficient condition (2.2) for the nonemptiness of $\mathcal{A}(R, S)$. This condition requires the checking of only $n - 1$ inequalities and one equality. A different condition was obtained by Ford and Fulkerson [8] by using a theorem in network flows. While this condition requires more inequalities than those in (2.2), it is nonetheless worthwhile in that it involves a quantity that arises in the investigation of certain parameters associated with $\mathcal{A}(R, S)$. We now discuss this condition of Ford and Fulkerson.

Let $R = (r_1, r_2, \ldots, r_m)$ and $S = (s_1, s_2, \ldots, s_n)$ be nonnegative integral vectors with elements summing to the same integer

$$\tau = r_1 + r_2 + \cdots + r_m = s_1 + s_2 + \cdots + s_n.$$

If $K \subseteq \{1, 2, \ldots, m\}$ and $L \subseteq \{1, 2, \ldots, n\}$, we define

$$t_{K,L} = |K||L| + \sum_{i \in \bar{K}} r_i - \sum_{j \in J} s_j.$$

The numbers $t_{K,L}$ are integers and we show below that they are nonnegative if $\mathcal{A}(R, S)$ is nonempty.

The *structure matrix* associated with R and S is the matrix

$$T = T(R, S) = [t_{kl}] \quad (k = 0, 1, \ldots, m; l = 0, 1, \ldots, n)$$

of size $m + 1$ by $n + 1$ where

$$t_{kl} = kl + \sum_{i=k+1}^{m} r_i - \sum_{j=1}^{l} s_j \quad (k = 0, 1, \ldots, m; l = 0, 1, \ldots, n). \qquad (2.4)$$

Under the assumption that R and S are nonincreasing,

$$t_{kl} = \min\{t_{K,L} : K \subseteq \{1, 2, \ldots, m\}, L \subseteq \{1, 2, \ldots, n\}, |K| = k, |L| = l\}.$$

Some of the elements of T are

$$t_{00} = \tau, t_{0n} = t_{m0} = 0, t_{1n} = n - r_1, t_{m1} = m - s_1, \text{ and } t_{mn} = mn - \tau.$$

We next observe that, if $\mathcal{A}(R, S)$ is nonempty, the elements of the structure matrix are nonnegative integers, by explaining their combinatorial significance. Let A be a matrix in $\mathcal{A}(R, S)$ and partition A as

$$\left[\begin{array}{c|c} A_1 & X \\ \hline Y & A_2 \end{array} \right],$$

where A_1 is of size k by l and where either or both of A_1 and A_2 may be empty. Then a straightforward calculation shows that

$$t_{kl} = N_0(A_1) + N_1(A_2),$$

the number $N_0(A_1)$ of 0's in A_1 plus the number $N_1(A_2)$ of 1's in A_2. The theorem of Ford and Fulkerson asserts that under the assumption that R and S are nonincreasing, the nonnegativity of T is also sufficient for the nonemptiness of $\mathcal{A}(R, S)$. It can be shown directly that the nonnegativity of T is equivalent to the condition that S is majorized by the conjugate R^* of R (see e.g. [17]). Rather than do this, we use the proof technique of Theorem 2.1.3 as given in [4]. This method of proof was earlier used in

[14]. A sequence $U = (u_1, u_2, \ldots, u_n)$ of integers is *nearly nonincreasing* provided

$$u_i \geq u_j - 1 \quad (1 \leq i < j \leq n).$$

Thus the sequence U is nearly nonincreasing if and only if there is a sequence $H = (h_1, h_2, \ldots, h_n)$ of 0's and 1's such that $U + H$ is nonincreasing. For example, $(5, 5, 4, 5, 5, 3, 4)$ is a nearly nonincreasing sequence. Michael [16] proved the theorem under the assumption that R and S are only nearly nonincreasing. We prove the theorem with no monotonicity assumption on R and with the assumption that S is nearly nonincreasing.

Theorem 2.1.4 *Let $R = (r_1, r_2, \ldots, r_m)$ and $S = (s_1, s_2, \ldots, s_n)$ be nonnegative integral vectors such that S is nearly nonincreasing and $r_1 + r_2 + \cdots + r_m = s_1 + s_2 + \cdots + s_n$. Then $\mathcal{A}(R, S)$ is nonempty if and only if*

$$t_{kl} \geq 0 \quad (k = 0, 1, \ldots, m; l = 0, 1, \ldots, n), \tag{2.5}$$

that is, the structure matrix T is a nonnegative matrix.

Proof. The necessity of (2.5) has already been demonstrated. We prove the sufficiency by contradiction. As in the proof of Theorem 2.1.3, we choose a counterexample with $m + n$ minimal, and for that $m + n$, with $\tau = r_1 + r_2 + \cdots + r_m$ minimal. Clearly $t_{00} = \tau > 0$ (otherwise take A to be the zero matrix O) and $t_{mn} > 0$ (otherwise take A to be the matrix J of all 1's).

Case 1: $t_{kl} = 0$ for some k and l with $(k, l) \neq (0, n)$ or $(m, 0)$. It is an immediate consequence that the structure matrices for $R_1 = (r_1 - l, \ldots, r_k - l)$, $S_1 = (s_{l+1}, \ldots, s_n)$ and for $R_2 = (r_{k+1}, \ldots, r_m)$, $S_2 = (s_1 - k, \ldots, s_l - k)$ satisfy the nonnegativity criteria (2.5). Since S_1 and S_2 are nearly monotone, by minimality there exist a $(0, 1)$-matrix A_1 with row sum vector R_1 and column sum vector S_1, and a $(0, 1)$-matrix A_2 with row sum vector R_2 and column sum vector S_2. Hence the matrix

$$\left[\begin{array}{c|c} J_{k,l} & A_1 \\ \hline A_2 & O \end{array} \right]$$

has row sums R and column sums S, a contradiction.

Case 2: $t_{kl} > 0$ for all k and l with $(k, l) \neq (0, n)$ or $(m, 0)$. By minimality, $r_m \geq 1$ and $s_n \geq 1$. Let $R' = (r_1, \ldots, r_{m-1}, r_m - 1)$, and let $S' = (s_1, \ldots, s_{n-1}, s_n - 1)$. Then S' is a nearly nonincreasing vector, and the structure matrix for R' and S' is nonnegative. By minimality again there is a matrix $A' = [a'_{ij}]$ with row sums R' and column sums S'. If $a'_{mn} = 0$, then changing this 0 to a 1 we obtain a matrix with row sums R and column sums S, a contradiction. Suppose that $a'_{mn} = 1$. Since $r_m - 1 \leq n - 1$, there is a q such that $a'_{mq} = 0$. Since S is nearly nonincreasing, $s_q \geq s_n - 1$ and hence there is a p such that $a'_{pq} = 1$ and $a'_{pn} = 0$.

Interchanging 0 with 1 in this 2 by 2 matrix gives a matrix A'' with row sums R' and column sums S', and we get a contradiction as before. $\square$

Assuming that R and S are nonincreasing, Fulkerson [9] gave conditions for the existence of a matrix $A = [a_{ij}]$ in $\mathcal{A}(R,S)$ such that $a_{ii} = 0$ $(i = 1, 2, \ldots, \min\{m,n\})$. These conditions also involve the elements of the structure matrix T. We obtain this theorem as a special case of Theorem 1.4.1.

Theorem 2.1.5 *Let* $R = (r_1, r_2, \ldots, r_m)$ *and* $S = (s_1, s_2, \ldots, s_n)$ *be non-increasing, nonnegative integral vectors satisfying* $r_1 + r_2 + \cdots + r_m = s_1 + s_2 + \cdots + s_n$. *There exists an* m *by* n $(0,1)$-*matrix* $A = [a_{ij}]$ *with row sum vector* R *and column sum vector* S *such that* $a_{ii} = 0$ $(i = 1, 2, \ldots, \min\{m,n\})$ *if and only if*

$$t_{kl} - \min\{k,l\} \geq 0 \quad (k = 0, 1, \ldots, m; l = 0, 1, \ldots, n). \qquad (2.6)$$

Proof. Define the m by n $(0,1)$-matrix $C = [c_{ij}]$ by $c_{ii} = 0$ only for $i = 1, 2, \ldots, \min\{m,n\}$. By Corollary 1.4.2 the matrix A specified in the theorem exists if and only if

$$\sum_{i \in I} \sum_{j \in J} c_{ij} + \sum_{i \notin I} r_i - \sum_{j \in J} s_j \geq 0 \qquad (2.7)$$

for all $I \subseteq \{1, 2, \ldots, m\}$ and $J \subseteq \{1, 2, \ldots, n\}$. Let $|I| = k$ and $|J| = l$. The sum $\sum_{i \in I} \sum_{j \in J} c_{ij} = kl - |I \cap J|$ is minimal when $|I \cap J|$ is maximal. Since R and S are nonincreasing, for fixed k and l the expression on the left side in (2.7) is smallest when $I = \{1, 2, \ldots, k\}$ and $J = \{1, 2, \ldots, l\}$ and in this case it has value

$$kl - \min\{k,l\} + \sum_{i=k+1}^{m} r_i - \sum_{j=1}^{l} s_j = t_{kl} - \min\{k,l\}.$$

$\square$

We close this section by recording the following inequalities that must be satisfied if there exists a matrix in $\mathcal{A}(R,S)$ where $R = (r_1, r_2, \ldots, r_m)$ and $S = (s_1, s_2, \ldots, s_n)$:

$$\sum_{i=1}^{m} r_i^2 + \sum_{j=1}^{n} s_j^2 \leq \tau \left(\max\{m,n\} + \frac{\tau}{\max\{m,n\}} \right),$$

$$\sum_{i=1}^{m} r_i^2 + \sum_{j=1}^{n} s_j^2 \leq \tau \max\{m,n\} + \sum_{i=1}^{\min\{m,n\}} r_i s_i.$$

The first of these inequalities is due to Khintchine [12] and the second is due to Matúš (see [15]).

2.2   Tournament Matrices and Landau's Theorem

A nonnegative matrix $A = [a_{ij}]$ of order n is a *generalized tournament matrix* provided

$$a_{ii} = 0 \quad (i = 1, 2, \ldots, n)$$

and

$$a_{ij} + a_{ji} = 1 \quad (1 \leq i < j \leq n),$$

that is, provided

$$A + A^{\mathrm{T}} = J_n - I_n.$$

Let $R = (r_1, r_2, \ldots, r_n)$ and $S = (s_1, s_2, \ldots, s_n)$ be the row sum vector and column sum vector of A, respectively. Then

$$r_i = \sum_{j \neq i} a_{ij} = \sum_{j \neq i}(1 - a_{ji}) = (n-1) - \sum_{j \neq i} a_{ji} = (n-1) - s_i.$$

Hence

$$r_i + s_i = n - 1 \quad (i = 1, 2, \ldots, n).$$

In particular, the column sum vector S of a generalized tournament matrix is determined by its row sum vector R. In addition, the sum of all the elements of a generalized tournament matrix of order n is

$$\sum_{i=1}^{n} r_i = \sum_{i=1}^{n} s_i = \sum_{i=1}^{n}\sum_{j=1}^{n} a_{ij} = \binom{n}{2},$$

an integer determined solely by n. Let I be a nonempty subset of $\{1, 2, \ldots, n\}$. Then the principal submatrix $A[I]$ of A is also a generalized tournament matrix, and hence

$$\sum_{i \in I} r_i \geq \binom{|I|}{2} \quad \text{and} \quad \sum_{i \in I} s_i \geq \binom{|I|}{2}. \tag{2.8}$$

If A is a generalized tournament matrix of order n, then so is PAP^{T} for each permutation matrix P of order n. Thus there is no loss in generality in assuming that the row sum vector R is nondecreasing, and hence that the column sum vector S is nonincreasing. If R is nondecreasing, then (2.8) is equivalent to

$$\sum_{i=1}^{k} r_i \geq \binom{k}{2} \quad \text{and} \quad \sum_{i=n-k+1}^{n} r_i \leq \binom{n}{2} - \binom{k}{2} \quad (k = 1, 2, \ldots, n).$$

A *tournament matrix* is a generalized tournament matrix each of whose elements equals 0 or 1. If $A = [a_{ij}]$ is a tournament matrix, then $a_{ij}a_{ji} = 0$ for all i and j. The digraph $D(A)$ of a tournament matrix A is called a *tournament.* We use the notation $T(A)$ for the tournament corresponding to the tournament matrix A. A tournament is a digraph obtained from the complete graph K_n with vertex set $\{x_1, x_2, \ldots, x_n\}$ by assigning an orientation to each of its edges.

A principal submatrix $A[I]$ of a tournament matrix A is also a tournament matrix and its digraph is a *subtournament* of the tournament $T(A)$. We can think of a tournament (or tournament matrix) as representing the outcomes of a round-robin tournament involving n teams with each team playing once against the other $n-1$ teams and with each game resulting in a winner and a loser (no ties allowed). An arc from the team represented by vertex x_i to the team represented by vertex x_j (a 1 in position (i,j) of A) signifies that in the tournament, x_i has beaten x_j. Thus r_i equals the number of wins of the team x_i; it is for this reason that r_i is called the *score* of team x_i and $R = (r_1, r_2, \ldots, r_n)$ is also called the *score vector* of a tournament. In a generalized tournament matrix $A = [a_{ij}]$, a_{ij} can be interpreted as the a priori probability that team x_i beats team x_j (whence the assumption that $a_{ij} + a_{ji} = 1$ for $i \neq j$); the row sum vector then gives the expected number of wins of each player.

Generalized tournament matrices and tournament matrices are related by the following theorem [19]. Let $\mathcal{T}_n$ and $\mathcal{T}_n^g$ denote the *set of all tournament matrices* and the *set of all generalized tournament matrices*, respectively, of order n. $\mathcal{T}_n^g$ is a convex set in real $n(n-1)$-dimensional space[2] $\Re^{n(n-1)}$.

Theorem 2.2.1 *Let n be a positive integer. Then $\mathcal{T}_n^g$ is a convex polytope whose extreme points are the tournament matrices $\mathcal{T}_n$.*

Proof. Let $A = [a_{ij}]$ be in $\mathcal{T}_n^g$. Let α be the number of 0's in A. Then $\alpha \leq n(n+1)/2$ with equality if and only if A is a tournament matrix. We argue by backwards induction on α. Suppose that $\alpha < n(n+1)/2$. Since $a_{ij} + a_{ji} = 1$ whenever $i \neq j$, there is a tournament matrix $T = [t_{ij}]$ such that for all $i \neq j$, $a_{ij} = 1$ implies that $t_{ij} = 1$. Let

$$c = \min\{a_{ij} : t_{ij} = 1, 1 \leq i, j \leq n\}.$$

Then $c > 0$ and $B = (A - cT)/(1 - c)$ is in $\mathcal{T}_n^g$ and has at least one more 0 than A. By induction B, and therefore A, is a convex combination of tournament matrices. A tournament matrix having only 0 and 1 as elements is clearly an extreme point of $\mathcal{T}_n^g$. □

Let $R = (r_1, r_2, \ldots, r_n)$ be a nonnegative vector with

$$r_1 + r_2 + \cdots + r_n = \binom{n}{2}.$$

---

[2]The matrices belong to $\Re^{n^2}$, but since the main diagonal of a generalized tournament matrix contains only 0's, it lies in a coordinate subspace of dimension $n(n-1)$.

We denote by

$$\mathcal{T}^{\mathrm{g}}(R)$$

the *class of all generalized tournament matrices with row sum vector R*, and by

$$\mathcal{T}(R)$$

the *class of all tournament matrices with row sum vector R*. Notice that $\mathcal{T}(R) \subseteq \mathcal{T}^{\mathrm{g}}(R)$.

Let A be a generalized tournament matrix in $\mathcal{T}^{\mathrm{g}}(R)$. The canonical form of A under simultaneous row and column permutations, as described in Theorem 1.3.2 but using the lower block triangular form, has the form

$$PAP^{\mathrm{T}} = \begin{bmatrix} A_1 & O & O & \ldots & O \\ J & A_2 & O & \ldots & O \\ J & J & A_3 & \ldots & O \\ \vdots & \vdots & \vdots & \ddots & \vdots \\ J & J & J & \ldots & A_t \end{bmatrix} \tag{2.9}$$

where all the blocks J below the diagonal are matrices of 1's, and the diagonal blocks $A_1, A_2, \ldots, A_t$ are the irreducible components of A. These irreducible components are themselves generalized tournament matrices. It is clear from (2.9) that, if $1 \le i < j \le t$, then the sum of the elements in a row of the matrix (2.9) that meets A_i is strictly greater than the sum of the elements in a row that meets A_j. It follows that the canonical form (2.9) can be determined simply by simultaneously permuting rows and columns so that the row sums are in nondecreasing order.

Landau [13] characterized the score vectors of tournaments. Moon [18] characterized the score vectors of generalized tournaments. We follow the short proof of Landau's theorem given by Mahmoodian [14] and Thomassen [21]. Another short proof is given by Bang and Sharp [2].

Theorem 2.2.2 *Let n be a positive integer and let $R = (r_1, r_2, \ldots, r_n)$ be a nondecreasing, nonnegative integral vector. Then $\mathcal{T}(R)$ is nonempty if and only if*

$$\sum_{i=1}^{k} r_i \ge \binom{k}{2} \quad (k = 1, 2, \ldots, n), \text{ with equality when } k = n. \tag{2.10}$$

Proof. We refer to (2.10) as *Landau's condition* and to the inequalities in (2.10) as *Landau's inequalities*. As already shown, Landau's condition is satisfied by a tournament matrix with row sum vector R. We prove the converse by contradiction. Choose a counterexample with n smallest and, for that n, with r_1 smallest. First suppose that there exists a k with $1 \le k < n$ such that

$$\sum_{i=1}^{k} r_i = \binom{k}{2}.$$

The minimality assumption on n implies that $R' = (r_1, r_2, \ldots, r_k)$ is the row sum vector of some tournament matrix A'. In addition,

$$\sum_{i=1}^{l}(r_{k+i} - k) = \sum_{i=1}^{k+l} r_i - \binom{k}{2} - kl$$

$$\geq \binom{k+l}{2} - \binom{k}{2} - kl$$

$$= \binom{l}{2} \quad (l = 1, 2, \ldots, n - k),$$

with equality when $l = n - k$. Hence the minimality assumption on n also implies that there exists a tournament matrix A'' with row sum vector $R'' = (r_{k+1} - k, r_{k+2} - k, \ldots, r_n - k)$. The matrix

$$\begin{bmatrix} A' & O_{k,n-k} \\ J_{n-k,k} & A'' \end{bmatrix}$$

is a tournament matrix with row sum vector R.

Now suppose that the inequalities (2.10) are strict inequalities for $k = 1, 2, \ldots, n-1$. Then $r_1 > 0$, and the vector $R''' = (r_1-1, r_2, \ldots, r_{n-1}, r_n+1)$ satisfies Landau's condition and so by the minimality assumption on r_1, there is a tournament matrix A with row sum vector R'''. Since $(r_n + 1) - (r_1 - 1) = (r_n - r_1) + 2 \geq 2$, there exists an integer p with $1 < p < n$ such that column p of A has a 0 in row 1 and a 1 in row n (and so row p of A has a 1 in column 1 and a 0 in column n). Interchanging the 0 and 1 in column p and in row p gives a tournament matrix in $\mathcal{T}(R)$.   □

We remark that Landau's condition (2.10) is equivalent to the majorization

$$(r_n, \ldots, r_2, r_1) \preceq (n - 1, \ldots, 2, 1, 0).$$

Fulkerson [10] reformulated Landau's condition in terms of what we shall call the normalized row sum vector of a tournament matrix or the *normalized score vector* of a tournament. Let $A = [a_{ij}]$ be a tournament matrix of order n with row sum vector $R = (r_1, r_2, \ldots, r_n)$. Because $a_{ij} + a_{ji} = 1$ for all $i \neq j$ and $a_{ii} = 0$ for all i, we obtain

$$\sum_{j=i+1}^{n} a_{ij} - \sum_{j=1}^{i-1} a_{ji} = r_i - (i - 1) \quad (i = 1, 2, \ldots, n). \qquad (2.11)$$

Conversely, given R, if the elements above the main diagonal of the tournament matrix $A = [a_{ij}]$ of order n satisfy (2.11), then A has row sum vector $R = (r_1, r_2, \ldots, r_n)$. The vector

$$H = (h_1, h_2, \ldots, h_n) = (r_1 - 0, r_2 - 1, \ldots, r_n - (n - 1)),$$

whose components are given by $r_i - (i-1)$ $(i = 1, 2, \ldots, n)$, is the *normalized row sum vector* of A. The vector R is nondecreasing if and only if $h_{i-1} - h_i \leq 1$ $(i = 2, 3, \ldots, n)$. Since $h_1 = r_1$, we have $h_1 \geq 0$.

The following theorem [10] is a restatement of Theorem 2.2.2 in terms of the normalized row sum vector.

Theorem 2.2.3 *Let $H = (h_1, h_2, \ldots, h_n)$ be a nonnegative vector satisfying*

$$h_1 \geq 0 \text{ and } h_{i-1} - h_i \leq 1 \quad (i = 2, 3, \ldots, n).$$

Then H is the normalized row sum vector of a tournament matrix if and only if

$$h_1 + h_2 + \cdots + h_k \geq 0 \quad (k = 1, 2, \ldots, n) \tag{2.12}$$

with equality for $k = n$. The row sum vector of such a tournament matrix is equal to $R = (r_1, r_2, \ldots, r_n)$ where $r_i = h_i + (i - 1)$ $(i = 1, 2, \ldots, n)$. $\square$

Let p and n be positive integers. Instead of a round-robin tournament in which each of n teams $x_1, x_2, \ldots, x_n$ plays one game against every other team, we may also consider a round-robin tournament in which each team plays p games against every other team. A nonnegative integral matrix $A = [a_{ij}]$ of order n is a *p-tournament matrix* provided

$$a_{ii} = 0 \quad (i = 1, 2, \ldots, n) \text{ and } a_{ij} + a_{ji} = p \quad (1 \leq i < j \leq n),$$

that is, provided

$$A + A^{\mathrm{T}} = p(J_n - I_n).$$

The element a_{ij} records the number of games in which team x_i defeats team x_j $(1 \leq i, j \leq n; i \neq j)$. We have

$$\sum_{i=1}^{n} \sum_{j=1}^{n} a_{ij} = \sum_{1 \leq i < j \leq n} (a_{ij} + a_{ji}) = \sum_{1 \leq i < j \leq n} p = p \binom{n}{2}.$$

We let $\mathcal{T}(R; p)$ denote the *class of all p-tournament matrices with a prescribed row sum vector* $R = (r_1, r_2, \ldots, r_n)$. Thus $\mathcal{T}(R; 1) = \mathcal{T}(R)$. The proof of Theorem 2.2.2 can easily be adapted to prove the following existence theorem for p-tournament matrices.

Theorem 2.2.4 *Let p and n be positive integers. Let $R = (r_1, r_2, \ldots, r_n)$ be a nondecreasing, nonnegative integral vector. Then $\mathcal{T}(R; p)$ is nonempty if and only if*

$$\sum_{i=1}^{k} r_i \geq p \binom{k}{2} \quad (k = 1, 2, \ldots, n), \text{ with equality when } k = n. \tag{2.13}$$

One can consider an even more general setting of a round-robin tournament in which there is a prescribed number p_{ij} of games to be played between each pair of teams $\{x_i, x_j\}$. Let $P = [p_{ij}]$ be a nonnegative integral matrix of order n with 0's on and below the main diagonal. Then a

P-tournament matrix is a nonnegative integral matrix $A = [a_{ij}]$ of order n satisfying

$$a_{ii} = 0 \quad (i = 1, 2, \ldots, n) \text{ and } a_{ij} + a_{ji} = p_{ij} \quad (1 \le i < j \le n),$$

equivalently, $A + A^{\mathrm{T}} = P + P^{\mathrm{T}}$. We have

$$\sum_{i=1}^{n} \sum_{j=1}^{n} a_{ij} = \sum_{1 \le i < j \le n} (a_{ij} + a_{ji}) = \sum_{1 \le i < j \le n} p_{ij}.$$

We denote the class of all P-tournament matrices with prescribed row sum vector $R = (r_1, r_2, \ldots, r_n)$ by $\mathcal{T}(R; P)$.

The following theorem is due to Cruse [7]. Since we shall not make use of this theorem, we do not include a proof.

Theorem 2.2.5 *Let n be a positive integer and let $P = [p_{ij}]$ be a nonnegative integral matrix of order n with 0's on and below its main diagonal. Let $R = (r_1, r_2, \ldots, r_n)$ be a nonnegative integral vector. Then $\mathcal{T}(R; P)$ is nonempty if and only if*

$$\sum_{i \in K} r_i \ge \sum_{i \in K} \sum_{j \in K} p_{ij} \quad (K \subseteq \{1, 2, \ldots, n\}), \tag{2.14}$$

with equality when $K = \{1, 2, \ldots, n\}$.

For generalized tournament matrices we have the following theorem of Moon [18].

Theorem 2.2.6 *Let n be a positive integer and let $R = (r_1, r_2, \ldots, r_n)$ be a nondecreasing, nonnegative vector. Then $\mathcal{T}^{\mathrm{g}}(R)$ is nonempty if and only if (2.10) holds.*

Proof. We have already verified that the conditions (2.10) are satisfied if there is a matrix in $\mathcal{T}^{\mathrm{g}}(R)$. Now suppose that the conditions (2.10) hold. We first observe the following. Let $S = (s_1, s_2, \ldots, s_n)$ where

$$s_i = n - 1 - r_i \quad (i = 1, 2, \ldots, n). \tag{2.15}$$

Suppose that there exists a matrix $A = [a_{ij}]$ of order n with row sum vector R and column sum vector S such that

$$a_{ii} = 0 \quad (i = 1, 2, \ldots, n) \text{ and } 0 \le a_{ij} \le 1 \quad (i, j = 1, 2, \ldots, n; i \ne j).$$

Let

$$B = [b_{ij}] = \frac{1}{2}(A - A^{\mathrm{T}} + J_n - I_n).$$

Then

$$b_{ii} = 0 \quad (i = 1, 2, \ldots, n),$$

$$0 \le b_{ij} = \frac{a_{ij} - a_{ji} + 1}{2} \le 1 \quad (i \ne j),$$

and

$$b_{ij} + b_{ji} = 1 \quad (i \ne j).$$

In addition, the sum of the elements in row i of B equals

$$\frac{r_i - s_i + n - 1}{2} = r_i.$$

Thus B is in $\mathcal{T}^{\mathrm{g}}(R)$. It thus suffices to show that (2.10) implies the existence of such a matrix A.

Let $C = [c_{ij}] = J_n - I_n$. Then we seek a matrix $A = [a_{ij}]$ of order n such that

$$0 \le a_{ij} \le c_{ij} \quad (i, j = 1, 2, \ldots, n),$$

$$\sum_{j=1}^{n} a_{ij} = r_i \quad (i = 1, 2, \ldots, n), \text{ and}$$

$$\sum_{i=1}^{n} a_{ij} = s_j \quad (j = 1, 2, \ldots, n).$$

Since R is nondecreasing and S is nonincreasing, it follows from Corollary 1.4.2 that such a matrix A exists if and only if

$$\sum_{i=k+1}^{n} \sum_{j=1}^{l} c_{ij} \ge \sum_{i=1}^{k} r_i - \sum_{j=1}^{l} s_j \ge 0 \quad (k = 1, 2, \ldots, n; j = 1, 2, \ldots, n). \quad (2.16)$$

First assume that $k \le l$. Then using (2.15) and the definition of C, we see that (2.16) is equivalent to

$$l(n - k) - (l - k) + \sum_{i=1}^{k} r_i - \sum_{j=1}^{l} (n - 1 - r_j) \ge 0,$$

that is,

$$-kl + k + \sum_{i=1}^{k} r_i + \sum_{i=1}^{l} r_i \ge 0. \quad (2.17)$$

But by the Landau inequalities,

$$-kl + k + \sum_{i=1}^{k} r_i + \sum_{i=1}^{l} r_i \ge -kl + k + \binom{k}{2} + \binom{l}{2}$$

$$= \frac{(l - k)(l - k + 1)}{2} \ge 0,$$

and hence (2.17) holds.

Now assume that $k \geq l + 1$. In a similar way we get that (2.16) is equivalent to

$$l(n-k) + \sum_{i=1}^{k} r_i - \sum_{j=1}^{l}(n-1-r_j) \geq 0,$$

that is,

$$-kl + l + \sum_{i=1}^{k} r_i + \sum_{j=1}^{l} r_j \geq 0. \tag{2.18}$$

Using Landau's inequalities again, we get

$$-kl + l + \sum_{i=1}^{k} r_i + \sum_{j=1}^{l} r_j \geq -kl + l + \binom{k}{2} + \binom{l}{2}$$

$$= \frac{(k-l)(k-(l+1))}{2} \geq 0,$$

and hence (2.18) holds. Thus the desired matrix exists and the proof of the theorem is complete. □

2.3 Symmetric Matrices

Let $R = (r_1, r_2, \ldots, r_n)$ be a nonnegative vector. Also let p be a positive integer. We define several classes of symmetric matrices with row sum vector R.

The class of all symmetric, nonnegative matrices with row sum vector R is denoted by $\mathcal{N}(R)$. If R is an integral vector, the class of all symmetric, nonnegative integral matrices of order n with row sum vector R is denoted by $\mathcal{Z}^+(R)$. The class $\mathcal{Z}^+(R; p)$ is the class of all symmetric, nonnegative integral matrices with row sum vector R, each of whose elements does not exceed p. The subclasses of $\mathcal{N}(R)$, $\mathcal{Z}^+(R)$, and $\mathcal{Z}^+(R; p)$ consisting of those matrices with only 0's on the main diagonal are denoted, respectively, by $\mathcal{N}(R)_0$, $\mathcal{Z}^+(R)_0$, and $\mathcal{Z}^+(R; p)_0$. If $p = 1$, we use $\mathcal{A}(R)$ instead of $\mathcal{Z}^+(R; 1)$, and $\mathcal{A}(R)_0$ instead of $\mathcal{Z}^+(R; 1)_0$. Thus $\mathcal{A}(R)$ is the set of adjacency matrices of graphs, with a loop permitted at each vertex, on n vertices with degree sequence $(r_1, r_2, \ldots, r_n)$, and $\mathcal{A}_0(R)$ is the set of adjacency matrices of (simple) graphs on n vertices with degree sequence R. The classes $\mathcal{N}(R)$ and $\mathcal{N}(R)_0$ are convex polytopes and will be investigated in Chapter 8.

Given a nonnegative vector $R \neq 0$, the diagonal matrix

$$\begin{bmatrix} r_1 & 0 & \cdots & 0 \\ 0 & r_2 & \cdots & 0 \\ \vdots & \vdots & \ddots & \vdots \\ 0 & 0 & \cdots & r_n \end{bmatrix}$$

is in $\mathcal{N}(R)$, and hence both $\mathcal{N}(R)$ and $\mathcal{Z}^+(R)$ are nonempty. The matrix $A = [a_{ij}]$ of order n with

$$a_{ij} = \frac{r_i r_j}{r_1 + r_2 + \cdots + r_n} \quad (i, j = 1, 2, \ldots, n)$$

is also a matrix in $\mathcal{N}(R)$, with all of its entries positive if R is a positive vector. If we insist on zero trace, then the classes $\mathcal{N}(R)_0$ and $\mathcal{Z}^+(R)_0$ may be empty. For instance, there is no symmetric, nonnegative matrix with zero trace with row sum vector $R = (1, 2)$.[3]

In investigating the nonemptiness of the above classes, there is no loss in generality in assuming that R is nonincreasing. This is because by replacing a symmetric matrix A with row sum vector R by $P^T A P$ for some permutation matrix P, we obtain a symmetric matrix with row sum vector RP, and PAP^T has zero trace if and only if A does.

Theorem 2.3.1 *Let $n \geq 2$ be an integer and let $R = (r_1, r_2, \ldots, r_n)$ be a nonincreasing, nonnegative vector. Then there exists a matrix in $\mathcal{N}(R)_0$ if and only if*

$$r_1 \leq r_2 + r_3 + \cdots + r_n. \tag{2.19}$$

Proof. First assume that there is a matrix in $\mathcal{N}(R)_0$. Then

$$r_1 = a_{12} + a_{13} + \cdots + a_{1n} = a_{21} + a_{31} + \cdots + a_{n1}$$
$$\leq r_2 + r_3 + \cdots + r_n.$$

Thus (2.19) holds. Now assume that (2.19) holds. If $n = 2$, then $r_1 = r_2$ and

$$\begin{bmatrix} 0 & r_1 \\ r_1 & 0 \end{bmatrix}$$

is the unique matrix in $\mathcal{N}(R)_0$. Now suppose that $n \geq 3$. It suffices to show that there is a nonnegative matrix B of order n with row and column sum vector equal to R and zero trace. This is because the matrix $(B + B^T)/2$ is then in $\mathcal{N}(R)_0$. We now apply Corollary 1.4.2 with $S = R$ and with $c_{ii} = 0$ $(i = 1, 2, \ldots, n)$ and $c_{ij} = +\infty$ (or a very large positive number) $(i, j = 1, 2, \ldots, n; i \neq j)$. By that corollary, since (1.6) is automatically satisfied unless $I = J = \{i\}$ for some i, such a matrix exists provided

$$0 \geq r_i - \sum_{j \neq i} r_j \quad (i = 1, 2, \ldots, n).$$

Since R is nonincreasing, this is equivalent to (2.19). $\qquad\qquad\square$

In case R is an integral vector, to guarantee a symmetric, nonnegative integral matrix with row sum vector R, we must assume that $r_1 + r_2 + \cdots + r_n$ is an even integer. The next theorem follows from Chungphaisan's theorem, Theorem 1.5.4, by choosing p to be a sufficiently large integer.

---

[3]Of course, there is no symmetric, nonnegative matrix of order 1 with zero trace and row sum vector $R = (1)$.

Theorem 2.3.2 *Let $n \geq 2$ be an integer and let $R = (r_1, r_2, \ldots, r_n)$ be a nonincreasing, nonnegative integral vector. Then there exists a matrix in $\mathcal{Z}^+(R)_0$ if and only if $r_1 + r_2 + \cdots + r_n$ is an even integer and the inequality (2.19) holds.* $\square$

Necessary and sufficient conditions for the nonemptiness of the classes $\mathcal{A}(R; p)$, $\mathcal{A}(R)$, $\mathcal{A}(R; p)_0$, and $\mathcal{A}(R)_0$ are given in Section 1.5 (Theorem 1.5.2, Corollary 1.5.3, Theorem 1.5.4, and Theorem 1.5.5, respectively).

References

[1] C.M. Bang and H. Sharp, An elementary proof of Moon's theorem on generalized tournaments, *J. Combin. Theory Ser. B*, **22** (1977), 299–301.

[2] C.M. Bang and H. Sharp, Score vectors of tournaments, *J. Combin. Theory Ser. B*, **26** (1979), 81–84.

[3] R.A. Brualdi, Matrices of zeros and ones with fixed row and column sum vectors, *Linear Algebra Appl.*, **33** (1980), 159–231.

[4] R.A. Brualdi, Short proofs of the Gale/Ryser and Ford/Fulkerson characterization of the row and column sum vectors of $(0,1)$-matrices, *Math. Inequalities Appl.*, **4** (2001), 157–159.

[5] R.A. Brualdi and H.J. Ryser, *Combinatorial Matrix Theory*, Cambridge U. Press, Cambridge, 1991.

[6] R.A. Brualdi and J. Shen, Landau's inequalities for tournament scores and a short proof of a theorem on transitive sub-tournaments, *J. Graph Theory*, (2001), 244–254.

[7] A.B. Cruse, On linear programming duality and Landau's characterization of tournament scores, preprint.

[8] L. R. Ford, Jr. and D. R. Fulkerson, *Flows in Networks*, Princeton U. Press, Princeton, 1962.

[9] D.R. Fulkerson, Zero–one matrices with zero trace, *Pacific J. Math.*, **10** (1960), 831–836.

[10] D.R. Fulkerson, Upsets in round robin tournaments, *Canad. J. Math.*, **17** (1965), 957–969.

[11] D. Gale, A theorem on flows in networks, *Pacific. J. Math.*, **7** (1957), 1073–1082.

[12] A. Khintchine, Über eine Ungleichung, *Mat. Sb.*, **39** (1932), 180–189.

[13] H. G. Landau, On dominance relations and the structure of animal societies. III. The condition for a score structure, *Bull. Math. Biophys.* **15** (1953), 143–148.

[14] E.S. Mahmoodian, A critical case method of proof in combinatorial mathematics, *Bull. Iranian Math. Soc.*, No. 8 (1978), 1L–26L.

[15] F. Matúš and A. Tuzar, Short proofs of Khintchine-type inequalities for zero–one matrices, *J. Combin. Theory. Ser. A*, **59** (1992), 155–159.

[16] T.S. Michael, The structure matrix and a generalization of Ryser's maximum term rank formula, *Linear Algebra Appl.*, **145** (1991), 21–31.

[17] L. Mirsky, *Transversal Theory*, Academic Press, New York, 1971.

[18] J.W. Moon, An extension of Landau's theorem on tournaments, *Pacific. J. Math.*, **13** (1963), 1343–1345.

[19] J.W. Moon, *Topics on Tournaments*, Holt, Rinehart, and Winston, New York, 1968.

[20] H. J. Ryser, *Combinatorial Mathematics*, Carus Math. Monograph #14, Math. Assoc. of America, 1963.

[21] C. Thomassen, Landau's characterization of tournament score sequences, *The Theory and Application of Graphs (Kalamazoo, Michigan, 1980)*, Wiley, New York, 1963, 589–591.

[22] Y. R. Wang, Characterizations of binary patterns and their projections, *IEEE Trans. Computers*, **C-24** (1975), 1032–1035.

3

The Class $\mathcal{A}(R, S)$ of (0,1)-Matrices

In this chapter we study the class $\mathcal{A}(R, S)$ of all $(0, 1)$-matrices with a prescribed row sum vector R and a prescribed column sum vector S. We study the structure of this class and the values (in particular, the extreme values) of certain combinatorial and linear algebraic parameters on it. In the next chapter we investigate the existence of matrices in $\mathcal{A}(R, S)$ with special combinatorial structures and the number of matrices in $\mathcal{A}(R, S)$. In Chapter 6 we study properties of a naturally defined graph whose vertices are the matrices in $\mathcal{A}(R, S)$.

3.1 A Special Matrix in $\mathcal{A}(R, S)$

Let $R = (r_1, r_2, \ldots, r_m)$ and $S = (s_1, s_2, \ldots, s_n)$ be nonnegative integral vectors. In Chapter 2 we gave two necessary and sufficient conditions for the nonemptiness of $\mathcal{A}(R, S)$. In this section we shall construct a matrix in $\mathcal{A}(R, S)$ which plays a special role in subsequent sections. In our investigations there is usually no loss in generality in assuming that R and S are nonincreasing vectors, and we generally make this assumption. This is because for permutation matrices P and Q of orders m and n, respectively,

$$\mathcal{A}(RP, SQ) = \{PAQ : A \in \mathcal{A}(R, S)\}.$$

As a consequence, combinatorial (and many other) properties of $\mathcal{A}(R, S)$ directly translate to properties of $\mathcal{A}(PR, SQ)$, and vice versa. We may also assume that the r_i and s_j are positive, since otherwise each matrix in $\mathcal{A}(R, S)$ has a row of all 0's or a column of all 0's.

Assume that $n \geq r_1 \geq r_2 \geq \cdots \geq r_m > 0$ and that $s_1 \geq s_2 \geq \cdots \geq s_n > 0$. Let $R^* = (r_1^*, r_2^*, \ldots, r_n^*)$ be the conjugate of R. Then R^* is a nonincreasing vector. There is a unique matrix $A(R, n)$ in $\mathcal{A}(R, R^*)$, and the 1's in each row of this matrix are in the leftmost positions.[1] Since R is nonincreasing, the 1's in each column of $A(R, n)$ occur in the topmost positions.

Now also assume that $S \preceq R^*$ so that $\mathcal{A}(R, S)$ is nonempty, although we shall not make use of this fact.[2] The construction in the algorithm below was used by both Ryser [59] and Gale [25] in their original derivations that $S \preceq R^*$ implies that $\mathcal{A}(R, S) \neq \oslash$, except that both Ryser and Gale constructed the columns according to decreasing column sums. While Gale placed the 1's in each column in those positions corresponding to the largest row sums, giving preference to the topmost positions in case of ties, it was Fulkerson and Ryser [23] who recognized that giving preference to the bottommost positions in case of ties leads to a matrix in $\mathcal{A}(R, S)$ with special properties.

Gale–Ryser Algorithm for Constructing a Matrix in $\mathcal{A}(R, S)$

(1) Begin with the matrix $A_n = A(R, n)$ of size m by n and an m by 0 empty matrix $\tilde{A}_n$.

(2) For $k = n, n - 1, n - 2, \ldots, 1$, do:

Shift to column k the final 1's in those s_k rows of A_k with the largest sum, with preference given to the lowest rows (those with the largest index) in case of ties. This results in a matrix

$$\left[A_{k-1} \middle| \tilde{A}_{n-k+1} \right],$$

where A_{k-1} has $k - 1$ columns.

Output $\tilde{A} = \tilde{A}_n$.

Theorem 3.1.1 *If $S \preceq R^*$, then the Gale–Ryser algorithm can be completed and the constructed matrix $\tilde{A}$ has row sum vector R and column sum vector S.*

Proof. If $n = 1$, then $\tilde{A} = A_1 = A(R, 1)$, since $R^* \preceq R$ implies that $s_1 = r_1 + r_2 + \cdots + r_m$. We now assume that $n > 1$. The assumption that $S \preceq R^*$ implies that

$$s_1 + s_2 + \cdots + s_{n-1} \leq r_1^* + r_2^* + \cdots + r_{n-1}^*,$$

$$s_1 + s_2 + \cdots + s_n = r_1^* + r_2^* + \cdots + r_n^*,$$

---

[1]The matrix $A(R, n)$ depends only on R and the number of its columns. As pointed out in the definition of conjugate given in Chapter 1, the number of components of R^* can be any integer $t \geq r_1$. We choose the number of columns to be the size n of S.

[2]Thus we will obtain an independent, algorithmic proof of the sufficient condition of Gale and Ryser for the nonemptiness of $\mathcal{A}(R, S)$.

and hence that

$$r_n^* \leq s_n. \tag{3.1}$$

We also have $s_1 \leq r_1^*$, $s_1 \geq s_n$, and hence

$$r_1^* \geq s_n. \tag{3.2}$$

It now follows from (3.1) and (3.2) that Step (1) of the Gale–Ryser algorithm can be carried out. By construction, the 1's in each row of the matrix A_{n-1} are in the leftmost positions, and A_{n-1} has a nonincreasing row sum vector $U = (u_1, u_2, \ldots, u_m)$ and a nonincreasing column sum vector equal to the conjugate $U^* = (u_1^*, u_2^*, \ldots, u_{n-1}^*)$ of U. Thus A_{n-1} equals the matrix $A(U, n-1)$. We now show that $(s_1, s_2, \ldots, s_{n-1}) \preceq U^*$.

We first observe that

$$s_1 + s_2 + \cdots + s_{n-1} = u_1^* + u_2^* + \cdots + u_{n-1}^*. \tag{3.3}$$

Let k be an integer with $1 \leq k \leq n-1$. Then

$$\sum_{i=1}^{k} u_i^* = \sum_{i=1}^{m} \min\{k, u_i\}$$

$$= \sum_{i=1}^{s_n} \min\{k, r_i - 1\} + \sum_{i=s_n+1}^{m} \min\{k, r_i\}. \tag{3.4}$$

We consider two cases.

Case 1: $r_{k+1}^* \geq s_n$. Since R is nonincreasing, we must have $r_i \geq k+1$ for $i = 1, 2, \ldots, s_n$. Therefore

$$\min\{k, r_i - 1\} = k = \min\{k, r_i\} \quad (i = 1, 2, \ldots, s_n),$$

and hence

$$\sum_{i=i}^{k} u_i^* = \sum_{i=1}^{m} \min\{k, r_i\} = \sum_{i=1}^{k} r_i^* \geq \sum_{i=1}^{k} s_i.$$

Case 2: $r_{k+1}^* < s_n$. Since the 1's in each column of $A(R, n)$ occur in the topmost positions, we have

$$r_i \geq k+1 \text{ and } \min\{k, r_i - 1\} = k = \min\{k, r_i\} \quad (i = 1, 2, \ldots, r_{k+1}^*),$$

and

$$r_i \leq k \text{ and } \min\{k, r_i - 1\} = \min\{k, r_i\} - 1 \quad (i = r_{k+1}^* + 1, \ldots, s_n).$$

It now follows from (3.4) that

$$\sum_{i=1}^{k} u_i^* = \sum_{i=1}^{m} \min\{k, r_i\} + r_{k+1}^* - s_n$$

$$= \sum_{i=1}^{k+1} r_i^* - s_n$$

$$\geq \sum_{i=1}^{k+1} s_i - s_n \quad (\text{since } S \preceq R^*)$$

$$\geq \sum_{i=1}^{k} s_i \quad (\text{since } S \text{ is nonincreasing}).$$

Using (3.3), we conclude that $(s_1, s_2, \ldots, s_{n-1}) \preceq U^*$, and the theorem follows. $\qquad \square$

Example. We illustrate the Gale–Ryser algorithm for $R = (4, 4, 3, 3, 2)$ and $S = (4, 3, 3, 3, 3)$. We have $R^* = (5, 5, 4, 2, 0)$. The following matrices $[A_i | \tilde{A}_{n-i}]$ are produced:

$$\begin{bmatrix} 1 & 1 & 1 & 1 & 0 \\ 1 & 1 & 1 & 1 & 0 \\ 1 & 1 & 1 & 0 & 0 \\ 1 & 1 & 1 & 0 & 0 \\ 1 & 1 & 0 & 0 & 0 \end{bmatrix}, \quad \left[\begin{array}{ccc|cc} 1 & 1 & 1 & 0 & 1 \\ 1 & 1 & 1 & 0 & 1 \\ 1 & 1 & 1 & 0 & 0 \\ 1 & 1 & 0 & 0 & 1 \\ 1 & 1 & 0 & 0 & 0 \end{array}\right], \quad \left[\begin{array}{ccc|cc} 1 & 1 & 0 & 1 & 1 \\ 1 & 1 & 0 & 1 & 1 \\ 1 & 1 & 0 & 1 & 0 \\ 1 & 1 & 0 & 0 & 1 \\ 1 & 1 & 0 & 0 & 0 \end{array}\right],$$

$$\left[\begin{array}{cc|ccc} 1 & 1 & 0 & 1 & 1 \\ 1 & 1 & 0 & 1 & 1 \\ 1 & 0 & 1 & 1 & 0 \\ 1 & 0 & 1 & 0 & 1 \\ 1 & 0 & 1 & 0 & 0 \end{array}\right], \quad \left[\begin{array}{c|cccc} 1 & 1 & 0 & 1 & 1 \\ 1 & 1 & 0 & 1 & 1 \\ 1 & 0 & 1 & 1 & 0 \\ 1 & 0 & 1 & 0 & 1 \\ 0 & 1 & 1 & 0 & 0 \end{array}\right], \quad \begin{bmatrix} 1 & 1 & 0 & 1 & 1 \\ 1 & 1 & 0 & 1 & 1 \\ 1 & 0 & 1 & 1 & 0 \\ 1 & 0 & 1 & 0 & 1 \\ 0 & 1 & 1 & 0 & 0 \end{bmatrix}.$$

The final matrix is the matrix $\tilde{A}$ in $\mathcal{A}(R, S)$. $\qquad \square$

The matrix $\tilde{A}$ has a simple characterization among all matrices in $\mathcal{A}(R, S)$. Let A be an m by n matrix. Let R_k denote the transpose of the row sum vector of the submatrix $A[\cdot, \{1, 2, \ldots, k\}]$ of A formed by its first k columns, $(k = 1, 2, \ldots, n)$. The matrix

$$M(A) = \begin{bmatrix} R_1 & R_2 & \cdots & R_n \end{bmatrix}$$

of size m by n is the *partial row sum matrix* of A. It is a consequence of the construction of $\tilde{A}$ that the columns of the partial row sum matrix of $\tilde{A}$ are nonincreasing vectors. The following theorem is from [6].

Theorem 3.1.2 *Let R and S be nonincreasing vectors for which $\mathcal{A}(R, S)$ is nonempty. Let A' be a matrix in $\mathcal{A}(R, S)$ and let*

$$M(A') = \begin{bmatrix} R'_1 & R'_2 & \cdots & R'_n \end{bmatrix}$$

be the partial row sum matrix of A'. Then $A' = \tilde{A}$ if and only if

(i) *each of the row sum vectors R'_i is nonincreasing, and*

(ii) $R'_i \preceq R_i$ $(i = 1, 2, \ldots, n)$ *for the partial row sum matrix*

$$M(A) = \begin{bmatrix} R_1 & R_2 & \cdots & R_n \end{bmatrix}$$

of each matrix A in $\mathcal{A}(R, S)$.

Proof. Let the partial row sum matrix of $\tilde{A}$ be

$$M(\tilde{A}) = \begin{bmatrix} \tilde{R}_1 & \tilde{R}_2 & \cdots & \tilde{R}_n \end{bmatrix}.$$

As already noted, each of the vectors $\tilde{R}_i$ is nonincreasing. We first show that $\tilde{R}_i \preceq R_i$ $(i = 1, 2, \ldots, n)$ by induction. First observe that $R_n = \tilde{R}_n$ so that $\tilde{R}_n \preceq R_n$ holds trivially. Suppose that i is an integer with $2 \leq i \leq n$ and that $\tilde{R}_i \preceq R_i$ holds. The vector R_{i-1} is obtained from R_i by reducing s_i of its components by 1. ¿From the construction of $\tilde{A}$ we see that a rearrangement of $\tilde{R}_{i-1}$ is obtained from $\tilde{R}_i$ by reducing the first s_i components of $\tilde{R}_i$ by 1. Hence it follows from Theorem 1.4.2 of Chapter 1 that $\tilde{R}_{i-1} \preceq R_{i-1}$, and this completes the induction.

Now suppose that R'_i is nonincreasing and that $R'_i \preceq R_i$ $(i = 1, 2, \ldots, n)$ for each matrix A in $\mathcal{A}(R, S)$. By above $\tilde{R}_i \preceq R'_i$ for each i. Since $\tilde{R}_i$ and R'_i are both monotone, it follows that $\tilde{R}_i = R'_i$ for $i = 1, 2, \ldots, n$ and hence that $A' = \tilde{A}$. $\square$

In contrast to the characterization of $\tilde{A}$ in Theorem 3.1.2 there need not exist a matrix in $\mathcal{A}(R, S)$ such that for each i with $1 \leq i \leq n$, the ith partial row sum vector is nonincreasing and majorizes the ith partial row vector of every matrix in $\mathcal{A}(R, S)$. The following example is from [6].

Example. Let $R = (3, 3, 3, 2, 1)$ and $S = (4, 2, 2, 2, 2)$. Let

$$A_1 = \begin{bmatrix} 1 & 1 & 1 & 0 & 0 \\ 1 & 1 & 1 & 0 & 0 \\ 1 & 0 & 0 & 1 & 1 \\ 1 & 0 & 0 & 1 & 0 \\ 0 & 0 & 0 & 0 & 1 \end{bmatrix} \text{ and } A_2 = \begin{bmatrix} 1 & 1 & 1 & 0 & 0 \\ 1 & 1 & 0 & 1 & 0 \\ 1 & 0 & 1 & 1 & 0 \\ 1 & 0 & 0 & 0 & 1 \\ 0 & 0 & 0 & 0 & 1 \end{bmatrix},$$

whose partial row sum matrices are, respectively,

$$M(A_1) = \begin{bmatrix} 1 & 2 & 3 & 3 & 3 \\ 1 & 2 & 3 & 3 & 3 \\ 1 & 1 & 1 & 2 & 3 \\ 1 & 1 & 1 & 2 & 2 \\ 0 & 0 & 0 & 0 & 1 \end{bmatrix} \text{ and } M(A_2) = \begin{bmatrix} 1 & 2 & 3 & 3 & 3 \\ 1 & 2 & 2 & 3 & 3 \\ 1 & 1 & 2 & 3 & 3 \\ 1 & 1 & 1 & 1 & 2 \\ 0 & 0 & 0 & 0 & 1 \end{bmatrix}.$$

Each of the column vectors of $M(A_1)$ and $M(A_2)$ is nonincreasing. We also observe that the third partial row sum vector of A_2 is majorized by the third partial row sum vector of A_1, and that the fourth partial row sum vector of A_1 is majorized by the fourth partial row sum vector of A_2.

Let B be any matrix in $\mathcal{A}(R, S)$ such that each partial row sum vector of A_1 is majorized by the corresponding partial row sum vector of B. Considering the third partial row sum vector of A_1 (i.e. the third column of $M(A_1)$), we conclude that the third partial row sum vector of B has at least two components equal to 3. Therefore, without loss of generality, we may assume that

$$B = \begin{bmatrix} 1 & 1 & 1 & 0 & 0 \\ 1 & 1 & 1 & 0 & 0 \\ * & 0 & 0 & * & * \\ * & 0 & 0 & * & * \\ * & 0 & 0 & * & * \end{bmatrix},$$

and then, using the fact that the third row contains exactly three 1's, we conclude that

$$B = \begin{bmatrix} 1 & 1 & 1 & 0 & 0 \\ 1 & 1 & 1 & 0 & 0 \\ 1 & 0 & 0 & 1 & 1 \\ * & 0 & 0 & * & * \\ * & 0 & 0 & * & * \end{bmatrix}.$$

Since the fourth partial row sum vector of B majorizes the fourth column of $M(A_1)$, we now see that $B = A_1$. In a similar way one can show that if B is any matrix in $\mathcal{A}(R, S)$ with the property that each partial row sum vector of A_2 is majorized by the corresponding partial row sum vector of B, then $B = A_2$. $\square$

3.2   Interchanges

In this section we take $R = (r_1, r_2, \ldots, r_m)$ and $S = (s_1, s_2, \ldots, s_n)$ to be nonnegative integral vectors such that $\mathcal{A}(R, S)$ is nonempty and show how, starting from any single matrix in $\mathcal{A}(R, S)$, the entire class $\mathcal{A}(R, S)$ can be generated by simple transformations. First we digress to introduce some additional simple ideas from the theory of graphs.

Let A and B be m by n matrices in $\mathcal{A}(R, S)$. The matrix $A - B$ is a $(0, \pm 1)$-matrix with all row and column sums equal to 0. This motivates the following discussion.

Let $C = [c_{ij}]$ be a $(0, \pm 1)$-matrix of size m by n. We define a *bipartite digraph* $\Gamma(C)$ as follows. We take two disjoint sets $X = \{x_1, x_2, \ldots, x_m\}$ and $Y = \{y_1, y_2, \ldots, y_n\}$ of m and n elements, respectively, and let $X \cup Y$ be the set of vertices of $\Gamma(C)$. In $\Gamma(C)$ there is an arc (x_i, y_j) from x_i to y_j provided $c_{ij} = 1$, and an arc from y_j to x_i provided $c_{ij} = -1$. There are no other arcs in $\Gamma(C)$. Now assume that the matrix C is *balanced*, that is, that each row and column sum of C equals 0. Then for each vertex, the number of arcs entering the vertex (its *indegree*) equals the number of arcs exiting the vertex (its *outdegree*). We call a digraph *balanced* provided the indegree of each vertex equals its outdegree.

Lemma 3.2.1 *Let D be a balanced digraph. Then the arcs of D can be partitioned into directed cycles. If D is also a bipartite digraph, then these directed cycles have even length.*

Proof. Since the indegree of a vertex of D equals its outdegree, starting at any vertex u of D and iteratively choosing arcs to construct a directed trail beginning at u, we eventually return to u and obtain a closed directed trail. The arcs of a closed directed trail can be partitioned into sets each of which determines a directed cycle. Removing the arcs of one (or all) of these directed cycles from D leaves a balanced digraph. Hence the first assertion follows by induction on the number of arcs of a balanced digraph. Every directed cycle of a bipartite digraph must have even length. □

Let C be a nonzero balanced $(0, \pm 1)$-matrix of size m by n. By Lemma 3.2.1, there are directed cycles $\gamma_1, \gamma_2, \ldots, \gamma_q$ partitioning the arcs of $\Gamma(C)$. We may view each directed cycle γ of $\Gamma(C)$ as a bipartite digraph with the same set of vertices as $\Gamma(C)$. Then γ is the bipartite digraph corresponding to a nonzero, balanced m by n $(0, \pm 1)$-matrix C_γ each of whose nonzero lines contains exactly two nonzero elements. We call such a matrix C_γ a *minimal balanced matrix*. Let $C_i = C_{\gamma_i}$ $(i = 1, 2, \ldots, q)$. Then

$$C = C_1 + C_2 + \cdots + C_q \tag{3.5}$$

where for $i \neq j$, the set of positions of the nonzero elements of C_i is disjoint from the set of positions of the nonzero elements of C_j. Such a decomposition (3.5) of C into minimal balanced matrices is called a *minimal balanced decomposition* of C.

A minimal balanced matrix has the property that its rows and columns can be permuted so that, for some integer $k \geq 2$, the resulting matrix $E = [e_{ij}]$ satisfies $e_{11} = e_{22} = \cdots = e_{kk} = 1$, and $e_{12} = \cdots = e_{k-1,k} = e_{k1} = -1$, and all other e_{ij} equal 0.

Example. The following example shows that a minimal balanced decomposition of a balanced matrix need not be unique. Consider the balanced matrix

$$C = \begin{bmatrix} 1 & -1 & -1 & 1 \\ 1 & 1 & -1 & -1 \\ -1 & 1 & 1 & -1 \\ -1 & -1 & 1 & 1 \end{bmatrix}.$$

Then

$$C = \begin{bmatrix} 1 & -1 & 0 & 0 \\ 0 & 1 & -1 & 0 \\ 0 & 0 & 1 & -1 \\ -1 & 0 & 0 & 1 \end{bmatrix} + \begin{bmatrix} 0 & 0 & -1 & 1 \\ 1 & 0 & 0 & -1 \\ -1 & 1 & 0 & 0 \\ 0 & -1 & 1 & 0 \end{bmatrix}$$

is a minimal balanced decomposition of C. Also

$$C = C_1 + C_2 + C_3 + C_4$$

is a minimal balanced decomposition, where

$$C_1 = \begin{bmatrix} 1 & -1 & 0 & 0 \\ 0 & 0 & 0 & 0 \\ -1 & 1 & 0 & 0 \\ 0 & 0 & 0 & 0 \end{bmatrix}, \quad C_2 = \begin{bmatrix} 0 & 0 & -1 & 1 \\ 0 & 0 & 0 & 0 \\ 0 & 0 & 1 & -1 \\ 0 & 0 & 0 & 0 \end{bmatrix},$$

$$C_3 = \begin{bmatrix} 0 & 0 & 0 & 0 \\ 1 & 0 & 0 & -1 \\ 0 & 0 & 0 & 0 \\ -1 & 0 & 0 & 1 \end{bmatrix}, \quad C_4 = \begin{bmatrix} 0 & 0 & 0 & 0 \\ 0 & 1 & -1 & 0 \\ 0 & 0 & 0 & 0 \\ 0 & -1 & 1 & 0 \end{bmatrix}.$$

$\square$

Let A and B be matrices in $\mathcal{A}(R, S)$. Applying the previous discussion to the balanced matrix $C = B - A$, we conclude that there exist minimal balanced matrices $C_1, C_2, \ldots, C_q$ whose nonzero elements are in pairwise disjoint positions, such that

$$B = A + C_1 + C_2 + \cdots + C_q.$$

A nonzero balanced matrix contains at least four nonzero elements. Balanced matrices $C = [c_{ij}]$ of size m by n with exactly four nonzero elements are obtained as follows. Choose distinct integers p and q with $1 \leq p, q \leq m$ and distinct integers k and l with $1 \leq k, l \leq n$, and set $c_{pk} = c_{ql} = 1$ and $c_{pl} = c_{qk} = -1$ and all other $c_{ij} = 0$. A balanced matrix with exactly four nonzero elements is called an *interchange matrix*. The matrices C_1, C_2, C_3, C_4 of order 4 in the previous example are interchange matrices. The negative of an interchange matrix is an interchange matrix.

Let A_1 and A_2 be matrices in $\mathcal{A}(R, S)$ such that $A_1 = A_2 + C$ where C is an interchange matrix. Then A_1 is obtained from A_2 by replacing a submatrix

$$A_2[\{p, q\}, \{k, l\}] = \begin{bmatrix} 1 & 0 \\ 0 & 1 \end{bmatrix}$$

of A_2 of order 2 with

$$\begin{bmatrix} 0 & 1 \\ 1 & 0 \end{bmatrix},$$

or vice versa. We say that A_1 is *obtained from A_2 by an interchange* or, more precisely, by a $(p, q; k, l)$-*interchange*. If A_1 is obtained from A_2 by a $(p, q; k, l)$-interchange, then A_2 is also obtained from A_1 by a $(p, q; k, l)$-interchange. A theorem of Ryser [59, 62] asserts that for each pair of matrices A, B in $\mathcal{A}(R, S)$, A can be obtained from B by a sequence of interchanges, that is, $A = B + E_1 + E_2 + \cdots + E_t$ where $E_1, E_2, \ldots, E_t$ are interchange matrices and $B + E_1 + \cdots + E_k$ is in $\mathcal{A}(R, S)$ for $i = 1, 2, \ldots, t$; these interchange matrices may have nonzero elements in overlapping positions. The proof of this assertion in [59] is inductive while that in [62] is obtained by showing that each matrix in $\mathcal{A}(R, S)$ can be obtained from the special matrix $\tilde{A}$ by a sequence of interchanges. We follow the direct constructive proof in [6].

Lemma 3.2.2 *Let $A = [a_{ij}]$ be a matrix in $\mathcal{A}(R, S)$. Let $C = [c_{ij}]$ be a minimal balanced matrix such that $A + C$ is in $\mathcal{A}(R, S)$, and let the number of nonzero elements of C be $2k$. Then $A + C$ can be obtained from A by a sequence of $k - 1$ interchanges.*

Proof. Let $B = A + C = [b_{ij}]$. We assume, without loss of generality, that $c_{11} = c_{22} = \cdots = c_{kk} = -1$ and $c_{12} = \cdots = c_{k-1,k} = c_{k1} = 1$. Thus

$$a_{11} = a_{22} = \cdots = a_{kk} = 1 \text{ and } a_{12} = \cdots = a_{k-1,k} = a_{k1} = 0,$$

and

$$b_{11} = b_{22} = \cdots = b_{kk} = 0 \text{ and } b_{12} = \cdots = b_{k-1,k} = b_{k1} = 1.$$

Since $a_{11} = 1$ and $a_{k1} = 0$, there exists an integer p with $2 \leq p \leq k$ such that

$$a_{11} = \cdots = a_{p-1,1} = 1 \text{ and } a_{p1} = 0.$$

The sequence of $(p - 1, p; 1, p)$-,$(p - 2, p - 1; 1, p - 1)$-,$\ldots, (1, 2; 1, 2)$-interchanges transforms A to a matrix A' in $\mathcal{A}(R, S)$ such that $B = A' + C'$ where C' is a minimal balanced matrix with exactly $2(k - p + 1)$ nonzero elements. Since $k - p + 1 \leq k - 1$, we may replace A by A' in the argument above and argue by induction that B can be obtained from A' by a sequence of $k - p + 1$ interchanges. Hence it follows that B can be obtained from A by a sequence of $(p - 1) + (k - p + 1) = k - 1$ interchanges $\qquad \square$

We now obtain a quantitative version of Ryser's theorem [6].

Theorem 3.2.3 *Let A and B be matrices in $\mathcal{A}(R, S)$, and let $B - A = C_1 + C_2 + \cdots + C_q$ be a minimal balanced decomposition of $B - A$. Let the number of nonzero elements in C_i equal $2k_i$ $(i = 1, 2, \ldots, q)$. Then B can be obtained from A by a sequence of $k_1 + k_2 + \cdots + k_q - q$ interchanges.*

Proof. The theorem follows from Lemma 3.2.2 by induction on the integer q. $\qquad \square$

Theorems 3.1.1 and 3.2.3 together imply a characterization of those pairs R and S for which $\mathcal{A}(R, S)$ contain a unique matrix [6, 62, 76].

Theorem 3.2.4 *Let $R = (r_1, r_2, \ldots, r_m)$ and $S = (s_1, s_2, \ldots, s_n)$ be nonincreasing, nonnegative integral vectors for which $\mathcal{A}(R, S)$ is nonempty. Then there is a unique matrix in $\mathcal{A}(R, S)$ if and only if S equals the conjugate R^* of R.*

Proof. First assume that $S = R^*$. Then the matrix $A(R, n)$ is in $\mathcal{A}(R, S)$. Since the 1's in $A(R, n)$ are in the leftmost and topmost positions, no interchange can be applied to $A(R, n)$. By Theorem 3.2.3 there can be no other matrix in $\mathcal{A}(R, S)$. Now assume that there is a unique matrix $B = [b_{ij}]$ in $\mathcal{A}(R, S)$. Suppose that in some row of B, say row p, a 0 precedes a 1. Let $b_{pk} = 0$ and $b_{pl} = 1$ where $k < l$. By hypothesis $s_k \geq s_l$ and hence there is a $q \neq p$ such that $b_{qk} = 1$ and $b_{ql} = 0$. An interchange now gives a matrix B' in $\mathcal{A}(R, S)$ with $B' \neq B$, contradicting the uniqueness of B. Hence the 1's in each row of B are in the leftmost positions implying that $B = A(R, n)$. □

Let A be an m by n $(0, 1)$-matrix with row sum vector R and column sum vector S. It follows from Theorem 3.2.4 that A is uniquely determined by its row sum vector and column sum vector[3] if and only if, after column permutations that ensure that S is nonincreasing, $S = R^*$.

The proofs of Lemma 3.2.2 and Theorem 3.2.3 provide an algorithm to transform a specified matrix A in $\mathcal{A}(R, S)$ to another specified matrix B in $\mathcal{A}(R, S)$. A minimal balanced decomposition of $B - A$ is easily found by a row and column scanning operation. Then the proof of Lemma 3.2.2 shows how to determine a sequence of interchange matrices that change A into B. To use this algorithm to transform A to B using the *smallest* number of interchanges we should find a minimal balanced decomposition of $B - A$ using the *largest* number of minimal balanced matrices.

If A and B are matrices of the same size, we define the (*combinatorial*) *distance*[4] $d(A, B)$ between A and B to be the number of nonzero elements of the matrix $A - B$, that is, the number of positions in which A and B differ. If A and B belong to $\mathcal{A}(R, S)$, we also define the *interchange distance*[5] $\iota(A, B)$ between A and B to be the smallest number of interchanges that transform A to B, and the *balanced separation* $q(A, B)$ between A and B to be the largest number of minimal balanced matrices in a minimal balanced decomposition of $A - B$. The following corollary is an immediate consequence of these definitions and Theorem 3.2.3.

Corollary 3.2.5 *If A and B are matrices in $\mathcal{A}(R, S)$, then*

$$\iota(A, B) \leq \frac{d(A, B)}{2} - q(A, B). \tag{3.6}$$

□

---

[3]Put another way, A can be uniquely *reconstructed* from its row and column sum vectors.

[4]This is a metric on the set of all real matrices of the same size.

[5]This is a metric on the class $\mathcal{A}(R, S)$.

A theorem of Walkup [76] (see also [6]) shows that equality holds in (3.6) and thus that the smallest number of interchanges that transforms a specified matrix A in $\mathcal{A}(R,S)$ to a specified matrix B in $\mathcal{A}(R,S)$ can be obtained by using a minimal cycle decomposition of $A - B$ as described in the proofs of Lemma 3.2.2 and Theorem 3.2.3.

Theorem 3.2.6 *Let A and B be matrices in $\mathcal{A}(R,S)$. Then*

$$\iota(A, B) = \frac{d(A, B)}{2} - q(A, B). \tag{3.7}$$

Proof. We prove the theorem by induction on the number $\iota = \iota(A, B)$ by showing that

$$\iota(A, B) \geq \frac{d(A, B)}{2} - q(A, B).$$

If $\iota = 1$, then $d(A, B) = 4$ and $q(A, B) = 1$. Hence (3.7) holds.

Assume that $\iota > 1$. There exist matrices $A_0 = A, A_1, \ldots, A_\iota = B$ in $\mathcal{A}(R,S)$ such that A_j can be obtained from A_{j-1} by an interchange $(j = 1, 2, \ldots, \iota)$. We have $\iota(A_1, B) = \iota - 1$, and hence by the induction hypothesis,

$$\iota - 1 = \iota(A_1, B) = \frac{d(A_1, B)}{2} - q(A_1, B).$$

Let $q_1 = q(A_1, B)$, and let $B - A_1 = C_1 + C_2 + \cdots + C_{q_1}$ be a minimal balanced decomposition of $B - A_1$. Let E be the interchange matrix $A_1 - A$. Let k be the number of matrices from among $C_1, C_2, \ldots, C_{q_1}$ that have a nonzero element in at least one of the four positions in which E has a nonzero element, and l be the number of such positions; we may assume that $C_1, \ldots, C_k$ are these minimal balanced matrices. We have $0 \leq k \leq l \leq 4$, since the nonzero positions of $C_1, C_2, \ldots, C_{q_1}$ are pairwise disjoint. The matrix $F = E + C_1 + \cdots + C_k$ is a balanced matrix and, if nonzero, is the sum of minimal balanced matrices $C_1' + C_2' + \cdots + C_{k'}'$. We have

$$B - A = F + C_{k+1} + \cdots + C_{q_1} = C_1' + \cdots + C_{k'}' + C_{k+1} + \cdots + C_{q_1},$$

the last being a minimal balanced decomposition of $B - A$ into $k' + (q_1 - k)$ minimal balanced matrices. Hence

$$q(A, B) \geq k' + (q_1 - k) = q(A_1, B) + (k' - k).$$

Case 1: $k = 0$. In this case $l = 0$, $F = E$, $d(A, B) = 4 + d(A_1, B)$, and $q(A, B) \geq q(A_1, B) + 1$. Thus

$$\frac{d(A, B)}{2} - q(A, B) \leq \frac{4 + d(A_1, B)}{2} - q(A_1, B) - 1$$
$$= 2 + (\iota - 1) - 1 = \iota.$$

Case 2: $k = 1$. We have that $1 \leq l \leq 4$ and

$$d(A, B) - d(A_1, B) = 3 - 1 = 2, \text{ if } l = 1,$$
$$d(A, B) - d(A_1, B) = 2 - 2 = 0, \text{ if } l = 2,$$
$$d(A, B) - d(A_1, B) = 1 - 3 = -2, \text{ if } l = 3,$$
$$d(A, B) - d(A_1, B) = 0 - 4 = -4, \text{ if } l = 4.$$

If $l = 1, 2$, or 3, we have $q(A, B) \geq q(A_1, B)$. If $l = 4$, we have $q(A, B) \geq q(A_1, B) - 1$. Hence either $d(A, B) \leq d(A_1, B) + 2$ and $q(A, B) \geq q(A_1, B)$, or $d(A, B) = d(A_1, B) - 4$ and $q(A, B) \geq q(A_1, B) - 1$. If $l = 1, 2$, or 3, we have

$$\frac{d(A, B)}{2} - q(A, B) \leq \frac{d(A_1, B) + 2}{2} - q(A_1, B)$$
$$= (\iota - 1) + 1 = \iota.$$

If $l = 4$, then

$$\frac{d(A, B)}{2} - q(A, B) \leq \frac{d(A_1, B) - 4}{2} - q(A_1, B) + 1$$
$$= (\iota - 1) - 2 + 1 = \iota - 2 \leq \iota.$$

Case 3: $k = 2$. We have $l = 2, 3$, or 4, $d(A, B) \leq d(A_1, B)$ and $q(A, B) \geq q(A_1, B) - 1$. Thus

$$\frac{d(A, B)}{2} - q(A, B) \leq \frac{d(A_1, B)}{2} - q(A_1, B) + 1$$
$$= (\iota - 1) + 1 = \iota.$$

Case 4: $k = 3$. Now $l = 3$ or 4, $d(A, B) \leq d(A_1, B) - 2$, and $q(A, B) \geq q(A_1, B) - 2$. Therefore

$$\frac{d(A, B)}{2} - q(A, B) \leq \frac{d(A_1, B) - 2}{2} - q(A_1, B) + 2$$
$$= (\iota - 1) - 1 + 2 = \iota.$$

Case 5: $k = 4$. Now $l = 4$, $d(A, B) = d(A_1, B) - 4$, and $q(A, B) \geq q(A_1, B) - 3$. Therefore

$$\frac{d(A, B)}{2} - q(A, B) \leq \frac{d(A_1, B) - 4}{2} - q(A_1, B) + 3$$
$$= (\iota - 1) - 2 + 3 = \iota.$$

This completes the induction, and Corollary 3.2.5 completes the proof. $\square$

By Theorem 3.2.3, given any two matrices A and B in a nonempty $\mathcal{A}(R, S)$ there exists a sequence of interchanges that transforms A into B. This motivates the introduction of the *interchange graph* $G(R, S)$ [6]. The vertices of $G(R, S)$ are the matrices in $\mathcal{A}(R, S)$. Two matrices A and B in $\mathcal{A}(R, S)$ are joined by an edge provided A differs from B by an interchange, equivalently, $A - B$ is an interchange matrix. Theorem 3.2.3 implies, in particular, that $G(R, S)$ is a connected graph. Interchange graphs are studied in Chapter 6.

3.3 The Structure Matrix $T(R, S)$

In this section we assume throughout that $R = (r_1, r_2, \ldots, r_m)$ and $S = (s_1, s_2, \ldots, s_n)$ are nonincreasing, nonnegative integral vectors satisfying

$$\tau = r_1 + r_2 + \cdots + r_m = s_1 + s_2 + \cdots + s_n. \tag{3.8}$$

Recall from Section 2.1 that the structure matrix associated with R and S is the matrix $T = T(R, S) = [t_{kl}]$ of size $m + 1$ by $n + 1$ where

$$t_{kl} = kl + \sum_{i=k+1}^{m} r_i - \sum_{j=1}^{l} s_j \quad (k = 0, 1, \ldots, m; l = 0, 1, \ldots, n). \tag{3.9}$$

By Theorem 2.1.4, $\mathcal{A}(R, S)$ is nonempty if and only if T is a nonnegative matrix. We have $t_{00} = \tau$. More generally, the entries of row and column 0 of T are

$$t_{0l} = \tau - \sum_{j=1}^{l} s_j = \sum_{j=l+1}^{n} s_j \quad (l = 0, 1, \ldots, n)$$

and

$$t_{k0} = \sum_{i=k+1}^{m} r_i = \tau - \sum_{i=1}^{k} r_i \quad (k = 0, 1, \ldots, m).$$

Thus for $k = 0, 1, \ldots, m$ and $l = 0, 1, \ldots, n$ we have

$$\begin{aligned} t_{kl} &= kl + \sum_{i=k+1}^{m} r_i - \sum_{j=1}^{l} s_j \\ &= kl + t_{k0} - (\tau - t_{0l}) \tag{3.10} \\[2mm] &= kl - (t_{00} - t_{k0} - t_{0l}). \tag{3.11} \end{aligned}$$

Equation (3.11) shows how to determine all of the elements of T from the $m + n + 1$ elements in row and column 0, and hence T is uniquely determined by these $m+n+1$ elements in row and column 0 of T. The elements of T also satisfy the simple recursive formulas [61]:

$$t_{k,l+1} = t_{kl} + k - s_{l+1} \quad (k=0,1,\ldots,m; l=0,1,\ldots n-1), \quad (3.12)$$

$$t_{k+1,l} = t_{kl} + l - r_{k+1} \quad (k=0,1,\ldots,m-1; l=0,1,\ldots,n). \quad (3.13)$$

These equations imply that

$$(t_{kl} - t_{k,l+1}) - (t_{k+1,l} - t_{k+1,l+1}) = 1$$

or, equivalently,

$$(t_{kl} - t_{k+1,l}) - (t_{k,l+1} - t_{k+1,l+1}) = 1$$

for $k=0,1,\ldots,m-1$ and $l=0,1,\ldots,n-1$, that is, consecutive differences of corresponding elements in a pair of adjacent rows of the structure matrix T differ by 1. A similar property holds for columns.

Example. The properties of a structure matrix discussed above are illustrated by the structure matrix

$$T = \begin{bmatrix} 12 & 9 & 6 & 3 & 1 & 0 \\ 8 & 6 & 4 & 2 & 1 & 1 \\ 5 & 4 & 3 & 2 & 2 & 3 \\ 2 & 2 & 2 & 2 & 3 & 5 \\ 0 & 1 & 2 & 3 & 5 & 8 \end{bmatrix}$$

when $R=(4,3,3,2)$ and $S=(3,3,3,2,1)$. $\qquad\qquad\qquad\qquad\square$

We note that (3.9) implies that the structure matrix can be factored as

$$T = E_{m+1} \begin{bmatrix} \tau & -s_1 & -s_2 & \cdots & -s_n \\ -r_1 & & & & \\ -r_2 & & & J_{m,n} & \\ \vdots & & & & \\ -r_m & & & & \end{bmatrix} E_n^{\mathrm{T}},$$

where E_k denotes the matrix of order k with 1's on and below the main diagonal and 0's elsewhere [63]. As observed by Sierksma and Sterken [69], since E_k is a nonsingular matrix, this factorization implies that the rank of T, over any field of characteristic 0, lies between 1 and 3. More specifically, neglecting the trivial cases where $\tau = 0$ or mn, the rank of T equals 2 if either R or S is a constant vector and equals 3 otherwise.

A sequence $c_0, c_1, \ldots, c_p$ of real numbers is called *log-convex* provided that

$$c_i + c_{i+2} \geq 2c_{i+1} \quad (i=0,1,\ldots,p-2),$$

that is, provided its *difference sequence* $\Delta(c)$ which is defined to be

$$(c_0 - c_1, c_1 - c_2, \ldots, c_{p-1} - c_p)$$

is a nonincreasing sequence. A log-convex sequence $c_0, c_1, \ldots, c_p$ is (*lower*) *unimodal*, that is, there exists an integer f with $0 \le f \le p$ such that $c_0 \ge c_1 \ge \cdots \ge c_f$ and $c_f \le c_{f+1} \le \cdots \le c_p$. The rows and columns of the structure matrix T above form log-convex sequences.

Theorem 3.3.1 *Let $R = (r_1, r_2, \ldots, r_m)$ and $S = (s_1, s_2, \ldots, s_n)$ be non-increasing, nonnegative vectors satisfying (3.8). Each row or column of the structure matrix $T(R, S) = [t_{ij}]$ forms a log-convex sequence.*

Proof. Using (3.12) we see that, for $0 \le i \le m$ and $0 \le j \le n - 2$,

$$\begin{aligned} t_{ij} + t_{i,j+2} &= (t_{i,j+1} - i + s_{j+1}) + (t_{i,j+1} + i - s_{j+2}) \\ &= 2t_{i,j+1} + (s_{j+1} - s_{j+2}) \\ &\ge 2t_{i,j+1}. \end{aligned}$$

Thus each row of $T(R, S)$ forms a log-convex sequence. In a similar way we show that each column of $T(R, S)$ forms a log-convex sequence. $\quad\square$

The smallest element in column 0 of the structure matrix is $t_{m0} = 0$. We now determine the smallest element in each of the other columns of the structure matrix [6].

Theorem 3.3.2 *Let $R = (r_1, r_2, \ldots, r_m)$ and $S = (s_1, s_2, \ldots, s_n)$ be non-increasing, nonnegative vectors satisfying (3.8). Let $R^* = (r_1^*, r_2^*, \ldots, r_n^*)$ be the conjugate sequence of R. The smallest element in column j of the structure matrix $T(R, S) = [t_{ij}]$ occurs in row r_j^*, and $t_{r_j^*, j} > t_{r_j^*+1, j}$ $(j = 1, 2, \ldots, n)$.*

Proof. Let $k = r_j^*$. By Theorem 3.3.1 the elements in column j of $T(R, S)$ form a log-convex, and so unimodal, sequence. It thus suffices to prove that

$$t_{k-1,j} - t_{kj} \ge 0, \text{ and } t_{kj} - t_{k+1,j} < 0.$$

By (3.13) we have

$$t_{k-1,j} - t_{kj} = r_k - j, \text{ and } t_{kj} - t_{k+1,j} = r_{k+1} - j.$$

By definition of the conjugate sequence, $k = r_j^*$ equals the number of terms of the nonincreasing sequence $(r_1, r_2, \ldots, r_m)$ which are greater than or equal to j. Hence $r_k \ge j$ and $r_{k+1} < j$. $\quad\square$

Since column 0 of the structure matrix is always nonnegative, Theorem 3.3.2 immediately implies the following result.

Corollary 3.3.3 *Under the assumptions in Theorem 3.3.2, the m by n structure matrix $T(R,S) = [t_{ij}]$ is a nonnegative matrix if and only if $t_{r_j^*,j} \geq 0$ $(j = 1, 2, \ldots, n)$.* □

Using (3.9) and the definition of the conjugate sequence, we calculate that

$$t_{r_j^*,j} = r_j^* j + (r_{r_j^*} + r_{r_j^*+1} + \cdots + r_m) - (s_1 + s_2 + \cdots + s_j)$$
$$= (r_1^* + r_2^* + \cdots + r_j^*) - (s_1 + s_2 + \cdots + s_j) \qquad (3.14)$$

It thus follows from Corollary 3.3.3 that if $r_1 + r_2 + \cdots + r_m = s_1 + s_2 + \cdots + s_n$, then the structure matrix $T(R,S)$ is nonnegative if and only if

$$(r_1^* + r_2^* + \cdots + r_j^*) - (s_1 + s_2 + \cdots + s_j) \geq 0 \quad (j = 1, 2, \ldots, n),$$

that is, if and only if $S \preceq R^*$. Thus we have obtained a direct proof of the equivalence of the Gale–Ryser (Theorem 2.1.1) and the Ford–Fulkerson (Theorem 2.1.2) condition for $\mathcal{A}(R,S)$ to be nonempty.

As remarked at the beginning of this section, the elements of the structure matrix $T = T(R,S)$ are determined by the $m + n + 1$ elements in row and column 0 of T. The formula (3.11) shows how to determine t_{kl} from t_{00}, t_{k0}, and t_{0l}. We now show how other sets of $m + n + 1$ elements of T uniquely determine T. Let $K_{m+1,n+1}$ be the *complete bipartite graph* whose set of vertices is $U \cup V$ where $U = \{u_0, u_1, \ldots, u_m\}$ and $V = \{v_0, v_1, \ldots, v_n\}$ are disjoint sets of sizes $m + 1$ and $n + 1$, respectively. The $(m+1)(n+1)$ edges of $K_{m+1,n+1}$ are the pairs $\{u_i, v_j\}$ with $0 \leq i \leq m$ and $0 \leq j \leq n$. The vertices in U correspond to the rows of T in the natural way, the vertices in V correspond to the columns of T, and the edges of $K_{m+1,n+1}$ correspond to the positions of T. Let G be a *spanning tree* of $K_{m+1,n+1}$. The vertex set of G is $U \cup V$, and G is a connected graph containing exactly $m + n + 1$ edges. A particular spanning tree is the spanning tree G_0 whose edges correspond to the positions in row 0 and column 0. Thus to say that the structure matrix T is determined by the elements in row and column 0 of T is to say that T is determined by the elements in the positions corresponding to the edges of the spanning tree G_0. We show that T is determined by the elements in the positions corresponding to any spanning tree [6].

Let p be an integer with $2 \leq p \leq \min\{m, n\}$. Let $i_1, i_2, \ldots, i_p$ be distinct integers between 0 and m, and let $j_1, j_2, \ldots, j_p$ be distinct integers between 0 and n. Let P be the set of positions

$$\{(i_1, j_1), (i_2, j_2), \ldots, (i_p, j_p), (i_1, j_2), \ldots, (i_{p-1}, j_p), (i_p, j_1)\}. \qquad (3.15)$$

Then P is the set of positions corresponding to a cycle of length $2p$ in $K_{m+1,n+1}$. Each cycle of $K_{m+1,n+1}$ is obtained in this way.

Lemma 3.3.4 *Let P be the set of positions (3.15) corresponding to a cycle of the complete bipartite graph $K_{m+1,n+1}$. Then*

$$\sum_{(i_k,j_l)\in P} (-1)^{k+l}(t_{i_k j_l} - i_k j_l) = 0. \tag{3.16}$$

Proof. By (3.9)

$$t_{i_k j_l} - i_k j_l = \sum_{i=i_k+1}^{m} r_i - \sum_{j=1}^{j_l} s_j \quad (k=1,2,\ldots,p; l=1,2\ldots,p).$$

Therefore

$$\sum_{(i_k,j_l)\in P} (-1)^{k+l}(t_{i_k j_l} - i_k j_l) = \sum_{(i_k,j_l)\in P} (-1)^{k+l}\left(\sum_{i=i_k+1}^{m} r_i - \sum_{j=1}^{j_l} s_j\right).$$

Since each row index i_k or column index j_l appears in P either twice or not at all, each of the terms $\sum_{i=i_k+1}^{m} r_i$ and $\sum_{j=1}^{j_l} s_j$ occurs twice in this summation and with opposite signs. Hence (3.16) follows. □

Let G be a spanning tree of $K_{m+1,n+1}$. Then each edge $\{u_i, v_j\}$ of $K_{m+1,n+1}$ which is not an edge of G belongs to a unique cycle γ_{ij} of $K_{m+1,n+1}$ each of whose other edges belongs to G. Let P_{ij} be the set of positions

$$\{(i_1,j_1),(i_2,j_2),\ldots,(i_p,j_p),(i_1,j_2),\ldots,(i_{p-1},j_p),(i_p,j_1)\} \tag{3.17}$$

corresponding to the cycle γ_{ij}. Let $P' = P \setminus \{(u_i, v_j)\}$ be obtained from P_{ij} be removing the position (u_i, v_j). From Lemma 3.3.4 we now obtain the following [6].

Theorem 3.3.5 *Let $R = (r_1, r_2, \ldots, r_m)$ and $S = (s_1, s_2, \ldots, s_n)$ be non-increasing, nonnegative vectors satisfying (3.8). Let G be a spanning tree of the complete bipartite graph $K_{m+1,n+1}$. Then the elements of the structure matrix $T(R,S)$ are determined by the elements of $T(R,S)$ in the positions corresponding to the edges of G. If (i,j) is a position of $T(R,S)$ that does not correspond to an edge of G, then using the notation above, we have*

$$t_{ij} = ij - \sum_{(i_k,j_l)\in P'} (-1)^{k+l}(t_{i_k j_l} - i_k j_l).$$

□

By using Theorem 3.3.5 a spanning tree of $K_{m+1,n+1}$ can be used to give other necessary and sufficient conditions for the nonemptiness of $\mathcal{A}(R,S)$ [6].

Finally we remark that in the case that $m = n$ the eigenvalues of the structure matrix are considered in [69], [19]. It is established in [19] that all the eigenvalues of T are real numbers and that the rank of T equals the number of its nonzero eigenvalues.

3.4 Invariant Sets

Again throughout this section, we assume that $R = (r_1, r_2, \ldots, r_m)$ and $S = (s_1, s_2, \ldots, s_n)$ are nonnegative integral vectors such that $\mathcal{A}(R,S)$ is nonempty. We do not generally assume that R and S are nonincreasing.

Let $A = [a_{ij}]$ be an m by n (0,1)-matrix. We denote by $\hat{A} = [\hat{a}_{ij}]$ the $(1, -1)$-matrix $2A - J_{m,n}$ obtained from A by replacing its 0's with -1's. The bipartite digraph $\Gamma(\hat{A})$ has vertex set $X \cup Y$ where $X = \{x_1, x_2, \ldots, x_m\}$ and $Y = \{y_1, y_2, \ldots, y_n\}$; for each x_i and y_j exactly one of the arcs (x_i, y_j) (when $\hat{a}_{ij} = 1$) and (y_j, x_i) (when $\hat{a}_{ij} = -1$) is present in $\Gamma(\hat{A})$.[6]

The class $\mathcal{A}(R,S)$ has the property that for each i with $1 \leq i \leq m$, row i of each matrix in $\mathcal{A}(R,S)$ contains the same number of 1's. A corresponding statement holds with row replaced by column. This invariance of the number of 1's in the set of positions of a row or of a column is the defining property of $\mathcal{A}(R,S)$. There may be other sets of positions, other than those obtained by unions of rows or unions of columns, which exhibit a similar invariance.

Example. Let $R = S = (3, 3, 1, 1)$. Then the structure matrix $T(R,S) = [t_{ij}]$ satisfies

$$t_{22} = 2 \cdot 2 + (r_3 + r_4) - (s_1 + s_2) = 4 + 2 - 6 = 0.$$

Thus if we partition matrices in $\mathcal{A}(R,S)$

$$\begin{bmatrix} A_{11} & A_{12} \\ A_{21} & A_{22} \end{bmatrix}$$

into four 2 by 2 matrices, then the number of 0's in A_{11} plus the number of 1's in A_{22} equals 0. Hence $A_{11} = J_2$ and $A_{22} = O_2$, and each matrix in $\mathcal{A}(R,S)$ has the form

$$\begin{bmatrix} J_2 & A_{12} \\ A_{21} & O_2 \end{bmatrix}.$$

Moreover, each of the matrices A_{12} and A_{21} equals one of

$$\begin{bmatrix} 1 & 0 \\ 0 & 1 \end{bmatrix} \quad \text{and} \quad \begin{bmatrix} 1 & 0 \\ 0 & 1 \end{bmatrix}.$$

In particular, the number of 1's in A_{12}, or A_{21}, is invariant over $\mathcal{A}(R,S)$.
□

We now turn to invariance properties for general classes $\mathcal{A}(R,S)$.

---

[6]If C is an m by n $(1, -1)$-matrix, then the directed bipartite graph $\Gamma(C)$ is called a *bipartite tournament*. A bipartite tournament is obtained from a complete bipartite graph $K_{m,n}$ by arbitrarily directing each of its edges. The digraph $\Gamma(\hat{A})$ is a bipartite tournament. We will discuss bipartite tournaments in Chapter 5.

Let $I \subseteq \{1, 2, \ldots, m\}$ and $J \subseteq \{1, 2, \ldots, n\}$. We say that (I, J) is an *invariant pair* for $\mathcal{A}(R, S)$ provided there exist integral vectors R_I and S_J such that for each matrix A in $\mathcal{A}(R, S)$, $A[I, J]$ belongs to $\mathcal{A}(R_I, S_L)$. We say that $I \times J$ is an *invariant set* for $\mathcal{A}(R, S)$ provided there exists an integer $\tau_{I \times J}$ such that each matrix in $\mathcal{A}(R, S)$ has the same number $\tau_{I \times J}$ of 1's in the positions of $I \times J$. Let $\bar{I}$ be the complement of I in $\{1, 2, \ldots, m\}$ and let $\bar{J}$ be the complement of J in $\{1, 2, \ldots, n\}$. If $(I, \{1, 2, \ldots, n\})$ is an invariant pair for $\mathcal{A}(R, S)$, then so is $(\bar{I}, \{1, 2, \ldots, n\})$. Similarly, $(\{1, 2, \ldots, m\}, \bar{J})$ is an invariant pair, if $(\{1, 2, \ldots, m\}, J)$ is. If $I \times J$ is an invariant set for $\mathcal{A}(R, S)$, then so are $\bar{I} \times J$, $I \times \bar{J}$, and $\bar{I} \times \bar{J}$. For each I and J, the sets

$$I \times \{1, 2, \ldots n\}, \ I \times \oslash, \ \{1, 2, \ldots, m\} \times J, \ \text{and} \ \oslash \times J$$

are invariant sets. Such sets are *trivial invariant sets*; all other invariant sets are nontrivial. In contrast, while $(I, \oslash)$ and $(\oslash, J)$ are invariant pairs, $(I, \{1, 2, \ldots, n\})$ and $(\{1, 2, \ldots, m\}, J)$ need not be invariant pairs when $I \neq \oslash, \{1, 2, \ldots, m\}$ and $J \neq \oslash, \{1, 2, \ldots, n\}$. The *trivial invariant pairs* are the pairs

$$(I, \oslash), (\oslash, J), \ \text{and} \ (\{1, 2, \ldots, m\}, \{1, 2, \ldots, n\})$$

for each I and J. If $I = \{i\}$ and $J = \{j\}$, then (I, J) is an invariant pair if and only if $I \times J$ is an invariant set, and in this case we say that (i, j) is an *invariant position*. We distinguish between an *invariant 1-position* and an *invariant 0-position* according as the element in position (i, j) of each matrix in $\mathcal{A}(R, S)$ is 1 or 0.

Recall that a digraph Γ is *strongly connected* provided that for each ordered pair of distinct vertices x and y there is a directed path in Γ from x to y. A *directed cut* of a digraph is a partition of its vertex set V into two nonempty sets V_1 and V_2 such that all arcs between V_1 and V_2 have their initial vertex in V_1 and terminal vertex in V_2. A digraph is strongly connected if and only it does not have a directed cut (see e.g. [13]). It follows easily that each arc of a strongly connected digraph is contained in a directed cycle. If a digraph Γ has the property that each arc is contained in a directed cycle and, considered as a graph by removing the directions of its arcs, is connected, then Γ is strongly connected.

The equivalence of (i), (ii), and (iii) in the following theorem is due to Ryser [59, 62] and Haber [30]; the equivalence of (i) and (iv) is due to Chen [17].

Theorem 3.4.1 *Let* $R = (r_1, r_2, \ldots, r_m)$ *and* $S = (s_1, s_2, \ldots, s_n)$ *be nonincreasing, nonnegative integral vectors such that* $\mathcal{A}(R, S)$ *is nonempty. Let* $A = [a_{ij}]$ *be a matrix in* $\mathcal{A}(R, S)$. *Then the following are equivalent.*

(i) $\mathcal{A}(R, S)$ *has an invariant position.*

(ii) *There exist integers* k *and* l *satisfying* $0 \leq k \leq m, 0 \leq l \leq n$, *and* $(k, l) \neq (0, n), (m, 0)$ *such that* $t_{kl} = 0$.

(iii) *There exist integers k and l satisfying $0 \leq k \leq m$ and $0 \leq l \leq n$ where $(k,l) \neq (0,n),(m,0)$ such that each of the positions in*

$$\{(i,j) : 1 \leq i \leq k, 1 \leq j \leq l\}$$

is an invariant 1-position of $\mathcal{A}(R,S)$, and each of the positions in

$$\{(i,j) : k+1 \leq i \leq m, \quad l+1 \leq j \leq n\}$$

is an invariant 0-position.

(iv) *The bipartite digraph $\Gamma(\hat{A})$ of $\hat{A} = 2A - J_{m,n}$ is not strongly connected.*

Proof. Let $\hat{A} = [\hat{a}_{ij}]$. First suppose that $\Gamma(\hat{A})$ is strongly connected. Consider any position (k,l) of A, and the arc α of $\Gamma(\hat{A})$ associated with that position. Since $\Gamma(\hat{A})$ is strongly connected, there is a directed cycle containing the arc α. Hence there is a minimal balanced matrix $C = [c_{ij}]$ such that $c_{kl} \neq 0$, and $c_{ij} \neq 0$ implies $c_{ij} = \hat{a}_{ij}$ $(i = 1, 2, \ldots, m; j = 1, 2, \ldots, n)$. The matrix $A - C$ is in $\mathcal{A}(R,S)$ and the element in position (k,l) is different from a_{kl}. Hence (k,l) is not an invariant position of $\mathcal{A}(R,S)$. Since (k,l) was arbitrary, $\mathcal{A}(R,S)$ does not have any invariant positions. Hence (i) implies (iv).

Now suppose that $\mathcal{A}(R,S)$ does not have any invariant positions. Consider any position (k,l) of A. Then there exists a matrix $B = [b_{ij}]$ in $\mathcal{A}(R,S)$ such that $a_{kl} \neq b_{kl}$. The matrix $A - B = [d_{ij}]$ is a nonzero balanced matrix, and $d_{ij} \neq 0$ implies that $d_{ij} = \hat{a}_{ij}$ $(i = 1, 2, \ldots, m; j = 1, 2, \ldots, n)$. Since $d_{kl} \neq 0$ and since C is a sum of minimal balanced matrices, there is a minimal balanced matrix C' which agrees with D in the (k,l) position. The matrix C corresponds to a directed cycle of $\Gamma(\hat{A})$. Hence each arc of $\Gamma(\hat{A})$ is an arc of a directed cycle. Since $\Gamma(\hat{A})$ is clearly connected (as a graph), $\Gamma(\hat{A})$ is strongly connected. Thus (iv) implies (i).

Let A be a matrix in $\mathcal{A}(R,S)$ and suppose that the digraph $\Gamma(\hat{A})$ with vertex set $X \cup Y$, $X = \{x_1, x_2, \ldots, x_m\}$ and $Y = \{y_1, y_2, \ldots, y_n\}$, is not strongly connected. Let $X_K \cup Y_{\bar{L}}$ be a cut of $\Gamma(\hat{A})$ where $X_K = \{x_i : i \in K\}$ and $Y_{\bar{L}} = \{y_j : j \in \bar{L}\}$. Let $X_{\bar{K}} = \{x_i : i \in \bar{K}\}$ and $Y_L = \{y_j : j \in L\}$. Then each arc between X_K and Y_L has initial vertex in X_K and terminal vertex in Y_L, and each arc between $Y_{\bar{L}}$ and $X_{\bar{K}}$ has initial vertex in $Y_{\bar{L}}$ and terminal vertex in $X_{\bar{K}}$. Hence $\hat{A}[K,L] = J$ and $\hat{A}[\bar{K},\bar{L}] = -J$ from which it follows that $A[K,L] = J$ and $A[\bar{K},\bar{L}] = O$. Thus $t_{K,L} = 0$ where $0 < |K| + |L| < m + n$, and hence $t_{kl} = 0$ where $k = |K|$ and $l = |L|$ and $0 < k + l < m + n$. We have

$$A[\{1, 2, \ldots, k\}, \{1, 2, \ldots, l\}] = J$$

and

$$A(\{1, 2, \ldots, k\}, \{1, 2, \ldots, l\}) = O.$$

It now follows that (iv) implies both (ii) and (iii). This argument can be reversed to show that each of (ii) and (iii) implies (iv). $\qquad\square$

The following theorem is from Chen [17] and is a weakened version of a theorem in [6].

Theorem 3.4.2 *Let R and S be nonnegative integral vectors such that $\mathcal{A}(R, S)$ is nonempty. Then $\mathcal{A}(R, S)$ has an invariant position if and only if $\mathcal{A}(R, S)$ has a nontrivial invariant pair.*

Proof. If A has an invariant position, then A has an invariant pair. Now assume that A does not have any invariant positions and consider the matrix $\hat{A} = [\hat{a}_{ij}] = 2A - J_{m,n}$. By Theorem 3.4.1 the digraph $\Gamma(\hat{A})$ with vertex set $X \cup Y$, $X = \{x_1, x_2, \ldots, x_m\}$ and $Y = \{y_1, y_2, \ldots, y_n\}$, is strongly connected.

Suppose $I \subseteq \{1, 2, \ldots, m\}$ and $J \subseteq \{1, 2, \ldots, n\}$ are such that $I \neq \varnothing, J \neq \varnothing$, and $(I, J) \neq (\{1, 2, \ldots, m\}, \{1, 2, \ldots, n\})$. Let $X_I = \{x_i : i \in I\}$ and $Y_J = \{y_j : j \in J\}$, and let $\Gamma(\hat{A}[I, J])$ be the subdigraph of $\Gamma(\hat{A})$ induced on the vertex set $X_I \cup Y_J$. Let $\Gamma(\hat{A})_{(I,J)}$ be the digraph obtained from $\Gamma(\hat{A})$ by removing the arcs of $\Gamma(\hat{A}[I, J])$. The vertex set of $\Gamma(\hat{A})_{(I,J)}$ is $X \cup Y$. First suppose that there is a directed cycle γ of $\Gamma(\hat{A})$ which contains both an arc of $\Gamma(\hat{A}[I, J])$ and an arc of $\Gamma(\hat{A})_{(I,J)}$. Let C be the minimal balanced matrix corresponding to γ. The matrix $B = A + C$ is in $\mathcal{A}(R, S)$. Moreover, some row (respectively, column) of $A[I, J]$ contains a different number of 1's than the corresponding row (respectively, column) of $B[I, J]$. Hence (I, J) is not an invariant pair. If either $\Gamma(\hat{A}[I, J])$ or $\Gamma(\hat{A})_{(I,J)}$ is not strongly connected, then since $\Gamma(\hat{A})$ is strongly connected, we easily obtain a directed cycle γ of the type above and conclude that (I, J) is not an invariant pair.

We may now assume that both $\Gamma(\hat{A}[I, J])$ and $\Gamma(\hat{A})_{(I,J)}$ are strongly connected. We take a directed cycle γ' in $\Gamma(\hat{A}[I, J])$, remove an arc (u, v) from γ, and determine a path in $\Gamma(\hat{A})_{(I,J)}$ from its vertex v to its vertex u. This gives a directed cycle γ which contains arcs of both $\Gamma(\hat{A}[I, J])$ and $\Gamma(\hat{A})_{(I,J)}$, and we conclude that (I, J) is not an invariant pair. $\square$

Assume that R and S are nonincreasing with $\mathcal{A}(R, S)$ nonempty. Let $T = T(R, S) = [t_{ij}]$ be the $m + 1$ by $n + 1$ structure matrix. Then $t_{0n} = t_{m0} = 0$. If $\mathcal{A}(R, S)$ does not have any invariant positions, then by Theorem 3.4.1 there are no more 0's in T. Otherwise, the matrix T has at least one other element equal to 0, and $\mathcal{A}(R, S)$ has at least one invariant position. It follows from Theorem 3.3.1 that the 0's of T occur consecutively in each row and in each column. From the combinatorial meaning of the elements of T we conclude that, if $t_{kl} = 0$ for integers k and l with $(k, l) \neq (0, n), (m, 0)$, then each matrix A in $\mathcal{A}(R, S)$ has a decomposition of the form

$$A = \begin{bmatrix} J_{k,l} & X \\ Y & O_{m-k,n-l} \end{bmatrix}.$$

At least one of the matrices $J_{k,l}$ and $O_{m-k,n-l}$ is nonvacuous and their positions within A are invariant positions of $\mathcal{A}(R, S)$. We also note that if

$0 \le i < k \le m$ and $0 \le j < l \le n$, then not both t_{ij} and t_{kl} can equal 0. Finally we note that, by Theorem 3.3.2, if column j of T contains a 0, then the last 0 in column j occurs in row r_j^* where $R^* = (r_1^*, r_2^*, \ldots, r_n^*)$ is the conjugate sequence of R. Similarly, if row i of T contains a 0, then the last 0 in row i occurs in column s_i^* where $S^* = (s_1^*, s_2^*, \ldots, s_m^*)$ is the conjugate sequence of S.

Let the 0's in T occupy the positions

$$(i_0, j_0) = (0, n), (i_1, j_1), \ldots, (i_{p-1}, j_{p-1}), (i_p, j_p) = (m, 0)$$

where $i_0 \le i_1 \le \cdots \le i_{p-1} \le i_p$, $j_0 \ge j_1 \ge \cdots \ge j_{p-1} \ge j_p$, and $(i_{t-1}, j_{t-1}) \ne (i_t, j_t)$ for $t = 1, 2, \ldots, p$. Let

$$I_t = \{i_{t-1} + 1, i_{t-1} + 2, \ldots, i_t\} \quad (t = 1, 2, \ldots, p).$$

and

$$J_t = \{j_t + 1, j_t + 2, \ldots, j_{t-1}\} \quad (t = 1, 2, \ldots, p).$$

Then $I_1, I_2, \ldots, I_p$ are pairwise disjoint sets satisfying $I_1 \cup I_2 \cup \cdots \cup I_p = \{1, 2, \ldots, m\}$. Similarly $J_1, J_2, \ldots, J_p$ are pairwise disjoint sets satisfying $J_1 \cup J_2 \cup \cdots \cup J_p = \{1, 2, \ldots, n\}$. One, but not both, of the sets in each pair I_t, J_t may be empty. We have

$$A[I_e, J_f] = \begin{cases} J, & \text{if } f > e, \\ O, & \text{if } e > f \end{cases} \quad (e, f = 1, 2, \ldots, t).$$

Let the matrix $A[I_t, J_t]$ have row sum vector R^t and column sum vector S^t $(t = 1, 2, \ldots, p)$. Then, since $A[I_t, J_t]$ is in $\mathcal{A}(R^{\mathrm{T}}, S^{\mathrm{T}})$, $\mathcal{A}(R^{\mathrm{T}}, S^{\mathrm{T}})$ is nonempty and has no invariant positions. This last assertion is a consequence of the observation that any invariant position of $\mathcal{A}(R^{\mathrm{T}}, S^{\mathrm{T}})$ would also be an invariant position of $\mathcal{A}(R, S)$. Thus the non-invariant positions of $\mathcal{A}(R, S)$ are precisely the positions occupied by the t matrices $A[I_1, J_1], A[I_2, J_2], \ldots, A[I_t, J_t]$.

Example. We illustrate the preceding discussion by taking

$$R = (10, 10, 9, 7, 6, 6, 5, 5, 2, 2, 1) \text{ and } S = (11, 9, 9, 8, 8, 5, 5, 3, 3, 1, 1).$$

The structure matrix $T(R, S)$ is

$$\begin{bmatrix}
63 & 52 & 43 & 34 & 26 & 18 & 13 & 8 & 5 & 2 & 1 & 0 \\
53 & 43 & 35 & 27 & 20 & 13 & 9 & 5 & 3 & 1 & 1 & 1 \\
43 & 34 & 27 & 20 & 14 & 8 & 5 & 2 & 1 & 0 & 1 & 2 \\
34 & 26 & 20 & 14 & 9 & 4 & 2 & 0 & 0 & 0 & 2 & 4 \\
27 & 20 & 15 & 10 & 6 & 2 & 1 & 0 & 1 & 2 & 5 & 8 \\
21 & 15 & 11 & 7 & 4 & 1 & 1 & 1 & 3 & 5 & 9 & 13 \\
15 & 10 & 7 & 4 & 2 & 0 & 1 & 2 & 5 & 8 & 13 & 18 \\
10 & 6 & 4 & 2 & 1 & 0 & 2 & 4 & 8 & 12 & 18 & 24 \\
5 & 2 & 1 & 0 & 0 & 0 & 3 & 6 & 11 & 16 & 23 & 30 \\
3 & 1 & 1 & 1 & 2 & 3 & 7 & 11 & 17 & 23 & 31 & 39 \\
1 & 0 & 1 & 2 & 4 & 6 & 11 & 16 & 23 & 30 & 39 & 48 \\
0 & 0 & 2 & 4 & 7 & 10 & 16 & 22 & 30 & 38 & 48 & 58
\end{bmatrix}.$$

The 0's in $T(R, S)$ are in positions
$$(0, 11), (2, 9), (3, 9), (3, 8), (3, 7), (4, 7), (6, 5),$$

$$(7, 5), (8, 5), (8, 4), (8, 3), (10, 1), (11, 1), (11, 0).$$
The pairs (I_t, J_t) both of whose sets are nonempty are

$$(I_1, J_1), \quad \text{where} \quad I_1 = \{1, 2\}, J_1 = \{10, 11\},$$
$$(I_6, J_6), \quad \text{where} \quad I_6 = \{5, 6\}, J_6 = \{5, 6\},$$
$$(I_{11}, J_{11}), \quad \text{where} \quad I_{11} = \{9, 10\}, J_{11} = \{2, 3\}.$$

Thus each matrix A in the class $\mathcal{A}(R, S)$ has the form

$$A = \begin{bmatrix} 1 & 1 & 1 & 1 & 1 & 1 & 1 & 1 & 1 & & \\ 1 & 1 & 1 & 1 & 1 & 1 & 1 & 1 & 1 & & \\ 1 & 1 & 1 & 1 & 1 & 1 & 1 & 1 & 1 & 0 & 0 \\ 1 & 1 & 1 & 1 & 1 & 1 & 1 & 0 & 0 & 0 & 0 \\ 1 & 1 & 1 & 1 & 1 & & & 0 & 0 & 0 & 0 \\ 1 & 1 & 1 & 1 & 1 & & & 0 & 0 & 0 & 0 \\ 1 & 1 & 1 & 1 & 1 & 0 & 0 & 0 & 0 & 0 & 0 \\ 1 & 1 & 1 & 1 & 1 & 0 & 0 & 0 & 0 & 0 & 0 \\ 1 & & & 0 & 0 & 0 & 0 & 0 & 0 & 0 & 0 \\ 1 & & & 0 & 0 & 0 & 0 & 0 & 0 & 0 & 0 \\ 1 & 0 & 0 & 0 & 0 & 0 & 0 & 0 & 0 & 0 & 0 \end{bmatrix}$$

where each of the submatrices $A[\{1, 2\}, \{10, 11\}]$, $A[\{5, 6\}, \{5, 6\}]$, and $A[\{9, 10\}, \{2, 3\}]$ has row and column sum vector equal to $(1, 1)$, and equals

$$\begin{bmatrix} 1 & 0 \\ 0 & 1 \end{bmatrix} \text{ or } \begin{bmatrix} 1 & 0 \\ 0 & 1 \end{bmatrix}.$$

There are $2^3 = 8$ matrices in this class $\mathcal{A}(R, S)$.   □

To conclude this section, we strengthen Theorem 3.4.2 by proving a theorem of Brualdi and Ross [11]. This theorem has been generalized by Anstee to network flows [2].

Theorem 3.4.3 *Let R and S be nonnegative integral vectors of sizes m and n, respectively, where $m, n \geq 2$. Assume that $\mathcal{A}(R, S)$ is nonempty. Then $\mathcal{A}(R, S)$ has an invariant position if and only if $\mathcal{A}(R, S)$ has a nontrivial invariant set.*

Proof. An invariant position gives a nontrivial invariant set. We now assume that $\mathcal{A}(R, S)$ does not have any invariant positions. Let I and J be sets with $\oslash \subset I \subset \{1, 2, \ldots, m\}$ and $\oslash \subset J \subset \{1, 2, \ldots, n\}$. Let A be a matrix in $\mathcal{A}(R, S)$. Then both $A[I, J]$ and $A(I, J)$ are nonvacuous matrices.

Since $\mathcal{A}(R,S)$ does not have any invariant positions, Theorem 3.4.1 implies that the digraph $\Gamma(\hat{A})$ is strongly connected. We show that $I \times J$ is not an invariant set, thus proving the theorem.

Let the vertex set of $\Gamma(\hat{A})$ be $X \cup Y$ where $X = \{x_1, x_2, \ldots, x_m\}$ and $Y = \{y_1, y_2, \ldots, y_n\}$. We let $X_I = \{x_i : i \in I\}$ and $Y_J = \{y_j : j \in J\}$. Suppose there is a directed cycle γ of $\Gamma(\hat{A})$ that contains an odd number of arcs joining a vertex in X_I and a vertex in Y_J. Let C be the minimal balanced matrix corresponding to γ. Then $A - C$ is in $\mathcal{A}(R,S)$ and the number of 1's in $(A-C)[I,J]$ is different from the number of 1's in $A[I,J]$ implying that $I \times J$ is not an invariant set. Thus we may assume that every directed cycle of $\Gamma(\hat{A})$ contains an even number of arcs joining a vertex in X_I and a vertex in Y_J. More generally, we may assume that every closed directed trail[7] of $\Gamma(\hat{A})$ contains an even number of arcs joining a vertex in X_I and a vertex in Y_J.

Consider an arc α joining a vertex in X_I and a vertex in Y_J and an arc β joining a vertex in $X_{\bar{I}}$ and a vertex in $Y_{\bar{J}}$. Since $\Gamma(\hat{A})$ is strongly connected, there is a path from the terminal vertex of α to the initial vertex of β, and a path from the terminal vertex of β to the initial vertex of α. The concatenation of these two paths gives a closed directed walk γ containing both of the arcs α and β; γ may have repeated vertices and repeated arcs. Suppose that γ has a repeated arc. Then γ properly contains a directed cycle δ. The directed cycle δ corresponds to a minimal balanced matrix C. Replacing the matrix A in $\mathcal{A}(R,S)$ with the matrix $B = A - C$ in $\mathcal{A}(R,S)$, and the strongly connected digraph $\Gamma(\hat{A})$ with the strongly connected digraph $\Gamma(\hat{B})$, we obtain a closed directed walk γ' containing α and β that has at least one less repeated arc than γ. Repeating this argument as necessary, we may assume that γ has no repeated arcs, that is, γ is a closed directed trail.

The closed directed trail γ contains an even number of arcs that join a vertex in X_I and a vertex in Y_J, and thus the line sums of $C[I,J]$ all equal 0. Let (y_j, x_i) be an arc of γ with y_j in Y_J and x_i in X_I, such that

$$(y_j, x_i), (x_i, y_{j_1}), (y_{j_1}, x_{i_1}), \ldots, (y_{i_k}, x_{i_k})$$

are arcs of γ where y_{j_1} is in $Y_{\bar{J}}$, and y_{i_k} and x_{i_k} are in $Y_{\bar{J}}$ and $X_{\bar{I}}$, respectively. Since $\Gamma(\hat{A})$ does not have any directed cycles that contain an odd number of arcs joining a vertex in X_I and a vertex in Y_J (and so none containing exactly one such arc), it follows that (y_j, x_{i_k}) is an arc of $\Gamma(\hat{A})$, that is, (x_{i_k}, y_j) is not an arc of $\Gamma(\hat{A})$. We may then shorten γ and obtain a closed directed trail that contains an odd number of arcs joining a vertex in X_I and a vertex in Y_J, a contradiction. $\square$

A notion of local invariance is discussed in [48].

---

[7]A directed trail may contain repeated vertices but not repeated arcs.

3.5   Term Rank

In this section we investigate the term rank $\rho(A)$ of matrices A in a nonempty class $\mathcal{A}(R, S)$. Since the term rank of a matrix is invariant under arbitrary permutations of its rows and of its columns, we make the general assumption that $R = (r_1, r_2, \ldots, r_m)$ and $S = (s_1, s_2, \ldots, s_n)$ are nonincreasing, nonnegative integral vectors. We recall from Chapter 1 that $\rho(A)$ equals the covering number $\kappa(A)$, the minimal number of rows and columns that contain all the 1's of A.   Let

$$\tilde{\rho} = \tilde{\rho}(R, S) = \min\{\rho(A) : A \in \mathcal{A}(R, S)\}$$

denote the *minimal term rank* of a matrix in $\mathcal{A}(R, S)$, and let

$$\bar{\rho} = \bar{\rho}(R, S) = \max\{\rho(A) : A \in \mathcal{A}(R, S)\}$$

denote the *maximal term rank* of a matrix in $\mathcal{A}(R, S)$. An interchange can change the covering number by at most 1. Hence if B is obtained from A by an interchange, $\rho(B) = \rho(A) \pm 1$. By Theorem 3.2.3, there exists a sequence of interchanges that transforms a matrix in $\mathcal{A}(R, S)$ of term rank $\tilde{\rho}$ into a matrix in $\mathcal{A}(R, S)$ of term rank $\bar{\rho}$. This establishes the following intermediate value theorem of Ryser [60].

Theorem 3.5.1 *Let R and S be nonnegative integral vectors such that $\mathcal{A}(R, S)$ is nonempty. For each integer ρ with $\tilde{\rho}(R, S) \leq \rho \leq \bar{\rho}(R, S)$, there exists a matrix in $\mathcal{A}(R, S)$ of term rank ρ.*   □

Assume that R and S are nonincreasing and $\mathcal{A}(R, S) \neq \varnothing$, and let $T(R, S) = [t_{kl}]$ be the structure matrix for R and S. Ryser [60, 61] obtained the remarkably simple formula

$$\bar{\rho} = \min\{t_{kl} + k + l : k = 0, 1, \ldots, m, l = 0, 1, \ldots, n\}.$$

To derive this formula, we first prove the following theorem of Haber [30] which establishes the existence of a special matrix in $\mathcal{A}(R, S)$ with term rank equal to $\bar{\rho}$.

Theorem 3.5.2 *Let $R = (r_1, r_2, \ldots, r_m)$ and $S = (s_1, s_2, \ldots, s_n)$ be non-increasing, nonnegative integral vectors such that $\mathcal{A}(R, S)$ is nonempty. Let $\bar{\rho} = \bar{\rho}(R, S)$. There exists a matrix A in $\mathcal{A}(R, S)$ such that $\rho(A) = \bar{\rho}$, and A has 1's in each of the positions $(1, \bar{\rho}), (2, \bar{\rho} - 1), \ldots, (\bar{\rho}, 1)$.*

Proof. We first show that there is a matrix in $\mathcal{A}(R, S)$ whose leading submatrix of order $\bar{\rho}$ has term rank equal to $\bar{\rho}$. Let $A = [a_{ij}]$ be any matrix in $\mathcal{A}(R, S)$ such that $\rho(A) = \bar{\rho}$ and choose a collection Λ of $\bar{\rho}$ 1's of A with no two on a line (more properly, positions occupied by 1's). Suppose p and q are such that $p \leq \bar{\rho} < q$, and row p does not contain any 1 of Λ but row q does. Let the 1 of Λ in row q be in column k. If $a_{pk} = 1$, then we may replace a_{qk} in Λ with a_{pk}. Assume that $a_{pk} = 0$. Since $r_p \geq r_q$, an interchange is available that replaces A with a matrix with term rank equal to $\bar{\rho}$ and replaces Λ with $(\Lambda \setminus \{(q,k)\}) \cup \{(p,k)\}$. In this way we obtain a matrix of term rank $\bar{\rho}$ with a Λ contained in the first $\bar{\rho}$ rows. A similar argument applied to columns gives us a matrix $A_{\bar{\rho}}$ of the form

$$A_{\bar{\rho}} = \begin{bmatrix} D & X \\ Y & O \end{bmatrix}, \tag{3.18}$$

where D in (3.18) is of order $\bar{\rho}$ and $\rho(D) = \bar{\rho}$. The submatrix of A complementary to D must be a zero matrix because the term rank of $A_{\bar{\rho}}$ cannot exceed $\bar{\rho}$. We now show how to obtain a matrix $A_{\bar{\rho}}$ with a collection of 1's Λ occupying the positions $\{(1,\bar{\rho}),(2,\bar{\rho}-1),\ldots,(\bar{\rho},1)\}$ of the *secondary diagonal* of D.

Suppose that Λ contains 1's in positions $(1,\bar{\rho}),(2,\bar{\rho}-1),\ldots,(k,\bar{\rho}-k+1)$ but not in position $(k+1,\bar{\rho}-k)$. The two 1's of Λ in row $k+1$ and column $\bar{\rho}-k$ determine a submatrix of order 2 equal to one of the matrices

$$\begin{bmatrix} 1 & 0 \\ 0 & 1 \end{bmatrix}, \begin{bmatrix} 1 & 0 \\ 1 & 1 \end{bmatrix}, \begin{bmatrix} 1 & 1 \\ 0 & 1 \end{bmatrix}, \begin{bmatrix} 1 & 1 \\ 1 & 1 \end{bmatrix}. \tag{3.19}$$

The 1's on the main diagonal of the matrices in (3.19) are 1's of Λ. In each case at most one interchange applied to $A_{\bar{\rho}}$ places 1's in positions corresponding to the secondary diagonal of the matrices in (3.19). This gives us a new Λ with 1's in positions $(1,\bar{\rho}),(2,\bar{\rho}-1),\ldots,(k,\bar{\rho}-k+1)$, $(k+1,\bar{\rho}-k)$. Continuing in this way we obtain a matrix of the form (3.18) with a Λ consisting of 1's from the secondary diagonal of D. $\square$

The combinatorial meaning of the elements of the structure matrix and the fact that the term rank of a matrix equals its covering number easily imply that each of the numbers $t_{kl}+k+l$ is an upper bound for $\bar{\rho}$. The proof of equality we give is taken from [12] with the simplifications introduced in [51]. The proof we give also makes use of the following preliminary theorem [49, 43, 42, 13].

Theorem 3.5.3 *Let* $R = (r_1, r_2, \ldots, r_m)$ *and* $S = (s_1, s_2, \ldots, s_n)$ *be nonnegative integral vectors. Let* $R' = (r'_1, r'_2, \ldots, r'_m)$ *and* $S' = (s'_1, s'_2, \ldots, s'_n)$ *be vectors where* $r'_i = r_i$ *or* $r_i - 1$ $(i = 1, 2, \ldots, m)$ *and* $s'_j = s_j$ *or* $s_j - 1$ $(j = 1, 2, \ldots, n)$. *If both* $\mathcal{A}(R,S)$ *and* $\mathcal{A}(R',S')$ *are nonempty, then there exist matrices* A *in* $\mathcal{A}(R,S)$ *and* B *in* $\mathcal{A}(R',S')$ *such that* $B \leq A$ *(entry-wise).*

Proof. Let $A = [a_{ij}]$ and $B = [b_{ij}]$ be matrices in $\mathcal{A}(R,S)$ and $\mathcal{A}(R',S')$ such that $A - B$ has the smallest number $h(A,B)$ of negative elements. We first note the following. Suppose that A and B have corresponding submatrices of order 2 of the form

$$A[\{p,q\},\{k,l\}] = \begin{bmatrix} 0 & 1 \\ 1 & a_{ql} \end{bmatrix} \text{ and } B[\{p,q\},\{k,l\}] = \begin{bmatrix} 1 & 0 \\ 0 & b_{ql} \end{bmatrix}.$$

If $b_{ql} = 1$, then a $(p,q;k,l)$-interchange applied to B gives a matrix C in $\mathcal{A}(R,S)$ with $h(A,C) < h(A,B)$. If $a_{ql} = 0$, then a $(p,q;k,l)$-interchange applied to A gives a matrix D in $\mathcal{A}(R,S)$ with $h(D,B) < h(A,B)$. Hence $a_{ql} = 1$ and $b_{ql} = 0$.

Suppose to the contrary that $h(A,B) > 0$. Choose p so that row p of $A - B$ contains the largest number t of negative elements of any row in A. Since $r_p' = r_p$ or $r_p - 1$, row p of $A - B$ contains u 1's where $u \geq t$. Let the 1's in row p of $A - B$ occur in columns $j_1, j_2, \ldots, j_u$. Choose j_0 so that the element in position (p, j_0) of $A - B$ equals -1. Since $s_{j_0} \geq s_{j_0}'$, there exists an integer q such that the element in the (q, j_0) position of $A - B$ equals 1, and hence $a_{qj_0} = 1$ and $b_{qj_0} = 0$. Applying the property established above, we obtain

$$a_{qj_0} = a_{qj_1} = \cdots = a_{qj_u} = 1 \text{ and } b_{qj_0} = b_{qj_1} = \cdots = b_{qj_u} = 0.$$

Since $r_q = r_q'$ or $r_q' + 1$, row q of $A - B$ contains at least u negative elements. Since $u \geq t$ it follows from the choice of p that $u = t$. Thus row q of $A - B$ contains $(u+1)$ 1's and u -1's. We now replace p with q in the preceding argument and obtain a contradiction. Hence $h(A,B) = 0$ and it follows that $B \leq A$. $\square$

We now prove Ryser's formula [60] for the maximal term rank.

Theorem 3.5.4 *Let R and S be nonincreasing, nonnegative integral vectors such that $\mathcal{A}(R,S)$ is nonempty. Then*

$$\bar{\rho}(R,S) = \min\{t_{kl} + k + l : k = 0,1,\ldots,m; l = 0,1,\ldots,n\}.$$

Proof. Let $T = [t_{ij}]$ be the structure matrix for $\mathcal{A}(R,S)$. Since $\mathcal{A}(R,S)$ is nonempty, T is a nonnegative matrix. Let ρ be an integer with $0 \leq \rho \leq \min\{m,n\}$. It follows from Theorems 3.5.2 and 3.5.3 that $\bar{\rho}$ equals the maximum integer ρ such that the class $\mathcal{A}(R^{(\rho)}, S^{(\rho)})$ is nonempty where $R^{(\rho)} = (r_1-1,\ldots,r_\rho-1,r_{\rho+1},\ldots,r_m)$ and $S^{(\rho)} = (s_1-1,\ldots,s_\rho-1,s_{\rho+1},\ldots,r_n)$. Since R and S are nonincreasing, $R^{(\rho)}$ and $S^{(\rho)}$ are nearly nonincreasing. Hence $\mathcal{A}(R^{(\rho)}, S^{(\rho)})$ is nonempty if and only if its structure matrix $T' = [t_{ij}']$ is nonnegative. A direct calculation establishes that

$$t_{kl}' = t_{kl} + \min\{\rho,k\} + \min\{\rho,l\} - \rho.$$

Thus $\mathcal{A}(R^{(\rho)}, S^{(\rho)})$ is nonempty if and only if

$$\rho \leq t_{kl} + \min\{\rho,k\} + \min\{\rho,l\} \quad (k = 0,1,\ldots,m; l = 0,1,\ldots,n). \quad (3.20)$$

If $k \geq \rho$ or $l \geq \rho$, then (3.20) trivially holds. If $k,l < \rho$, then (3.20) holds if and only if $\rho \leq t_{kl} + k + l$. The theorem now follows. $\square$

The following decomposition theorem for the class $\mathcal{A}(R,S)$ is due to Ryser [60].

Theorem 3.5.5 *Let $R = (r_1, r_2, \ldots, r_m)$ and $S = (s_1, s_2, \ldots, s_n)$ be nonincreasing, nonnegative integral vectors such that $\mathcal{A}(R, S)$ is nonempty. Then there exist integers e and f with $0 \le e \le m$ and $0 \le f \le n$ such that each matrix A in $\mathcal{A}(R, S)$ has a decomposition of the form*

$$A = \begin{bmatrix} A_{11} & A_{12} \\ A_{21} & A_{22} \end{bmatrix} \qquad (3.21)$$

where A_{11} is of size e by f and the number of 0's in A_{11} plus the number of 1's in A_{22} equals $\bar{\rho}(R, S) - (e + f)$.

Proof. By Theorem 3.5.4 there exist integers e and f such that $\bar{\rho}(R, S) = t_{ef} + e + f$. Since for each matrix A in $\mathcal{A}(R, S)$, t_{ef} equals the number of 0's in the upper left e by f submatrix of A plus the number of 1's in the lower right $m - e$ by $n - f$ submatrix, the theorem follows.   □

If in (3.21) we take A to be the matrix $A_{\bar{\rho}}$ of Theorem 3.5.2, then A_{11} is a matrix of all 1's and A_{22} has exactly $\bar{\rho}(R, S) - (e + f)$ 1's, namely those 1's on the secondary diagonal of the leading submatrix of order $\bar{\rho}$ of $A_{\bar{\rho}}$ which lie in A_{22}.

Let nonincreasing, nonnegative integral vectors $R = (r_1, r_2, \ldots, r_m)$ and $S = (s_1, s_2, \ldots, s_n)$ be given with $r_1 + r_2 + \cdots + r_m = s_1 + s_2 + \cdots + s_n$. According to Theorems 3.5.4 and 3.5.2, we may construct a matrix in $\mathcal{A}(R, S)$ with $\rho(A) = \bar{\rho}(R, S)$ as follows.

(i) First we compute the elements of the structure matrix $T(R, S) = [t_{kl}]$ and check that they are nonnegative (if so, then $\mathcal{A}(R, S)$ is nonempty).

(ii) We then compute the elements of the $m+1$ by $n+1$ matrix $T'(R, S) = [t_{kl} + k + l]$ and determine p and q such that $t_{pq} + p + q$ is minimal (by Theorem 3.5.4, $\bar{\rho}(R, S) = t_{pq} + p + q$).

(iii) Let $\bar{\rho} = \bar{\rho}(R, S)$. A matrix in $\mathcal{A}(R, S)$ exists with 1's in positions $(1, \bar{\rho}), (2, \bar{\rho} - 1), \ldots, (\bar{\rho}, 1)$. Let

$$R' = (r_1 - 1, r_2 - 1, \ldots, r_{\bar{\rho}} - 1, r_{\bar{\rho}+1}, \ldots, r_m)$$

and

$$S' = (s_1 - 1, s_2 - 1, \ldots, s_{\bar{\rho}} - 1, s_{\bar{\rho}+1}, \ldots, s_n).$$

A matrix A' in $\mathcal{A}(R', S')$ with 0's in positions $(1, \bar{\rho}), (2, \bar{\rho} - 1), \ldots, (\bar{\rho}, 1)$ may be constructed using known algorithms from network flow theory [21, 1]. By replacing the 0's of A' in positions $(1, \bar{\rho}), (2, \bar{\rho} - 1), \ldots, (\bar{\rho}, 1)$ with 1's we obtain a matrix A in $\mathcal{A}(R, S)$ satisfying $\rho(A) = \bar{\rho}$.

Another algorithm for constructing a matrix in $\mathcal{A}(R, S)$ of maximal term rank $\bar{\rho}(R, S)$ is given by Haber [30].

We now turn to a formula for the minimal term rank $\tilde{\rho} = \tilde{\rho}(R, S)$. A formula for $\tilde{\rho}$ was first derived by Haber [31]. This formula and its proof were subsequently simplified in [7]. We begin with the following covering lemma [31, 7] which, in particular, implies the existence of a special matrix in $\mathcal{A}(R, S)$ with term rank $\tilde{\rho}$.

Lemma 3.5.6 *Let $R = (r_1, r_2, \ldots, r_m)$ and $S = (s_1, s_2, \ldots, s_n)$ be nonincreasing, nonnegative integral vectors such that the class $\mathcal{A}(R, S)$ is nonempty. Let e and f be integers with $0 \le e \le m$ and $0 \le f \le n$. If there is a matrix in $\mathcal{A}(R, S)$ all of whose 1's are contained in the union of a set of e rows and f columns, then there is a matrix in $\mathcal{A}(R, S)$ all of whose 1's are contained in the union of its first e rows and first f columns.*

Proof. Let A be a matrix in $\mathcal{A}(R, S)$ all of whose 1's are contained in rows $i_1, i_2, \ldots, i_e$ and columns $j_1, j_2, \ldots, j_f$. Suppose there is an integer p such that $p \ne i_1, i_2, \ldots, i_e$ and $p < i_s$ for some s. Since R is nonincreasing, $r_p \ge r_{i_s}$. All the 1's in row p of A are contained in columns $j_1, j_2, \ldots, j_f$. If all the 1's in row i_s are also contained in these columns, then we may replace i_s by p without changing the matrix A. Otherwise, let $k_1, k_2, \ldots, k_t$ be the columns different from $j_1, j_2, \ldots, j_f$ that contain 1's in row i_s of A. Then row p of A has 0's in these columns, and since $r_p \ge r_{i_s}$, there are columns $q_1, q_2, \ldots, q_t$ such that row p has 1's in these columns and row i_s has 0's in these columns. Now t interchanges give a matrix in $\mathcal{A}(R, S)$ all of whose 1's are contained in rows $i_1, \ldots, i_{s-1}, i_{s+1}, \ldots, i_e, p$ and columns $j_1, j_2, \ldots, j_f$. Repeating as necessary, we obtain a matrix in $\mathcal{A}(R, S)$ all of whose 1's are contained in the first e rows and columns $j_1, j_2, \ldots, j_f$. Applying our argument to columns, we complete the proof. $\square$

The following corollary is an immediate consequence of Lemma 3.5.6.

Corollary 3.5.7 *Let R and S be nonincreasing, nonnegative integral vectors such that $\mathcal{A}(R, S)$ is nonempty. Then*

$$\tilde{\rho}(R, S) = \min\{e + f\}$$

where the minimum is taken over all integers e and f with $0 \le e \le m$ and $0 \le f \le n$ such that there exists a matrix in $\mathcal{A}(R, S)$ of the form

$$\begin{bmatrix} X & Y \\ Z & O_{m-e, n-f} \end{bmatrix}. \tag{3.22}$$

$\square$

Let $T = T(R, S) = [t_{ij}]$ be the $m+1$ by $n+1$ structure matrix associated with the nonincreasing, nonnegative vectors R and S. Define a matrix $\Phi = \Phi(R, S) = [\phi_{kl}]$ $(k = 0, 1, \ldots, m; l = 0, 1, \ldots, n)$ by

$$\phi_{kl} = \min\{t_{i_1, l+j_2} + t_{k+i_2, j_1} + (k - i_1)(l - j_1)\}$$

where the minimum is taken over all integers i_1, i_2, j_1, j_2 such that

$$0 \le i_1 \le k \le k + i_2 \le m \text{ and } 0 \le j_1 \le l \le l + j_2 \le n.$$

Theorem 3.5.8 *Let* $R = (r_1, r_2, \ldots, r_m)$ *and* $S = (s_1, s_2, \ldots, s_n)$ *be non-increasing, nonnegative integral vectors such that* $\mathcal{A}(R, S)$ *is nonempty. Then, with the above notation,*

$$\tilde{\rho}(R, S) = \min\{e + f : \phi_{ef} \geq t_{ef}, 0 \leq e \leq m, 0 \leq f \leq n\}.$$

Proof. Let e and f be integers with $0 \leq e \leq m$ and $0 \leq f \leq n$. We first determine when there exists a matrix in $\mathcal{A}(R, S)$ of the form (3.22). Let

$$C = [c_{ij}] = \begin{bmatrix} J_{e,f} & J_{e,n-f} \\ J_{m-e,f} & O_{m-e,n-f} \end{bmatrix}.$$

By Corollary 1.4.2 there exists a matrix in $\mathcal{A}(R, S)$ of the form (3.22) if and only if

$$\sum_{i \in I, j \in J} c_{ij} \geq \sum_{j \in J} s_j - \sum_{i \notin I} r_i \quad (I \subseteq \{1, 2, \ldots, m\}, J \subseteq \{1, 2, \ldots, n\}). \quad (3.23)$$

For $I \subseteq \{1, 2, \ldots, m\}$ and $J \subseteq \{1, 2, \ldots, n\}$, we write $I = I_1 \cup I_2$ where $I_1 \subseteq \{1, 2, \ldots, e\}$ and $I_2 \subseteq \{e + 1, e + 2, \ldots, m\}$, and we write $J = J_1 \cup J_2$ where $J_1 \subseteq \{1, 2, \ldots, f\}$ and $J_2 \subseteq \{f + 1, f + 2, \ldots, n\}$. We agree to take complements of $I_1, I_2, J_1,$ and J_2 with respect to their supersets $\{1, 2, \ldots, e\}, \{e + 1, e + 2, \ldots, m\}, \{1, 2, \ldots, f\},$ and $\{f + 1, f + 2, \ldots, n\}$, respectively. Then (3.23) is equivalent to

$$|I_1||J_1| + |I_1||J_2| + |I_2||J_1| \geq \sum_{j \in J_1 \cup J_2} s_j - \sum_{i \in \bar{I}_1 \cup \bar{I}_2} r_i, \quad (3.24)$$

holding for all $I_1 \subseteq \{1, 2, \ldots, e\}, I_2 \subseteq \{e + 1, e + 2, \ldots, m\}$ and for all $J_1 \subseteq \{1, 2, \ldots, f\}, J_2 \subseteq \{f + 1, f + 2, \ldots, n\}$. Let $k_1 = |I_1|, k_2 = |I_2|, l_1 = |J_1|$, and $l_2 = |J_2|$. Since R and S are nonincreasing, it follows that (3.23) is equivalent to

$$k_1 l_1 + k_1 l_2 + k_2 l_1 \geq \sum_{j=1}^{l_1} s_j + \sum_{j=f+1}^{f+l_2} s_j - \sum_{i=e-k_1+1}^{e} r_i - \sum_{i=m-k_2+1}^{m} r_i, \quad (3.25)$$

holding for all integers k_1, k_2 with $0 \leq k_1 \leq e \leq e + k_2 \leq m$ and all integers l_1, l_2 with $0 \leq l_1 \leq f \leq f + l_2 \leq n$. The inequality in (3.25) is equivalent to

$$t(K, L) - k_2 l_2 \geq 0 \quad (3.26)$$

where $K_1 = \{1, 2, \ldots, k_1\}, K_2 = \{e + 1, e + 2, \ldots, e + k_2\}, K = K_1 \cup K_2,$ $L_1 = \{1, 2, \ldots, l_1\}, L_2 = \{f + 1, f + 2, \ldots, f + l_2\},$ and $L = L_1 \cup L_2$. Let

A be any matrix in $\mathcal{A}(R,S)$ and partition A according to the diagram

	L_1	$\bar{L}_1$	L_2	$\bar{L}_2$
K_1	(0)		(0)	
$\bar{K}_1$		(1)		(1)
K_2	(0)		$-(1)$	
$\bar{K}_2$		(1)		(1)

.

Then (3.26) counts 0's and 1's in submatrices of A as shown above, where $-(1)$ means the negative of the number of 1's of $A[K_2, L_2]$. On the other hand, the expression $t_{i_1, f+j_2} + t_{e+i_2, j_1}$ counts 0's and 1's as shown below:

	L_1	$\bar{L}_1$	L_2	$\bar{L}_2$
K_1	2(0)	(0)	(0)	
$\bar{K}_1$	(0)		(1)	
K_2	(0)		(1)	
$\bar{K}_2$		(1)	(1)	2(1)

.

In this matrix, 2(0) means twice the number of 0's in $A[K_1, L_1]$, and 2(1) means twice the number of 1's in $A[\bar{K}_2, \bar{L}_2]$. It now follows that

$$t_{i_1, f+j_2} + t_{e+i_2, j_1} - (t(K,L) - k_2 l_2) \qquad (3.27)$$

counts 0's and 1's in submatrices of A as indicated in the matrix diagram

	L_1	$\bar{L}_1$	L_2	$\bar{L}_2$
K_1	(0)	(0)		
$\bar{K}_1$	(0)	$-(1)$		
K_2			(1)	(1)
$\bar{K}_2$			(1)	(1)

Hence (3.27) equals $t_{ef} - (e - k_1)(f - l_1)$. Therefore (3.26) holds if and only if $\phi_{ef} \geq t_{ef}$ all e and f, and the theorem follows now from Corollary 3.5.7. □

Let nonincreasing, nonnegative integral vectors $R = (r_1, r_2, \ldots, r_m)$ and $S = (s_1, s_2, \ldots, s_n)$ be given with $r_1 + r_2 + \cdots + r_m = s_1 + s_2 + \cdots + s_n$. According to Corollary 3.5.7 and Theorem 3.5.8 and its proof, we may construct a matrix in $\mathcal{A}(R,S)$ with $\rho(A) = \tilde{\rho}(R,S)$ as follows.

(i) First we compute the elements of the structure matrix $T(R,S) = [t_{kl}]$ and check that they are nonnegative (if so, then $\mathcal{A}(R,S)$ is nonempty).

(ii) We then compute the elements of the $m+1$ by $n+1$ matrix $\Phi(R,S) = [\phi_{kl}]$ and determine e and f such that $e + f$ is maximal subject to $\phi_{ef} \geq t_{ef}$ (by Theorem 3.5.8, $\tilde{\rho}(R,S) = e + f$).

(iii) A matrix in $\mathcal{A}(R,S)$ exists whose lower left $m - e$ by $n - f$ submatrix is the zero matrix $O_{m-e,n-f}$. Such a matrix has term rank equal to $\tilde{\rho}(R,S)$ and may be constructed using known algorithms from network flows [21].

There are instances of classes $\mathcal{A}(R,S)$ for which the maximal term rank equals the minimal term rank, that is, the term rank is an invariant of the class. If S is the conjugate R^* of R, this is surely the case because $\mathcal{A}(R,R^*)$ contains a unique matrix. Let k be an integer with $1 \leq k \leq n - 1$. A nontrivial class for which $\tilde{\rho} = \bar{\rho}$ is the class of all $(0,1)$-matrices of order n with exactly k 1's in each row and column. It is a consequence of Theorem 1.2.1 that each matrix in this class has term rank equal to n.

The following theorem of Ryser [60] gives sufficient conditions for the minimal term rank to be different from the maximal term rank.

Theorem 3.5.9 *Let R and S be nonincreasing, positive integral vectors such that $\mathcal{A}(R,S)$ is nonempty. Assume that $\mathcal{A}(R,S)$ has no invariant 1-positions and that $\bar{\rho}(R,S) < \min\{m,n\}$. Then $\tilde{\rho}(R,S) < \bar{\rho}(R,S)$.*

Proof. Let $T = [t_{ij}]$ be the structure matrix for R and S. By Theorem 3.5.4 there exist integers e and f satisfying $0 \le e \le m$ and $0 \le f \le n$ such that $\bar{\rho}(R, S) = t_{ef} + e + f$. Since $\bar{\rho}(R, S) < \min\{m, n\}$, we have $e < m$ and $f < n$. We have $t_{0f} + 0 + f \ge n$ because S is a positive vector. Hence $e > 0$ and similarly $f > 0$. Let A be a matrix in $\mathcal{A}(R, S)$ with $\rho(A) = \bar{\rho}(R, S) = t_{ef} + e + f$. Since $t_{ef} = \rho(A) - (e + f)$, it follows from Theorem 1.2.1 that $A[\{e+1, e+2, \ldots, m\}, \{f+1, f+2, \ldots, n\}]$ has exactly t_{ef} 1's and hence that $A[\{1, 2, \ldots, e\}, \{1, 2, \ldots, f\}] = J_{ef}$. By hypothesis, the position $(1,1)$ of A is not an invariant 1-position. Hence there is a matrix $B = [b_{ij}]$ in $\mathcal{A}(R, S)$ with $b_{11} = 0$. This implies that $B[\{e+1, e+2, \ldots, m\}, \{f+1, f+2, \ldots, n\}]$ contains fewer than t_{ef} 1's, and hence its covering number is strictly less than $t_{ef} + e + f = \bar{\rho}(R, S)$ By Theorem 1.2.1, $\rho(B) < \bar{\rho}(R, S)$ and the theorem follows. □

The following theorem gives a characterization of those classes $\mathcal{A}(R, S)$ with a constant term rank. The proof, which we omit, can be found in [6].

Theorem 3.5.10 *Let $R = (r_1, r_2, \ldots, r_m)$ and $S = (s_1, s_2, \ldots, s_n)$ be nonincreasing, nonnegative integral vectors such that $\mathcal{A}(R, S)$ is nonempty, and let $T = [t_{ij}]$ be the $m + 1$ by $n + 1$ structure matrix for R and S. Then $\tilde{\rho}(R, S) = \bar{\rho}(R, S)$ if and only if there exist integers e and f with $0 \le e \le m$ and $0 \le f \le n$ such that*

(i) *$t_{ef} = 0$, and*

(ii) *$\tilde{\rho}(R', S') = e$ and $\tilde{\rho}(R'', S'') = f$ where*
$R' = (r_1 - f, r_2 - f, \ldots, r_e - f)$ and $S' = (s_{f+1}, s_{f+2}, \ldots, s_n)$, and
$R'' = (r_{e+1}, r_{e+2}, \ldots, r_m)$ and $S'' = (s_1 - e, s_2 - e, \ldots, s_f - e)$.

We conclude this section by determining as in [6], given m and n, for what integers p and q there exists a nonempty class $\mathcal{A}(R, S)$ with $\tilde{\rho}(R, S) = p$ and $\bar{\rho}(R, S) = q$.

Lemma 3.5.11 *The minimum term rank $\tilde{\rho}$ and maximum term rank $\bar{\rho}$ of a nonempty class $\mathcal{A}(R, S)$ satisfy*

$$\bar{\rho} \le \begin{cases} \tilde{\rho} + \frac{\tilde{\rho}^2}{4}, & \text{if } \tilde{\rho} \text{ is even,} \\[2mm] \tilde{\rho} + \frac{\tilde{\rho}^2 - 1}{4}, & \text{if } \tilde{\rho} \text{ is odd.} \end{cases}$$

Proof. We may assume that R and S are nonincreasing. By Corollary 3.5.7 there exist nonnegative integers e and f with $\tilde{\rho} = e + f$ and a matrix A in $\mathcal{A}(R, S)$ all of whose 1's are contained in the union of the first e rows and first f columns. Let $T = [t_{ij}]$ be the structure matrix for R and S. Since A has no 1's in its lower right $m - e$ by $n - f$ submatrix, $t_{ef} \le ef$. Using Theorem 3.5.4, we obtain

$$\bar{\rho} \le t_{ef} + e + f \le ef + e + f = ef + \tilde{\rho},$$

and the lemma follows. □

Theorem 3.5.12 *Let m and n be positive integers, and let p and q be nonnegative integers. There exist nonnegative integral vectors R and S such that $\mathcal{A}(R,S)$ is nonempty, $\tilde{\rho}(R,S) = p$, and $\bar{\rho}(R,S) = q$ if and only if*

$$p \leq q \leq \begin{cases} \min\left\{p + \frac{p^2}{4}, m, n\right\}, & \text{if } p \text{ is even,} \\[2mm] \min\left\{p + \frac{p^2-1}{4}, m, n\right\}, & \text{if } p \text{ is odd.} \end{cases} \tag{3.28}$$

Proof. The term rank of an m by n matrix cannot exceed m or n. Lemma 3.5.11 now implies that (3.28) holds if $\mathcal{A}(R,S)$ is nonempty, $\tilde{\rho}(R,S) = p$, and $\bar{\rho}(R,S) = q$.

Now suppose that (3.28) holds. First assume that p is even, and let $p = 2a$. Let $b = q - p$ so that $b \leq \min\{a^2, m-2a, n-2a\}$. Let $u = m - 2a - b$ and $v = n - 2a - b$. Define an m by n matrix by

$$A = \begin{bmatrix} J_{a,a} & J_{a,a} & X & J_{a,v} \\ J_{a,a} & O & O & O \\ Y & O & I_b & O \\ J_{u,a} & O & O & O \end{bmatrix},$$

where X is an a by b (0,1)-matrix with exactly one 0 in each column and $p_i \leq a$ 0's in row i ($1 \leq i \leq a$), and Y is a b by a (0,1)-matrix with exactly one 0 in each row and $q_j \leq a$ 0's in column j ($1 \leq j \leq a$). The p_i and q_j are chosen so that there exists a $(0,1)$-matrix U with row sum vector $(p_1, p_2, \ldots, p_a)$ and column sum vector $(q_1, q_2, \ldots, q_a)$. Such matrices X and Y exist, because $b \leq a^2$. We may choose X to have a nonincreasing row sum vector and Y to have a nonincreasing column sum vector. The matrix A then has a nonincreasing row sum vector R and a nonincreasing column sum vector S. Let $T = [t_{ij}]$ be the $m+1$ by $n+1$ structure matrix for R and S. The term rank and covering number of A obviously satisfy $\rho(A) \geq 2a + b$ and $\kappa(A) \leq 2a + b$. By Theorem 1.2.1, $\rho(A) = 2a + b$. From A we also see that $t_{aa} + a + a = 2a + b$. By Theorem 3.5.4, $\bar{\rho}(R,S) \leq 2a+b$, and hence $\bar{\rho}(R,S) = 2a + b = q$.

Suppose there is a matrix B in $\mathcal{A}(R,S)$ with $\rho(B) < p$. Then for some e and f with $e + f < p = 2a$, there exist e rows and f columns of B which together contain all the 1's of B. By symmetry we may assume that $e < a$. Since each column sum of B is at least a, we must have $f = n$. This gives $p > n$, contradicting (3.28). Hence $\rho(B) \geq p$ and $\tilde{\rho}(R,S) \geq p$. By our choice of X and Y to imply the existence of U, it follows that there exists a matrix C of the form

$$\begin{bmatrix} Z & J_{a,n-a} \\ J_{m-a,a} & O \end{bmatrix}$$

in $\mathcal{A}(R,S)$. By Theorem 1.2.1 again, $\rho(C) \leq 2a = p$ and hence $\rho(C) = p$, and we conclude that $\tilde{\rho}(R,S) = p$. This completes the proof when p is even. A very similar proof is possible in case p is odd. $\qquad\square$

3.6   Widths and Multiplicities

Let A be a $(0,1)$-matrix of size m by n and let the row sum vector of A be $R = (r_1, r_2, \ldots, r_m)$. Let α be an integer with $0 \le \alpha \le \min\{r_1, r_2, \ldots, r_m\}$. We recall from Chapter 1 that the α-width of A is the smallest integer $\epsilon_\alpha(A)$ such that there exists $J \subseteq \{1, 2, \ldots, n\}$ with $|J| = \epsilon_\alpha(A)$ for which the full-rowed submatrix $E = A[\cdot, J]$ has at least α 1's in each row. The α-width satisfies

$$\alpha \le \epsilon_\alpha(A) \le n.$$

The submatrix F of E (and so of A) consisting of all rows of E with exactly α 1's is a critical α-submatrix of A. A critical α-submatrix of A is nonempty and has no zero columns. The minimum number of rows in an α-critical submatrix of A is the α-multiplicity $\delta_\alpha(A)$ of A.

Now let the column sum vector of A be $S = (s_1, s_2, \ldots, s_n)$. Let

$$\tilde{\epsilon}_\alpha = \tilde{\epsilon}_\alpha(R, S) = \min\{\epsilon_\alpha(B) : B \in \mathcal{A}(R, S)\}$$

denote the *minimal α-width* of a matrix in $\mathcal{A}(R, S)$, and let

$$\bar{\epsilon}_\alpha = \bar{\epsilon}_\alpha(R, S) = \max\{\epsilon_\alpha(B) : B \in \mathcal{A}(R, S)\}$$

denote the *maximal α-width* of a matrix in $\mathcal{A}(R, S)$. We also let

$$\delta_\alpha = \delta_\alpha(R, S) = \min\{\delta_\alpha(A) : B \in \mathcal{A}(R, S), \epsilon_\alpha(B) = \tilde{\epsilon}_\alpha(R, S)\}$$

denote the *minimal α-multiplicity* of a matrix in $\mathcal{A}(R, S)$ of width $\epsilon_\alpha(R, S)$.

An interchange can change the α-width of a matrix by at most 1. Hence if B is obtained from A by an interchange, then $\epsilon_\alpha(B) = \epsilon_\alpha(A) \pm 1$. By Theorem 3.2.3 there exists a sequence of interchanges that transforms a matrix in $\mathcal{A}(R, S)$ of α-width $\tilde{\epsilon}_\alpha$ into a matrix in $\mathcal{A}(R, S)$ of α-width $\bar{\epsilon}_\alpha$. This establishes the following intermediate value theorem of Fulkerson and Ryser [22].

Theorem 3.6.1 *Let R and S be nonnegative integral vectors such that $\mathcal{A}(R, S)$ is nonempty. For each integer ϵ with $\tilde{\epsilon}_\alpha(R, S) \le \epsilon \le \bar{\epsilon}_{\alpha(R,S)}$ there exists a matrix in $\mathcal{A}(R, S)$ of α-width ϵ.* $\square$

For the remainder of this section, we assume without loss of generality that both R and S are nonincreasing vectors. We next show that the special matrix $\tilde{A}$ in $\mathcal{A}(R, S)$, obtained by applying the Gale–Ryser algorithm in Section 3.1 and characterized in Theorem 3.1.2, has the property that $\epsilon_\alpha(\tilde{A})$ equals the minimal α-width of $\mathcal{A}(R, S)$ for each α and that this number can be read off from the last row of $\tilde{A}$. First we prove the following lemma.

Lemma 3.6.2 *Let* $R = (r_1, r_2, \ldots, r_m)$ *and* $S = (s_1, s_2, \ldots, s_n)$ *be non-increasing, nonnegative integral vectors and assume that the class* $\mathcal{A}(R,S)$ *is nonempty. Let* α *be any integer with* $0 \leq \alpha \leq r_m$. *Then there exists a matrix* B *in* $\mathcal{A}(R,S)$ *such that* $\epsilon_\alpha(B) = \tilde{\epsilon}(R,S)$, *where the submatrix* G *formed by the first* $\tilde{\epsilon}_\alpha(R,S)$ *columns of* B *has at least* α *1's in each row, and such that* $\delta_\alpha(B) = \delta_\alpha(R,S)$, *where the number of rows of* G *with exactly* α *1's equals* $\delta_\alpha(R,S)$.

Proof. Let $A = [a_{ij}]$ be a matrix in $\mathcal{A}(R,S)$ such that $\epsilon_\alpha(A) = \tilde{\epsilon}_\alpha$ and $\delta_\alpha(A) = \delta_\alpha$. Let $E = A[\cdot, K]$ be a submatrix of A such that $|K| = \tilde{\epsilon}_\alpha$, each row sum of E is at least α, and E has δ_α rows with exactly α 1's. Suppose that there is an integer j such that $1 \leq j \leq \tilde{\epsilon}_\alpha$ and $j \notin K$. Then there exists an integer k such that $k \in K$ and $\tilde{\epsilon}_\alpha < k \leq n$. First assume that column j of A contains a 1 in each row in which column k contains a 1. Then we replace k in K with j and E with $F = A[\cdot, K \setminus \{k\} \cup \{j\}]$. Now assume that there are $t \geq 1$ rows of A that have a 0 in column j and a 1 in column k. Since S is nonincreasing, there are at least t rows of A that have a 1 in column j and a 0 in column k. A sequence of t interchanges now gives a matrix A' in $\mathcal{A}(R,S)$ such that $F = A'[\cdot, K \setminus \{k\} \cup \{j\}]$ has at least α 1's in each row. Moreover, in both cases, since $s_j \geq s_k$, the number of rows of F with exactly α 1's is at most the number for E and hence equals δ_α. We may repeat this argument until we finally obtain a matrix B with the properties specified in the theorem. $\qquad \square$

We are now able to prove the following theorem of Fulkerson and Ryser [23] which shows, in particular, that the special matrix $\tilde{A}$ in $\mathcal{A}(R,S)$ is an instance of the matrix B in Lemma 3.6.2.

Theorem 3.6.3 *Let* $R = (r_1, r_2, \ldots, r_m)$ *and* $S = (s_1, s_2, \ldots, s_n)$ *be non-increasing, nonnegative integral vectors such that* $\mathcal{A}(R,S)$ *is nonempty. Let* α *be any integer with* $1 \leq \alpha \leq r_m$. *Then*

$$\tilde{\epsilon}_\alpha(R,S) = \epsilon_\alpha(\tilde{A}) \text{ and } \delta_\alpha(R,S) = \delta_\alpha(\tilde{A}).$$

Moreover, the α*th 1 in row* m *of* $\tilde{A}$ *occurs in column* $\tilde{\epsilon}_\alpha(R,S)$ *of* $\tilde{A}$ *and* $\delta_\alpha(R,S)$ *equals the number of rows of* $\tilde{A}[\cdot, \{1, 2, \ldots, \tilde{\epsilon}_\alpha(R,S)\}]$ *that contain exactly* α *1's.*

Proof. Let B be the matrix of Lemma 3.6.2, and let the row sum vector of the first k columns of B be R_k ($k = 1, 2, \ldots, n$). Let $\tilde{R}_k$ be the row sum vector of the first k columns of $\tilde{A}$ ($k = 1, 2, \ldots, n$). By Theorem 3.1.2, each of these partial row sum vectors $\tilde{R}_k$ is nonincreasing and $\tilde{R}_k \preceq R_k$ ($k = 1, 2, \ldots, n$). Let $R_{\tilde{\epsilon}_\alpha} = (u_1, u_2, \ldots, u_m)$, and let $\tilde{R}_{\tilde{\epsilon}_\alpha} = (v_1, v_2, \ldots, v_m)$. Since $\tilde{R}_{\tilde{\epsilon}_\alpha} \preceq R_{\tilde{\epsilon}_\alpha}$ and each component of $R_{\tilde{\epsilon}_\alpha}$ is at least α, we have $v_i \geq \alpha$ ($i = 1, 2, \ldots, m$). This implies that $v_m = \alpha$ and that the αth 1 in row m of $\tilde{A}$ occurs in column $\tilde{\epsilon}_\alpha$, for otherwise each of the rows of the first $\tilde{\epsilon}_\alpha - 1$ columns of $\tilde{A}$ would contain at least α 1's. If $\tilde{R}_{\tilde{\epsilon}_\alpha}$ had $t > \delta_\alpha$ components equal to α, then since $\tilde{R}_{\epsilon_\alpha} \preceq R_{\epsilon_\alpha}$, so would $R_{\tilde{\epsilon}_\alpha}$, a contradiction. Thus $(v_1, v_2, \ldots, v_m)$ has at most δ_α components equal to α and hence exactly δ_α components equal to α. □

Example. We illustrate the properties of the matrix $\tilde{A}$ given in Theorem 3.6.3 by taking $R = (4,4,4,3,3,3)$ and $S = (5,4,4,3,3,2)$. Using the Gale–Ryser algorithm given in Section 3.1, we obtain

$$\tilde{A} = \begin{bmatrix} 1 & 1 & 1 & 0 & 1 & 0 \\ 1 & 1 & 0 & 1 & 0 & 1 \\ 1 & 1 & 0 & 1 & 0 & 1 \\ 1 & 0 & 1 & 1 & 0 & 0 \\ 1 & 0 & 1 & 0 & 1 & 0 \\ 0 & 1 & 1 & 0 & 1 & 0 \end{bmatrix}.$$

Using Theorem 3.6.3, we obtain the following table from $\tilde{A}$:

width	multiplicity
$\tilde{\epsilon}_1 = 2$	$\delta_1 = 3$
$\tilde{\epsilon}_2 = 3$	$\delta_2 = 5$
$\tilde{\epsilon}_3 = 5$	$\delta_3 = 5$

□

Theorem 3.6.3 gives a very simple algorithm for determining the widths and multiplicities of a class $\mathcal{A}(R,S)$. After constructing the matrix $\tilde{A}$ using the Gale–Ryser algorithm, the minimal width $\tilde{\epsilon}_\alpha$ is evident from the last row of $\tilde{A}$, and the multiplicity δ_α equals $m-k$ where k is the smallest integer such that rows $k+1, k+2, \ldots, m$ of $\tilde{A}[\cdot, \{1, 2, \ldots, \tilde{\epsilon}_\alpha\}]$ contain exactly α 1's. Fulkerson and Ryser [22] have, in addition, given a formula for the minimal α-width based on the structure matrix of a class $\mathcal{A}(R,S)$.[8]

Let $R = (r_1, r_2, \ldots, r_m)$ and $S = (s_1, s_2, \ldots, s_n)$ be nonincreasing, nonnegative integral vectors such that $\mathcal{A}(R,S)$ is nonempty, and let $T = [t_{kl}]$ be the structure matrix for R and S. Let $\epsilon, e,$ and f be integers with $0 \le e \le m$ and $0 \le \epsilon \le f \le n$, and let

$$N(\epsilon, e, f) = (r_{e+1} + r_{e+2} + \cdots + r_m) - (s_{\epsilon+1} + s_{\epsilon+2} + \cdots + s_f) - e(f - \epsilon).$$

Let A be a matrix in $\mathcal{A}(R,S)$ and suppose that A is partitioned as

$$A = \begin{bmatrix} * & Y & * \\ X & * & Z \end{bmatrix},$$

where X is an $m - e$ by ϵ matrix, Y is an e by $f - \epsilon$ matrix, and Z is an $m - e$ by $n - f$ matrix. We have

$$N(\epsilon, e, f) = N_1(X) + N_0(Y) + N_1(Z),$$

---

[8]In fact, this formula came first [22], before the special role played by the matrix $\tilde{A}$ in determining the minimal widths was observed [23].

the number of 1's in X and Z plus the number of 0's in Y. An easy calculation shows that

$$N(0, e, f) = t_{ef} \text{ and } N(\epsilon, e, f) = t_{ef} + t_{e0} - t_{e\epsilon};$$

thus the numbers $N(\epsilon, e, f)$ are generalizations of the basic invariants t_{ij} of $\mathcal{A}(R,S)$.

Theorem 3.6.4 *Let R and S be nonincreasing, nonnegative integral vectors such that $\mathcal{A}(R,S)$ is nonempty. Let α be an integer with $1 \le \alpha \le r_m$. Then the minimal α-width $\tilde{\epsilon}_\alpha(R,S)$ of $\mathcal{A}(R,S)$ is given by*

$$\min\{\epsilon : N(\epsilon, e, f) \ge \alpha(m - e) \text{ for all } e, f \text{ with } 0 \le e \le m, \epsilon \le f \le n\}. \tag{3.29}$$

Proof. We first show that the matrix $\tilde{A}$ has the form

$$\begin{bmatrix} G & K & * \\ F & & \\ E & * & L \end{bmatrix}, \tag{3.30}$$

where E has size δ_α by $\tilde{\epsilon}_\alpha$ with exactly α 1's in each row, F has exactly $\alpha + 1$ 1's in each row, G has at least $\alpha + 1$ 1's in each row, and K and L are all 1's and all 0's matrices, respectively. Some of the matrices G, F, K and L may be vacuous.

It follows from Theorems 3.1.2 and 3.6.3 that $\tilde{A}$ has the form (3.30) where E has the specified size, and the numbers of 1's in the rows of the matrices F and G have the required properties. This does not yet define the matrices F and G, since G may have some rows with exactly $\alpha + 1$ 1's. We denote by H the matrix formed by G and F. Since each row of H contains at least $\alpha + 1$ 1's and each row of E contains exactly α 1's, it follows that for each row of H and each row of E there is a 1 above a 0. If to the right of column $\tilde{\epsilon}_\alpha$ there were a 0 above a 1 in the corresponding rows, then an interchange would contradict the minimality of δ_α. Hence there is no such 0 above a 1, that is, in any column j with $j > \tilde{\epsilon}_\alpha$, a 1 in one of the last δ_α positions implies that there are only 1's in the first $m - \delta_\alpha$ positions. Since the column sums are nonincreasing, this implies that $\tilde{A}$ has the form (3.30) with the properties as stated. Each of the first $\tilde{\epsilon}_\alpha$ columns of $\tilde{A}$ must contain strictly more than $m - \delta_\alpha$ 1's, for otherwise we may use interchanges confined to these columns and create a column of 0's in E, contradicting the minimality of $\tilde{\epsilon}_\alpha$.

Considering $\tilde{A}$ and using the fact that the numbers $N(\epsilon, e, f)$ depend only on R and S, we see that the minimum ϵ' in (3.29) is at most $\tilde{\epsilon}_\alpha(R,S)$.

Suppose that for some fixed ϵ, $N(\epsilon, e, f) \ge \alpha(m - e)$ for all e and f as restricted in (3.29) where $\epsilon < \tilde{\epsilon}_\alpha$. In (3.30), let G have e' rows and let

$K = J$ have $f' - \tilde{\epsilon}_\alpha$ columns. Then $0 \le e' < m$ and $\epsilon < \tilde{\epsilon}_\alpha \le f' \le n$. We have

$$N(\epsilon, e', f') = N(\tilde{\epsilon}_\alpha, e', f') + N_0(T) - N_1(U), \qquad (3.31)$$

where T is the submatrix at the intersection of rows $1, 2, \ldots, e'$ and columns $\epsilon + 1, \epsilon + 2, \ldots, \tilde{\epsilon}_\alpha$ of $\tilde{A}$, and U is the submatrix at the intersection of rows $e' + 1, e' + 2, \ldots, m$ and columns $\epsilon + 1, \epsilon + 2, \ldots, \tilde{\epsilon}_\alpha$ of $\tilde{A}$. Since each of the first $\tilde{\epsilon}_\alpha$ columns of $\tilde{A}$ contains strictly more than $m - \delta_\alpha$ 1's, we have

$$N_1(U) - N_0(T) + e'(\tilde{\epsilon}_\alpha - \epsilon) = s_{\epsilon+1} + \cdots + s_{\tilde{\epsilon}_\alpha} > (m - \delta_\alpha)(\tilde{\epsilon}_\alpha - \epsilon),$$

and so

$$N_1(U) - N_0(T) > m - (\tilde{\epsilon}_\alpha + \delta_\alpha).$$

We also have that

$$N(\tilde{\epsilon}_\alpha, e', f') = (\alpha + 1)(m - e') - \delta_\alpha.$$

It now follows that

$$\begin{aligned} N(\epsilon, e', f') &= (\alpha + 1)(m - e') - \delta_\alpha + N_0(T) - N_1(U) \\ &< (\alpha + 1)(m - e') - \delta_\alpha - m + e' + \delta_\alpha \\ &= \alpha(m - e'). \end{aligned}$$

This contradiction completes the proof. $\qquad\qquad\square$

We now turn our attention to the maximal width $\bar{\epsilon}_\alpha(R, S)$ of a nonempty class $\mathcal{A}(R, S)$. Here the results are much less complete and consist primarily of special cases and upper bounds, often of a technical nature. Evidence for the difficulty of the maximum width problem was provided by Ryser [62] who showed that the problem of determining the 1-width of certain regular classes of matrices is equivalent to the existence problem for finite projective planes.

We recall (see [13], [62] for more details) that a finite projective plane of order $n \ge 2$ is equivalent to a $(0, 1)$-matrix A of order $n^2 + n + 1$ with exactly $n + 1$ 1's in each row and column such that

$$A^{\mathrm{T}}A = nI_{n^2+n+1} + J_{n^2+n+1}. \qquad (3.32)$$

The matrix A is the line–point incidence matrix of the plane. The *complementary matrix* $B = J_{n^2+n+1} - A$ has exactly n^2 1's in each row and column and satisfies

$$B^{\mathrm{T}}B = nI_{n^2+n+1} + n(n - 1)J_{n^2+n+1}. \qquad (3.33)$$

The problem of determining for what integers n there exists a finite projective plane of order n is a major unsettled combinatorial problem. We now establish Ryser's theorem [62].

Theorem 3.6.5 *Let $n \geq 2$ be an integer, and let $v = n^2 + n + 1$. Let $K_v = (n^2, n^2, \ldots, n^2)$ be the vector of size v each of whose components equals n^2. The 1-width $\epsilon_1(B)$ of matrices B in $\mathcal{A}(K_v, K_v)$ that are complements of incidence matrices of projective planes of order n equals 3, and all other matrices in $\mathcal{A}(K_v, K_v)$ have 1-width equal to 2. Thus $\bar{\epsilon}_1(K_v, K_v) = 3$ if there exists a projective plane of order n, and $\bar{\epsilon}_1(K_v, K_v) = 2$ otherwise.*

Proof. Let B be a matrix in $\mathcal{A}(K_v, K_v)$. Let m be the minimum value and let $\bar{m}$ be the average value of the elements of $B^{\mathrm{T}}B$ that are not on its main diagonal. Using the fact that each row and column of B contains n^2 1's, we compute the average value to be

$$\bar{m} = \frac{n^2(n^2 - 1)}{n^2 + n} = n(n - 1).$$

The fact that each column of B is of size $n^2 + n + 1$ and contains exactly n^2 1's implies that $m \geq n(n - 1) - 1$, and hence that $m = n(n - 1)$ or $n(n - 1) - 1$.

First assume that $m = n(n - 1)$. Then each element of $B^{\mathrm{T}}B$ not on its main diagonal equals $n(n - 1)$ and B is the complement of the incidence matrix of a projective plane of order n. Moreover, since each pair of distinct columns has exactly $m = n(n - 1)$ 1's in the same rows, each v by 2 submatrix of B has exactly one row of 0's. We conclude that in this case, $\epsilon_1(B) = 3$. Now assume that $m = n(n - 1) - 1$. Then B is not the complement of the incidence matrix of a projective plane and there is a pair of columns that has no row of 0's. Hence in this case $\epsilon_1(B) = 2$. The theorem now follows. $\square$

Theorem 3.6.5 implies that the determination of the maximal 1-width $\bar{\epsilon}_1$ is likely to be a computationally difficult problem. Considerable effort has gone into obtaining upper bounds for $\bar{\epsilon}_1$. These results are usually quite technical and, as a result, we limit our discussion of them here and refer to the original papers for more details. One of the first in a sequence of papers is that of Gel'fond [26]. For k and n integers with $0 \leq k \leq n$, we use the more concise notation $\mathcal{A}(n, k)$ for the class of all $(0, 1)$-matrices of order n with exactly k 1's in each row and column and $\bar{\epsilon}_\alpha(n, k)$ for the maximal α-width of the class $\mathcal{A}(n, k)$.

Theorem 3.6.6 *Let n and k be integers with $1 \leq k \leq n$. Let A be a $(0, 1)$-matrix of order n with exactly k 1's in each row and column. Then*

$$\epsilon_1(A) \leq \left\lfloor \frac{k}{2k - 1} n \right\rfloor \text{ and hence } \bar{\epsilon}_1(n, k) \leq \left\lfloor \frac{k}{2k - 1} n \right\rfloor.$$

Proof. Consider a $J \subseteq \{1, 2, \ldots, n\}$ of smallest cardinality $r = |J|$ such that $E = A[\cdot, J]$ has no row of all 0's. Since r is minimal, each of the r columns of E contains a 1 which is the only 1 in its row. The corresponding r rows of the matrix $F = A[\cdot, \bar{J}]$ each contain $k - 1$ 1's. Hence F contains at least $(k-1)r$ 1's. On the other hand, since each column of F contains k 1's, F contains exactly $(n-r)k$ 1's. Therefore, $(k-1)r \leq (n-r)k$ and the theorem follows.   □

The following corollary is from Fulkerson and Ryser [24] (see also [72]).

Corollary 3.6.7 *Let n be an integer with $n \geq 2$. Then*

$$\bar{\epsilon}_1(n,2) = \left\lfloor \frac{2n}{3} \right\rfloor.$$

Proof. By Theorem 3.6.6 we have $\bar{\epsilon}_1(n,2) \leq \lfloor 2n/3 \rfloor$. We now construct matrices in $\mathcal{A}(n,2)$ with 1-width equal $\lfloor 2n/3 \rfloor$. The matrices

$$A_2 = J_2, \ A_3 = \begin{bmatrix} 1 & 1 & 0 \\ 1 & 0 & 1 \\ 0 & 1 & 1 \end{bmatrix}, \text{ and } A_4 = J_2 \oplus J_2$$

work for $n = 2, 3$, and 4, respectively. Now let $n \geq 5$ and write $n = 3k + a$ where $a = 0, 1$, or 2. We then take A of the form

$$A_3 \oplus A_3 \oplus \cdots \oplus A_3 \quad (a = 0), \ A_3 \oplus \cdots \oplus A_3 \oplus A_4 \quad (a = 1),$$

$$\text{and } A_3 \oplus \cdots \oplus A_3 \oplus A_2 \quad (a = 2)$$

to obtain a matrix with 1-width equal to $\lfloor 2n/3 \rfloor$.   □

Henderson and Dean [34] and Tarakanov [73, 74] (see also [35], [70], [71]) have obtained an estimate for the 1-width of an arbitrary $(0,1)$-matrix in terms of the minimal row sum and maximal column sum. We quote the following theorem from [74] without proof.

Theorem 3.6.8 *Let A be an m by n $(0,1)$-matrix with at least r 1's in each row and at most s 1's in each column, Then*

$$\epsilon_1(A) \leq n \left(1 - (s - m/n) \frac{(r-1)^{s-1}(s-1)!}{(2r-1)(3r-2)\dots(sr-(s-1))} \right).$$

For integers m, n, r, and s with $0 \leq r \leq n, 0 \leq s \leq m$, and $mr = ns$, we use the notation $\mathcal{A}(m, n; r, s)$ to denote the class $\mathcal{A}(R, S)$ with constant row sum vector $R = (r, r, \dots, r)$ and constant column sum vector $S = (s, s, \dots, s)$. The maximal 1-width of such a class is denoted by $\bar{\epsilon}_1(m, n; r, s)$. The constraints on m, n, r, and s imply that this class is nonempty. Tarakanov [74] also obtained the following results for these classes.

Theorem 3.6.9 *The following hold:*

(i) $\bar{\epsilon}_1(m, n : 2, s) = \left\lfloor \frac{2m}{s+1} \right\rfloor$;

$$\text{(ii) } \bar{\epsilon}_1(m,n;r,3) \leq \begin{cases} \left\lfloor \frac{5s-4}{3(3s-2)}n \right\rfloor, & \text{if } s \text{ is even}, \\ \\ \left\lfloor \frac{5s-3}{3(3s-1)}n \right\rfloor, & \text{if } s \text{ is odd}; \end{cases}$$

$$\text{(iii) } \bar{\epsilon}_1(m,n;r,4) \leq \begin{cases} \left(1 - \frac{(5s-3)^2-1}{4(3s-2)(4s-3)}\right) n, & \text{if } s \text{ is even}, \\ \\ \left(1 - \frac{(5s-2)^2+s-4}{4(3s-1)(4s-3)}\right) n, & \text{if } s \text{ is odd}. \end{cases}$$

Corollary 3.6.10 *The following hold:*

(i) [34, 74] $\bar{\epsilon}_1(n,3) = \left\lfloor \frac{n}{2} \right\rfloor$;

(ii) [34, 74] $\bar{\epsilon}_1(n,4) \leq \left\lfloor \frac{29}{65}n \right\rfloor$;

(iii) [34, 74] $\bar{\epsilon}_1(m,n;3,4) = \left\lfloor \frac{5}{12}m \right\rfloor = \left\lfloor \frac{5}{9}n \right\rfloor$.

Information about the 1-widths of classes of matrices connected with combinatorial designs is contained in [24], [57], [68]; information about the distribution of 1-widths can be found in [42].

We conclude this section by quoting the following estimate of Leont'ev [43] for the maximal 1-width of the class of matrices of a prescribed size having a constant number of 1's in each row but with no restrictions on the columns.

Theorem 3.6.11 *Let $\mathcal{B}(m,n;k)$ denote the class of all m by n $(0,1)$-matrices with exactly k 1's in each row. Then*

$$\max\{\epsilon_1(A) : A \in \mathcal{B}(m,n;k)\} \leq \frac{n}{k} \ln \frac{mk}{n} + \frac{n}{k} + 1.$$

3.7   Trace

Let $A = [a_{ij}]$ be a matrix of size m by n. The *trace* of A is defined by

$$\sigma(A) = \sum_{i=1}^{\min\{m,n\}} a_{ii}.$$

We refer to the positions $\{(1,1),(2,2),\ldots,(p,p)\}$, $p = \min\{m,n\}$, as the *main diagonal* of A. The trace of a matrix is not in general invariant under arbitrary permutations of its rows and columns but, in case $m = n$, the trace of a matrix is invariant under simultaneous permutations of its rows and columns. In this section we investigate the trace of matrices in a nonempty class $\mathcal{A}(R,S)$, and derive formulas for the maximal and minimal trace attained by matrices in $\mathcal{A}(R,S)$. The results in this section are taken from Ryser [61].

Let
$$\tilde{\sigma} = \tilde{\sigma}(R,S) = \min\{\sigma(A) : A \in \mathcal{A}(R,S)\}$$
denote the *minimal trace* of a matrix in $\mathcal{A}(R,S)$, and let
$$\bar{\sigma} = \bar{\sigma}(R,S) = \max\{\sigma(A) : A \in \mathcal{A}(R,S)\}$$
denote the *maximal trace* of a matrix in $\mathcal{A}(R,S)$. In determining $\tilde{\sigma}$ and $\bar{\sigma}$ we make the assumption that R and S are nonincreasing. This does impose a restriction but this is necessary in order to obtain the uncomplicated conclusions to be described. We also assume that R and S are positive vectors and this is without loss of generality. Our assumptions imply that $\bar{\sigma}(R,S) \geq 1$, since if A is a matrix with a 0 in the (1,1)-position, an interchange is available to move a 1 to that position.

The matrices
$$\begin{bmatrix} 1 & 0 \\ 0 & 1 \end{bmatrix} \text{ and } \begin{bmatrix} 0 & 1 \\ 1 & 0 \end{bmatrix}$$

demonstrate that, unlike the term rank, a single interchange may alter the trace by 2, but clearly by no more than 2. Ryser [61] has characterized those classes $\mathcal{A}(R,S)$ for which the trace takes on all integral intermediate values between $\tilde{\sigma}$ and $\bar{\sigma}$.

We first show that any trace realizable in $\mathcal{A}(R,S)$ is realizable by a matrix in $\mathcal{A}(R,S)$ with the 1's in the initial positions on the main diagonal [61].

Theorem 3.7.1 *Let R and S be nonincreasing, positive integral vectors. If $\mathcal{A}(R,S)$ contains a matrix of trace t, then $\mathcal{A}(R,S)$ contains a matrix of trace t with the 1's in the initial positions on the main diagonal.*

Proof. Let $A = [a_{ij}]$ be a matrix in $\mathcal{A}(R,S)$ of trace t with
$$a_{11} = a_{22} = \cdots = a_{k-1,k-1} = 1 \text{ and } a_{kk} = 0,$$
where $k \leq t$. To establish the theorem, we show that it is possible to transform A by interchanges into a matrix of trace t with 1's in the initial k positions on the main diagonal.

There exists an integer $l > t$ such that $a_{ll} = 1$. We consider four cases according to whether (a_{kl}, a_{lk}) equals $(0,0), (1,0), (0,1)$, or $(1,1)$. We treat only the case $(a_{kl}, a_{lk}) = (0,0)$, because the other cases can be handled by similar arguments. Since R and S are nonincreasing, $a_{kl} = a_{lk} = 0, a_{ll} = 1$, and $k < l$ imply that there exist integers u and v such that $a_{uk} = 1, a_{ul} = 0$ and $a_{kv} = 1, a_{lv} = 0$. An $(l, u; k, l)$-interchange followed by a $(k, l; l, v)$-interchange puts a 0 in position (l, l) and a 1 in position (k, k) and leaves unaltered all other elements on the main diagonal. The theorem now follows. $\square$

To obtain a formula for the maximal trace, we first show the existence of a matrix of maximal trace in $\mathcal{A}(R,S)$ of very special form.

Theorem 3.7.2 *Let R and S be nonincreasing, positive integral vectors of sizes m and n, respectively, such that $\mathcal{A}(R,S)$ is nonempty. Assume that the maximal trace $\bar{\sigma}$ for $\mathcal{A}(R,S)$ satisfies $\bar{\sigma} \neq \min\{m,n\}$. Then there exist integers e and f with $0 < e \leq \bar{\sigma}$ and $0 < f \leq \bar{\sigma}$, and there exists a matrix $A_{\bar{\sigma}}$ in $\mathcal{A}(R,S)$ of the form*

$$A_{\bar{\sigma}} = \begin{bmatrix} J_{e,f} & * & * \\ * & O' & O \\ * & O & O_{m-\bar{\sigma},n-\bar{\sigma}} \end{bmatrix}, \tag{3.34}$$

where O' is a matrix of size $\bar{\sigma} - e$ by $\bar{\sigma} - f$ having 1's in the corresponding main diagonal positions of $A_{\bar{\sigma}}$ and 0's elsewhere.

Proof. Let $R = (r_1, r_2, \ldots, r_m)$ and $S = (s_1, s_2, \ldots, s_n)$. By Theorem 3.7.1 there exists a matrix A in $\mathcal{A}(R,S)$ with $\bar{\sigma}$ 1's in the initial positions of the main diagonal. Because R and S are nonincreasing, the submatrix of A of size $m - \bar{\sigma}$ by $n - \bar{\sigma}$ in the lower right corner is a zero matrix, for otherwise a single interchange would increase the trace. We now claim that the matrix A of trace $\bar{\sigma}$ may be taken to be of the form

$$A = \begin{bmatrix} S_1 & * & U \\ * & * & O \\ V & O & O_{m-\bar{\sigma},n-\bar{\sigma}} \end{bmatrix} \tag{3.35}$$

where U has at least one 1 in each row and V has at least one 1 in each column. To see this, suppose that rows k and l with $1 \leq k < l \leq \bar{\sigma}$ have the property that row k has all 0's in its last $n - \bar{\sigma}$ positions while row l has at least one 1 in its last $n - \bar{\sigma}$ positions. If $r_k > r_l$, then we may apply an interchange that puts a 0 in one of the last $n - \bar{\sigma}$ positions of row k and does not shift a 1 off the main diagonal. If $r_k = r_l$, we may have to apply an interchange that shifts a 1 off the main diagonal, but then a second interchange is available within the first $\bar{\sigma}$ columns that restores this 1 on the main diagonal. In this way we get the matrices U and the zero matrix below it in (3.35). In a similar way we get the matrices V and the zero matrix O to its right; if we have to shift a 1 off the main diagonal, then a second interchange within the first $\bar{\sigma}$ rows and columns restores it or else an interchange would move a 1 to a position occupied by $O_{m-\bar{\sigma},n-\bar{\sigma}}$ contradicting the maximality of $\bar{\sigma}$. Thus we obtain a matrix of the form (3.35).

Let the matrix S_1 be of size e' by f'. Then $S_1 = J_{e',f'}$, for otherwise we could increase $\bar{\sigma}$ by interchanges. A matrix A of the form given in (3.35) with e' and f' fixed will be called *special* (in $\mathcal{A}(R,S)$) and the 1's on its main diagonal called the *essential* 1's of A. Without loss of generality, we assume that $e' \leq f'$.

We now consider the further structure of special matrices by considering initial columns of all 1's of the matrix labeled with an $*$ to the right of S_1

in (3.35). This leads to a special A of the form

$$A^* = \left[\begin{array}{c|c|c|c} S_1 & S_2 & Y & U \\ & X & Z & O \\ \hline V & & O & O \end{array} \right] \tag{3.36}$$

where $S_2 = J_{e',f^*-f'}$ with $f^* - f'$ maximal in A^*. Among all special A in $\mathcal{A}(R,S)$ we select A^* whose corresponding $f^* - f'$ is minimal. If $f^* - f' > 0$, then S_2 is a matrix of all 1's that appears in all of the special matrices in $\mathcal{A}(R,S)$. If Y (and hence Z) is empty, then (3.34) holds. We now assume that Y is nonempty and hence the matrix A^* has a 0 in the first column of Y. If the matrix Z does not contain an unessential 1, then again (3.34) holds. Now assume that there is an unessential 1 in position (s,t) of Z where s is chosen to be maximal. If $t = 1$, then a 0 occurs in column t of Y. Suppose that $t \neq 1$ and let position $(s,1)$ of Z contain a 0 or an essential 1. If column t of Y contains only 1's, then we may perform an interchange using unessential 1's and the 0 in column 1 of Y to obtain a 0 in column t of Y. Thus we may now require our matrix A^* to have a 0 in column t of Y, but not in column 1 any more. This now implies that row s of X has only 1's, for otherwise either an interchange gives a special A^* with a 0 in S_2, contradicting the minimality of $f^* - f'$, or a sequence of two interchanges place a 1 in the zero matrix in the lower right corner, contradicting the maximality of $\bar{\sigma}$.

We now show that rows $1, 2, \ldots, s - 1$ of X contain only 1's, and this implies, invoking the maximality of s, that we have a matrix of the required form (3.34).

Suppose that X contains a 0 in a position (u,v) where $u < s$. Then we may apply an interchange involving this 0 and the 1 in position (s,v) of X. If this interchange does not involve an essential 1, then a second interchange involving the unessential 1 in position (s,t) of Z puts a 0 in either S_1 or S_2, and this leads to a contradiction as before. Thus we suppose that the interchange involving the 0 in position (u,v) and the 1 in position (s,v) involves an essential 1. We consider two cases. First suppose that $s \leq f^* - e'$. Then a second interchange involving the unessential 1 in position (s,t) of Z restores trace $\bar{\sigma}$ and puts a 0 in either S_1 or S_2, and this gives us a contradiction again. Now suppose that $s > f^* - e'$. Then the second interchange involving the unessential 1 in position (s,t) of Z puts a 0 in either S_1 or S_2 but leaves the trace at $\bar{\sigma} - 1$. But then a third interchange involving rows u and s of Z restores trace $\bar{\sigma}$. Thus we have a contradiction in this case as well. Thus rows $1, 2, \ldots, s - 1$ of X contain only 1's, and the proof is complete. $\square$

We now obtain a remarkably simple formula for maximal trace involving the elements of the structure matrix.

Theorem 3.7.3 *Let R and S be nonincreasing, positive integral vectors such that $\mathcal{A}(R,S)$ is nonempty, and let $T = [t_{ij}]$ be the structure matrix for*

$\mathcal{A}(R,S)$. *Then*

$$\bar{\sigma}(R,S) = \min\{t_{kl} + \max\{k,l\} : 0 \le k \le m; 0 \le l \le n\}. \qquad (3.37)$$

Proof. Let A be a matrix of maximal trace $\bar{\sigma} = \bar{\sigma}(R,S)$. Let

$$B = A[\{1,2,\ldots,k;1,2,\ldots,l\}]$$

and let

$$C = A[\{k+1,k+2,\ldots,m\};\{l+1,l+2,\ldots,n\}].$$

Then, as in Section 2.1,

$$\begin{aligned}
t_{kl} + \max\{k,l\} &= N_0(B) + N_1(C) + \max\{k,l\} \\
&\ge N_1(C) + \max\{k,l\} \\
&\ge (\bar{\sigma} - \max\{k,l\}) + \max\{k,l\} \\
&= \bar{\sigma}. \qquad (3.38)
\end{aligned}$$

First suppose that $\bar{\sigma} \ne \min\{m,n\}$. Then taking A to be the matrix $A_{\bar{\sigma}}$ of Theorem 3.7.2, and taking $k = e$ and $l = f$, we obtain that $N_0(B) = 0$ and $N_1(C) = \bar{\sigma} - \max\{e,f\}$, and we have equality throughout (3.38) for $k = e$ and $l = f$. If $\bar{\sigma} = m$, equality is attained for $k = m$ and $l = 0$. If $\bar{\sigma} = n$, then equality is attained for $k = 0$ and $l = n$. $\square$

We now consider the minimal trace $\tilde{\sigma} = \tilde{\sigma}(R,S)$.

Theorem 3.7.4 *Let R and S be nonincreasing, positive integral vectors of sizes m and n, respectively, such that $\mathcal{A}(R,S)$ is nonempty. Let u and v be the maximal integers such that the first u rows and first v columns of all matrices in $\mathcal{A}(R,S)$ are composed entirely of 1's. Assume that $\tilde{\sigma} \ne \max\{u,v\}$. Then there exists a matrix $A_{\tilde{\sigma}}$ of trace $\tilde{\sigma}$ in $\mathcal{A}(R,S)$ of the form*

$$A_{\tilde{\sigma}} = \begin{bmatrix} J_{\tilde{\sigma}} & J_{\tilde{\sigma},s} & * \\ J_{t,\tilde{\sigma}} & W & * \\ * & * & O \end{bmatrix}, \qquad (3.39)$$

where W is a t by s matrix with 0's in the main diagonal positions of $A_{\tilde{\sigma}}$ and 1's in all other positions. Here s and t are nonnegative integers.

Proof. By adding zero rows or zero columns, we may assume without loss of generality that $n = m$. Let A be a matrix in $\mathcal{A}(R,S)$. Let $\hat{A} = J_n - A$ be the complement of A. Then $\hat{A}$ determines a class $\mathcal{A}(\hat{R},\hat{S})$, and we let C be a matrix of maximal trace $\bar{\sigma}_c$ in this class. We have $\tilde{\sigma} = n - \bar{\sigma}_c$. Since $\tilde{\sigma} \ne \max\{u,v\}$, we have that $\bar{\sigma}_c \ne n - \max\{u,v\}$. Applying Theorem 3.7.2 to the submatrix of C of size $n - u$ by $n - v$ in its lower right corner and observing that the row and column sums of the matrices in the class

$\mathcal{A}(\hat{R}, \hat{S})$ are in nondecreasing order, we see that C can be taken to have the form

$$
C = \begin{bmatrix} O_{u,v} & O & * \\ O & \begin{array}{|c|c|} \hline O & O \\ \hline O & P \\ \hline \end{array} & * \\ * & * & J_{e',f'} \end{bmatrix},
\tag{3.40}
$$

where P is of size $\bar{\sigma}_c - e'$ by $\bar{\sigma}_c - f'$ and has 1's in the main diagonal positions of C and 0's in all other positions. Taking the complement of C and deleting any added zero rows or columns, we obtain a matrix in $\mathcal{A}(R, S)$ of the form specified in (3.39. $\qquad\square$

We now obtain a formula for minimal trace.

Theorem 3.7.5 *Let R and S be nonincreasing, positive integral vectors of sizes m and n, respectively, such that $\mathcal{A}(R, S)$ is nonempty. Then*

$$
\tilde{\sigma}(R, S) = \max\{\min\{k, l\} - t_{kl} : 0 \le k \le m; 0 \le l \le n\}.
\tag{3.41}
$$

Proof. By Theorem 3.7.1 there is a matrix A in $\mathcal{A}(R, S)$ of trace $\tilde{\sigma}$ with $\tilde{\sigma}$ 1's in the initial positions on the main diagonal. As in the proof of Theorem 3.7.3, let

$$
B = A[\{1, 2, \ldots, k\}, \{1, 2, \ldots, l\}]
$$

and let

$$
C = A[\{k+1, k+2, \ldots, m\}, \{l+1, l+2, \ldots, n\}].
$$

Then

$$
\begin{aligned}
\min\{k, l\} - t_{kl} &= \min\{k, l\} - N_0(B) - N_1(C) \\
&\le \min\{k, l\} - N_0(B) \\
&\le \min\{k, l\} - (\min\{k, l\} - \tilde{\sigma}) \\
&= \tilde{\sigma}.
\end{aligned}
\tag{3.42}
$$

Suppose first that $\tilde{\sigma} \ne \max\{u, v\}$ where the matrices in $\mathcal{A}(R, S)$ have precisely u rows and v columns consisting entirely of 1's. Then considering the matrix $A_{\tilde{\sigma}}$ in (3.39) of Theorem 3.7.4, we have $N_0(B) = \min\{k, l\}$ and $N_1(C) = 0$. Thus in this case we have a k and l for which equality holds throughout (3.42). If $\tilde{\sigma} = u$, then equality is attained for $k = \tilde{\sigma}$ and $l = n$. If $\tilde{\sigma} = v$, then equality is attained for $k = m$ and $l = \tilde{\sigma}$. $\qquad\square$

3.8 Chromatic Number

Let A be a $(0, 1)$-matrix of size m by n whose row sum vector is $R = (r_1, r_2, \ldots, r_m)$. We recall from Chapter 1 that for a positive integer t, a t-coloring of A is a partition of $\{1, 2, \ldots, n\}$ into t sets $I_1, I_2, \ldots, I_t$ with the property that for each i such that $r_i \ge 2$, the ith row sum of at least two of the matrices $A[\cdot, I_i]$ is nonzero. The sets $I_1, I_2, \ldots, I_t$ are the color classes of the t-coloring. The chromatic number $\gamma(A)$ of A is the smallest integer t such that A has a t-coloring.

Now let $S = (s_1, s_2, \ldots, s_n)$ be the column sum vector of A. Let

$$\tilde{\gamma} = \tilde{\gamma}(R, S) = \min\{\gamma(A) : A \in \mathcal{A}(R, S)\}$$

denote the *minimal chromatic number* of a matrix in $\mathcal{A}(R, S)$, and let

$$\bar{\gamma} = \tilde{\gamma}(R, S) = \max\{\gamma(A) : A \in \mathcal{A}(R, S)\}$$

denote the *maximal chromatic number* of a matrix in $\mathcal{A}(R, S)$. The chromatic number of a matrix is invariant under arbitrary permutations of its rows and columns. Hence without loss of generality we may assume that R and S are nonincreasing. The following example and theorem are from [9].

Example. Let $R = (2, 2, 2, 2, 2, 2)$ and $S = (3, 3, 3, 3)$. Let

$$A = \begin{bmatrix} 1 & 1 & 0 & 0 \\ 1 & 1 & 0 & 0 \\ 1 & 0 & 0 & 1 \\ 0 & 1 & 1 & 0 \\ 0 & 0 & 1 & 1 \\ 0 & 0 & 1 & 1 \end{bmatrix}.$$

The partition $\{1, 3\}, \{2, 4\}$ of $\{1, 2, 3, 4\}$ shows that $\gamma(A) = 2$. Let

$$B = \begin{bmatrix} 1 & 1 & 0 & 0 \\ 1 & 0 & 1 & 0 \\ 1 & 0 & 0 & 1 \\ 0 & 1 & 1 & 0 \\ 0 & 1 & 0 & 1 \\ 0 & 0 & 1 & 1 \end{bmatrix}$$

be the matrix obtained from A by a $(2, 5; 2, 3)$-interchange. For each pair of distinct columns, there exists a row of B that has both of its 1's in those two columns. This implies that $\gamma(B) = 4$. Hence an interchange can alter the chromatic number by 2, but no more than 2 according to the following theorem. $\square$

Theorem 3.8.1 *Let B be obtained from a matrix A by a single interchange. Then $\gamma(B) = \gamma(A) \pm 2$.*

Proof. Let $I_1, I_2, \ldots, I_{\gamma(A)}$ be a $\gamma(A)$-coloring of A. Let B be obtained from A by a $(p, q; k, l)$-interchange. Let k be in I_u and let l be in I_v. Then replacing I_u by the pair $I_u \backslash \{k\}$ and $\{k\}$, and replacing I_v the pair by $I_v \backslash \{l\}$ and $\{l\}$, we obtain a $(\gamma(A) + 2)$-coloring of B. Hence $\gamma(B) \leq \gamma(A) + 2$.

Since A can be obtained from B by a single interchange, we also have $\gamma(A) \leq \gamma(B) + 2$. $\qquad\qquad\qquad\qquad\qquad\qquad\qquad\qquad\qquad\qquad\qquad\quad\square$

If a matrix has at most one 1 in each row, then its chromatic number equals 1; if there are at least two 1's in some row, then its chromatic number is greater than 1. The next theorem shows that the minimal chromatic number of a class cannot be more than 3.

Theorem 3.8.2 *Let $\mathcal{A}(R,S)$ be nonempty. Then $\tilde{\gamma}(R,S) \leq 3$. In addition, $\tilde{\gamma}(R,S) = 2$ if $r_i \geq 3 \quad (i = 1, 2, \ldots, m)$.*

Proof. The theorem is trivially true if $n \leq 2$. We assume that $n \geq 3$ and that R and S are nonincreasing. Without loss of generality we also assume that $r_i \geq 2 \quad (i = 1, 2, \ldots, m)$

First suppose that $r_i \geq 3 \quad (i = 1, 2, \ldots, m)$. We show that there is a 2-coloring of the special matrix $\tilde{A} = [a_{ij}]$, and hence $\gamma(\tilde{A}) = 2$. Let the row sum vector of the m by q submatrix $\tilde{A}[\cdot, \{1, 2, \ldots, q\}]$ of $\tilde{A}$ be $R^{[q]} = (r_1^{[q]}, r_2^{[q]}, \ldots, r_m^{[q]}) \, (q = 1, 2, \ldots, n)$. By the construction of $\tilde{A}$, $R^{[q]}$ is a nonincreasing vector for each q. Let k be the largest integer such that

$$a_{ik} + a_{i,k+1} + \cdots + a_{in} \geq 1 \quad (i = 1, 2, \ldots, m). \tag{3.43}$$

There exists an integer p with $1 \leq p \leq m$ such that $a_{pk} = 1$ and $a_{pj} = 0 (j = k + 1, \ldots, n)$. We have $r_p^{[k]} = r_p \geq 3$, and hence that $r_i^{[k]} \geq 3$ for $i = 1, 2, \ldots, p$. We now show that $r_i^{[k]} \geq 2$ for $i = p + 1, p + 2, \ldots, m$.

Suppose that $r_u^{[k]} \leq 1$ for some integer u with $p + 1 \leq u \leq m$. Since $r_u \geq 3$, at least two of $a_{u,k+1}, a_{u,k+2}, \ldots, a_{un}$ equal 1. Let a_{uv} be the first among these that equals 1. Since $r_p \geq r_u$, it follows that $r_p^{[v]} > r_u^{[v]}$. Since $a_{pv} = 0$ and $a_{uv} = 1$, we contradict the construction of $\tilde{A}$. Thus $r_i^{[k]} \geq 2$ for $i = p + 1, p + 2, \ldots, m$. It now follows that $\{1, 2, \ldots, k-1\}, \{k, k+1, \ldots, n\}$ is a 2-coloring of $\tilde{A}$. Hence $2 \leq \tilde{\gamma}(R,S) \leq \gamma(\tilde{A}) \leq 2$, and $\tilde{\gamma}(R,S) = 2$.

Now suppose that some r_i equals 2. We show that $\tilde{A}$ has a 3-coloring. Let p be the largest integer such that $r_{m-p+1} = 2$; thus $r_i > r_{m-p+1}$ for $i = 1, 2, \ldots, m - p$. Let k be the largest integer such that $a_{m-p+1,k} = 1$. Then the rule of construction of $\tilde{A}$ implies that (3.43) holds with equality for $i = m - p + 1$. In particular, we have $r_{m-p+1}^{[k]} = 1$. Suppose that

$$r_i^{[k]} = a_{i1} + a_{i2} + \cdots + a_{ik} = 0.$$

If $i \leq m - p$, then we contradict the monotonicity of $R^{[k]}$. If $m - p + 1 < i \leq m$, then since $r_{m-p+1} = r_i = 2$, we contradict the rule of construction of $\tilde{A}$. Thus

$$a_{i1} + a_{i2} + \cdots + a_{ik} \geq 1 \quad (i = 1, 2, \ldots, m). \tag{3.44}$$

Inequalities (3.43) and (3.44) now imply $I_1 = \{1, 2, \ldots, k-1\}, I_2 = \{k\}, I_3 = \{k+1, k+2, \ldots, n\}$ is a 3-coloring of $\tilde{A}$. Therefore $\tilde{\gamma}(R,S) \leq \gamma(\tilde{A}) \leq 3$. $\square$

The following example shows that $\tilde{A}$ need not have chromatic number equal to $\tilde{\gamma}(R, S)$ [9].

Example. Let $R = (2, 2, 2, 2, 2, 2, 2, 2)$ and $S = (5, 4, 3, 3, 1)$. Then

$$\tilde{A} = \begin{bmatrix} 1 & 1 & 0 & 0 & 0 \\ 1 & 0 & 1 & 0 & 0 \\ 1 & 0 & 1 & 0 & 0 \\ 1 & 0 & 1 & 0 & 0 \\ 1 & 0 & 0 & 1 & 0 \\ 0 & 1 & 0 & 1 & 0 \\ 0 & 1 & 0 & 1 & 0 \\ 0 & 1 & 0 & 0 & 1 \end{bmatrix}.$$

One verifies that $\tilde{A}$ does not have a 2-coloring and hence, as shown in the proof of Theorem 3.8.2, $\gamma(\tilde{A}) = 3$. The matrix

$$A = \begin{bmatrix} 1 & 1 & 0 & 0 & 0 \\ 1 & 1 & 0 & 0 & 0 \\ 1 & 1 & 0 & 0 & 0 \\ 1 & 1 & 0 & 0 & 0 \\ 1 & 0 & 0 & 1 & 0 \\ 0 & 0 & 1 & 1 & 0 \\ 0 & 0 & 1 & 1 & 0 \\ 0 & 0 & 1 & 0 & 1 \end{bmatrix}$$

in $\mathcal{A}(R, S)$ has a 2-coloring, namely $I_1 = \{1, 3\}, I_2 = \{2, 4, 5\}$, and hence $\tilde{\gamma}(R, S) = \gamma(R, S) = 2$. $\qquad\qquad\qquad\qquad\qquad\qquad\qquad\qquad\qquad\qquad\qquad\square$

If $R = (1, 1, \ldots, 1)$, then $\tilde{\gamma}(R, S) = 1$. If $R \neq (1, 1, \ldots, 1)$, then by Theorem 3.8.2, $\tilde{\gamma}(R, S) = 2$ or 3. Using the theory of network flows, Brualdi and Manber [9] characterized those classes $\mathcal{A}(R, S)$ for which $\tilde{\gamma}(R, S) = 2$. We only state the result and refer to the original paper for its proof. For a class $\mathcal{A}(R, S)$ of m by n matrices, we let $\rho_i(R)$ equal the number of components of R equal to $i + 1$ $(i = 0, 1, \ldots, m - 1)$, and for a subset I of $\{1, 2, \ldots n\}$, we let $\sigma_k(I; S)$ be the sum of the k largest components of S among those indexed by I $(k = 1, 2, \ldots, |I|)$.

Theorem 3.8.3 *Let $\mathcal{A}(R, S)$ be a nonempty class having row sum vector $R = (r_1, r_2, \ldots, r_m)$ with $r_i \geq 2$ $(i = 1, 2, \ldots, m)$. Then $\tilde{\gamma}(R, S) = 2$ if and only if there is a partition I_1, I_2 of $\{1, 2, \ldots, n\}$ such that*

$$\sigma_k(I_2, S) \leq mk - \sum_{i=1}^{k-1}(k - i)\rho_i(R) \quad (k = 1, 2, \ldots, |I_2|)$$

and

$$\sigma_k(I_1, S) \leq mk - \sum_{i=1}^{k-1}(k - i)\rho_i(R) \quad (k = 1, 2, \ldots, |I_1|).$$

The determination of the maximal chromatic number $\bar{\gamma}(R, S)$ is a more intractable problem. We have the following elementary result.

Theorem 3.8.4 *Let $\mathcal{A}(R, S)$ be a nonempty class having row sum vector $R = (r_1, r_2, \ldots, r_m)$, and let $r = \min\{r_i : i = 1, 2, \ldots, m\} \geq 2$. Then*

$$\bar{\gamma}(R, S) \leq \left\lfloor \frac{n}{r-1} \right\rfloor.$$

Proof. Let A be a matrix in $\mathcal{A}(R, S)$. We partition $\{1, 2, \ldots, n\}$ in any way into $t = \lfloor n/(r-1) \rfloor$ sets, each with at most $r - 1$ elements, and obtain a t-coloring of A. $\qquad \square$

We now turn briefly to the strong chromatic number. As discussed in Chapter 1, a strong t-coloring of an m by n (0,1)-matrix A is a partition of $\{1, 2, \ldots, n\}$ into t sets $I_1, I_2, \ldots, I_t$ with the property that each row sum of the matrix $A[\cdot, I_k]$ is at most 1 $(k = 1, 2, \ldots, p)$. The strong chromatic number $\gamma_s(A)$ of A equals the smallest t such that A has a strong t-coloring. If the number of 1's in each row of A is at most 2, then the strong chromatic number is identical to the chromatic number.

If the class $\mathcal{A}(R, S)$ is nonempty, then we let

$$\tilde{\gamma}_s(R, S) = \min\{\gamma_s(A) : A \in \mathcal{A}(R, S)\},$$

and

$$\bar{\gamma}_s(R, S) = \max\{\gamma_s(A) : A \in \mathcal{A}(R, S)\}.$$

Clearly we have $r \leq \gamma_s(A) \leq n$ where r is the maximal row sum of A. The following theorem is proved in a way similar to the proof of Theorem 3.8.1. The example that we used to show that an interchange can alter the chromatic number by 2 also shows, since the rows sums were all equal to 2, that an interchange can alter the strong chromatic number by 2.

Theorem 3.8.5 *Let B be obtained from a matrix A by a single interchange. Then $\gamma_s(B) = \gamma_s(A) \pm 2$.* $\qquad \square$

We now obtain tight bounds for $\tilde{\gamma}_s(R, S)$ [9].

Theorem 3.8.6 *Let $\mathcal{A}(R, S)$ be a nonempty class with row sum vector $R = (r_1, r_2, \ldots, r_m)$, and let $r = \max\{r_i : i = 1, 2, \ldots, m\} \geq 2$. Then*

$$r \leq \tilde{\gamma}(R, S) \leq 2r - 1.$$

Proof. The lower bound has been observed above. We prove the upper bound by showing by induction on r that the special matrix $\tilde{A}$ satisfies $\gamma_s(\tilde{A}) \leq 2r - 1$. If $r = 1$, then $\gamma_s(\tilde{A}) = 1$. Now let $r > 1$, and let k be the largest integer such that $a_{pk} + a_{p,k+1} + \cdots + a_{pn} \geq 2$ for all p with $1 \leq p \leq m$. Then each of the row sums of the matrices $\tilde{A}[\cdot, \{k\}]$ and

$\tilde{A}[\cdot, \{k+1, \ldots, n\}]$ is at most 1. The matrix $\tilde{A}[\cdot, \{1, 2, \ldots, k-1\}]$ is the special matrix of a class $\mathcal{A}(R', S')$ with maximal row sum at most $r-1$. Hence by induction there is a strong t-coloring of $\tilde{A}[\cdot, \{1, 2, \ldots, k-1\}]$ for some t with $t \leq 2r-3$. It follows that $\tilde{A}$ has a strong $(t+2)$-coloring where $t+2 \leq 2r-1$. □

Example. In this example we show that the bounds in Theorem 3.8.6 can be attained for infinitely many values of r. Let m and n be positive integers with $n = mr$, and let $R = (r, r, \ldots, r)$ and $S = (1, 1, \ldots, 1)$. The matrix

$$\begin{bmatrix} I_m & I_m & \cdots & I_m \end{bmatrix}$$

of size m by mr is in $\mathcal{A}(R, S)$ and it has strong chromatic number equal to r; in fact each matrix A in $\mathcal{A}(R, S)$ satisfies $\gamma_s(A) = r$. Hence $\tilde{\gamma}_s(R, S) = r$.

Now let p be a positive integer, and let $m = 4p - 2$, and $n = 2p - 1$. Let $R = (p, p, \ldots, p)$ and $S = (2p, 2p, \ldots, 2p)$ be vectors of size m and n, respectively. If C is a circulant matrix of order $2p - 1$ with p 1's in each row and column, then

$$\begin{bmatrix} C \\ C \end{bmatrix}$$

is in $\mathcal{A}(R, S)$. Let A be any matrix in $\mathcal{A}(R, S)$. Since $2(2p) > m$, each m by 2 submatrix of A has a row with two 1's. Hence the only strong coloring of A is the one into sets of size 1. We conclude that $\gamma_s(A) = n = 2p - 1$ for all A and hence that $\tilde{\gamma}_s(R, S) = 2p - 1$. □

The following theorem gives a characterization of the minimal strong chromatic number of a nonempty class $\mathcal{A}(R, S)$ [9].

Theorem 3.8.7 *Let $R = (r_1, r_2, \ldots, r_m)$ and $S = (s_1, s_2, \ldots, s_n)$ be positive integral vectors such that $\mathcal{A}(R, S)$ is nonempty. If $I_1, I_2, \ldots, I_t$ is a partition of $\{1, 2, \ldots, n\}$, let*

$$s'_p = \sum_{j \in I_p} s_j \ (p = 1, 2, \ldots, t) \ and \ S' = (s'_1, s'_2, \ldots, s'_t).$$

Then $\tilde{\gamma}_s(R, S)$ equals the smallest integer t such that there exists a partition $I_1, I_2, \ldots, I_t$ of $\{1, 2, \ldots, n\}$ satisfying

$$|K||L| + \sum_{i \in \bar{K}} r_i - \sum_{j \in L} s'_j \geq 0 \quad (K \subseteq \{1, 2, \ldots, m\}, L \subseteq \{1, 2, \ldots, t\}). \quad (3.45)$$

Proof. Suppose that $I_1, I_2, \ldots, I_t$ is a strong t-coloring of a matrix $A = [a_{ij}]$ in $\mathcal{A}(R, S)$. Let $A' = [a'_{ij}]$ be the m by t matrix defined by

$$a'_{ij} = \sum_{k \in I_j} a_{ik} \quad (i = 1, 2, \ldots, m; j = 1, 2, \ldots, t).$$

Then A' is a (0,1)-matrix and belongs to $\mathcal{A}(R, S')$. Conversely, given a matrix A' in $\mathcal{A}(R, S)$ we can reverse this procedure and obtain a (not unique in general) matrix in $\mathcal{A}(R, S)$ with a strong t-coloring. The theorem now follows from Theorem 2.1.4. □

3.9 Discrepancy

Let m and n be positive integers, and let $R = (r_1, r_2, \ldots, r_m)$ and $S = (s_1, s_2, \ldots, s_n)$ be nonincreasing, nonnegative integral vectors such that $\mathcal{A}(R, S)$ is nonempty. Let $R^* = (r_1^*, r_2^*, \ldots, r_n^*)$ be the conjugate of R and, as in Section 3.1, let $A(R, n)$ be the unique matrix in the class $\mathcal{A}(R, R^*)$. In each row of $A(R, n)$ the 1's are in the leftmost positions.

The *discrepancy* of matrix was defined in Brualdi and Sanderson [14] as a metric to be used in wildlife ecology for measuring the nestedness of species composition on a collection of sites under investigation.[9] Let A be a matrix in $\mathcal{A}(R, S)$. Then A can be gotten from $A(R, n)$ by shifting 1's to the right along rows. Let k_i equal the number of positions in row i in which $A(R, n)$ has a 1 but A has a zero ($i = 1, 2, \ldots, m$). Then A can be obtained from $A(R, n)$ by shifting

$$d(A) = k_1 + k_2 + \cdots + k_m$$

1's and this number is called the *discrepancy* of A. The discrepancy of A equals zero if and only if A is the matrix $A(R, n)$, equivalently, A is the unique matrix in the class $\mathcal{A}(R, S)$ it determines. As usual we let

$$\tilde{d}(R, S) = \min\{d(A) : A \in \mathcal{A}(R, S)\}$$

denote the *minimal discrepancy* of a matrix in $\mathcal{A}(R, S)$ and

$$\bar{d}(R, S) = \max\{d(A) : A \in \mathcal{A}(R, S)\}$$

denote the *maximal discrepancy* of a matrix in $\mathcal{A}(R, S)$.

Example. Let $R = S = (2, 2, 1)$. Then

$$A(R, 3) = \begin{bmatrix} 1 & 1 & 0 \\ 1 & 1 & 0 \\ 1 & 0 & 0 \end{bmatrix}.$$

The matrices

$$\begin{bmatrix} 1 & 1 & 0 \\ 0 & 1 & 1 \\ 1 & 0 & 0 \end{bmatrix} \quad \text{and} \quad \begin{bmatrix} 1 & 0 & 1 \\ 1 & 1 & 0 \\ 0 & 1 & 0 \end{bmatrix}$$

have discrepancy $\tilde{d}(R, S) = 1$ and $\bar{d}(R, S) = 2$, respectively. $\qquad\square$

An interchange can change the discrepancy of a matrix by at most 1, and hence the following theorem holds.

---

[9]For instance, as described in [14], species distributions on islands in an archipelago or on isolated mountain tops – *sites* for brevity – have been found to exhibit patterns of nestedness whereby species compositions on less rich sites are proper subsets of those on richer sites. Forming a *site–species incidence matrix* A, one generates a class $\mathcal{A}(R, S)$ where S gives the species richness on sites and R gives the site-richness of species. Assume that the sites and species have been ordered so that R and S are nonincreasing vectors. Then perfect nestedness would mean that S is the conjugate of R^*, that is, $A = A(R, n)$. But perfect nestedness rarely occurs in field observations, so a measure of nestedness is used to test the hypothesis that the species exhibit patterns of nestedness.

Theorem 3.9.1 *Let R and S be nonincreasing, nonnegative integral vectors such that $\mathcal{A}(R, S)$ is nonempty. For each integer d with $\tilde{d}(R, S) \leq d \leq \bar{d}(R, S)$, there exists a matrix in $\mathcal{A}(R, S)$ of discrepancy d.* $\qquad\square$

In [14] a lower bound was given for the discrepancy of a matrix in $\mathcal{A}(R, S)$. This lower bound was shown to be the minimal discrepancy by Brualdi and Shen [15], and we now turn to a derivation of this formula.

Let $X = (x_1, x_2, \ldots, x_n)$ be a vector of size n. Then we let $X^+ = \sum_{i=1}^{n} x_i^+$ where for a real number a, $a^+ = \max\{0, a\}$. If $Y = (y_1, y_2, \ldots, y_n)$ is also a vector of vectors of size n, then

$$(X - Y)^+ = \sum_{i=1}^{n} \max\{0, x_i - y_i\}.$$

Notice that, in general, $(X - Y)^+ \neq (Y - X)^+$. If A is a matrix in $\mathcal{A}(R, S)$ we clearly have

$$d(A) \geq (R^* - S)^+ = \sum_{i=1}^{n} \max\{0, r_i^* - s_j\}. \tag{3.46}$$

Before showing that there is a matrix A in $\mathcal{A}(R, S)$ for which equality holds in this inequality, we prove the following lemma.

Lemma 3.9.2 *Let $S = (s_1, s_2, \ldots, s_n)$ and $T = (t_1, t_2, \ldots, t_n)$ be nonincreasing, nonnegative integral vectors such that $S \preceq T$. Then there exist $k = (T - S)^+ + 1$ nonincreasing, nonnegative integral vectors $S^{(i)} = (s_1^{(i)}, s_2^{(i)}, \ldots, s_n^{(i)})$ $(i = 1, 2, \ldots, k)$, such that*

(i) $S = S^{(1)} \preceq S^{(2)} \preceq \cdots \preceq S^{(k)} = T$, *and*

(ii) $(S^{(i+1)} - S^{(i)})^+ = 1$ $(i = 1, 2, \ldots, k - 1)$.

Proof. The lemma clearly holds when $k = 1$ or 2, and we now assume that $k \geq 3$ and proceed by induction. Let $S^{(1)} = S$. We have $S \neq T$ and there exists a smallest index l_0 satisfying $s_{l_0} > t_{l_0}$. If $l_0 \leq n - 1$, then either $s_{l_0} > s_{l_0+1}$, or $s_{l_0+1} = s_{l_0} > t_{l_0} \geq t_{l_0+1}$. Thus there exists a smallest index $l_1 \geq l_0$ satisfying $s_{l_1} > t_{l_1}$, and satisfying $s_{l_1} > s_{l_1+1}$ if $l_1 \leq n - 1$. Therefore

$$\begin{cases} s_i > t_i, & \text{if } l_0 \leq i \leq l_1, \\ s_i \leq t_i, & \text{if } i \leq l_0 - 1. \end{cases}$$

Since $S \preceq T$, we have $s_1 \leq t_1$ and $l_0 > 1$. Let l_2 be the smallest index i satisfying $1 \leq i < l_0$ and $s_i < t_i$. (Such an i exists since $S \preceq T$ but $S \neq T$.) Since $S \preceq T$,

$$\sum_{i=1}^{l_2} t_i = \sum_{i=1}^{l_2-1} t_i + t_{l_2} > \sum_{i=1}^{l_2-1} s_i + s_{l_2} = \sum_{i=1}^{l_2} s_i.$$

Let S_2 be defined by

$$
s_j^{(2)} = \begin{cases} s_j - 1, & \text{if } j = l_1, \\ s_j + 1, & \text{if } j = l_2, \\ s_j, & \text{otherwise.} \end{cases}
$$

We have

$$
\sum_{i=1}^{l} t_i = \sum_{i=1}^{l_2} t_i + \sum_{i=l_2+1}^{l} t_i > \sum_{i=1}^{l_2} s_i + \sum_{i=l_2+1}^{l} s_i = \sum_{i=1}^{l} s_i \quad (l_2 \leq l \leq l_0 - 1),
$$

$$(3.47)$$

and

$$
\sum_{i=1}^{l} t_i = \sum_{i=1}^{l_1} t_i - \sum_{i=l+1}^{l_1} t_i > \sum_{i=1}^{l_1} s_i - \sum_{i=l+1}^{l_1} s_i = \sum_{i=1}^{l} s_i \quad (l_0 \leq l \leq l_1 - 1). \quad (3.48)
$$

Since $S \preceq T$, it follows from (3.47) and (3.48) that $S_2 \preceq T$. By the choices of l_1 and l_2, we have

$$
s_{l_1}^{(2)} = s_{l_1} - 1 \geq s_{l_1+1} = s_{l_1+1}^{(2)} \quad \text{if } l_1 + 1 \leq n, \text{ and}
$$

$$
s_{l_2-1}^{(2)} = s_{l_2-1} \geq t_{l_2-1} \geq t_{l_2} \geq s_{l_2} + 1 = s_{l_2}^{(2)} \quad \text{if } l_2 - 1 \geq 1.
$$

Thus S_2 is nonincreasing. It is now easily verified that $S_1 \preceq S_2$, $(S_2 - S_1)^+ = 1$, and $(T - S_2)^+ = k - 2$. The lemma now follows by induction upon replacing S with $S^{(2)}$. $\qquad\square$

Theorem 3.9.3 *Let $R, S,$ and S' be nonincreasing, nonnegative integral vectors such that $S \preceq S'$ and $S \neq S'$. Let A' be a matrix in $\mathcal{A}(R, S')$. Then there is a matrix in $\mathcal{A}(R, S)$ which can be obtained from A' by shifting at most $(S' - S)^+$ 1's in rows.*

Proof. Because of Lemma 3.9.2 we may assume that $(S' - S)^+ = 1$. Let $S = (s_1, s_2, \ldots, s_n)$ and $S' = (s_1', s_2', \ldots, s_n')$. Since $S \preceq S'$, there are integers p and q such that $p < q$ and

$$
s_i' = \begin{cases} s_i - 1, & \text{if } i = q, \\ s_i + 1, & \text{if } i = p, \\ s_i, & \text{otherwise.} \end{cases}
$$

In particular, we have $s_p' = s_p + 1 \geq s_q + 1 = s_q' + 2$, and so column p of A' contains at least two more 1's than column q of A'. Thus some row of A' contains a 1 in column p and a 0 in column q. By shifting this 1 from column p to column q we obtain a matrix in $\mathcal{A}(R, S)$. $\qquad\square$

The next theorem gives a formula for minimal discrepancy.

Theorem 3.9.4 *Let R and S be nonincreasing, nonnegative integral vectors such that $\mathcal{A}(R,S)$ is nonempty. Then*

$$\tilde{d}(R,S) = (R^* - S)^+.$$

Proof. Since $\mathcal{A}(R,S)$ is nonempty, $S \preceq R^*$. We apply Theorem 3.9.3 with $S' = R^*$ and with $A' = A(R,n)$, and obtain a matrix A in $\mathcal{A}(R,S)$ by shifting at most $(R^* - S)^+$ 1's in rows. Hence $d(A) \le (R^* - S)^+$. The reverse inequality is a consequence of (3.46), and the lemma follows.   □

An explicit formula for the maximal discrepancy $\bar{d}(R,S)$ is not known and appears to be very difficult. An algorithm to compute the maximal discrepancy is given in [15].

3.10   Rank

We denote the rank over the real field of a matrix A by $\nu(A)$. If A is a $(0,1)$-matrix, then $\nu(A)$ equals the rank of A over any field of characteristic 0. In contrast to the other parameters considered in this chapter, rank is not a purely combinatorial parameter, although it is invariant under arbitrary permutations of rows and columns. For a nonempty class $\mathcal{A}(R,S)$ we define

$$\tilde{\nu} = \tilde{\nu}(R,S) = \min\{\nu(A) : A \in \mathcal{A}(R,S)\}$$

to be the *minimal rank* of a matrix in $\mathcal{A}(R,S)$, and

$$\bar{\nu} = \tilde{\nu}(R,S) = \max\{\nu(A) : A \in \mathcal{A}(R,S)\}$$

to be the *maximal rank* of a matrix in $\mathcal{A}(R,S)$.

The quantities $\tilde{\nu}(R,S)$ and $\bar{\nu}(R,S)$ appear to be very complicated functions of R and S, and we do not expect that there are formulas for these quantities like the formulas for minimal and maximal term rank or trace. An interchange, however, can alter the rank by only 1.

Theorem 3.10.1 *Let A be a $(0,1)$-matrix and let B be obtained from the matrix by a single interchange. Then $\nu(B) = \nu(A) \pm 1$.*

Proof. Let B be obtained from A by, say, a $(1,2;1,2)$-interchange. Let C be the matrix obtained from A by adding column 1 to column 2. Then $\nu(C) = \nu(A)$. Let C' be the matrix obtained from B by adding column 1 to column 2. Then $\nu(C') = \nu(B)$. Since C and C' differ only in column 1, $\nu(C') = \nu(C) \pm 1$ and hence $\nu(B) = \nu(A) \pm 1$.   □

Since a matrix in $\mathcal{A}(R,S)$ can be gotten from any other matrix in $\mathcal{A}(R,S)$ by a sequence of interchanges, the following corollary is immediate from Theorem 3.10.1.

Corollary 3.10.2 *Let R and S be nonnegative integral vectors such that $\mathcal{A}(R, S)$ is nonempty. For each integer t with $\tilde{\nu}(R, S) \leq t \leq \bar{\nu}(R, S)$, there exists a matrix A in $\mathcal{A}(R, S)$ such that $\nu(A) = t$.* □

In order to obtain some information about $\tilde{\nu}(R, S)$ and $\bar{\nu}(R, S)$, we need to place restrictions on R and S. A natural restriction is to consider *regular matrices* – matrices with constant row and column sums.

Let n and k be integers with $0 \leq k \leq n$. Let $R = S = (k, k, \ldots, k)$ be a constant vector of size n. We consider the class $\mathcal{A}(R, S)$ of all $(0, 1)$-matrices of order n with exactly k 1's in each row and column. This class is nonempty and, as before, we denote it more simply as $\mathcal{A}(n, k)$. We also use the simpler notations

$$\tilde{\nu}(n, k) = \min\{\nu(A) : A \in \mathcal{A}(n, k)\},$$

and

$$\bar{\nu}(n, k) = \max\{\nu(A) : A \in \mathcal{A}(n, k)\}$$

for the minimal and maximal ranks of matrices in this class. Recall that if A is an m by n matrix, then the complement of A is the matrix $A^c = J_{m,n} - A$. The matrix A is a $(0, 1)$-matrix if and only if A^c is. The *complementary class* of a class $\mathcal{A}(R, S)$ is the class

$$\mathcal{A}^c(R, S) = \{A^c : A \in \mathcal{A}(R, S)\}.$$

The row sum and column sum vectors of the complementary class are denoted by R^c and S^c, respectively. We also let 1^n denote the vector of all 1's of size n.

The following lemma is due to Ryser (see [10]).

Lemma 3.10.3 *Let A be an m by n matrix. Then $\nu(A) = \nu(A^c)$ if and only if the vector 1^n is a linear combination of the rows of A exactly when it is a linear combination of the rows of A^c.*

Proof. Let $\widehat{A}$ be the $m + 1$ by n matrix obtained from A by appending the vector 1^n as a last row. We have $\nu(A) = \nu(\widehat{A})$ if and only if 1^n is a linear combination of the rows of A. Similarly, $\nu(A^c) = \nu(\widehat{A^c})$ if and only if 1^n is a linear combination of the rows of A^c. The theorem is thus equivalent to the assertion that $\nu(A) = \nu(\widehat{A})$ if and only if $\nu(A^c) = \nu(\widehat{A^c})$. The matrix $\widehat{A^c}$ can be obtained from $\widehat{A}$ by elementary row operations and hence $\nu(\widehat{A^c}) = \nu(\widehat{A})$. It follows that $\nu(A) = \nu(\widehat{A})$ if and only if $\nu(A^c) = \nu(\widehat{A^c})$. □

Corollary 3.10.4 *Let A be an m by n matrix in the class $\mathcal{A}(R, S)$ where $S = (k, k, \ldots, k)$ where $1 \leq k \leq m - 1$. Then $\nu(A) = \nu(A^c)$, and hence $\tilde{\nu}(R, S) = \tilde{\nu}(R^c, S^c)$ and $\bar{\nu}(R, S) = \bar{\nu}(R^c, S^c)$.*

Proof. The sum of the rows of A equals S and hence 1^n is a linear combination of the rows of A. Similarly, the sum of the rows of A^c is $S^c = (m - k, m - k, \ldots, m - k)$ and 1^n is also a linear combination of the rows of A^c. By Lemma 3.10.3, $\nu(A) = \nu(A^c)$ and the corollary follows. $\square$

The following theorem is an immediate consequence of Corollary 3.10.4.

Theorem 3.10.5 *Let n and k be positive integers with $0 \leq k \leq n$. Then*

$$\tilde{\nu}(n,0) = \bar{\nu}(n,0) = 0,$$
$$\tilde{\nu}(n,n) = \bar{\nu}(n,n) = 1,$$
$$\tilde{\nu}(n,k) = \tilde{\nu}(n,n-k) \text{ if } 1 \leq k \leq n-1, \text{ and}$$
$$\bar{\nu}(n,k) = \bar{\nu}(n,n-k) \text{ if } 1 \leq k \leq n-1.$$

Moreover, a matrix A in $\mathcal{A}(n,k)$ satisfies $\nu(A) = \tilde{\nu}(n,k)$ (respectively, $\nu(A) = \bar{\nu}(n,k)$) if and only if the complementary matrix A^c in $\mathcal{A}(n, n-k)$ satisfies $\nu(A^c) = \tilde{\nu}(n, n-k)$ (respectively, $\nu(A^c) = \bar{\nu}(n, n-k)$) $\square$

The next theorem, proved independently by Newman [54] and Houck and Paul [37], asserts that, except for two trivial cases and one exceptional case, there always exists a nonsingular matrix in $\mathcal{A}(n,k)$. We follow the proof in [54]. First we recall the following basic facts about circulant matrices.

The *circulant matrix* of order n determined by $(c_0, c_1, \ldots, c_{n-1})$ is the matrix

$$\mathrm{Circ}_n = \mathrm{Circ}(c_0, c_1, \ldots, c_{n-1})$$
$$= c_0 I_n + c_1 C_n + c_1 C_n^2 + \cdots + c_{n-1} C_n^{n-1}$$
$$= \begin{bmatrix} c_0 & c_1 & \cdots & c_{n-1} \\ c_{n-1} & c_0 & \cdots & c_{n-2} \\ \vdots & \vdots & \ddots & \vdots \\ c_1 & c_2 & \cdots & c_0 \end{bmatrix}$$

obtained from $(c_0, c_1, \ldots, c_{n-1})$ by cyclically shifting rows one position to the right. Here C_n is the full cycle permutation matrix of order n corresponding to the permutation $(2, 3, \ldots, n, 1)$. The eigenvalues of Circ_n are the numbers

$$\psi(\epsilon_n^l) \quad (l = 1, 2, \ldots, n) \tag{3.49}$$

where

$$\psi(\lambda) = c_0 + c_1 \lambda^1 + \cdots + c_{n-1} \lambda^{n-1} \quad \text{and} \quad \epsilon_n = \exp(2\pi i/n).$$

Thus Circ_n is nonsingular if and only if each of the n numbers in (3.49) is nonzero.

The following lemma is in [54] but the proof is taken from [20].

Lemma 3.10.6 *Let A be a matrix of order n such that $AJ_n = kJ_n$ where $k \neq 0$. Then $\det(J_n + A) = (n+k)\det(A)/k$ and $\det(J_n - A) = (-1)^{n-1}(n-k)\det(A)/k$.*

Proof. Since $AJ_n = kJ_n$ we have

$$J_n + A = \frac{1}{k}AJ_n + A = A\left(\frac{1}{k}J_n + I_n\right),$$

and hence

$$\det(J_n + A) = \det(A)\det\left(\frac{1}{k}J_n + I_n\right).$$

Since

$$\det\left(\frac{1}{k}J_n + I_n\right) = \frac{n+k}{k},$$

the first assertion follows. The second assertion in the theorem is an elementary consequence of the first. $\qquad\square$

Theorem 3.10.7 *Let n and k be positive integers with $0 \leq k \leq n$. Then*

$$\bar{\nu}(n,k) = \begin{cases} 0 & \text{if } k = 0, \\ 1 & \text{if } k = n, \\ 3 & \text{if } k = 2 \text{ and } n = 4, \\ n & \text{otherwise.} \end{cases} \tag{3.50}$$

Proof. That $\bar{\nu}(n,0) = 0$ and $\bar{\nu}(n,n) = 1$ has already been noted in Theorem 3.10.5. Up to row and column permutations, there are only two matrices in $\mathcal{A}(4,2)$, namely, the matrices

$$J_2 \oplus J_2 \quad \text{and} \quad \begin{bmatrix} 1 & 1 & 0 & 0 \\ 0 & 1 & 1 & 0 \\ 0 & 0 & 1 & 1 \\ 1 & 0 & 0 & 1 \end{bmatrix}$$

of ranks 2 and 3, respectively. Hence $\bar{\nu}(4,2) = 3$. We now assume that $0 < k < n$ and $(n,k) \neq (4,2)$ and, to complete the proof, construct a nonsingular matrix in $\mathcal{A}(n,k)$. If $k = 1$, then I_n is a nonsingular matrix in $\mathcal{A}(n,1)$. If $k = 2$, then $I_n + C_n$ is nonsingular for $n \geq 3$ with n odd, and $I_n + (C_3 \oplus C_{n-3})$ is nonsingular for $n \geq 4$ and even. If $k = n-1$, then the circulant matrix $J_n - I_n$ is nonsingular for $n \geq 2$. Now suppose that $k = n-2$ and $n \geq 5$. If n is odd, then $J_n - (I_n + C_n)$ is in $\mathcal{A}(n, n-2)$ and by Lemma 3.10.6, its determinant equals $n - 2 > 0$. If n is even, then $J_n - (I_n + (C_3 \oplus C_{n-3}))$ is in $\mathcal{A}(n, n-2)$ and Lemma 3.10.6 implies that its determinant equals $4 - 2n < 0$.

It remains to prove the theorem for $3 \leq k \leq n-3$. First assume that $3 \leq k \leq (n/2) + 1$. Consider the circulant matrix $A_k = I_n + C_n + C_n^3 +$

$C_n^5 + \cdots + C_n^{2k-3}$ in $\mathcal{A}(n,k)$. The matrix A_k is singular if and only if there exists an nth root of unity $\zeta = \epsilon_n^j$ such that

$$1 + \zeta + \zeta^3 + \cdots + \zeta^{2k-3} = 0. \tag{3.51}$$

If $\zeta = 1$, then (3.51) implies the contradiction $k = 0$. If $\zeta = -1$, then (3.51) implies the contradiction $k = 2$. Now suppose that $\zeta \neq \pm 1$. Thus (3.51) is equivalent to

$$1 + \zeta\frac{1 - \zeta^{2k-2}}{1 - \zeta^2} = 0. \tag{3.52}$$

Taking complex conjugates in (3.52) and equating the result with (3.52), we obtain $\zeta^{2k-2} = 1$. Putting this in (3.52) gives the contradiction $1 = 0$. Thus A_k is a nonsingular matrix.

Now assume that $(n/2) + 1 < k \leq n - 3$ so that $3 \leq n - k < (n/2) - 1$. Then A_{n-k} is a nonsingular matrix, $J_n - A_{n-k}$ is in $\mathcal{A}(n,k)$, and by Lemma 3.10.6

$$\det(J_n - A_{n-k}) = (-1)^{n-1}(n-k)\det(A_{n-k})/k \neq 0.$$

Hence $J_n - A_{n-k}$ is a nonsingular matrix completing the proof. $\square$

We now turn to the determination of the minimal rank $\tilde{\nu}(n,k)$, which is substantially more complicated and, as a consequence, the results are incomplete.

First we obtain a lower bound for $\tilde{\nu}(n,k)$. In doing so we shall make use of the following lemma.

Lemma 3.10.8 *Let k and n be integers with $1 \leq k \leq n$. Let*

$$A = \begin{bmatrix} A_1 \\ \hline A_2 \end{bmatrix} = \begin{bmatrix} A_{11} & A_{12} \\ \hline A_2 \end{bmatrix} \tag{3.53}$$

be a matrix in $\mathcal{A}(n,k)$, where $\nu(A) = \nu(A_1)$ and where A_{12} is nonvacuous, has no zero rows, and exactly one 1 in each column. Then A_1 has no zero columns and k divides n.

Proof. The rows of A_2 are linear combinations of the rows of A_1. If A_1 has a zero column, then so does A, contradicting $k \geq 1$. Let the row vectors of A_1 be $x_1, x_2, \ldots, x_s$. Each of the vectors $x_1, x_2, \ldots, x_s$ has a 1 in a position belonging to A_{12}. Each row vector of A_2 is a linear combination of rows of A_1, and the assumptions on A_{12} imply that in such a linear combination, all the coefficients equal 0 or 1. Since all the rows of A contain exactly k 1's, we conclude that exactly one coefficient equals 1. Thus each row of A_2 is also a row of A_1. Since each column of A_{12} contains exactly one 1, each row of A_1 occurs $k - 1$ times as a row of A_2, and hence n is a multiple of k. $\square$

The following result is from [10].

Theorem 3.10.9 *For each integer k with $1 \leq k \leq n$, we have*

$$\tilde{\nu}(n, k) \geq \lceil n/k \rceil,$$

with equality if and only if k divides n. In addition, if k divides n and A is a matrix in $\mathcal{A}(n, k)$ satisfying $\nu(A) = n/k$, then the rows and columns of A can be permuted to give the matrix $J_k \oplus J_k \oplus \cdots \oplus J_k$ in $\mathcal{A}(n, k)$.

Proof. Let A be a matrix in $\mathcal{A}(n, k)$. Assume to the contrary that $\nu(A) = t < \lceil n/k \rceil$. Then $tk < n$ and every t by n submatrix of A has a zero column. Since there exists a t by n submatrix of A whose rows span all the rows of A, this implies that A has a zero column, contradicting $k \geq 1$. Hence $\nu(n, k) \geq \lceil n/k \rceil$.

First suppose that $k | n$. Then the matrix $J_k \oplus J_k \oplus \cdots \oplus J_k$ of order n is in $\mathcal{A}(n, k)$ and has rank n/k. Thus $\nu(n, k) = n/k = \lceil n/k \rceil$. Let A be any matrix in $\mathcal{A}(n, k)$ with $\nu(A) = n/k$. Then there is an n/k by n submatrix A_1 of A whose rows span all the rows of A. By Lemma 3.10.8, A_1 has no zero columns and so A_1 has exactly one 1 in each row. It follows that each row of A equals some row of A_1 and that, after row and column permutations, $A = J_k \oplus J_k \oplus \cdots \oplus J_k$.

Now suppose that $\nu(n, k) = \lceil n/k \rceil$, and let A be a matrix in $\mathcal{A}(n, k)$ of rank $l = \nu(n, k)$. Let A_1 be an l by n submatrix of A of rank l. Define subsets $S_1, S_2, \ldots, S_l$ of $\{1, 2, \ldots, n\}$ recursively by

$$S_i = \{t : a_{it} = 1 \text{ and } a_{jt} = 0 \text{ for } j = 1, 2, \ldots, i-1\} \quad (i = 1, 2, \ldots, l).$$

We have $n = \sum_{i=1}^{l} |S_i|$ and $|S_i| \leq k$ $(i = 1, 2, \ldots, l)$. If $|S_p| = 0$ for some p, then

$$n = \sum_{j=1}^{l} |S_j| \leq k(l-1) < n,$$

a contradiction. Thus $|S_j| > 0$ for $i = 1, 2, \ldots, l$. For each i with $1 \leq i \leq l$, let $A_1^i = A_1[\{1, \ldots, i-1, i+1, \ldots, l\}, S_i]$. The number of 1's in A_1^i is at most

$$\sum_{j=1, j \neq i}^{l} |S_j| = (l-1)k - \sum_{j=1, j \neq i}^{l} |S_j|$$

$$= \left\lceil \frac{n}{k} \right\rceil k - k - n + |S_i|$$

$$\leq (k-1) - k + |S_i|$$

$$= |S_i| - 1.$$

Therefore A_1^i contains a zero column and thus $A_1[\cdot, S_i]$ has a column with exactly one 1. Hence A_1 has a submatrix of order l equal to a permutation matrix. Without loss of generality we assume that

$$A = \left[\begin{array}{c} A_1 \\ \hline A_2 \end{array}\right] = \left[\begin{array}{c|c} A_{11} & I_l \\ \hline & A_2 \end{array}\right]$$

By Lemma 3.10.8, k divides n.   □

We now apply Theorem 3.10.9 with $k = 2$.

Theorem 3.10.10 *Let $n \geq 2$ be an integer. Then*

$$\tilde{\nu}(n,2) = \begin{cases} \frac{n}{2} & \text{if } n \text{ is even,} \\[2mm] \frac{n+3}{2} & \text{if } n \text{ is odd.} \end{cases}$$

In addition, a matrix in $\mathcal{A}(n,2)$ has rank equal to $\tilde{\nu}(n,2)$ if and only if its rows and columns can be permuted to the form

$$J_2 \oplus J_2 \oplus \cdots \oplus J_2, \quad \text{if } n \text{ is even, and} \tag{3.54}$$

$$J_2 \oplus \cdots \oplus J_2 \oplus (J_3 - I_3), \quad \text{if } n \text{ is odd.} \tag{3.55}$$

Proof. Because of Theorem 3.10.9 we need only consider the case of n odd.

Let $n = 2p + 1$. By Theorem 3.10.9, $\tilde{\nu}(n,2) \geq \lceil n/2 \rceil + 1 = (n+3)/2$. The matrix (3.55) of order n belongs to $\mathcal{A}(n,2)$ and has rank equal to $(p-1) + 3 = p + 2 = (n+3)/2$. Hence $\tilde{\nu}(n,2) = (n+3)/2$. Now let A be a matrix in $\mathcal{A}(n,2)$ with $\nu(A) = (n+3)/2 = p + 2$. We may assume that A has the form (3.53) where A_1 has $p + 2$ rows and has rank equal to $p + 2$. The number of 1's in A_1 equals $2(p+2)$. Since A_1 cannot have a zero column, three columns of A_1 contain two 1's and the other $2p - 2$ columns each contain one 1. Thus we may take the matrix A_{11} in (3.53) to have 3 columns and the matrix A_{12} to have $2p - 2$ columns. Since 2 does not divide n, it follows from Lemma 3.10.8 that A_{12} has a zero row and hence that A_{11} has a row containing two 1's. This fact, along with the fact that the rows of A_1 are linearly independent, implies that the nonzero rows of the matrix A_1, after row and column permutations, form a submatrix equal to one of the three matrices

$$\begin{bmatrix} 1 & 1 & 0 \\ 1 & 0 & 0 \\ 0 & 1 & 0 \\ 0 & 0 & 1 \\ 0 & 0 & 1 \end{bmatrix}, \begin{bmatrix} 1 & 1 & 0 \\ 1 & 0 & 1 \\ 0 & 1 & 0 \\ 0 & 0 & 1 \end{bmatrix}, \text{ and } \begin{bmatrix} 0 & 1 & 1 \\ 1 & 0 & 1 \\ 1 & 1 & 0 \end{bmatrix}. \tag{3.56}$$

We take this submatrix in the upper left corner of A_1. We show that the first two submatrices lead to a contradiction.

First suppose the 5 by 3 submatrix in (3.56) occurs. Let u and v be the columns of A_1 that contain a 1 in rows 4 and 5, respectively. There is a row of A_2 that has a 1 in column u and this row is a linear combination of the rows of A_2. In this linear combination the coefficient of row 4 must be 1, and since column 3 of A_2 is a zero column, the coefficient of row 5 must be -1. This implies that column u has a negative element, a contradiction.

Now suppose that the 4 by 3 submatrix in (3.56) occurs. Arguing as above, there is a row of A_2 which is a linear combination of the rows of A_1 where the coefficients of rows $1, 2, 3, 4$ are, respectively, $-1, 1, 1, -1$. This gives again the contradiction that A_2 has a negative element.

Thus we have $A = E \oplus F$ where E is the matrix $J_3 - I_3$ of order 3 in (3.56) and F is a matrix in $\mathcal{A}(2p - 2, 2)$ of rank $p - 1$. Applying Theorem 3.10.9, we complete the proof. □

Theorems 3.10.5 and 3.10.10 imply the following corollary.

Corollary 3.10.11 *Let $n \geq 2$ be an integer. Then*

$$\tilde{\nu}(n, n - 2) = \begin{cases} \frac{n}{2} & \textit{if } n \textit{ is even,} \\ \\ \frac{n+3}{2} & \textit{if } n \textit{ is odd.} \end{cases}$$

In addition, a matrix A in $\mathcal{A}(n, 2)$ has rank equal to $\tilde{\nu}(n, n-2)$ if and only the rows and columns of the complementary matrix $J_n - A$ can be permuted to give the matrix in equation (3.54) or (3.55). □

The proof of the following theorem is more complicated and we refer to [10].

Theorem 3.10.12 *Let $n \geq 3$ be an integer. Then*

$$\tilde{\nu}(n, 3) = \begin{cases} \frac{n}{3} & \textit{if } n \textit{ is divisible by 3,} \\ \\ \lfloor \frac{n}{3} \rfloor + 3 & \textit{otherwise.} \end{cases} \tag{3.57}$$

Let n and k be integers with $1 \leq k \leq n-1$. In [10] a construction is given for a special matrix A^* in $\mathcal{A}(n, k)$. This construction arises by combining the matrices of the form $J_k \oplus J_k \oplus \cdots \oplus J_k$ which achieve the minimal rank when $k|n$ and their complements $(J_k \oplus J_k \oplus \cdots \oplus J_k)^c$ in $\mathcal{A}(n, n - k)$ which achieve the minimal rank also when $k|n$. As a result, the matrix A^* has small rank but it need not have minimal rank. We use the notation $\bigoplus_p X$ to denote the direct sum of p copies of a matrix X; the matrix $\bigoplus_0 X$ is an empty matrix with no rows or columns. The construction of A^* proceeds recursively as follows.

(I) Let $n = qk + r$ where $q \geq 1$ and $0 \leq r < k$. If $r = 0$, set $m = q$. If $q \geq 2$ and $r > 0$, set $m = q - 1$. If $q = 1$ and $r > 0$, set $m = 0$. Now replace n with $n - mk = k + r$. Note that if $n - mk \neq 0$, then $k < n - mk = k + r < 2k$. Let $A^* = (\bigoplus_m J_k) \oplus B$ where B is vacuous if $m = q$, and is determined in (II) otherwise.

(II) Let $k < n < 2k$, and let $n = p(n-k) + s$ where $0 \le s < n-k$ and hence $p \ge 2$. If $s = 0$, let $l = p$. Otherwise, set $l = p-1$. Let $B = ((\bigoplus_l J_{n-k}) \oplus C^c)^c$ where C is determined as A^* in (I) with k replaced by s.

We alternate between (I) and (II) until we eventually obtain a vacuous matrix. The resulting matrix A^* is in $\mathcal{A}(n,k)$ and its rank $\nu^*(n,k)$ is readily determined and is an upper bound for $\tilde{\nu}(n,k)$. We note that for $n \ge 2k$, the construction of A^* implies that $\nu^*(n,k) = \nu^*(n-k,k) + 1$.

Example. We illustrate this construction by taking $n = 11$ and $k = 4$. We get $A^* = J_4 \oplus B$, $B = (J_4 \oplus C^c)^c$ and $C = \bigoplus_4 J_1 = I_4$. Hence

$$A^* = \left[\begin{array}{c|cc} J_4 & \multicolumn{2}{c}{O_{4,7}} \\ \hline \multirow{2}{*}{$O_{7,4}$} & O_3 & J_{3,4} \\ & J_{4,3} & I_4 \end{array} \right].$$

We have $\nu^*(11,4) = 6$. Since $\lceil 11/4 \rceil = 3$ and 11 is not a multiple of 4, we have $5 \le \tilde{\nu}(11,4) \le 6$. □

The upper bound for $\tilde{\nu}(n,k) \le \lfloor \frac{n}{k} \rfloor + k$ was proved in [56], [29]; as shown in the next theorem, $\nu^*(n,k)$ is a better upper bound.

Theorem 3.10.13 *Let n and k be integers with $1 \le k \le n$. Then*

$$\tilde{\nu}(n,k) \le \nu^*(n,k) \le \left\lfloor \frac{n}{k} \right\rfloor + k.$$

Proof. We need only show the inequality

$$\nu^*(n,k) \le \left\lfloor \frac{n}{k} \right\rfloor + k. \tag{3.58}$$

This inequality holds when k divides n, in particular when $k = 1$, since then $\nu^*(n,k) = n/k$. If $n \ge 2k$, then it follows from the construction of A^* that $\nu^*(n,k) = 1 + \nu^*(n-k,k)$. By induction, $\nu^*(n-k,k) \le \lfloor (n-k)/k \rfloor + k$, and (3.58) follows.

Now assume that $k < n < 2k$. The construction of A^* now effectively begins with step (II) (the first step produces an empty matrix). Let $n = q(n-k) + r$ where $0 \le r < n-k < k$. The assumption $k < n < 2k$ implies that $q \ge 2$. If $q > 2$, then $\nu^*(n,k) = 1 + \nu^*(n-k,k)$. By induction, $\nu^*(n-k,k) \le \lfloor \frac{n-k}{k} \rfloor + k$ and (3.58) follows again. We now take $q = 2$, so that

$$n = 2(n-k) + r \text{ and } k = (n-k) + r.$$

We now have $\nu^*(n,k) = 1 + \nu^*(k,r)$ where by induction

$$\nu^*(k,r) \le \left\lfloor \frac{k}{r} \right\rfloor + r.$$

Hence

$$\nu^*(n,k) \leq 1 + \left\lfloor \frac{k}{r} \right\rfloor + r \leq \left\lfloor \frac{n}{k} \right\rfloor + k.$$

The last inequality can be checked by elementary calculation (see [56] for details).   □

While $\nu^*(n,k)$ is an upper bound for $\tilde{\nu}(n,k)$, equality need not hold [10].

Example. Let $n = 9$ and $k = 4$. By Theorem 3.10.9, $\tilde{\nu}(9,4) \geq 4$. An easy calculation shows that $\nu^*(9,4) = 6$. The matrix

$$\begin{bmatrix}
1 & 1 & 1 & 1 & 0 & 0 & 0 & 0 & 0 \\
1 & 1 & 1 & 1 & 0 & 0 & 0 & 0 & 0 \\
1 & 1 & 1 & 1 & 0 & 0 & 0 & 0 & 0 \\
0 & 0 & 0 & 0 & 1 & 1 & 1 & 1 & 0 \\
0 & 0 & 0 & 0 & 1 & 1 & 1 & 1 & 0 \\
0 & 0 & 0 & 0 & 0 & 1 & 1 & 1 & 1 \\
0 & 0 & 0 & 0 & 0 & 1 & 1 & 1 & 1 \\
0 & 0 & 1 & 1 & 1 & 0 & 0 & 0 & 1 \\
1 & 1 & 0 & 0 & 1 & 0 & 0 & 0 & 1
\end{bmatrix}$$

in $\mathcal{A}(9,4)$ has rank 5. It follows that $\tilde{\nu}(n,4) < \nu^*(n,4)$ for every $n \geq 9$ with $n \equiv 1 \mod 4$.   □

It follows from Theorem 3.10.10 that $\tilde{\nu}(n,2) = \nu^*(n,2)$ and that a matrix A in $\mathcal{A}(n,2)$ satisfies $\nu(A) = \tilde{\nu}(n,2)$ if and only if its rows and columns can be permuted to give the matrix A^* in $\mathcal{A}(n,2)$. By Theorem 3.10.12, $\tilde{\nu}(n,3) = \nu^*(n,3)$. Our final example shows that the constructed matrix A^* in $\mathcal{A}(n,3)$ need not be, up to row and column permutations, the only matrix in $\mathcal{A}(n,3)$ of minimal rank [10].

Example. Let $n = 7$ and $k = 3$. Then $\tilde{\nu}(7,3) = 5$. We have $A^* = J_3 \oplus (J_4 - I_4)$ and $\nu^*(7,3) = \nu(A^*) = 5$. The matrix

$$A = \begin{bmatrix}
1 & 1 & 1 & 0 & 0 & 0 & 0 \\
1 & 1 & 1 & 0 & 0 & 0 & 0 \\
1 & 1 & 0 & 1 & 0 & 0 & 0 \\
0 & 0 & 1 & 1 & 1 & 0 & 0 \\
0 & 0 & 0 & 1 & 0 & 1 & 1 \\
0 & 0 & 0 & 0 & 1 & 1 & 1 \\
0 & 0 & 0 & 0 & 1 & 1 & 1
\end{bmatrix}$$

is in $\mathcal{A}(7,3)$ and has rank equal to 5. Since A does not have a submatrix equal to J_3, its rows and columns cannot be permuted to give the matrix A^*.   □

Another point of view is considered in [38]. For r a positive integer, let $\mathcal{R}_r$ equal the set of all numbers $\frac{k}{n}$ for which there exists a matrix in $\mathcal{A}(n,k)$

with rank equal to r. The ratios k/n and r are invariant under taking the Kronecker product of a matrix A with a square matrix J_m of all 1's.

Theorem 3.10.14 *Let r be a positive integer. Then*

$$|\mathcal{R}_r| < 2^{r^2};$$

in particular, $\mathcal{R}_r$ is a finite set.

Proof. Let A be a matrix in $\mathcal{A}(n, k)$, and let r be the rank of A. Since $AJ_n = kJ_n$, the all 1's vector 1^n is in the column space of A. Hence there exist an n by r submatrix A' of A and a column vector $\alpha = (a_1, a_2, \ldots, a_r)^{\mathrm{T}}$ such that $A'\alpha = 1^n$. Summing the elements of $A'\alpha$ and of 1^n, and using the fact that each column of A' contains k 1's, we get

$$k \sum_{i=1}^{r} a_i = n, \quad \text{equivalently,} \quad \frac{k}{n} = \frac{1}{\sum_{i=1}^{r} a_i}.$$

Let A'' be a nonsingular submatrix of order r of A'. Then we also have $A''\alpha = 1^r$, and this equation determines $\sum_{i=1}^{r} a_i$ and thus k/n. Since there are strictly less than 2^{r^2} nonsingular $(0,1)$-matrices of order r, the theorem follows.   □

In addition to the (real) rank of a $(0,1)$-matrix, there are two other ranks of some combinatorial significance. Let A be an m by n $(0,1)$-matrix. The *nonnegative integer rank* $z(A)$ of A is the smallest integer t for which there exist $(0,1)$-matrices U and V of sizes m by t and t by n, respectively, satisfying[10]

$$A = UV. \tag{3.59}$$

Since $A = UV$ implies that $\nu(A) \le t$, we have $\nu(A) \le z(A)$.

Let $A = UV$ where U and V are $(0,1)$-matrices of sizes m by t and t by n, respectively. Let the columns of U be the m by 1 matrices $U^1, U^2, \ldots, U^{\mathrm{T}}$, and let the rows of V be $V_1, V_2, \ldots, V_t$. Then

$$A = UV = \sum_{i=1}^{\mathrm{T}} U^i V_i. \tag{3.60}$$

The m by n matrix $U^i V_i$ is a $(0,1)$-matrix whose 1's form an a_i by b_i submatrix of A, where a_i is the number of 1's in U^i and b_i is the number of 1's in V_i. Each of the bipartite graphs $\mathrm{BG}(U^i V_i)$ is, except for possible vertices meeting no edges, a complete bipartite graph, and is a *biclique* of $\mathrm{BG}(A)$. Thus from (3.60) we conclude that $z(A)$ *is the smallest integer t*

---

[10]In general, in defining the nonnegative integer rank of a nonnegative integral matrix A, the factors U and V are nonnegative integral matrices. Since A is a $(0,1)$-matrix, the factor U, or V, in a factorization $A = UV$ can only have elements greater than 1 when U, or V, has a zero column, or a zero row, respectively. In a factorization with t minimal, U and V must be $(0,1)$-matrices.

such that the edges of $BG(A)$ can be partitioned into t bicliques; equivalently the smallest integer t such that the 1's of A can be partitioned into t submatrices of all 1's.

If instead of using real arithmetic in (3.59) and (3.60), we use *Boolean arithmetic* where $1 + 1 = 1$, then we obtain the *Boolean rank* $b(A)$ of A. Thus we see that $b(A)$ is the smallest integer t such that the edges of $BG(A)$ can be covered by t bicliques;[11] equivalently, the smallest integer t such that the 1's of A can be covered by t submatrices of all 1's.

For all $(0,1)$-matrices A of size m by n, we have

$$\max\{\nu(A), b(A)\} \leq z(A) \leq \min\{m, n\}.$$

The following example is from [53].

Example. The matrix $J_7 - I_7$ is nonsingular and hence has rank equal to 7. Using Boolean arithmetic, we also have

$$J_7 - I_7 = \begin{bmatrix} 1 & 1 & 1 & 0 & 0 \\ 0 & 1 & 1 & 0 & 1 \\ 1 & 0 & 0 & 0 & 1 \\ 0 & 0 & 1 & 1 & 1 \\ 0 & 1 & 0 & 1 & 1 \\ 1 & 1 & 0 & 1 & 0 \\ 0 & 1 & 1 & 1 & 0 \end{bmatrix} \begin{bmatrix} 0 & 1 & 0 & 1 & 1 & 0 & 1 \\ 0 & 0 & 1 & 1 & 0 & 0 & 0 \\ 0 & 0 & 0 & 0 & 1 & 1 & 0 \\ 1 & 1 & 1 & 0 & 0 & 0 & 0 \\ 1 & 0 & 0 & 0 & 0 & 1 & 1 \end{bmatrix}.$$

Hence $b(J_7 - I_7) \leq 5$; in fact, $b(J_7 - I_7) = 5$. In particular, $b(J_7 - I_7) < \nu(A)$. More generally, it is shown in [16] that

$$b(J_n - I_n) = \min\left\{ k : \binom{k}{\lfloor k/2 \rfloor} \geq n \right\}.$$

Since the right-hand side is strictly less than n for $n > 4$, we have $b(J_n - I_n) < \nu(J_n - I_n) = n$ for $n > 4$.

Now consider the matrix $I_n + C_n$ where, as before, C_n is the cyclic permutation matrix corresponding to the cyclic permutation $(2, 3, \ldots, n-1, 1)$. The matrix $I_n + C_n$ is nonsingular if n is odd and has rank $n - 1$ if n is even. If $n > 2$, then clearly $b(I_n + C_n) = n$. Hence $\nu(I_n + C_n) < b(I_n + C_n)$ for all even $n \geq 4$.

This example shows that the Boolean rank may be strictly less than the rank, and it may also be strictly greater. □

Let n and k be integers with $1 \leq k \leq n$. Let $\tilde{b}(n, k)$ and $\bar{b}(n, k)$ denote the *minimal and maximal Boolean ranks*, respectively, of a matrix in $\mathcal{A}(n, k)$. Also let $\tilde{z}(n, k)$ and $\bar{z}(n, k)$ denote the *minimal and maximal nonnegative integer ranks*, respectively, of a matrix in $\mathcal{A}(n, k)$. We have

---

[11]That is, each edge belongs to at least one biclique.

$\tilde{b}(n,1) = \bar{b}(n,1) = n$ and $\tilde{b}(n,n) = \bar{b}(n,n) = 1$. Also it follows from the above example that

$$\tilde{b}(n,n-1) = \bar{b}(n,n-1) = \min\left\{k : \binom{k}{\lfloor k/2 \rfloor} \geq n\right\}.$$

We also have $\tilde{z}(n,1) = \bar{z}(n,1) = n$ and $\tilde{z}(n,n) = \bar{z}(n,n) = 1$. In addition, since $J_n - I_n$ is a nonsingular matrix, it follows that $\tilde{z}(n,n-1) = \bar{z}(n, n-1) = n$.

The following bounds for $\tilde{b}(n,k)$ and $\tilde{z}(n,k)$ are due to Gregory *et al.* [29]. The equality case in the lower bound on the left is due to Hefner and Lundgren [32].

Theorem 3.10.15 *Let n and k be integers with $n \geq k \geq 1$. Then*

$$\left\lceil \frac{n}{k} \right\rceil \leq \tilde{b}(n,k) \leq \tilde{z}(n,k) \leq \left\lfloor \frac{n}{k} \right\rfloor + k, \tag{3.61}$$

with equality on the far left if and only if k is a divisor of n.

Proof. Let $n = qk + r$ where r is an integer with $0 \leq r < k$. Consider the circulant matrix $A(k, k-r) = \text{Circ}_k(1, 1, \ldots, 1, 0 \ldots, 0)$ of order k with $k - r$ 1's in each row and column. Thus $A(k, k-r) = I_k + C_k + \cdots + C_k^{k-r-1}$ is in $\mathcal{A}(k, k-r)$, and the matrix

$$B(r+k, k) = \begin{bmatrix} O_{r,r} & J_{r,k} \\ J_{k,r} & A(k, k-r) \end{bmatrix}$$

is in $\mathcal{A}(r+k, k)$. We can partition the 1's of $B(r+k, k)$ into an r by k submatrix of 1's and k submatrices of 1's consisting of one row and hence $z(B(r+k, k)) \leq k + 1$. The matrix $A = J_k \oplus \cdots \oplus J_k \oplus B(r, k)$ ($q-1$ J_k's) is in $\mathcal{A}(n, k)$ and satisfies $z(A) \leq q - 1 + k + 1 = q + k$. Hence $z(n,k) \leq q + k$ and the inequality on the far right in (3.61) holds.

Let A be a matrix in $\mathcal{A}(n, k)$. Then A contains exactly kn 1's and a submatrix of all 1's of A contains at most k^2 1's. Hence $k^2 b(A) \geq kn$ and $b(A) \geq n/k$ implying the inequality on the far left in (3.61). If k is a divisor of n, then the matrix $J_k \oplus J_k \oplus \cdots \oplus J_k$ (n/k J_k's) has Boolean rank n/k. Now assume that k is not a divisor of n. Let $J^1, J^2, \ldots, J^p$ be a collection of p submatrices of 1's of A which together contain all the 1's of A. We assume without loss of generality that $J^1 = J_{k-a,k-b}$, where $0 \leq a, b < k$, is maximal in the sense that there is no submatrix of 1's of A that properly contains J^1, that J^1 contains the largest number of 1's of all the matrices $J^1, J^2, \ldots, J^p$, and that J^1 is in the upper left corner of A. We distinguish three cases.

Case 1: $a, b \geq 1$. One of the matrices $J^2, \ldots, J^p$ covers a nonempty set of 1's in A to the right of J^1 and another covers a nonempty set of 1's below J^1. Together these two matrices cover at most $(a+b)k$ 1's not already covered by J^1. The remaining $p - 3$ matrices each cover at most $(k - a)(k - b)$ 1's of A. Thus the number of 1's covered by $J^1, J^2, \ldots, J^p$ is at most

$$(p - 2)(k - a)(k - b) + (a + b)k. \tag{3.62}$$

If $a, b \leq k/2$, then (3.62) is at most

$$(p - 2)k^2 + k^2 = (p - 1)k^2.$$

If $a \geq k/2$ and $b \leq k/2$ (or $a \leq k/2$ and $b \geq k/2$), then (3.62) is at most

$$(p - 2)\frac{k^2}{2} + \frac{3}{2}k^2 \leq (p - 1)k^2.$$

Finally, if $a, b \geq k/2$, then in fact, the number of 1's covered by the matrices $J^1, J^2, \ldots, J^p$ is at most

$$p(k - a)(k - b) \leq m\frac{k^2}{4} \leq (p - 1)k^2.$$

Hence

$$(p - 1)k^2 \geq nk,$$

and we have $p \geq \lceil n/k \rceil + 1$.

Case 2: $a \geq 1$ and $b = 0$ (or, $a = 0$ and $b \geq 1$). The maximality property of J^1 implies that at least two of the matrices $J^2, \ldots, J^p$ are used to cover 1's in A below J^1. Each of these submatrices covers at most ak 1's not covered by J^1. The number of 1's of A covered by $J^1, J^2, \ldots, J^p$ is at most

$$(p - 2)(k - a)k + 2ak \leq (m - 2)k^2 + k^2 = (p - 1)k^2, \text{ if } a \leq k/2,$$

$$pk(k - a) \leq p\frac{k^2}{2} \leq (p - 1)k^2, \text{ if } a \geq k/2.$$

Hence as in Case 1, $p \geq \lceil n/k \rceil + 1$.

Case 3: $a = b = 0$. Let $J^1, J^2, \ldots, J^q$ equal J_k where, since k is not a divisor of n, $1 \leq q < p$. Then, up to permutations of rows and columns,

$$A = J_k \oplus \cdots \oplus J_k \oplus B \quad (q \ J_k\text{'s}),$$

where B is a matrix in $\mathcal{A}(n - qk, k)$. Applying Case 1 or Case 2 to B, we conclude that $J^{q+1}, \ldots, J^p$ cover at most $(p - q - 1)k^2$ 1's of B, and hence $J^1, J^2, \ldots, J^p$ cover at most $qk^2 + (p - q - 1)k^2 = (p - 1)k^2$ 1's of A. Again as in case 1, we have $p \geq \lceil n/k \rceil + 1$. $\square$

Example. [10, 32] Let q be a positive integer, let k be an positive even integer, and let $n = qk + (k/2)$. Let $A = J_k \oplus J_k \oplus A'$ $(q - 1 \ J_k\text{'s})$ where

$$A' = \begin{bmatrix} O_{k/2} & J_{k/2} & J_{k/2} \\ J_{k/2} & O_{k/2} & J_{k/2} \\ J_{k/2} & J_{k/2} & O_{k/2} \end{bmatrix}.$$

Then A is in $\mathcal{A}(n, k)$ and it follows easily that

$$\nu(A) = b(A) = z(A) = q + 2 = \left\lfloor \frac{n}{k} \right\rfloor + 2 = \left\lceil \frac{n}{k} \right\rceil + 1.$$

Hence Theorems 3.10.9 and 3.10.15 are best possible when k is even.

Theorem 3.10.16 *Let $n \geq 2$ be an integer. Then*

$$\tilde{b}(n, 2) = \tilde{z}(n, 2) = \begin{cases} \frac{n}{2} & \text{if } n \text{ is even,} \\[2mm] \frac{n+3}{2} & \text{if } n \text{ is odd.} \end{cases}$$

Proof. By Theorem 3.10.15, we have

$$\tilde{z}(n, 2) \geq \tilde{b}(n, 2) \geq \left\lceil \frac{n}{2} \right\rceil$$

with equality if and only if n is even. Taking $k = 2$ in the previous example we complete the proof. $\qquad\square$

In [29] the following formulas are also proved; the formula in (3.63) is the same as that for $\tilde{\nu}(n, 3)$ quoted in Theorem 3.10.12.

Theorem 3.10.17 *Let n be an integer. If $n \geq 3$, then*

$$\tilde{b}(n, 3) = \tilde{z}(n, 3) = \begin{cases} \frac{n}{3}, & \text{if } n \text{ is divisible by 3,} \\[2mm] \left\lfloor \frac{n}{3} \right\rfloor + 3, & \text{otherwise.} \end{cases} \tag{3.63}$$

Let a_n equal 0, 3, 2, or 4 according as $n \equiv 0$, 1, 2, or 3 mod 4. If $n > 5$, then

$$\tilde{b}(n, 4) = \tilde{z}(n, 4) = \left\lceil \frac{n}{4} \right\rceil + a_n. \tag{3.64}$$

In addition, $\tilde{b}(5, 4) = \tilde{z}(5, 4) = 5$.

To conclude this section we note that, generalizing a theorem of Longstaff [46], Odlyzko [55] has determined the maximal number of distinct rows in a (0,1)-matrix with n columns and all row sums equal to k. This number equals

$$\binom{n-1}{k}, \qquad \text{if } 1 \leq k < n/2,$$

$$2\binom{n}{(n-2)/2}, \quad \text{if } k = n/2, \text{ and}$$

$$\binom{n-1}{m-1}, \qquad \text{if } n/2 < k \leq n.$$

3.11   Permanent

Let $A = [a_{ij}]$ be a matrix of size m by n with $m \leq n$. Recall from Section 1.3 that the permanent of A is defined by

$$\mathrm{per}(A) = \sum a_{1i_1} a_{2i_2} \ldots a_{mi_m}, \tag{3.65}$$

where the summation extends over all sequences $i_1, i_2, \ldots, i_m$ with $1 \leq i_1 < i_2 < \cdots < i_m \leq n$. The permanent is invariant under arbitrary permutations of rows and columns. There is a nonzero term in the summation (3.65) if and only if the term rank $\rho(A)$ of A equals m. According to Theorem 1.3.1, $\rho(A) = m$ if and only if A does not have a zero submatrix $O_{k,l}$ with $k + l = n + 1$.

There is an extensive literature on the permanent, much of it concerned with lower and upper bounds. In this section we briefly review results for the permanent as they pertain to the class $\mathcal{A}(R, S)$. A fuller treatment of the permanent is given in Chapter 7 of the book [13]. With a few exceptions, we refer to [13] for proofs.

For a nonempty class $\mathcal{A}(R, S)$ we define

$$\widetilde{\mathrm{per}}(R, S) = \min\{\mathrm{per}(A) : A \in \mathcal{A}(R, S)\}$$

to be the *minimal permanent* of a matrix in $\mathcal{A}(R, S)$, and

$$\overline{\mathrm{per}}(R, S) = \max\{\mathrm{per}(A) : A \in \mathcal{A}(R, S)\}$$

to be the *maximal permanent* of a matrix in $\mathcal{A}(R, S)$. It follows that $\widetilde{\mathrm{per}}(R, S) = 0$ if and only if $\tilde{\rho}(R, S) < m$, and that $\overline{\mathrm{per}}(R, S) > 0$ if and only if $\bar{\rho}(R, S) = m$.

The best lower bound for the nonzero permanent of a (0,1)-matrix in terms of its row sum vector is due to Ostrand (see [13]).

Theorem 3.11.1 *Let A be an m by n (0, 1)-matrix with $\mathrm{per}(A) > 0$. Let the row sum vector of A be $R = (r_1, r_2, \ldots, r_m)$ where R is nonincreasing. Then*

$$\mathrm{per}(A) \geq \prod_{i=1}^{m} \max\{1, r_i - m + i\}.$$

If we assume that $m = n$, then we obtain the following inequalities for $\mathcal{A}(R, S)$.

Corollary 3.11.2 *Let $R = (r_1, r_2, \ldots, r_n)$ and $S = (s_1, s_2, \ldots, s_n)$ be nonincreasing integral vectors of size n such that $\mathcal{A}(R, S)$ is nonempty. If*

$\bar{\rho}(R,S) = n$, then

$$\overline{\mathrm{per}}(R,S) \geq \max \left\{ \prod_{i=1}^{n} \max\{1, r_i - n + i\}, \prod_{i=1}^{n} \max\{1, s_i - n + i\} \right\}.$$

If $\tilde{\rho}(R,S) = n$, then

$$\widetilde{\mathrm{per}}(R,S) \geq \max \left\{ \prod_{i=1}^{n} \max\{1, r_i - n + i\}, \prod_{i=1}^{n} \max\{1, s_i - n + i\} \right\}.$$

$\square$

The best upper bound for the permanent of a (0,1)-matrix in terms of its row sum vector was conjectured by Minc and proved first by Brégman [5] and later more simply by Schrijver [64] (see also [13]).

Theorem 3.11.3 *Let A be a $(0,1)$-matrix of order n with row sum vector $R = (r_1, r_2, \ldots, r_n)$. Then*

$$\mathrm{per}(A) \leq \prod_{i=1}^{n} (r_i!)^{1/r_i}.$$

Moreover, equality holds if and only if $n = mk$ for some integers m and k, and the rows and columns of A can be permuted to give $J_k \oplus J_k \oplus \cdots \oplus J_k$ (m J_k's).

Corollary 3.11.4 *Let $R = (r_1, r_2, \ldots, r_n)$ and $S = (s_1, s_2, \ldots, s_n)$ be integral vectors such that $\mathcal{A}(R,S)$ is nonempty. Then*

$$\overline{\mathrm{per}}(R,S) \leq \min \left\{ \prod_{i=1}^{n} (r_i!)^{1/r_i}, \prod_{i=1}^{n} (s_i!)^{1/s_i} \right\}.$$

$\square$

The inequality in Theorem 3.11.3 relies on the assumption that A is a (0,1)-matrix.[12] A weaker inequality, valid for all nonnegative matrices A, is $\mathrm{per} A \leq r_1 r_2 \ldots r_n$. Combining this with the corresponding inequality for column sums we get

$$\mathrm{per}(A) \leq \min \left\{ \prod_{i=1}^{n} r_i, \prod_{i=1}^{n} s_i \right\}. \tag{3.66}$$

Jurkat and Ryser [39, 13] improved (3.66) obtaining an inequality that is not merely the minimum of one for row sums and one for column sums as in (3.66). Because of its strong connection to the class $\mathcal{A}(R,S)$, we give the proof. The following lemma is easily proved by induction.

---

[12] E.g. if A is the matrix of order 1 with unique element equal to 2, then $\mathrm{per} A = 2 \not\leq 2^{1/2}$.

Lemma 3.11.5 *Let $(a_1, a_2, \ldots, a_n)$ and $(b_1, b_2, \ldots, b_n)$ be nonincreasing vectors of nonnegative numbers. Then*

$$\prod_{i=1}^{n} \min\{a_i, b_{j_i}\} \leq \prod_{i=1}^{n} \min\{a_i, b_i\}$$

for each permutation $j_1, j_2, \ldots, j_n$ of $\{1, 2, \ldots, n\}$. □

Theorem 3.11.6 *Let $A = [a_{ij}]$ be a nonnegative matrix of order n with nonincreasing row sum vector $R = (r_1, r_2, \ldots, r_n)$ and nonincreasing column sum vector $S = (s_1, s_2, \ldots, s_n)$. Then*

$$\text{per}(A) \leq \prod_{i=1}^{n} \min\{r_i, s_i\}. \tag{3.67}$$

Proof. The inequality (3.67) holds if $n = 1$. We assume that $n > 1$ and proceed by induction on n. Without loss of generality we assume that $r_n \leq s_n$. Using the Laplace expansion of the permanent on row n we obtain

$$\text{per}(A) = \sum_{j=1}^{n} a_{nj} \text{per}(A(n, j)).$$

Let the row and column sum vectors of $A(n, j)$ be $R^j = (r_1^j, r_2^j, \ldots, r_{n-1}^j)$ and $S^j = (s_1^j, \ldots, s_{j-1}^j, s_{j+1}^j, \ldots, s_n^j)$, respectively. There are permutations $i_1, i_2, \ldots, i_{n-1}$ of $\{1, 2, \ldots, n-1\}$ and $k_1, k_2, \ldots, k_{n-1}$ of $\{1, \ldots, j-1, j+1, \ldots, n\}$ such that

$$r_{i_1}^j \geq r_{i_2}^j \geq \cdots \geq r_{i_{n-1}}^j \quad \text{and} \quad s_{k_1}^j \geq s_{k_2}^j \geq \cdots \geq s_{k_{n-1}}^j.$$

We have $r_{i_t}^j \leq r_{i_t}$ and $s_{k_t}^j \leq s_{k_t}$ for $t = 1, 2, \ldots, n-1$ and $j = 1, 2, \ldots, n$. Taking these inequalities into account and applying the induction hypothesis, we obtain for each $j = 1, 2, \ldots, n$

$$\text{per}(A(n, j)) \leq \prod_{t=1}^{n-1} \min\{r_{i_t}, s_{k_t}\}$$

$$\leq \prod_{t=1}^{j-1} \min\{r_t, s_t\} \prod_{t=j}^{n-1} \min\{r_t, s_{t+1}\} \quad \text{(by Lemma 3.11.5)}$$

$$\leq \prod_{t=1}^{n-1} \min\{r_t, s_t\} \quad \text{(since S is nonincreasing).}$$

Therefore

$$\mathrm{per}(A) = \sum_{j=1}^{n} a_{nj} \mathrm{per}(A(n,j))$$

$$\leq \sum_{j=1}^{n} \left(a_{nj} \prod_{t=1}^{n-1} \min\{r_t, s_t\} \right)$$

$$= \left(\sum_{j=1}^{n} a_{nj} \right) \prod_{t=1}^{n-1} \min\{r_t, s_t\}$$

$$= r_n \prod_{t=1}^{n-1} \min\{r_t, s_t\}$$

$$= \prod_{t=1}^{n} \min\{r_t, s_t\} \quad (\text{since } r_n \leq s_n).$$

$\square$

Corollary 3.11.7 *Let* $R = (r_1, r_2, \ldots, r_n)$ *and* $S = (s_1, s_2, \ldots, s_n)$ *be nonincreasing, nonnegative integral vectors such that* $\mathcal{A}(R, S)$ *is nonempty. Then*

$$\overline{\mathrm{per}}(R, S) \leq \prod_{i=1}^{n} \min\{r_i, s_i\}.$$

$\square$

More information about $\widetilde{\mathrm{per}}(R, S)$ and $\overline{\mathrm{per}}(R, S)$ can be obtained by restricting attention to a class $\mathcal{A}(n, k)$ of regular matrices. We use the simpler notations $\widetilde{\mathrm{per}}(n, k)$ and $\overline{\mathrm{per}}(n, k)$ for the minimal and maximal permanent, respectively, in the class $\mathcal{A}(n, k)$.

From Theorem 3.11.3 we obtain the following.

Theorem 3.11.8 *Let* k *and* n *be integers with* $1 \leq k \leq n$. *Then*

$$\overline{\mathrm{per}}(n, k) \leq (k!)^{n/k}$$

with equality if and only if k *is a divisor of* n. *If* k *is a divisor of* n, *then a matrix* A *in* $\mathcal{A}(n, k)$ *satisfies* $\mathrm{per} A = (k!)^{n/k}$ *if and only if its rows and columns can be permuted to give* $J_k \oplus J_k \oplus \cdots \oplus J_k$ (n/k J_k's). $\square$

We summarize the other known results about $\overline{\mathrm{per}}(n, k)$ in the following theorem. Many of these results are proved in [13]. The case $k = n$ is trivial because $\mathcal{A}(n, n)$ contains the unique matrix J_n whose permanent equals $n!$ The cases $k = 1$ and $k = n - 1$ are also trivial because, up to row and column permutations, the class $\mathcal{A}(n, 1)$ contains the unique matrix I_n whose permanent equals 1, and the class $\mathcal{A}(n, n - 1)$ contains the unique matrix $J_n - I_n$ whose permanent equals the nth derangement number D_n. Theorem 3.11.8 determines $\overline{\text{per}}(n, k)$ if k is a divisor of n.

As before, let C_n denote the cyclic permutation matrix of order n having 1's in positions $(1, 2), (2, 3), \ldots, (n - 1, n), (n, 1)$.

Theorem 3.11.9 *The following hold.*

(i) [8] $\overline{\text{per}}(n, 2) = 2^{\lceil n/2 \rceil}$. *A matrix A in $\mathcal{A}(n, 2)$ achieves the maximal permanent if and only if its rows and columns can be permuted to give*

$$J_2 \oplus J_2 \oplus \cdots \oplus J_2 \quad (n \text{ even}), \text{ and } J_2 \oplus \cdots \oplus J_2 \oplus (J_3 - I_3) \ (n \text{ odd}).$$

(ii) [50] *If $n \equiv 1 \bmod 3$, then*

$$\overline{\text{per}}(n, 3) = 6^{(n-4)/3} 9;$$

a matrix achieving the maximal permanent is

$$J_3 \oplus \cdots \oplus J_3 \oplus (J_4 - I_4).$$

(iii) [50] *If $n \equiv 2 \bmod 3$, then*

$$\overline{\text{per}}(n, 3) = \begin{cases} 13 \text{ if } n = 5, \\ 6^{(n-8)/3} 9^2 \text{ if } n \geq 8; \end{cases}$$

a matrix achieving the maximal permanent is $J_5 - I_5 - C_5$ if $n = 5$ and

$$J_3 \oplus \cdots \oplus (J_4 - I_4) \oplus (J_4 - I_4), \text{ if } n \geq 8.$$

(iv) [3, 4] *If $n \equiv 1 \bmod 4$, then*

$$\overline{\text{per}}(n, 4) = 24^{(n-5)/4} 44;$$

a matrix achieving the maximal permanent is

$$J_4 \oplus \cdots \oplus J_4 \oplus (J_5 - I_5).$$

(v) [8, 49] *The matrices achieving the maximal permanent in $\mathcal{A}(n, n - 2)$ are those whose rows and columns can be permuted to give*

$$J_n - (J_2 \oplus J_2 \oplus \cdots \oplus J_2), \text{ if } n \text{ is even and } n \geq 8,$$

$$J_n - (J_2 \oplus \cdots \oplus J_2 \oplus (J_3 - I_3)), \text{ if } n \text{ is odd and } n \geq 9,$$

$$J_5 - (I_5 + C_5), \text{ if } n = 5,$$

$$J_6 - ((J_3 - I_3) \oplus (J_3 - I_3)), \text{ if } n = 6,$$

$$J_7 - ((I_5 + C_5) \oplus J_2) \text{ or } J_7 - ((I_3 + C_3) \oplus J_2 \oplus J_2), \text{ if } n = 7.$$

(There are in general no simple formulas for these maximal permanents.)

(vi) [49] *If $k \geq 3$ and k is a divisor of n with $n/k \geq 5$, then the matrices achieving the maximal permanent in $\mathcal{A}(n, n-k)$ are those matrices whose rows and columns can be permuted to give*

$$J_n - (J_k \oplus J_k \oplus \cdots \oplus J_k).$$

(vii) [77] *Up to permutations of rows and columns, each matrix attaining the maximal permanent in $\mathcal{A}(n,k)$ is a direct sum of matrices of bounded order of which fewer than k differ from J_k.*

It has been conjectured in [49] that, with the exception of a finite number, possibly two, of special cases, if k is a divisor of n, then the matrices achieving the maximal permanent in $\mathcal{A}(n, n-k)$ are those matrices whose rows and columns can be permuted to give $J_n - (J_k \oplus J_k \oplus \cdots \oplus J_k)$. The two known exceptions are $n = 6, k = 2$ (see (v) in Theorem 3.11.9), and $n = 9, k = 3$ where the matrix

$$\begin{bmatrix}
0 & 1 & 0 & 0 & 1 & 1 & 1 & 1 & 1 \\
0 & 0 & 1 & 0 & 1 & 1 & 1 & 1 & 1 \\
0 & 0 & 0 & 1 & 1 & 1 & 1 & 1 & 1 \\
1 & 0 & 0 & 0 & 1 & 1 & 1 & 1 & 1 \\
1 & 1 & 1 & 1 & 0 & 1 & 1 & 0 & 0 \\
1 & 1 & 1 & 1 & 0 & 0 & 0 & 1 & 1 \\
1 & 1 & 1 & 1 & 0 & 0 & 0 & 1 & 1 \\
1 & 1 & 1 & 1 & 1 & 1 & 0 & 0 & 0 \\
1 & 1 & 1 & 1 & 1 & 0 & 1 & 0 & 0
\end{bmatrix}$$

has permanent equal to $12\,108$ while $J_9 - (J_3 \oplus J_3 \oplus J_3)$ has permanent equal to $12\,096$.

We now turn to the minimal permanent $\widetilde{\mathrm{per}}(n,k)$. We have the trivial cases $\widetilde{\mathrm{per}}(n,1) = 1$, $\widetilde{\mathrm{per}}(n,2) = 2$, with $\mathrm{per}(I_n + C_n) = 2$, and $\widetilde{\mathrm{per}}(n, n-1) = D_n$, the nth derangement number, and $\widetilde{\mathrm{per}}(n,n) = n!$ In addition, Henderson [33] proved that if n is even, $\widetilde{\mathrm{per}}(n, n-2) = U_n$, the nth ménage number, and if n is odd, $\widetilde{\mathrm{per}}(n,k) = U_n - 1$; the permanent of the matrix $J_n - (I_n + C_n)$ equals U_n, and if n is odd, the permanent of the matrix $J_n - ((I_{(n-1)/2} + C_{(n-1)/2}) \oplus (I_{(n+1)/2} + C_{(n+1)/2}))$ equals $U_n - 1$. The cases of equality were completed settled by McKay and Wanless [49] and we summarize in the following theorem.

Theorem 3.11.10 *Let $n \geq 2$ be an integer. If n is even, then $\widetilde{\mathrm{per}}(n, n-2) = U_n$, and if n is odd, $\widetilde{\mathrm{per}}(n,k) = U_n - 1$. The matrices in $\mathcal{A}(n, n-2)$ achieving the maximal permanent are:*

$J_n - (I_n + C_n), n \leq 4$;
$J_n - (I_t + C_t) \oplus (I_{t+1} + C_{t+1}), n = 2t + 1 \geq 4$;
$J_6 - (I_6 + C_6)$ or $J_6 - (I_4 + C_4) \oplus J_2$ or $J_6 - (J_2 \oplus J_2 \oplus J_2), n = 6$;
$J_n - (I_n + C_n)$ or $J_n - (I_{t-1} + C_{t-1}) \oplus (I_{t+1} + C_{t+1}), n = 2t \geq 8$.

We summarize the remaining known facts in the following theorem. Let

$$\theta_k = \lim_{n \to \infty} \widetilde{\mathrm{per}}(n,k)^{1/n}.$$

Also let $\mathcal{Z}^+(n,k)$ denote the *class of all nonnegative integral matrices with all row and column sums equal to k*. We also let

$$\widetilde{\mathrm{per}}^*(n,k) = \min\{\mathrm{per}(A) : A \in \mathcal{Z}^+(n,k)\}, \text{ and}$$
$$\theta_k^* = \lim_{n \to \infty} \widetilde{\mathrm{per}}^*(n,k)^{1/n}.$$

The numbers θ_k and θ_k^* give the best exponential lower bound $\mathrm{per}(A) \geq \theta_k^n$ for matrices A in $\mathcal{A}(n,k)$ and $\mathcal{Z}^+(n,k)$, respectively, for n sufficiently large, and thus provide an exponential lower bound for $\widetilde{\mathrm{per}}(n,k)$ and $\widetilde{\mathrm{per}}(n,k)$.

Theorem 3.11.11 *The following hold:*

(i) [75] $\widetilde{\mathrm{per}}^*(n,k) \geq 6 \left(\frac{4}{3}\right)^{n-3} \geq \left(\frac{4}{3}\right)^n$ *for all n and k with $n \geq k \geq 3$;*

(ii) [67, 65, 66] $\theta_k^* = \frac{(k-1)^{k-1}}{k^{k-2}}$ *for all $k \geq 3$;*

(iii) [78] $\theta_k = \frac{(k-1)^{k-1}}{k^{k-2}}$ *for all $k \geq 3$;*

(iv) $\widetilde{\mathrm{per}}^*(n,k) \geq n!(k/n)^n$, *and so $\theta_k \geq k/e$ for $k \geq 4$.*

The last statement (iv) follows by applying the Egoryčev–Falikman solution of the van der Waerden conjecture for the minimal value of the permanent of a doubly stochastic matrix of order n to the doubly stochastic matrix $(1/k)A$. It gives the exponential lower bound $(k/e)^n$ which is not as good as that given by (ii). This conjecture and its solution will be discussed in Chapter 9. There remain many conjectures and open problems on permanents [18].

3.12   Determinant

In this section we investigate the determinant over a nonempty class $\mathcal{A}(R,S)$ of square matrices with row sum vector $R = (r_1, r_2, \ldots, r_n)$ and column sum vector $S = (s_1, s_2, \ldots, s_n)$. We assume that R and S are positive integral vectors, for otherwise every matrix in $\mathcal{A}(R,S)$ has determinant equal to zero. We define

$$\widetilde{\det}(R,S) = \min\{|\det(A)| : A \in \mathcal{A}(R,S)\}$$

to be the *minimal determinant*, in absolute value, of a matrix in $\mathcal{A}(R,S)$, and

$$\overline{\det}(R,S) = \max\{|\det(A)| : A \in \mathcal{A}(R,S)\}$$

to be the *maximal determinant*, in absolute value, of a matrix in $\mathcal{A}(R,S)$. Since interchanging two rows of a matrix changes the sign of its determinant, these extreme values are attained by matrices in $\mathcal{A}(R,S)$ with nonnegative determinants. We have $\widetilde{\det}(R,S) > 0$ if and only if the minimal rank satisfies $\tilde{\nu}(R,S) = n$. We also have $\overline{\det}(R,S) > 0$ if and only if the maximal rank satisfies $\bar{\nu}(R,S) = n$. Both the minimal and maximal determinants are very complicated functions of R and S, and as a result we will not have very much to say about them for general R and S. When $\bar{\nu}(R,S) = n$ but $\tilde{\nu}(R,S) < n$, in place of $\widetilde{\det}(R,S)$ one can also consider

$$\widetilde{\det}^{\,\neq 0}(R,S) = \min\{|\det(A)| : A \in \mathcal{A}(R,S) \text{ and } \det(A) \neq 0\},$$

the *minimal nonzero determinant*, in absolute value, of a matrix in $\mathcal{A}(R,S)$, equivalently the minimal determinant of the subclass of nonsingular matrices in $\mathcal{A}(R,S)$. Again, this extreme value is attained by a matrix with a positive determinant. We have

$$\widetilde{\det}(R,S) = \begin{cases} 0 & \text{if } \tilde{\nu}(R,S) < n, \\ \widetilde{\det}^{\,\neq 0}(R,S) & \text{if } \tilde{\nu}(R,S) = n. \end{cases}$$

As usual for the regular classes $\mathcal{A}(n,k)$ we use the abbreviations

$$\overline{\det}(n,k), \ \widetilde{\det}(n,k), \text{ and } \widetilde{\det}^{\,\neq 0}(n,k).$$

To illustrate these ideas and their difficulty, we prove the following simple theorem for the regular class $\mathcal{A}(n,2)$.

Theorem 3.12.1 *Let $n \geq 3$ be an integer. Then*

$$\overline{\det}(n,2) = 2^{q(n)},$$

where $q(n)$ is the largest number of parts in a partition of n into odd parts of size 3 or greater.

Proof. Each matrix in $\mathcal{A}(n,2)$ is, up to row and column permutations, equal to a matrix of the form

$$(I_{k_1} + C_{k_1}) \oplus (I_{k_2} + C_{k_2}) \oplus \cdots \oplus (I_{k_t} + C_{k_t}) \tag{3.68}$$

where $k_1, k_2, \ldots, k_n$ is a partition of n into parts of size 2 or greater and C_j denotes the full cycle permutation matrix of order j. The determinant of the matrix (3.68) is nonzero if and only if none of the parts $k_1, k_2, \ldots, k_t$ is even, and when this happens, its determinant equals 2^{T}. The theorem now follows. $\square$

The following corollary is contained in [20] and follows directly from Theorem 3.12.1.

Corollary 3.12.2 *Let $n \geq 3$ be an integer. Then*

$$\overline{\det}(n, 2) = \begin{cases} 2^{\lfloor n/3 \rfloor}, & \text{if } n \equiv 0 \text{ or } 2 \bmod 3, \\ 2^{\lfloor n/3 \rfloor - 1}, & \text{if } n \equiv 1 \bmod 3. \end{cases}$$

A general theorem of Ryser [58] concerning the maximal determinant of a $(0,1)$-matrix of order n with a specified number of 1's furnishes an upper bound for $\overline{\det}(n, k)$ whose proof we omit.

Theorem 3.12.3 *Let n and k be integers with $1 < k < n$, and let $\lambda = k(k-1)/(n-1)$. Then*

$$\overline{\det}(n, k) \leq k(k - \lambda)^{(n-1)/2}.$$

If for some matrix $A \in \mathcal{A}(n, k)$, $|\det(A)|$ equals the upper bound in Theorem 3.12.3, then A is the incidence of a combinatorial configuration known as a (v, k, λ)-design.

We now consider $\widetilde{\det}^{\neq 0}(n, k)$, defined according to Theorem 3.10.7, when $1 \leq k \leq n-1$ and $(n, k) \neq (4, 2)$.[13] It was conjectured by Newman [54] (see also [27]) and proved by Li *et al.* [45] that $\widetilde{\det}^{\neq 0}(n, k) = k \gcd\{n, k\}$.

The proof we give is taken from [45]. We first note the following immediate corollary of Lemma 3.10.6.

Corollary 3.12.4 *Let n and k be integers with $1 \leq k \leq n - 1$. Then*

$$\overline{\det}(n, n - k) = \frac{n - k}{k} \overline{\det}(n, k).$$

In particular, a formula for $\overline{\det}(n, n - 2)$ can be obtained from the formula given in Corollary 3.12.2. □

The following lemma is from [54]. *The set of all integral matrices of order n with row and column sums k is denoted by $\mathcal{Z}(n, k)$.* The set of nonnegative matrices in $\mathcal{Z}(n, k)$ is the set $\mathcal{Z}^+(n, k)$ previously defined. We thus have $\mathcal{A}(n, k) \subseteq \mathcal{Z}^+(n, k) \subseteq \mathcal{Z}(n, k)$.

Lemma 3.12.5 *Let n and k be integers with $1 \leq k \leq n$, and let A be a matrix in $\mathcal{Z}(n, k)$. Then $\det(A)$ is an integral multiple of $k \gcd\{n, k\}$.*

Proof. The matrix B obtained from A by adding rows $2, 3, \ldots, n$ to row 1 and then columns $2, 3, \ldots, n$ to column 1 has the same determinant as A. The first row and the transpose of the first column of B both equal

---

[13] We neglect the trivial case $n = k = 1$.

$[nk\ k\ \ldots\ k]$. Hence each term in the determinant expansion of B contains either nk or k^2 as a factor, and hence contains $k\gcd\{n,k\}$ as a factor. □

Thus to prove the conjecture it is only necessary to show that there is a matrix in $\mathcal{A}(n,k)$ with determinant equal to $k\gcd\{n,k\}$. In fact, such a matrix can be found in the subclass $\mathcal{A}'(n,k)$ of $\mathcal{A}(n,k)$ to be defined now.

Let n and k be positive integers with $2 \le k < n$. Let $\mathcal{A}'(n,k)$ denote the set of all matrices A in $\mathcal{A}(n,k)$ for which the following two properties hold:

(i) the submatrix $A[\{n-k+2, n-k+3, \ldots, n\}, \{1,2,\ldots,k-1\}]$ of order $k-1$ has 1's on and below its main diagonal (i.e. the elements in the lower left corner of A form a triangle of 1's of size $k-1$), and

(ii) the submatrix $A[\{k+2, k+3, \ldots, n\}, \{k, k+1, \ldots, n-2\}]$ of order $n-k-1$ has 0's on and below its main diagonal (i.e. deleting the first $k-1$ columns of A results in a matrix whose elements in the lower left corner form a triangle of 0's of size $n-k-1$).

The following lemma can be verified by calculation.

Lemma 3.12.6 *The matrix*

$$\begin{bmatrix} 0 & 1 & 0 & 1 & 0 & 0 \\ 0 & 1 & 1 & 0 & 0 & 0 \\ 0 & 0 & 1 & 1 & 0 & 0 \\ 1 & 0 & 0 & 0 & 1 & 0 \\ 0 & 0 & 0 & 0 & 1 & 1 \\ 1 & 0 & 0 & 0 & 0 & 1 \end{bmatrix}$$

in $\mathcal{A}'(6,2)$ has determinant equal to -2^2, and thus has the minimal determinant in absolute value of all matrices in $\mathcal{A}(6,2)$. □

The construction in the following lemma is due to Grady [27]. It proves the conjecture when $(n,k) = (2k,k), k \ne 2$. As before C_n denotes the permutation matrix of order n corresponding to the cyclic permutation $(2,3,\ldots,n,1)$

Lemma 3.12.7 *Let $k > 2$ be an integer, and let*

$$Q = \begin{bmatrix} -I_{k-1} & O_{k-1,2} & I_{k-1} \\ O_{2,k-1} & O_2 & O_{2,k-1} \\ I_{k-1} & O_{k-1,2} & -I_{k-1} \end{bmatrix}.$$

Then the matrix

$$A = C_{2k}^{-1} + I_{2k} + C_{2k} + \cdots + C_{2k}^{k-2} + Q$$

is in $\mathcal{A}'(2k,k)$ and $\det(A) = -k^2$.

Proof. We note that the matrix A can be obtained from $C_{2k}^{-1} + I_{2k} +$ $C_{2k} + \cdots + C_{2k}^{k-2}$ by a sequence of $k-1$ "disjoint" interchanges. It is straightforward to verify that the matrix A belongs to $\mathcal{A}'(2k, k)$. It follows from Lemma 3.10.6 that $\det(A) = -\det(J_{2k} - A)$. Thus it suffices to show that $\det(J_{2k} - A) = k^2$. We sketch the proof of this calculation leaving some elementary details to the interested reader.

We begin with the partition

$$J_{2k} - A = \begin{bmatrix} E & F \\ G & H \end{bmatrix}$$

of $J_{2k} - A$ into submatrices of order k, and using elementary row and column operations,[14] we obtain

$$\det(J_{2k} - A) = (-1)^k \det(G) \det(F - EG^{-1}H).$$

Straightforward computations give

$$G^{-1} = \begin{bmatrix} 1 & -1 & 0 & 0 & \cdots & 0 & 0 & 1 \\ 0 & 1 & -1 & 0 & \cdots & 0 & 0 & 0 \\ 0 & 0 & 1 & -1 & \cdots & 0 & 0 & 0 \\ 0 & 0 & 0 & 1 & \cdots & 0 & 0 & 0 \\ \vdots & \vdots & \vdots & \vdots & \ddots & \vdots & \vdots & \vdots \\ 0 & 0 & 0 & 0 & \cdots & 1 & -1 & 0 \\ 0 & 0 & 0 & 0 & \cdots & 0 & 1 & -1 \\ 0 & 0 & 0 & 0 & \cdots & 0 & 0 & 1 \end{bmatrix}.$$

Computing EG^{-1} and $(EG^{-1})H$ column by column results in

$$EG^{-1} = \begin{bmatrix} 1 & -1 & 0 & 0 & \cdots & 0 & 0 & 2 \\ 0 & 1 & -1 & 0 & \cdots & 0 & 0 & 0 \\ 1 & -1 & 1 & -1 & \cdots & 0 & 0 & 1 \\ 1 & 0 & -1 & 1 & \cdots & 0 & 0 & 1 \\ \vdots & \vdots & \vdots & \vdots & \ddots & \vdots & \vdots & \vdots \\ 1 & 0 & 0 & 0 & \cdots & 1 & -1 & 0 \\ 1 & 0 & 0 & 0 & \cdots & -1 & 1 & 0 \\ 1 & 0 & 0 & 0 & \cdots & 0 & -1 & 1 \end{bmatrix},$$

---

[14]First interchange the first k columns with the last k columns. Then subtract EG^{-1} times the last k rows from the first k rows. The matrix $F - EG^{-1}H$ obtained in the upper left corner is the *Schur complement* of G.

and

$$EG^{-1}H = \begin{bmatrix} 2 & 1 & 2 & 2 & 2 & \cdots & 2 & 0 & 3 \\ -1 & 1 & -1 & 0 & 0 & \cdots & 0 & 0 & 0 \\ 1 & -1 & 2 & 0 & 1 & \cdots & 1 & 0 & 2 \\ 0 & 1 & -1 & 2 & 0 & \cdots & 1 & 0 & 2 \\ 0 & 0 & 1 & -1 & 2 & \cdots & 1 & 0 & 2 \\ \vdots & \vdots & \vdots & \vdots & \vdots & \ddots & \vdots & \vdots & \vdots \\ 0 & 0 & 0 & 0 & 0 & \cdots & 2 & -1 & 2 \\ 0 & 0 & 0 & 0 & 0 & \cdots & -1 & 1 & 1 \\ 0 & 0 & 0 & 0 & 0 & \cdots & 1 & -1 & 2 \end{bmatrix}.$$

The above gives

$$F - EG^{-1}H = \begin{bmatrix} -1 & -1 & -1 & -1 & -1 & \cdots & -1 & 1 & 3 \\ 2 & 0 & 1 & 1 & 1 & \cdots & 1 & 1 & 1 \\ -1 & 2 & -1 & 0 & 0 & \cdots & 0 & 1 & -1 \\ 0 & -1 & 2 & -1 & 0 & \cdots & 0 & 1 & -1 \\ 0 & 0 & -1 & 2 & -1 & \cdots & 0 & 1 & -1 \\ \vdots & \vdots & \vdots & \vdots & \vdots & \ddots & \vdots & \vdots & \vdots \\ 0 & 0 & 0 & 0 & 0 & \cdots & -1 & 1 & -1 \\ 0 & 0 & 0 & 0 & 0 & \cdots & 2 & 0 & -1 \\ 0 & 0 & 0 & 0 & 0 & \cdots & -1 & 2 & -1 \end{bmatrix}.$$

Since $\det(G) = 1$, it remains to evaluate $\det(F - EG^{-1}H)$. We do this by adding rows $2, 3, \ldots, k$ to row 1, and adding columns $1, 2, \ldots, k-1$ to column k. This gives a matrix X whose only nonzero element in the first row is k in column $k-1$ and whose only nonzero element in the last column is k in row 2. In addition, the submatrix of X of order $k-2$ in the lower left corner is a triangular matrix with -1's on its main diagonal. Hence $\det(X) = (-1)^k k^2$, and we have

$$\det(A) = -\det(J_{2k} - A) = (-1)^{k+1} \det(G) \det(F - EG^{-1}H)$$
$$= (-1)^{k+1}(1)(-1)^k k^2 = -k^2,$$

as was to be proved.   □

In the next lemma, a procedure is described for generating matrices in classes $\mathcal{A}'(n, k)$.

Lemma 3.12.8 *Let A be a matrix in $\mathcal{A}'(n, k)$ for some n and k.*

(i) *Let A^* be the matrix obtained from $J_n - A$ by permuting columns so that columns $k, k+1, \ldots, n-2$ come first. Then A^* is in $\mathcal{A}(n, n-k)$ and $|\det(A^*)| = (n-k)|\det(A)|/k$.*

(ii) *Let*

$$A^* = \begin{bmatrix} \hat{A} & B \\ C & D \end{bmatrix}$$

where $\hat{A}$ is obtained from A by replacing its triangle of 1's of size $k-1$ in its lower left corner with 0's, B is a $(0,1)$-matrix of size n by k all of whose 1's form a triangle of size $k-1$ in its lower left corner, C is a $(0,1)$-matrix of size k by n all of whose 1's form a triangle of size $k-1$ in its lower left corner, and D is a $(0,1)$-matrix of size k by k with 1's on and above its main diagonal and 0's elsewhere. Then A^* is in $\mathcal{A}'(n+k,k)$ and $\det(A^*) = \det(A)$.

Proof. Assertion (i) is readily verified using Lemma 3.10.6. That the matrix A^* in assertion (ii) is in $\mathcal{A}'(n+k,k)$ follows by direct observation; the second conclusion in (ii) is a consequence of the observations that $A = \hat{A} - BD^{-1}C$ and $\det(A^*) = \det(D)\det(\hat{A} - BD^{-1}C).$[15] $\qquad\square$

The following theorem [45] is the main result of this section.

Theorem 3.12.9 *Let n and k be positive integers with $1 \le k \le n-1$ and $(n,k) \ne (4,2)$. Then*

$$\widetilde{\det}^{\ne 0}(n,k) = k \gcd\{n,k\}.$$

Moreover, there exists a matrix in $\mathcal{A}'(n,k)$ whose determinant in absolute value equals $k \gcd\{n,k\}$.

Proof. Because of Lemma 3.12.5, it suffices to show the existence of a matrix A in $\mathcal{A}'(n,k)$ for which $|\det(A)| = k \gcd\{n,k\}$. If $k = 1$, we may take $A = I_n$. Now assume that $k \ge 2$ and $(n,k) \ne (4,2)$. We use Lemma 3.12.8 to construct the desired matrix A using the following algorithm.

(1) Begin with $r = 1$ and $(n_r, k_r) = (n, k)$.

(2) While $n_r \ne 2k_r$, do:

   (i) Set
   $$(n_{r+1}, k_{r+1}) = \begin{cases} (n_r - k_r, k_r) & \text{if } n_r > 2k_r, \\ (n_r, n_r - k_r) & \text{if } n_r < 2k_r. \end{cases}$$

   (ii) Replace r with $r + 1$.

It follows from the Euclidean algorithm that after a finite number of steps the algorithm above terminates with $(n_r, k_r) = (2d, d)$ where $d = \gcd\{n, k\}$. First assume that $d > 2$. By Lemma 3.12.7 there is a matrix A_r in $\mathcal{A}'(2d, d)$ with $|\det(A_r)| = d^2$. Working backwards from r to 1 using Lemma 3.12.8, we construct matrices $A_{r-1}, \ldots, A_2, A_1$ such that A_i is in $\mathcal{A}'(n_i, k_i)$ and $|\det(A_i)| = dk_i$ $(i = r, \ldots, 2, 1)$. Thus A_1 is in $\mathcal{A}'(n, k)$ and $|\det(A_1)| = dk$.

Now assume that $d = 2$. Then $(n_r, k_r) = (4, 2)$ and $(n_{r-1}, k_{r-1}) = (6, 2)$. Working backwards starting from the matrix A_{r-1} in Lemma 3.12.6 with $|\det(A_{r-1})| = 2^2$, we complete the proof as above. $\qquad\square$

In computing the minimal nonzero determinant in Theorem 3.12.9, the restriction on $(0,1)$-matrices can be relaxed.

---

[15]See the proof of Lemma 3.12.7 and the footnote concerning the Schur complement.

Corollary 3.12.10 *Let n and k be positive integers. Then*

$$\min\{\det(A) : A \in \mathcal{Z}^+(n,k) \text{ and } \det(A) \neq 0\}$$

equals $k\gcd\{n,k\}$.

Proof. By Lemma 3.12.5 and Theorem 3.12.9 the conclusion holds if $1 \leq k \leq n$ unless $(n,k) = (4,2)$ or $k = n$. If $(n,k) = (4,2)$ then $(J_3 - I_3) \oplus [2]$ is a matrix in $\mathcal{Z}^+(4,2)$ with determinant equal to 4. If $k = n$, then the matrix $(I_{n-1} + J_{n-1}) \oplus [n]$ is a matrix in $\mathcal{Z}^+(n,n)$ with determinant equal to n^2.

Now assume that $k > n$ and let $k = qn + k'$ for some integers q and k' with $q \geq 1$ and $1 \leq k' < n$. Then $\gcd\{n,k\} = \gcd\{n,k'\}$ and as above there exists a matrix A in $\mathcal{Z}^+(n,k')$ with $\det(A) = k\gcd\{n,k\}$. Using Lemma 3.10.6, we see that the matrix $qJ_n + A$ is in $\mathcal{Z}^+(n,k)$ and has determinant equal to $k\gcd\{n,k\}$. $\qquad\square$

An alternative proof of Theorem 3.12.9 is given by Grady and Newman [28]. In particular, they show that the circulant $E_{n,k} = I_n + C_n + C_n^2 + \cdots + C_n^{k-1}$ in $\mathcal{A}(n,k)$ has rank equal to $n - \gcd\{n,k\} + 1$ and that a matrix in $\mathcal{A}(n,k)$ with determinant equal to $\pm k\gcd\{n,k\}$ can be found whose interchange-distance to $E_{n,k}$ equals $\gcd\{n,k\} - 1$. Since an interchange can increase the rank of a matrix by at most 1, $E_{n,k}$ can be no closer to a matrix with such a determinant. Some results and conjectures concerning the possible values of the determinant of matrices in $\mathcal{A}(n,k)$ are presented in [44] and also in [28].

References

[1] R.K. Ahujo, T.L. Magnanti, and J.B. Orlin, *Network Flows: Theory, Algorithms, and Applications*, Prentice-Hall, Englewood Cliffs, 1993.

[2] R.P. Anstee, Invariant sets of arcs in network flow problems, *Discrete Appl. Math.*, **13** (1986), 1–7.

[3] V.I. Bol′shakov, Upper values of a permanent in Γ_n^k (in Russian), *Combinatorial analysis, No. 7*, Moscow, 92–118 and 164–165 (1986).

[4] V.I. Bol′shakov, The spectrum of the permanent on Γ_N^k (in Russian), *Proceedings of the All-Union Seminar on Discrete Mathematics and Its Applications, Moscow (1984)*, Moskov. Gos. Univ. Mekh.-Mat. Fak., Moscow, 65–73, 1986.

[5] L.M. Brégman, Certain properties of nonnegative matrices and their permanents, *Dokl. Akad. Nauk SSSR*, **211** (1973), 27–30 (*Soviet Math. Dokl.*, **14** [1973], 194–230.

[6] R.A. Brualdi, Matrices of zeros and ones with fixed row and column sum vectors, *Linear Algebra Appl.*, **33** (1980), 159–231.

[7] R.A. Brualdi, On Haber's minimum term rank formula, *Europ. J. Combin.*, **2** (1981), 17–20.

[8] R.A. Brualdi, J.L. Goldwasser, and T.S. Michael, Maximum permanents of matrices of zeros and ones, *J. Combin. Theory, Ser. A*, **47** (1988), 207–245.

[9] R.A. Brualdi and R. Manber, Chromatic number of classes of matrices of 0's and 1's, *Discrete Math.*, **50** (1984), 143–152.

[10] R.A. Brualdi, R. Manber, and J.A. Ross, On the minimum rank of regular classes of matrices of zeros and ones, *J. Combin. Theory, Ser. A*, **41** (1986), 32–49.

[11] R.A. Brualdi and J.A. Ross, Invariants sets for classes of matrices of zeros and ones, *Proc. Amer. Math. Soc.*, **80** (1980), 706–710.

[12] R.A. Brualdi and J.A. Ross, On Ryser's maximum term rank formula, *Linear Algebra Appl.*, **29** (1980), 33–38.

[13] R.A. Brualdi and H.J. Ryser, *Combinatorial Matrix Theory*, Cambridge U. Press, Cambridge, 1991.

[14] R.A. Brualdi and J.G. Sanderson, Nested species subsets, gaps, and discrepancy, *Oecologia*, **119** (1999), 256–264.

[15] R.A. Brualdi and J. Shen, Dicrepancy of matrices of zeros and ones, *Electron. J. Combin.*, **6** (1999), #R15.

[16] D. de Caen, D.A. Gregory, and N.J. Pullman, The Boolean rank of zero–one matrices, *Proceedings of the Third Caribbean Conference on Combinatorics and Computing*, Barbados (1981), 169–173.

[17] W.Y.C. Chen, Integral matrices with given row and column sums, *J. Combin. Theory, Ser. A*, **2** (1992), 153–172.

[18] G.S. Cheon and I.M. Wanless, An update on Minc's survey of open problems involving permanents, *Linear Algebra Appl.*, **403** (2005), 314–342.

[19] R. Ermers and B. Polman, On the eigenvalues of the structure matrix of matrices of zeros and ones, *Linear Algebra Appl.*, **95** (1987), 17–41.

[20] S. Fallat and P. van den Driessche, Maximum determinant of $(0, 1)$-matrices with certain constant row and column sums, *Linear Multilin. Alg.*, **42** (1997), 303–318.

[21] L. R. Ford, Jr. and D. R. Fulkerson, *Flows in Networks*, Princeton U. Press, Princeton, 1962.

[22] D.R. Fulkerson and H.J. Ryser, Widths and heights of (0,1)-matrices, *Canad. J. Math.*, **13** (1961), 239–255.

[23] D.R. Fulkerson and H.J. Ryser, Multiplicities and minimal widths for (0,1)-matrices, *Canad. J. Math.*, **14** (1962), 498–508.

[24] D.R. Fulkerson and H.J. Ryser, Width sequences for special classes of (0,1)-matrices, *Canad. J. Math.*, **15** (1963), 371–396.

[25] D. Gale, A theorem on flows in networks, *Pacific. J. Math.*, **7** (1957), 1073–1082.

[26] A.O. Gel'fond, Some combinatorial properties of (0,1)-matrices, *Math. USSR – Sbornik*, **4** (1968), 1.

[27] M. Grady, Research problems: combinatorial matrices having minimal non-zero determinant, *Linear Multilin. Alg.*, **35** (1993), 179–183.

[28] M. Grady and M. Newman, The geometry of an interchange: minimal matrices and circulants, *Linear Algebra Appl.*, **262** (1997), 11–25.

[29] D.A. Gregory, K.F. Jones, J.R. Lundgren, and N.J. Pullman, Biclique coverings of regular bigraphs and minimum semiring ranks of regular matrices, *J. Combin. Theory, Ser. B*, **51** (1991), 73–89.

[30] R.M. Haber, Term rank of (0, 1)-matrices, *Rend. Sem. Mat. Padova*, **30** (1960), 24–51.

[31] R.M. Haber, Minimal term rank of a class of (0,1)-matrices, *Canad. J. Math.*, **15** (1963), 188–192.

[32] K.A.S. Hefner and J.R. Lundgren, Minimum matrix rank of k-regular (0, 1)-matrices, *Linear Algebra Appl.*, **133** (1990), 43–52.

[33] J.R. Henderson, Permanents of (0, 1)-matrices having at most two 0's per line, *Canad. Math. Bull.*, **18** (1975), 353–358.

[34] J.R. Henderson and R.A. Dean, the 1-width of (0,1)-matrices having constant row sum 3, *J. Combin. Theory, Ser. A*, **16** (1974), 355–370.

[35] J.R. Henderson and R.A. Dean, A general upper bound for 1-widths, *J. Combin. Theory, Ser. A*, **18** (1975), 236–238.

[36] G.T. Herman and A. Kuba, Discrete tomography: a historical overview, *Discrete Tomography: Foundations, Algorithms, and Applications*, ed. G.T. Herman and A. Kuba, Birkhäuser, Boston, 1999, 3–34.

[37] D.J. Houck and M.E. Paul, Nonsingular 0–1 matrices with constant row and column sums, *Linear Algebra Appl.*, **50** (1978), 143–152.

[38] L.K. Jorgensen, Rank of adjacency matrices of directed (strongly) regular graphs, *Linear Algebra Appl.*, **407** (2005), 233–241.

[39] W.B. Jurkat and H.J. Ryser, Term ranks and permanents of nonnegative matrices, *J. Algebra*, **5** (1967), 342–357.

[40] D.J. Kleitman and D.L. Wang, Algorithms for constructing graphs and digraphs with given valencies and factors, *Discrete Math.*, **6** (1973), 79–88.

[41] S. Kundu, The k-factor conjecture is true, *Discrete Math.*, **6** (1973), 367–376.

[42] C.W.H. Lam, The distribution of 1-widths of (0, 1)-matrices, *Discrete Math.*, **20** (1970), 109–122.

[43] V.K. Leont'ev, An upper bound for the α-height of (0, 1)-matrices, *Mat. Zametki*, **15** (1974), 421–429.

[44] C.K. Li, J.S.-J. Lin and L. Rodman, Determinants of certain classes of zero–one matrices with equal line sums, *Rocky Mountain J. Math.*, **29** (1999), 1363–1385.

[45] C.K. Li, D.D. Olesky, D.P. Stanford, and P. Van den Driessche, Minimum positive determinant of integer matrices with constant row and column sums, *Linear Multilin. Alg.*, **40** (1995), 163–170.

[46] W.E. Longstaff, Combinatorial solution of certain systems of linear equations involving $(0, 1)$-matrices, *J. Austral. Math. Soc. Ser. A*, **23** (1977), 266–274.

[47] L. Lovász, Valencies of graphs with 1-factors, *Period. Math. Hung.*, **5** (1974), 149–151.

[48] K. McDougal, Locally invariant positions of $(0, 1)$ matrices, *Ars Combin.*, **44** (1996), 219–224.

[49] B.D. McKay and I.M. Wanless, Maximising the permanent of $(0, 1)$-matrices and the number of extensions of Latin rectangles, *Elecron. J. Combin.*, **5** (1998), #11.

[50] D. Merriell, The maximum permanent in Γ_n^k, *Linear Multilin. Alg.* **9** (1980), 81–91.

[51] T.S. Michael, The structure matrix and a generalization of Ryser's maximum term rank formula, *Linear Algebra Appl.*, **145** (1991), 21–31.

[52] L. Mirsky, *Transversal Theory*, Academic Press, New York, 1971.

[53] S.D. Monson, N.J. Pullman, and R. Rees, A survey of clique and biclique coverings and factorizations of $(0, 1)$-matrices, *Bull. Inst. Combin. Appl.*, **14** (1995), 17–86.

[54] M. Newman, Combinatorial matrices with small determinants, *Canad. J. Math.*, **30** (1978), 756–762.

[55] A.M. Odlyzko, On the ranks of some $(0,1)$-matrices with constant row sums, *J. Austral. Math. Soc., Ser. A*, **31** (1981), 193–201.

[56] N.J. Pullman and M. Stanford, Singular $(0, 1)$-matrices with constant row and column sums, *Linear Algebra Appl.*, **106** (1988), 195–208.

[57] M. Raghavachari, Some properties of the widths of the incidence matrices of balanced incomplete block designs through partially balanced ternary designs, *Saṅkhyā, Ser. B*, **37** (1975), 211–219.

[58] H.J. Ryser, Maximal determinants in combinatorial investigations, *Canad. J. Math.*, **8** (1956), 245–249.

[59] H.J. Ryser, Combinatorial properties of matrices of zeros and ones, *Canad. J. Math.*, **9** (1957), 371–377.

[60] H.J. Ryser, The term rank of a matrix, *Canad. J. Math.*, **10** (1957), 57–65.

[61] H.J. Ryser, Traces of matrices of zeros and ones, *Canad. J. Math.*, **12** (1960), 463–476.

[62] H. J. Ryser, *Combinatorial Mathematics*, Carus Math. Monograph #14, Math. Assoc. of America, Washington, 1963.

[63] H.J. Ryser, Matrices of zeros and ones in combinatorial mathematics, *Recent Advances in Matrix Theory*, ed. by H. Schneider, U. Wisconsin Press, Madison, 103–124, 1964.

[64] A. Schrijver, A short proof of Minc's conjecture, *J. Combin. Theory, Ser. A*, **25** (1978), 80–83.

[65] A. Schrijver, Bounds on permanent and the number of 1-factors and 1-factorizations of bipartite graphs, *Surveys in Combinatorics* (Southampton 1983), London Math Soc. Lecture Notes 82, Cambridge U. Press, Cambridge, 1983, 107–134.

[66] A. Schrijver, Counting 1-factors in regular bipartite graphs, *J. Combin. Theory, Ser. B*, **72** (1998), 122–135.

[67] A. Schrijver and W.G. Valiant, On lower bounds for permanents, *Indag. Math.*, **42** (1980), 425–427.

[68] S.M. Shah and C.C. Gujarathi, The α-width of the incidence matrices of incomplete block designs, *Saṅkhyā, Ser. B*, **46** (1984), 118–121.

[69] G. Sierksma and E. Sterken, The structure matrix of $(0, 1)$-matrices: its rank, trace and eigenvalues. An application to econometric models, *Linear Algebra Appl.*, **83** (1986), 151–166.

[70] S.K. Stein, Two combinatorial covering problems, *J. Combin. Theory, Ser. A*, **16** (1974), 391–397.

[71] V.E. Tarakanov, Correlations of maximal depths of classes of square $(0, 1)$-matrices for different parameters, *Mat. Sb. (N.S.)*, **77** (119) (1968), 59–70.

[72] V.E. Tarakanov, The maximal depth of a class of $(0,1)$-matrices, *Math. Sb. (N.S.)*, **75** (117) (1968), 4–14.

[73] V.E. Tarakanov, The maximum depth of arbitary classes of $(0,1)$-matrices and some of its applications (in Russian), *Math. Sb. (N.S.)*, **92**(134) (1973), 472–490. Translated in *Math. USSR – Sbornik*, **21** (1973), 467–484.

[74] V.E. Tarakanov, The depth of (0,1)-matrices with identical row and identical column sums (in Russian), *Mat. Zametki*, **34** (1983), 463–476. Translated in *Math Notes.*, **34** (1983), 718–725.

[75] M. Voorhoeve, On lower bounds for permanents of certain $(0, 1)$-matrices, *Indag. Math.*, **41** (1979), 83–86.

[76] D.W. Walkup, Minimal interchanges of (0,1)-matrices and disjoint circuits in a graph, *Canad. J. Math.*, **17** (1965), 831–838.

[77] I.M. Wanless, Maximising the permanent and complementary permanent of $(0, 1)$-matrices with constant line sums, *Discrete Math.*, **205** (1999), 191–205.

[78] I.M. Wanless, Extremal properties for binary matrices, preprint.

4

More on the Class $\mathcal{A}(R, S)$ of (0,1)-Matrices

In this chapter we continue our study of the class $\mathcal{A}(R, S)$ of $(0, 1)$-matrices. Our primary focus is on the evaluation of the number of matrices in $\mathcal{A}(R, S)$ and the existence of matrices in $\mathcal{A}(R, S)$ with special combinatorial structures. In Chapter 6 we study properties of the interchange graph $G(R, S)$.

4.1 Cardinality of $\mathcal{A}(R, S)$ and the RSK Correspondence

Let m and n be fixed positive integers, and let $R = (r_1, r_2, \ldots, r_m)$ and $S = (s_1, s_2, \ldots, s_n)$ be nonnegative integral vectors satisfying

$$r_i \leq n \quad (i = 1, 2, \ldots, m), \quad s_j \leq m \ (j = 1, 2, \ldots, n), \text{ and}$$

$$r_1 + r_2 + \cdots + r_m = s_1 + s_2 + \cdots + s_n.$$

We let

$$\kappa(R, S) = |\mathcal{A}(R, S)|$$

denote the *number of matrices in the class* $\mathcal{A}(R, S)$. Let $x_1, x_2, \ldots, x_m$ and $y_1, y_2, \ldots, y_n$ be $m + n$ variables. Let $\mathbf{x} = (x_1, x_2, \ldots, x_m)$ and $\mathbf{y} = (y_1, y_2, \ldots, y_n)$. By the Gale–Ryser theorem of Chapter 2, $\kappa(R, S) \neq 0$ if and only if the nonincreasing rearrangement of S is majorized by the conjugate R^* of R. The *generating function for the numbers* $\kappa(R, S)$ is $g_{m,n}(\mathbf{x}; \mathbf{y}) = g_{m,n}(x_1, x_2, \ldots, x_m; y_1, y_2, \ldots, y_n)$ where

$$g_{m,n}(\mathbf{x}; \mathbf{y}) = \sum_{R,S} \kappa(R, S) x_1^{r_1} x_2^{r_2} \ldots x_m^{r_m} y_1^{s_1} y_2^{s_2} \ldots y_n^{s_n}.$$

Since $\kappa(R, S) = \kappa(RP, SQ)$ for all permutation matrices P of order m and Q of order n, there is a redundancy built into this generating function.

The following theorem has been proved in [25] and in [46], [34]. It is also implicit in Knuth's derivation of identities of Littlewood [32] and is used in [28], [44].

Theorem 4.1.1 *Let m and n be positive integers. Then*

$$g_{m,n}(\mathbf{x}; \mathbf{y}) = \prod_{i=1}^{m} \prod_{j=1}^{n} (1 + x_i y_j).$$

Proof. We have

$$\prod_{i=1}^{m} \prod_{j=1}^{n} (1 + x_i y_j) = \sum_{\{a_{ij}=0 \text{ or } 1: 1 \le i \le m, 1 \le j \le n\}} \prod_{i=1}^{m} \prod_{j=1}^{n} (x_i y_j)^{a_{ij}} \qquad (4.1)$$

where the summation extends over all 2^{mn} $(0,1)$-matrices $A = [a_{ij}]$ of size m by n. Collecting terms, we see that the coefficient of

$$x_1^{r_1} x_2^{r_2} \dots x_m^{r_m} y_1^{s_1} y_2^{s_2} \dots y_n^{s_n}$$

on the right side of (4.1) equals $|\mathcal{A}(R, S)|$.   □

By replacing some of the terms $(1 + x_i y_j)$ in (4.1) with a 1 or with $x_i y_j$'s we get the generating function for the number of matrices in $\mathcal{A}(R, S)$ having 0's or 1's in prescribed places.

Let $a_{n,k} = |\mathcal{A}(n, k)|$ equal the number of $(0, 1)$-matrices with exactly k 1's in each row and column ($1 \le k \le n$). We have $a_{n,1} = a_{n,n-1} = n!$ and $a_{n,n} = 1$. In the next corollary we evaluate $a_{n,2}$, [1] and [25, pp. 235–236].

Corollary 4.1.2 *For $n \ge 2$ we have*

$$a_{n,2} = \frac{1}{4^n} \sum_{j=0}^{n} (-1)^j (2n - 2j)! \, j! \binom{n}{j}^2 2^j.$$

Proof. According to Theorem 4.1.1, $a_{n,2}$ is the coefficient of the term $(x_1 y_1)^2 (x_2 y_2)^2 \dots (x_n y_n)^2$ in

$$\prod_{i=1}^{n} \prod_{j=1}^{n} (1 + x_i y_j).$$

Thus $a_{n,2}$ equals the coefficient of $x_1^2 x_2^2 \dots x_n^2$ in each of the expressions

$$\left(\sum_{1 \le i < j \le n} x_i x_j \right)^n,$$

$$\left(\frac{1}{2} \right)^n ((x_1 + x_2 + \dots + x_n)^2 - (x_1^2 + x_2^2 + \dots + x_n^2))^n,$$

$$\left(\frac{1}{2} \right)^n \sum_{j=0}^{n} (-1)^j \binom{n}{j} (x_1^2 + x_2^2 + \dots + x_n^2)^j (x_1 + x_2 + \dots + x_n)^{2(n-j)},$$

and thus equals

$$\left(\frac{1}{2}\right)^n \sum_{j=0}^{n} (-1)^j \binom{n}{j} \frac{n!}{(n-j)!} \frac{(2n-2j)!}{2^{n-j}}.$$

The corollary now follows.   □

Corollary 4.1.2 implies that the exponential-like generating function

$$g(t) = \sum_{n \geq 0} a_{n,2} \frac{t^n}{n!^2}$$

satisfies

$$g(t) = \frac{e^{-t/2}}{\sqrt{1-t}}.$$

From Theorem 4.1.1 one can also deduce a formula for $a_{n,3}$ (see [25, p. 236]).

An argument similar to that used in the proof of Theorem 4.1.1 determines the generating function $g'_{m,n}(\mathbf{x}; \mathbf{y})$ for $\kappa'(R, S) = |\mathcal{Z}^+(R, S)|$, the number of nonnegative integral matrices with row sum vector R and column sum vector S.

Theorem 4.1.3 *Let m and n be positive integers. Then*

$$g'_{m,n}(\mathbf{x}; \mathbf{y}) = \prod_{i=1}^{m} \prod_{j=1}^{n} \frac{1}{1 - x_i y_j} = \prod_{i=1}^{m} \prod_{j=1}^{n} \sum_{a_{ij}=0}^{\infty} (x_i y_j)^{a_{ij}}.$$

□

There is a remarkable correspondence between matrices and pairs of combinatorial constructs known as Young tableaux that enables one to evaluate the number of matrices in $\mathcal{A}(R, S)$ in terms of the so-called Kostka numbers, to which we now turn our attention. We rely heavily on the excellent treatments by Fulton [28] and Stanley [44].

A Young diagram is a geometric realization of a partition of a nonnegative integer.[1] More precisely, let $\lambda = (\lambda_1, \lambda_2, \ldots, \lambda_m)$ be a partition of the nonnegative integer τ. Recall that this means that $\lambda_1, \lambda_2, \ldots, \lambda_m$ are positive integers satisfying $\lambda_1 \geq \lambda_2 \geq \cdots \geq \lambda_m$ and $\tau = \lambda_1 + \lambda_2 + \cdots + \lambda_m$. The *Young diagram* corresponding to λ is the collection of τ boxes arranged with λ_i boxes in row i which have been left-justified $(i = 1, 2, \ldots, m)$.

Example. If $\lambda = (5, 4, 4, 2, 1)$ is a partition of 16, then the corresponding Young diagram is

---

[1] The integer 0 has exactly one partition, namely the empty partition with no parts.

As in this example, we usually draw an m by λ_1 array of boxes and, to indicate that they are not present, put the symbol $\times$ in those boxes that are not part of the Young diagram. $\square$

As is customary, we identify a partition λ and its Young diagram. A *Young tableau of shape* λ is obtained from a Young diagram λ by putting a positive integer in each of its boxes, subject to the following conditions:

   (i) the sequence $a_1, a_2, a_3, \ldots$ of numbers in each row is nondecreasing,
       $a_1 \leq a_2 \leq a_3 \leq \cdots$;

   (ii) the sequence $b_1, b_2, b_3, \ldots$ of numbers in each column is strictly increasing, $b_1 < b_2 < b_3 < \cdots$.

A Young tableau T of shape λ where λ is a partition of the positive integer τ is called *standard* provided its elements are the numbers $1, 2, \ldots, \tau$ with each number occurring once. A Young tableau T whose elements are restricted to a set X of positive integers is a *Young tableau based on* X.

Example. A Young tableau of shape $\lambda = (5, 4, 4, 2, 1)$ is

1	2	2	3	5
2	3	4	6	$\times$
4	5	7	7	$\times$
6	6	$\times$	$\times$	$\times$
7	$\times$	$\times$	$\times$	$\times$

This tableau is based on $\{1, 2, \ldots, 7\}$ (or any set containing $\{1, 2, \ldots, 7\}$). A standard Young tableau of shape λ is

1	2	3	6	7
4	5	9	10	$\times$
8	12	14	15	$\times$
11	16	$\times$	$\times$	$\times$
13	$\times$	$\times$	$\times$	$\times$

$\square$

We now describe a basic algorithm, called *row-insertion* or *row-bumping*, due to Schensted [43] (see also [28], [44]) for inserting a positive integer k into a given Young tableau T. This algorithm works row by row starting from the top row and working down as needed, and produces a new Young tableau $T \leftarrow k$ with one more box than T and with the additional element k.

Row-insertion of k into T proceeds as follows.

(i) If k is at least as large as the last element in row 1, then a new box is added at the end of row 1 and k is put in this box.

(ii) Otherwise, determine the leftmost element k_1 in row 1 that is strictly greater than k, replace k_1 with k, and repeat with the second row and k replaced with the bumped element k_1. (Recursively, the first step results in a Young tableau T_1, and we do the row-insertion $T_1' \leftarrow k_1$ where T_1' is the tableau obtained by deleting row 1 of T_1.)

(iii) We continue as in steps (i) and (ii) until either we put an element at the end of a row (step (i) applies), or we arrive at the last row in which case the element bumped from the last row is put in a new last row as its only element.

(iv) $T \leftarrow k$ is the final tableau obtained by this process.

We observe that there is one filled box in $T \leftarrow k$ that is not a filled box in T, and that it is a *corner box*, that is, the rightmost box in some row and simultaneously the bottommost box in some column. We denote this corner box by $b_k(T)$ and the element inserted in $b_k(T)$ (the last bumped element) by $\iota_k(T)$; thus $\iota_k(T)$ is in box $b_k(T)$.

Example. Let

$$
T = \begin{array}{|c|c|c|c|c|}
\hline
1 & 1 & 1 & 2 & 3 \\
\hline
2 & 2 & 4 & 4 & \times \\
\hline
3 & 4 & 5 & \times & \times \\
\hline
4 & \times & \times & \times & \times \\
\hline
\end{array}.
$$

We insert 2 into T to get

$$
T \leftarrow 2 = \begin{array}{|c|c|c|c|c|}
\hline
1 & 1 & 1 & 2 & \mathbf{2} \\
\hline
2 & 2 & \mathbf{3} & 4 & \times \\
\hline
3 & 4 & \mathbf{4} & \times & \times \\
\hline
4 & \mathbf{5} & \times & \times & \times \\
\hline
\end{array}.
$$

The box $b_2(T)$ is the box in the fourth row and second column; the element $\iota_2(T)$ is 5. □

Define the *insertion-chain* of $T \leftarrow k$ to be the set $\mathrm{IC}(T \leftarrow k)$ consisting of those boxes where an element is bumped from a row and the new box $b_k(T)$ containing the last bumped element. In the above example, $\mathrm{IC}(T \leftarrow 2)$ consists of the boxes containing bold elements.

Lemma 4.1.4 *If T is a Young tableau and k is a positive integer, then the result $T \leftarrow k$ of row-inserting k in T is also a Young tableau. In addition, starting in the first row, the insertion-chain $\mathrm{IC}(T \leftarrow k)$ moves down and possibly to the left.*

Proof. By construction the rows of $T \leftarrow k$ are nondecreasing. Suppose that an element a bumps an element b in a row. Then, by definition of a Young tableau, the element c directly below b in the next row, if there is one, is strictly greater than b. Thus when b is inserted in this row, b goes into either the same column or one to its left. Moreover, the element immediately above b in its new position is strictly smaller than b.   □

Given the Young tableau $T \leftarrow k$ and the location of the box $b_k(T)$ filled in $T \leftarrow k$ but not in T, the Young tableau T can be recovered by reversing the steps in row-insertion. In fact, we now describe an algorithm, called *row-extraction*, that takes a Young tableau T' and an element k' occupying a corner box of the diagram, and produces a Young tableau T'', denoted $k' \leftarrow T'$, with one filled box less than T'. Moreover, if T' equals $T \leftarrow k$, then $\iota_k(T) \leftarrow (T \leftarrow k)$ equals T.[2]

Row-extraction for a Young tableau T' beginning with an element k' in a box of T' proceeds as follows.

  (i) If k' is in the rightmost box of row 1, then T'' is obtained from T' by emptying the box containing k'.

 (ii) Otherwise, there is a box immediately above k' in the next row and it contains an integer strictly less than k'. We determine the element k'_1 in this row which is strictly less than k' and is farthest to the right. We empty the box containing k'_1, and replace k'_1 in this box with k'. We repeat with k' replaced with the bumped element k'_1.

(iii) We continue as in steps (i) and (ii) until we reach the top row and bump an element from it.

(iv) $k' \leftarrow T'$ is the final tableau T'' obtained by this process.

The algorithm for row-extraction is that for row-insertion done in reverse, and hence we get the following.

Lemma 4.1.5 *Let T' be a Young tableau, let k' be the element in a specified corner box of T', and let k be the element bumped from the first row in the row-extraction $k' \leftarrow T'$. Then $k' \leftarrow T'$ is a Young tableau that satisfies*

$$(k' \leftarrow T') \leftarrow k \ = \ T'.$$

In addition,

$$\iota_k(T) \leftarrow (T \leftarrow k) \ = \ T$$

for every Young tableau T and positive integer k.   □

---

[2] The notation $k' \leftarrow T'$ is, of course, ambiguous since there may be more than one box containing the element k'. We always have in mind a particular box containing k' when we use it.

Example. Consider the Young tableau

$$
T' = \begin{array}{|c|c|c|c|}
\hline
1 & 1 & 2 & 3 \\
\hline
1 & 3 & 3 & \times \\
\hline
2 & 4 & 4 & \times \\
\hline
3 & 5 & \times & \times \\
\hline
\end{array}
$$

and the element 5 in the lower right corner box. Then

$$
5 \leftarrow T' = \begin{array}{|c|c|c|c|}
\hline
1 & 1 & \mathbf{3} & \mathbf{3} \\
\hline
1 & 3 & \mathbf{4} & \times \\
\hline
2 & 4 & \mathbf{5} & \times \\
\hline
3 & \times & \times & \times \\
\hline
\end{array}
$$

with the element 2 bumped from the first row. Row-inserting the element 2 we recover T': $(5 \leftarrow T') \leftarrow 2) = T'$. □

We now compare the insertion-chains for two successive row-insertions. Let b and c be two boxes in a tableau (or diagram) where b occupies row i and column p, and c occupies row j and column q. Then b is *strictly-left* of c provided $i = j$ and $p < q$; b is *weakly-left* of b' provided $i = j$ and $p \le q$. Similar terminology applies when replacing "left" with "right," "above," or "below." Moreover, some combinations are possible. For instance, b is strictly-left of and weakly-below c means that $p < q$ and $i \ge j$. Now let I and I' be two sets of boxes in a tableau (or diagram), each containing at most one box in a row. Then I is *strictly-left* (respectively, *weakly-left*) of I' provided in each row that contains a box of I', I also has a box that is strictly-left (respectively, weakly-left) of the box of I'.

Lemma 4.1.6 *Let T be a Young tableau, and let k and l be positive integers. Let $b = b_k(T)$ be the new box resulting from the row insertion $T \leftarrow k$, and let $c = b_l(T \leftarrow k)$ be the new box resulting from the row insertion $(T \leftarrow k) \leftarrow l$.*
If $k \le l$, then

(i) $IC(T \leftarrow k)$ *is strictly-left of* $IC((T \leftarrow k) \leftarrow l)$,

(ii) b *is strictly-left of and weakly-below c, and*

(iii) $|IC((T \leftarrow k) \leftarrow l)| \le |IC(T \leftarrow k)|$.

If $k > l$, then

(iv) $IC((T \leftarrow k) \leftarrow l)$ *is weakly-left of* $IC(T \leftarrow k)$,

(v) c *is weakly-left of and strictly-below b, and*

(vi) $|IC((T \leftarrow k)| < |IC(T \leftarrow k) \leftarrow l)|$.

Proof. First assume that $k \leq l$. Consider first the case in which k is placed in a new box at the end of row 1. Then k is at least as large as each element in row 1 of T. Since $k \leq l$, l is at least as large as each element in row 1 of $T \leftarrow k$, and is inserted in a new box at the end of row 1 of $T \leftarrow k$. Hence all the conclusions hold in this case. Now consider the case that k bumps k_1 from row 1 in the row-insertion $T \leftarrow k$. If l is inserted in a new box at the end of row 1 of $T \leftarrow k$ then, since an insertion-chain never moves right, all conclusions hold in this case. Now suppose that l bumps l_1 in the row-insertion $(T \leftarrow k) \leftarrow l$. Since $k \leq l$, the box containing the element l_1 in row 1 of $T \leftarrow k$ that is bumped by l must lie strictly-right of the box in which k has been inserted in $T \leftarrow k$. In particular, $k_1 \leq l_1$. We now continue the argument with k_1 and l_1 in place of k and l. The only fact to check is that $\mathrm{IC}(T \leftarrow k)$ does not end before $\mathrm{IC}((T \leftarrow k) \leftarrow l)$. But this follows as in the first case, for if $I(T \leftarrow k)$ ends by placing an element at the end of a (possibly new) row, then $\mathrm{IC}((T \leftarrow k) \leftarrow l)$ ends at that row also, if it had not already ended earlier. It follows that b is strictly-left of c. Also b is in the same row as c if the two row-insertions end in the same row and, since the insertion-chains never move right, b is strictly-left of and strictly-below c, otherwise.

Now assume that $k > l$. The argument here is very similar. Consider first the case that k is placed in a new box at the end of row 1. Then in $(T \leftarrow k) \leftarrow l$, l bumps l_1 where either $l_1 = k$ or l_1 is an element in a box strictly to the left of k. Since l is not placed at the end of row 1 in $T \leftarrow k$, the insertion-chain continues to the next row. Hence c is weakly-left of and strictly-below b, and all conclusions hold in this case. Now consider the case that k bumps k_1 from row 1. Since $k_1 > k > l$, l bumps either the element k that has been inserted in row 1 of $T \leftarrow k$, or one strictly-left of it. In either case, $k_1 > l_1$ and we proceed with k_1 and l_1 in place of k and l. As above, the insertion-chain for $(T \leftarrow k) \leftarrow l$ must continue at least one row beyond the insertion-chain for $T \leftarrow k$. $\square$

The row-insertion algorithm is the basis for a remarkable bijective correspondence between nonnegative integral matrices and pairs of Young tableaux of the same shape. This correspondence is called the *Robinson–Schensted–Knuth correspondence*, or *RSK correspondence* [42, 43, 10]. We refer to Fulton [28] and Stanley [44] for some of the historical background for this correspondence, and its important connections with symmetric functions and representation theory.

A *generalized permutation array* is a two-rowed array

$$\Upsilon = \begin{pmatrix} i_1 & i_2 & \cdots & i_p \\ j_1 & j_2 & \cdots & j_p \end{pmatrix} \tag{4.2}$$

where the pairs $\begin{pmatrix} i_k \\ j_k \end{pmatrix}$ are in lexicographic order. Thus $i_1 \leq i_2 \leq \cdots \leq i_p$ and, in addition, $j_k \leq j_l$ whenever $i_k = i_l$ and $k \leq l$. The array Υ is a *permutation array* provided $\{i_1, i_2, \ldots, i_p\} = \{j_1, j_2, \ldots, j_p\} = \{1, 2, \ldots, p\}$.

Let $A = [a_{ij}]$ be a matrix of size m by n whose elements are nonnegative integers. Let the sum of the elements of A be $\tau = \sum_{i,j} a_{ij}$. With A we can associate a 2 by τ *generalized permutation array* $\Upsilon(A)$ in which the pair $\binom{i}{j}$ appears a_{ij} times $(i = 1, 2 \ldots, m; j = 1, 2, \ldots, n)$. Conversely, given a 2 by τ generalized permutation array Υ with elements from the first row taken from $\{1, 2, \ldots, m\}$ and elements from the second row taken from $\{1, 2, \ldots, n\}$, we can reverse this construction to obtain an m by n nonnegative integral matrix A such that $\Upsilon(A) = \Upsilon$. As a result, we may identify a nonnegative integral matrix with a generalized permutation array. A zero matrix corresponds to an empty 2 by 0 array.

Example. Let

$$A = \begin{bmatrix} 2 & 1 & 0 & 0 \\ 0 & 2 & 1 & 2 \\ 3 & 1 & 0 & 0 \end{bmatrix}. \tag{4.3}$$

Then

$$\Upsilon_A = \begin{pmatrix} 1 & 1 & 1 & 2 & 2 & 2 & 2 & 2 & 3 & 3 & 3 & 3 \\ 1 & 1 & 2 & 2 & 2 & 3 & 4 & 4 & 1 & 1 & 1 & 2 \end{pmatrix}. \tag{4.4}$$

□

Let m and n be positive integers, and let Υ_n^m denote the set of all generalized permutation arrays in which the elements in row 1 are taken from $\{1, 2, \ldots, m\}$ and the elements in row 2 are taken from $\{1, 2, \ldots, n\}$. Summarizing, we have the following.

Lemma 4.1.7 *The correspondence $A \to \Upsilon(A)$ is a one-to-one correspondence between nonnegative integral matrices of size m by n and generalized permutation arrays in Υ_n^m.* □

Let m and n be positive integers. The RSK correspondence gives a bijection between nonnegative integral matrices of size m by n (or generalized permutation arrays in Υ_n^m) and pairs $(\mathcal{P}, \mathcal{Q})$ of Young tableaux of the same shape such that $\mathcal{P}$ is based on $\{1, 2, \ldots, n\}$ and $\mathcal{Q}$ is based on $\{1, 2, \ldots, m\}$. Under this correspondence, the zero matrix (equivalently, the empty 2 by 0 generalized permutation array) corresponds to a pair of empty tableaux (with shape equal to that of the empty partition of 0).

Consider a generalized permutation array Υ of size 2 by p as given in (4.2). We recursively construct a pair $(\mathcal{P}, \mathcal{Q}) = (\mathcal{P}_\Upsilon, \mathcal{Q}_\Upsilon)$ of Young tableaux of the same shape as follows.

(i) Begin with a pair of empty tableaux $(\mathcal{P}_0, \mathcal{Q}_0)$.

(ii) Suppose that k is an integer with $0 \le k < p$ and that $(\mathcal{P}_k, \mathcal{Q}_k)$ has been defined. Then

   (a) $\mathcal{P}_{k+1} = \mathcal{P}_k \leftarrow j_{k+1}$, and

(b) $\mathcal{Q}_{k+1}$ is obtained from $\mathcal{Q}_k$ by inserting i_{k+1} in the box that is filled in $\mathcal{P}_{k+1}$ but not in $\mathcal{P}_k$.

(iii) Set $(\mathcal{P}_\Upsilon, \mathcal{Q}_\Upsilon)$ equal to $(\mathcal{P}_p, \mathcal{Q}_p)$.

The correspondence

$$\Upsilon \xrightarrow{\text{RSK}} (\mathcal{P}_\Upsilon, \mathcal{Q}_\Upsilon)$$

is the *RSK correspondence* between generalized permutation arrays and pairs of Young tableaux of the same shape. The tableau $\mathcal{P}$ is called the *insertion tableau* of the RSK correspondence, and $\mathcal{Q}$ is called the *recording tableau*. The RSK correspondence

$$A \xrightarrow{\text{RSK}} (\mathcal{P}_A, \mathcal{Q}_A)$$

between nonnegative integral matrices and pairs of Young tableaux of the same shape is the composite correspondence

$$A \to \Upsilon_A \xrightarrow{\text{RSK}} (\mathcal{P}_{\Upsilon_A}, \mathcal{Q}_{\Upsilon_A}).$$

Let $R = (r_1, r_2, \ldots, r_m)$ and $S = (s_1, s_2, \ldots, s_n)$ be the row and column sum vectors of A. Then the integer i appears exactly r_i times in $\mathcal{Q}_A$ ($i = 1, 2, \ldots, m$), and the integer j appears exactly s_j times in $\mathcal{P}_A$ ($j = 1, 2, \ldots, n$).

Example. Let

$$A = \begin{bmatrix} 1 & 0 & 2 \\ 2 & 1 & 0 \\ 2 & 1 & 0 \end{bmatrix}$$

with corresponding generalized permutation array

$$\Upsilon_A = \begin{pmatrix} 1 & 1 & 1 & 2 & 2 & 2 & 3 & 3 & 3 \\ 1 & 3 & 3 & 1 & 1 & 2 & 1 & 1 & 2 \end{pmatrix}.$$

Applying the algorithm for the RSK correspondence, we get

$$A \xrightarrow{\text{RSK}} \left(\begin{array}{c} \begin{array}{|c|c|c|c|c|c|} \hline 1 & 1 & 1 & 1 & 1 & 2 \\ \hline 2 & 3 & \times & \times & \times & \times \\ \hline 3 & \times & \times & \times & \times & \times \\ \hline \end{array} \end{array}, \begin{array}{c} \begin{array}{|c|c|c|c|c|c|} \hline 1 & 1 & 1 & 2 & 3 & 3 \\ \hline 2 & 2 & \times & \times & \times & \times \\ \hline 3 & \times & \times & \times & \times & \times \\ \hline \end{array} \end{array} \right).$$

□

We now show that the RSK correspondence is a bijection.

Theorem 4.1.8 *Let m and n be positive integers. The RSK correspondence*

$$A \xrightarrow{\text{RSK}} (\mathcal{P}_A, \mathcal{Q}_A)$$

is a bijection between nonnegative integral matrices A of size m by n and ordered pairs $(\mathcal{P}, \mathcal{Q})$ of Young tableaux of the same shape with $\mathcal{P}$ based on

$\{1, 2, \ldots, n\}$ and $\mathcal{Q}$ based on $\{1, 2, \ldots, m\}$. *In this correspondence, the row sum vector of A is $R = (r_1, r_2, \ldots, r_m)$ where r_i is the number of times the integer i appears in $\mathcal{Q}_A$ ($i = 1, 2, \ldots, m$), and the column sum vector is $S = (s_1, s_2, \ldots, s_n)$ where s_j is the number of times the integer j appears in $\mathcal{P}_A$ ($j = 1, 2, \ldots, n$). The matrix A is a permutation matrix of order n if and only if both $\mathcal{P}_A$ and $\mathcal{Q}_A$ are standard tableaux based on $\{1, 2, \ldots, n\}$.*

Proof. Let A be a nonnegative integral matrix of size m by n. It follows from the construction for the RSK correspondence that $\mathcal{P}_A$ is a Young tableau based on $\{1, 2, \ldots, n\}$ and that the column sum vector of A is as specified in the theorem. It also follows from this construction that, provided $\mathcal{Q}_A$ is a Young tableaux, it has the same shape as $\mathcal{P}_A$ and is based on $\{1, 2, \ldots, m\}$, and the row sum vector of A is as specified in the theorem. The assertion about permutation matrices is also an immediate consequence of the other assertions in the theorem. Thus it remains to show that

(i) $\mathcal{Q}_A$ is a Young tableau,

(ii) the RSK correspondence as specified in the theorem is

    (a) injective, and

    (b) surjective.

(i) By definition of Υ_A the elements of $\mathcal{Q}_A$ are inserted in nondecreasing order. By definition of row-insertion as used in the construction of $\mathcal{P}_A$, the elements of $\mathcal{Q}_A$ are recursively inserted in a new box which is at the end of some row and at the bottom of some column. ¿From these two facts we conclude that the elements in each row and in each column of $\mathcal{Q}_A$ are nondecreasing. It remains to show that no two equal elements occur in the same column of $\mathcal{Q}_A$. But this is a consequence of property (i) of the insertion-chain given in Lemma 4.1.6 which implies that equal elements of $\mathcal{Q}_A$ are inserted strictly from left to right (and weakly from bottom to top).

(ii)(a) By Lemma 4.1.5 we can reverse the steps in the construction of $\mathcal{P}_A$ provided we know the order in which new boxes are filled. This order is determined by recursively choosing the rightmost largest element (necessarily in a corner box) of $\mathcal{Q}_A$.

(ii)(b) Let $(\mathcal{P}, \mathcal{Q})$ be a pair of Young tableaux of the same shape with $\mathcal{P}$ based on $\{1, 2, \ldots, n\}$ and $\mathcal{Q}$ based on $\{1, 2, \ldots, m\}$. Since the largest integer in a Young tableau occurs in boxes that occupy positions strictly from left to right and weakly from bottom to top, we can apply the procedure in (iia) to $(\mathcal{P}, \mathcal{Q})$. We need to show that if e occurs in boxes b_1 and b_2 with b_1 strictly-left of and weakly-below b_2, then the element e_1 bumped from the first row in applying row-extraction starting with e in b_2 is at least as large as the element e_2 bumped from row 1 in applying row-extraction starting with e in box b_1. Let b_1 and b_2 occupy positions (i_1, j_1) and (i_2, j_2), respectively, where $i_1 \geq i_2$ and $j_1 < j_2$. It follows as in the first part of the proof of Lemma 4.1.6 that the element bumped in row p by the row-extraction starting from e in b_1 lies strictly-left of the element bumped from row p by the row-extraction starting with e in b_2 ($p = i_1, i_1 - 1, \ldots, 1$). Hence $e_1 \leq e_2$. $\qquad\square$

The RSK algorithm has a remarkable symmetry property. Since we shall not make use of this property, we refer to Knuth [32], Fulton [28], or Stanley [44] for a proof.

Theorem 4.1.9 *Let A be an m by n nonnegative integral matrix. If $A \xrightarrow{\mathrm{RSK}}$ $(\mathcal{P}_A, \mathcal{Q}_A)$, then $A^{\mathrm{T}} \xrightarrow{\mathrm{RSK}} (\mathcal{Q}_A, \mathcal{P}_A)$. Equivalently, $\mathcal{P}_{A^{\mathrm{T}}} = \mathcal{Q}_A$ and $\mathcal{Q}_{A^{\mathrm{T}}} = \mathcal{P}_A$. The matrix A is symmetric if and only if $\mathcal{P}_A = \mathcal{Q}_A$.*

Before proceeding with our development, we mention one application of the RSK correspondence to Schur functions. Let λ be a partition of an integer into at most m parts. Let $x_1, x_2, \ldots, x_m$ be m commuting variables, and let $\mathbf{x} = (x_1, x_2, \ldots, x_m)$. Corresponding to each Young tableau T of shape λ based on $\{1, 2, \ldots, m\}$ there is a monomial

$$\mathbf{x}^T = \prod_{i=1}^{m} x_i^{a_i} \quad (\text{where } i \text{ occurs } a_i \text{ times in } T).$$

The *Schur polynomial* corresponding to λ is

$$s_\lambda(\mathbf{x}) = s_\lambda(x_1, x_2, \ldots, x_m) = \sum_T \mathbf{x}^T$$

where the summation extends over all Young tableaux of shape λ based on $\{1, 2, \ldots, m\}$. A basic fact is that Schur polynomials are symmetric in the variables $x_1, x_2, \ldots, x_m$ and they form an additive basis for the ring of symmetric polynomials.

Let $y_1, y_2, \ldots, y_n$ be another set of n commuting variables, and let $\mathbf{y} = (y_1, y_2, \ldots, y_n)$. By Theorem 4.1.3 the generating function $g'_{m,n}(\mathbf{x}; \mathbf{y})$ for the number $\kappa'(R, S)$ of nonnegative integral matrices with row sum vector R and column sum vector S satisfies

$$g'_{m,n}(\mathbf{x}; \mathbf{y}) = \prod_{i=1}^{m} \prod_{j=1}^{n} \frac{1}{1 - x_i y_j} = \prod_{i=1}^{m} \prod_{j=1}^{n} \sum_{a_{ij}=0}^{\infty} (x_i y_j)^{a_{ij}}; \qquad (4.5)$$

the coefficient of $x^{r_1} x^{r_2} \ldots x^{r_m} y^{s_1} y^{s_2} \ldots y^{s_n}$ in $g'_{m,n}(\mathbf{x}; \mathbf{y})$ is $\kappa'(R, S)$. On the other hand, the coefficient of $x^{r_1} x^{r_2} \ldots x^{r_m} y^{s_1} y^{s_2} \ldots y^{s_n}$ in the summation

$$\sum_\lambda s_\lambda(\mathbf{x}) s_\lambda(\mathbf{y}),$$

where the summation extends over all partitions λ, equals the number of pairs $(\mathcal{P}, \mathcal{Q})$ of Young tableaux of the same shape such that i occurs r_i times in $\mathcal{P}$ ($i = 1, 2, \ldots, m$), and j occurs s_j times in $\mathcal{Q}$ ($j = 1, 2, \ldots, n$). Applying Theorem 4.1.8 we obtain a proof of the following theorem.

Theorem 4.1.10 *The generating function $g'_{m,n}(\mathbf{x};\mathbf{y})$ for the number of nonnegative integral matrices with row sum vector R and column sum vector S satisfies*

$$g'_{m,n}(\mathbf{x};\mathbf{y}) = \sum_\lambda s_\lambda(\mathbf{x})s_\lambda(\mathbf{y}). \qquad (4.6)$$

$\square$

The identity (4.6), which, because of (4.5), asserts that

$$\prod_{i=1}^{m}\prod_{j=1}^{n}\frac{1}{1-x_iy_j} = \sum_\lambda s_\lambda(\mathbf{x})s_\lambda(\mathbf{y}),$$

is known as the *Cauchy identity*.

Let λ be a partition of a nonnegative integer. A Young tableau T of shape λ based on $\{1,2,\ldots,m\}$ is of *type* $\mu = (\mu_1,\mu_2,\ldots,\mu_m)$ provided i occurs exactly μ_i times in T ($i = 1,2,\ldots,m$). When μ is a partition, that is, $\mu_1 \geq \mu_2 \geq \cdots \geq \mu_m$, the number of Young tableaux of shape λ and of type μ is the *Kostka number* $K_{\lambda,\mu}$. The following property of Kostka numbers follows readily.

Corollary 4.1.11 *If λ and μ are partitions of an integer τ, then $K_{\lambda,\mu} \neq 0$ if and only if $\mu \preceq \lambda$.* $\square$

As a corollary to Theorem 4.1.8 or Theorem 4.1.10 we obtain an evaluation of the number $\kappa'(R,S)$ of nonnegative integral matrices with row sum vector R and column sum vector S in terms of the Kostka numbers.

Corollary 4.1.12 *Let m, n, and τ be positive integers, and let nonincreasing, nonnegative integral vectors $R = (r_1, r_2, \ldots, r_m)$ and $S = (s_1, s_2, \ldots, s_n)$ be given with*

$$\tau = r_1 + r_2 + \cdots + r_m = s_1 + s_2 + \cdots + s_n.$$

Then the number of nonnegative integral matrices with row sum vector R and column sum vector S is

$$\kappa'(R,S) = \sum_\lambda K_{\lambda,R}K_{\lambda,S}$$

where the summation extends over all partitions of τ. $\square$

Row-insertion is an algorithm for inserting an integer into a Young tableau. It proceeds row by row in such a way as to maintain the nondecreasing property of the elements in a row (bumping the first, from right to left, possible element in a row) and the strictly increasing property of the elements in a column. It is natural to consider column-insertion whereby the first, from the bottom up, possible element in a column is bumped to the next column. Since the columns are strictly increasing from top to bottom, this means bumping the topmost element, which is at least as large as the element to be inserted. If one does this recursively as for row-insertion arriving at a $\mathcal{P}$ and a $\mathcal{Q}$ as in the RSK correspondence, then while $\mathcal{P}$ will be a Young tableau, $\mathcal{Q}$ need not be a Young tableau.

Example. Let

$$A = \begin{bmatrix} 1 & 1 \\ 1 & 0 \end{bmatrix},$$

where

$$\Upsilon_A = \begin{pmatrix} 1 & 1 & 2 \\ 1 & 2 & 1 \end{pmatrix}.$$

Then column bumping as described above leads to

$$\mathcal{P} = \begin{array}{|c|c|} \hline 1 & 1 \\ \hline 2 & \times \\ \hline \end{array} \quad \text{and} \quad \mathcal{Q} = \begin{array}{|c|c|} \hline 1 & 2 \\ \hline 1 & \times \\ \hline \end{array}.$$

While $\mathcal{P}$ is a Young tableau, $\mathcal{Q}$ is not. But notice that if $\mathcal{Q}$ is transposed, the result is a Young tableau.[3] This will always be the case if A is a $(0, 1)$-matrix.   □

We now confine our attention to $(0, 1)$-matrices and describe a bijective correspondence due to Knuth [32] between $(0, 1)$-matrices and certain pairs of Young tableaux.

Column-insertion $T \uparrow k$ of an integer k into a Young tableau T proceeds as follows.

(i)   If k is strictly larger than the last element in column 1, then a new box is added at the end of column 1 and k is put in this box.

(ii) Otherwise, determine the topmost element k_1 in column 1 that is at least as large as k, replace this element with k, and repeat with the second column and k replaced with the bumped element k_1. (Recursively, the first step results in a Young tableau T_1, and we do the column-insertion $T_1' \uparrow k_1$ where T_1' is the tableau obtained by deleting column 1 of T_1.)

(iii) We continue as in steps (i) and (ii) until either we put an element at the end of a column (step (i) applies), or we arrive at the last column in which case the element bumped from the last column is put in a new last column as its only element.

(iv) $T \uparrow k$ is the final tableau obtained by this process.

---

[3]There is an alternative. As shown in Fulton [28], column-insertion (starting with the rightmost element in the second row of a generalized permutation array) and a procedure called *sliding* lead to a pair of Young tableaux $(\mathcal{P}, \mathcal{Q})$ that is identical to that obtained by the RSK correspondence.

Corresponding to Lemma 4.1.6 we have the following lemma. We define the insertion-chain $\mathrm{IC}(T \uparrow k)$ in a way similar to the insertion-chain $\mathrm{IC}(T \leftarrow k)$.

Lemma 4.1.13 *Let T be a Young tableau, and let k and l be positive integers. Let $c = c_k(T)$ be the new box resulting from the column-insertion $T \uparrow k$, and let $d = c_l(T \uparrow k)$ be the new box resulting from the column insertion $(T \uparrow k) \uparrow l$.*
If $k < l$, then

(i) $\mathrm{IC}((T \uparrow k) \uparrow l)$ *is strictly-below* $\mathrm{IC}(T \uparrow k)$,

(ii) d *is strictly-below and weakly-left of c, and*

(iii) $|\mathrm{IC}(T \uparrow k) \uparrow l)| \leq |\mathrm{IC}(T \uparrow k)|$.
If $k \geq l$, then

(iv) $\mathrm{IC}((T \uparrow k) \uparrow l)$ *is weakly-above* $\mathrm{IC}(T \uparrow k)$,

(v) d *is weakly-above and strictly-right of c, and*

(vi) $|\mathrm{IC}(T \uparrow k)| < |\mathrm{IC}(T \uparrow k) \uparrow l)|$.

Proof. The proof is very similar to the proof of Lemma 4.1.6 and consequently is omitted. □

Generalized permutation arrays corresponding to $(0,1)$-matrices are characterized by the property that there is no repeated pair $\binom{i}{j}$. We thus consider generalized permutation arrays

$$\Upsilon = \begin{pmatrix} i_1 & i_2 & \cdots & i_p \\ j_1 & j_2 & \cdots & j_p \end{pmatrix} \tag{4.7}$$

with distinct pairs $\binom{i_k}{j_k}$ in lexicographic order. We recursively construct a pair $(\mathcal{S}, \mathcal{R}) = (\mathcal{S}_\Upsilon, \mathcal{R}_\Upsilon)$ of Young tableaux of *conjugate shapes*, that is, of shapes corresponding to conjugate partitions. Thus a box in position (p,q) is filled in $\mathcal{S}$ if and only if the box in the *conjugate position* (q,p) is filled in $\mathcal{R}$.

(i) Begin with a pair of empty tableaux $(\mathcal{S}_0, \mathcal{R}_0)$.

(ii) Suppose that k is an integer with $0 \leq k < p$ and that $(\mathcal{S}_k, \mathcal{R}_k)$ has been defined. Then

   (a) $\mathcal{S}_{k+1} = \mathcal{S}_k \uparrow j_{k+1}$, and

   (b) $\mathcal{R}_{k+1}$ is obtained from $\mathcal{R}_k$ by inserting i_{k+1} in the box that is conjugate to the box filled in $\mathcal{S}_{k+1}$ but not in $\mathcal{S}_k$.

(iii) Set $(\mathcal{S}_\Upsilon, \mathcal{R}_\Upsilon)$ equal to $(\mathcal{S}_p, \mathcal{R}_p)$.

The correspondences

$$\Upsilon \xrightarrow{\text{K}} (\mathcal{S}_\Upsilon, \mathcal{R}_\Upsilon)$$

and

$$A \xrightarrow{\text{K}} (\mathcal{S}_A, \mathcal{R}_A)$$

defined by

$$A \to \Upsilon_A \xrightarrow{\text{K}} (\mathcal{S}_{\Upsilon_A}, \mathcal{R}_{\Upsilon_A})$$

are, respectively, the *Knuth correspondence* between generalized permutation arrays with no repeated pair and pairs of Young tableaux of conjugate shapes and the *Knuth correspondence* between $(0, 1)$-matrices and pairs of Young tableaux of conjugate shapes.

Theorem 4.1.14 *Let m and n be positive integers. The Knuth correspondence*

$$A \xrightarrow{\text{K}} (\mathcal{S}_A, \mathcal{R}_A)$$

is a bijection between m by n $(0, 1)$-matrices A and ordered pairs $(\mathcal{S}, \mathcal{R})$ of Young tableaux of conjugate shapes where the tableau $\mathcal{S}$ is based on $\{1, 2, \ldots, n\}$ and the tableau $\mathcal{R}$ is based on $\{1, 2, \ldots, m\}$. In this correspondence, the row sum vector of A is $R = (r_1, r_2, \ldots, r_m)$ where r_i is the number of times the integer i appears in $\mathcal{R}_A$ $(i = 1, 2, \ldots, m)$ and the column sum vector is $S = (s_1, s_2, \ldots, s_n)$ where s_j is the number of times the integer j appears in $\mathcal{S}_A$ $(j = 1, 2, \ldots, n)$. The matrix A is a permutation matrix of order n if and only if both $\mathcal{S}_A$ and $\mathcal{R}_A$ are standard tableaux based on $\{1, 2, \ldots, n\}$.

Proof. The proof is very similar to the proof of Theorem 4.1.8 using Lemma 4.1.13 in place of Lemma 4.1.6. As a result we only make some comments concerning its proof. The construction used in column-insertion guarantees that $\mathcal{S}$ is a Young tableau. To verify that $\mathcal{R}$ is a Young tableau, we refer to the generalized permutation array Υ_A with no repeated pair. The elements of $\mathcal{R}$ are inserted in nondecreasing order. By Lemma 4.1.13, at each step within a row of A the new box in the construction of $\mathcal{S}$ is strictly-below and weakly-left of the previous new box; hence in $\mathcal{R}$ the new box is strictly-right of and weakly-below the previous new box. Thus no two equal elements are inserted in the same column of $\mathcal{R}$, and it follows that $\mathcal{R}$ is also a Young tableau.

The inverse of the Knuth correspondence is based on *column-extraction* which is similar to row-extraction. Let k' be an element in a corner box of a column of a Young tableau T'. A new Young tableau T is obtained as follows.

(i) If k' is in the last box of column 1, then T is obtained from T' by emptying this box.

(ii) Otherwise, there is a box immediately to the left of k' in the next column and it contains an integer which does not exceed k'. Determine the integer k_1' which is lowest in the column and does not exceed k'. We empty the box containing k_1', and replace k_1' with k'. We repeat with k' replaced by the bumped element k_1'.

(iii) We continue as in steps (i) and (ii) until we reach the first column and bump an element from it.

(iv) T' is the final tableau obtained by this process.

To obtain the inverse of the Knuth correspondence we apply column-extraction to the insertion tableau. To do this we have to specify how to choose the elements k' to extract at each stage. This is done by choosing the rightmost largest element in the recording tableau, and then extracting the element in the conjugate position of the insertion tableau. □

Example. We illustrate how to determine the generalized permutation array with no repeated pairs (and so the $(0, 1)$-matrix) corresponding to a pair of Young tableaux of conjugate shape. Consider the pair $(\mathcal{S}, \mathcal{R})$ of Young tableaux

$$
\mathcal{S} =
\begin{array}{|c|c|c|c|}
\hline
1 & 1 & 1 & 2 \\
\hline
2 & 2 & 3 & \times \\
\hline
3 & \times & \times & \times \\
\hline
4 & \times & \times & \times \\
\hline
\end{array}
, \quad
\mathcal{R} =
\begin{array}{|c|c|c|c|}
\hline
1 & 1 & 1 & 3 \\
\hline
2 & 2 & \times & \times \\
\hline
3 & 4 & \times & \times \\
\hline
4 & \times & \times & \times \\
\hline
\end{array}
.
$$

Applying the column-extraction algorithm, we obtain the generalized permutation array

$$
\begin{pmatrix}
1 & 1 & 1 & 2 & 2 & 3 & 3 & 4 & 4 \\
1 & 2 & 3 & 2 & 3 & 1 & 4 & 1 & 2
\end{pmatrix}
$$

and corresponding $(0, 1)$-matrix

$$
\begin{bmatrix}
1 & 1 & 1 & 0 \\
0 & 1 & 1 & 0 \\
1 & 0 & 0 & 1 \\
1 & 1 & 0 & 0
\end{bmatrix}.
$$

□

Analogous to Corollary 4.1.12 we have the following evaluation of $\kappa(R, S)$ in terms of Kostka numbers.

Corollary 4.1.15 *Let m, n, and τ be positive integers, and let nonincreasing, nonnegative integral vectors $R = (r_1, r_2, \ldots, r_m)$ and $S = (s_1, s_2, \ldots, s_n)$ satisfy*

$$
\tau = r_1 + r_2 + \cdots + r_m = s_1 + s_2 + \cdots + s_n.
$$

Then the number of $(0, 1)$-matrices with row sum vector R and column sum vector S is

$$
\kappa(R, S) = \sum_{\lambda} K_{\lambda, R} K_{\lambda^*, S}
$$

where the summation extends over all partitions λ of τ. □

The preceding corollary furnishes another proof of the Gale–Ryser theorem in Chapter 2.

Corollary 4.1.16 *Let m, n, and τ be positive integers, and let nonincreasing, nonnegative integral vectors $R = (r_1, r_2, \ldots, r_m)$ and $S = (s_1, s_2, \ldots, s_n)$ satisfy*

$$\tau = r_1 + r_2 + \cdots + r_m = s_1 + s_2 + \cdots + s_n.$$

There exists a matrix in $\mathcal{A}(R, S)$ if and only if $S \preceq R^*$.

Proof. By Corollary 4.1.15, $\mathcal{A}(R, S)$ is nonempty if and only if there exists a partition λ of τ such that $K_{\lambda, R} \neq 0$ and $K_{\lambda^*, S} \neq 0$. Using Corollary 4.1.11, we now see that $\mathcal{A}(R, S)$ is nonempty if and only if there exists a partition λ of τ such that $R \preceq \lambda$ and $S \preceq \lambda^*$, equivalently, $S \preceq \lambda^* \preceq R^*$. The latter is equivalent to $S \preceq R^*$. $\qquad\square$

From Theorem 4.1.1 and Corollary 4.1.15 we obtain the so-called *dual Cauchy identity.*

Theorem 4.1.17 *The generating function $g_{m,n}(\mathbf{x}; \mathbf{y})$ for the number of $(0, 1)$-matrices with row sum vector R and column sum vector S satisfies*

$$g_{m,n}(\mathbf{x}; \mathbf{y}) = \prod_{i=1}^{m} \prod_{j=1}^{n} (1 + x_i y_j) = \sum_{\lambda} s_\lambda(\mathbf{x}) s_{\lambda^*}(\mathbf{y}). \qquad (4.8)$$

$\qquad\square$

From Corollary 4.1.15 we know that

$$\kappa(R, S) = \sum_{\lambda} K_{\lambda, R} K_{\lambda^*, S}, \qquad (4.9)$$

equivalently

$$\kappa(R, S) = \sum_{\lambda} K_{\lambda^*, R} K_{\lambda, S}, \qquad (4.10)$$

where the summations extend over all partitions λ of τ. Since $K_{\lambda^*, R} K_{\lambda, S} \neq 0$ if and only if $R \preceq \lambda^*$ and $S \preceq \lambda$, equivalently, $S \preceq \lambda \preceq R^*$, we have

$$\kappa(R, S) = \sum_{S \preceq \lambda \preceq R^*} K_{\lambda^*, R} K_{\lambda, S}. \qquad (4.11)$$

We conclude from (4.11) that there exists a matrix in $\mathcal{A}(R, S)$ whose insertion tableau has shape λ (and thus whose recording tableau has shape λ^*) if and only if $S \preceq \lambda \preceq R^*$.

Since the Knuth correspondence is a bijection, given a pair $(\mathcal{P}, \mathcal{Q})$ of Young tableaux of conjugate shapes λ and λ^*, respectively, where $\mathcal{P}$ has content S (so $S \preceq \lambda$) and $\mathcal{Q}$ has content R (so $R \preceq \lambda^*$ and hence $\lambda \preceq R^*$), we can invert the correspondence as already explained and obtain a matrix A in $\mathcal{A}(R, S)$ whose insertion tableau has shape λ. Given a partition λ with $S \preceq \lambda \preceq R^*$, it is of interest to have a direct algorithm that constructs a matrix A_λ in $\mathcal{A}(R,S)$ whose insertion tableau has shape λ that does not depend on first constructing the pair $(\mathcal{P}, \mathcal{Q})$. We now describe such a construction in the two extreme cases where $\lambda = S$ and $\lambda = R^*$ [10], and relate these algorithms to Ryser's algorithm for constructing a matrix in $\mathcal{A}(R,S)$ as given in Chapter 3.

Let $R = (r_1, r_2, \ldots, r_m)$ and $S = (s_1, s_2, \ldots, s_n)$ be two nonincreasing, positive integral vectors which are both partitions of the same integer τ. We assume that $S \preceq R^*$ so that $\mathcal{A}(R,S)$ is nonempty. We first consider the case where $\lambda = R^*$ and show that a modification of Ryser's algorithm, but not Ryser's algorithm itself, works.

Algorithm for an Insertion Tableau of Shape $\lambda = R^*$

(1) Begin with the m by n matrix $A(R;n)$ with row sum vector R and column sum vector R^*.

(2) Shift s_n of the last 1's in s_n rows of $A(R;n)$ to column n, choosing those 1's in the rows with the largest sums but, in the case of ties, giving preference to the *topmost* rows.

(3) The matrix left in columns $1, 2, \ldots, n-1$ of $A(R;n)$ is a matrix $A(R'';n-1)$ whose row sum vector R'' is determined by R and the 1's chosen to be shifted. (In general, unlike the vector R' in Ryser's algorithm, the vector R'' will not be nonincreasing.) We now repeat with $A(R'';n-1)$ in place of $A(R;n)$ and s_{n-1} in place of s_n. We continue like this until we arrive at a matrix $\widetilde{A'}$ in $\mathcal{A}(R,S)$.

This algorithm must terminate with a matrix in $\mathcal{A}(R,S)$, because it is a permuted form of Ryser's algorithm.

Example. Let $R = (4,4,3,3,2)$ and $S = (4,3,3,3,3)$, and let $\lambda = R^* = (5,5,4,2)$. Applying the above algorithm, we get

$$
\begin{bmatrix}
1 & 1 & 1 & 1 & 0 & \| & 4 \\
1 & 1 & 1 & 1 & 0 & \| & 4 \\
1 & 1 & 1 & 0 & 0 & \| & 3 \\
1 & 1 & 1 & 0 & 0 & \| & 3 \\
1 & 1 & 0 & 0 & 0 & \| & 2
\end{bmatrix}
\rightarrow
\begin{bmatrix}
1 & 1 & 1 & 0 & 1 & \| & 3 \\
1 & 1 & 1 & 0 & 1 & & 3 \\
1 & 1 & 0 & 0 & 1 & & 2 \\
1 & 1 & 1 & 0 & 0 & & 3 \\
1 & 1 & 0 & 0 & 0 & & 2
\end{bmatrix}
$$

$$
\rightarrow
\begin{bmatrix}
1 & 1 & 0 & 1 & 1 & \| & 2 \\
1 & 1 & 0 & 1 & 1 & & 2 \\
1 & 1 & 0 & 0 & 1 & & 2 \\
1 & 1 & 0 & 1 & 0 & & 2 \\
1 & 1 & 0 & 0 & 0 & \| & 2
\end{bmatrix}
\rightarrow
\begin{bmatrix}
1 & 0 & 1 & 1 & 1 & \| & 1 \\
1 & 0 & 1 & 1 & 1 & & 1 \\
1 & 0 & 1 & 0 & 1 & & 1 \\
1 & 1 & 0 & 1 & 0 & & 2 \\
1 & 1 & 0 & 0 & 0 & \| & 2
\end{bmatrix}
$$

$$
\rightarrow
\begin{bmatrix}
0 & 1 & 1 & 1 & 1 & \| & 0 \\
1 & 0 & 1 & 1 & 1 & \| & 1 \\
1 & 0 & 1 & 0 & 1 & \| & 1 \\
1 & 1 & 0 & 1 & 0 & \| & 1 \\
1 & 1 & 0 & 0 & 0 & \| & 1
\end{bmatrix}
\rightarrow
\begin{bmatrix}
0 & 1 & 1 & 1 & 1 & \| & 0 \\
1 & 0 & 1 & 1 & 1 & \| & 0 \\
1 & 0 & 1 & 0 & 1 & \| & 0 \\
1 & 1 & 0 & 1 & 0 & \| & 0 \\
1 & 1 & 0 & 0 & 0 & \| & 0
\end{bmatrix}.
$$

Thus the matrix constructed by this algorithm is

$$\widetilde{A'} = \begin{bmatrix} 0 & 1 & 1 & 1 & 1 \\ 1 & 0 & 1 & 1 & 1 \\ 1 & 0 & 1 & 0 & 1 \\ 1 & 1 & 0 & 1 & 0 \\ 1 & 1 & 0 & 0 & 0 \end{bmatrix},$$

and its corresponding generalized permutation array is

$$\begin{pmatrix} 1 & 1 & 1 & 1 & 2 & 2 & 2 & 2 & 3 & 3 & 3 & 4 & 4 & 4 & 5 & 5 \\ 2 & 3 & 4 & 5 & 1 & 3 & 4 & 5 & 1 & 3 & 5 & 1 & 2 & 4 & 1 & 2 \end{pmatrix}.$$

Applying the Knuth correspondence to construct the corresponding insertion tableau $\mathcal{P}$ and showing only the results of inserting the five sets of column indices corresponding to the 1's in each row, we get

$$
\begin{array}{l}
2 \\ 3 \\ 4 \\ 5
\end{array}
\rightarrow
\begin{array}{ll}
1 & 2 \\ 3 & 3 \\ 4 & 4 \\ 5 & 5
\end{array}
\rightarrow
\begin{array}{lll}
1 & 1 & 2 \\ 3 & 3 & 3 \\ 4 & 4 & 5 \\ 5 & 5
\end{array}
\rightarrow
\begin{array}{llll}
1 & 1 & 1 & 2 \\ 2 & 3 & 3 & 3 \\ 4 & 4 & 4 & 5 \\ 5 & 5
\end{array}
\rightarrow
\begin{array}{lllll}
1 & 1 & 1 & 1 & 2 \\ 2 & 2 & 3 & 3 & 3 \\ 4 & 4 & 4 & 5 \\ 5 & 5
\end{array} .
$$

Thus, in our usual notation,

$$\mathcal{P} = \begin{array}{|c|c|c|c|c|}
\hline
1 & 1 & 1 & 1 & 2 \\
\hline
2 & 2 & 3 & 3 & 3 \\
\hline
4 & 4 & 4 & 5 & \times \\
\hline
5 & 5 & \times & \times & \times \\
\hline
\times & \times & \times & \times & \times \\
\hline
\end{array}$$

which has shape $R^* = (5, 5, 4, 2)$. The recording tableau $\mathcal{Q}$, having shape $R^{**} = R$ and content R, must be

$$\begin{array}{|c|c|c|c|c|}
\hline
1 & 1 & 1 & 1 & \times \\
\hline
2 & 2 & 2 & 2 & \times \\
\hline
3 & 3 & 3 & \times & \times \\
\hline
4 & 4 & 4 & \times & \times \\
\hline
5 & 5 & \times & \times & \times \\
\hline
\end{array} .$$

$\square$

The next lemma contains an important property of the constructed matrix $\widetilde{A'}$.

Lemma 4.1.18 *Let q be an integer with $1 \le q \le n$. Let $\widetilde{A'}_q$ be the m by $n - q + 1$ submatrix of $\widetilde{A'}$ formed by columns $q, q+1, \ldots, n$. Then the row sums of $\widetilde{A'}_q$ are nonincreasing.*

Proof. Let p be an integer with $2 \leq p \leq m$. The assertion in the lemma is that the number of 1's in row $p-1$ of $\widetilde{A'}$ beginning with column q is at least as large as the number of 1's in row p beginning with column q. We argue by contradiction. Let $l \geq q$ be the largest index such that beginning with column l, row p contains strictly more 1's than row $p-1$. Then there are a 0 in position $(p-1,l)$ and a 1 in position (p,l), and rows $p-1$ and p are identical starting with column $l+1$. Let rows $p-1$ and p contain e 1's starting with column $l+1$. Since $r_{p-1} \geq r_p$, we have $r_{p-1} - e \geq r_p - e$ and we contradict our rules for construction of $\widetilde{A'}$ which would have us put a 1 in column l of row $p-1$ before putting a 1 in column l of row p. This contradiction completes the proof. $\qquad \square$

The property of $\widetilde{A'}$ given in Lemma 4.1.18 should be contrasted with the property of the matrix $\widetilde{A}$ constructed by Ryser's algorithm; the matrix $\widetilde{A}$ has the property that for each q the row sums of the m by q submatrix of $\widetilde{A}$ formed by its first q columns are nonincreasing.

Before verifying that the algorithm has the stated property, we need a lemma concerning the permutation array corresponding to a $(0,1)$-matrix. This lemma contains an important result of Greene [29] which generalized a theorem of Schensted [43]. We defer the proof to an appendix-section at the end of this chapter.

Lemma 4.1.19 *Let*

$$\Upsilon = \begin{pmatrix} i_1 & i_2 & \cdots & i_\tau \\ j_1 & j_2 & \cdots & j_\tau \end{pmatrix}$$

be a generalized permutation array with distinct pairs $\binom{i_k}{j_k}$ in lexicographic order. Let $\mathcal{P}$ be the insertion tableau that results by applying the column bumping operation to Υ. The number of occupied boxes in the first k columns of $\mathcal{P}$ equals the maximal number of terms in a subsequence of $j_1, j_2, \ldots, j_\tau$ that is the union of k strictly increasing subsequences $(k = 1, 2, \ldots)$.

Theorem 4.1.20 *The matrix $\widetilde{A'}$ in $\mathcal{A}(R,S)$ constructed by the algorithm has corresponding insertion tableau of shape R^*.*

Proof. Consider the generalized permutation array

$$\Upsilon_{\widetilde{A'}} = \begin{pmatrix} i_1 & i_2 & \cdots & i_\tau \\ j_1 & j_2 & \cdots & j_\tau \end{pmatrix}$$

corresponding to $\widetilde{A'}$. By Lemma 4.1.19, in order for the insertion tableau to have shape R^*, the maximal number of terms in a subsequence of the sequence of column indices $j_1, j_2, \ldots, j_\tau$ that is the disjoint union of k strictly increasing subsequences must equal $r_1 + r_2 + \cdots + r_k$ $(k = 1, 2, \ldots, m)$. In fact, we shall show that a much stronger property holds: the maximal number of terms in a subsequence of the sequence of column indices that is the disjoint union of k strictly increasing subsequences is attained by the initial subsequence of $j_1, j_2, \ldots, j_\tau$ consisting of the $r_1 + r_2 + \cdots + r_k$ column indices of the 1's in the first k rows ($k = 1, 2, \ldots, m$).

First consider the case $k = 1$. We want to show that a longest strictly increasing subsequence of $j_1, j_2, \ldots, j_\tau$ is given by the column indices of the 1's in row 1 of $\widetilde{A'}$. A strictly increasing subsequence σ of $j_1, j_2, \ldots, j_\tau$ corresponds to a collection of 1's in $\widetilde{A'}$ with the property that the column index of the last 1 in row i is strictly less than the column index of the first 1 in row $i + 1$ ($i = 1, 2, \ldots, m - 1$). We refer to these 1's as 1's of σ. Let the largest row index containing a 1 of σ be $p > 1$ and let the first 1 of σ in row p be in column q.

By Lemma 4.1.18 the number of 1's in row $p - 1$ of $\widetilde{A'}$ beginning with column q is at least as large as the number of 1's in row p beginning with column q. We may now replace the 1's of σ in row p with the 1's of $\widetilde{A'}$ in row $p - 1$ beginning with column q. This shows that there is a longest strictly increasing subsequence of $j_1, j_2, \ldots, j_\tau$ contained in rows $1, 2, \ldots, p - 1$. Arguing inductively, we conclude that there is a longest strictly increasing subsequence corresponding to 1's in row 1. Hence the longest strictly increasing subsequence has length r_1.

Generalizing to $k > 1$, we show how a subsequence σ of the sequence of column indices which is the disjoint union of k strictly increasing subsequences $\sigma^1, \sigma^2, \ldots, \sigma^k$ can be transformed to a subsequence of the same or greater length, but taken from the first k rows of $\widetilde{A'}$. This implies that the longest such σ has length $r_1 + r_2 + \cdots + r_k$. Let p be the largest row index containing a 1 of σ. The sequence σ^i is a union of sequences $\sigma_1^i, \sigma_2^i, \ldots, \sigma_p^i$, where σ_j^i is a (possibly empty) sequence of column indices corresponding to 1's in row i of $\widetilde{A'}$ ($i = 1, 2, \ldots, k; j = 1, 2, \ldots, p$). We may assume that each σ_j^i contains all the 1's in row i of $\widetilde{A'}$ that lie between the first and last 1 of σ_j^i. We may choose our notation so that $\sigma^1 = \sigma_1^1, \sigma_2^1, \ldots, \sigma_p^1$, that is, the 1's of σ_j^1 are to the left of the 1's of σ_{j+1}^1 ($j = 1, 2, \ldots, p - 1$). We show that we may assume that this property holds for each σ^i, that is,

$$\sigma^i = \sigma_1^i, \sigma_2^i, \ldots, \sigma_p^i \quad (i = 1, 2, \ldots, k),$$

so that the 1's of each σ_j^i are to the left of the 1's of σ_{j+1}^i ($i = 1, 2, \ldots, k; j = 1, 2, \ldots, p-1$). This is because if, for instance, $\sigma^2 = \sigma_{j_1}^2, \sigma_{j_2}^2, \ldots, \sigma_{j_p}^2$, where $j_1, j_2, \ldots, j_p$ is a permutation of $\{1, 2, \ldots, p\}$, then we may replace σ_t^1 with $\sigma_{j_t}^1$ without changing the size of σ or the property that σ is the disjoint union of k strictly increasing subsequences. We may now repeatedly use Lemma 4.1.18 as in the case of $k = 1$ above to transform σ into a new sequence σ' such that the size of σ' is at least that of σ and σ' is the disjoint union of k strictly increasing subsequences, with the first from row 1, the second from row 2, ..., and the last from row k. $\qquad \square$

We now consider the case where $\lambda = S$ and give an algorithm that constructs a matrix in $\mathcal{A}(R, S)$ whose insertion tableau has shape $\lambda = S$.

In this case Ryser's algorithm with the rows and columns interchanged works. This algorithm starts with the m by n $(0,1)$-matrix $A'(S;m)$ with column sum vector S and row sum vector S^* whose 1's are in the initial (topmost) positions in each column.

Algorithm for an Insertion Tableau of Shape S

(1) Begin with the m by n matrix $A'(S;m)$ with row sum vector S^* and column sum vector S.

(2) Shift r_m of the last 1's in r_m columns of $A'(S;m)$ to row m, choosing those 1's in the columns with the largest sums but, in the case of ties, giving preference to the *rightmost* columns.

(3) The matrix left in rows $1, 2, \ldots, m-1$ of $A'(S;m)$ is a matrix $A'(S^1; m-1)$ with column sum vector S^1 determined by S and the 1's chosen to be shifted. (By choice of 1's to be shifted, the vector S^1 is nondecreasing.) We now repeat with $A'(S^1; m-1)$ in place of $A'(S;m)$ and r_{m-1} in place of r_m. We continue like this until we arrive at a matrix $\widetilde{A''}$ in $\mathcal{A}(R,S)$.

This algorithm must terminate with a matrix in $\mathcal{A}(R,S)$ because it is just the "transpose" of Ryser's algorithm.

Example. As in the preceding example, we use $R = (4,4,3,3,2)$ and $S = (4,3,3,3,3)$. We have $S^* = (5,5,5,1,0)$. Applying the algorithm we get

$$
\begin{bmatrix}
1 & 1 & 1 & 1 & 1 \\
1 & 1 & 1 & 1 & 1 \\
1 & 1 & 1 & 1 & 1 \\
1 & 0 & 0 & 0 & 0 \\
0 & 0 & 0 & 0 & 0 \\
\hline
4 & 3 & 3 & 3 & 3
\end{bmatrix}
\rightarrow
\begin{bmatrix}
1 & 1 & 1 & 1 & 1 \\
1 & 1 & 1 & 1 & 1 \\
1 & 1 & 1 & 1 & 0 \\
0 & 0 & 0 & 0 & 0 \\
1 & 0 & 0 & 0 & 1 \\
\hline
3 & 3 & 3 & 3 & 2
\end{bmatrix}
$$

$$
\rightarrow
\begin{bmatrix}
1 & 1 & 1 & 1 & 1 \\
1 & 1 & 1 & 1 & 1 \\
1 & 0 & 0 & 0 & 0 \\
0 & 1 & 1 & 1 & 0 \\
1 & 0 & 0 & 0 & 1 \\
\hline
3 & 2 & 2 & 2 & 2
\end{bmatrix}
\rightarrow
\begin{bmatrix}
1 & 1 & 1 & 1 & 1 \\
1 & 1 & 1 & 0 & 0 \\
1 & 0 & 0 & 1 & 1 \\
0 & 1 & 1 & 1 & 0 \\
1 & 0 & 0 & 0 & 1 \\
\hline
2 & 2 & 2 & 1 & 1
\end{bmatrix}
$$

$$
\rightarrow
\begin{bmatrix}
1 & 1 & 1 & 1 & 0 \\
1 & 1 & 1 & 0 & 1 \\
1 & 0 & 0 & 1 & 1 \\
0 & 1 & 1 & 1 & 0 \\
1 & 0 & 0 & 0 & 1 \\
\hline
1 & 1 & 1 & 1 & 0
\end{bmatrix}
\rightarrow
\begin{bmatrix}
1 & 1 & 1 & 1 & 0 \\
1 & 1 & 1 & 0 & 1 \\
1 & 0 & 0 & 1 & 1 \\
0 & 1 & 1 & 1 & 0 \\
1 & 0 & 0 & 0 & 1 \\
\hline
0 & 0 & 0 & 0 & 0
\end{bmatrix}.
$$

Thus the matrix constructed by this algorithm is

$$\widetilde{A''} = \begin{bmatrix} 1 & 1 & 1 & 1 & 0 \\ 1 & 1 & 1 & 0 & 1 \\ 1 & 0 & 0 & 1 & 1 \\ 0 & 1 & 1 & 1 & 0 \\ 1 & 0 & 0 & 0 & 1 \end{bmatrix},$$

and its corresponding generalized permutation array is

$$\begin{pmatrix} 1 & 1 & 1 & 1 & 2 & 2 & 2 & 2 & 3 & 3 & 3 & 4 & 4 & 4 & 5 & 5 \\ 1 & 2 & 3 & 4 & 1 & 2 & 3 & 5 & 1 & 4 & 5 & 2 & 3 & 4 & 1 & 5 \end{pmatrix}.$$

The corresponding insertion tableau $\mathcal{P}$ (as before we only show the results of inserting the five sets of column indices corresponding to the 1's in each row) is

$$
\begin{array}{c}
1 \\ 2 \\ 3 \\ 4
\end{array}
\rightarrow
\begin{array}{cc}
1 & 1 \\ 2 & 2 \\ 3 & 3 \\ 4 & \\ 5 &
\end{array}
\rightarrow
\begin{array}{ccc}
1 & 1 & 1 \\ 2 & 2 & \\ 3 & 3 & \\ 4 & 4 & \\ 5 & 5 &
\end{array}
\rightarrow
\begin{array}{ccc}
1 & 1 & 1 \\ 2 & 2 & 2 \\ 3 & 3 & 3 \\ 4 & 4 & 4 \\ 5 & 5 &
\end{array}
\rightarrow
\begin{array}{cccc}
1 & 1 & 1 & 1 \\ 2 & 2 & 2 & \\ 3 & 3 & 3 & \\ 4 & 4 & 4 & \\ 5 & 5 & 5 &
\end{array},
$$

or in our usual notation

1	1	1	1	×
2	2	2	×	×
3	3	3	×	×
4	4	4	×	×
5	5	5	×	×

which has shape and content equal to $S = (4, 3, 3, 3, 3)$. The recording tableau equals

1	1	1	1	2
2	2	2	3	3
3	4	4	4	5
5	×	×	×	×
×	×	×	×	×

and has shape $S^* = (5, 5, 5, 1)$.   □

As in the example, if a matrix in $\mathcal{A}(R, S)$ has insertion tableau $\mathcal{P}$ of shape $\lambda = S$, then since the content of $\mathcal{P}$ is S and since the columns are strictly increasing, the insertion tableau must have s_1 1's in row 1, s_2 2's in row 2, s_3 3's in row 3, and so on.

Theorem 4.1.21 *The matrix $\widetilde{A''}$ in $\mathcal{A}(R, S)$ constructed by the algorithm has corresponding insertion tableau of shape S.*

Proof. Let A be a matrix in $\mathcal{A}(R,S)$ with corresponding generalized permutation array Υ_A. In column bumping, an integer k is inserted in column 1 in a row numbered $j_1 \leq k$, the element k_1 bumped is inserted in column 2 in a row numbered $j_2 \leq j_1$ (since the rows are nondecreasing), and so on. Thus the insertion path moves strictly to the right and never down. In order that the insertion tableau have shape and content S, the sequence of column indices $j_1, j_2, \ldots, j_\tau$ from the second row of Υ_A must have the property:

(i) where t_i is the number of integers equal to i in any initial segment $j_1, j_2, \ldots, j_p$ $(i = 1, 2, \ldots, n)$, we have $t_1 \geq t_2 \geq \cdots \geq t_n$.

In terms of the matrix A this property is equivalent to:

(ii) the column sums of each p by n submatrix of A are nonincreasing $(p = 1, 2, \ldots, m)$.

Since the matrix $\widetilde{A}$ has the property that the row sums of each m by q initial submatrix are nonincreasing, and since the algorithm constructs a matrix $\widetilde{A}''$ like $\widetilde{A}$ but with rows and columns interchanged, it follows that the column sums of each q by n initial submatrix of $\widetilde{A}''$ are nonincreasing. Hence the insertion tableau of $\widetilde{A}''$ has shape $\lambda = S$. $\qquad\square$

The matrices $\widetilde{A}'$ and $\widetilde{A}''$ in $\mathcal{A}(R,S)$ whose insertion tableaux have shapes $\lambda = R^*$ and $\lambda = S$, respectively, are not the only matrices with these insertion tableaux. For example, let $R = S = (3,2,1,1)$. Then each of the matrices

$$\widetilde{A}'' = \begin{bmatrix} 1 & 1 & 1 & 0 \\ 1 & 0 & 0 & 1 \\ 0 & 1 & 0 & 0 \\ 1 & 0 & 0 & 0 \end{bmatrix} \quad \text{and} \quad A = \begin{bmatrix} 1 & 1 & 1 & 0 \\ 1 & 1 & 0 & 0 \\ 0 & 0 & 0 & 1 \\ 1 & 0 & 0 & 0 \end{bmatrix}$$

has insertion tableau

1	1	1	×
2	2	×	×
3	×	×	×
4	×	×	×

of shape $\lambda = S$. The matrix $\widetilde{A}''$ has recording tableau

1	1	1	2
2	3	×	×
4	×	×	×
×	×	×	×

and the matrix A has recording tableau

1	1	1	3
2	2	×	×
4	×	×	×
×	×	×	×

We conclude this section with some remarks about other results concerning the number of matrices in nonempty classes $\mathcal{A}(R,S)$ where $R = (r_1, r_2, \ldots, r_m)$ and $S = (s_1, s_2, \ldots, s_n)$. Wang [50] gave a very complex formula for $|\mathcal{A}(R,S)|$ involving a sum of products of binomial coefficients. By introducing different binomial coefficient products, Wang and Zhang [51] reduced the number $2^n - 2n$ of terms in the summation by half. Using this formula, they computed the cardinality of $\mathcal{A}(n, k)$ for some n and k. The following computations are some of those given in [51]:

$$|\mathcal{A}(n, k)| = \begin{cases} 90 & \text{if } n = 4 \text{ and } k = 2, \\ 2\,040 & \text{if } n = 5 \text{ and } k = 2, \\ 67\,950 & \text{if } n = 6 \text{ and } k = 2, \\ 3\,110\,940 & \text{if } n = 7 \text{ and } k = 2, \\ 297\,200 & \text{if } n = 6 \text{ and } k = 3, \\ 68\,938\,800 & \text{if } n = 7 \text{ and } k = 3. \end{cases}$$

Wang and Zhang [52] also investigated the number of normal matrices in $\mathcal{A}(n, 2)$.

We next derive a lower bound for $|\mathcal{A}(n, k)|$ due to Wei [49]. Let $S^{(0)} = (s_1^{(0)}, s_2^{(0)}, \ldots, s_n^{(0)})$ and $S = (s_1, s_2, \ldots, s_n)$ be two nonincreasing, nonnegative integral vectors such that $S \preceq S^{(0)}$. Assume that $S \neq S^{(0)}$. There exists a smallest index p with $1 < p \leq n$ such that $s_p > s_p^{(0)}$, and then a largest index $q < p$ such that $s_q < s_q^{(0)}$. In particular, $s_i = s_i^{(0)}$ for $q < i < p$. We define a new vector $S^{(1)} = (s_1^{(1)}, s_2^{(1)}, \ldots, s_n^{(1)})$ by:

(i) If $s_q^{(0)} - s_q \geq s_p - s_p^{(0)}$, then

$$s_j^{(1)} = \begin{cases} s_q^{(0)} - (s_p - s_p^{(0)}), & \text{if } j = q, \\ s_p^{(0)} + (s_p - s_p^{(0)}) = s_p, & \text{if } j = p \\ s_j^{(0)}, & \text{otherwise.} \end{cases}$$

(ii) If $s_q^{(0)} - s_q < s_p - s_p^{(0)}$, then

$$s_j^{(1)} = \begin{cases} s_q^{(0)} - (s_q^{(0)} - s_q) = s_q, & \text{if } j = q, \\ s_p^{(0)} + (s_q^{(0)} - s_q), & \text{if } j = p \\ s_j^{(0)}, & \text{otherwise.} \end{cases}$$

It is straightforward to check that

$$S \preceq S^{(1)} \preceq S^{(0)}.$$

In addition, the number of components in which S and $S^{(1)}$ agree is at least one more than the number in which S and $S^{(0)}$ agree. Let

$$w(S, S^{(0)}) = \begin{pmatrix} s_q^{(0)} - s_p^{(0)} \\ \min\{s_q^{(0)} - s_q, s_p - s_p^{(0)}\} \end{pmatrix}, \qquad (4.12)$$

a binomial coefficient. Let $A^{(0)} = [a_{ij}]$ be a matrix in $\mathcal{A}(R, S^{(0)})$ for some vector R. Consider the m by 2 submatrix of A determined by columns q and p. At least $s_q^{(0)} - s_p^{(0)}$ rows of this submatrix are of the form $[1 \; 0]$. If we exchange the 1 for the 0 in $\min\{s_q^{(0)} - s_q, s_p - s^{(0)}\}$ of these rows, then we obtain a matrix $A^{(1)}$ in $\mathcal{A}(R, S^{(1)})$. By the choice of p and q and the nonincreasing property of $S^{(0)}$ and of S, we have

$$s_q^{(0)} - s_p^{(0)} > \begin{cases} s_q^{(0)} - s_p \geq s_q^{(0)} - s_q, \\ s_q - s_p^{(0)} \geq s_p - s_p^{(0)}. \end{cases}$$

Thus the number of matrices $A^{(1)}$ obtained in this way equals $w(S, S^{(0)})$ as given in (4.12). We now replace the majorization $S \preceq S^{(0)}$ with the majorization $S \preceq S^{(1)}$ and repeat. In a finite number t of steps we obtain

$$S = S^{(t)} \preceq S^{(t-1)} \preceq \cdots \preceq S^{(1)} \preceq S^{(0)} \tag{4.13}$$

and matrices

$$A^{(0)}, A^{(1)}, \ldots, A^{(t)} \tag{4.14}$$

with $A^{(i)} \in \mathcal{A}(R, S^{(i)})$ for $i = 1, 2, \ldots, t$. If $S = S^{(0)}$, then $t = 0$. The sequence of vectors $S^{(0)}, S^{(1)}, \ldots, S^{(t-1)}, S^{(t)} = S$ satisfying (4.13) is called the *descent chain* from $S^{(0)}$ to S. Its *weight* is defined to be

$$w(S, S^{(0)}) = \prod_{i=1}^{t-1} w(S, S^{(i)}).$$

Theorem 4.1.22 *Let $R = (r_1, r_2, \ldots, r_m)$ and $S = (s_1, s_2, \ldots, s_n)$ be nonincreasing, nonnegative integral vectors such that $S \preceq R^*$. Let (4.13) be the descent chain from $S^{(0)} = R^*$ to S. Then*

$$|\mathcal{A}(R,S)| \geq w(S, R^*).$$

Proof. Let $A = [a_{ij}]$ be a matrix in $\mathcal{A}(R, R^*)$. Since $S \preceq R^*$, there exist a descent chain (4.13) from $S^{(0)}$ to S, and matrices (4.14) with $A^{(t)} \in \mathcal{A}(R, S^{(t)})$. The theorem will be proved once we show that all the matrices $A^{(t)}$ obtained in this way are distinct. But this follows from the fact that in passing from $A^{(i)}$ to $A^{(i-1)}$ in the sequence (4.14), at least one more column of the final matrix $A^{(t)}$ in $\mathcal{A}(R, S)$ is determined. Hence the number of matrices $A^{(t)}$ obtained in this way is $w(S, S^{(0)}) = w(S, R^*)$. $\qquad\square$

Example. [49] Let

$$R = (7, 5, 5, 4, 4, 3, 2, 2, 2, 1) \quad \text{and} \quad S = (7, 6, 4, 4, 4, 4, 4).$$

Then $R^* = (9,8,6,5,3,1,1)$ and $S \preceq R^*$. The descent chain from R^* to S is

$$S^{(0)} = (9,8,6,5,3,1,1), \quad w(S,S^{(0)}) = \binom{2}{1} = 2,$$

$$S^{(1)} = (9,8,6,4,4,1,1), \quad w(S,S^{(1)}) = \binom{5}{2} = 10,$$

$$S^{(2)} = (9,8,4,4,4,3,1), \quad w(S,S^{(2)}) = \binom{5}{1} = 5,$$

$$S^{(3)} = (9,7,4,4,4,4,1), \quad w(S,S^{(3)}) = \binom{6}{1} = 6,$$

$$S^{(4)} = (9,6,4,4,4,4,2), \quad w(S,S^{(4)}) = \binom{7}{2} = 21,$$

$$S^{(5)} = (7,6,4,4,4,4,4).$$

By Theorem 4.1.22,

$$|\mathcal{A}(R,S)| \geq 2 \cdot 10 \cdot 5 \cdot 6 \cdot 21 = 12\,600.$$

$\square$

The following corollary is a straightforward consequence of Theorem 4.1.22.

Corollary 4.1.23 *Let n and k be positive integers with k a divisor of n. Then*

$$|\mathcal{A}(n,k)| \geq \frac{(n!)^k}{(k!)^n}.$$

$\square$

The lower bound in Theorem 4.1.22 has been extended by Wan [45] to the subclass of $\mathcal{A}(R,S)$ consisting of those matrices with at most one prescribed 1 in each column.

McKay [39] derived the asymptotic estimate, for $0 < \epsilon < 1/6$,

$$|\mathcal{A}(n,k)| \sim \frac{(nk)!}{(k!)^{2n}} \exp\left(\frac{(k-1)^2}{2} + O\left(\frac{k^3}{n}\right)\right),$$

uniformly for $1 \leq k \leq \epsilon n$. An improvement of this result is contained in [40]. In [39], [40], and [20] one may find other asymptotic results for $|\mathcal{A}(R,S)|$ valid with certain restrictions. Henderson [30] surveys the literature on the number and enumeration of matrices in $\mathcal{A}(R,S)$, and carries out the enumeration in some special cases that have arisen in applications. An enumeration by rank of matrices over rings with prescribed row and column sums is studied in [7]. Enumeration is also treated in [50, 51, 52].

4.2   Irreducible Matrices in $\mathcal{A}(R, S)$

Let A be a square matrix of order n. We recall from Section 1.3 that A is irreducible provided there does not exist a permutation matrix P such that

$$PAP^{\mathrm{T}} = \begin{bmatrix} A_1 & A_{12} \\ O & A_2 \end{bmatrix}$$

where A_1 and A_2 are square matrices of order at least 1. A matrix that is not irreducible is reducible. By Theorem 1.3.1 the matrix A is irreducible if and only if the digraph $D(A)$ determined by its nonzero elements is strongly connected. In this section we investigate the existence of irreducible matrices in a class $\mathcal{A}(R, S)$. We also consider the nonexistence of reducible matrices in $\mathcal{A}(R, S)$, equivalently, classes which contain only irreducible matrices.

Since the property of being irreducible is not in general invariant under arbitrary row and column permutations, we cannot without loss of generality assume that both R and S are nonincreasing vectors. We can assume that one of R and S is nonincreasing since, if Q is a permutation matrix, the matrix QAQ^{T} is irreducible whenever A is. A trivial necessary condition for $\mathcal{A}(R, S)$ to contain an irreducible matrix is that both R and S are positive vectors, and we generally make this assumption.

We begin with three elementary lemmas.

Lemma 4.2.1 *Let* $A = A_1 \oplus A_2$ *where* A_1 *and* A_2 *are irreducible* $(0, 1)$-*matrices of orders* m *and* n, *respectively. Let* A' *be obtained from* A *by a* $(p, q; k, l)$-*interchange where* $1 \leq p, k \leq m < q, l \leq m + n$. *Then* A' *is also irreducible.*

Proof. Let $D_1 = D(A_1)$ and $D_2 = D(A_2)$. Since A_1 and A_2 are irreducible matrices, D_1 and D_2 are strongly connected digraphs. The digraph $D(A')$ is obtained from $D(A)$ by removing the arc from vertex p to vertex k and the arc from vertex q to vertex l, and inserting arcs from p to l and q to k. Since D_1 is strongly connected, there is a path γ_1 from k to p; since D_2 is strongly connected, there is a path γ_2 from l to q. Combining the arc (p, l), the path γ_1, and the arc (q, k), we obtain a path in $D(A')$ from p to k. Similarly, using γ_2 we obtain a path from q to l. It now follows that $D(A')$ is strongly connected and hence that A' is irreducible. $\qquad\square$

The following lemma can be proved in a similar way.

Lemma 4.2.2 *Let* A_1 *and* A_2 *be irreducible* $(0, 1)$-*matrices of orders* m *and* n, *respectively. Let*

$$A = [a_{ij}] = \begin{bmatrix} A_1 & X \\ O & A_2 \end{bmatrix}.$$

Assume that row p of the matrix X is not a zero row. Assume also that there is an integer q with $1 \leq q \leq m$ such that that $a_{pq} = 1$. Then the matrix obtained from A by replacing a_{pq} with a 0 and some 0 in column q of O with a 1 is also irreducible.   □

The following lemma will be used implicitly in the proof of Theorem 4.2.4.

Lemma 4.2.3 *Let*

$$X = \begin{bmatrix} X_1 & X_{12} & \ldots & X_{1k} \\ X_{21} & X_2 & \ldots & X_{2k} \\ \vdots & \vdots & \ddots & \vdots \\ X_{k1} & X_{k2} & \ldots & X_k \end{bmatrix}$$

where $X_1, X_2, \ldots, X_k$ are irreducible matrices. Then the number of irreducible components of X is at most k.

Proof. Let D be the digraph with vertex set $\{1, 2, \ldots, k\}$ in which there is an arc from i to j if and only if $i \neq j$ and $X_{ij} \neq O$. The matrices $X_1, X_2, \ldots, X_k$ are the irreducible components of A exactly when D has no directed cycles. If D has a directed cycle γ, then the matrices $X_1, X_2, \ldots, X_k$ that correspond to vertices of γ are part of a larger irreducible submatrix of A.   □

The following theorem is proved in [9]. Recall that if $R = (r_1, r_2, \ldots, r_n)$ and $S = (s_1, s_2, \ldots, s_n)$ are nonnegative integral vectors, then

$$t_{K,L} = |K||L| + \sum_{i \notin K} r_i - \sum_{j \in J} s_j \quad (K, L \subseteq \{1, 2, \ldots, n\}).$$

Theorem 4.2.4 *Let $n \geq 2$ be an integer, and let $R = (r_1, r_2, \ldots, r_n)$ and $S = (s_1, s_2, \ldots, s_n)$ be positive integral vectors such that R is nonincreasing and $\mathcal{A}(R,S)$ is nonempty. Then the following assertions are equivalent.*

(i) *There exists an irreducible matrix in $\mathcal{A}(R,S)$.*

(ii) *There exist an irreducible matrix in $\mathcal{A}(R,S)$ whose trace is the minimal trace $\tilde{\sigma}(R,S)$ possible for a matrix in $\mathcal{A}(R,S)$.*

(iii) $\sum_{i=1}^{k} r_i < \sum_{i=1}^{k} s_i + \sum_{i=k+1}^{n} \min\{k, s_i\}$ $(k = 1, 2, \ldots, n-1)$.

(iv) $t_{K,L} > 0$ $(\varnothing \neq K, L \subset \{1, 2, \ldots, n\}, K \cap L = \varnothing)$.

Proof. Assertion (ii) clearly implies (i). Assume that (i) holds, and let A be an irreducible matrix in $\mathcal{A}(R,S)$. Let K and L be disjoint, nonempty subsets of $\{1, 2, \ldots, n\}$. Suppose that $t_{K,L} = 0$. Let $k = |K|$ and $l = |L|$ so that $k > 0, l > 0$, and $k + l \leq n$. Then, recalling the combinatorial meaning of $t_{K,L}$ and using the disjointness of K and L, we see that there exists a permutation matrix P such that

$$PAP^{\mathrm{T}} = \left[\begin{array}{c|cc} X & * & J_{k,l} \\ \hline O_{n-k,k} & O_{n-k,n-(k+l)} & * \end{array}\right].$$

The matrix PAP^{T} is reducible and thus so is A. This contradiction shows that (i) implies (iv).

Now assume that (iv) holds, and let k be an integer with $1 \le k \le n-1$. Suppose that

$$\sum_{i=1}^{k} r_i \ge \sum_{i=1}^{k} s_i + \sum_{i=k+1}^{n} \min\{k, s_i\}. \tag{4.15}$$

Let A be a matrix in $\mathcal{A}(R,S)$. Then in (4.15) equality holds and $s_i \le k$ $(i = 1, 2, \ldots, k)$. If $s_i \le k$ for $i = k+1, k+2, \ldots, n$, then (4.15) would imply that $\sum_{i=1}^{k} r_i = \sum_{i=1}^{n} s_i$ and hence $r_n = 0$, contradicting our assumption that R is a positive vector. Thus the set $L = \{i : s_i > k, k+1 \le i \le n\}$ is nonempty. Let $K = \{1, 2, \ldots, k\}$. Then (4.15) implies that $A[K, L] = J$ and $A(K, L) = O$. Therefore, $t_{K,L} = N_0(A[K, L]) + N_1(A(K, L)) = 0$. Since K and L are nonempty with $K \cap L = \varnothing$, we contradict (iv). Hence (iv) implies (iii).

Finally, we assume that (iii) holds and deduce (ii), completing the proof of the theorem. Note that the monotonicity assumption on R implies that (iii) holds for every set of k components of R. Let A be a matrix in $\mathcal{A}(R,S)$ of minimal trace $\tilde{\sigma}(R,S)$ having the smallest number t of irreducible components. Our goal is to show that $t = 1$. Suppose to the contrary that $t > 1$. Let Q be a permutation matrix such that

$$QAQ^{\mathrm{T}} = \begin{bmatrix} A_1 & A_{12} & \cdots & A_{1t} \\ O & A_2 & \cdots & A_{2t} \\ \vdots & \vdots & \ddots & \vdots \\ O & O & \cdots & A_t \end{bmatrix}, \tag{4.16}$$

where A_i is an irreducible matrix of order n_i $(i = 1, 2, \ldots, t)$. In that which follows, the irreducible components $A_1, A_2, \ldots, A_t$ of A which are zero matrices of order 1 will be called *trivial*, and all others will be called *nontrivial*. Since R and S are positive vectors, A_1 and A_t are nontrivial components. It follows from Lemma 4.2.1 that the following property holds:

(*) if A_i and A_j are nontrivial components with $i < j$, then $A_{ij} = J$.

Therefore, if A does not have any trivial components, QAQ^{T} has the form

$$\begin{bmatrix} A_1 & J \\ O & A' \end{bmatrix},$$

contradicting (iii) with $k = n_1$. Hence A has at least one trivial component. Choose $p \geq 1$ so that $A_1, \ldots, A_p$ are nontrivial and A_{p+1} is trivial. By (*), $A_{ij} = J$ for i and j satisfying $1 \leq i < j \leq p$. Let

$$
X = \begin{bmatrix}
A_{1,p+1} & A_{1,p+2} & \cdots & A_{1t} \\
A_{2,p+1} & A_{2,p+2} & \cdots & A_{2t} \\
\vdots & \vdots & \ddots & \vdots \\
A_{p,p+1} & A_{p,p+2} & \cdots & A_{pt}
\end{bmatrix},
$$

the matrix in the upper right corner of QAQ^{T}. If $X = J$, then we contradict (iii) with $k = n_1 + \cdots + n_p$. Consider the first column of X that contains a 0. This 0 must lie in a column of A that contains a trivial component A_u, for otherwise we may apply Lemma 4.2.1 and contradict the minimality of t. Let the row of A containing this 0 meet the component A_r. We have $p < u < t$ and $1 \leq r \leq p$. Suppose that A_{lu} contains a 1 for some l with $p < l < u$. Then it follows from Lemma 4.2.2 that we may apply an interchange which decreases the number of components and does not increase the trace, contradicting the minimality assumption on t. Hence $A_{lu} = O$ for each l with $p < l < u$. This implies that we may assume that $u = p + 1$. In a similar way, we see that we may assume that all of the columns of X that contain a 0 lie in columns of A that contain trivial components and that these columns precede the columns of X that contain only 1's. Thus QAQ^{T} has the form

$$
\left[
\begin{array}{ccccc|c|c}
A_1 & J & \cdots & J & & & \\
O & A_2 & \cdots & J & & & \\
\vdots & \vdots & \ddots & \vdots & & E_1 & X_1 \\
O & O & \cdots & A_p & & & \\
\hline
& & O & & & B_1 & B_{12} \\
\hline
& & O & & & O & B_2
\end{array}
\right]
$$

where B_1 is a matrix with 0's on and below the main diagonal and E_1 has a 0 in each column. We may assume that the zero rows of B_1 come last, since this can be achieved by simultaneous row and column permutations without changing the form of B_1.

Now consider the rows of B_1 that contain a 1 above the main diagonal. The intersection of these rows (in B_{12}) with the columns of nontrivial components contained in B_2 must contain only 1's; otherwise two interchanges decrease the number of irreducible components without increasing the trace. Thus the 0's in B_{12} occur in columns that contain trivial components. Arguing as above, we may assume that these trivial components come next.

Let C_1 be the submatrix of B_2 with these trivial components on its main diagonal. We may assume that the zero rows of C_1 come last. We now argue with C_1 as we did with B_1. But we also observe that the zero rows of B_1 and the columns of C_1 corresponding to the zero rows intersect in a zero matrix; otherwise a sequence of two interchanges decreases the number of fully indecomposable components without increasing the trace. We continue like this and eventually obtain a matrix which, after simultaneous row and column permutations, has the form

$$\begin{bmatrix} F & * & * & J_{u,l} \\ O & D & * & J_{v,l} \\ O & O & O_t & * \\ O & O & O & G \end{bmatrix}$$

where F is a matrix of order u, D is a matrix of order v with 0's on and below its main diagonal, O_t is a zero matrix of order t, and G is a matrix of order l. Now we contradict (iii) with $k = u + v$. Thus $t = 1$ and the proof is complete. □

Theorem 4.2.4 has some important corollaries.

Corollary 4.2.5 *Let* $n \geq 2$ *be an integer, and let* $R = (r_1, r_2, \ldots, r_n)$ *and* $S = (s_1, s_2, \ldots, s_n)$ *be positive integral vectors such that* $\mathcal{A}(R,S)$ *is nonempty. If all of the matrices in* $\mathcal{A}(R,S)$ *are reducible, then there are an integer* k *with* $1 \leq k \leq n - 1$ *and a permutation matrix* P *such that* PAP^{T} *has the form*

$$PAP^{\mathrm{T}} = \begin{bmatrix} A_1 & A_{12} \\ O & A_2 \end{bmatrix} \quad (A_1 \text{ of order } k) \tag{4.17}$$

for every matrix A *in* $\mathcal{A}(R,S)$.

Proof. Let P be a permutation matrix such that RP is a nonincreasing vector. By Theorem 4.2.4, there exists an integer k with $1 \leq k \leq n-1$ such that (iii) does not hold. Every matrix A in $\mathcal{A}(R,S)$ satisfies (4.17) for this k and P. □

If R and S are both nonincreasing, then the mere nonemptiness of $\mathcal{A}(R,S)$ guarantees the existence of an irreducible matrix.

Corollary 4.2.6 *Let* $R = (r_1, r_2, \ldots, r_n)$ *and* $S = (s_1, s_2, \ldots, s_n)$ *be nonincreasing, positive integral vectors such that* $\mathcal{A}(R,S)$ *is nonempty. Then there exists an irreducible matrix in* $\mathcal{A}(R,S)$.

Proof. Since each matrix of order 1 is irreducible, we assume that $n \geq 2$. Suppose (iii) of Theorem 4.2.4 does not hold. Then there is an integer k with $1 \leq k \leq n-1$ such that

$$\sum_{i=1}^{k} r_i = \sum_{i=1}^{k} s_i + \sum_{i=k+1}^{n} \min\{k, s_i\}. \tag{4.18}$$

Let A be a matrix in $\mathcal{A}(R,S)$. Then (4.18) implies that $s_i \leq k$, for $i = 1, 2, \ldots, k$, and hence since S is nonincreasing that $s_i \leq k$ for $i = 1, 2, \ldots, n$. Now (4.18) implies that $\sum_{i=1}^{k} r_i = \sum_{i=1}^{n} s_i$. Since $k < n$, we conclude that $r_n = 0$, contradicting the positivity of R. $\square$

We now characterize when $\mathcal{A}(RP, SQ)$ contains an irreducible matrix for all permutation matrices P and Q of order n. Since irreducibility is invariant under simultaneous row and column permutations, we may assume that $P = I_n$.

Corollary 4.2.7 *Let $R = (r_1, r_2, \ldots, r_n)$ and $S = (s_1, s_2, \ldots, s_n)$ be nonincreasing, positive integral vectors such that $\mathcal{A}(R,S)$ is nonempty. Let $T = [t_{kl}]$ be the structure matrix for R and S. The following are equivalent.*

(a) *There exists an irreducible matrix in $\mathcal{A}(R, SQ)$ for each permutation matrix Q of order n.*

(b) *There exists an irreducible matrix in $\mathcal{A}(R, \overrightarrow{S})$ where $\overrightarrow{S}$ is the nondecreasing rearrangement $(s_n, s_{n-1}, \ldots, s_1)$ of R.*

(c) *We have $t_{kl} > 0$ for all positive integers k and l with $k + l \leq n$.*

Proof. Clearly, (a) implies (b). First assume that (c) holds. Let K and L be disjoint, nonempty subsets of $\{1, 2, \ldots, n\}$ with $|K| = k$ and $|L| = l$. Then k and l are positive integers with $k + l \leq n$. Let Q be a permutation matrix and let $S' = SQ = (s_1', s_2', \ldots, s_n')$. Let

$$t'_{K,L} = |K||L| + \sum_{i \notin K} r_i - \sum_{j \in L} s_j'.$$

Then

$$t'_{K,L} \geq kl + \sum_{i=k+1}^{n} r_i - \sum_{j=1}^{l} s_j = t_{kl} > 0,$$

where equality holds if $S' = \overrightarrow{S}$, $K = \{1, 2, \ldots, k\}$, and $L = \{n - l + 1, n - l + 2, \ldots, n\}$. Now Theorem 4.2.4 implies that (a) and (b) hold. Thus (c) implies both (a) and (b). From the above calculation and Theorem 4.2.4 we also get that (b) implies (c), and hence the corollary holds. $\square$

In a similar way one proves the following corollary.

Corollary 4.2.8 *Let $R = (r_1, r_2, \ldots, r_n)$ and $S = (s_1, s_2, \ldots, s_n)$ be nonincreasing, positive integral vectors such that there exists a matrix in $\mathcal{A}(R,S)$ with trace equal to 0. Let $T = [t_{kl}]$ be the structure matrix for R and S, and let $\overrightarrow{S}$ be the nondecreasing rearrangement of S. Then there exists an irreducible matrix in $\mathcal{A}(R, \overrightarrow{S})$ with trace equal to 0 if and only if $t_{kl} > 0$ for all positive integers k and l with $k + l \leq n$.* $\square$

In Chapter 2 we characterized those row and column sum vectors R and S for which there exists a matrix in $\mathcal{A}(R,S)$ with zero trace (equivalently, for which the minimal trace $\tilde{\sigma}(R,S)$ equals 0). Such matrices correspond to digraphs without directed loops the indegree sequence of whose vertices is R and the outdegree sequence is S. The following theorem [9] is an immediate consequence of Theorem 4.2.4; the equivalence of (ii) and (iii) in the theorem is due to Beineke and Harary [5].

Theorem 4.2.9 *Let $n \geq 2$ be an integer, and let $R = (r_1, r_2, \ldots, r_n)$ and $S = (s_1, s_2, \ldots, s_n)$ be positive integral vectors such that R is nonincreasing and there exists a matrix in $\mathcal{A}(R,S)$ with trace equal to 0. Then the following assertions are equivalent.*

(i) *There exists an irreducible matrix in $\mathcal{A}(R,S)$ with zero trace.*

(ii) *There exists a strongly connected digraph with indegree sequence R and outdegree sequence S without any directed loops.*

(iii) $\sum_{i=1}^{k} r_i < \sum_{i=1}^{k} s_i + \sum_{i=k+1}^{n} \min\{k, s_i\}$ $(k = 1, 2, \ldots, n-1)$.

(iv) $t_{K,L} > 0$ $(\varnothing \neq K, L \subset \{1, 2, \ldots, n\}, K \cap L = \varnothing)$.

                                                               $\square$

4.3   Fully Indecomposable Matrices in $\mathcal{A}(R,S)$

We recall from Chapter 1 that a square matrix A of order n is fully indecomposable provided either $n = 1$ and A is not a zero matrix, or $n \geq 2$ and there do not exist permutation matrices P and Q such that

$$PAQ = \begin{bmatrix} A_1 & A_{12} \\ O & A_2 \end{bmatrix}$$

where A_1 and A_2 are square matrices of order at least 1. A matrix that is not fully indecomposable is partly decomposable. The term rank of a fully indecomposable matrix A of order n equals n, and hence there exist permutation matrices P' and Q' such that $P'AQ'$ has only nonzero elements on its main diagonal; in fact, by Theorem 1.3.3, any nonzero element of A can be brought to the main diagonal of such a $P'AQ'$ for some choice of permutation matrices P' and Q'. A matrix A of order n that satisfies this latter property, equivalently, for which every nonzero element belongs to a nonzero diagonal, is said to have total support. A zero matrix has total support. If a nonzero matrix A has total support, then there exist permutation matrices U and V such that UAV is a direct sum of fully indecomposable matrices. By Theorem 1.3.7, a square matrix with total support is fully indecomposable if and only the bipartite graph $\mathrm{BG}(A)$ determined by its nonzero elements is connected. Finally we note that a fully indecomposable matrix of order $n \geq 2$ has at least two nonzero elements in each row and column.

The connection between fully indecomposable matrices and irreducible matrices is provided by the following lemma [17, 19] which, for later use, we state in terms of $(0, 1)$-matrices.

Lemma 4.3.1 *Let A be a $(0, 1)$-matrix of n. Then A is fully indecomposable if and only if there exist permutation matrices P and Q such that PAQ has only 1's on its main diagonal and PAQ is irreducible.*

Proof. The lemma holds trivially for $n = 1$. Now let $n \geq 2$. Then A is fully indecomposable if and only if there do not exist nonempty subsets K and L of $\{1, 2, \ldots, n\}$ with $|K| + |L| = n$ such that $A[K, L] = O$. For irreducible, we have also to include the condition $K \cap L = \emptyset$ (thus $K \cup L = \{1, 2, \ldots, n\}$). The lemma now follows from the fact that a fully indecomposable matrix of order n has term rank equal to n and, if K and L are not disjoint, then a_{pp} is an element of $A[K, L]$ for each p in $K \cap L$.   □

We remark that the condition in the theorem can be replaced with the condition: PAQ has only 1's on its main diagonal and $PAQ - I_n$ is irreducible. This is because the property of being irreducible is independent of the elements on the main diagonal. The following theorem is from [9]. The assumption that R and S are nonincreasing is without loss of generality.

Theorem 4.3.2 *Let $n \geq 2$ be an integer, and let $R = (r_1, r_2, \ldots, r_n)$ and $S = (s_1, s_2, \ldots, s_n)$ be nonincreasing, positive integral vectors such that $\mathcal{A}(R, S)$ is nonempty. Let $T = [t_{kl}]$ be the structure matrix for R and S. There exists a fully indecomposable matrix in $\mathcal{A}(R, S)$ if and only if*

$$t_{kl} + k + l \geq n \quad (k, l = 0, 1, \ldots, n) \tag{4.19}$$

with equality only if $k = 0$ or $l = 0$.

Proof. Since $\mathcal{A}(R, S)$ is nonempty, (4.19) is automatically satisfied if $k + l \geq n$. First suppose A is a fully indecomposable matrix in $\mathcal{A}(R, S)$. Then $\rho(A) = n$ and hence the maximal term rank $\bar{\rho}(R, S) = n$. By Ryser's formula given in Theorem 3.5.4 the inequality (4.19) holds. Suppose there exist integers k and l with $1 \leq k, l \leq n$ such that

$$t_{kl} + k + l = n. \tag{4.20}$$

Then $k + l \leq n$, and we partition A as

$$\begin{bmatrix} A_1 & X \\ Y & A_2 \end{bmatrix}$$

where A_1 is a k by l matrix. Then (4.20) implies that

$$N_1(A_2) \leq N_0(A_1) + N_1(A_2) = t_{kl} = n - (k + l).$$

Let A' be the matrix of order $n-1$ obtained from A by deleting row and column 1. The nonzero elements of A' are contained in

$$(k-1) + (l-1) + N_1(A_2) \le n-2$$

rows and columns. Hence the term rank of A' satisfies $\rho(A') \le n-2$, and by Theorem 1.2.1 A' has a p by q zero submatrix with $p+q = (n-1)+1 = n$. This contradicts the full indecomposability of A. Therefore (4.19) holds.

Now suppose that (4.19) holds. By Theorem 3.5.4 again, the maximal term rank satisfies $\bar{\rho}(R,S) = n$. Let $\overrightarrow{S} = RP = (s_n, s_{n-1}, \ldots, s_1)$ be the nondecreasing rearrangement of S. It follows from Theorem 3.5.5 there exists a matrix B in $\mathcal{A}(R, \overrightarrow{S})$ with only 1's on the main diagonal. Let $U = (r_1-1, r_2-1, \ldots, r_n-1)$ and $V = (s_1-1, s_2-1, \ldots, s_n-1)$, and let $T' = [t'_{kl}]$ be the structure matrix for U and V. Let $\overrightarrow{V} = (s_n-1, s_{n-1}-1, \ldots, s_1-1)$ be the nondecreasing rearrangement of V. The matrix $B - I_n$ is a matrix in $\mathcal{A}(U, \overrightarrow{V})$ with zero trace. We calculate that

$$t'_{kl} = t_{kl} + k + l - n \quad (k, l = 0, 1, \ldots, n).$$

Hence $t'_{kl} \ge 0$ for all k and l and hence when $k+l \le n$. We now apply Corollary 4.2.8 and conclude that there is an irreducible matrix A in $\mathcal{A}(U, \overrightarrow{V})$ with trace zero. By Lemma 4.3.1, $A+I$ is a fully indecomposable matrix in $\mathcal{A}(R, \overrightarrow{S})$, and therefore $(A + I_n)P$ is a fully indecomposable matrix in $\mathcal{A}(R, S)$. $\square$

We now record some corollaries of Theorem 4.3.2. From Corollary 4.2.5 and the proof of Theorem 4.3.2, we obtain the following property of classes that do not contain a fully indecomposable matrix.

Corollary 4.3.3 *Let $n \ge 2$ be an integer, and let $R = (r_1, r_2, \ldots, r_n)$ and $S = (s_1, s_2, \ldots, s_n)$ be positive integral vectors such that $\mathcal{A}(R,S)$ is nonempty. If all the matrices in $\mathcal{A}(R,S)$ are partly decomposable, then there are an integer k with $1 \le k \le n-1$ and permutation matrices P and Q such that for every matrix A in $\mathcal{A}(R,S)$,*

$$PAQ = \begin{bmatrix} A_1 & A_{12} \\ O & A_2 \end{bmatrix} \quad (A_1 \text{ of order } k).$$

$\square$

In order that a nonzero matrix of order n be of total support, all row and column sums must be positive. The next corollary characterizes classes $\mathcal{A}(R,S)$ which contain a matrix of total support and is a consequence of Theorem 4.3.2.

Corollary 4.3.4 *Let $n \ge 2$ be an integer, and let $R = (r_1, r_2, \ldots, r_n)$ and $S = (s_1, s_2, \ldots, s_n)$ be nonincreasing, positive integral vectors. Let p and q be integers satisfying $r_1 \ge \cdots \ge r_p > 1 = r_{p+1} = \cdots = r_n$ and*

$s_1 \geq \cdots \geq s_q > 1 = s_{q+1} = \cdots = s_n$. *There exists a matrix in $\mathcal{A}(R,S)$ having total support if and only if $p = q$, and $R' = (r_1, r_2, \ldots, r_p)$ and $S' = (s_1, s_2, \ldots, s_p)$ are such that $\mathcal{A}(R',S')$ is nonempty and the structure matrix $T' = T(R',S') = [t'_{kl}]$ of order p satisfies*

$$t'_{kl} + k + l \geq p \quad (k, l = 0, 1, \ldots, p),$$

with equality only if $k = 0$ or $l = 0$. $\square$

We now characterize those classes $\mathcal{A}(R,S)$ which contain only fully indecomposable matrices. We require the following lemma.

Lemma 4.3.5 *Let $R = (r_1, r_2, \ldots, r_m)$ and $S = (s_1, s_2, \ldots, s_n)$ be nonincreasing, nonnegative integral vectors such that $\mathcal{A}(R,S)$ is nonempty. Let k and l be positive integers for which there exists a matrix in $\mathcal{A}(R,S)$ containing the zero matrix $O_{k,l}$ as a submatrix. Then there exists a matrix A in $\mathcal{A}(R,S)$ such that*

$$A[\{m-k+1, m-k+2, \ldots, m\}, \{n-l+1, n-l+2, \ldots, n\}] = O_{k,l}.$$

Proof. Let A be a matrix in $\mathcal{A}(R,S)$ for which there exists $K \subseteq \{1, 2, \ldots, m\}$ and $L \subseteq \{1, 2, \ldots, n\}$ with $|K| = k$, $|L| = l$, and $A[K, L] = O$. Let $C = [c_{ij}]$ be the $(0,1)$-matrix of size m by n such that $c_{ij} = 0$ if and only if $i \in K$ and $j \in L$. Let $C = [c'_{ij}]$ be the $(0,1)$-matrix of size m by n such that $c'_{ij} = 0$ if and only if $m - k + 1 \leq i \leq m$ and $n - l + 1 \leq j \leq n$. By Corollary 1.4.2,

$$\sum_{i \in I, j \in J} c_{ij} - \sum_{j \in J} s_j + \sum_{i \notin I} r_i \geq 0 \quad (I \subseteq \{1, 2, \ldots, m\}, J \subseteq \{1, 2, \ldots, n\}).$$

$$(4.21)$$

Let $I' \subseteq \{1, 2, \ldots, m\}$ and $J' \subseteq \{1, 2, \ldots, n\}$. There exist $I \subseteq \{1, 2, \ldots, m\}$ and $J \subseteq \{1, 2, \ldots, n\}$ such that $|I \cap K| = |I' \cap \{m-k+1, m-k+2, \ldots, m\}|$ and $|J \cap L| = |J' \cap \{n-l+1, n-l+2, \ldots, n\}|$ and hence

$$\sum_{i \in I', j \in J'} c'_{ij} = \sum_{i \in I, j \in J} c_{ij}.$$

Since R and S are nonincreasing, for this I and J we must have

$$\sum_{j \in J'} s_j \geq \sum_{j \in J} s_j \text{ and } \sum_{i \notin I'} r_i \leq \sum_{i \notin I} r_i.$$

Hence

$$\sum_{i \in I', j \in J'} c'_{ij} - \sum_{j \in J'} s_j + \sum_{i \notin I'} r_i \geq \sum_{i \in I, j \in J} c_{ij} - \sum_{j \in J} s_j + \sum_{i \notin I} r_i \geq 0$$

for all $I' \subseteq \{1, 2 \ldots, m\}$ and $J' \subseteq \{1, 2, \ldots, n\}$. By Corollary 1.4.2 again, there exists a matrix in $\mathcal{A}(R,S)$ whose lower right k by l submatrix is a zero matrix. $\square$

We recall the matrix $\Phi = \Phi(R, S) = [\phi_{kl}]$ of size $m+1$ by $n+1$ defined in Chapter 3 by

$$\phi_{kl} = \min\{t_{i_1, l+j_2} + t_{k+i_2, j_1} + (k-i_1)(l-j_1)\} \quad (0 \le k \le m, 0 \le l \le n)$$

where the minimum is taken over all integers i_1, i_2, j_1, j_2 such that

$$0 \le i_1 \le k \le k + i_2 \le m \text{ and } 0 \le j_1 \le l \le l + j_2 \le n.$$

This matrix was introduced in order to obtain Theorem 3.5.8. In the proof of that theorem we showed that there exists a matrix in $\mathcal{A}(R, S)$ whose lower right $m - e$ by $n - f$ submatrix is a zero matrix if and only if $\phi_{ef} \ge t_{ef}$.

Theorem 4.3.6 *Let $n \ge 2$ be an integer, and let $R = (r_1, r_2, \ldots, r_n)$ and $S = (s_1, s_2, \ldots, s_n)$ be nonincreasing, positive integral vectors such that $\mathcal{A}(R, S)$ is nonempty. Let $T = [t_{ij}]$ be the $m + 1$ by $n + 1$ structure matrix for R and S. Then every matrix in $\mathcal{A}(R, S)$ is fully indecomposable if and only if*

$$\phi_{n-k,k} < t_{n-k,k} \quad (k = 1, 2, \ldots, n).$$

Proof. Applying Lemma 4.3.5 we see that every matrix in $\mathcal{A}(R, S)$ is fully indecomposable if and only if there does not exist a matrix in $\mathcal{A}(R, S)$ which has a k by $n - k$ zero submatrix in its lower right corner for any $k = 1, 2 \ldots, n - 1$. By the discussion preceding the theorem, this holds if and only if $\phi_{n-k,k} < t_{n-k,k}$ for $k = 1, 2, \ldots, n$, and the theorem follows. $\square$

The final result in this section concerns the existence of an indecomposable matrix in a class $\mathcal{A}(R, S)$. Recall from Chapter 1 that an m by n matrix is indecomposable provided there do not exist permutation matrices P and Q such that $PAQ = A_1 \oplus A_2$ where A_1 is a k by l matrix and $0 < k + l < m + n$. An indecomposable $(0,1)$-matrix must have positive row and column sums. The property of a matrix being indecomposable is invariant under arbitrary row and column permutations.

Theorem 4.3.7 *Let $R = (r_1, r_2, \ldots, r_m)$ and $S = (s_1, s_2, \ldots, s_n)$ be nonincreasing, positive integral vectors such that $\mathcal{A}(R, S)$ is nonempty. There exists an indecomposable matrix in $\mathcal{A}(R, S)$ if and only if*

$$\sum_{i=1}^{m} r_i \ge m + n - 1. \tag{4.22}$$

Proof. If there is an indecomposable matrix in $\mathcal{A}(R, S)$, then the bipartite graph BG is connected. Since this graph has $m + n$ vertices, it must have at least $m + n - 1$ edges, that is, $\sum_{i=1}^{m} r_i \ge m + n - 1$ and (4.22) holds.

Now suppose that (4.22) holds. Consider a matrix A in $\mathcal{A}(R, S)$ whose bipartite graph BG(A) has the smallest number t of connected components.

If $t = 1$, then A is indecomposable. Suppose to the contrary that $t > 1$. Since R and S are positive vectors, there exist permutation matrices P and Q such that

$$PAQ = A_1 \oplus A_2 \oplus \cdots \oplus A_t$$

where A_i is a nonvacuous indecomposable matrix of size m_i by n_i ($i = 1, 2, \ldots, t$). Equation (4.22) implies that A_i contains at least $m_i + n_i$ 1's for at least one i. We assume that $i = 1$. The matrix A_1 contains at least one 1 whose replacement with 0 results in an indecomposable matrix. An interchange applied to $A_1 \oplus A_2$ involving such a 1 of A_1 and any 1 of A_2 results in a matrix B for which it is readily verified that $\mathrm{BG}(B)$ is connected. Hence B is indecomposable and $\mathcal{A}(R,S)$ contains a matrix A' for which $\mathrm{BG}(A')$ has $t - 1$ connected components, a contradiction. $\quad\square$

Some sufficient conditions for every matrix in a class $\mathcal{A}(R,S)$ to be indecomposable are given in Cho *et al.* [24].

4.4 $\mathcal{A}(R,S)$ and $\mathcal{Z}^+(R,S)$ with Restricted Positions

In this section we focus on the existence of matrices in a class $\mathcal{A}(R,S)$ or, more generally, $\mathcal{Z}^+(R,S)$ in which certain restrictions have been placed on the elements of the matrix. In previous chapters we have already given two results of this type. Under the assumption that R and S are nonincreasing vectors, Theorem 2.1.5 gives necessary and sufficient conditions for the existence of a matrix in $\mathcal{A}(R,S)$ each of whose elements on the main diagonal equals 0. In Theorem 3.5.3 it is shown that if $\mathcal{A}(R,S)$ and $\mathcal{A}(R',S')$ are both nonempty where the components of R' and S' are equal to or one less than the corresponding components of R and S, respectively, then there exists a matrix A in $\mathcal{A}(R,S)$ that "contains" a matrix in $\mathcal{A}(R',S')$. In the first theorem local restrictions are placed on a matrix in $\mathcal{A}(R,S)$; in the second theorem there are global restrictions. In this section we shall generalize both of these theorems.

Corollary 1.4.2 (also see below) gives a general existence theorem with local restrictions (lower[4] and upper bounds) on each element. The number of inequalities that need to be verified is exponential and equals 2^{m+n}. Anstee [2] generalized Theorem 2.1.5 and showed n inequalities suffice for the existence of a matrix in $\mathcal{A}(R,S)$ where at most one element in each column is restricted to be 0. This theorem was further generalized by Chen [22] and Nam [41]. We shall prove a special case of Nam's theorem in its general form for nonnegative integral matrices.

The following lemma of Nam [41] shows that the 2^{m+n} inequalities in Corollary 1.4.2 are equivalent to 2^n inequalities. In the theorem that follows

---

[4] In Corollary 1.4.2 the lower bounds all equal 0. It is a routine matter to replace the 0's by integers l_{ij} satisfying $0 \le l_{ij} \le c_{ij}$ ($i = 1, 2, \ldots, m; j = 1, 2, \ldots, n$).

we show that with certain restrictions on the upper bounds c_{ij}, the number of inequalities that need to be checked can be reduced to a linear number.

If $C = [c_{ij}]$ is an m by n nonnegative integral matrix, and R and S are nonnegative integral vectors of sizes m and n, respectively, then we let

$$\mathcal{Z}_C^+(R,S) = \{A : A \in \mathcal{Z}^+(R,S), A \le C\}$$

be the set of all nonnegative integral matrices with row and column sum vector R and S, respectively, whose elements are bounded by the corresponding elements of C. If C is the matrix $J_{m,n}$ of all 1's, then $\mathcal{Z}_{J_{m,n}}^+(R,S) = \mathcal{A}(R,S)$. If C is an arbitrary m by n (0,1)-matrix, then $\mathcal{Z}_C^+(R,S)$ is the set $\mathcal{A}_C(R,S)$ of matrices in $\mathcal{A}(R,S)$ having 0's in all those positions in which C has 0's.

Lemma 4.4.1 *Let $C = [c_{ij}]$ be a nonnegative integral matrix of size m by n. Let $R = (r_1, r_2, \ldots, r_m)$ and $S = (s_1, s_2, \ldots, s_n)$ be nonnegative integral vectors satisfying*

$$r_1 + r_2 + \cdots + r_m = s_1 + s_2 + \cdots + s_n.$$

There exists a matrix $A = [a_{ij}]$ in $\mathcal{Z}_C^+(R,S)$ if and only if

$$\sum_{i=1}^{m} \left(r_i - \sum_{j \in J} c_{ij} \right)^+ \le \sum_{j \notin J} s_j \quad (J \subseteq \{1, 2, \ldots, n\}). \qquad (4.23)$$

Proof. By Corollary 1.4.2 there exists a matrix $A = [a_{ij}]$ in $\mathcal{Z}_C^+(R,S)$ if and only if for all $I \subseteq \{1, 2, \ldots, m\}$ and $J \subseteq \{1, 2, \ldots, n\}$,

$$\sum_{i \in I} \sum_{j \in J} c_{ij} \ge \sum_{i \in I} r_i - \sum_{j \notin J} s_j. \qquad (4.24)$$

First assume that A exists and thus that (4.24) holds. Let $J \subseteq \{1, 2, \ldots, n\}$. Then with $I = \{i : r_i \ge \sum_{j \in J} c_{ij}\}$, we have

$$\sum_{i=1}^{m} \left(r_i - \sum_{j \in J} c_{ij} \right)^+ = \sum_{i \in I} r_i - \sum_{i \in I} \sum_{j \in J} c_{ij}$$
$$\le \sum_{j \notin J} s_j,$$

and (4.23) holds.

Now assume that (4.23) holds. Then for all $I \subseteq \{1, 2, \ldots, m\}$ and $J \subseteq \{1, 2, \ldots, n\}$, we have

$$\sum_{i \in I} r_i - \sum_{i \in I} \sum_{j \in J} c_{ij} \le \sum_{i=1}^{m} \left(r_i - \sum_{j \in J} c_{ij} \right)^+$$
$$\le \sum_{j \notin J} s_j$$

and (4.24) holds. $\qquad \square$

The following theorem[5] is due to Nam [41].

Theorem 4.4.2 *Let* $C = [c_{ij}]$ *be a nonnegative integral matrix of size* m *by* n *satisfying*

$$\sum_{i=1}^{m} (c_{ij} - c_{ik})^+ \leq s_j - s_k + 1 \quad (1 \leq j < k \leq n). \qquad (4.25)$$

Let $R = (r_1, r_2, \ldots, r_m)$ *and* $S = (s_1, s_2, \ldots, s_n)$ *be nonnegative integral vectors satisfying*

$$r_1 + r_2 + \cdots + r_m = s_1 + s_2 + \cdots + s_n. \qquad (4.26)$$

There exists a matrix $A = [a_{ij}]$ *in* $\mathcal{Z}_C^+(R, S)$ *if and only if*

$$\sum_{i=1}^{m} \left(r_i - \sum_{j=1}^{h} c_{ij} \right)^+ \leq \sum_{j=h+1}^{n} s_j \quad (h = 1, 2, \ldots, n). \qquad (4.27)$$

Proof. First we remark that we have not assumed that S is a nonincreasing vector. However, if for some j and k with $j < k$ we have $s_j < s_k$ then $s_j - s_k + 1 \leq 0$, and in order for (4.25) to be satisfied, $s_j = s_k - 1$. Thus S is nearly nonincreasing. If the desired matrix A exists, then by Lemma 4.4.1, (4.27) holds. We now assume that (4.27) holds and prove the existence of the matrix A.

Suppose to the contrary that $\mathcal{Z}_C^+(R, S)$ is empty. Taking $h = n$ in (4.27), we obtain

$$\sum_{i=1}^{m} \left(r_i - \sum_{j=1}^{n} c_{ij} \right)^+ \leq 0$$

and hence that

$$\sum_{j=1}^{n} c_{ij} \geq r_i \quad (i = 1, 2, \ldots, n).$$

Hence there exists a nonnegative integral matrix $A = [a_{ij}]$ of size m by n such that A has row sum vector R and $A \leq C$. Let $S' = (s'_1, s'_2, \ldots, s'_n)$ be the column sum vector of A'. If $S' = S$, then A is in $\mathcal{Z}_C^+(R, S)$. Otherwise, there is an integer k with $1 \leq k \leq n$ such that $s'_j = s_j$ for $j < k$ and $s'_k \neq s_k$. We choose A, firstly, so that k is as large as possible and, secondly, so that $|s'_k - s_k|$ is as small as possible. We distinguish two cases according as $s'_k > s_k$ or $s'_k < s_k$.

---

[5]In [41], a proof is given for a more general theorem. We are indebted to Yunsun Nam for providing us with a proof of this special case.

Case 1: $s_k' > s_k$. It follows from (4.26) that there exists an integer $l > k$ such that $s_l' < s_l$. Let

$$I = \{i : a_{ik} > 0, 1 \le i \le m\}.$$

We have

$$a_{il} = c_{il} \quad (i \in I),$$

for otherwise for some i in I we could replace a_{ik} by $a_{ik} - 1$ and a_{il} by $a_{il} + 1$ and contradict the choice of A. Hence

$$\begin{aligned}
\sum_{i=1}^{m} (c_{ik} - c_{il})^+ &\ge \sum_{i \in I} (c_{ik} - c_{il}) \\
&\ge \sum_{i \in I} (a_{ik} - a_{il}) \\
&\ge s_k' - s_l' \\
&\ge s_k + 1 - (s_l - 1) \\
&= s_k - s_l + 2,
\end{aligned}$$

contradicting (4.25).

Case 2: $s_k' < s_k$. Let i be an integer with $1 \le i \le m$. Column k is called *shiftable from row* i provided there exists in A a sequence of an odd number of distinct positions

$$(i_1, j_1), (i_1, j_2), (i_2, j_2), \ldots, (i_{t-1}, j_{t-1}), (i_{t-1}, j_t), (i_t, j_t) \qquad (4.28)$$

such that $j_1 = k$ and $i_t = i$, and

$$a_{i_p j_p} < c_{i_p j_p} \quad (p = 1, 2, \ldots, t) \text{ and } a_{i_p j_{p+1}} > 0 \quad (p = 1, 2, \ldots, t-1).$$

Let

$$I = \{i : \text{column } k \text{ is not shiftable from row } i \text{ in } A, 1 \le i \le m\},$$

and

$$\bar{I} = \{i : \text{column } k \text{ is shiftable from row } i \text{ in } A, 1 \le i \le m\}.$$

By taking $t = 1$ in the definition of shiftable we see that

$$a_{ik} = c_{ik} \quad (i \in I). \qquad (4.29)$$

Suppose that $a_{il} > 0$ for some $i \in \bar{I}$ and $l > k$. Since column k is shiftable from row i, then tacking on the position (i_t, j_{t+1}) to (4.28) where $i_t = i$ and $j_{t+1} = l$, and adding and subtracting 1 at these positions in A, we increase column sum k of A by 1 without changing the other chosen properties of A, and hence contradict the choice of A. We conclude that

$$a_{il} = 0 \quad (i \in \bar{I} \text{ and } l > k). \qquad (4.30)$$

Now suppose that

$$a_{pq} < c_{pq}, \quad \text{for some } p \text{ in } I \text{ and } q < k. \tag{4.31}$$

Since p is in I, it follows that

$$a_{iq} = 0, \quad \text{for all } i \text{ in } \overline{I}; \tag{4.32}$$

otherwise, column k is shiftable from row i and hence from row p, contradicting the fact that p is in I. Calculating we obtain

$$
\begin{aligned}
\sum_{i=1}^{m} (c_{iq} - c_{ik})^+ &\geq \sum_{i \in I} (c_{iq} - c_{ik}) \\
&\geq \sum_{i \in I} (a_{iq} - a_{ik}) + 1 \text{ (by (4.29) and (4.31))} \\
&\geq s'_q - s'_k + 1 \text{ (by (4.32))} \\
&> s_q - s_k + 1,
\end{aligned}
$$

contradicting (4.25). This contradiction and (4.29) now imply

$$a_{ij} = c_{ij} \quad (i \in I, j \leq k). \tag{4.33}$$

Calculating again, we obtain

$$
\begin{aligned}
\sum_{i=1}^{m} \left(r_i - \sum_{j=1}^{k} c_{ij} \right)^+ &\geq \sum_{i \in I} \left(r_i - \sum_{j=1}^{k} c_{ij} \right) \\
&= \sum_{i \in I} \left(r_i - \sum_{j=1}^{k} a_{ij} \right) \text{ (by (4.33))} \\
&= \sum_{j=k+1}^{n} s'_j \text{ (by (4.30))} \\
&> \sum_{j=k+1}^{n} s_j.
\end{aligned}
$$

The last inequality follows from the facts that

$$\sum_{j=1}^{n} s_j = \sum_{i=1}^{m} r_i = \sum_{j=1}^{n} s'_j$$

and

$$s'_j = s_j \quad (j = 1, 2, \ldots, k), \quad \text{and } s'_k < s_k.$$

We have thus contradicted the inequalities (4.27) with $h = k$ and the proof is now complete. $\square$

Chen [22] showed that the inequalities (4.27) are also necessary and sufficient for the existence of a matrix in $\mathcal{Z}_C^+(R, S)$ if the assumption (4.25) is replaced by the assumption

$$m\Delta_C - \sum_{i=1}^m c_{ik} \le s_j - s_k + 1 \quad (1 \le j < k \le n), \tag{4.34}$$

where Δ_C is the largest element of C. In this condition each column of C is required to be close to the constant column vector of size m each of whose elements equals Δ_C. The conditions (4.25) only require that corresponding elements in columns be close to one another, and hence (4.25) is less restrictive than (4.34).

Theorem 4.4.2 contains as a special case a theorem of Anstee [2] which in turn is a generalization of Theorem 2.1.5. We call a $(0, 1)$-matrix C *column-restricted* provided it has at most one 0 in each column. The number of 0's in a row is unrestricted. The complement of a permutation matrix is a column-restricted matrix as is the matrix

$$\begin{bmatrix} 1 & 0 & 0 & 1 & 0 & 1 \\ 0 & 1 & 1 & 1 & 1 & 1 \\ 1 & 1 & 1 & 0 & 1 & 1 \\ 1 & 1 & 1 & 1 & 1 & 1 \end{bmatrix}. \tag{4.35}$$

If $S = (s_1, s_2, \ldots, s_n)$ is a nondecreasing, nonnegative integral vector and $C = [c_{ij}]$ is a column-restricted $(0, 1)$-matrix of size m by n, then C satisfies the inequalities (4.25) in Theorem 4.4.2.

Let $C = [c_{ij}]$ be a column-restricted $(0, 1)$-matrix of size m by n, and let $(r_1', r_2', \ldots, r_m')$ be the row sum vector of C. Let $R = (r_1, r_2, \ldots, r_m)$ and $S = (s_1, s_2, \ldots, s_n)$ be nonnegative integral vectors where $r_i \le r_i'$ $(i = 1, 2, \ldots, m)$, and S is nonincreasing. Consider the class $\mathcal{A}_C(R, S)$ of all $(0, 1)$-matrices A which have 0's in at least those positions in which C has 0's. Let $A_C = A_C(R, n)$ be the m by n $(0, 1)$-matrix with row sum vector R whose 1's are as far to the left as possible subject to the condition that $A_C \le C$. The column sum vector $R_C^* = (r_1^*, r_2^*, \ldots, r_n^*)$ of A_C is the *C-restricted conjugate of R*. If $C = J_{m,n}$, then $R_{J_{m,n}}^* = R^*$, the (ordinary) conjugate of R.

Example. Let C be the column-restricted matrix in (4.35), and let $R = (2, 4, 4, 4)$. Then

$$A_C(R, 6) = \begin{bmatrix} 1 & 0 & 0 & 1 & 0 & 0 \\ 0 & 1 & 1 & 1 & 1 & 0 \\ 1 & 1 & 1 & 0 & 1 & 0 \\ 1 & 1 & 1 & 1 & 0 & 0 \end{bmatrix}$$

and $R_C^* = (3, 3, 3, 3, 2, 0)$.   □

We observe that for each integer $h = 0, 1, \ldots, n$,

$$\sum_{i=1}^{m}\left(r_i - \sum_{j=1}^{h} c_{ij}\right)^{+} = \sum_{j=h+1}^{n} r_i^*.$$

Thus in this case (4.27) is equivalent to

$$\sum_{j=h+1}^{n} r_i^* \leq \sum_{j=h+1}^{n} s_j \quad (h = 0, 1, \ldots, n-1).$$

This implies the following corollary due to Anstee [2].

Corollary 4.4.3 *Let $C = [c_{ij}]$ be a column-restricted $(0,1)$-matrix of size m by n. Let $R = (r_1, r_2, \ldots, r_m)$ and $S = (s_1, s_2, \ldots, s_n)$ be nonnegative integral vectors, with S nonincreasing, such that $r_1 + r_2 + \cdots + r_m = s_1 + s_2 + \cdots + s_n$. Let $R_C^* = (r_1^*, r_2^*, \ldots, r_n^*)$ be the C-restricted conjugate of R. Then $\mathcal{A}_C(R, S)$ is nonempty if and only if*

$$\sum_{j=1}^{h} s_j \leq \sum_{j=1}^{h} r_j^* \quad (h = 1, 2, \ldots, n). \tag{4.36}$$

$\square$

Corollary 4.4.3 can also be formulated in the following way.

Corollary 4.4.4 *Let $C = [c_{ij}]$ be a column-restricted $(0,1)$-matrix of size m by n with row sum vector $R_C = (r_1', r_2', \ldots, r_m')$ and column sum vector $S_C = (s_1', s_2', \ldots, s_n')$. Let $R = (r_1, r_2, \ldots, r_m)$ and $S = (s_1, s_2, \ldots, s_n)$ be nonnegative integral vectors, such that $r_1 + r_2 + \cdots + r_m = s_1 + s_2 + \cdots + s_n$, $R_C \leq R$, $S_C \leq S$, and $S - S_C$ is nonincreasing. Let $(R - R_C)_C^* = (r_1^*, r_2^*, \ldots, r_n^*)$ be the C-restricted conjugate of $R - R_C$. Then $\mathcal{A}_C(R, S)$ is nonempty if and only if*

$$\sum_{j=1}^{h} (s_j - s_j') \leq \sum_{j=1}^{h} r_j^* \quad (h = 1, 2, \ldots, n). \tag{4.37}$$

$\square$

The following corollary is also contained in [2].

Corollary 4.4.5 *Let R and S be nonnegative integral vectors of size n, and assume that R and S are nonincreasing. If there exists a matrix in $\mathcal{A}_{J_n - I_n}(R, S)$, then for every permutation matrix P of order n, there exists a matrix in $\mathcal{A}_{J_n - P}(R, S)$.*

Proof. We recall that the all 1's vector of size n is denoted by 1^n. Because R is nonincreasing, we have

$$\sum_{j=1}^{h} r_j^* \le \sum_{j=1}^{h} r_j^{\#} \quad (h = 1, 2, \ldots, n)$$

where $(r_1^*, r_2^*, \ldots, r_n^*)$ and $(r_1^{\#}, r_2^{\#}, \ldots, r_n^{\#})$ are the I_n-restricted conjugate and P-restricted conjugates of $R-1^n$ and $S-1^n$, respectively. The corollary now follows from Corollary 4.4.4. $\qquad\square$

Corollary 4.4.5 should be contrasted with Theorem 3.5.2 which implies that if for some permutation matrix P of order n, there is a matrix A in $\mathcal{A}(R,S)$ with $A \ge P$, then there exists a matrix B in $\mathcal{A}(R,S)$ with $A \ge E_n$, where E_n is the permutation matrix whose 1's lie on the back-diagonal. In this regard, McDougal [38] has investigated for each $t \le \bar{\rho}(R,S)$, the positions occupied by t 1's, no two from the same row or column, in matrices in $\mathcal{A}(R,S)$.

Corollary 4.4.3 contains the Gale–Ryser theorem as a special case. Moreover, by taking $C = [c_{ij}]$ to be the m by n $(0,1)$-matrix whose only 0's are $c_{ii} = 0$ for $i = 1, 2, \ldots, \min\{m, n\}$, we obtain a linear number of conditions for the existence of a $(0,1)$-matrix with zero trace and prescribed row and column sums. These conditions, first discovered by Fulkerson [27], are different from the exponential number given in Theorem 2.1.5.

We now extend the results on interchanges in Section 3.2 to classes $\mathcal{Z}_C^+(R,S)$. Let $D = [d_{ij}]$ be an integral matrix of size m by n. Let $X = \{x_1, x_2, \ldots, x_m\}$ and $Y = \{y_1, y_2, \ldots, y_n\}$ be disjoint sets of m and n elements, respectively. The *bipartite multidigraph* $\Gamma(D)$ has vertex set $X \cup Y$ and $|d_{ij}|$ arcs from x_i to y_j or y_j to x_i depending on whether $d_{ij} > 0$ or $d_{ij} < 0$. We regard two arcs in $\Gamma(D)$ with the same initial vertex and same terminal vertex as different for the purposes of our discussion below. Assume that the matrix D is *balanced*, that is, every row or column sum equals 0. Then the multidigraph $\Gamma(D)$ is *balanced*, meaning that the indegree of each vertex (the number of arcs entering the vertex) of $\Gamma(D)$ equals its outdegree (the number of arcs exiting the vertex). As in Lemma 3.2.1, the arcs of a balanced, bipartite multidigraph $\Gamma(D)$ can be partitioned into directed cycles of even length.

Now let A and B be matrices in $\mathcal{Z}_C^+(R,S)$. Then $D = A - B$ is a balanced matrix and $\Gamma(D)$ is a balanced bipartite multidigraph. There exist minimal balanced matrices $D_1, D_2, \ldots, D_q$, corresponding to directed cycles $\gamma_1, \gamma_2, \ldots, \gamma_q$ in $\Gamma(D)$, such that

$$A = B + D_1 + D_2 + \cdots + D_q$$

where $B + D_1 + D_2 + \cdots + D_k$ is in $Z_C^+(R,S)$ for each $k = 1, 2, \ldots, q$. Let the directed cycle γ_i corresponding to the minimal balanced matrix D_i be

$$x_{i_1}, y_{j_1}, x_{i_2}, y_{j_2}, \ldots, x_{i_p}, y_{j_p}, x_{i_1}$$

where $x_{i_1}, x_{i_2}, \ldots, x_{i_p}$ and $y_{j_1}, y_{j_2}, \ldots, y_{j_p}$ are all distinct. Then γ_i and D_i are called *chordless with respect to* C provided $c_{i_k j_l} = 0$ whenever $l \not\equiv k, k-1 \bmod p$. The following theorem is due to Anstee [4] (see also Chen [22]).

Theorem 4.4.6 *Let A and B be matrices in a class $\mathcal{Z}_C^+(R, S)$. Then there exist minimal balanced matrices $D_1, D_2, \ldots, D_t$ which are chordless with respect to C such that*

$$A = B + D_1 + D_2 + \cdots + D_t$$

and

$$B + D_1 + \cdots + D_i \in \mathcal{Z}_C^+(R, S) \quad (i = 1, 2, \ldots, t).$$

Proof. By arguing by induction it suffices to prove the theorem in the special case where $D = A - B$ is a minimal balanced matrix corresponding to a directed cycle γ. This special case we also argue by induction on the length of γ. If D is not chordless, then $D = D_1 + D_2$ where D_1 and D_2 are minimal balanced matrices corresponding to directed cycles of length strictly less than that of γ, and $A + D_1$ is in $\mathcal{Z}_C^+(R, S)$. The theorem now follows by induction. $\qquad\square$

If the matrix C has no zeros, then the chordless directed cycles have length 4, and hence correspond to the interchanges used in our study of $\mathcal{A}(R, S)$, namely, subtracting 1 from the two diagonal positions of a submatrix of order 2 and adding 1 to the other two positions, or vice versa. In particular, we get the following corollary of Theorem 4.4.6.

Corollary 4.4.7 *If A and B are matrices in the same nonempty class $\mathcal{Z}^+(R, S)$, then A can be gotten from B by a sequence of interchanges with all intermediary matrices in $\mathcal{Z}^+(R, S)$.* $\qquad\square$

Let A_1 and A_2 be $(0, 1)$-matrices in a column-restricted class $\mathcal{A}_C(R, S)$. Then A_1 can be obtained from A_2 by a *triangle interchange* provided A_1 differs from A_2 only in a submatrix of order 3 which, in A_1, has the form

$$P \begin{bmatrix} 1 & 0 & * \\ * & 1 & 0 \\ 0 & * & 1 \end{bmatrix} Q$$

and, in A_2, has the form

$$P \begin{bmatrix} 0 & 1 & * \\ * & 0 & 1 \\ 1 & * & 0 \end{bmatrix} Q$$

for some permutation matrices P and Q of order 3, or vice versa. In a column-restricted class, chordless directed cycles can only be of lengths 4 and 6. Theorem 4.4.6 immediately implies the following corollary [3].

Corollary 4.4.8 *Let $C = [c_{ij}]$ be a column-restricted $(0,1)$-matrix of size m by n. Let $R = (r_1, r_2, \ldots, r_m)$ and $S = (s_1, s_2, \ldots, s_n)$ be nonnegative integral vectors such that $\mathcal{A}_C(R,S)$ is nonempty. Then A can be obtained from B by a sequence of interchanges and triangle interchanges with all intermediary matrices in $\mathcal{A}_C(R,S)$.*   □

We now turn to a theorem of Anstee [3] which gives a simple characterization of those classes $\mathcal{A}(R,S)$, with R and S nonincreasing, that contain a triangle[6] of 0's. The emphasis here is on "simple characterization" as Corollary 1.4.2 provides a characterization that requires the checking of an exponential number of inequalities. We give a simple proof which is a natural extension of the proof of the Gale–Ryser theorem given in Section 2.1.

Let m and n be positive integers, and let q be an integer with $0 \le q \le \min\{m, n\}$. Let $L = L(m, n; q) = [l_{ij}]$ be the (0,1)-matrix of size m by n such that L has a triangle of 0's in its lower right corner with sides of size q:

$$l_{ij} = \begin{cases} 0, & \text{if } i + j \ge m + n - q + 1, \\ 1, & \text{otherwise.} \end{cases}$$

The row sum vector of L is $R_L = (a_1, a_2, \ldots, a_m)$ where $a_i = n - (i - (m - q))^+$ $(i = 1, 2, \ldots, m)$, and the column sum vector is $S_L = (b_1, b_2, \ldots, b_n)$ where $b_j = m - (j - (n - q))^+$ $(j = 1, 2, \ldots, n)$.

Example. If $m = 5, n = 7$, and $q = 3$, then

$$L(5, 7; 3) = \begin{bmatrix} 1 & 1 & 1 & 1 & 1 & 1 & 1 \\ 1 & 1 & 1 & 1 & 1 & 1 & 1 \\ 1 & 1 & 1 & 1 & 1 & 1 & 0 \\ 1 & 1 & 1 & 1 & 1 & 0 & 0 \\ 1 & 1 & 1 & 1 & 0 & 0 & 0 \end{bmatrix}.$$

The row sum vector of $L(5, 7; 3)$ is $(7, 7, 6, 5, 4)$ and its column sum vector is $(5, 5, 5, 5, 4, 3, 2)$. If $m = 5, n = 7$, and $q = 6$, then the row and column sum vectors of

$$L(5, 7; 6) = \begin{bmatrix} 1 & 1 & 1 & 1 & 1 & 0 & 0 \\ 1 & 1 & 1 & 1 & 0 & 0 & 0 \\ 1 & 1 & 1 & 0 & 0 & 0 & 0 \\ 1 & 1 & 0 & 0 & 0 & 0 & 0 \\ 1 & 0 & 0 & 0 & 0 & 0 & 0 \end{bmatrix}$$

are $(5, 4, 3, 2, 1)$ and $(5, 4, 3, 2, 1, 0, 0)$, respectively.   □

---

[6]This triangle may actually be a right trapezoid, but we will refer to it more simply as a triangle.

Theorem 4.4.9 *Let $R = (r_1, r_2, \ldots, r_m)$ and $S = (s_1, s_2, \ldots, s_n)$ be non-increasing, nonnegative integral vectors. Let q be an integer with $0 \leq q \leq \max\{m, n\}$ and consider the matrix $L(m, n; q)$. There exists a matrix A in $\mathcal{A}(R, S)$ with $A \leq L(m, n; q)$ if and only if*

$$S \preceq R^*, \ R \leq R_L \text{ (entrywise), and } S \leq S_L \text{ (entrywise).} \qquad (4.38)$$

Proof. The last two inequalities in (4.38) are clearly necessary conditions for the existence of a matrix $A \in \mathcal{A}(R, S)$ with $A \leq L(m, n; q)$. Since $r_1 \leq a_1 = n - (1 - (m - q))^+$ implies $r_1 \leq n$, it follows from Theorem 2.1.3 that the conditions (4.38) are also necessary conditions for the existence of a matrix $A \in \mathcal{A}(R, S)$ with $A \leq L(m, n; q)$. We prove the converse by contradiction. We choose a counterexample with $m + n$ minimal, and for that m and n with $r_1 + r_2 + \cdots + r_m$ minimal. The minimality implies that $r_m \geq 1$ and $s_n \geq 1$.

Case 1: $\sum_{i=1}^{l} s_i = \sum_{i=1}^{l} r_i^*$ for some l with $0 < l < n$. Let

$$R_1 = (r_1', \ldots, r_m') = (\min\{l, r_1\}, \ldots, \min\{l, r_m\}) \text{ and } S_1 = (s_1, \ldots, s_l),$$

and let

$$R_2 = (r_1 - r_1', \ldots r_m - r_m'), \text{ and } S_2 = (s_{l+1}, \ldots, s_n).$$

Also let L_1 be the m by l matrix formed by the first l columns of $L(m, n; q)$, and let L_2 be the m by $n - l$ matrix formed by the last $n - l$ columns. It follows from the minimality assumption that there exist matrices A_1 in $\mathcal{A}(R_1, S_1)$ with $A_1 \leq L_1$ and A_2 in $\mathcal{A}(R_2, S_2)$ with $A_2 \leq L_2$. The matrix

$$\begin{bmatrix} A_1 & A_2 \end{bmatrix}$$

has row sum vector R and column sum vector S, and satisfies $A \leq L(m, n; q)$, a contradiction.

Case 2: $\sum_{i=1}^{l} s_i < \sum_{i=1}^{l} r_i^*$ for all l with $0 < l < n$. Let

$$R' = (r_1, \ldots, r_{m-q-1}, r_{m-q}-1, r_{m-q+1}, \ldots, r_m), \ S' = (s_1, \ldots, s_{n-1}, s_n-1).$$

Then the assumption in this case implies that $S' \preceq R'^*$.[7] By minimality again there is an m by n matrix $A' = [a_{ij}']$ with row sum vector R', column sum vector S', satisfying $A' \leq L(m, n; q)$. If $a_{m-q,n}' = 0$, then changing this 0 to a 1 we obtain a matrix with row sum vector R and column sum vector S satisfying $A \leq L(m, n; q)$, a contradiction. Suppose that $a_{m-q,n}' = 1$. Since $s_n' = s_n - 1 \leq m - q - 1$, there is an i with $i \leq m - q$ such that $a_{in}' = 0$. Since $r_i' = r_i \geq r_{m-q} = r_{m-q}'$, there is a j such that $a_{ij}' = 1$ and $a_{m-q,j}' = 1$. Interchanging 0 with 1 in this 2 by 2 matrix gives a

---

[7]The vector R' is not necessarily nonincreasing but this doesn't affect the construction of its conjugate vector R'^*.

matrix A'' with row sum vector R' and column sum vector S'. Moreover, this interchange does not put a 1 in a position in which $L(m,n;q)$ has a 0. Hence $A'' \leq L(m,n;q)$, and we get a contradiction as before.   □

Example. [2] Without the monotonicity assumption, Theorem 4.4.9 may fail. For example, let $R = S = (0,1)$. Then the conditions given in Theorem 4.4.9 are satisfied with $m = n = 2$ and $q = 1$, but there does not exist a matrix in $\mathcal{A}(R,S)$ with $A \leq L(2,2;1)$.   □

The following corollary is also from [2]. Again we denote by E_n the permutation matrix of order n with 1's in positions $(1,n),(2,n-1),\ldots,(n,1)$. Let $L_n = L(n,n;n-1)$, the $(0,1)$-matrix of order n with 1's on and above the back-diagonal and 0's elsewhere.

Corollary 4.4.10 *Let $R = (r_1, r_2, \ldots, r_n)$ and $S = (s_1, s_2, \ldots, s_n)$ be nonincreasing, nonnegative integral vectors. Let $R' = (r_1 - 1, r_2 - 1, \ldots, r_n - 1)$ and $S' = (s_1 - 1, s_2 - 1, \ldots, s_n - 1)$. There exists a matrix A in $\mathcal{A}(R,S)$ such that*

$$E_n \leq A \leq L_n \qquad (4.39)$$

if and only if $\mathcal{A}(R', S')$ is nonempty, and $r_i, s_i \leq n - i + 1$ $(i = 1, 2, \ldots, n)$.

Proof. If there is a matrix A in $\mathcal{A}(R,S)$ satisfying (4.39), then since $A - E_n$ is in $\mathcal{A}(R', S')$ and $A - E_n \leq L_n - E_n$, the stated conditions are clearly satisfied. Now assume these conditions hold. The vectors R' and S' are nonincreasing and their components satisfy $r_i', s_i' \leq n - i$ $(i = 1, 2, \ldots, n)$. Hence by Theorem 4.4.9 there exists a matrix A' in $\mathcal{A}(R', S')$ with $A' \leq L_n - E_n$. The matrix $A = A' + E_n$ is in $\mathcal{A}(R,S)$ and satisfies (4.39).   □

We now generalize the matrices $L(m,n;q)$ to allow for more general restrictions on the positions of the 1's in a matrix and discuss some results of Brualdi and Dahl [11, 12]. Let $C = [c_{ij}]$ be an m by n $(0,1)$-matrix with the property that for all k and j with $1 \leq k < j \leq n$, either (i) column k and column j do not have 1's in the same row (their inner product is 0), or (ii) for each row in which column j has a 1, column k also has a 1 in that same row (the inner product of column k and column j equals the number of 1's in row j). Let

$$P_j = \{i : c_{ij} = 1\} \quad (1 \leq j \leq n),$$

the set of indices of those rows in which C has a 1 in column j. Thus conditions (i) and (ii) say that for $k < j$ either $P_k \cap P_j = \varnothing$ or $P_j \subseteq P_k$. Since we are not going to require any monotonicity assumptions on the row and column sum vectors R and S, no additional generality is obtained by replacing these conditions with the condition that for each k and j with $k \neq j$, either $P_k \cap P_j = \varnothing$, or $P_k \subseteq P_j$ or $P_j \subseteq P_k$. A special case occurs when (ii) always holds. Thus the matrix C for this special case can be specified by a partition $\lambda = (\lambda_1, \lambda_2, \ldots, \lambda_m)$ of a nonnegative integer

$\lambda_1 + \lambda_2 + \cdots + \lambda_m$ where $n \geq \lambda_1 \geq \lambda_2 \geq \cdots \geq \lambda_m \geq 0$ and $c_{ij} = 1$ if and only if $1 \leq j \leq \lambda_i$ $(i = 1, 2, \ldots, m)$. The 1's of this C are in the pattern of the boxes of a Young diagram. To display the dependency of C on this partition, we write C_λ in place of C, and we say that C is a *Young pattern*.[8] In general, we call C a *generalized Young pattern*.

Now let $R = (r_1, r_2, \ldots, r_m)$ and $S = (s_1, s_2, \ldots, s_n)$ be nonnegative integral vectors without any monotonicity assumptions, and let $C = [c_{ij}]$ be an m by n generalized Young pattern. The structure of C allows us to generalize certain basic results for $\mathcal{A}(R, S)$ to $\mathcal{A}_C(R, S)$.

We first show that there is a canonical matrix $\widetilde{A}_C$ in $\mathcal{A}_C(R, S)$ that equals the canonical matrix $\widetilde{A}$ in the special case that $C = J_{m,n}$.

Lemma 4.4.11 *Assume that $C = [c_{ij}]$ is a generalized Young pattern and that $\mathcal{A}_C(R, S)$ is nonempty. Then there exists a matrix in $\mathcal{A}_C(R, S)$ such that the 1's in column n appear in those rows allowed by C for which the corresponding component of R is largest, giving preference to the bottommost positions in case of ties.*

Proof. Let $A = [a_{ij}]$ be in $\mathcal{A}_C(R, S)$. Suppose there exist i and k such that $c_{in} = 1$, $a_{in} = 0$, $a_{kn} = 1$ (and so $c_{kn} = 1$), and $r_i > r_k$. Since $r_i > r_k$ there is an integer $j < n$ such that $a_{ij} = 1$ (and hence $c_{ij} = 1$) and $a_{kj} = 0$. Since C is a generalized Young pattern, we have $c_{kj} = 1$ also. Hence an interchange results in a matrix in $\mathcal{A}_C(R, S)$ which has a 1 in position (i, n). A similar argument works in case $r_i = r_k$, $a_{in} = 1$, $a_{kn} = 0$, $i < k$, and $c_{kn} = 1$. Repeating these arguments we obtain a matrix of the type specified in the theorem.   □

Let B be the matrix in the nonempty class $\mathcal{A}_C(R, S)$ whose existence is proved in Lemma 4.4.11. Deleting column n of B, we may apply the same argument to the resulting matrix. Continuing recursively, we obtain a matrix $\widetilde{A}_C$ such that the last column of the leading m by k submatrix of A_C has the property that its last column satisfies the properties in the lemma for each $k = 1, 2, \ldots, n$ with respect to its row sum vector. The matrix $\widetilde{A}_C$ is uniquely determined and is called the *canonical matrix* in the class $\mathcal{A}_\lambda(R, S)$.

It follows from Lemma 4.4.11 that the following algorithm, generalizing that of Ryser, constructs the canonical matrix in the class $\mathcal{A}_C(R, S)$ when the class is nonempty. If the class is empty, then the algorithm must fail.

Algorithm for Constructing a Matrix in $\mathcal{A}_C(R, S)$

Let $R = (r_1, r_2, \ldots, r_m)$ and $S = (s_1, s_2, \ldots, s_n)$ be nonnegative integral vectors, and let C be a generalized Young pattern.

---

[8]If $R = (r_1, r_2, \ldots, r_m)$ is a nonincreasing, nonnegative integral vector and $r_1 \leq n$, then the matrix $A(R; n)$ as defined in Chapter 3 is a Young pattern and every Young pattern arises in this way.

(1) Begin with the m by n matrix $A_n = A(R,n)$ with row sum vector R for which the 1's in each row are in the initial positions.

(2) For $k = n, n-1, n-2, \ldots, 1$, do:

   Shift to column k the final 1's in the s_k rows of A_k that are permitted by C to contain a 1 in column k and that have the largest sum, with preference given to the lowest rows in case of ties. This results in a matrix

$$\left[\, A_{k-1} \,\middle|\, \widetilde{A}_{C,n-k+1} \,\right].$$

(3) Output $\widetilde{A}_C = \widetilde{A}_{C,n}$.

The recursive application of the interchanges as applied in Lemma 4.4.11 implies that every matrix in $\mathcal{A}_C(R,S)$ can be transformed to the canonical matrix by a sequence of interchanges. Hence we get the following.

Corollary 4.4.12 *Let A and B be matrices in the class $\mathcal{A}_C(R,S)$. Then there is a sequence of interchanges that transforms A to B where every intermediary matrix also belongs to $\mathcal{A}_C(R,S)$.* □

We now consider the question of uniqueness of a matrix in $\mathcal{A}_C(R,S)$. Such questions are of importance in discrete tomography: if $\mathcal{A}_C(R,S)$ contains a unique matrix A, then this means that the $(0,1)$-matrix A is uniquely reconstructable from R, S, and the set of allowable positions for 1's as given by C.

To faciltate our discussion, we consider the bipartite digraph $\Gamma_C(A)$ associated with an m by n $(0,1)$-matrix $A = [a_{ij}]$ with vertices $\{x_1, x_2, \ldots, x_m\}$ and $\{y_1, y_2, \ldots, y_n\}$ in which there is an arc from x_i to y_j if $a_{ij} = 1$ (and so $c_{ij} = 1$) and an arc from y_j to x_i if $a_{ij} = 0$ and $c_{ij} = 1$. When $C = J_{m,n}$, this is the same digraph as considered in Section 3.4.

Theorem 4.4.13 *Let $C = [c_{ij}]$ be a generalized Young pattern, and let A be a matrix in $\mathcal{A}_C(R,S)$. The following are equivalent.*

(i) *A is the unique matrix in $\mathcal{A}_C(R,S)$.*

(ii) *The bipartite digraph $\Gamma_C(A)$ does not have any directed cycles.*

(iii) *The bipartite digraph $\Gamma_C(A)$ does not have any directed cycles of length 4.*

(iv) *Starting with A and recursively deleting rows and columns that have all 0's, or have all 1's, in positions in which C has 1's, one obtains an empty matrix.*

Proof. A directed cycle in $\Gamma_C(A)$ corresponds to an alternating sequence of 0's and 1's with the property that if these 1's and 0's are interchanged in A the result is a matrix A' in $\mathcal{A}_C(R,S)$ with $A' \neq A$. Conversely, if there is a matrix $A' \neq A$ in $\mathcal{A}_C(R,S)$, then $A - A'$ is a balanced nonzero matrix and it follows that $\Gamma_C(A)$ has a directed cycle. Thus (i) and (ii) are equivalent. In view of Corollary 4.4.12, (i) and (iii) are equivalent, since a directed cycle of length 4 in $\Gamma_C(A)$ corresponds to an interchange from A to a different matrix in $\mathcal{A}_C(R,S)$. In terms of the digraph $\Gamma_C(A)$, the procedure in (iv) recursively deletes vertices of indegree or outdegree zero. Such vertices cannot be in a directed cycle. If all vertices are deleted there is no directed cycle in $\Gamma_C(A)$. Otherwise, one determines a subdigraph in which each vertex has positive indegree and outdegree and hence contains a directed cycle. □

Note that in Theorem 4.4.13, for the equivalence of (i), (ii), and (iv) to hold, C may be any $(0,1)$-matrix. The recursive algorithm described in (iii) may be used in the reverse to determine if $\mathcal{A}_C(R,S)$ contains a unique matrix and to find that matrix. This algorithm may be described as follows.

Reconstruction Algorithm for Uniqueness in $\mathcal{A}_C(R,S)$

(1) Start with the matrix $A = [a_{ij}]$ such that $a_{ij} = 0$ for all (i,j) for which $c_{ij} = 0$, whose other entries are unspecified.

(2) Locate a line in A such that its line sum, 0's fixed by C, and previously determined entries, force all remaining entries in that line to be all 0's or all 1's.

(3) Update A and return to (2).

(4) If A is completely specified, then A is the unique matrix in $\mathcal{A}_C(R,S)$. Otherwise, $\mathcal{A}(R,S)$ either is empty or contains more than one matrix.

Further discussion of the classes $\mathcal{A}_C(R,S)$ can be found in [11], [12]. In [11] the special classes $\mathcal{A}_C(R,S)$, where C corresponds to a partition λ of the form $(m,m,\ldots,m,k,k,\ldots,k)$, are investigated. For these special classes, the analog of the structure matrix for $\mathcal{A}(R,S)$ is studied, and a different algorithm is given for construction of a matrix in $\mathcal{A}_C(R,S)$.

We conclude this section with a very different kind of restriction. Let $R = (r_1, r_2, \ldots, r_m)$ and $S = (s_1, s_2, \ldots, s_n)$ be nonnegative integral vectors of lengths m and n, respectively. Suppose that $R = R' + R''$ and $S = S' + S''$ where R', R'', S', S'' are nonnegative integral vectors. Then $\mathcal{A}(R,S)$ has an $(R', S'; R'', S'')$-*joint realization* provided there exist matrices A in $\mathcal{A}(R,S)$, A' in $\mathcal{A}(R',S')$, and A'' in $\mathcal{A}(R'',S'')$ such that $A = A' + A''$. Necessary conditions for such a joint realization are that each of the classes $\mathcal{A}(R,S)$, $\mathcal{A}(R',S')$, and $\mathcal{A}(R'',S'')$ is nonempty. In general, these conditions are not sufficient to guarantee a joint realization.

Example. [21] Let $R = (2,3,1)$ and $S = (3,1,2)$. Then $R = R' + R''$ where $R' = (2,1,0)$ and $R'' = (0,2,1)$, and $S = S' + S''$ where $S' = (2,1,0)$ and $S'' = (1,0,2)$. The unique matrices in the classes $\mathcal{A}(R,S)$, $\mathcal{A}(R',S')$, and $\mathcal{A}(R'',S'')$ are, respectively,

$$A = \begin{bmatrix} 1 & 0 & 1 \\ 1 & 1 & 1 \\ 1 & 0 & 0 \end{bmatrix}, A' = \begin{bmatrix} 1 & 1 & 0 \\ 1 & 0 & 0 \\ 0 & 0 & 0 \end{bmatrix}, \text{ and } A'' = \begin{bmatrix} 0 & 0 & 0 \\ 1 & 0 & 1 \\ 0 & 0 & 1 \end{bmatrix}.$$

Since $A \neq A' + A''$, $\mathcal{A}(R,S)$ does not have an $(R',S';R'',S'')$-joint realization.

Now let $R = S = (3,4,1,2)$. Then $R = R' + R''$ and $S = S' + S''$ where $R' = S' = (3,2,1,0)$ and $R'' = S'' = (0,2,0,2)$ Again, there are unique matrices A, A', and A'' in each of the three classes, namely,

$$\begin{bmatrix} 1 & 1 & 0 & 1 \\ 1 & 1 & 1 & 1 \\ 0 & 1 & 0 & 0 \\ 1 & 1 & 0 & 0 \end{bmatrix}, \begin{bmatrix} 1 & 1 & 1 & 0 \\ 1 & 1 & 0 & 0 \\ 1 & 0 & 0 & 0 \\ 0 & 0 & 0 & 0 \end{bmatrix}, \text{ and } \begin{bmatrix} 0 & 0 & 0 & 0 \\ 0 & 1 & 0 & 1 \\ 0 & 0 & 0 & 0 \\ 0 & 1 & 0 & 1 \end{bmatrix},$$

respectively. Since $A \neq A' + A''$, there is no joint realization. □

We now show that if the row sum vector R'' is nearly constant, then the necessary conditions for joint realization are sufficient. This is a theorem of Anstee [2] who generalized a theorem of Brualdi and Ross [18] in which it was also assumed that S'' is nearly constant. We follow a proof of Chen and Shastri [23].

Theorem 4.4.14 *Let $R = (r_1, r_2, \ldots, r_m)$ and $S = (s_1, s_2, \ldots, s_n)$ be nonnegative integral vectors. Let $(k_1, k_2, \ldots, k_m)$ be an integral vector such that for some nonnegative integer k, $k \leq k_i \leq k+1$ $(i = 1, 2, \ldots, m)$. Let $R' = (r_1 - k_1, r_2 - k_2, \ldots, r_m - k_m)$, and let $S' = (s'_1, s'_2, \ldots, s'_n)$ be a nonnegative integral vector such that $s'_j \leq s_j$ $(j = 1, 2, \ldots, n)$. Then $\mathcal{A}(R,S)$ has an $(R', S'; R - R', S - S')$-joint realization if and only if both $\mathcal{A}(R,S)$ and $\mathcal{A}(R',S')$ are nonempty.*

Proof. That the stated conditions are necessary is clear and so we turn to the sufficiency. Recall that the complement of a $(0,1)$-matrix X of size m by n is the matrix $X^c = J_{m,n} - X$.

Suppose that $\mathcal{A}(R,S)$ and $\mathcal{A}(R',S')$ are nonempty. If there are matrices A in $\mathcal{A}(R,S)$ and A' in $\mathcal{A}(R',S')$ such that $B = A^c + A'$ is a $(0,1)$-matrix, then $A = A' + B^c$ where B^c is in $\mathcal{A}(R - R', S - S')$, and thus $\mathcal{A}(R,S)$ has an $(R', S'; R - R', S - S')$-joint realization. So let A and A' be matrices in $\mathcal{A}(R,S)$ and $\mathcal{A}(R',S')$, respectively, such that $B = [b_{ij}] = A^c + A'$ has the fewest number of 2's. Suppose $b_{pk} = 2$. Then there exists a q such that $b_{qk} = 0$. Since the row sum vector of B is $(n - k_1, n - k_2, \ldots, n - k_m)$ where $k_i = k$ or $k + 1$, there exists a j such that $b_{pj} < b_{qj}$. Applying a

$(p,q:k,j)$-interchange to either A^c or A', we reduce the number of 2's in B. This contradiction completes the proof.   □

Joint realizability is also discussed in [36]. Chen and Shastri [23] have proved that when joint realization is not possible, certain restrictions apply.

Theorem 4.4.15 *Let R, S, R', S', R'', and S'' be nonnegative integral vectors such that $R = R' + R''$ and $S = S' + S''$. Assume that each of the classes $\mathcal{A}(R,S), \mathcal{A}(R', S')$, and $\mathcal{A}(R'', S'')$ is nonempty. If $\mathcal{A}(R,S)$ does not have an $(R', S'; R'', S'')$-realization, then there exist matrices A in $\mathcal{A}(R,S)$, A' in $\mathcal{A}(R', S')$, and A'' in $\mathcal{A}(R'', S'')$ such that, up to row and column permutations, each of A' and A'' contains*

$$
\begin{bmatrix} 1 & 1 & 0 \\ 1 & 0 & 0 \\ 0 & 0 & 0 \end{bmatrix} \quad or \quad \begin{bmatrix} 1 & 1 & 0 & 0 \\ 1 & 1 & 0 & 0 \\ 0 & 0 & 0 & 0 \\ 0 & 0 & 0 & 0 \end{bmatrix}
$$

as a submatrix, and A contains

$$
\begin{bmatrix} 1 & 1 & 1 \\ 1 & 0 & 0 \\ 1 & 0 & 0 \end{bmatrix} \quad or \quad \begin{bmatrix} 1 & 1 & 1 & 1 \\ 1 & 1 & 1 & 1 \\ 1 & 1 & 0 & 0 \\ 1 & 1 & 0 & 0 \end{bmatrix}
$$

as a submatrix.

In a subsequent chapter we shall consider symmetric matrices and tournament matrices in $\mathcal{A}(R,S)$ and other combinatorially defined classes.

We conclude this section with the following sufficient condition of McDougal [37] for a class $\mathcal{A}(R,S)$ of $(0,1)$-matrices of order n to contain an *asymmetric matrix* $A = [a_{ij}]$, that is, one satisfying $a_{ij}a_{ji} = 0$ $(i, j = 1, 2, \ldots, n)$.

Theorem 4.4.16 *Let $R = (r_1, r_2, \ldots, r_n)$ and $S = (s_1, s_2, \ldots, s_n)$ be nonnegative integral vectors such that $\mathcal{A}(R,S)$ is nonempty. Let k be an integer with $1 \le k \le n - 1$ such that $r_i + s_i = k - 1$ or k for each $i = 1, 2, \ldots, n$. Then $\mathcal{A}(R,S)$ contains an asymmetric matrix.*

4.5   The Bruhat Order on $\mathcal{A}(R,S)$

The Bruhat order on the set S_n of permutations of order n is defined by: if τ and π are permutations of $\{1, 2, \ldots, n\}$, then π is less than τ in the *Bruhat order*, denoted $\pi \preceq_B \tau$, provided either $\pi = \tau$ or π can be obtained from τ by a sequence of transformations on permutations of the form:

(*) if $i < j$ and $a_i > a_j$, then $a_1 \ldots a_i \ldots a_j \ldots a_n$ is replaced with $a_1 \ldots a_j \ldots a_i \ldots a_n$.

A transformation of this type reduces the number of inversions in a permutation. The identity permutation $12 \ldots n$ is the unique minimal element in the Bruhat order on S_n, and the unique maximal element is the permutation $n(n-1) \ldots 1$.

Each permutation $i_1 i_2 \ldots i_n$ of $\{1, 2, \ldots, n\}$ corresponds to a permutation matrix $P = [p_{kl}]$ of order n where $p_{kl} = 1$ if and only if $(k, l) = (t, i_t)$ for some t. The Bruhat order on the set $\mathcal{P}_n$ of permutation matrices of order n is defined by $P \preceq_{\mathrm{B}} Q$ if and only if the corresponding permutations π and τ satisfy $\pi \preceq_{\mathrm{B}} \tau$. Let

$$L_2 = \begin{bmatrix} 0 & 1 \\ 1 & 0 \end{bmatrix}.$$

In terms of permutation matrices, the transformation (*) replaces a submatrix of order 2 equal to L_2 by the identity matrix I_2 of order 2:

$$L_2 \to I_2.$$

Thus the transformation (*) in terms of permutation matrices is one of the two types of interchanges possible, and we call it an *inversion-reducing interchange*.

There are alternative, but equivalent, ways to define the Bruhat order on S_n. One of these is in terms of the *Gale order* (see e.g. [6]) on subsets of a fixed size of $\{1, 2, \ldots, n\}$. Let k be an integer with $1 \leq k \leq n$, and let $X = \{a_1, a_2, \ldots, a_k\}$ and $Y = \{b_1, b_2, \ldots, b_k\}$ be subsets of $\{1, 2, \ldots, n\}$ of size k where $a_1 < a_2 < \cdots < a_k$ and $b_1 < b_2 < \cdots < b_k$. In the Gale order, $X \leq_{\mathrm{G}} Y$ if and only if $a_1 \leq b_1, a_2 \leq b_2, \ldots, a_k \leq b_k$. For $\tau = i_1 i_2 \ldots i_n \in S_n$, let $\tau[k] = \{i_1, i_2, \ldots, i_k\}$.

Lemma 4.5.1 *Let τ and π be permutations in S_n, Then*

$$\tau \leq_{\mathrm{B}} \pi \text{ if and only if } \tau[k] \leq_{\mathrm{G}} \pi[k] \quad (k = 1, 2, \ldots, n).$$

$\square$

We prove this lemma in an equivalent form in terms of permutation matrices.

Let $A = [a_{ij}]$ be an m by n matrix, and let Σ_A denote the m by n matrix whose (k, l)-entry is

$$\sigma_{kl}(A) = \sum_{i=1}^{k} \sum_{j=1}^{l} a_{ij} \quad (1 \leq k \leq m; 1 \leq l \leq n),$$

the sum of the entries in the leading k by l submatrix of A. The following lemma, equivalent to Lemma 4.5.1, is classical. Our proof is taken from [35].

Lemma 4.5.2 *If $P = [p_{ij}]$ and $Q = [q_{ij}]$ are permutation matrices of order n, then $P \leq_{\mathrm{B}} Q$ if and only if $\Sigma_P \geq \Sigma_Q$ (entrywise order).*

Proof. If $P \leq_{\mathrm{B}} Q$, then it follows easily that $\Sigma_P \geq \Sigma_Q$. Now assume that $P \neq Q$ and $\Sigma_P \geq \Sigma_Q$. Let (k,l) be the lexicographically first position in which P and Q differ. Thus $p_{ij} = q_{ij}$ if $i < k$, $p_{kj} = q_{kj}$ if $j < l$, and $p_{kl} = 1$ and $q_{kl} = 0$; in particular, $\sigma_{kl}(P) > \sigma_{kl}(Q)$. Let (r,s) with $r + s$ maximal be such that $\sigma_{ij}(P) > \sigma_{ij}(Q)$ for all i and j with $k \leq i < r$ and $l \leq j < s$; the integers r and s are not necessarily unique. Suppose that the submatrix $P[\{k+1, k+2, \dots, r\}, \{l+1, l+2, \dots, s\}]$ is a zero matrix. The maximality of $r+s$ implies that $\sigma_{is}(P) = \sigma_{is}(Q)$ and $\sigma_{rj}(P) = \sigma_{rj}(Q)$ for some i and j with $k \leq i \leq r-1$ and $l \leq j \leq s-1$. But then

$$\sigma_{rs}(P) = \sigma_{is}(P) + \sigma_{rj}(P) - \sigma_{ij}(P)$$
$$< \sigma_{is}(Q) + \sigma_{rj}(Q) - \sigma_{ij}(Q) = \sigma_{rs}(Q),$$

a contradiction. Hence there exist u and v with $k+1 \leq u \leq r$ and $l+1 \leq v \leq s$ such that $p_{uv} = 1$. We may choose (u,v) with $u+v$ minimal implying that $p_{ij} = 0$ for all i and j with $k+1 \leq i \leq u$ and $l+1 \leq j \leq v$ with the exception of $i = u$ and $j = v$. Thus

$$P[\{k, k+1, \dots, u\}, \{l, l+1, \dots, v\}] = \begin{bmatrix} 1 & 0 & \dots & 0 & 0 \\ 0 & & & & 0 \\ \vdots & & O & & \vdots \\ 0 & & & & 0 \\ 0 & 0 & \dots & 0 & 1 \end{bmatrix}.$$

Let P' be the permutation matrix obtained from P by the $I_2 \to L_2$ interchange involving the two 1's in positions (k,l) and (u,v). We then have

$$\sigma_{ij}(P') = \begin{cases} \sigma_{ij}(P) - 1 \geq \sigma_{ij}(Q), & \text{if } k \leq i \leq u-1 \text{ and } l \leq j \leq v-1, \\ \sigma_{ij}(P) = \sigma_{ij}(Q), & \text{otherwise.} \end{cases}$$

Hence P' is a permutation matrix such that P is obtained from P' by an $L_2 \to I_2$ interchange with $P \prec_{\mathrm{B}} P' \preceq_{\mathrm{B}} Q$. Replacing P with P' we complete the proof by induction. $\square$

The Bruhat order on permutation matrices can be extended to matrix classes $\mathcal{A}(R,S)$ in two natural ways [15, 13]. Let A_1 and A_2 be matrices in $\mathcal{A}(R,S)$. Then A_1 precedes A_2 in the *Bruhat order*, written $A_1 \preceq_{\mathrm{B}} A_2$, provided, in the entrywise order, $\Sigma_{A_1} \geq \Sigma_{A_2}$. The matrix A_1 precedes A_2 in the *secondary Bruhat order*, written $A_1 \preceq_{\widehat{\mathrm{B}}} A_2$, provided A_1 can be obtained from A_2 by a sequence of inversion-reducing interchanges. It follows that the Bruhat order is a refinement of the secondary Bruhat order, that is, $A_1 \preceq_{\widehat{\mathrm{B}}} A_2$ implies that $A_1 \preceq_{\mathrm{B}} A_2$. By Lemma 4.5.2, the secondary Bruhat order agrees with the Bruhat order on the permutation matrices of order n, that is, on the class $\mathcal{A}(R,S)$ where $R = S = (1, 1, \dots, 1)$.

Since the Bruhat order is a refinement of the secondary Bruhat order, we immediately obtain the following.

Corollary 4.5.3 *Let A be a matrix in $\mathcal{A}(R,S)$ that is minimal in the Bruhat order. Then no submatrix of A of order 2 equals L_2.* $\square$

Example. Let $R = S = (2,2,2,2,2)$. Then both of

$$A = \begin{bmatrix} 1 & 1 & 0 & 0 & 0 \\ 1 & 1 & 0 & 0 & 0 \\ 0 & 0 & 1 & 1 & 0 \\ 0 & 0 & 1 & 0 & 1 \\ 0 & 0 & 0 & 1 & 1 \end{bmatrix}, \text{ with } \Sigma_A = \begin{bmatrix} 1 & 2 & 2 & 2 & 2 \\ 2 & 4 & 4 & 4 & 4 \\ 2 & 4 & 5 & 6 & 6 \\ 2 & 4 & 6 & 7 & 8 \\ 2 & 4 & 6 & 8 & 10 \end{bmatrix},$$

and

$$B = \begin{bmatrix} 1 & 1 & 0 & 0 & 0 \\ 1 & 0 & 1 & 0 & 0 \\ 0 & 1 & 1 & 0 & 0 \\ 0 & 0 & 0 & 1 & 1 \\ 0 & 0 & 0 & 1 & 1 \end{bmatrix}, \text{ with } \Sigma_B = \begin{bmatrix} 1 & 2 & 2 & 2 & 2 \\ 2 & 3 & 4 & 4 & 4 \\ 2 & 4 & 6 & 6 & 6 \\ 2 & 4 & 6 & 7 & 8 \\ 2 & 4 & 6 & 8 & 10 \end{bmatrix},$$

are minimal matrices of $\mathcal{A}(R,S)$ in the Bruhat order. $\square$

We now describe the cover relation for the secondary Bruhat order on a class $\mathcal{A}(R,S)$ [13]. If A_1 and A_2 are distinct matrices in $\mathcal{A}(R,S)$ with $A_1 \preceq_{\widehat{B}} A_2$, then A_2 *covers* A_1 in the secondary Bruhat order provided there does not exist a matrix C in $\mathcal{A}(R,S)$ with $C \neq A_1, A_2$ such that $A_1 \preceq_{\widehat{B}} C \preceq_{\widehat{B}} A_2$.

Theorem 4.5.4 *Let $A = [a_{ij}]$ be a matrix in $\mathcal{A}(R,S)$ where $A[\{i,j\}, \{k,l\}] = L_2$. Let A' be the matrix obtained from A by the $L_2 \to I_2$ interchange that replaces $A[\{i,j\},\{k,l\}] = L_2$ with I_2. Then A covers A' in the secondary Bruhat order on $\mathcal{A}(R,S)$ if and only if*

(i) $a_{pk} = a_{pl} \quad (i < p < j)$,

(ii) $a_{iq} = a_{jq} \quad (k < q < l)$,

(iii) $a_{pk} = 0$ *and* $a_{iq} = 0$ *imply* $a_{pq} = 0 \quad (i < p < j, k < q < l)$,

(iv) $a_{pk} = 1$ *and* $a_{iq} = 1$ *imply* $a_{pq} = 1 \quad (i < p < j, k < q < l)$.

Proof. We first show that if any of the conditions (i)–(iv) does not hold, then A does not cover A'. First suppose that $a_{pk} = 0$ and $a_{pl} = 1$ for some p with $i < p < j$. Then a sequence of two interchanges replaces A with A' where the matrix C resulting from the first interchange satisfies $A' \preceq_{\widehat{B}} C \preceq_{\widehat{B}} A$. Hence A does not cover A' in this case. A similar argument works when $a_{pk} = 1$ and $a_{pl} = 0$, and when $a_{iq} = 0$ and $a_{jq} = 1$, or $a_{iq} = 1$ and $a_{jq} = 0$ $(k < q < l)$.

Now assume (i) and (ii), and suppose that for some q with $k < q < l$, we have $a_{pk} = 0$, $a_{iq} = 0$ but $a_{pq} = 1$. In this case, a sequence of three interchanges replaces A with A' and we obtain distinct matrices C_1 and C_2, with neither A nor A' equal to C_1 or C_2, such that $A' \preceq_{\widehat{B}} C_1 \preceq_{\widehat{B}} C_2 \preceq_{\widehat{B}} A$. For example, and showing only the submatrix $A[\{i, i+1, \ldots, j\}, \{k, k+1, \ldots, l\}]$,

$$
\begin{bmatrix}
0 & 0 & 0 & 1 & 1 & 1 \\
0 & & & & & 0 \\
0 & 1 & & & & 0 \\
1 & & & & & 1 \\
1 & & & & & 1 \\
1 & 0 & 0 & 1 & 1 & 0
\end{bmatrix}
\rightarrow
\begin{bmatrix}
0 & 0 & 0 & 1 & 1 & 1 \\
0 & & & & & 0 \\
1 & 0 & & & & 0 \\
1 & & & & & 1 \\
1 & & & & & 1 \\
0 & 1 & 0 & 1 & 1 & 0
\end{bmatrix}
$$

$$
\rightarrow
\begin{bmatrix}
1 & 0 & 0 & 1 & 1 & 0 \\
0 & & & & & 0 \\
0 & 0 & & & & 1 \\
1 & & & & & 1 \\
1 & & & & & 1 \\
0 & 1 & 0 & 1 & 1 & 0
\end{bmatrix}
\rightarrow
\begin{bmatrix}
1 & 0 & 0 & 1 & 1 & 0 \\
0 & & & & & 0 \\
0 & 1 & & & & 0 \\
1 & & & & & 1 \\
1 & & & & & 1 \\
0 & 0 & 0 & 1 & 1 & 1
\end{bmatrix}
$$

where unspecified entries don't change. Hence A does not cover A' in this case. A similar argument works when $a_{pk} = 1$, $a_{iq} = 1$, and $a_{pq} = 0$ ($i < p < j, k < q < l$).

For the converse, suppose that (i)–(iv) hold. Since A and A' agree outside of the contiguous submatrix $A[\{i, i+1, \ldots, j\}, \{k, k+1, \ldots, l\}]$, and since the row and column sum vectors of $A[\{i, i+1, \ldots, j\}, \{k, k+1, \ldots, l\}]$, and $A'[\{i, i+1, \ldots, j\}, \{k, k+1, \ldots, l\}]$ are the same, it follows that $\sigma_{uv}(A) = \sigma_{uv}(A')$ for all u, v for which one of the following holds: (a) $u < i$, (b) $v < k$, (c) $u \geq j$ and (d) $v \geq l$.

We now claim that if there is a matrix $C \neq A, A'$ with $A' \preceq_{\widehat{B}} C \preceq_{\widehat{B}} A$, then every sequence of $L_2 \to I_2$ interchanges that transforms A into C takes place entirely within $A[\{i, i+1, \ldots, j\}, \{k, k+1, \ldots, l\}]$. Assume that a sequence of $L_2 \to I_2$ interchanges transforms A into C where at least one of these interchanges involves an entry of A outside of $A[\{i, i+1, \ldots, j\}, \{k, k+1, \ldots, l\}]$. First assume that one of these $L_2 \to I_2$ interchanges involves an entry a_{pq} where either $p < i$ or $q > l$. Choose such an entry with p minimal, and for this p with q maximal. Since A and A' agree outside of their submatrices determined by rows $i, i+1, \ldots, j$ and columns $k, k+1, \ldots, l$, either $a_{pq} = a'_{pq} = 0$ or $a_{pq} = a'_{pq} = 1$. First suppose that $a_{pq} = a'_{pq} = 0$. Then in our sequence of $L_2 \to I_2$ interchanges, $a_{pq} = 0$ changes to a 1 and then back to a 0. But, for our choice of p and q, the first change is impossible in $L_2 \to I_2$ interchanges. Now suppose that $a_{pq} = a'_{pq} = 1$. Then in our sequence of $L_2 \to I_2$ interchanges, $a_{pq} = 1$ changes to a 0 and then back to a 1. This second change is impossible for our choice of p and q. A similar contradiction results if $p > j$ or $q < k$, and our claim holds.

The conditions (i)–(iv) imply that with row and column permutations $A[\{i, i+1, \ldots, j\}, \{k, k+1, \ldots, l\}]$ can be brought to the form

$$\begin{bmatrix} J & J & X \\ J & L_2 & O \\ Y & O & O \end{bmatrix}, \tag{4.40}$$

where J designates a matrix of all 1's and where $A[\{i, j\}, \{k, l\}]$ is transformed into the displayed matrix L_2. Row and column permutations can change an $L_2 \to I_2$ interchange into an $I_2 \to L_2$ interchange, or vice versa, but otherwise do not affect the existence of interchanges in submatrices of order 2. In the class $\mathcal{A}(R'S'$ determined by the matrix (4.40), the J's represent invariant 1's and the O's represent invariant 0's. This implies that the only $L_2 \to I_2$ interchange in $A[\{i, i+1, \ldots, j\}, \{k, k+1, \ldots, l\}]$ that can change any of its corner entries is the corner one itself. Thus we conclude that the only sequence of $L_2 \to I_2$ interchanges that transforms A to A' is the single $L_2 \to I_2$ interchange involving all the corner entries of $A[\{i, i+1, \ldots, j\}, \{k, k+1, \ldots, l\}]$. Hence A covers A' in the secondary Bruhat order on $\mathcal{A}(R, S)$. $\square$

A characterization of the cover relation for the Bruhat order on $\mathcal{A}(R, S)$ is not known. Since the Bruhat order is a refinement of the secondary Bruhat order, it follows that if A covers A' in the Bruhat order but A does not cover A' in the secondary Bruhat order, then A and A' are incomparable in the secondary Bruhat order.

It was conjectured in [15] that a matrix in $\mathcal{A}(R, S)$ was minimal in the Bruhat order if it did not contain a submatrix equal to L_2, that is, if no inversion-reducing interchange was possible. Implicit was the conjecture that the secondary Bruhat order agrees with the Bruhat order on all classes $\mathcal{A}(R, S)$. This conjecture was shown to be false in [13] even in the case where R and S are constant vectors.

Example. Let $R = S = (3, 3, 3, 3, 3, 3)$. Consider the three matrices in $\mathcal{A}(R, S) = \mathcal{A}(6, 3)$ defined by

$$A = \begin{bmatrix} 1 & 0 & 0 & 0 & 1 & 1 \\ 1 & 0 & 1 & 1 & 0 & 0 \\ 1 & 1 & 0 & 1 & 0 & 0 \\ 0 & 0 & 0 & 1 & 1 & 1 \\ 0 & 1 & 1 & 0 & 1 & 0 \\ 0 & 1 & 1 & 0 & 0 & 1 \end{bmatrix}, \quad C = \begin{bmatrix} 0 & 0 & 0 & 1 & 1 & 1 \\ 1 & 0 & 1 & 1 & 0 & 0 \\ 1 & 1 & 0 & 1 & 0 & 0 \\ 1 & 0 & 0 & 0 & 1 & 1 \\ 0 & 1 & 1 & 0 & 1 & 0 \\ 0 & 1 & 1 & 0 & 0 & 1 \end{bmatrix},$$

and

$$D = \begin{bmatrix} 0 & 0 & 0 & 1 & 1 & 1 \\ 1 & 1 & 0 & 1 & 0 & 0 \\ 1 & 0 & 1 & 1 & 0 & 0 \\ 1 & 0 & 0 & 0 & 1 & 1 \\ 0 & 1 & 1 & 0 & 1 & 0 \\ 0 & 1 & 1 & 0 & 0 & 1 \end{bmatrix}.$$

By calculation we obtain that

$$
\Sigma_A = \begin{bmatrix} 1 & 1 & 1 & 1 & 2 & 3 \\ 2 & 2 & 3 & 4 & 5 & 6 \\ 3 & 4 & 5 & 7 & 8 & 9 \\ 3 & 4 & 5 & 8 & 10 & 12 \\ 3 & 5 & 7 & 10 & 13 & 15 \\ 3 & 6 & 9 & 12 & 15 & 18 \end{bmatrix}, \quad \Sigma_C = \begin{bmatrix} 0 & 0 & 0 & 1 & 2 & 3 \\ 1 & 1 & 2 & 4 & 5 & 6 \\ 2 & 3 & 4 & 7 & 8 & 9 \\ 3 & 4 & 5 & 8 & 10 & 12 \\ 3 & 5 & 7 & 10 & 13 & 15 \\ 3 & 6 & 9 & 12 & 15 & 18 \end{bmatrix},
$$

and

$$
\Sigma_D = \begin{bmatrix} 0 & 0 & 0 & 1 & 2 & 3 \\ 1 & 2 & 2 & 4 & 5 & 6 \\ 2 & 3 & 4 & 7 & 8 & 9 \\ 3 & 4 & 5 & 8 & 10 & 12 \\ 3 & 5 & 7 & 10 & 13 & 15 \\ 3 & 6 & 9 & 12 & 15 & 18 \end{bmatrix}.
$$

Thus

$$
\Sigma_A > \Sigma_D > \Sigma_C
$$

and hence

$$
A \prec_{\mathrm{B}} D \prec_{\mathrm{B}} C.
$$

In fact, it follows from Theorem 4.5.4 that C covers both D and A in the secondary Bruhat order. This implies that D and A are incomparable in the secondary Bruhat order. We conclude that the Bruhat order and secondary Bruhat order are different on $\mathcal{A}(6, 3)$.   □

Making use of three lemmas, we now show that the Bruhat order and secondary Bruhat order are identical on each of the classes $\mathcal{A}(n, 2)$ [13] generalizing the classical result on $\mathcal{A}(n, 1)$. For integers k and l with $k \leq l$, we use the notation $[k, l]$ to denote the set $\{k, k+1, \ldots, l\}$.

Lemma 4.5.5 *Let A and C be matrices in $\mathcal{A}(R, S)$ with $A \prec_{\mathrm{B}} C$, and let i and j be integers with $\sigma_{ij}(A) > \sigma_{ij}(C)$. Let s and t be integers with $s + t$ maximal such that*

$$
(r, c) \in [i, s-1] \times [j, t-1] \text{ implies } \sigma_{rc}(A) > \sigma_{rc}(C).
$$

Then there exists $(i_0, j_0) \in [i+1, s] \times [j+1, t]$ with $a_{i_0 j_0} = 1$.

Proof. By the maximality of $s + t$, there exist a $k \in [i, s-1]$ and an $l \in [j, t-1]$ such that $\sigma_{kt}(A) = \sigma_{kt}(C)$ and $\sigma_{sl}(A) = \sigma_{sl}(C)$. If $a_{rc} = 0$ for every $(r, c) \in [i+1, s] \times [j+1, t]$, then we calculate that

$$
\begin{aligned}
\sigma_{st}(A) &= \sigma_{sl}(A) + \sigma_{kt}(A) - \sigma_{kl}(A) \\
&< \sigma_{sl}(A) + \sigma_{kt}(A) - \sigma_{kl}(C) \\
&= \sigma_{sl}(C) + \sigma_{kt}(C) - \sigma_{kl}(C) \\
&\leq \sigma_{st}(C),
\end{aligned}
$$

a contradiction to the hypothesis that $A \prec_{\mathrm{B}} C$.   □

If M is a square matrix of order n, then M^{at} denotes the *antitranspose* of M, that is, the matrix obtained from M by flipping about the antidiagonal running from position $(1, n)$ down to position $(n, 1)$.

Lemma 4.5.6 *Let A and C be matrices in $\mathcal{A}(n, k)$. Then for every $(i, j) \in [1, n-1] \times [1, n-1]$,*

$$\sigma_{ij}(A) \geq \sigma_{ij}(C) \Leftrightarrow \sigma_{n-i,n-j}(A^{\mathrm{at}}) \geq \sigma_{n-i,n-j}(C^{\mathrm{at}}).$$

Proof. We observe that

$$nk = \sigma_{ij}(A) + (ik - \sigma_{ij}(A)) + (jk - \sigma_{ij}(A)) + \sigma_{n-i,n-j}(A^{\mathrm{at}}),$$

and thus

$$\sigma_{ij}(A) = (ik + jk - nk) + \sigma_{n-i,n-j}(A^{\mathrm{at}}).$$

The same inequality holds with A replaced with C. Therefore $\sigma_{ij}(A) \geq \sigma_{ij}(C)$ is equivalent to

$$(ij + jk - nk) + \sigma_{n-i,n-j}(A^{\mathrm{at}}) \geq (ij + jk - nk) + \sigma_{n-i,n-j}(C^{\mathrm{at}}),$$

and this is equivalent to $\sigma_{n-i,n-j}(A^{\mathrm{at}}) \geq \sigma_{n-i,n-j}(C^{\mathrm{at}})$. $\square$

Lemma 4.5.7 *Let A and C be matrices in $\mathcal{A}(n, k)$ with $A \prec_{\mathrm{B}} C$, and let i and j be integers with $\sigma_{ij}(A) > \sigma_{ij}(C)$. Let s and t be integers with $s+t$ minimal such that*

$$(r, c) \in [s+1, i] \times [t+1, j] \ \text{implies} \ \sigma_{rc}(A) > \sigma_{rc}(C).$$

Then there exists $(i_0, j_0) \in [s+1, i] \times [t+1, j]$ with $a_{i_0 j_0} = 1$.

Proof. Let $i' = n - i$ and let $j' = n - j$. Since $\sigma_{ij}(A) > \sigma_{ij}(C)$, by Lemma 4.5.6 we have $\sigma_{i'j'}(A^{\mathrm{at}}) > \sigma_{i'j'}(C^{\mathrm{at}})$. Let $s' = n - s$ and $t' = n - t$. Let $(r, c) \in [i', s'-1] \times [j', t'-1]$. Then $n - r \in [n - (s'-1), n - i'] = [s+1, i]$ and $n - c \in [n - (t'-1), n - j'] = [t+1, j]$, and thus $\sigma_{n-r,n-c}(A) > \sigma_{n-r,n-c}(C)$. By Lemma 4.5.6, this implies that $\sigma_{rc}(A^{\mathrm{at}}) > \sigma_{rc}(C^{\mathrm{at}})$. We conclude that

$$(r, c) \in [i', s'-1] \times [j', t'-1] \Rightarrow \sigma_{rc}(A^{\mathrm{at}}) > \sigma_{rc}(B^{\mathrm{at}}). \tag{4.41}$$

Finally, by the minimality of $s+t$, there exist a $k \in [s, i]$ and an $l \in [t, j]$ such that $\sigma_{kt}(A) = \sigma_{kt}(C)$ and $\sigma_{sl}(A) = \sigma_{sl}(C)$. Let $k' = n - k$ and $l' = n - l$. By Lemma 4.5.6, we have that $\sigma_{k't'}(A^{\mathrm{at}}) = \sigma_{k't'}(C^{\mathrm{at}})$ and $\sigma_{s'l'}(A^{\mathrm{at}}) = \sigma_{s'l'}(C^{\mathrm{at}})$. This implies that $s' + t'$ is maximal with property (4.41).

We may now apply Lemma 4.5.5 to A^{at} and C^{at} to conclude that there exists $(i_0', j_0') \in [i'+1, s'] \times [j'+1, t']$ such that the (i_0', j_0')-entry of A^{at} is 1.

Let $i_0 = n - i_0' + 1$ and $j_0 = n - j_0' + 1$. Then the (i_0, j_0)-entry of A is 1. Further, since $i' + 1 \le i_0' \le s'$,

$$i = n - (i' + 1) + 1 \ge n - i_0' + 1 = i_0 \ge n - s' + 1 = s + 1,$$

so that $i_0 \in [s + 1, i]$. Similarly, since $j' + 1 \le j_0' \le t'$,

$$j = n - (j' + 1) + 1 \ge n - j_0' + 1 = j_0 \ge n - t' + 1 = t + 1,$$

and hence $j_0 \in [t + 1, j]$. $\square$

Theorem 4.5.8 *Let A and C be two matrices in $\mathcal{A}(n, 2)$. Then $A \prec_B C$ if and only if $A \prec_{\widehat{B}} C$. In other words the Bruhat order and secondary Bruhat order are the same on $\mathcal{A}(n, 2)$.*

Proof. We know that $A \prec_{\widehat{B}} C$ implies that $A \prec_B C$. So we need only prove that if $A \prec_B C$ then $A \prec_{\widehat{B}} C$. Because the Bruhat order is a refinement of the secondary Bruhat order, it suffices to show this under the assumption that C covers A. So assume that C covers A in the Bruhat order.

Since $A \prec_B C$, there is a position (i, j) such that $a_{ij} = 1$ while $\sigma_{ij}(A) > \sigma_{ij}(C)$ (the lexicographically first position in which Σ_A and Σ_B differ, for example, is easily seen to be such a position). We choose such a position (i, j) with $i + j$ as large as possible.

Applying Lemma 4.5.5, we choose $(i_0, j_0) \in [i + 1, n] \times [j + 1, n]$ such that $a_{i_0 j_0} = 1$ and for any $(r, c) \in [i, i_0 - 1] \times [j, j_0 - 1]$, $\sigma_{rc}(A) > \sigma_{rc}(B)$. We consider three cases.

Case 1: $a_{i_0 j} = a_{i j_0} = 0$. In this case, an $I_2 \to L_2$ interchange that replaces $A[\{i, i_0\}, \{j, j_0\}] = I_2$ with L_2 results in a matrix D with $A \prec_{\widehat{B}} D$. Now, since for $(r, c) \in [1, n] \times [1, n]$

$$\sigma_{rc}(D) = \begin{cases} \sigma_{rc}(A) - 1, & \text{if } (r, c) \in [i, i_0 - 1] \times [j, j_0 - 1], \\ \sigma_{rc}(A), & \text{otherwise,} \end{cases}$$

and since $\sigma_{rc}(A) > \sigma_{rc}(C)$ for any $(r, c) \in [i, i_0 - 1] \times [j, j_0 - 1]$, then $A \prec_B C$ implies $D \preceq_B C$. Since C covers A in the Bruhat order, we conclude that in fact $D = C$ and hence $A \prec_{\widehat{B}} C$.

Case 2: $a_{i_0 j} = 1$. Because $i_0 > i$, by the maximality condition on i and j we know that $\sigma_{i_0 j}(A) = \sigma_{i_0 j}(C)$. We also know that $\sigma_{i_0 - 1, j}(A) > \sigma_{i_0 - 1, j}(C)$.

We claim that $\sigma_{i_0 - 1, j - 1}(A) > \sigma_{i_0 - 1, j - 1}(C)$. Suppose to the contrary that this is not so, that is, $\sigma_{i_0 - 1, j - 1}(A) = \sigma_{i_0 - 1, j - 1}(C)$. By calculation we get

$$1 + \sigma_{i_0 - 1, j}(A) + \sigma_{i_0, j - 1}(A) - \sigma_{i_0 - 1, j - 1}(A) = \sigma_{i_0 j}(A)$$

where

$$\sigma_{i_0 j}(A) = \sigma_{i_0 j}(C)$$
$$= c_{i_0 j} + \sigma_{i_0 - 1, j}(C) + \sigma_{i_0, j - 1}(C) - \sigma_{i_0 - 1, j - 1}(C)$$
$$\le 1 + \sigma_{i_0 - 1, j}(C) + \sigma_{i_0, j - 1}(C) - \sigma_{i_0 - 1, j - 1}(C),$$

and this implies that

$$\sigma_{i_0-1,j}(A) + \sigma_{i_0,j-1}(A) \le \sigma_{i_0-1,j}(C) + \sigma_{i_0,j-1}(C),$$

and so

$$0 < \sigma_{i_0-1,j}(A) - \sigma_{i_0-1,j}(C) \le \sigma_{i_0,j-1}(C) - \sigma_{i_0,j-1}(A).$$

Thus

$$\sigma_{i_0,j-1}(C) > \sigma_{i_0,j-1}(A),$$

a contradiction.

Therefore, $\sigma_{i_0-1,j-1}(A) > \sigma_{i_0-1,j-1}(C)$, and we can apply Lemma 4.5.7 to obtain $(i_1, j_1) \in [1, i_0 - 1] \times [1, j - 1]$ such that $a_{i_1 j_1} = 1$ and for any $(r, c) \in [i_1, i_0 - 1] \times [j_1, j - 1]$, $\sigma_{rc}(A) > \sigma_{rc}(C)$. We consider two possibilities.

First assume that $i_1 = i$. In this case we have already identified two positions, namely (i, j) and (i_1, j_1), in row $i_1 = i$ that are occupied by 1's in A. Since any row of A contains exactly two 1's, it follows that $a_{ij_0} = 0$. Similarly, since we have identified 1's in A at positions (i_0, j) and (i_0, j_0), it follows that $a_{i_0 j_1} = 0$. Finally, note that, in addition,

$$(r, c) \in [i_1, i_0 - 1] \times [j_1, j_0 - 1] \Rightarrow \sigma_{rc}(A) > \sigma_{rc}(C).$$

As in Case 1, an $I_2 \to L_2$ interchange results in a matrix D such that $A \prec_{\mathrm{B}} D \preceq_{\mathrm{B}} C$, implying $C = D$ and hence $A \prec_{\widehat{B}} C$.

Now assume that $i_1 \ne i$. In this case as well, we have identified two positions, namely (i, j) and (i_0, j), in column j that are occupied by 1's in A. Since any column of A contains exactly two 1's, it follows that $a_{i_1 j} = 0$. Similarly, since we have identified 1's in A at positions (i_0, j) and (i_0, j_0), it follows that $a_{i_0 j_1} = 0$. Finally, note that, in addition,

$$(r, c) \in [i_1, i_0 - 1] \times [j_1, j - 1] \Rightarrow \sigma_{rc}(A) > \sigma_{rc}(C).$$

Therefore, as above, an $I_2 \to L_2$ interchange results in a matrix D that must equal C and we conclude that $A \prec_{\widehat{B}} C$.

Case 3: $a_{i_0 j} = 1$. The proof here is completely symmetric to the proof in Case 2 (or we could apply Case 2 to the transpose matrices). $\qquad \square$

Let A be a matrix in $\mathcal{A}(R, S)$ which is minimal in the Bruhat order. Let $A^c = J_{m,n} - A$ be the complement of A. Let R^c and S^c be, respectively, the row and column sum vectors of A^c. Then $\Sigma_{A^c} = \Sigma_{J_{m,n}} - \Sigma_A$, and hence A^c is a maximal matrix in the class $\mathcal{A}(R^c, S^c)$. It follows that properties of minimal matrices in the Bruhat order carry over under complementation to properties of maximal matrices in the Bruhat order. In particular, from Corollary 4.5.3 we get the following property of maximal matrices.

Corollary 4.5.9 *Let A be a matrix in $\mathcal{A}(R,S)$ that is maximal in the Bruhat order. Then no submatrix of A of order 2 equals I_2.*   □

Example. Let $R = S = (2,2,2,2,2)$. Then $R^c = S^c = (3,3,3,3,3)$. The matrix in $\mathcal{A}(R^c, S^c) = \mathcal{A}(5,3)$ given by

$$\begin{bmatrix} 1 & 1 & 1 & 0 & 0 \\ 1 & 1 & 1 & 0 & 0 \\ 1 & 0 & 0 & 1 & 1 \\ 0 & 1 & 0 & 1 & 1 \\ 0 & 0 & 1 & 1 & 1 \end{bmatrix}$$

is minimal in the Bruhat order. Thus the matrix

$$\begin{bmatrix} 0 & 0 & 0 & 1 & 1 \\ 0 & 0 & 0 & 1 & 1 \\ 0 & 1 & 1 & 0 & 0 \\ 1 & 0 & 1 & 0 & 0 \\ 1 & 1 & 0 & 0 & 0 \end{bmatrix}$$

in $\mathcal{A}(R,S)$ is maximal in the Bruhat order.

In general there are many minimal matrices in a nonempty class $\mathcal{A}(R,S)$. We now give a Ryser-like algorithm for determining a minimal matrix under the assumption that R and S are nonincreasing vectors, and then specialize it for classes with constant row and column sums [15].

**Algorithm to Construct a Minimal Matrix
in the Bruhat Order on $\mathcal{A}(R,S)$**

Let $R = (r_1, r_2, \ldots, r_m)$ and $S = (s_1, s_2, \ldots, s_n)$ be nonincreasing, positive integral vectors with $S \preceq R^*$. Let $\overline{A}$ be the unique matrix in $\mathcal{A}(R, R^*)$.

(1) Rewrite R by grouping together its components of equal value:

$$R = (\underbrace{a_1, \ldots, a_1}_{p_1}, \underbrace{a_2, \ldots, a_2}_{p_2}, \ldots, \underbrace{a_k, \ldots, a_k}_{p_k})$$

where $a_1 > a_2 > \cdots > a_k$, and $p_1, p_2, \ldots, p_k > 0$.

(2) Determine nonnegative integers $x_1, x_2, \ldots, x_k$ satisfying $x_1 + x_2 + \cdots + x_k = s_n$ where $x_k, x_{k-1}, \ldots, x_1$ are maximized in turn *in this order* subject to $(s_1, s_2, \ldots, s_{n-1}) \preceq R_1^*$ where $R_1 = R_{(x_1, x_2, \ldots, x_k)}$ is the vector

$$(\underbrace{a_1, \ldots, a_1, \overbrace{a_1 - 1, \ldots, a_1 - 1}^{x_1}}_{p_1}, \ldots, \underbrace{a_k, \ldots, a_k \overbrace{a_k - 1, \ldots, a_k - 1}^{x_k}}_{p_k}).$$

(3) Shift $s_n = x_1 + x_2 + \cdots + x_k$ 1's to the last column as specified by those rows whose sums have been diminished by 1. (Thus the last column consists of $p_1 - x_1$ 0's followed by x_1 1's, $p_2 - x_2$ 0's followed by x_2 1's, ..., $p_k - x_k$ 0's followed by x_k 1's.)

(4) Proceed recursively, returning to step (1), with the components of R replaced with those of R_1, and the components of S replaced with those of $S_1 = (s_1, s_2, \ldots, s_{n-1})$

We refer to this algorithm as the *minimal-matrix algorithm*.

Example. Let $R = (4, 4, 3, 3, 2, 2)$ and $S = (4, 4, 3, 3, 3, 1)$. Then $R^* = (6, 6, 4, 2, 0, 0)$. Starting with the matrix $\overline{A}$ in $\mathcal{A}(R, R^*)$ and applying the minimal-matrix algorithm, we get

$$
\begin{bmatrix}
1 & 1 & 1 & 1 & 0 & 0 \\
1 & 1 & 1 & 1 & 0 & 0 \\
1 & 1 & 1 & 0 & 0 & 0 \\
1 & 1 & 1 & 0 & 0 & 0 \\
1 & 1 & 0 & 0 & 0 & 0 \\
1 & 1 & 0 & 0 & 0 & 0
\end{bmatrix}
\rightarrow
\begin{bmatrix}
1 & 1 & 1 & 1 & 0 & 0 \\
1 & 1 & 1 & 1 & 0 & 0 \\
1 & 1 & 1 & 0 & 0 & 0 \\
1 & 1 & 1 & 0 & 0 & 0 \\
1 & 1 & 0 & 0 & 0 & 0 \\
1 & 0 & 0 & 0 & 0 & 1
\end{bmatrix}
$$

$$
\rightarrow
\begin{bmatrix}
1 & 1 & 1 & 1 & 0 & 0 \\
1 & 1 & 1 & 1 & 0 & 0 \\
1 & 1 & 1 & 0 & 0 & 0 \\
1 & 1 & 0 & 0 & 1 & 0 \\
1 & 0 & 0 & 0 & 1 & 0 \\
0 & 0 & 0 & 0 & 1 & 1
\end{bmatrix}
\rightarrow
\begin{bmatrix}
1 & 1 & 1 & 1 & 0 & 0 \\
1 & 1 & 1 & 1 & 0 & 0 \\
1 & 1 & 1 & 0 & 0 & 0 \\
1 & 1 & 0 & 0 & 1 & 0 \\
0 & 0 & 0 & 1 & 1 & 0 \\
0 & 0 & 0 & 0 & 1 & 1
\end{bmatrix}.
$$

Since no more shifting can be done, we can now stop. The resulting matrix has no submatrix equal to L_2, and it is straightforward to verify that it is a minimal matrix in its class $\mathcal{A}(R, S)$.

Theorem 4.5.10 *Let R and S be nonincreasing positive vectors such that the class $\mathcal{A}(R, S)$ of m by n matrices is nonempty. Then the minimal-matrix algorithm constructs a minimal matrix in $\mathcal{A}(R, S)$ in the Bruhat order.*

Proof. We prove the theorem by induction on n. If $n = 1$, there is a unique matrix in $\mathcal{A}(R, S)$, and the theorem holds trivially. Assume that $n > 1$ and let the matrix constructed by the minimal-matrix algorithm be $A = [a_{ij}]$. Let R_1 be defined as in the algorithm. Let $P = [p_{ij}]$ be a matrix in $\mathcal{A}(R, S)$ such that $P \preceq_B A$. Let $u = (u_1, u_2, \ldots, u_m)^\mathrm{T}$ and $v = (v_1, v_2, \ldots, v_m)^\mathrm{T}$ be, respectively, the last columns of A and P. First suppose that $u = v$. Then the matrices A' and P' obtained by deleting the last column of A and P, respectively, belong to the same class $\mathcal{A}(R', S')$, and $P' \preceq_B A'$. Since A' is constructed by the minimal-matrix algorithm, it now follows from the inductive assumption that $P' = A'$ and hence $P = A$.

Now suppose that $u \neq v$. We may assume that the last column of P consists of $p_1 - y_1$ 0's followed by y_1 1's, ..., $p_k - y_k$ 0's followed by y_k 1's where $y_1, y_2, \ldots, y_k$ are nonnegative integers satisfying $y_1 + y_2 + \cdots + y_k = s_n$. Otherwise, the last column of P contains a 1 above a 0 in two rows with equal sums, and then P contains a submatrix equal to L_2. An inversion-reducing interchange then replaces P with Q where $Q \preceq_B P \preceq_B A$.

The row sum vector $R_{(y_1, y_2, \ldots, y_k)}$ of the matrix P' obtained by deleting the last column of P is nonincreasing. Since $P \in \mathcal{A}(R,S)$, we have $(s_1, s_2, \ldots, s_{n-1}) \preceq R^*_{(y_1, y_2, \ldots, y_k)}$. The choice of $x_1, x_2, \ldots, x_k$ implies that

$$y_1 + \cdots + y_j \leq x_1 + \cdots + x_j \quad (j = 1, 2, \ldots, k) \qquad (4.42)$$

with equality for $j = k$. Let q be the smallest integer such that $u_q \neq v_q$. Then it follows from (4.42) that $u_q = 0$ and $v_q = 1$. We calculate that

$$\sum_{i=1}^{q} \sum_{j=1}^{n-1} p_{ij} = r_1 + \cdots + r_q - \sum_{j=1}^{q-1} v_j - 1$$

$$= r_1 + \cdots + r_q - \sum_{j=1}^{q-1} u_j - 1$$

$$= r_1 + \cdots + r_q - \sum_{j=1}^{q} u_j - 1$$

$$= \sum_{i=1}^{q} \sum_{j=1}^{n-1} a_{ij} - 1,$$

contradicting the fact that $P \preceq_B A$. The theorem now follows easily by induction. $\qquad \square$

Let k be an integer with $1 \leq k \leq n$ and let $R = S = (k, k, \ldots, k)$, the n-vector of k's. Then $\mathcal{A}(R,S) = \mathcal{A}(n,k)$. The class $\mathcal{A}(n,1)$ is the set of permutation matrices of order n. The minimal-matrix algorithm simplifies for the class $\mathcal{A}(n,k)$.

Algorithm to Construct a Minimal Matrix
in the Bruhat Order on $\mathcal{A}(n,k)$

(1) Let $n = qk + r$ where $0 \leq r < k$.

(2) If $r = 0$, then $A = J_k \oplus \cdots \oplus J_k$ (q J_k's) is a minimal matrix.

(3) Else, $r \neq 0$.

   (a) If $q \geq 2$, let

$$A = X \oplus \underbrace{J_k \oplus \cdots \oplus J_k}_{q-1}, \text{ where } X \text{ has order } k + r,$$

   and replace n with $k + r$.

(b) Else, $q = 1$, and let

$$A = \left[\begin{array}{c|c} J_{r,k} & O_k \\ \hline X & J_{k,r} \end{array}\right] \quad (X \text{ has order } k),$$

and replace k with $k - r$ and n with k.

(c) Proceed recursively with the current values of n and k to determine X.

Example. Let $n = 18$ and $k = 11$. The algorithm constructs the following minimal matrix in $\mathcal{A}(n, k)$.

$$\left[\begin{array}{c|c} \begin{array}{c} J_{7,11} \\ \hline \begin{array}{c|c} J_{3,4} & O_3 \\ \hline I_4 & J_{4,3} \\ \hline O_{4,7} & J_4 \end{array} \end{array} \begin{array}{c} O_{7,4} \end{array} & \begin{array}{c} O_7 \\ \hline J_{11,7} \end{array} \end{array}\right].$$

Here we first construct (using $18 = 1 \cdot 11 + 7$)

$$\left[\begin{array}{c|c} J_{7,11} & O_7 \\ \hline X & J_{11,7} \end{array}\right].$$

Then to construct the matrix X of order 11 with $k = 11 - 7 = 4$ (using $11 = 2 \cdot 4 + 3$), we construct

$$\left[\begin{array}{c|c} Y & O_{7,4} \\ \hline O_{4,7} & J_4 \end{array}\right].$$

Then to construct the matrix Y of order $4 + 3 = 7$ with $k = 4$ (using $7 = 1 \cdot 4 + 3$), we construct

$$\left[\begin{array}{c|c} J_{3,4} & O_3 \\ \hline Z & J_{4,3} \end{array}\right].$$

Finally to construct the matrix Z of order 4 with $k = 4 - 3 = 1$ (using $4 = 4 \cdot 1 + 0$), we construct

$$Z = I_1 \oplus I_1 \oplus I_1 \oplus I_1 = I_4.$$

$\square$

The identity matrix I_n is the only minimal matrix in $\mathcal{A}(n, 1)$. In [15] the minimal matrices in $\mathcal{A}(n, 2)$ and $\mathcal{A}(n, 3)$ are characterized. We give this characterization but refer to [15] for details. Let F_n denote the matrix of order n with 0's in positions $(1, n), (2, n - 2), \ldots, (n, 1)$ and 0's elsewhere.

Theorem 4.5.11 *Let n be an integer with $n \geq 2$. Then a matrix in $\mathcal{A}(n, 2)$ is a minimal matrix in the Bruhat order if and only if it is the direct sum of matrices equal to J_2 and F_3.*

Let

$$V = \begin{bmatrix} 1 & 1 & 1 & 0 & 0 \\ 1 & 1 & 1 & 0 & 0 \\ 1 & 0 & 0 & 1 & 1 \\ 0 & 1 & 0 & 1 & 1 \\ 0 & 0 & 1 & 1 & 1 \end{bmatrix}.$$

For $i \geq 1$, let U_i be the matrix in $\mathcal{A}(i+6,3)$ of the form

$$\begin{bmatrix} 1 & 1 & 1 & 0 & & & & & \cdots & \\ 1 & 1 & 1 & 0 & & & & & \cdots & \\ 1 & 1 & 0 & 1 & & & & & \cdots & \\ 0 & 0 & 1 & 1 & 1 & & & & & \\ & & & & \ddots & \ddots & \ddots & & & \\ & & & & & 1 & 1 & 1 & 0 & 0 \\ & & & & & & 1 & 0 & 1 & 1 \\ & & & & & & 0 & 1 & 1 & 1 \\ & & & & & & 0 & 1 & 1 & 1 \end{bmatrix}.$$

Thus

$$U_1 = \begin{bmatrix} 1 & 1 & 1 & 0 & 0 & 0 & 0 \\ 1 & 1 & 1 & 0 & 0 & 0 & 0 \\ 1 & 1 & 0 & 1 & 0 & 0 & 0 \\ 0 & 0 & 1 & 1 & 1 & 0 & 0 \\ 0 & 0 & 0 & 1 & 0 & 1 & 1 \\ 0 & 0 & 0 & 0 & 1 & 1 & 1 \\ 0 & 0 & 0 & 0 & 1 & 1 & 1 \end{bmatrix}.$$

Theorem 4.5.12 *Let n be an integer with $n \geq 3$. Then a matrix in $\mathcal{A}(n,3)$ is a minimal matrix in the Bruhat order if and only if it is the direct sum of matrices equal to J_3, F_4, V, V^{T}, and U_i ($i \geq 1$).*

Already for $k = 4$, it seems that there is no simple characterization of minimal matrices in $\mathcal{A}(n,4)$ [15].

In [13] it is proved that a class $\mathcal{A}(n,k)$ contains a unique minimal element in the Bruhat order (and so in the secondary Bruhat order) if and only if $k = 0, 1, n-1, n$ or $n = 2k$. The unique minimal matrix in the class $\mathcal{A}(2k,k)$ is $J_k \oplus J_k$.

4.6 The Integral Lattice $\mathcal{L}(R,S)$

Throughout this section $R = (r_1, r_2, \ldots, r_m)$ and $S = (s_1, s_2, \ldots, s_n)$ are nonincreasing, positive integral vectors and the class $\mathcal{A}(R,S)$ is nonempty. The *integral lattice generated by* $\mathcal{A}(R,S)$ is the set $\mathcal{L}(R,S)$ consisting of all integer linear combinations

$$\left\{ \sum_{A \in \mathcal{A}(R,S)} c_A A : c_A \in \mathcal{Z} \right\}$$

of matrices in $\mathcal{A}(R,S)$. If $X = [x_{ij}]$ and $Y = [y_{ij}]$ are two real matrices of size m by n, then

$$X \circ Y = \sum_{i=1}^{m} \sum_{j=1}^{n} x_{ij} y_{ij}$$

is their *inner product*. The *dual lattice* of $\mathcal{L}(R,S)$ is the set

$$\mathcal{L}^*(R,S) = \{X : X \circ A \in Z \text{ for all } A \in \mathcal{L}(R,S)\}$$

of all *rational* m by n matrices X whose inner product with each matrix in $\mathcal{L}(R,S)$ (equivalently, $\mathcal{A}(R,S)$) is an integer. The third set we consider is the *linear span of* $\mathcal{A}(R,S)$,

$$\mathcal{LS}(R,S) = \left\{ \sum_{A \in \mathcal{A}(R,S)} r_A A : r_A \text{ real} \right\}$$

consisting of all real linear combinations of the matrices in $\mathcal{A}(R,S)$. We also consider the set $\mathcal{LS}_Z(R,S)$ of integral matrices in $\mathcal{LS}(R,S)$. Clearly, $\mathcal{L}(R,S) \subseteq \mathcal{LS}_Z(R,S)$.

In view of (iii) of Theorem 3.4.1, there is no essential loss in generality in assuming that $\mathcal{A}(R,S)$ does not have any invariant positions. For, if $\mathcal{A}(R,S)$ has invariant positions, then there exist integers k and l with $0 \leq k \leq m$, $0 \leq l \leq n$, and $(k,l) \neq (0,n),(m,0)$ such that each matrix in $\mathcal{A}(R,S)$ has the form

$$\begin{bmatrix} J_{k,l} & * \\ * & O_{m-k,n-l} \end{bmatrix}.$$

Thus the study of the integral lattice $\mathcal{L}(R,S)$ reduces to the study of two smaller integral lattices.

We first characterize $\mathcal{LS}(R,S)$ [14, 33, 16]. To do this we make use of the following lemma.

Lemma 4.6.1 *Let $X = [x_{ij}]$ be an m by n real matrix. Then X is a linear combination of matrices in $\mathcal{A}(R,S)$ with nonnegative coefficients if and only if there is a real number q such that*

(i) *the row and column sum vectors of X are qR and qS, respectively, and*

(ii) $0 \leq x_{ij} \leq q$ $(1 \leq i \leq m, 1 \leq j \leq n)$.

Proof. Let $\mathcal{LS}_{\geq 0}(R,S)$ be the set of nonnegative linear combinations of matrices in $\mathcal{A}(R,S)$. The matrices in $\mathcal{LS}_{\geq 0}(R,S)$ clearly satisfy (i) and (ii) in the theorem. Now let X be a real matrix satisfying (i) and (ii) for some real number q. Since $X \in \mathcal{LS}_{\geq 0}(R,S)$ if and only if $(1/q)X \in \mathcal{LS}_{\geq 0}(R,S)$, we assume that $q = 1$. The set of such matrices with row sum vector R, column sum vector S, and entries between 0 and 1 forms a convex polytope

$\mathcal{P}$.[9] We now show that the extreme points of $\mathcal{P}$ are the matrices in $\mathcal{A}(R,S)$ and thus that each matrix in $\mathcal{P}$ is a nonnegative linear combination, with sum of the coefficients equal to 1, of matrices in $\mathcal{A}(R,S)$.

If $X \in \mathcal{P}$ is a (0,1)-matrix, then $X \in \mathcal{A}(R,S)$. Otherwise X has an entry strictly between 0 and 1. Let G be the bipartite graph with vertices partitioned into sets $U = \{u_1, u_2, \ldots, u_m\}$ and $W = \{w_1, w_2, \ldots, w_n\}$ and with edges consisting of all pairs $\{u_i, w_j\}$ for which $0 < x_{ij} < 1$. The graph G clearly has an (even) cycle. Let C be the $(0, 1, -1)$-matrix with 1's and -1's alternating in the positions corresponding to the edges of the cycle. Then there is a positive number ϵ such that $X \pm \epsilon C \in \mathcal{P}$. But then

$$X = \frac{1}{2}(X + \epsilon C) + \frac{1}{2}(X - \epsilon C),$$

and hence X is not an extreme point of $\mathcal{P}$. Thus the extreme points of $\mathcal{P}$ are the matrices in $\mathcal{A}(R,S)$.   □

Corollary 4.6.2 *An m by n real matrix X can be expressed as a nonnegative integer linear combination of matrices in $\mathcal{A}(R,S)$ if and only if X is a nonnegative integral matrix, and there exists a nonnegative integer q such that X has row sum vector qR, column sum vector qS, and entries at most equal to q.*

Proof. Let X be a matrix satisfying the conditions given in the corollary. It follows from Lemma 4.6.1 that there is a matrix $A \in \mathcal{A}(R,S)$ such that X has positive entries in at least those positions in which A has 1's. Then $X - A$ satisfies the conditions given in the corollary with q replaced by $q - 1$. It now follows inductively that $X \in \mathcal{LS}_{\mathcal{Z}}(R,S)$. The converse is obvious.   □

We define the special integral matrix $J_{R,S}$ to be the sum of all the matrices in $\mathcal{A}(R,S)$. The matrix $J_{R,S}$ has row and column sum vectors qR and qS, respectively, where $q = \kappa(R,S)$ is the number of matrices in $\mathcal{A}(R,S)$.

Theorem 4.6.3 *Assume that $\mathcal{A}(R,S)$ has no invariant positions. Then an m by n real matrix X belongs to $\mathcal{LS}(R,S)$ if and only if there is a real number q such that the row and column sum vectors of X are qR and qS, respectively.*

Proof. Each matrix in $\mathcal{LS}(R,S)$ satisfies the conditions given in the theorem. Now let X be a real matrix satisfying these conditions for some real number q. Since $\mathcal{A}(R,S)$ has no invariant positions, $J_{R,S}$ is a positive matrix. Hence there exists an integer d such that the matrix $X' = X + dJ_{R,S}$ is a nonnegative matrix. Since X belongs to $\mathcal{LS}(R,S)$ if and only if X' does, the theorem now follows from Lemma 4.6.1.   □

---

[9]Convex polytopes of matrices are considered in Chapter 8.

We now characterize the lattice $\mathcal{L}(R, S)$.

Theorem 4.6.4 *Assume that $\mathcal{A}(R, S)$ has no invariant positions. Then $\mathcal{L}(R, S)$ consists of all the m by n integral matrices X for which X has row sum vector qR and column sum vector qS for some integer q.*

Proof. A matrix X in $\mathcal{L}(R, S)$ clearly satisfies the conditions given in the theorem. Now let $X = [x_{ij}]$ be a matrix satisfying these conditions. Since $\mathcal{A}(R, S)$ has no invariant positions, the matrix $J_{R,S}$ is a positive matrix in $\mathcal{L}(R, S)$. There exists a positive integer p such that $X + pJ_{R,S}$ is a nonnegative matrix, and it follows that we may assume that X is a nonnegative matrix. If $x_{ij} \leq q$ for all i and j, then by Corollary 4.6.2, $X \in \mathcal{L}(R, S)$. Since $X = cJ_{R,S} - (cJ_{R,s} - X)$ for each integer c, it suffices to show that the matrix $G_c = cJ_{R,s} - X$ is in $\mathcal{L}(R, S)$ for some positive integer c. Let t be the number of matrices in $\mathcal{A}(R, S)$. Since $\mathcal{A}(R, S)$ has no invariant positions, each entry of $J_{R,S} = [e_{ij}]$ satisfies $e_{ij} < t$. We may choose c large enough so that the entries of $G_c = [g_{ij}]$ satisfy

$$\begin{cases} g_{ij} \geq 0 \, (1 \leq i \leq m, 1 \leq j \leq n), \\ \sum_{j=1}^{n} g_{ij} = (ct - q)r_i \, (1 \leq i \leq m), \\ \sum_{i=1}^{m} g_{ij} = (ct - q)s_j \, (1 \leq j \leq n). \end{cases}$$

Since the e_{ij} satisfy $e_{ij} < t$, we have for c large enough that

$$q + c(e_{ij} - t) \leq 0 \leq x_{ij},$$

and hence

$$0 \leq g_{ij} = ce_{ij} - x_{ij} \leq ct - q \, (1 \leq i \leq m, 1 \leq j \leq n).$$

Since the row and column sum vectors of G are $(ct - q)R$ and $(ct - q)S$, respectively, Corollary 4.6.2 now implies that G, and hence X, belongs to $\mathcal{L}(R, S)$ $\qquad\square$

Example. Theorem 4.6.4 is no longer true if we drop the assumption that there do not exist invariant positions. Let $R = S = (3, 3, 1, 1)$. Then there are exactly four matrices in $\mathcal{A}(R, S)$, one of which is

$$\begin{bmatrix} 1 & 1 & 1 & 0 \\ 1 & 1 & 0 & 1 \\ 1 & 0 & 0 & 0 \\ 0 & 1 & 0 & 0 \end{bmatrix},$$

and they all have a J_2 in the upper left corner and an O_2 in the lower right corner. The matrix

$$\begin{bmatrix} 3 & 0 & 0 & 0 \\ 0 & 3 & 0 & 0 \\ 0 & 0 & 1 & 0 \\ 0 & 0 & 0 & 1 \end{bmatrix}$$

satisfies the conditions in Theorem 4.6.4 with $q = 1$ but is clearly not in $\mathcal{L}(R, S)$. This example can be generalized [16].   □

Corollary 4.6.5 *Assume that $\mathcal{A}(R, S)$ has no invariant positions and that the components of R are relatively prime. Then*

$$\mathcal{L}(R, S) = \mathcal{LS}_Z(R, S).$$

Proof. Let $R = (r_1, r_2, \ldots, r_m)$ where $r_1, r_2, \ldots, r_m$ are relatively prime integers. Since $\mathcal{L}(R, S) \subseteq \mathcal{LS}_Z(R, S)$, we need only show the reverse containment. Let X be an integral matrix in $\mathcal{LS}(R, S)$ with row sum vector qR and column sum vector qS for some integer q. Since X is an integral matrix, qR is an integral vector. The relative primeness of $r_1, r_2, \ldots, r_m$ now implies that q is an integer. From Theorem 4.6.4 we now conclude that X is in $\mathcal{L}(R, S)$.   □

Example. If $r_1, r_2, \ldots, r_m$ are not relatively prime, the conclusion of Corollary 4.6.5 need not hold. For example, the identity matrix I_3 satisfies

$$I_3 = \frac{1}{2} \begin{bmatrix} 1 & 1 & 0 \\ 0 & 1 & 1 \\ 1 & 0 & 1 \end{bmatrix} + \frac{1}{2} \begin{bmatrix} 1 & 0 & 1 \\ 1 & 1 & 0 \\ 0 & 1 & 1 \end{bmatrix} - \frac{1}{2} \begin{bmatrix} 0 & 1 & 1 \\ 1 & 0 & 1 \\ 1 & 1 & 0 \end{bmatrix},$$

but $I_3 \notin \mathcal{LS}_Z(R, S)$.   □

We now determine a basis for $\mathcal{L}(R, S)$. Let $\mathcal{E}_{m,n}$ equal the set of $mn - m - n + 1$ m by n $(0, 1, -1)$-matrices $C_{i,j}$ that have a 1 in positions (i, j) and $(1, n)$, a -1 in positions $(1, j)$ and (i, n), $2 \le i \le m, 1 \le j \le n - 1$, and 0's everywhere else. Let $\mathcal{C}_{m,n}$ be the set of all m by n $(0, 1, -1)$-matrices all of whose entries are zero except for a submatrix of order 2 of the form

$$\pm \begin{bmatrix} 1 & -1 \\ -1 & 1 \end{bmatrix}.$$

Lemma 4.6.6 *Assume that $\mathcal{A}(R, S)$ has no invariant positions. Then*

$$\mathcal{E}_{m,n} \subseteq \mathcal{L}(R, S).$$

Proof. Let R' be obtained from R by decreasing each of r_1 and r_i by 1, and let S' be obtained from S by decreasing s_j and s_n by 1. We first show that $\mathcal{A}(R', S')$ is nonempty. Since S is nonincreasing, it follows that S' is nearly nonincreasing. Hence by Theorem 2.1.4, there exists a matrix A in $\mathcal{A}(R, S)$ if and only if the structure matrix $T(R', S') = [t'_{kl}]$ is nonnegative. Since $\mathcal{A}(R, S)$ has no invariant positions, it follows from Theorem 3.4.1 that the entries of the structure matrix $T(R, S) = [t_{kl}]$ are positive unless $(k, l) = (m, 0)$ or $(n, 0)$. It is now easy to verify that $T(R', S')$ is a nonnegative matrix and hence $\mathcal{A}(R', S')$ is nonempty. By Theorem 4.4.14 we conclude that there are matrices $A \in \mathcal{A}(R,S)$ and $A' \in \mathcal{A}(R',S')$ such that $A' \leq A$ (componentwise). The matrix $A - A'$ has all 0's outside the submatrix

$$(A - A')[\{1,i\}, \{j,n\}] = \begin{bmatrix} 1 & 0 \\ 0 & 1 \end{bmatrix} \quad \text{or} \quad \begin{bmatrix} 0 & 1 \\ 1 & 0 \end{bmatrix}.$$

Applying the corresponding interchange to A results in a matrix B for which $A - B = \pm C_{i,j}$. Hence $C_{i,j} \in \mathcal{L}(R,S)$. $\qquad\square$

Example. Let $R = S = (3,2,1,1)$. Then $\mathcal{A}(R,S) \neq \varnothing$. Let $T(R,S) = [t_{ij}]$ be the corresponding structure matrix. Then $t_{22} = 1$. Hence each matrix A in $\mathcal{A}(R,S)$ has either three or four 1's in its submatrix $A[\{1,2\}, \{1,2\}]$. Hence there is no matrix in $\mathcal{A}(R,S)$ to which an interchange can be applied in its leading submatrix of order 2.[10] $\qquad\square$

Corollary 4.6.7 *Assume that $\mathcal{A}(R,S)$ has no invariant positions. Then the dimension of the lattice $\mathcal{L}(R,S$ equals $mn - m - n + 2$.*

Proof. The $mn - m - n + 1$ matrices in $\mathcal{E}_{m,n}$ are linearly independent matrices in $\mathcal{L}(R,S)$. Moreover it is easy to check that each matrix in $\mathcal{C}(m,n)$ is the sum of matrices in $\mathcal{E}_{m,n}$ and thus belongs to $\mathcal{L}(R,S)$. Let A be any matrix in $\mathcal{A}(R,S)$. It now follows from the interchange theorem (see Theorem 3.2.3) that every matrix in $\mathcal{A}(R,S)$ is the sum of A and an integer linear combination of the matrices in $\mathcal{E}_{m,n}$, and hence every matrix in $\mathcal{L}(R,S)$ is an integer linear combination of A and the matrices in $\mathcal{E}_{m,n}$. The corollary now follows. $\qquad\square$

We now consider the dual lattice $\mathcal{L}^*(R,S)$, and define the set

$$\mathcal{L}^\circ(R,S) = \{X : X \circ B = 0 \text{ for all } B \in \mathcal{L}(R,S)\}$$

of all *rational* m by n matrices X whose inner product with each matrix in $\mathcal{L}(R,S)$ (equivalently, $\mathcal{A}(R,S)$) equals 0. The set $\mathcal{L}^\circ(R,S)$ is a lattice and is called the *lattice of $\mathcal{L}(R,S)$-orthogonal matrices*. Clearly, $\mathcal{L}^\circ(R,S) \subseteq \mathcal{L}^*(R,S)$. Of course, $\mathcal{L}^*(R,S)$ also contains the set $\mathcal{Z}^{m,n}$ of integral m by n matrices. Matrices in $\mathcal{L}^*(R,S)$ are characterized in the next theorem.

Theorem 4.6.8 *Let $R = (r_1, r_2, \ldots, r_m)$ and $S = (s_1, s_2, \ldots, s_n)$ be positive integral vectors such that $\mathcal{A}(R,S) \neq \varnothing$, and assume that $\mathcal{A}(R,S)$ has no invariant positions. Let d be the greatest common divisor of the integers $r_1, r_2, \ldots, r_m, s_1, s_2, \ldots, s_n$. Then*

$$\mathcal{L}^*(R,S) = \mathcal{L}^\circ(R,S) + \frac{1}{d} \mathcal{Z}^{m,n},$$

---

[10]This is contrary to the assertion on page 188 of [16] that $\mathcal{C}_{m,n} \subseteq \mathcal{L}(R,S)$, making the proof there of Theorem 2 incomplete. That assertion is true with the added assumption that the entries of the structure matrix $T(R,S) = [t_{ij}]$ satisfy $t_{ij} \geq 2$ for $(i,j) \neq (0,n)$ or $(m,0)$.

that is, every matrix in $\mathcal{L}^*(R,S)$ *can be represented as the sum of an* $\mathcal{L}^\circ(R,S)$-*orthogonal matrix and* $1/d$ *times an integral matrix.*

Proof. Let $X = [x_{ij}]$ be a (rational) matrix in $\mathcal{L}^*(R,S)$. Since $X \circ B = 0$ for all $B \in \mathcal{L}(R,S)$, we have that

$$X \circ C = 0, \quad C \in \mathcal{E}_{m,n}. \tag{4.43}$$

There exist rational numbers $a_1, a_2, \ldots, a_m, b_1, b_2, \ldots, b_n$ such that $x_{ij} = a_i + b_j$ for all i and j; in fact, using (4.43), we can take $a_1 = b_n = x_{1n}/2$, and then $a_i = x_{in} - x_{1n}/2$ for $i = 2, \ldots, n$ and $b_j = x_{1j} - x_{1n}/2$ for $j = 1, 2, \ldots, n-1$. We now conclude that

$$X \circ A = q, \quad \text{where } q = a_1 r_1 + \cdots + a_m r_m + b_1 s_1 + \cdots + b_n s_n.$$

The number q is an integer since $X \in \mathcal{L}^*(R,S)$. By definition of d we can write

$$qd = e_1 r_1 + \cdots e_m r_m + f_1 s_1 + \cdots + f_n s_n$$

where $e_1, \ldots, e_m, f_1, \ldots, f_n$ are integers. Let $P = [p_{ij}]$ be the m by n integral matrix defined by $p_{ij} = e_i + f_j$ for all i and j. Then $(1/d)P \circ A = q$ for all $A \in \mathcal{A}(R,S)$. Hence

$$(X - (1/d)P) \circ A = 0, \quad (A \in \mathcal{A}(R,S)).$$

Thus $X - (1/d)P$ is a matrix Q in $\mathcal{L}^\circ(R,S)$ and $X = Q + (1/d)P$. $\qquad\square$

Corollary 4.6.9 *Let* $R = (r_1, r_2, \ldots, r_m)$ *and* $S = (s_1, s_2, \ldots, s_n)$ *be positive integral vectors such that* $\mathcal{A}(R,S) \neq \oslash$. *Assume that* $\mathcal{A}(R,S)$ *has no invariant positions and that* $r_1, \ldots, r_m, s_1, \ldots, s_n$ *are relatively prime. Then every matrix in* $\mathcal{L}^*(R,S)$ *is the sum of an* $\mathcal{L}(R,S)$-*orthogonal matrix and an integral matrix.* $\qquad\square$

Example. Let $R = S = (2,2,2)$. The matrix

$$X = \begin{bmatrix} 1/2 & 1/2 & 1/2 \\ 0 & 0 & 0 \\ 0 & 0 & 0 \end{bmatrix}$$

is in $\mathcal{L}(R,S)$, as it has inner product 1 with each matrix in $\mathcal{A}(R,S)$. The matrix X cannot be written as the sum of an $\mathcal{L}(R,S)$ and an integral matrix, as is easily verified. Hence Corollary 4.6.9 is not true in general without the relatively prime assumption. $\qquad\square$

4.7 Appendix

In this appendix we give a proof of Greene's theorem [29], Lemma 4.1.19, which we restate below as a theorem. We rely heavily also on proofs given in [28] for the row-bumping operation and nondecreasing subsequences, and [44] for permutation arrays (as done in [29]).

Theorem 4.7.1 *Let*

$$\Upsilon = \begin{pmatrix} i_1 & i_2 & \cdots & i_\tau \\ j_1 & j_2 & \cdots & j_\tau \end{pmatrix}$$

be a generalized permutation array with distinct pairs $\begin{pmatrix} i_k \\ j_k \end{pmatrix}$ in lexicographic order. Let $\mathcal{P}$ be the insertion tableau that results by applying the column insertion algorithm to Υ, that is, to the sequence $j_1, j_2, \ldots, j_\tau$. The number of occupied boxes in the first k columns of $\mathcal{P}$ equals the maximal number of terms in a subsequence of $j_1, j_2, \ldots, j_\tau$ that is the union of k strictly increasing subsequences $(k = 1, 2, \ldots)$.

The particular case of $k = 1$, due to Schensted [43], asserts that the length of the longest strictly increasing subsequence of $j_1, j_2, \ldots, j_\tau$ equals the number of occupied boxes in the first column of $\mathcal{P}$. The other columns of $\mathcal{P}$ have no direct interpretation, although their union, as stated in the theorem, does.

Example. Consider the sequence $\sigma = 2, 3, 6, 1, 4, 5$. Column insertion gives the following Young tableau:

$$
\begin{array}{|c|c|}
\hline
1 & 2 \\
\hline
3 & 6 \\
\hline
4 & \times \\
\hline
5 & \times \\
\hline
\end{array}
\qquad (4.44)
$$

The length of the longest strictly increasing subsequence of σ is 4, attained uniquely by $2, 3, 4, 5$, and the maximal combined length of the union of two strictly increasing subsequence is 6, attained uniquely by the union of $2, 3, 6$ and $1, 4, 5$. Now consider the sequence $\sigma' = 2, 6, 1, 3, 4, 5$ obtained by listing the elements of the Young tableau (4.44) column by column starting from the last and going down each column. Then column insertion applied to σ' also results in the Young tableau (4.44). In contrast to σ, for σ' we can obtain the union of two strictly increasing sequences of combined length 6 by taking the union of the subsequence $1, 3, 4, 5$ of length 4 with the subsequence $2, 6$ of length 2. $\qquad \square$

For a sequence σ of integers and a positive integer k, let $\ell(\sigma, k)$ equal the maximal combined length of the disjoint union of k strictly increasing subsequences of σ. If T is a Young tableau, then we define the *word* of T to be the sequence $w(T)$ of integers obtained by reading the elements of T column by column starting from the last column and going down each column, as in the above example. The tableau T is uniquely reconstructable from its word $w(T)$, by breaking $w(T)$ at those entries that are greater than or equal to the next, with the pieces then forming the columns of T. It follows from the method of constructing $w(T)$ and the fact that the elements in each row of T are nondecreasing that each strictly increasing subsequence of $w(T)$ consists of numbers from T obtained by proceeding in order from top to bottom never going to the right, thus taking numbers from different rows. Since the columns of T are strictly increasing, it follows that $\ell(\sigma,1)$ equals the number of nonempty rows of T, and more generally, that $\ell(\sigma,k)$ equals the number of occupied boxes in the first k columns of T. We thus have the following lemma.

Lemma 4.7.2 *Let T be a Young tableau of shape $\lambda = (\lambda_1, \lambda_2, \ldots, \lambda_m)$. Let $p = \lambda_1$, and let $\lambda^* = (\lambda_1^*, \lambda_2^*, \ldots, \lambda_p^*)$ be the conjugate of λ. Then*

$$\ell(w(T), k) = \lambda_1^* + \lambda_2^* + \cdots + \lambda_p^* \quad (1 \le k \le p).$$

$\square$

We now investigate when two sequences have the same insertion tableau.

Example. The sequences $2, 1, 3$ and $2, 3, 1$ have insertion tableau

1	2
3	×

,

and thus the words of their corresponding insertion tableaux are identical. The sequences $3, 1, 2$ and $1, 3, 2$ have the same insertion tableau

1	3
2	×

.

In language to be defined below, $2, 1, 3$ and $2, 3, 1$ are Knuth equivalent, as are $3, 1, 2$ and $1, 3, 2$. $\square$

A *Knuth transformation* of a sequence σ is a transformation of one of the following types:

$$\mathrm{K}_1 : \ldots, a, c, b, \ldots \longleftrightarrow \ldots, c, a, b, \ldots \text{ provided } a < b \le c,$$

and

$$\mathrm{K}_2 : \ldots, b, a, c, \ldots \longleftrightarrow \ldots, b, c, a, \ldots \text{ provided } a \le b < c.$$

Thus in a Knuth transformation a pair of adjacent integers can be interchanged provided there is an integer, between them as specified above, located directly on either side. By definition, the inverse of a Knuth transformation is a Knuth transformation. Two sequences are *Knuth equivalent* provided each can be obtained from the other by a sequence of Knuth transformations. Since the inverse of a Knuth transformation is also a Knuth transformation, Knuth equivalence is indeed an equivalence relation.

Lemma 4.7.3 *Let σ and σ' be Knuth equivalent sequences. Then for all k we have*

$$\ell(\sigma, k) = \ell(\sigma', k).$$

Proof. We need to consider the two cases:

(i) $\sigma = \pi, a, c, b, \tau$ and $\sigma' = \pi, c, a, b, \tau$ $(a < b \le c)$, and

(ii) $\sigma = \pi, b, a, c, \tau$ and $\sigma' = \pi, b, c, a, \tau$ $(\le b < c)$,

where π and τ are arbitrary sequences.

In both cases we have $\ell(\sigma, k) \ge \ell(\sigma', k)$ for all k, since any collection of k strictly increasing disjoint subsequences of σ' is also a collection of strictly increasing disjoint subsequences of σ'. This is because in (i) the order relations $a < b, a < c$ in the subsequence a, c, b of σ hold while only the order relation $a < b$ holds in the subsequence c, a, b of σ', while in (ii) it's $b < c, a < c$ for σ and $b < c$ for σ'. We now show that we also have $\ell(\sigma', k) \ge \ell(\sigma, k)$. We do this by showing that if we have k pairwise disjoint strictly increasing subsequences of σ', then there are k pairwise disjoint strictly increasing subsequences of σ of the same total length. This is certainly true unless one of the subsequences for σ contains a and c, that is, is of the form θ, a, c, ϕ where θ and ϕ are subsequences of σ. If none of the k subsequences for σ uses b, then in (i) we may substitute a, b for a, c while in (ii) we may substitute b, c for a, c, and get the same combined length in σ'. Now assume that one of the k subsequences for σ is of the form θ', b, ϕ'. Then we can replace θ, a, c, ϕ and θ', b, ϕ' with θ, a, b, ϕ and θ, c, ϕ in case (i), and by θ, b, c, ϕ and θ, a, ϕ' in case (ii). Since no other subsequences need be changed, the combined length is the same. Hence $\ell(\sigma', k) \ge \ell(\sigma, k)$, and so $\ell(\sigma, k) = \ell(\sigma', k)$. □

We require one more lemma.

Lemma 4.7.4 *Let σ and σ' be Knuth equivalent sequences. Let $\hat{\sigma}$ and $\hat{\sigma}'$ be obtained by removing the first occurrence of the largest integer from each of σ and σ', respectively. Then $\hat{\sigma}$ and $\hat{\sigma}'$ are Knuth equivalent. A similar conclusion holds when the first occurrence of the smallest integer is removed.*

Proof. By induction, it suffices to prove the lemma when σ' is obtained from σ by one Knuth transformation, that is, $\sigma = \pi, a, c, b, \tau$ and $\sigma' = \pi, c, a, b, \tau$, where $a < b \le c$, or $\sigma = \pi, b, a, c, \tau$ and $\sigma' = \pi, b, c, a, \tau$, where $a \le b < c$. If the integer removed is not c (it cannot be a or b), then $\hat{\sigma}$ and $\hat{\sigma}'$ are clearly Knuth equivalent. If c is removed (in the place shown), then $\hat{\sigma}$ and $\hat{\sigma}'$ are equal. □

We now come to the main result.

Theorem 4.7.5 *Every sequence σ of positive integers is Knuth equivalent to the word of a unique Young tableau, namely, the Young tableau constructed by applying the column-insertion algorithm to σ.*

Proof. First we establish that every sequence σ of positive integers is Knuth equivalent to the word of the Young tableau constructed by the column-insertion algorithm. This follows from the fact that the construction of a Young tableau from σ by column-insertion can be broken down into a sequence of Knuth transformations.[11] To see this, consider a Young tableau T and the insertion of an integer x into the tableau producing the new Young tableau $T \uparrow x$. The assertion is that $w(T \uparrow x)$ is Knuth equivalent to $w(T), x$ (the word of T with x appended at the right). Let column 1 of T be the strictly increasing sequence (from top to bottom) $a_1 < a_2 < \cdots < a_p$, thereby comprising the subword of $w(T)$ consisting of the last p integers of $w(T)$. If $x > a_p$, then x is placed at the end of column 1, and in this case $w(T \uparrow x)$ equals $w(T), x$. If $x = a_p$, then since $a_p = x$ is bumped, an operation K_1 replaces $a_1, a_2, \ldots, a_{p-1}, x, x$ with $a_1, a_2, \ldots, a_{p-2}, x, a_{p-1}, x$. If $x \leq a_{p-1} < a_p$, an operation K_2 replaces $a_1, a_2, \ldots, a_p, x$ with $a_1, a_2, \ldots, a_{p-1}, x, a_p$. We continue like this until the integer bumped to the second column in the insertion algorithm arrives at the position in $w(T)$ right before a_1 and hence immediately to the left of the sequence in $w(T)$ corresponding to the subword of $w(T)$ corresponding to column 2 of T. Thus the word of $T \uparrow x$ can be gotten from the word $w(T), x$ by a sequence of Knuth transformations. Since Knuth transformations are invertible, the word $w(T), x$ can also be gotten from $w(T \uparrow x)$ by Knuth transformations.

To complete the proof let T be a Young tableau such that σ is Knuth equivalent to $w(T)$. We show by induction on the length of the sequence σ that T is uniquely determined. This is obvious if the length of σ is 1. Assume that the length of σ is greater than 1. It follows from Lemmas 4.7.2 and 4.7.3 that the shape of T is uniquely determined by σ. Let the largest integer in σ be p. Let $\hat{\sigma}$ be the sequence obtained from σ by removing the first occurrence of p. Let $\hat{T}$ be the Young tableau obtained from T by removing p from its occurrence in the rightmost position of T. Using the notation in Lemma 4.7.4 we have $w(\hat{T})$ equals the word $\widehat{w(T)}$ obtained from $w(T)$ by removing the first occurrence of p. By Lemma 4.7.4, $\hat{\sigma}$ is Knuth equivalent to $w(\hat{T})$. By induction $\hat{T}$ is the unique Young tableau whose word is Knuth equivalent to $\hat{\sigma}$. Since the Young diagrams corresponding to T and $\hat{T}$ are identical except for one box, we have that T is obtained from $\hat{T}$ by inserting the integer p in that box. $\square$

Theorem 4.7.1 is now an immediate consequence of Lemmas 4.7.2 and 4.7.3 and Theorem 4.7.5.

---

[11] Knuth [32] defined the transformations that now bear his name as a way to frame the insertion or bumping operation.

References

[1] H. Anand, V.C. Dumir, and H. Gupta, A combinatorial distribution problem, *Duke Math. J.*, **33** (1966), 757–769.

[2] R.P. Anstee, Properties of a class of (0, 1)-matrices covering a given matrix, *Canad. J. Math*, **34** (1982), 438–453.

[3] R.P. Anstee, Triangular (0, 1)-matrices with prescribed row and column sums, *Discrete Math.*, **40** (1982), 1–10.

[4] R.P. Anstee, The network flows approach for matrices with given row and column sums, *Discrete Math.*, **44** (1983), 125–138.

[5] L.W. Beineke and F. Harary. Local restrictions for various classes of directed graphs, *J. London Math. Soc.*, **40** (1965), 87–95.

[6] A.V. Borovik, I.M. Gelfand, and N. White, *Coxeter Matroids*, Birkhäuser, Boston–Basle–Berlin, 2003.

[7] J.V. Brawley and L. Carlitz, Enumeration of matrices with prescribed row and column sums, *Linear Algebra Appl.*, **6** (1973), 165–174.

[8] R.A. Brualdi, Matrices of 0's and 1's with total support, *J. Combin. Theory, Ser. A*, **28** (1980), 249–256.

[9] R.A. Brualdi, Matrices of zeros and ones with fixed row and column sum vectors, *Linear Algebra Appl.*, **33** (1980), 159–231.

[10] R.A. Brualdi, Algorithms for constructing (0,1)-matrices with prescribed row and column sum vectors, *Discrete Math.*, to appear.

[11] R.A. Brualdi and G. Dahl, Matrices of zeros and ones with given line sums and a zero block, *Linear Algebra Appl.*, **371** (2003), 191–207.

[12] R.A. Brualdi and G. Dahl, Constructing (0,1)-matrices with given line sums and certain fixed zeros, to appear.

[13] R.A. Brualdi and L. Deaett, More on the Bruhat order for (0, 1)-matrices, preprint 2005.

215

[14] R.A. Brualdi, D.J. Hartfiel, and S.-G. Hwang, On assignment functions, *Linear Multilin. Alg.*, **19** (1986), 203–219.

[15] R.A. Brualdi and S.-G. Hwang, A Bruhat order for the class of (0,1)-matrices with row sum vector R and column sum vector S, *Electron. J. Linear Algebra*, **12** (2004), 6–16.

[16] R.A. Brualdi and B. Liu, A lattice generated by (0,1)-matrices, *Ars Combin.*, **31** (1991), 183–190.

[17] R.A. Brualdi, S.V. Parter, and H. Schneider, The diagonal equivalence of a nonnegative matrix to a stochastic matrix, *J. Math. Anal. Appl.*, 16 (1966), 31–50.

[18] R.A. Brualdi and J.A. Ross, On Ryser's maximum term rank formula, *Linear Algebra Appl.*, **29** (1980), 33–38.

[19] R.A. Brualdi and H.J. Ryser, *Combinatorial Matrix Theory*, Cambridge U. Press, Cambridge, 1991.

[20] E.R. Canfield and B.D. McKay, Asymptotic enumeration of 0–1 matrices with constant row and column sums, *Electron. J. Combin.*, to appear.

[21] Y.C. Chen, A counterexample to a conjecture of Brualdi and Anstee (in Chinese), *J. Math. Res. Exposition*, **6** (1986), 68.

[22] Y.C. Chen, Integral matrices with given row and column sums, *J. Combin. Theory, Ser. A*, **2** (1992), 153–172.

[23] Y.C. Chen and A. Shastri, On joint realization of (0, 1)-matrices, *Linear Algebra Appl.*, **112** (1989), 75–85.

[24] H.H. Cho, C.Y. Hong, S.-R. Kim, C.H. Park, and Y. Nam, On $\mathcal{A}(R, S)$ all of whose members are indecomposable, *Linear Algebra Appl.*, **332/334** (2001), 119–129.

[25] L. Comtet, *Advanced Combinatorics*, Reidel, Dordrecht, 1974.

[26] L. R. Ford, Jr. and D. R. Fulkerson, *Flows in Networks*, Princeton U. Press, Princeton, 1962.

[27] D.R. Fulkerson, Zero–one matrices with zero trace, *Pacific J. Math.*, **10** (1960), 831–836.

[28] W. Fulton, *Young Tableaux*, London Math. Soc. Student Texts 35, Cambridge U. Press, Cambridge, 1997.

[29] C. Greene, An extension of Schensted's theorem, *Advances Math.*, **14** (1974), 254–265.

[30] J.A. Henderson, Jr., On the enumeration of rectangular $(0, 1)$-matrices, *J. Statist. Comput. Simul.*, **51** (1995), 291–313.

[31] D. Jungnickel and M. Leclerc, A class of lattices, *Ars Combin.*, **26** (1988), 243–248.

[32] D.E. Knuth, Permutation matrices and generalized Young tableaux, *Pacific J. Math*, **34** (1970), 709–727.

[33] J.L. Lewandowski, C.L. Liu, and J.W.S. Liu, An algorithmic proof of a generalization of the Birkhoff–von Neumann theorem, *J. Algorithms*, **7** (1986), 323–330.

[34] Q. Li and H.H. Wan, Generating function for cardinalities of several (0,1)-matrix classes, preprint.

[35] P. Magyar, Bruhat order for two flags and a line, *J. Algebraic Combin.*, **21** (2005), 71–101.

[36] K. McDougal, The joint realization of classes of $(0, 1)$ matrices with given row and column sum vectors, Proceedings of the Twenty-Third Southeastern International Conference on Combinatorics, Graph Theory, and Computing (Boca Raton, FL, 1992), *Cong. Numer.*, **91** (1992), 201–216.

[37] K. McDougal, On asymmetric (0,1)-matrices with given row and column vectors, *Discrete Math.*, **137** (1995), 377–381.

[38] K. McDougal, A generalization of Ryser's theorem on term rank, *Discrete Math.*, **170** (1997), 283–288.

[39] B.D. McKay, Asymptotics for 0–1 matrices with prescribed line sums, *Enumeration and Design*, ed. D. M. Jackson and S.A. Vanstone, Academic Press, New York, 1984, 225–238.

[40] B.D. McKay and X. Wang, Asymptotic enumeration of 0–1 matrices with equal row sums and equal column sums, *Linear Algebra Appl.*, **373** (2003), 273–287.

[41] Y. Nam, Integral matrices with given row and column sums, *Ars Combin.*, **52** (1999), 141–151.

[42] G. deB. Robinson, On the representations of the symmetric group, *Amer. J. Math.*, **60** (1938), 745–760.

[43] C. Schensted, Longest increasing and decreasing subsequences, *Canad. J. Math.*, **13** (1961), 179–191.

[44] R.P. Stanley, *Enumerative Combinatorics Vol. II*, Cambridge Studies in Adv. Math., Cambridge U. Press, Cambridge, 1998.

[45] H.H. Wan, Structure and cardinality of the class $\mathcal{A}(R, S)$ of $(0, 1)$-matrices, *J. Math. Res. Exposition*, **4** (1984), 87–93.

[46] H.H. Wan, Cardinal function $f(R, S)$ of the class $\mathcal{A}(R, S)$ and its nonzero point set, *J. Math. Res. Exposition*, **5** (1985), 113–116.

[47] H.H Wan, Cardinality of a class of $(0, 1)$-matrices covering a given matrix, *J. Math. Res. Exposition*, **6** (1984), 33–36.

[48] H.H. Wan, Generating functions and recursion formula for $|\mathcal{A}(R, S)|$ (in Chinese), *Numerical Mathematics: A Journal of Chinese Universities*, **6** (4), 319–326.

[49] W.D. Wei, The class $\mathcal{A}(R, S)$ of (0,1)-matrices, *Discrete Math.*, **39** (1982), 301–305.

[50] B.Y. Wang, Precise number of (0,1)-matrices in $\mathcal{A}(R, S)$, *Scientia Sinica, Ser. A*, 1 (1988), 1–6.

[51] B.Y. Wang and F. Zhang, On the precise number of (0,1)-matrices in $\mathcal{A}(R, S)$, *Discrete Math.*, **187** (1998), 211–220.

[52] B.Y. Wang and F. Zhang, On normal matrices of zeros and ones with fixed row sum, *Linear Algebra Appl.*, **275**–**276** (1998), 517–526.

5

The Class $\mathcal{T}(R)$ of Tournament Matrices

In this chapter we study the class $\mathcal{T}(R)$ of tournament matrices with a prescribed row sum vector R, equivalently, tournaments with score vector R. We show that Landau's inequalities for R from Chapter 2 can be individually strengthened, although collectively the two sets of inequalities are equivalent. These strengthened inequalities are used to show the existence of a special "half-transitive" tournament matrix in $\mathcal{T}(R)$. We also investigate the values of certain natural combinatorial parameters over the class $\mathcal{T}(R)$ and the number of matrices in $\mathcal{T}(R)$. As with the class $\mathcal{A}(R,S)$, there is a naturally defined graph associated with $\mathcal{T}(R)$, and this graph is studied in the next chapter.

5.1 Algorithm for a Matrix in $\mathcal{T}(R)$

Let $R = (r_1, r_2, \ldots, r_n)$ be a nonnegative integral vector such that

$$r_1 \leq r_2 \leq \cdots \leq r_n, \tag{5.1}$$

and

$$r_1 + r_2 + \cdots + r_n = \binom{n}{2}. \tag{5.2}$$

In Section 2.2 we proved that (5.1), (5.2), and the Landau inequalities

$$\sum_{i=1}^{k} r_i \geq \binom{k}{2} \quad (k = 1, 2, \ldots, n-1) \tag{5.3}$$

are necessary and sufficient for the existence of a tournament matrix with row sum vector R. Ryser [24] gave a construction for a matrix $\widetilde{A}$ in $\mathcal{T}(R)$

which is the analog of the matrix $\widetilde{A}$ in $\mathcal{A}(R,S)$ constructed in Section 3.1. Let $S = (s_1, s_2, \ldots, s_n)$, where $s_i = n - 1 - r_i$, $(i = 1, 2, \ldots, n)$, be the column sum vector of a matrix in $\mathcal{T}(R)$. From (5.1) we get that $s_1 \geq s_2 \geq \cdots \geq s_n$.

Ryser's Algorithm for Constructing a Matrix in $\mathcal{T}(R)$

(1) Begin with the zero matrix $A^{(n)} = O$ of order n.

(2) In the $n - 1$ nondiagonal positions of column n of A^n, insert $s_n = (n-1) - r_n$ 1's in those positions corresponding to the s_n largest row sums, giving preference to the topmost positions in case of ties. Next insert 1's in row n in the complementary transposed positions thereby achieving the row sum r_n. The result is a matrix $A^{(n-1)}$ of order n whose leading principal submatrix of order $n - 1$ is a zero matrix, and whose nth row and column sums are r_n and s_n, respectively. Let the vectors $R^{(n-1)} = (r_1^{(n-1)}, r_2^{(n-1)}, \ldots, r_{n-1}^{(n-1)})$ and $S^{(n-1)} = (s_1^{(n-1)}, s_2^{(n-1)}, \ldots, s_{n-1}^{(n-1)})$ be obtained from R and S by subtracting, respectively, the row sum vector and column sum vector of $A^{(n-1)}$ and then deleting the last component (equal to 0). We continue recursively.

(3) For $k = n - 1, n - 2, \ldots, 2$, do:

In the nondiagonal positions of column k of the principal submatrix $A^{(k)}[\{1, 2, \ldots, k\}] = O$ of $A^{(k)}$, insert $s_k^{(k)}$ 1's in those positions corresponding to the $s_k^{(k)}$ largest row sums, giving preference to the topmost positions in case of ties. Next insert 1's in the complementary transposed positions in row k of $A^{(k)}$ thereby achieving row sum r_k. This results in a matrix $A^{(k-1)}$, and we let $R^{(k-1)} = (r_1^{(k-1)}, r_2^{(k-1)}, \ldots, r_{k-1}^{(k-1)})$ and $S^{(k-1)} = (s_1^{(k-1)}, s_2^{(k-1)}, \ldots, s_{k-1}^{(k-1)})$ be the vectors obtained from R and S by subtracting, respectively, the row sum vector and column sum vector of $A^{(k-1)}$ and then deleting the last component.

(4) Output $\widetilde{A} = A^{(1)}$.

Theorem 5.1.1 *Let $R = (r_1, r_2, \ldots, r_n)$ be a nonnegative integral vector satisfying (5.1), (5.2), and (5.3). Then the Ryser algorithm can be completed and the constructed matrix $\widetilde{A}$ is a tournament matrix with row sum vector R.*

Proof. It follows from (5.1), (5.2), and (5.3) that $r_n \leq n - 1$ and then that $0 \leq s_n \leq n - 1$. Thus step (2) of the algorithm can be carried out. It now suffices to show that for each $k = n - 1, n - 2, \ldots, 2$, the components of the vector $R^{(k)} = (r_1^{(k)}, r_2^{(k)}, \ldots, r_k^{(k)})$ are nondecreasing and satisfy the

Landau conditions

$$r_1^{(k)} + r_2^{(k)} + \cdots + r_k^{(k)} = \binom{k}{2} \tag{5.4}$$

and

$$r_1^{(k)} + r_2^{(k)} + \cdots + r_e^{(k)} \geq \binom{e}{2} \quad (e = 1, 2, \ldots, k - 1). \tag{5.5}$$

Arguing by induction, we need only verify (5.4) and (5.5) for $k = n - 1$.

For each $i = 1, 2, \ldots, n - 1$, $r_i^{(n-1)}$ is equal to r_i or $r_i - 1$. Since the 1's in column n are placed in those rows with the largest row sums, giving preference to the topmost positions in case of ties, the vector $R^{(n-1)}$ is a nondecreasing, nonnegative integral vector. We have

$$\sum_{i=1}^{n-1} r_i^{(n-1)} = \sum_{i=1}^{n} r_i - (r_n + s_n) = \binom{n}{2} - (n-1) = \binom{n-1}{2}.$$

Suppose that there exists an integer p with $1 \leq p < n - 1$ such that

$$\sum_{i=1}^{p} r_i^{(n-1)} < \binom{p}{2}, \tag{5.6}$$

and assume p is the smallest such integer. By minimality,

$$\sum_{i=1}^{p-1} r_i^{(n-1)} \geq \binom{p-1}{2}. \tag{5.7}$$

Inequalities (5.6) and (5.7) together imply that $r_p^{(n-1)} \leq p - 1$ and hence

$$r_p \leq p. \tag{5.8}$$

Let the number of 1's in the first p positions of column n of $A^{(n-1)}$ be t. If $t = 0$, then (5.2) implies that

$$\sum_{i=1}^{p} r_i^{(n-1)} = \sum_{i=1}^{p} r_i \geq \binom{p}{2}$$

contradicting (5.6). Hence $t > 0$. Let $l = p$ if there are no 0's in positions $p+1, p+2, \ldots, n-1$ of column n of $A^{(n)}$; otherwise, let l be the position of the bottommost 0 in these positions of column n. Step (2) of the algorithm implies that

$$r_p = r_{p+1} = \cdots = r_l. \tag{5.9}$$

From the definitions of p and l, column n contains at least $t + (n-1) - l$ 1's, and hence

$$r_n = n - 1 - s_n \leq l - t. \tag{5.10}$$

Using (5.6), (5.8), (5.9), and (5.10), we get

$$
\binom{n}{2} = \sum_{i=1}^{n} r_i
$$

$$
= \sum_{i=1}^{p} r_i^{(n-1)} + t + \sum_{i=p+1}^{l} r_i + \sum_{i=l+1}^{n-1} r_i + r_n
$$

$$
< \binom{p}{2} + \sum_{i=p+1}^{l} r_i + \sum_{i=l+1}^{n-1} r_i + l
$$

$$
\leq \binom{n}{2}.
$$

This contradiction completes the verification of the validity of Ryser's algorithm.   □

Example. If $R = (1, 2, 2, 2, 4, 4)$ and $S = (4, 3, 3, 3, 1, 1)$. Then the tournament matrix constructed from Ryser's algorithm is

$$
\tilde{A} = \begin{bmatrix}
0 & 0 & 1 & 0 & 0 & 0 \\
1 & 0 & 0 & 0 & 1 & 0 \\
0 & 1 & 0 & 1 & 0 & 0 \\
1 & 1 & 0 & 0 & 0 & 0 \\
1 & 0 & 1 & 1 & 0 & 1 \\
1 & 1 & 1 & 1 & 0 & 0
\end{bmatrix}.
$$

     □

5.2   Basic Properties of Tournament Matrices

We first observe that the rank of a generalized tournament matrix of order n is either $n - 1$ or n [10].

Theorem 5.2.1 *Let A be a generalized tournament matrix of order n. Then the rank of A is at least $n - 1$.*

     Proof. The matrix A satisfies the equation $A + A^{\mathrm{T}} = J_n - I_n$. Let B be the n by $n + 1$ matrix obtained from A by appending a column vector $e = 1^n$ of all 1's. Suppose that for some real column vector x of size n, $x^{\mathrm{T}} B = 0$. Then $x^{\mathrm{T}} e = 0$ and $x^{\mathrm{T}} A = 0$, and hence $A^{\mathrm{T}} x = 0$. Since A is a generalized tournament matrix, we have

$$
x^{\mathrm{T}}(J_n - I_n)x = x^{\mathrm{T}}(A + A^{\mathrm{T}})x = (x^{\mathrm{T}} A)x + x^{\mathrm{T}}(A^{\mathrm{T}} x) = 0.
$$

Since $x^{\mathrm{T}} J_n x = (x^{\mathrm{T}} J_n)x = 0$, we conclude that $-x^{\mathrm{T}} x = 0$ and hence x is a zero vector. Therefore the rank of B equals n and the rank of A is at least $n - 1$.   □

     It is to be noted that it was not necessary to invoke the nonnegativity of A in the proof of Theorem 5.2.1. Since the term rank of a matrix is always at least as large as its rank, we obtain the following corollary.

Corollary 5.2.2 *The term rank of a generalized tournament matrix of order n equals $n-1$ or n.*   □

By definition a tournament has no directed cycles of lengths 1 or 2. A *transitive tournament* is a tournament that has no directed cycles of length 3, and hence no directed cycles of any length. A *transitive tournament matrix* is a tournament matrix A whose associated tournament $T(A)$ is transitive. Thus a tournament matrix $A = [a_{ij}]$ is transitive if and only if

$$a_{ij} + a_{jk} + a_{ki} \geq 1 \quad (i, j, k \text{ distinct}).$$

Since we must also have

$$a_{ik} + a_{kj} + a_{ji} \geq 1,$$

and since $a_{rs} + a_{sr} = 1$ for $r \neq s$, it follows that a transitive tournament matrix $A = [a_{ij}]$ satisfies

$$2 \geq a_{ij} + a_{jk} + a_{ki} \geq 1 \quad (i, j, k \text{ distinct}).$$

The digraph of an irreducible matrix is strongly connected, and a strongly connected digraph with more than one vertex has a directed cycle. Hence the irreducible components of a transitive tournament matrix are zero matrices of order 1. It follows that a tournament matrix A is transitive if and only if there is a permutation matrix P such that

$$PAP^{\mathrm{T}} = L_n - I_n$$

where L_n is the triangular $(0,1)$-matrix of order n with 1's on and below the main diagonal. Equivalently, a tournament matrix of order n is transitive if and only if its row sum vector is a permutation of the vector $(0, 1, 2, \ldots, n-1)$.

The next theorem implies that whether or not a tournament matrix in $\mathcal{T}(R)$ is irreducible is determined solely by R.

Theorem 5.2.3 *Let n be a positive integer and let $R = (r_1, r_2, \ldots, r_n)$ be a nondecreasing, nonnegative integral vector. The following are equivalent.*

(i) *There exists an irreducible matrix in $\mathcal{T}(R)$.*

(ii) *$\mathcal{T}(R)$ is nonempty and every matrix in $\mathcal{T}(R)$ is irreducible.*

(iii) *$\sum_{i=1}^{k} r_i \geq \binom{k}{2}$ $(k = 1, 2, \ldots, n)$, with equality if and only if $k = n$.*

Proof. Let A be any matrix in $\mathcal{T}(R)$. Suppose that A is reducible. Then there exists a permutation matrix P of order n such that

$$PAP^{\mathrm{T}} = \begin{bmatrix} A_1 & O \\ A_{21} & A_2 \end{bmatrix}, \tag{5.11}$$

where A_1 is a tournament matrix of order k with $1 \leq k \leq n - 1$. Since PAP^{T} is also a tournament matrix it follows that $A_{21} = J$. This implies that each of the first k row sums of PAP^{T} does not exceed $k - 1$, and each of the last $n - k$ row sums is at least equal to k. Since R is assumed to be nondecreasing, this implies that (5.11) holds with $P = I_n$, and also that $\sum_{i=1}^{k} r_i = \binom{k}{2}$.

Conversely, suppose that for some integer k with $1 \leq k \leq n-1$, we have

$$\sum_{i=1}^{k} r_i = \binom{k}{2}.$$

Since $A[\{1, 2, \ldots, k\}]$ is a tournament matrix of order k for each A in $\mathcal{T}(R)$, each matrix A in $\mathcal{T}(R)$ has the form

$$\begin{bmatrix} A_1 & O \\ J_{n-k,k} & A_2 \end{bmatrix},$$

and hence A is reducible. The theorem now follows from Landau's theorem. $\qquad\square$

If R is a vector satisfying (iii) in Theorem 5.2.3, then every tournament with score vector R is strongly connected, and if R does not satisfy these conditions, then no tournament with score vector R is strongly connected. A nondecreasing, nonnegative integral vector $R = (r_1, r_2, \ldots, r_n)$ is called a *strong score vector* or a *strong row sum vector* provided R satisfies the conditions (iii) in Theorem 5.2.3.

A *directed Hamilton cycle* in a digraph with n vertices is a directed cycle of length n, that is, a directed cycle that passes through all the vertices. The following theorem is due to Camion [12].

Theorem 5.2.4 *Every strongly connected tournament T of order $n \geq 3$ has a directed Hamilton cycle.*

Proof. Since $n \geq 3$, T has a directed cycle

$$\gamma : x_1 \rightarrow x_2 \rightarrow \cdots \rightarrow x_m \rightarrow x_1,$$

and we choose γ so that its length m is as large as possible. Suppose that $m < n$. If there is a vertex y not on γ for which there are both an edge from y to a vertex of γ and an edge from a vertex of γ to y, then there exists an integer i such that (x_i, y) and (y, x_{i+1}) are edges of T (subscripts interpreted modulo m). But then

$$x_1 \rightarrow \cdots \rightarrow x_i \rightarrow y \rightarrow x_{i+1} \rightarrow \cdots \rightarrow x_m \rightarrow x_1$$

is a directed cycle of length $m + 1$, contradicting the maximality property of γ. Thus for each vertex y not on γ, either all edges between y and the vertices of γ go out of y (such vertices y are put in a set U) or they all go into y (these vertices are put in a set V). There must be an edge from a vertex v in V to a vertex u in U, for otherwise T is not strongly connected. But then

$$v \to u \to x_1 \to x_2 \to \cdots \to x_m \to v$$

is a directed cycle of length $m + 2$, again contradicting the maximality of γ. Hence $m = n$.   □

If A is a tournament matrix of order n whose corresponding tournament has a directed Hamilton cycle, then there is a permutation matrix P such that

$$C_n \le PAP^{\mathrm{T}}$$

where C_n denotes the cyclic permutation matrix with 1's in each of the positions $(1,2), (2,3), \ldots, (n{-}1, n)$, and $(n, 1)$. In particular, the term rank of A equals n. If, however, the tournament corresponding to A is transitive, then there is a permutation matrix P such that $PAP^{\mathrm{T}} = L_n - I_n$ where, as before, L_n is the triangular $(0,1)$-matrix with 1's on and below the main diagonal. In this case the term rank of A is $n - 1$. Using the canonical form under simultaneous row and column permutations (see Section 2.2), we see again that the term rank of a tournament matrix equals $n - 1$ or n. The term rank equals $n - 1$ if and only if A has an irreducible component of order 1 [24].

The following lemma describes an elementary property of tournaments.

Lemma 5.2.5 *Let* $R_n = (r_1, r_2, \ldots, r_n)$ *be the score vector of a tournament* T. *Then the number of 3-cycles of* T *is*

$$k_3(R_n) = \binom{n}{3} - \sum_{i=1}^{n} \binom{r_i}{2}.$$

In particular, the number of 3-cycles of a tournament depends only on its score vector.

Proof. Let the vertices of T be $p_1, p_2, \ldots, p_n$ where p_i has outdegree r_i. There are $\binom{r_i}{2}$ pairs $\{p_k, p_l\}$ of distinct vertices such that there are edges from p_i to p_k and p_l. Each such triple $\{p_i, p_k, p_l\}$ induces a transitive subtournament of order 3. Every other triple induces a subtournament of order 3 which is a 3-cycle. Hence the number of 3-cycles of T equals

$$\binom{n}{3} - \sum_{i=1}^{n} \binom{r_i}{2}.$$

□

We now investigate edge-disjoint 3-cycles in tournaments, and prove a theorem of Gibson [17].

Theorem 5.2.6 *Let $R_n = (r_1, r_2, \ldots, r_n)$ be a strong score vector with $n \geq 3$. Then there exists a tournament T in $\mathcal{T}(R_n)$ with at least $\lfloor (n-1)/2 \rfloor$ edge-disjoint 3-cycles. For the strong score vector $\hat{R}_n = (1, 1, 2, 3, \ldots, n - 3, n - 2, n - 2)$, each matrix in $\mathcal{T}(\hat{R}_n)$ has at most $\lfloor (n-1)/2 \rfloor$ edge-disjoint 3-cycles.*

Proof. If $n = 3$ or 4, the theorem is easily verified. We assume that $n \geq 5$ and proceed by induction on n. Let T be in $\mathcal{T}(R_n)$. By Theorem 5.2.4, T contains a directed Hamilton cycle

$$\gamma : x_1 \to x_2 \to \cdots \to x_n \to x_1.$$

Reading the subscripts on the x_i modulo n, we distinguish three cases.

Case 1: $x_i \to x_{i+3}$ and $x_{j+3} \to x_j$ are edges of T for some i, j with $1 \leq i < j \leq n$. The assumptions of this case imply that for some $k = 1, 2, \ldots, n$ we have $x_k \to x_{k+3}$ and $x_{k+4} \to x_{k+1}$. First suppose that $x_{k+3} \to x_{k+1}$ is an edge of T. Then

$$x_{k+1} \to x_{k+2} \to x_{k+3} \to x_{k+1}$$

is a 3-cycle in T. The subtournament T' of T of order $n - 2$ obtained by deleting vertices x_{k+1} and x_{k+2} has a directed Hamilton cycle and hence has a strong score vector R_{n-2}. Applying the inductive assumption to $\mathcal{T}(R_{n-2})$ we conclude that $\mathcal{T}(R_{n-2})$ contains a tournament T_1 with at least $\lfloor (n-3)/2 \rfloor$ edge-disjoint 3-cycles. Replacing T' in T with T_1 we obtain a tournament in $\mathcal{T}(R_n)$ with at least

$$\left\lfloor \frac{n-3}{2} \right\rfloor + 1 = \left\lfloor \frac{n-1}{2} \right\rfloor$$

edge-disjoint 3-cycles. If instead $x_{k+1} \to x_{k+3}$ is an edge in T, a similar argument can be applied to the tournament in $\mathcal{T}(R_n)$ obtained from T by reversing the directions of the edges of the directed Hamilton cycle.

Case 2: $x_i \to x_{i+3}$ is an edge of T for each $i = 1, 2, \ldots, n$. First suppose that $x_i \to x_{i+2}$ is an edge of T for $i = 1, 2, \ldots, n$. Then

$$x_1 \to x_3 \to x_4 \to \cdots \to x_{n-3} \to x_{n-1} \to x_1 \ (n \text{ even})$$

or

$$x_1 \to x_3 \to x_5 \to \cdots \to x_{n-2} \to x_n$$
$$\to x_2 \to x_4 \to \cdots \to x_{n-3} \to x_{n-1} \to x_1 \ (n \text{ odd})$$

is a directed cycle in T. Let T^* be the tournament in $\mathcal{T}(R_n)$ obtained from T by reversing the edges of this cycle. Then in T', $x_i \to x_{i+3}$ is an edge for $i = 1, 2, \ldots, n$ and $x_3 \to x_1$ is also an edge in T'. Then we may proceed with T^* as in Case 1 by considering the subtournament of T^* of order $n - 2$ obtained by deleting the vertices x_2 and x_3 and obtain a tournament in $\mathcal{T}(R_n)$ with at least $\lfloor (n-1)/2 \rfloor$ edge-disjoint cycles.

Case 3: $x_{i+3} \to x_i$ is an edge of T for each $i = 1, 2, \ldots, n$. In this case we may apply the argument in Case 2 to the tournament in $\mathcal{T}(R_n)$ obtained by reversing the edges of the directed Hamilton cycle γ.

The proof of the theorem is now complete.   □

Finally we discuss invariant positions for a nonempty tournament class $\mathcal{T}(R_n)$. An *invariant position* for $\mathcal{T}(R_n)$ is a position (i, j) with $i \neq j$ such that every matrix in $\mathcal{T}(R_n)$ has a 1 in that position (an *invariant 1-position*) or every matrix in $\mathcal{T}(R_n)$ has a 0 in that position (an *invariant 0-position*). For $i \neq j$, (i, j) is an invariant 1-position if and only if (j, i) is an invariant 0-position. Assume that R_n is a nondecreasing vector of size n. First suppose that $n > 1$ and every matrix in $\mathcal{T}(R_n)$ is irreducible (see Theorem 5.2.3). Let A be a matrix in $\mathcal{T}(R_n)$ and let T be its associated tournament. Then T is a strongly connected digraph and hence each edge is contained in a directed cycle. Reversing the direction of edges on a directed cycle of T gives a tournament T' whose corresponding tournament matrix is in $\mathcal{T}(R_n)$. Hence $\mathcal{T}(R_n)$ has no invariant positions. Now suppose that every matrix in $\mathcal{T}(R_n)$ is reducible (again see Theorem 5.2.3). Then every matrix in $\mathcal{T}(R_n)$ is of the form

$$A = \begin{bmatrix} A_1 & O_{k,n-k} \\ J_{n-k,k} & A_2 \end{bmatrix}$$

where $1 \leq k \leq n - 1$. The positions in A occupied by the 1's of $J_{k,n-k}$ are invariant 1-positions, and the positions occupied by the 0's of $O_{k,n-k}$ are invariant 0-positions. Therefore, the class $\mathcal{T}(R_n)$ does not have invariant positions if and only if some matrix, and hence every matrix, in $\mathcal{T}(R_n)$ is irreducible.

5.3   Landau's Inequalities

According to Theorem 2.2.2, if n is a positive integer and $R = (r_1, r_2, \ldots, r_n)$ is a nondecreasing, nonnegative integral vector, then the class $\mathcal{T}(R)$ of tournament matrices with row sum vector R is nonempty if and only if the Landau conditions are satisfied:

$$\sum_{i=1}^{k} r_i \geq \binom{k}{2} \quad (k = 1, 2, \ldots, n), \tag{5.12}$$

with equality when $k = n$. Since $r_1 \leq r_2 \leq \cdots \leq r_n$, the above inequalities are equivalent to

$$\sum_{i \in I} r_i \geq \binom{|I|}{2} \quad (I \subseteq \{1, 2, \ldots, n\}), \tag{5.13}$$

with equality when $I = \{1, 2, \ldots, n\}$.

The following theorem of Brualdi and Shen [9] shows that the row sum vector of a tournament matrix satisfies inequalities that are individually stronger than the inequalities (5.13), although collectively the two sets of inequalities are equivalent.

Theorem 5.3.1 *Let n be a positive integer and let $R = (r_1, r_2, \ldots, r_n)$ be a nondecreasing, nonnegative integral vector. Then $\mathcal{T}(R)$ is nonempty if and only if*

$$\sum_{i \in I} r_i \geq \frac{1}{2} \sum_{i \in I} (i - 1) + \frac{1}{2} \binom{|I|}{2} \quad (I \subseteq \{1, 2, \ldots, n\}), \tag{5.14}$$

with equality when $I = \{1, 2, \ldots, n\}$.

Proof. Since

$$\frac{1}{2} \sum_{i \in I} (i - 1) \geq \frac{1}{2} \sum_{i=1}^{|I|} (i - 1) = \frac{1}{2} \binom{|I|}{2},$$

the inequalities (5.14) imply the inequalities in (5.13) and equality holds when $I = \{1, 2, \ldots, n\}$. Hence there exists a matrix in $\mathcal{T}(R)$.

Now suppose that $\mathcal{T}(R)$ is nonempty so that R satisfies Landau's inequalities (5.13). For each subset J of $\{1, 2, \ldots, n\}$ we define

$$f(J) = \sum_{i \in J} r_i - \frac{1}{2} \sum_{i \in J} (i - 1) - \frac{1}{2} \binom{|J|}{2}.$$

We choose a subset I of $\{1, 2, \ldots, n\}$ of minimal cardinality such that

$$f(I) = \min\{f(J) : J \subseteq \{1, 2, \ldots, n\}\}.$$

Suppose that $I \neq \{i : 1 \leq i \leq |I|\}$. Then there exist $i \notin I$ and $j \in I$ such that $j = i + 1$. For this i and j we have $r_i \leq r_j$. The minimality assumptions imply that

$$r_j - \frac{1}{2}(j + |I| - 2) = f(I) - f(I \setminus \{j\}) < 0$$

and

$$r_i - \frac{1}{2}(i + |I| - 1) = f(I \cup \{i\}) - f(I) \geq 0.$$

These inequalities imply the contradiction

$$\frac{1}{2}(i + |I| - 1) \leq r_i \leq r_j < \frac{1}{2}(j + |I| - 2) = \frac{1}{2}(i + |I| - 1).$$

Hence $I = \{1, 2, \ldots, |I|\}$.

Therefore

$$f(I) = \sum_{i=1}^{|I|} r_i - \frac{1}{2}\sum_{i=1}^{|I|}(i-1) - \frac{1}{2}\binom{|I|}{2} = \sum_{i=1}^{|I|} r_i - \binom{|I|}{2} \geq 0,$$

where the last inequality follows from Landau's inequalities. By the choice of the subset I, $f(J) \geq f(I) \geq 0$ for all $J \subseteq \{1,2,\ldots,n\}$; moreover $f(I) = 0$ when $I = \{1,2,\ldots,n\}$.   □

Example. Let n be a positive integer. If n is odd, then the vector $R^{(n)} = ((n-1)/2, (n-1)/2, \ldots, (n-1)/2)$ of size n is the row sum vector of a tournament matrix of order n. Such tournament matrices are called *regular tournament matrices* and their column sum vectors also equal $R^{(n)}$. Examples of regular tournament matrix of order n are the circulant matrices

$$\text{Circ}_n(0,1,\ldots,1,0,\ldots,0) = C_n + C_n^2 + \cdots + C_n^{(n-1)/2}$$

and

$$\text{Circ}_n(0,0,\ldots,0,1,\ldots,1) = C_n^{(n+1)/2} + C_n^{(n+3)/2} + \cdots + C_n^{n-1}$$

determined by the indicated vectors with $(n-1)/2$ 1's. For the row sum vector $R^{(n)}$, equality holds in (5.14) for all I of the form $I = \{n-k+1, n-k+2, \ldots, n\}$.

If n is even, then $R^{(n)} = ((n-2)/2, \ldots, (n-2)/2, n/2, \ldots, n/2)$, where $(n-2)/2$ and $n/2$ each occur $n/2$ times, is the row sum vector of a tournament matrix, called a *near-regular tournament matrix*. An example of a near-regular tournament matrix of order n is

$$\begin{bmatrix} L_{n/2} & L_{n/2}^{\mathrm{T}} \\ I_{n/2} + L_{n/2}^{\mathrm{T}} & L_{n/2} \end{bmatrix},$$

where $L_{n/2}$ is the lower triangular matrix of order $n/2$ with 0's on and above the main diagonal and 1's below the main diagonal. When $n = 8$, this matrix is

$$\begin{bmatrix} 0 & 0 & 0 & 0 & 0 & 1 & 1 & 1 \\ 1 & 0 & 0 & 0 & 0 & 0 & 1 & 1 \\ 1 & 1 & 0 & 0 & 0 & 0 & 0 & 1 \\ 1 & 1 & 1 & 0 & 0 & 0 & 0 & 0 \\ 1 & 1 & 1 & 1 & 0 & 0 & 0 & 0 \\ 0 & 1 & 1 & 1 & 1 & 0 & 0 & 0 \\ 0 & 0 & 1 & 1 & 1 & 1 & 0 & 0 \\ 0 & 0 & 0 & 1 & 1 & 1 & 1 & 0 \end{bmatrix}.$$

□

Theorem 5.3.1 implies an upper bound for row sums of a tournament matrix [9].

Corollary 5.3.2 *Let n be a positive integer and let $R = (r_1, r_2, \ldots, r_n)$ be a nondecreasing, nonnegative integral vector. Then $\mathcal{T}(R)$ is nonempty if and only if*

$$\sum_{i \in I} r_i \leq \frac{1}{2} \sum_{i \in I} (i-1) + \frac{1}{4} |I| \, (2n - |I| - 1) \quad (I \subseteq \{1, 2, \ldots, n\}),$$

with equality if $I = \{1, 2, \ldots, n\}$.

Proof. Let $J = \{1, 2, \ldots, n\} \setminus I$. Since

$$\sum_{i \in I} r_i + \sum_{j \in J} r_i = \binom{n}{2},$$

$$\sum_{i \in I} r_i \leq \frac{1}{2} \sum_{i \in I} (i-1) + \frac{1}{4} |I| \, (2n - |I| - 1),$$

if and only if

$$\sum_{i \in J} r_i = \binom{n}{2} - \sum_{i \in I} r_i \geq \frac{1}{2} \sum_{i \in J} (i-1) + \frac{1}{2} \binom{|J|}{2}.$$

Corollary 5.3.2 now follows from Theorem 5.3.1.   □

Theorem 5.3.1 and Corollary 5.3.2 can be regarded as generalizations of the following corollary due to Landau [20].

Corollary 5.3.3 *Let $R = (r_1, r_2, \ldots, r_n)$ be a nondecreasing vector such that $\mathcal{T}(R)$ is nonempty. Then*

$$\frac{i-1}{2} \leq r_i \leq \frac{n+i-2}{2} \quad (i = 1, 2, \ldots, n).$$

Proof. The corollary follows immediately from Theorem 5.3.1 and Corollary 5.3.2 by taking $I = \{i\}$.   □

In [9] the implications of instances of equality in (5.14) for the structure of tournaments in a class $\mathcal{T}(R)$ are investigated.

5.4   A Special Matrix in $\mathcal{T}(R)$

Ao and Hansen [3] and Guiduli, Gyárfás, Thomassé, and Weidl [18] independently proved the existence of a "half-transitive" tournament matrix in a nonempty class $\mathcal{T}(R)$. A very simple proof of this theorem was given by Brualdi and Shen [9] based on the strengthened Landau inequalities (5.14). This proof makes use of the Ford–Fulkerson criteria for the nonemptiness of a class of $(0, 1)$-matrices with a prescribed row and column sum vector. Before stating this theorem we briefly discuss bipartite tournaments.

Let m and n be positive integers and consider the complete bipartite graph $K_{m,n}$ with bipartition $\{U, V\}$ where $U = \{u_1, u_2, \ldots, u_m\}$ and $V = \{v_1, v_2, \ldots, v_n\}$. A *bipartite tournament of size m by n* [5] is a digraph D obtained from $K_{m,n}$ by arbitrarily assigning a direction to each of its edges. We can think of the bipartite tournament D as representing the outcomes of a competition between two teams U and V in which each player on team U competes once against each player on team V. An edge from player u_i to player v_j means that u_i beats v_j; if the edge between u_i and v_j is from v_j to u_i, then player v_j beats player u_i (u_i loses to v_j). Again no ties are allowed. There are two score vectors associated with a bipartite tournament. The score vector of team U is $R_U = (r_1, r_2, \ldots, r_m)$ where r_i equals the number of players in V beaten by u_i ($i = 1, 2, \ldots, m$). The score vector of team V is $R_V = (r'_1, r'_2, \ldots, r'_n)$ where r'_j equals the number of players in U beaten by v_j ($j = 1, 2, \ldots, n$). Let $C = [c_{ij}]$ be the $(1, -1)$-matrix of size m by n such that $c_{ij} = 1$ if u_i beats v_j and $c_{ij} = -1$ if v_j beats u_i ($i = 1, 2, \ldots, m; j = 1, 2, \ldots, n$). Then the bipartite digraph $\Gamma(C)$, as defined in Section 3.2, equals T. Conversely, given an m by n $(1, -1)$-matrix C, we can define a bipartite tournament D such that D equals $\Gamma(C)$. It follows that there is a bijection between bipartite tournaments D of size m by n and $(1, -1)$-matrices C of size m by n. Under this correspondence the score vector R_U of D (for team U) gives the number of 1's in each row of C and the score vector R_V of D (for team V) gives the number of -1's in each column of C. We can replace the -1's in C by 0's and obtain a bijection between bipartite tournaments D of size m by n and $(0, 1)$-matrices A of size m by n. Under this correspondence, the score vector R_U is the row sum vector R of A and the score vector R_V is $(m, m, \ldots, m) - S$ where S is the column sum vector of A.

We denote the vector $(m, m, \ldots, m)$ of size n by $m^{(n)}$.

Theorem 5.4.1 *Let m, n be positive integers, and let $R = (r_1, r_2, \ldots, r_m)$ and $R' = (r'_1, r'_2, \ldots, r'_n)$ be nonnegative integral vectors. Let $S = m^{(n)} - R'$. Then the following are equivalent.*

(i) *There exists a bipartite tournament with score vectors R and R'.*

(ii) *There exists a $(0, 1)$-matrix in $\mathcal{A}(R, S)$.*

(iii) $\sum_{i \in K} r_i + \sum_{j \in L} r'_j \geq |K||L|$ *for all $K \subseteq \{1, 2, \ldots, m\}$ and $L \subseteq \{1, 2, \ldots, n\}$, with equality if $K = \{1, 2, \ldots, m\}$ and $L = \{1, 2, \ldots, n\}$.*

Proof. By the discussion preceding the statement of the theorem, (i) and (ii) are equivalent. Let $S = (s_1, s_2, \ldots, s_n)$. By Theorem 2.1.4 and the discussion preceding that theorem (since we are not assuming that the vectors R and R' are nonincreasing), (ii) holds if and only if

$$|K||L| + \sum_{i \in \overline{K}} r_i - \sum_{j \in L} s_j \geq 0 \quad (K \subseteq \{1,2,\ldots,m\}, L \subseteq \{1,2,\ldots,n\}), \quad (5.15)$$

with equality when $K = \oslash$ and $L = \{1,2,\ldots,n\}$ (that is, $r_1 + r_2 + \cdots + r_m = s_1 + s_2 + \cdots + s_m$). Interchanging K and $\overline{K}$ and replacing s_j by $m - r'_j$, (5.15) is equivalent to (iii). $\square$

A transitive tournament matrix of order n with a nondecreasing row sum vector has row sum vector $(0,1,2,\ldots,n-1)$ and equals the $(0,1)$-matrix L_n of order n with 0's on and above the main diagonal and 1's below the main diagonal.

The following theorem shows that a nonempty tournament class $\mathcal{T}(R)$ always contains a "half-transitive" tournament matrix.

Theorem 5.4.2 *Let $R = (r_1, r_2, \ldots, r_n)$ be a nondecreasing, nonnegative integral vector such that $\mathcal{T}(R)$ is nonempty. Then there exists a tournament with score vector R such that the subtournaments on both the even-indexed vertices and the odd-indexed vertices are transitive. Equivalently, there exists a tournament matrix A in $\mathcal{T}(R)$ such that*

$$A[\Theta_n] = L_{\lfloor (n-1)/2 \rfloor + 1} \text{ and } A[\Pi_n] = L_{\lfloor n/2 \rfloor}, \quad (5.16)$$

where $\Theta_n = \{1, 3, 5, \ldots, 2\lfloor (n-1)/2 \rfloor + 1\}$ and $\Pi_n = \{2, 4, 6, \ldots, 2\lfloor n/2 \rfloor\}$.

Proof. By Corollary 5.3.3,

$$r_i \geq \left\lfloor \frac{i-1}{2} \right\rfloor \quad (i = 1, 2, \ldots, n).$$

Let

$$S = (r_i - (i-1)/2 : i = 1, 3, 5, \ldots, 2\lfloor (n-1)/2 \rfloor + 1)$$

and

$$S' = (r_i - (i-2)/2 : i = 2, 4, 6, \ldots, 2\lfloor n/2 \rfloor).$$

Suppose there exists a bipartite tournament with score vectors S and S', equivalently, by Theorem 5.4.1, a $(0,1)$-matrix B with row sum vector S and column sum vector $(\lfloor n/2 \rfloor, \lfloor n/2 \rfloor, \ldots, \lfloor n/2 \rfloor) - S'$. Setting $A[\Theta_n, \Pi_n] = B$ and $A[\Pi_n, \Theta_n] = J - B^{\mathrm{T}}$, we obtain using (5.16) a tournament matrix in $\mathcal{T}(R)$. We now show the existence of such a matrix B.

Since $\mathcal{T}(R)$ is nonempty, it follows from Theorem 5.3.1 that for all $K \subseteq \Theta_n$ and $L \subseteq \Pi_n$,

$$\sum_{i \in K}(r_i - (i-2)/2) + \sum_{i \in L}(r_i - (i-1)/2)$$

$$= \sum_{i \in K \cup L} r_i - \frac{1}{2}\sum_{i \in K \cup L}(i-1) + \frac{1}{2}\sum_{i \in L} 1$$

$$\geq \frac{1}{2}\binom{|K \cup L|}{2} + \frac{1}{2}|L|$$

$$= \frac{1}{2}\binom{|K| + |L|}{2} + \frac{1}{2}|L|$$

$$\geq |K||L|,$$

with equality when $K = \Theta_n$ and $L = \Pi_n$. Hence by Theorem 5.4.1 again, the matrix B exists and the theorem follows. $\qquad\square$

Theorem 5.4.2 combined with the Gale–Ryser algorithm of Section 3.1 provides another algorithm for the construction of a matrix in a nonempty tournament class $\mathcal{T}(R)$. The algorithm in Section 5.1 for constructing a tournament matrix with row sum vector R requires the determination of elements in roughly $n^2/2$ positions. By Theorem 5.4.2 we may reduce this number to $n^2/4$. We illustrate this in the next example.

Example. Let $R = (1, 3, 4, 4, 4, 5, 5, 5, 7, 7)$. Then R satisfies Landau's condition so that, according to Theorem 5.4.2, there exists a matrix in $\mathcal{T}(R)$ of the form

$$A = \begin{bmatrix}
0 & * & 0 & * & 0 & * & 0 & * & 0 & * \\
\diamond & 0 & \diamond & 0 & \diamond & 0 & \diamond & 0 & \diamond & 0 \\
1 & * & 0 & * & 0 & * & 0 & * & 0 & * \\
\diamond & 1 & \diamond & 0 & \diamond & 0 & \diamond & 0 & \diamond & 0 \\
1 & * & 1 & * & 0 & * & 0 & * & 0 & * \\
\diamond & 1 & \diamond & 1 & \diamond & 0 & \diamond & 0 & \diamond & 0 \\
1 & * & 1 & * & 1 & * & 0 & * & 0 & * \\
\diamond & 1 & \diamond & 1 & \diamond & 1 & \diamond & 0 & \diamond & 0 \\
1 & * & 1 & * & 1 & * & 1 & * & 0 & * \\
\diamond & 1 & \diamond & 1 & \diamond & 1 & \diamond & 1 & \diamond & 0
\end{bmatrix}.$$

We now seek a matrix in $\mathcal{A}(S, S')$ where $S = (1, 3, 2, 2, 3)$ and $S' = (2, 2, 2, 3, 2)$. Note that neither S nor S' is nonincreasing or nondecreasing. We now apply the Gale–Ryser algorithm[1] to construct $\widetilde{A}$ in $\mathcal{A}(S, S')$ obtaining

$$\begin{bmatrix} 1 & 0 & 0 & 0 & 0 \\ 0 & 1 & 0 & 1 & 1 \\ 0 & 0 & 1 & 1 & 0 \\ 1 & 0 & 1 & 0 & 0 \\ 0 & 1 & 0 & 1 & 1 \end{bmatrix}.$$

We now set

$$A[\{1, 3, 5, 7, 9\}, \{2, 4, 6, 8, 10\}] = \widetilde{A}$$

and

$$A[\{2, 4, 6, 8, 10\}, \{1, 3, 5, 7, 9\}] = J_5 - \widetilde{A}^{\mathrm{T}}$$

to get

$$A = \begin{bmatrix} 0 & 1 & 0 & 0 & 0 & 0 & 0 & 0 & 0 & 0 \\ 0 & 0 & 1 & 0 & 1 & 0 & 0 & 0 & 1 & 0 \\ 1 & 0 & 0 & 1 & 0 & 0 & 0 & 1 & 0 & 1 \\ 1 & 1 & 0 & 0 & 1 & 0 & 1 & 0 & 0 & 0 \\ 1 & 0 & 1 & 0 & 0 & 1 & 0 & 1 & 0 & 0 \\ 1 & 1 & 1 & 1 & 0 & 0 & 0 & 0 & 1 & 0 \\ 1 & 1 & 1 & 0 & 1 & 1 & 0 & 0 & 0 & 0 \\ 1 & 1 & 0 & 1 & 0 & 1 & 1 & 0 & 0 & 0 \\ 1 & 0 & 1 & 1 & 1 & 0 & 1 & 1 & 0 & 1 \\ 1 & 1 & 0 & 1 & 1 & 1 & 1 & 1 & 0 & 0 \end{bmatrix}$$

in $\mathcal{T}(R)$. $\square$

It has been conjectured in [9] that if $n_1, n_2, \ldots, n_t$ is any sequence of integers such that $\sum_{i=1}^{\mathrm{T}} n_i = n$ and $n_i \leq (n+1)/2$ for all i, then the nonemptiness of the class $\mathcal{T}(R)$ implies the existence of a tournament with score vector R whose vertex set V can be partitioned into t pairwise disjoint subsets U_i of cardinality n_i, such that the subtournament on U_i is transitive $(i = 1, 2, \ldots, t)$. Theorem 5.4.2 established the conjecture for $t = 2$. The validity of the conjecture for $t = 3$ implies its validity for all $t \geq 3$. The conjecture has been verified for regular tournaments in [9], and has been settled recently in [2] with a very complicated proof.

5.5   Interchanges

In this section we take $R = (r_1, r_2, \ldots, r_n)$ to be a nonnegative integral vector such that $\mathcal{T}(R)$ is nonempty. As with a class of $(0, 1)$-matrices with prescribed row and column sum vectors, we show that $\mathcal{T}(R)$ can be generated, starting with any single matrix in $\mathcal{T}(R)$, by simple transformations.

Let $A = [a_{ij}]$ and $B = [b_{ij}]$ be tournament matrices in $\mathcal{T}(R)$ with associated tournaments $T(A)$ and $T(B)$, respectively, on the vertex set

---

[1] In doing so, we would first rearrange S and S' to get nonincreasing vectors.

$X = \{x_1, x_2, \ldots, x_n\}$. The $(0, 1)$-matrix $C = [c_{ij}] = (A - B)^+$ of order n defined by

$$c_{ij} = (a_{ij} - b_{ij})^+ \quad (i, j = 1, 2, \ldots, n),$$

is the *positive difference* of A and B.

The digraph $D(C)$ is the digraph obtained from $T(A)$ by removing those edges that are also edges of $T(B)$. Recall that a digraph is *balanced* provided, for each vertex, the number of edges entering the vertex (its *indegree*) equals the number of edges exiting the vertex (its *outdegree*). The edges of a balanced digraph can be partitioned in at least one way into directed cycles.

The following lemma and theorem are from Brualdi and Li [8].

Lemma 5.5.1 *The digraph $D(C)$ of the positive difference C of two matrices A and B in $\mathcal{T}(R)$ is a balanced digraph.*

Proof. Let i be an integer with $1 \leq i \leq n$, and let row i of A and row i of B have p_i 1's in common positions. Since row i of A and of B contains r_i 1's, row i of C contains $r_i - p_i$ 1's, and row i of A and B have $n - 1 - (2r_i - p_i)$ 0's in common positions off the main diagonal. Thus the columns i of A and B have 1's in $(n - 1) - (2r_i - p_i)$ common positions off the main diagonal. Since column i of A contains $(n - 1) - r_i$ 1's, the number of 1's in column i of C is

$$(n - 1) - r_i - ((n - 1) - (2r_i - p_i)) = r_i - p_i.$$

Thus row and column i of C each contain $r_i - p_i$ 1's. Since this is true for each i, $D(C)$ is balanced. $\square$

If D is a digraph with a vertex set $X = \{x_1, x_2, \ldots, x_n\}$, we let $\mathcal{B}(D)$ denote the set of all balanced digraphs with vertex set X whose edges are a subset of the edges of D. Note that the digraph with vertex set X and no edges belongs to $\mathcal{B}(D)$.

Theorem 5.5.2 *Let R be a nonnegative integral vector such that $\mathcal{T}(R)$ is nonempty. Let A be a tournament matrix in $\mathcal{T}(R)$ and let $T = T(A)$. There is a bijection between $\mathcal{B}(T)$ and $\mathcal{T}(R)$. Thus the number of tournament matrices in $\mathcal{T}(R)$ equals the number of balanced digraphs in $\mathcal{B}(T)$ (independent of which tournament matrix A is chosen in $\mathcal{T}(R)$).*

Proof. By Lemma 5.5.1, $f(B) = D((A - B)^+)$ defines a function $f : \mathcal{T}(R) \to \mathcal{B}(T)$. For a balanced digraph D in $\mathcal{B}(T)$, define $g(D)$ to be the digraph obtained from $T(A)$ by reversing the direction of all edges in D. Then $g(D)$ is a tournament and, since D is balanced, its corresponding tournament matrix $h(D)$ is in $\mathcal{T}(R)$. Thus $h : \mathcal{B}(T) \to \mathcal{T}(R)$ is a function with $h = f^{-1}$. $\square$

Ryser [24] introduced the notion of a *double interchange* in a tournament class $\mathcal{T}(R)$, namely a sequence consisting of an $(i_1, i_2; j_1, j_2)$-interchange followed by a $(j_1, j_2; i_1, i_2)$-interchange, and showed that any two matrices in

$\mathcal{T}(R)$ can be gotten from one another by a sequence of such double interchanges. Double interchanges are of three types and their result is the replacement of a principal submatrix of order 3 or 4 by another as given below:

$$
\begin{bmatrix}
0 & 0 & 1 & * \\
1 & 0 & * & 0 \\
0 & * & 0 & 1 \\
* & 1 & 0 & 0
\end{bmatrix}
\leftrightarrow
\begin{bmatrix}
0 & 1 & 0 & * \\
0 & 0 & * & 1 \\
1 & * & 0 & 0 \\
* & 0 & 1 & 0
\end{bmatrix},
\tag{5.17}
$$

$$
\begin{bmatrix}
0 & * & 0 & 1 \\
* & 0 & 1 & 0 \\
1 & 0 & 0 & * \\
0 & 1 & * & 0
\end{bmatrix}
\leftrightarrow
\begin{bmatrix}
0 & * & 1 & 0 \\
* & 0 & 0 & 1 \\
0 & 1 & 0 & * \\
1 & 0 & * & 0
\end{bmatrix},
\tag{5.18}
$$

and

$$
\begin{bmatrix}
0 & 0 & 1 \\
1 & 0 & 0 \\
0 & 1 & 0
\end{bmatrix}
\leftrightarrow
\begin{bmatrix}
0 & 1 & 0 \\
0 & 0 & 1 \\
1 & 0 & 0
\end{bmatrix}.
\tag{5.19}
$$

A double interchange applied to a tournament matrix results in a tournament matrix with the same row sum vector. Double interchanges of the types (5.17) and (5.18) leave unchanged the number of 1's above the main diagonal, and the number of 1's below the main diagonal. We call double interchanges of the types (5.17) and (5.18) *four-way double interchanges*. Double interchanges of the type (5.19) are called Δ-*interchanges*. Fulkerson [14] and Moon [23] observed that a four-way double interchange can be accomplished by a sequence of two Δ-interchanges.

We refer to a (p, q, r)-Δ-interchange if the principal submatrices of order 3 in (5.19) lie in rows (and columns) $p, q,$ and r. A Δ-interchange reverses the direction of a directed cycle of length 3 in the associated tournament T and is called a Δ-*reversal* of T [21]. A Δ-interchange does not change the row sum vector and hence any matrix obtained from a tournament matrix in $\mathcal{T}(R)$ by a sequence of Δ-interchanges also belongs to $\mathcal{T}(R)$.

The following theorem is from Brualdi and Li [8].

Theorem 5.5.3 *Let A and B be tournament matrices in $\mathcal{T}(R)$ and let $C = (A - B)^+$. Let the edges of the balanced digraph $D(C)$ be partitioned into q directed cycles $\gamma_1, \gamma_2, \ldots, \gamma_q$ of lengths $k_1, k_2, \ldots, k_q$, respectively. Then B can be obtained from A by a sequence of $(\sum_{i=1}^{q} k_i) - 2q$ Δ-interchanges.*

Proof. Let $T = T(A)$ the tournament associated with A. By induction, it suffices to show that the tournament T' obtained from T by reversing the edges of a cycle

$$\gamma : y_1 \rightarrow y_2 \rightarrow y_3 \rightarrow \cdots \rightarrow y_k \rightarrow y_1$$

of length k can be obtained from T by $(k - 2)$ Δ-reversals. We prove this by induction on k. If $k = 3$, there is nothing to prove. Suppose $k > 3$.

There is an edge in T between y_1 and y_3. First assume that this edge is from y_3 to y_1. Then $y_1 \to y_2 \to y_3 \to y_1$ is a directed cycle of length 3. Reversing this directed cycle reverses the direction of the edge (y_3, y_1) and creates a directed cycle γ' of length $k - 1$. By induction the edges of γ' can be reversed using $(k-1) - 2$ Δ-reversals. These Δ-reversals restore the direction of the edge between y_1 and y_3, and hence we obtain D'. A similar argument works if the edge between y_1 and y_3 is from y_1 to y_3. $\square$

Let A and B be tournament matrices in a class $\mathcal{T}(R)$. The *combinatorial (tournament) distance*[2] $d_t(A, B)$ between A and B is the number of positions in which A has a 1 and B has a 0 (equivalently, since A and B are tournament matrices, the number of positions in which B has a 1 and A has a 0). We have that $d_t(A, B)$ equals the number of 1's in $(A - B)^+$. If D is a balanced digraph, then we define the *cycle packing number* $q(D)$ to be the maximal number of directed cycles into which the edges of D can be partitioned.[3] We let $q(A, B)$ equal $q(D)$ where D is the balanced digraph $D((A - B)^+)$. Finally, we define the Δ-*interchange distance* $\iota_\Delta(A, B)$ between the tournament matrices A and B in $\mathcal{T}(R)$ to be the smallest number of Δ-interchanges that transform A into B. The following formula for the Δ-interchange distance is due to Brualdi and Li [8].

Theorem 5.5.4 *Let A and B be tournament matrices in $\mathcal{T}(R)$. Then*

$$\iota_\Delta(A, B) = d_t(A, B) - 2q(A, B). \tag{5.20}$$

Proof. By Theorem 5.5.3, $\iota_\Delta(A, B) \le d_t(A, B) - 2q(A, B)$. We prove that equality hold by induction on $\iota_\Delta(A, B)$. If $\iota_\Delta(A, B) = 1$, then (5.20) holds. Let d be an integer with $d \ge 1$ and assume that (5.20) holds for each pair of tournament matrices A and B with Δ-interchange distance d. Now assume that $\iota_\Delta(A, B) = d + 1$. There exists a sequence of tournament matrices $A = A_1, A_2, \ldots, A_d, A_{d+1} = B$ in $\mathcal{T}(R)$ such that A_{i+1} can be obtained from A_i by a Δ-interchange $(i = 1, 2, \ldots, d)$. Then $\iota_\Delta(A, A_d) = d$ and by the induction hypothesis, the edges of the balanced digraph $D((A - A_d)^+)$ can be partitioned into p directed cycles, where

$$p = q(A, A_d) \text{ and } d = \iota_\Delta(A, A_d) = d_t(A, A_d) - 2p. \tag{5.21}$$

The matrix B can be obtained from A_d by a Δ-interchange. Let l be the number of 1's in $(A_d - A)^+$ that are moved by this Δ-interchange. Then

$$d_t(A, B) = d_t(A, A_d) + 3 - 2l. \tag{5.22}$$

---

[2]Not to be confused with the combinatorial distance between two $(0, 1)$-matrices of the same size defined in Section 3.2.

[3]In Section 3.2 we defined the balanced separation $q(A, B)$ of two matrices $A, B \in \mathcal{A}(R, S)$; $q(A, B)$ is the same as the cycle packing number of the balanced digraph $\Gamma(A - B)$.

The edges of $D((A-B)^+)$ contain $p+1-l$ pairwise edge-disjoint directed cycles, and hence

$$q(A,B) \geq p+1-l. \tag{5.23}$$

Using (5.21), (5.22), and (5.23), we now obtain

$$\begin{aligned} d_{\mathrm{t}}(A,B) - 2q(A,B) &\leq d_{\mathrm{t}}(A,A_d) + 3 - 2l - 2(p+1-l) \\ &= d_{\mathrm{t}}(A,A_d) - 2p + 1 = d + 1 \\ &= \iota_\Delta(A,B). \end{aligned}$$

Thus $\iota_\Delta(A,B) \geq d_{\mathrm{t}}(A,B) - 2q(A,B)$ and the proof is complete. $\qquad\square$

Theorem 5.5.3 motivates the introduction of the Δ-*interchange graph* $G_\Delta(R)$ whose vertices are the tournament matrices in $\mathcal{T}(R)$. There is an edge in $G_\Delta(R)$ between two matrices A and B in $\mathcal{T}(R)$ provided A can be obtained from B by a single Δ-interchange (and so B can be obtained from A by a single Δ-interchange). By Theorem 5.5.3, $G_\Delta(R)$ is a connected graph. The Δ-interchange graph arises again in Section 5.8 and is studied in more detail in Chapter 6.

5.6 Upsets in Tournaments

Let $R = (r_1, r_2, \ldots, r_n)$ be a nondecreasing, integer vector such that $\mathcal{T}(R)$ is nonempty, and let $A = [a_{ij}]$ be a tournament matrix in $\mathcal{T}(R)$. Then A represents the outcomes of a round-robin tournament amongst teams $p_1, p_2, \ldots, p_n$ where p_i defeats p_j if and only if $a_{ij} = 1$ $(i, j = 1, 2, \ldots, n;$ $i \neq j)$. Since

$$r_1 \leq r_2 \leq \cdots \leq r_n,$$

if $i < j$, then the record of team p_i is worse than that of team p_j (or at least no better than) and we say that p_i is *inferior* to p_j and p_j is *superior* to p_i. In the case that $r_i = r_j$, whether p_i is inferior or superior to p_j depends solely on the order in which the teams have been listed. An *upset* occurs whenever an inferior team defeats a superior team. The upsets are in one-to-one correspondence with the 1's above the main diagonal of A. The *total number of upsets* is

$$v(A) = \sum_{i>j} a_{ij}.$$

We have $v(A) = 0$ if and only if A is a transitive tournament matrix with only 0's on and above its main diagonal. If T is the tournament associated with A, then we set $v(T) = v(A)$.

A Δ-interchange increases or decreases the number of upsets by 1, in particular cannot leave the number of upsets fixed. This gives the following result.

Lemma 5.6.1 *Let A be a tournament matrix and let B be obtained from A by a Δ-interchange. Then $v(B) = v(A) \pm 1$.* □

Let
$$\tilde{v}(R) = \min\{v(A) : A \in \mathcal{T}(R)\}$$
be the minimal number of upsets of a tournament matrix with row sum vector R, and let
$$\bar{v}(R) = \max\{v(A) : A \in \mathcal{T}(R)\}$$
be the maximal number of upsets of a tournament matrix with row sum vector R.

Matrices in $\mathcal{T}(R)$ with the minimal number of upsets have a special property [24]. If $A = [a_{ij}]$ is a matrix of order n, then the *initial part* and *terminal part* of row i are, respectively,

$$(a_{i1}, a_{i2}, \ldots, a_{i,i-1}) \text{ and } (a_{i,i+1}, a_{i,i+2}, \ldots, a_{in}) \quad (i = 1, 2, \ldots, n).$$

Note that row 1 has an empty initial part and row n has an empty terminal part.

Theorem 5.6.2 *Let $A = [a_{ij}]$ be a matrix in $\mathcal{T}(R)$ with the minimal number of upsets. Then*

$$a_{21} = a_{32} = \cdots = a_{n,n-1} = 1 \text{ and } a_{12} = a_{23} = \cdots = a_{n-1,n} = 0.$$

Proof. Suppose to the contrary that there exists an i such that $a_{i,i-1} = 0$ and thus $a_{i-1,i} = 1$. Suppose there exists a j such that in the terminal parts of rows $i - 1$ and i we have $a_{i-1,j} = 0$ and $a_{ij} = 1$. Then we may perform an $(i - 1, i, j)$-Δ-interchange and obtain a matrix with fewer upsets than A, contradicting the choice of A. Since $r_{i-1} \leq r_i$, this implies that there exists a j such that in the initial parts of rows $i - 1$ and i we have $a_{i-1,j} = 0$ and $a_{ij} = 1$. Now a $(j, i - 1, i)$-Δ-interchange leads to a contradiction of the choice of A. □

Ryser [24] derived a formula for the minimal number of upsets $v(R)$ by inductively showing the existence of a special matrix in $\mathcal{T}(R)$ with the minimal number of upsets. Fulkerson [14] gave an algorithm to construct such a matrix, and we now turn to the description and verification of this algorithm. The description of the algorithm is in terms of the normalized row sum vector. Recall from Section 2.2 that for a nondecreasing, nonnegative integral vector $R = (r_1, r_2, \ldots, r_n)$, the normalized row sum vector is an integral vector $H = (h_1, h_2, \ldots, h_n)$ defined by

$$h_i = r_i - (i - 1) \quad (i = 1, 2, \ldots, n) \tag{5.24}$$

that satisfies the conditions

$$h_1 \geq 0 \text{ and } h_{i-1} - h_i \leq 1 \quad (i = 2, 3, \ldots, n), \tag{5.25}$$

and
$$h_1 + h_2 + \cdots + h_k \geq 0 \quad (k = 1, 2, \ldots, n) \tag{5.26}$$
with equality for $k = n$. We call the conditions (5.25) and (5.26) the *normalized Landau conditions*. Since $h_n = r_n - (n-1)$, it follows that $-h_n$ is the number of 1's in column n of a tournament matrix with row sum vector R.

Example. Let R be the nondecreasing vector $(1, 2, 3, 3, 3, 6, 6, 6, 6)$ of size $n = 9$. Then $H = (1, 1, 1, 0, -1, 1, 0, -1, -2)$. Since the partial sums of H are nonnegative with the total sum equal to 0, H is the normalized row sum vector of a tournament matrix, that is, R is the row sum vector of a tournament matrix.   □

Fulkerson's Algorithm I for a Minimal Upset Tournament in $\mathcal{T}(R)$

(1) Begin with the zero matrix $A^{(n)} = O$ of order n and $H^{(n)} = H$.

(2) Insert $-h_n$ 1's in the last column of $A^{(n)}$ according to the following rule.

   (i) Find the first member of the last consecutive subsequence of *positive* components of $H^{(n)}$ and insert 1's consecutively downward until either $-h_n$ 1's have been inserted or this subsequence of positive components has been exhausted. If the latter case holds, then
      (a) Find the first member of the next-to-last consecutive subsequence of positive components, and insert 1's as in (i).
      (b) Continue like this until $-h_n$ 1's have been inserted.
   (ii) Insert 1's in the complementary transposed positions in the last row of A_n, thereby achieving row sum r_n.
   (iii) Let $H^{(n-1)} = (h_1^{(n-1)}, h_2^{(n-1)}, \ldots, h_{n-1}^{(n-1)})$ be defined by

$$h_i^{(n-1)} = \begin{cases} h_i - 1, & \text{if column } n \text{ has a 1 in position } i, \\ h_i, & \text{otherwise.} \end{cases}$$

(3) Replace $H^{(n)}$ by $H^{(n-1)}$ and repeat until columns $n, n-1, \ldots, 2$ and rows $n, n-1, \ldots, 2$ of $A^{(n)}$ have been filled in.

(4) Output $A_{\tilde{v}} = A^{(n)}$.

The formula for $\tilde{v}(R)$ in the next theorem is due to Ryser [24].

Theorem 5.6.3 *Let* $R = (r_1, r_2, \ldots, r_n)$ *be a nondecreasing integral vector satisfying Landau's conditions. Let* $H = (h_1, h_2, \ldots, h_n)$ *be defined by* $h_i = r_i - (i-1)$ $(i = 1, 2, \ldots, n)$. *Then Fulkerson's Algorithm* I *can be applied and the constructed matrix* $A_{\tilde{v}}$ *is a tournament matrix in* $\mathcal{T}(R)$ *with the minimal number*

$$\tilde{v}(R) = \sum_{i=1}^{n} (r_i - (i-1))^+ \tag{5.27}$$

of upsets.

Proof. The theorem is trivial if $n = 1$, so we assume that $n \geq 2$. Since R satisfies the Landau conditions, the vector H satisfies the normalized Landau conditions (5.25) and (5.26). These conditions imply that at least $-h_n$ of the first $n-1$ components of H are positive. Hence step (2) of the algorithm can be carried out to fill in column n and row n as described. To verify that the construction can be completed, it suffices to show that $H^{(n-1)}$ satisfies the normalized Landau conditions. That

$$h_1^{(n-1)} \geq 0 \text{ and that } h_{i-1}^{(n-1)} - h_i^{(n-1)} \leq 1 \quad (i = 2, 3, \ldots, n)$$

are immediate consequences of the procedure for inserting 1's in column n as described in step (2) of the algorithm. Also since $h_1 + h_2 + \cdots + h_n = 0$ and $-h_n$ of $h_1, h_2, \ldots, h_{n-1}$ are decreased by 1, it follows that

$$h_1^{(n-1)} + h_2^{(n-1)} + \cdots + h_{n-1}^{(n-1)} = 0. \tag{5.28}$$

Suppose to the contrary that there is an integer k with $1 \leq k \leq n-2$ such that

$$h_1^{(n-1)} + h_2^{(n-1)} + \cdots + h_k^{(n-1)} < 0. \tag{5.29}$$

We choose k to be the largest integer in the interval $1 \leq k \leq n-2$ satisfying (5.29), and this implies that

$$h_k^{(n-1)} < 0 \text{ and } h_{k+1}^{(n-1)} \geq 0. \tag{5.30}$$

There exist nonnegative integers p and q with $p + q = -h_n$ such that

$$h_1^{(n-1)} + h_2^{(n-1)} + \cdots + h_k^{(n-1)} = h_1 + h_2 + \cdots + h_k - p \tag{5.31}$$

and

$$h_{k+1}^{(n-1)} + h_{k+2}^{(n-1)} + \cdots + h_{n-1}^{(n-1)} = h_{k+1} + h_{k+2} + \cdots + h_{n-1} - q.$$

From (5.28) and (5.29) we now get

$$h_{k+1}^{(n-1)} + h_{k+2}^{(n-1)} + \cdots + h_{n-1}^{(n-1)} > 0,$$

and hence

$$h_{k+1} + h_{k+2} + \cdots + h_{n-1} > 0. \tag{5.32}$$

Now using the normalized Landau conditions (5.25), and (5.30), and (5.32), we conclude that the subsequence $h_{k+1}, h_{k+2}, \ldots, h_n$ has at least $-h_{n-1} + 1 \geq -h_n$ positive terms. But then the construction in the algorithm implies that $q = -h_n$ and $p = 0$. Using (5.29) and (5.31), we obtain the contradiction

$$h_1 + h_2 + \cdots + h_k < 0$$

to the normalized Landau conditions (5.26). This contradiction completes the verification of the algorithm.

Next we prove the number of 1's above the main diagonal of $A_{\tilde{v}}$ is $\sum_{i=1}^{n} ((r_i - (i-1))^+$. This is trivial for $n = 1$. In the recursive application of step (2) of the algorithm, the number of positive components of $H^{(i)}$ which are reduced by 1 equals the number of 1's placed above the main diagonal of $A_{\tilde{v}}$. Since all positive components are reduced to 0,

$$v(A_{\tilde{v}}) = \sum_{i=1}^{n} ((r_i - (i-1))^+ .$$

Since every matrix A in $\mathcal{T}(R)$ clearly satisfies

$$v(A) \geq \sum_{i=1}^{n} ((r_i - (i-1))^+ ,$$

(5.27) holds   □

Let A be any tournament matrix in $\mathcal{T}(R)$ with the minimal number of upsets for its row sum vector R. A consequence of the formula (5.27) is that row i of A has $(r_i - (i-1))^+$ 1's in its terminal part $(i = 1, 2, \ldots, n)$. Moreover, if a team defeats a superior team, then it beats all teams inferior to it, that is, no team upsets it. In terms of A this means that, if row i of A contains an upset, then column i contains no upsets. This implies that either there are $\lfloor n/2 \rfloor$ rows that contain all the 1's above the main diagonal of A or there are $\lfloor n/2 \rfloor$ columns with this property. Finally, since $h_{i-1} \leq h_i + 1$, it follows that $h_{i-1}^+ \leq h_i^+ + 1$, and hence that row $i-1$ contains at most one more 1 in its terminal part than row i $(i = 2, 3, \ldots, n)$.

Example. [14] We illustrate Fulkerson's Algorithm I with the vectors $R = (1, 2, 3, 3, 3, 6, 6, 6, 6)$ and $H = (1, 1, 1, 0, -1, 1, 0, -1, -2)$ from the previous example. We write the vectors $H^{(9)}, H^{(8)}, \ldots, H^{(1)}$ as column vectors to facilitate the construction of the matrix $A_{\tilde{v}}$. Applying the algorithm we obtain

$$H^{(i)} \text{ for } i = 9, 8, \ldots, 1$$

9	8	7	6	5	4	3	2	1
1	0	0	0	0	0	0	0	0
1	1	0	0	0	0	0	0	
1	1	1	1	1	0	0		
0	0	0	0	0	0			
−1	−1	−1	−1	−1				
1	0	0	0					
0	0	0						
−1	−1							
−2								

and

$$A_{\tilde{v}} = \begin{bmatrix} 0 & 0 & 0 & 0 & 0 & 0 & 0 & 0 & 1 \\ 1 & 0 & 0 & 0 & 0 & 0 & 0 & 1 & 0 \\ 1 & 1 & 0 & 0 & 1 & 0 & 0 & 0 & 0 \\ 1 & 1 & 1 & 0 & 0 & 0 & 0 & 0 & 0 \\ 1 & 1 & 0 & 1 & 0 & 0 & 0 & 0 & 0 \\ 1 & 1 & 1 & 1 & 1 & 0 & 0 & 0 & 1 \\ 1 & 1 & 1 & 1 & 1 & 1 & 0 & 0 & 0 \\ 1 & 0 & 1 & 1 & 1 & 1 & 1 & 0 & 0 \\ 0 & 1 & 1 & 1 & 1 & 0 & 1 & 1 & 0 \end{bmatrix},$$

where $v(A_{\tilde{v}}) = \tilde{v}(R) = 4$. $\square$

Brualdi and Li [8] have characterized strong row sum vectors R for which there is a unique tournament matrix A in $\mathcal{T}(R)$ with $v(A) = \tilde{v}(R)$. We record this theorem without proof.

Theorem 5.6.4 *Let R be a nondecreasing integral vector of size n with R strong. Then there exists a unique matrix A in $\mathcal{T}(R)$ with $v(A) = \tilde{v}(R)$ if and only if for some positive integer k satisfying $2k + 1 \leq n$,*

$$R = (k, k, \ldots, k, k+1, k+2, \ldots, n-k+1, n-k+1, \ldots, n-k-1)$$

with $k + 1$ initial k's and $k + 1$ terminal $(n - k - 1)$'s. For such a vector R, $\tilde{v}(R) = 1 + 2 + \cdots + k$ and the unique matrix in $\mathcal{T}(R)$ with this number of upsets is the matrix A all of whose 1's above the main diagonal are contained in the submatrix $A[\{1, 2, \ldots, k\}, \{n - k + 1, n - k + 2, \ldots, n\}]$ of order k, and $A[\{1, 2, \ldots, k\}, \{n - k + 1, n - k + 2, \ldots, n\}]$ is the triangular matrix of order k with 1's on and above its main diagonal.

Example. If we take $n = 9$ and $k = 3$, then in Theorem 5.6.4, $R = (3, 3, 3, 3, 4, 5, 5, 5, 5)$, and the unique matrix in $\mathcal{T}(R)$ with $\tilde{v}(R) = 6$ upsets is

$$\begin{bmatrix} 0 & 0 & 0 & 0 & 0 & 0 & 1 & 1 & 1 \\ 1 & 0 & 0 & 0 & 0 & 0 & 0 & 1 & 1 \\ 1 & 1 & 0 & 0 & 0 & 0 & 0 & 0 & 1 \\ 1 & 1 & 1 & 0 & 0 & 0 & 0 & 0 & 0 \\ 1 & 1 & 1 & 1 & 0 & 0 & 0 & 0 & 0 \\ 1 & 1 & 1 & 1 & 1 & 0 & 0 & 0 & 0 \\ 0 & 1 & 1 & 1 & 1 & 1 & 0 & 0 & 0 \\ 0 & 0 & 1 & 1 & 1 & 1 & 1 & 0 & 0 \\ 0 & 0 & 0 & 1 & 1 & 1 & 1 & 1 & 0 \end{bmatrix}.$$

$\square$

No simple closed formula for the maximal number of upsets $\bar{v}(R)$ is known and such a formula may not be possible.[4] Fulkerson [14] has described an algorithm that constructs a tournament matrix in $\mathcal{T}(R)$ with

---

[4]For an explanation in terms of network flow theory, see [14].

the maximal number of upsets, equivalently, with the minimal number of
1's below the main diagonal. The description of the algorithm uses what
we shall call here the skew row sum vector.

Let $R = (r_1, r_2, \ldots, r_n)$ be a nondecreasing, nonnegative integral vector.
Then the *skew row sum vector* of a matrix $A = [a_{ij}]$ in $\mathcal{T}(R)$ is the integral
vector $G = (g_1, g_2, \ldots, g_n)$ defined by

$$g_i = r_i - (n - i) \quad (i = 1, 2, \ldots, n).$$

Since $0 \le r_1 \le r_2 \le \cdots \le r_n$, we have

$$-(n - 1) \le g_1 < g_2 < \cdots < g_n. \tag{5.33}$$

We also have

$$\sum_{j<i} a_{ij} - \sum_{j>i} a_{ij} = g_i \quad (i = 1, 2, \ldots, n).$$

In terms of the components of G, Landau's conditions are equivalent to

$$\sum_{i=1}^{k} g_k \ge k(k - n) \quad (k = 1, 2, \ldots, n),$$

with equality for $k = n$. If the column sum vector of A is $S = (s_1, s_2, \ldots, s_n)$,
then

$$g_i = -(s_i - (i - 1)) \quad (i = 1, 2, \ldots, n),$$

and thus G is the negative of the normalized column sum vector. In particular, $s_1 = -g_1$.

Fulkerson's Algorithm II for a Maximal Upset Tournament in $\mathcal{T}(R)$

(1) Begin with the zero matrix $A^{(1)} = O$ of order n and $G^{(1)} = G$.

(2) Insert $-g_1$ 1's in column 1 of $A^{(1)}$ in rows corresponding to the $-g_1$
largest components of G, with preference given to the topmost rows
(those with smallest index) in case of ties. Then insert 1's in the
complementary transposed positions of row 1, thereby achieving row
sum r_1. The result is a matrix $A^{(2)}$. Define the vector $G^{(2)}$ of size
$n - 1$ by $G^{(2)} = (g_2^{(2)}, g_3^{(2)}, \ldots, g_n^{(2)})$, where

$$g_i^{(2)} = \begin{cases} g_i - 1, & \text{if a 1 has been inserted in row } i, \\ g_i, & \text{otherwise.} \end{cases}$$

We continue recursively.

(3) For $k = 2, 3 \ldots, n - 1$, do:

Insert $-g_k^{(k)}$ 1's in column k of $A^{(k)}$ in rows corresponding to the $-g_k^{(k)}$
largest components of $G^{(k)}$, with preference given to the topmost rows

in case of ties. Insert 1's in the complementary transposed positions of row k of $A^{(k)}$ obtaining a matrix $A^{(k+1)}$. Then define the vector $G^{(k+1)}$ of size $n - k + 1$ by $G^{(k+1)} = (g_{k+1}^{(k+1)}, g_{k+2}^{(k+1)}, \ldots, g_n^{(k+1)})$ where

$$
g_i^{(k+1)} = \begin{cases} g_i^{(k)} - 1, & \text{if a 1 has been inserted in row } i, \\ g_i^{(k)}, & \text{otherwise.} \end{cases}
$$

(4) Output $A_{\bar{v}} = A^{(n)}$.

Theorem 5.6.5 *Let $R = (r_1, r_2, \ldots, r_n)$ be a nondecreasing integral vector satisfying Landau's condition. Let $G = (g_1, g_2, \ldots, g_n)$ be defined by $g_i = r_i - (n-i)$ $(i = 1, 2, \ldots, n)$. Then Fulkerson's Algorithm II can be completed and the constructed matrix $A_{\bar{v}}$ is a tournament matrix in $\mathcal{T}(R)$ with the maximal number of upsets.*

Proof. Before proving the theorem we remark that the rule for breaking ties in the algorithm maintains the nondecreasing property of the components of the vectors $G^{(k)}$ but the strictly increasing property (5.33) of the skew normalized row sum vector $G^{(1)} = G$ is not maintained in general.[5] In the inductive argument given below, we only make use of the nondecreasing property of the components of the skew normalized row sum vector.

Let $A = [a_{ij}]$ be any tournament matrix in $\mathcal{T}(R)$ with $v(A) = \bar{v}(R)$. We show that there is a sequence of four-way double interchanges that transforms A into another tournament matrix A' in $\mathcal{T}(R)$ where the first column (and thus first row) of A' coincides with that of $A_{\bar{v}}$ and where $v(A') = v(A)$. Arguing inductively, we then conclude that there is a sequence of four-way double interchanges that transforms A into $A_{\bar{v}}$ which does not change the number of upsets, and hence that $A_{\bar{v}}$ has the maximal number of upsets in $\mathcal{T}(R)$.

Let the last 1 in column 1 of A occur in row e. Suppose that $e < n$. Then

$$
a_{e1} = 1 \text{ and } a_{e+1,1} = 0, \tag{5.34}
$$

and

$$
\sum_{j<e} a_{ej} - \sum_{j>e} a_{je} = g_e \leq g_{e+1} = \sum_{j<e+1} a_{e+1,j} - \sum_{j>e+1} a_{j,e+1}. \tag{5.35}
$$

Formulas (5.34) and (5.35) imply that there is an integer l satisfying

$$
e < l \leq n, \ a_{le} = 1, \text{ and } a_{l,e+1} = 0, \tag{5.36}
$$

or

$$
1 \leq l < e, \ a_{el} = 0, \text{ and } a_{e+1,l} = 1. \tag{5.37}
$$

---

[5]This implies that the subtournaments $A_{\bar{v}}[\{k+1, k+2, \ldots, n\}]$ need not have nondecreasing row sums. This is to be contrasted with the construction of the minimal upset matrix $A_{\underline{v}}$ whose subtournaments $A_{\underline{v}}[\{1, 2, \ldots, k\}]$ have nondecreasing row sums.

First assume that (5.36) holds. If $j = e+1$, then a $(1, e, e+1)$-Δ-interchange increases the number of 1's above the main diagonal, contradicting the maximal property of A. Hence $j > e + 1$, and a four-way interchange in rows and columns $1, e, e + 1, j$ lowers the 1 in position $(e, 1)$ to position $(e + 1, 1)$, without changing the number of 1's above the main diagonal. Similar reasoning gives the same conclusion if (5.37) holds. We may repeat this argument and obtain a matrix in $\mathcal{T}(R)$ with the maximal number of 1's above the main diagonal and with the 1's in column 1 occupying the bottommost positions. We remark that the same argument as above allows us to move up 1's in column 1 to higher rows in case of equal row sums, since under this circumstance we have $g_e = g_{e+1}$. Thus we can obtain a matrix A' whose first column and first row agree with those of $A_{\bar{v}}$ and with $v(A') = v(A_{\bar{v}})$.   □

Example. [14] We illustrate Fulkerson's Algorithm II with the vectors $R = (1, 2, 3, 3, 3, 6, 6, 6, 6)$ and $G = (-7, -5, -3, -2, -1, 3, 4, 5, 6)$. We write the vectors $G^{(1)}, G^{(2)}, \ldots, G^{(9)}$ as column vectors to facilitate the construction of the matrix $A_{\bar{v}}$. Applying the algorithm we obtain

$$G^{(i)} \text{ for } i = 1, 2, \ldots, 9$$

9	8	7	6	5	4	3	2	1
								-7
							-5	-5
						-4	-4	-3
					-3	-3	-3	-2
				-3	-3	-3	-2	-1
			-1	0	0	1	2	3
		0	0	0	1	2	3	4
	0	0	0	1	2	3	4	5
0	0	0	1	2	3	4	5	6

and

$$A_{\bar{v}} = \begin{bmatrix}
0 & 1 & 0 & 0 & 0 & 0 & 0 & 0 & 0 \\
0 & 0 & 1 & 1 & 0 & 0 & 0 & 0 & 0 \\
1 & 0 & 0 & 1 & 1 & 0 & 0 & 0 & 0 \\
1 & 0 & 0 & 0 & 1 & 1 & 0 & 0 & 0 \\
1 & 1 & 0 & 0 & 0 & 0 & 1 & 0 & 0 \\
1 & 1 & 1 & 0 & 1 & 0 & 1 & 1 & 0 \\
1 & 1 & 1 & 1 & 0 & 0 & 0 & 1 & 1 \\
1 & 1 & 1 & 1 & 1 & 0 & 0 & 0 & 1 \\
1 & 1 & 1 & 1 & 1 & 1 & 0 & 0 & 0
\end{bmatrix},$$

where $v(A_{\bar{v}}) = \bar{v}(R) = 13$.   □

5.7   Extreme Values of $\tilde{v}(R)$ and $\bar{v}(R)$

In this section we consider the extreme values of $\tilde{v}(R_n)$ and $\bar{v}(R_n)$ as R_n ranges over all nondecreasing row sum vectors $(r_1, r_2, \ldots, r_n)$ of size n for which $\mathcal{T}(R_n)$ is nonempty. In studying the minimum value of $\tilde{v}(R_n)$, we assume that R_n is a strong row sum vector. Recall that this means that

$$\sum_{i=1}^{k} r_i \geq \binom{k}{2} + 1 \quad (k = 1, 2, \ldots, n-1) \text{ and } \sum_{i=1}^{n} r_i = \binom{n}{2}. \qquad (5.38)$$

Otherwise, if $R_n = (0, 1, 2, \ldots, n-1)$, the row sum vector of a transitive tournament matrix, then $\tilde{v}(R_n) = 0$. We could, but need not, assume that R_n is strong in studying the maximal value as well. By Theorem 5.2.3, if R_n satisfies (5.38), then every tournament matrix in $\mathcal{T}(R_n)$ is irreducible.

Let

$$\tilde{v}_n^{\min} = \min\{\tilde{v}(R_n)) : \mathcal{T}(R_n) \neq \oslash, \text{ and } R_n \text{ nondecreasing and strong}\}$$

be the minimal number of upsets in an irreducible tournament matrix of order n with a nondecreasing strong row sum vector R. Also let

$$\tilde{v}_n^{\max} = \max\{\tilde{v}(R_n) : \mathcal{T}(R_n) \neq \oslash, \text{ and } R_n \text{ nondecreasing}\}$$

be the maximal number t for which there exists a nonempty class $\mathcal{T}(R_n)$ with a nondecreasing row sum vector for which every tournament matrix in $\mathcal{T}(R_n)$ has at least t upsets. We generally assume that $n > 2$, since otherwise the only possible row sum vectors are $R_1 = (0)$ and $R_2 = (0, 1)$.

Most of the material in this section is taken from Brualdi and Li [7, 8]. We first consider the minimal upset number $\tilde{v}^{\min}$. Let

$$\hat{R}_n = (1, 1, 2, 3, \ldots, n-3, n-2, n-2) \quad (n \geq 2).$$

If $n = 2$, then $\hat{R}_2$ is taken to be $(1, 0)$, and it is to be noted that $\hat{R}_2$ is not nondecreasing. For R_n with $n \geq 3$, each of the inequalities in (5.38) holds with equality. In terms of majorization, this implies, since R_n is nondecreasing, that $\hat{R}_n$ is a strong row sum vector and

$$R_n \preceq \hat{R}_n \text{ for each strong row sum vector } R_n.$$

Theorem 5.7.1 *Let $n \geq 3$ be an integer. Then $\tilde{v}_n^{\min} = 1$. Moreover, a nondecreasing, strong row sum vector R_n satisfies $\tilde{v}(R_n) = 1$ if and only if $R_n = \hat{R}_n$.*

     Proof. An irreducible tournament matrix of order $n \geq 3$ satisfies $v(A) \geq 1$, and hence $\tilde{v}_n \geq 1$. The tournament matrix $\hat{A}_n$ obtained from the transitive tournament matrix L_n with row sum vector $(0, 1, 2, \ldots, n-1)$ by interchanging the 1 in position $(n, 1)$ with the 0 in position $(1, n)$ has row sum vector equal to $\hat{R}_n$ and satisfies $v(\hat{A}_n) = 1$. Hence $\tilde{v}(R_n) = \tilde{v}_n^{\min} = 1$.

Suppose that R_n is a strong row sum vector with $\tilde{v}(R_n) = 1$. Then a tournament matrix in $\mathcal{T}(R_n)$ with $v(A) = 1$ has exactly one 1 above the main diagonal. Since this matrix is irreducible, this 1 must be in position $(1, n)$, for otherwise row 1 or column n of A contains only 0's. It now follows that $R_n = \hat{R}_n$. □

Example. In this example we completely describe the tournament matrices in $\mathcal{T}(\hat{R}_n)$. The matrix $\hat{A}_n$ is defined in the proof of Theorem 5.7.1 for $n \geq 3$. We define $\hat{A}_2$ to be the unique matrix

$$\begin{bmatrix} 0 & 1 \\ 0 & 0 \end{bmatrix}$$

in $\mathcal{T}(\hat{R}_2)$. Let A be a tournament matrix in $\mathcal{T}(\hat{R}_n)$. Since the first row sum equals 1, A has exactly one 1 in row 1, say in column $k_1 \geq 2$. If $k_1 = n$, then $A = \hat{A}_n$. Otherwise, $k_1 < n$ and the leading principal submatrix of order k_1 of A satisfies

$$A[\{1, 2, \ldots, k_1\}] = \hat{A}_{k_1}.$$

The principal submatrix $A[\{k_1, k_1+1, \ldots, n\}]$ belongs to $\mathcal{T}(\hat{R}_{n-k_1+1})$. Proceeding inductively we see that there exist integers $k_1, k_2, \ldots, k_p$, satisfying

$$k_i \geq 2 \quad (i = 1, 2, \ldots, p) \text{ and } k_1 + k_2 + \cdots + k_p - (p - 1) = n,$$

such that

$$A = \hat{A}_{k_1} * \hat{A}_{k_2} * \cdots * \hat{A}_{k_p}. \tag{5.39}$$

Here for two tournament matrices A and B of orders k and l, respectively, $C = A * B$ is the tournament matrix of order $k + l - 1$ satisfying $C[\{1, 2, \ldots, k\}] = A$, $C[\{k, k + 1, \ldots, k + l - 1\}] = B$, $A[\{1, 2, \ldots, k - 1\}, \{k + 1, k + 2, \ldots, k + l - 1\}] = O_{k-1,l-1}$, and $A[\{k + 1, k + 2, \ldots, k + l - 1\}, \{1, 2, \ldots, k - 1\}] = J_{l-1,k-1}$. The 1's above the main diagonal in the matrix (5.39) occur in positions

$$(1, k_1), (k_1, k_1 + k_2 - 1), \ldots, (k_1 + k_2 + \cdots + k_{p-1} - (p - 2), n).$$

For example,

$$\hat{A}_3 * \hat{A}_3 = \begin{bmatrix} 0 & 0 & 1 & 0 & 0 \\ 1 & 0 & 0 & 0 & 0 \\ 0 & 1 & 0 & 0 & 1 \\ 1 & 1 & 1 & 0 & 0 \\ 1 & 1 & 0 & 1 & 0 \end{bmatrix}.$$

□

In determining the maximal value of $\tilde{v}_n$, we shall make use of the normalized row sum vector $H_n = (h_1, h_2, \ldots, h_n)$ of a tournament class $\mathcal{T}(R)$.

Recall that this vector is defined by $h_i = r_i - (i-1)$ $(i = 1, 2, \ldots, n)$ and satisfies the normalized Landau conditions

$$h_1 \geq 0 \text{ and } h_{i-1} - h_i \leq 1 \quad (i = 2, 3, \ldots, n),$$

and

$$h_1 + h_2 + \cdots + h_k \geq 0 \quad (k = 1, 2, \ldots, n)$$

with equality for $k = n$. We let

$$H^+ = \{j : h_j > 0, 1 \leq j \leq n\} \text{ and } H^- = \{j : h_j < 0, 1 \leq j \leq n\}.$$

Because the sum of the normalized row sums equals 0, $H^+ = \emptyset$ if and only if $H^- = \emptyset$. Hence both H^+ and H^- are nonempty unless $R_n = (0, 1, 2, \ldots, n-1)$ in which case $\mathcal{T}(R_n)$ contains only the transitive tournament matrix of order n with no upsets.

With this notation, we now prove two lemmas which will be used to determine $\tilde{v}_n^{\max}$.

Lemma 5.7.2 *Let j and k be integers with $1 \leq j < k \leq n$ such that $h_j > 0$ and $h_k < 0$. Then there exists an integer i with $j < i < k$ such that $h_i = 0$.*

Proof. We have $h_{j+1} \geq h_j - 1 \geq 0$. Since $h_k < 0$ and $k > j$, for some i with $j < i < k$, h_i takes the value 0. $\square$

Corollary 5.7.3 $|H^+| + |H^-| \leq n - 1$. $\square$

Lemma 5.7.4 *Assume that $|H^+| = p$ and $|H^-| = q$. Then*

$$\sum_{i \in H^+} h_i \leq \binom{p+1}{2} \text{ and } \sum_{j \in H^-} h_j \geq -\binom{q+1}{2}.$$

Equality holds on the left if and only if H^+ is a set of p consecutive integers $\{k, k+1, \ldots, k+p-1\}$ and $h_k = p, h_{k+1} = p-1, \ldots, h_{k+p-1} = 1$. Equality holds on the right if and only if H^- is a set of q consecutive integers $\{l, l+1, \ldots, l+q-1\}$ and $h_l = -1, h_{l+1} = -2, \ldots, h_{l+q-1} = -q$.

Proof. We have $h_n^+ = 0$, and $h_{i-1} \leq h_i + 1$ $(i = 2, 3, \ldots, n)$. We first use these inequalities to conclude that if j is the largest integer in H^+, then $h_j = 1$, and if k is the smallest integer in H^-, then $h_k = -1$. Now the inequalities imply the conclusions of the lemma. $\square$

We now determine the maximal value for the minimal number of upsets in a class $\mathcal{T}(R_n)$.

Theorem 5.7.5 *Let $n \geq 3$ be an integer. Then*

$$\tilde{v}_n^{\max} = \binom{\lfloor \frac{n+1}{2} \rfloor}{2}. \tag{5.40}$$

Let $R_n = (r_1, r_2, \ldots, r_n)$ be a nondecreasing, nonnegative integral vector such that $\mathcal{T}(R_n)$ is nonempty. Then $\tilde{v}(R_n) = \tilde{v}_n^{\max}$ if and only if

$$n \text{ is odd and } R_n = ((n-1)/2, (n-1)/2, \ldots, (n-1)/2),$$

or

$$n \text{ is even and } r_1 + r_2 + \cdots + r_{n/2} = \frac{n}{2}\left(\frac{n}{2} - 1\right).$$

Proof. First assume that n is odd and that $R_n = ((n-1)/2, (n-1)/2, \ldots, (n-1)/2)$. Then

$$\begin{aligned}
\tilde{v}(R_n) &= \sum_{i=1}^{n} ((n-1)/2 - (i-1))^+ \\
&= 1 + 2 + \cdots + (n-1)/2 \\
&= \binom{\frac{n+1}{2}}{2}.
\end{aligned}$$

Now assume that n is even and $r_1 + r_2 + \cdots + r_{n/2} = (n/2)((n/2) - 1)$. Then

$$\begin{aligned}
\tilde{v}(R_n) &= \sum_{i=1}^{n} (r_i - (i-1))^+ \\
&\geq \sum_{i=1}^{n/2} (r_i - (i-1))^+ \\
&\geq \sum_{i=1}^{n/2} r_i - \sum_{i=1}^{n/2} (i-1) \\
&= \frac{n}{2}\left(\frac{n}{2} - 1\right) - \frac{1}{2}\frac{n}{2}\left(\frac{n}{2} - 1\right) \\
&= \frac{1}{2}\frac{n}{2}\left(\frac{n}{2} - 1\right) \\
&= \binom{\frac{n}{2}}{2} = \binom{\lfloor \frac{n+1}{2} \rfloor}{2}.
\end{aligned}$$

We now show that $\tilde{v}(R_n)$ is bounded above by the value given in (5.40) which then implies that (5.40) holds and that those R_n described in the theorem satisfy $\tilde{v}(R_n) = \tilde{v}_n^{\max}$.

Assume to the contrary that for some R_n we have

$$\tilde{v}(R_n) = \sum_{i \in H^+} h_i > \binom{\lfloor \frac{n+1}{2} \rfloor}{2} = 1 + 2 + \cdots + \left\lfloor \frac{n-1}{2} \right\rfloor. \tag{5.41}$$

By Lemma 5.7.4, $|H^+| \geq \lfloor (n+1)/2 \rfloor$ and by Corollary 5.7.3, $|H^-| \leq \lfloor (n-2)/2 \rfloor$. From Lemma 5.7.4, we get

$$\sum_{i \in H^-} h_i \geq -\binom{\lfloor \frac{n}{2} \rfloor}{2}. \tag{5.42}$$

Combining (5.41) and (5.42) we get

$$\sum_{i=1}^{n} h_i = \sum_{i \in H^+} h_i + \sum_{i \in H^-} h_i > \binom{\lfloor \frac{n+1}{2} \rfloor}{2} - \binom{\lfloor \frac{n}{2} \rfloor}{2} \geq 0.$$

This contradicts the normalized Landau conditions.

It remains to show that if $\mathcal{T}(R_n)$ is nonempty and $\tilde{v}(R_n) = \tilde{v}_n^{\max}$ as given in (5.40), then R_n has the form given in the theorem.

First suppose that n is odd, and $\tilde{v}(R_n) = \tilde{v}_n^{\max}$. By Lemma 5.7.4, we have $|H^+| \geq \lfloor (n-1)/2 \rfloor$. Arguing as above we conclude that $|H^+| = (n-1)/2$ and $|H^-| \leq (n-1)/2$. Lemma 5.7.4 and the normalized Landau conditions imply that $|H^-| = (n-1)/2$. Since H^+ and H^- are sets of consecutive integers by Lemma 5.7.4, and $1 \notin H^-$ and $n \notin H^+$, Lemma 5.7.2 implies that

$$H^+ = \left\{ 1, 2, \ldots, \frac{n-1}{2} \right\} \text{ and } H^- = \left\{ \frac{n+3}{2}, \frac{n+3}{2} + 1, \ldots, n \right\}.$$

Thus $h_{(n+1)/2} = 0$ and $r_{(n+1)/2} = (n-1)/2$. Since R_n is nondecreasing, we have $r_i \leq (n-1)/2$ for $i = 1, 2, \ldots, (n-1)/2$. Since $\tilde{v}(R_n) = \tilde{v}_n^{\max}$, this implies that $r_i = (n-1)/2$ for $i = 1, 2, \ldots, (n-1)/2$, and then $R_n = ((n-1)/2, (n-1)/2, \ldots, (n-1)/2)$.

Now suppose that n is even, and $\tilde{v}(R_n) = \tilde{v}_n^{\max}$. First assume that $r_1 = 0$ and hence $h_1 = 0$. Let A be a matrix in $\mathcal{T}(R_n)$ with $v(A) = \tilde{v}(R_n) = \tilde{v}_n^{\max}$. Let A' be the tournament matrix of odd order $n-1$ obtained by deleting row 1 and column 1 of A. Then $v(A') = \tilde{v}_{n-1}^{\max}$ and by what we have already proved, the row sum vector R' of A' is the vector $((n-2)/2, (n-2)/2, \ldots, (n-2)/2)$. Hence $R_n = (0, n/2, n/2, \ldots, n/2)$ and

$$r_1 + r_2 + \cdots + r_{n/2} = \left(\frac{n}{2} - 1 \right) \frac{n}{2}, \tag{5.43}$$

as required.

Now assume that $r_1 = h_1 \geq 1$. From Lemma 5.7.4 we get that $|H^+| \geq (n/2) - 1$, and using Corollary 5.7.3 and Lemma 5.7.4 as before we see that $|H^+| = (n/2) - 1$ or $n/2$. Suppose that $|H^+| = (n/2) - 1$. Since $h_1 \geq 1$, it follows from Lemma 5.7.4 that $H^+ = \{1, 2, \ldots, (n/2) - 1\}$ and $r_1 = r_2 = \cdots = r_{(n/2)-1} = (n/2) - 1$. If $r_{n/2} > (n/2) - 1$, then we obtain the contradiction that $\sum_{i=1}^{n} r_i > \binom{n}{2}$. Hence $r_{n/2} = (n/2) - 1$ and we satisfy (5.43) again. Now suppose that $|H^+| = n/2$. Then $|H^-| \leq (n-2)/2$

and

$$\sum_{i \in H^-} h_i = -\sum_{i \in H^+} h_i$$
$$= -\tilde{v}_n$$
$$= -\binom{\frac{n}{2}}{2}$$
$$= -1 - 2 - \cdots - \frac{n-2}{2}.$$

Lemma 5.7.4 now implies that $|H^-| = (n-2)/2$ and that H^- is a set of $(n-2)/2$ consecutive integers. Now $|H^+| = n/2$, $1 \in H^+$, $n \notin H^+$, and Lemma 5.7.2 imply that H^+ is the set of $n/2$ consecutive integers equal to $\{1, 2, \ldots, n/2\}$. Thus

$$\sum_{i=1}^{n/2} r_i = \sum_{i=1}^{n/2} h_i + \sum_{i=1}^{n/2} (i-1)$$
$$= 2 \binom{\frac{n}{2}}{2}$$
$$= \frac{n}{2} \left(\frac{n}{2} - 1 \right).$$

The proof of the theorem is now complete.   □

Another formulation of (5.40) is

$$\max_{\{R_n : \mathcal{T}(R_n) \neq \varnothing\}} \min_{A \in \mathcal{T}(R_n)} v(A) = \binom{\lfloor \frac{n+1}{2} \rfloor}{2}.$$

Example. If n is even there are many vectors R_n that satisfy the conditions for equality in Theorem 5.7.5. For example, if $n = 8$, those nondecreasing vectors R_8 for which $\mathcal{T}(R_8)$ is nonempty and $\tilde{v}(R_8) = \tilde{v}_8^{\max} = 6$ are

$$(0, 4, 4, 4, 4, 4, 4, 4), (3, 3, 3, 3, 4, 4, 4, 4), (2, 3, 3, 4, 4, 4, 4, 4),$$

$$(1, 3, 4, 4, 4, 4, 4, 4), \text{ and } (3, 3, 3, 3, 3, 4, 4, 5).$$

The near-regular vector $R^{(n)} = ((n-2)/2, \ldots, (n-2)/2, n/2, \ldots, n/2)$ ($(n-2)/2$ and $n/2$ each occur $n/2$ times) always satisfies $\tilde{v}(R^{(n)}) = \tilde{v}_n^{\max}$.   □

Now let

$$\bar{v}_n^{\min} = \min\{\bar{v}(R_n)) : \mathcal{T}(R_n) \neq \varnothing \text{ and } R_n \text{ nondecreasing and strong}\}$$

be the minimal number t for which there exists a nonempty class $\mathcal{T}(R_n)$ with a nondecreasing strong row sum vector R_n such that every tournament matrix in $\mathcal{T}(R_n)$ has at most t upsets, and let

$$\bar{v}_n^{\max} = \max\{\bar{v}(R_n) : \mathcal{T}(R_n) \neq \varnothing \text{ and } R_n \text{ nondecreasing}\}$$

be the maximal number of upsets in any tournament matrix of order n with a nondecreasing row sum vector R_n. In the remainder of this section we determine $\bar{v}_n^{\min}$ and $\bar{v}_n^{\max}$ and characterize the row sum vectors that attain these extreme values.

The following lemma is used to determine $\bar{v}_n^{\min}$. It asserts that for each nondecreasing strong score vector R_n, there exists a tournament in $\mathcal{T}(R_n)$ with a directed Hamilton cycle that passes through the players in order of increasing superiority, closing with the best player beating the poorest player. Recall that Theorem 5.2.4 asserts the existence of a directed Hamilton cycle in every strong tournament of order $n \geq 3$.

Lemma 5.7.6 *Let $R_n = (r_1, r_2, \ldots, r_n)$ be a nondecreasing, strong row sum vector of size $n \geq 3$. Then there exists a tournament matrix $A = [a_{ij}]$ in $\mathcal{T}(R_n)$ such that*

$$a_{12} = a_{23} = \cdots = a_{n-1,n} = a_{n1} = 1. \tag{5.44}$$

Proof. Let k be the maximal integer for which there exists a matrix $A = [a_{ij}]$ in $\mathcal{T}(R_n)$ with $a_{12} = a_{23} = \cdots = a_{k-1,k} = 1$. First assume that $k = n$. If $a_{n1} = 1$, then A satisfies the conclusion of the theorem. Otherwise, $a_{n1} = 0$ and so $a_{1n} = 1$. Since $r_1 \leq r_n$, for some i with $1 < i < n$, there is a $(1, i, n)$-Δ-interchange that produces the required matrix.

Now assume that $k < n$. Then $a_{k,k+1} = 0$ and $a_{k+1,k} = 1$. First suppose that there is an integer j with $j > k$ such that $a_{kj} = 1$ and hence $a_{jk} = 0$. Since $r_k \leq r_{k+1}$, there is a four-way double interchange which replaces $a_{k,k+1}$ with a 1 and leaves unchanged all the elements in the leading principal submatrix of A of order k. This contradicts the maximality property of k. Thus the terminal part of row k of A contains only 0's. Since R_n is strong, $X = A[\{1, 2, \ldots, k\}, \{k+1, k+2, \ldots, n\}] \neq O$. Let j be the last row of X that contains a 1. Then $j < k$ and $a_{jl} = 1$ for some $l > k$. We have $a_{j,j+1} = a_{jl} = 1$ and $a_{j+1,j+1} = a_{j+1,l} = 0$. Since $r_j \leq r_{j+1}$, there exists an integer $i \neq j - 1$ such that $a_{ji} = 0$ and $a_{j+1,i} = 1$ and hence a double interchange which places a 1 in row $j + 1$ of X and does not change the 1's in positions $(1, 2), (2, 3), \ldots, (k-1, k)$. Repeating this argument, we can place a 1 in row k of X and, as above, contradict the maximality of k. It follows that $k = n$ and the theorem holds. $\qquad \square$

Corollary 5.7.7 *Let R_n be a nondecreasing, strong row sum vector of size $n \geq 3$. Then $\bar{v}(R_n) - \tilde{v}(R_n) \geq n - 2$.*

Proof. Let $A = [a_{ij}]$ be the matrix in Lemma 5.7.6. Then $a_{12} = a_{23} = \cdots = a_{n-1,n} = a_{n1} = 1$ and $a_{21} = a_{32} = \cdots = a_{n,n-1} = a_{1n} = 0$. Interchanging these 1's and 0's we obtain a matrix A' in $\mathcal{T}(R_n)$ satisfying $v(A) - v(A') = n - 2$. $\qquad \square$

Recall that $\hat{R}_n$ denotes the strong score vector $(1, 1, 2, 3, \ldots, n - 3, n - 2, n - 2)$ $(n \geq 2)$.

Theorem 5.7.8 *Let $n \geq 3$ be an integer. Then*

$$\bar{v}_n^{\min} = n - 1. \tag{5.45}$$

Moreover, a nondecreasing, strong score vector R_n satisfies $\bar{v}(R_n) = n - 1$ if and only if $R_n = \hat{R}_n$.

Proof. By Lemma 5.7.6, if R_n is a nondecreasing, strong score vector with $n \geq 3$, then there exists a matrix $A = [a_{ij}] \in \mathcal{T}(R_n)$ satisfying (5.44) and so with $v(A) \geq n - 1$. Hence $\bar{v}_n^{\min} \geq n - 1$. If $v(A) = n - 1$, then $a_{ij} = 0$ for $3 \leq i + 2 \leq j \leq n$ and $R_n = \hat{R}_n$; indeed $A = \hat{A}_2 * \hat{A}_2 * \cdots * \hat{A}_2$ where there are $n - 1$ matrices

$$\hat{A}_2 = \begin{bmatrix} 0 & 1 \\ 0 & 0 \end{bmatrix} \in \mathcal{T}(\hat{R}_2).$$

Since $\bar{v}(\hat{R}_n) = n - 1$, the theorem follows. $\qquad\square$

Example. Let $n = 8$ and $R_8 = (1, 1, 3, 3, 4, 4, 6, 6)$. By the formula (5.27) for the minimal number of upsets, we have $\tilde{v}(R_8) = 2$. Applying Fulkerson's algorithm for a tournament matrix in $\mathcal{T}(R_8)$ with the maximal number of upsets we see that $\bar{v}(R_8) = 8$. Hence $\bar{v}(R_8) - \tilde{v}(R_8) = 6 = n - 2$, and we have equality in Corollary 5.7.7. Since $\tilde{v}(\hat{R}_n) = 1$ and $\bar{v}(\hat{R}_n) = n - 1$, we also have equality for $\hat{R}_n$ whenever $n \geq 3$. $\qquad\square$

In the following theorem we determine $\bar{v}_n^{\max}$ and show that only regular and near-regular row sum vectors attain the maximum.

Theorem 5.7.9 *Let $n \geq 3$ be an integer. Then*

$$\bar{v}_n^{\max} = \binom{n}{2} - \binom{\lfloor \frac{n+2}{2} \rfloor}{2}. \tag{5.46}$$

Let $R_n = (r_1, r_2, \ldots, r_n)$ be a nondecreasing, nonnegative integral vector such that $\mathcal{T}(R_n)$ is nonempty. Then $\bar{v}(R_n) = \bar{v}_n^{\max}$ if and only if

$$n \text{ is odd and } R_n = ((n-1)/2, (n-1)/2, \ldots, (n-1)/2),$$

or

$$n \text{ is even and } R_n = ((n-2)/2. \ldots, (n-2)/2, n/2, \ldots, n/2).$$

Proof. We consider n odd and n even separately.

Case 1: n is odd. Let

$$a = r_{(n+3)/2} + r_{(n+5)/2} + \cdots + r_n, \text{ and } b = r_2 + r_3 + \cdots + r_{(n+1)/2}.$$

Since R_n is nondecreasing, we have

$$r_1 + b + a = \binom{n}{2}, a \geq b, \text{ and hence } r_1 + 2a \geq \binom{n}{2}.$$

Thus

$$a \geq \frac{n(n-1)}{4} - \frac{r_1}{2}.$$

By Landau's conditions, we have $r_1 \leq (n-1)/2$ with equality if and only if $R_n = ((n-1)/2, (n-1)/2, \ldots, (n-1)/2)$. Therefore

$$a \geq \frac{(n-1)^2}{4}, \qquad (5.47)$$

with equality if and only if $R_n = ((n-1)/2, (n-1)/2, \ldots, (n-1)/2)$.

Now let $A = [a_{ij}]$ be a matrix in $T(R_n)$. Rows i of T with $(n+3)/2 \leq i \leq n$ contain in total at most

$$1 + 2 + \cdots + \frac{n-3}{2} = \binom{\frac{n-1}{2}}{2}$$

1's in their terminal parts. Let a' be the number of 1's in the initial parts in rows i with $(n+3)/2 \leq i \leq n$. Then using (5.47) we see that

$$a' \geq a - \binom{\frac{n-1}{2}}{2} \geq \binom{\frac{n+1}{2}}{2},$$

with equality only if $R_n = ((n-1)/2, (n-1)/2, \ldots, (n-1)/2)$. Thus the number $v(A)$ of 1's above the main diagonal of A satisfies

$$v(A) \leq \binom{n}{2} - a' \leq \binom{n}{2} - \binom{\frac{n+1}{2}}{2},$$

again with equality only if $R_n = ((n-1)/2, (n-1)/2, \ldots, (n-1)/2)$. Therefore, the expression in (5.46) is an upper bound for $\bar{v}^{\max}$. Equality holds for the regular tournament matrix

$$A = \mathrm{Circ}_n(0, 1, \ldots, 1, 0, \ldots, 0) = C_n + C_n^2 + \cdots + C_n^{(n-1)/2}.$$

Thus when n is odd, (5.46) holds and $\bar{v}(R_n) = \bar{v}^{\max}$ if and only if $R_n = ((n-1)/2, (n-1)/2, \ldots, (n-1)/2)$.

Case 2: n is even. Let

$$c = r_1 + r_2 + \cdots + r_{n/2} \text{ and } d = r_{(n/2)+1} + r_{(n/2)+2} + \cdots + r_n.$$

Suppose that $c > (n/2)(n-2)/2$. Since R_n is nondecreasing, $r_{n/2} \geq n/2$ and thus $r_i \geq n/2$ $((n/2) \leq i \leq n)$. Hence $d \geq (n/2)^2$ and we have

$$\binom{n}{2} = c + d > \frac{n}{2}\left(\frac{n-2}{2}\right) + \left(\frac{n}{2}\right)^2 = \binom{n}{2}.$$

This contradiction implies that

$$c \leq \frac{n}{2}\left(\frac{n-2}{2}\right) \text{ and } d \geq \binom{n}{2} - \frac{n}{2}\left(\frac{n-2}{2}\right) = \frac{n^2}{4}. \qquad (5.48)$$

Now let $A = [a_{ij}]$ be a matrix in $\mathcal{T}(R_n)$. The number of 1's in total in the initial parts of the last $n/2$ rows of A is at least

$$d - \left(1 + 2 + \cdots + \left(\frac{n}{2} - 1\right)\right) \geq \frac{n^2}{4} - \binom{\frac{n}{2}}{2} = \binom{\frac{n+2}{2}}{2}.$$

As in the odd case we conclude that

$$v(A) \leq \binom{n}{2} - \binom{\frac{n+2}{2}}{2}. \tag{5.49}$$

Thus the expression in (5.46) is an upper bound for $\bar{v}^{\max}$. Equality holds for the near-regular tournament matrix

$$\begin{bmatrix} L_{n/2}^{\mathrm{T}} & L_{n/2} \\ (I_{n/2} + L_{n/2})^{\mathrm{T}} & L_{n/2}^{\mathrm{T}} \end{bmatrix}$$

where $L_{n/2}$ is the lower triangular matrix of order $n/2$ with 0's on and above the main diagonal and 1's below the main diagonal. Thus when n is even, (5.46) holds and $\bar{v}(R_n) = \bar{v}_n^{\max}$ where $R_n = ((n-2)/2, \ldots, (n-2)/2, n/2, \ldots, n/2)$.

Now suppose that equality holds in (5.49). Then equality holds for c and d in (5.48) and A has the form

$$\begin{bmatrix} L_{n/2}^{\mathrm{T}} & X \\ Y & L_{n/2}^{\mathrm{T}} \end{bmatrix} \tag{5.50}$$

Since $r_{n/2} \geq n/2$, it follows from (5.50) that $r_n = n/2$. Since $d = n^2/4$, we also have $r_i = n/2$ $((n/2) + 1 \leq i \leq n)$. From (5.50) we conclude that $r_1 \geq (n-2)/2$. Since $c = (n/2)(n-2)/2$, we also have $r_i = (n-2)/2$ $(1 \leq i \leq n/2)$. Thus we have $R_n = ((n-2)/2, \ldots, (n-2)/2, n/2, \ldots, n/2)$. Therefore when n is even, (5.46) holds and $\bar{v}(R_n) = \bar{v}^{\max}$ if and only if $R_n = ((n-2)/2, \ldots, (n-2)/2, n/2, \ldots, n/2)$. $\qquad\qquad\square$

In [13] a parameter $s(T)$ called the *Slater index* of a tournament T is defined as the minimum number of edges whose reversal gives a transitive tournament, and some connections between this index and the number of upsets of T are derived. In investigating upsets, we have assumed that the teams are ordered from worst score to best score. By reversing the upset edges we obtain a transitive tournament with the teams in this same order. Thus $s(T) \leq v(T)$. In determining the Slater index, the order of the teams may change in the resulting transitive tournament.

5.8   Cardinality of $\mathcal{T}(R)$

The determination of the number

$$|\mathcal{T}(R)| \quad (\mathcal{T}(R) \neq \oslash)$$

of tournament matrices with a given nondecreasing row sum vector R is of considerable difficulty. In this section we determine the generating function for these numbers $|\mathcal{T}(R)|$ that exhibits a surprising connection with Vandermonde matrices. We also obtain some estimates for the number of tournament matrices with special row sum vectors.

Let p and n be positive integers. Recall that a nonnegative integral matrix $A = [a_{ij}]$ of order n is a p-tournament matrix provided

$$a_{ii} = 0 \ (i = 1, 2, \ldots, n) \text{ and } a_{ij} + a_{ji} = p \quad (1 \le i < j \le n),$$

that is, provided

$$A + A^{\mathrm{T}} = p(J_n - I_n).$$

We have

$$\sum_{i=1}^{n}\sum_{j=1}^{n} a_{ij} = \sum_{1 \le i < j \le n} (a_{ij} + a_{ji}) = \sum_{1 \le i < j \le n} p = p\binom{n}{2}.$$

Recall also that the class of all p-tournament matrices with a given row sum vector R is denoted by

$$\mathcal{T}(R; p).$$

If $p = 1$, then a 1-tournament matrix is just an ordinary tournament matrix. In the next section we study in more detail the class $\mathcal{T}(R; 2)$ of 2-tournament matrices with row sum vector R. In determining the generating function for the number of tournament matrices, we more generally determine the generating function for the number of p-tournament matrices.

Let $x_1, x_2, \ldots, x_n$ be n independent complex variables. The *Vandermonde matrix* based on $x_1, x_2, \ldots, x_n$ is the matrix

$$V_n(x_1, x_2, \ldots, x_n) = \begin{bmatrix} 1 & 1 & \cdots & 1 \\ x_1 & x_2 & \cdots & x_n \\ \vdots & \vdots & \ddots & \vdots \\ x_1^{n-1} & x_2^{n-1} & \cdots & x_n^{n-1} \end{bmatrix}.$$

We consider the rows of $V_n(x_1, x_2, \ldots, x_n)$ to be indexed by $0, 1, \ldots, n-1$. The determinant of a Vandermonde matrix is a classical formula. We give Gessel's [15] derivation using tournaments.

Lemma 5.8.1 $\quad \det(V_n(x_1, x_2, \ldots, x_n)) = \prod_{1 \le i < j \le n}(x_j - x_i)$.

Proof. Let T_n be a tournament of order n with vertex set $\{1, 2, \ldots, n\}$. Define the weight $w(i, j)$ of an edge (i, j) of T_n to be $-x_i$ if $i < j$ and x_i if $i > j$. Then define the weight $w(T_n)$ of T_n to be the product of the weights of all of its edges. By identifying each term in the following product with a tournament T_n, we see that

$$\prod_{1 \le i < j \le n}(x_j - x_i) = \sum_{T_n} w(T_n), \tag{5.51}$$

where the sum is over all tournaments T_n of order n. Each factor $(x_j - x_i)$ of the product in (5.51) corresponds to an edge between x_i and x_j. In expanding this product, we identify x_j in $(x_j - x_i)$ with an edge from x_j to x_i and $-x_i$ with an edge from x_i to x_j. For instance, if $n = 4$, then the product of the underlined terms in

$$(\underline{x_4} - x_1)(\underline{x_3} - x_1)(x_2\underline{-x_1})(x_4\underline{-x_2})(\underline{x_3} - x_2)(\underline{x_4} - x_3)$$

is

$$x_4 x_3(-x_1)(-x_2)(x_3)(x_4) = x_1 x_2 x_3^2 x_4^2,$$

and this term corresponds to the tournament matrix

$$\begin{bmatrix} 0 & 1 & 0 & 0 \\ 0 & 0 & 0 & 1 \\ 1 & 1 & 0 & 0 \\ 1 & 0 & 1 & 0 \end{bmatrix}$$

with row sum vector $R_4 = (1, 1, 2, 2)$. Let R_n be the row sum vector of T_n. Then the sum on the right in (5.51) can also be written as

$$\sum_{T_n} w(T_n) = \sum_{\{R_n : \mathcal{T}(R_n) \neq \varnothing\}} \sum_{T_n \in \mathcal{T}(R_n)} w(T_n).$$

The weight of a tournament in $\mathcal{T}(R_n)$ equals

$$\pm x_1^{r_1} x_2^{r_2} \dots x_n^{r_n}.$$

Suppose $\mathcal{T}(R_n) \neq \varnothing$ and R_n is not the score vector of a transitive tournament. Let T_n be a tournament with score vector R_n. Then T_n has a 3-cycle and, applying a Δ-interchange, we obtain a tournament T_n' for which $w(T_n') = -w(T_n)$. This implies that the Δ-interchange graph $G_\Delta(R_n)$, defined at the end of Section 5.5, is a bipartite graph and, since by Lemma 5.2.5, each tournament in $\mathcal{T}(R_n)$ has the same number of 3-cycles, $G_\Delta(R_n)$ is a graph with all vertices of the same degree. Hence the number of tournaments with weight $x_1^{r_1} x_2^{r_2} \dots x_n^{r_n}$ equals the number with weight $-x_1^{r_1} x_2^{r_2} \dots x_n^{r_n}$, and thus

$$\sum_{T_n \in \mathcal{T}(R_n)} w(T_n) = 0$$

if R_n is not the score vector of a transitive tournament. Since a transitive tournament is uniquely determined by its score vector, we have

$$\sum_{T_n} w(T_n) = \sum_{T_n \text{ transitive}} w(T_n).$$

The score vectors of transitive tournaments of order n are the vectors $(i_1, i_2, \dots, i_n)$ that are permutations of $\{0, 1, \dots, n - 1\}$. The weight of the transitive tournament with score vector $(i_1, i_2, \dots, i_n)$ is

$$\text{sign}(i_1, i_2, \dots, i_n) x_1^{i_1} x_2^{i_2} \dots x_n^{i_n}$$

where $\text{sign}(i_1, i_2, \ldots, i_n)$ is the sign of the permutation $(i_1, i_2, \ldots, i_n)$ of $\{0, 1, \ldots, n-1\}$.   □

The above argument was generalized in [6]; additional related work is in [11].

Yutsis [26] determined the generating function for the number of p-tournament matrices of order n with score vector R_n. Let

$$g_n^p(x_1, x_2, \ldots, x_n) = \sum_{R_n = (r_1, r_2, \ldots, r_n)} |\mathcal{T}(R_n; p)| x_1^{r_1} x_2^{r_2} \ldots x_n^{r_n}$$

where the sum is over all nonnegative integral vectors $R_n = (r_1, r_2, \ldots, r_n)$ with $r_1 + r_2 + \cdots + r_n = p\binom{n}{2}$.

Theorem 5.8.2 *Let n and p be positive integers. Then*

$$g_n^p(x_1, x_2, \ldots, x_n) = \frac{\det(V_n(x_1^{p+1}, x_2^{p+1}, \ldots, x_n^{p+1}))}{\det(V_n(x_1, x_2, \ldots, x_n))}.$$

Proof. We have

$$g_n^p(x_1, x_2, \ldots, x_n) = \sum_{R_n = (r_1, r_2, \ldots, r_n)} |\mathcal{T}(R_n; p)| x_1^{r_1} x_2^{r_2} \ldots x_n^{r_n}$$

$$= \prod_{1 \le i < j \le n} \left(\sum_{k=0}^{p} x_i^{p-k} x_j^{k} \right)$$

$\quad (x_i^{p-k} x_j^{k}$ corresponds to the choice of $p - k$
$\quad$ in position (i, j) and k in position $(j, i))$

$$= \prod_{1 \le i < j \le n} \frac{x_j^{p+1} - x_i^{p+1}}{x_j - x_i}$$

$$= \frac{\prod_{1 \le i < j \le n}(x_j^{p+1} - x_i^{p+1})}{\prod_{1 \le i < j \le n}(x_j - x_i)}$$

$$= \frac{\det(V_n(x_1^{p+1}, x_2^{p+1}, \ldots, x_n^{p+1}))}{\det(V_n(x_1, x_2, \ldots, x_n))}.$$

(by two applications of Lemma 5.8.1).

□

Let $R_n = (r_1, r_2, \ldots, r_n)$ be a nonnegative integral vector with $r_1 + r_2 + \cdots + r_n = p\binom{n}{2}$. We now show a bijection between the set $\mathcal{T}(R_n; p)$ of

p-tournament matrices with row sum vector R_n and the set $\mathcal{Y}_{\lambda_{n,p}}(R_n)$ of Young tableaux of shape

$$\lambda_{n,p} = (p(n-1), p(n-2), \ldots, p)$$

based on the set $\{1, 2, \ldots, n\}$ in which i occurs r_i times $(i = 1, 2, \ldots, n)$. This bijection due to Yutsis [26] is based on the row-insertion algorithm and the RSK correspondence developed in Section 4.3.

Yutsis' Algorithm for Associating a Young Tableau in $\mathcal{Y}_{n,p}(R_n)$ with a p-Tournament Matrix in $\mathcal{T}(R_n, p)$

Let $R_n = (r_1, r_2, \ldots, r_n)$, and let $A = [a_{ij}]$ be a p-tournament matrix in $\mathcal{T}(R_n; p)$.

(1) If $n = 1$, set Y_1 equal to the empty tableau.

(2) Otherwise, for $k = 2, 3, \ldots, n$, do:

   (i) From right to left, row-insert

$$\underbrace{1, 1, \ldots, 1}_{a_{1k}}, \underbrace{2, 2, \ldots, 2}_{a_{2k}}, \ldots, \underbrace{k-1, k-1, \ldots, k-1}_{a_{k-1,k}}$$

   in Y_{k-1}.

   (ii) Add new squares to the tableau Y_k in order to obtain a Young diagram of shape $\lambda_{k,p}$ and insert k in each of the new squares.

   (iii) Set Y_k equal to the resulting tableau.

(3) Output $Y_A = Y_n$.

Before proving that this algorithm does produce a Young tableau in $\mathcal{Y}_{n,p}(R_n)$, we give some examples of the algorithm

Example. First consider the transitive tournament matrix

$$A = \begin{bmatrix} 0 & 0 & 0 & 0 \\ 1 & 0 & 0 & 0 \\ 1 & 1 & 0 & 0 \\ 1 & 1 & 1 & 0 \end{bmatrix}$$

with $p = 1$ and row sum vector $R_4 = (0, 1, 2, 3)$. Then

$$Y_1 = \varnothing, \quad Y_2 = \begin{bmatrix} 2 \end{bmatrix}, \quad Y_3 = \begin{array}{|c|c|} \hline 2 & 3 \\ \hline 3 & \times \\ \hline \end{array}, \quad \text{and } Y_A = Y_4 = \begin{array}{|c|c|c|} \hline 2 & 3 & 4 \\ \hline 3 & 4 & \times \\ \hline 4 & \times & \times \\ \hline \end{array}.$$

Now consider the transitive tournament matrix

$$\begin{bmatrix} 0 & 1 & 1 & 1 \\ 0 & 0 & 1 & 1 \\ 0 & 0 & 0 & 1 \\ 0 & 0 & 0 & 0 \end{bmatrix}$$

with row sum vector $R_4 = (3, 2, 1, 0)$. Then

$$Y_1 = \varnothing, \ Y_2 = \boxed{1}, \ Y_3 = \begin{array}{|c|c|} \hline 1 & 1 \\ \hline 2 & \times \\ \hline \end{array}, \ \text{and} \ Y_A = Y_4 = \begin{array}{|c|c|c|} \hline 1 & 1 & 1 \\ \hline 2 & 2 & \times \\ \hline 3 & \times & \times \\ \hline \end{array}.$$

In general, the p-tournament matrix of order 4 with p in every position above the main diagonal gives the Young tableau with $3p$ 1's in row 1, $2p$ 2's in row 2, and p 3's in row 3.

Finally, consider the 2-tournament matrix

$$A = \begin{bmatrix} 0 & 1 & 2 & 0 \\ 1 & 0 & 1 & 2 \\ 0 & 1 & 0 & 1 \\ 2 & 0 & 1 & 0 \end{bmatrix}$$

with row sum vector $R_4 = (3, 4, 2, 3)$. Then

$$Y_1 = \varnothing, \ Y_2 = \boxed{1 \ | \ 2}, \ Y_3 = \begin{array}{|c|c|c|c|} \hline 1 & 1 & 1 & 3 \\ \hline 2 & 2 & \times & \times \\ \hline \end{array},$$

and

$$Y_A = Y_4 = \begin{array}{|c|c|c|c|c|c|} \hline 1 & 1 & 1 & 2 & 2 & 4 \\ \hline 2 & 2 & 3 & 3 & \times & \times \\ \hline 4 & 4 & \times & \times & \times & \times \\ \hline \end{array}.$$

$\square$

Theorem 5.8.3 *Let n and p be positive integers with $R_n = (r_1, r_2, \ldots, r_n)$ be a nonnegative integral vector with $r_1 + r_2 + \cdots + r_n = p\binom{n}{2}$. Then Yutsis' algorithm determines a bijection*

$$A \longrightarrow Y_A$$

between p-tournament matrices in $\mathcal{T}(R_n; p)$ and Young tableaux in $\mathcal{Y}_{\lambda_{n,p}}(R_n)$. In particular,

$$|\mathcal{T}(R_n; p)| = |\mathcal{Y}_{\lambda_{n,p}}(R_n)|.$$

Proof. Given a p-tournament matrix $A = [a_{ij}]$ in $\mathcal{T}(R_n; p)$, since the algorithm is successive row-insertion, it certainly produces a Young tableau Y_A.

We first show by induction on n that Y_A contains r_i i's for each $i = 1, 2, \ldots, n$. This is trivial for $n = 1$. Assume $n > 1$. Let $A' = A[\{1, 2, \ldots, n-1\}]$ and let the row sum vector of A' be $R' = (r'_1, r'_2, \ldots, r'_{n-1})$. By the induction hypothesis, the number of times i occurs in $Y_{A'}$ equals r'_i ($i = 1, 2, \ldots, n - 2$). Since Y_A is obtained from $Y_{A'}$ by row-inserting a_{in} i's for $i = n - 1, n - 2, \ldots, 1$, it follows that Y_A contains r_i i's for each

$i = 1, 2, \ldots, n - 1$. After that, n is row-inserted $p(n - 1) - \sum_{i=1}^{n-1} a_{in} = r_n$ times. Hence n also occurs r_n times in Y_A.

We next show that Y_A has shape $\lambda_{n,p}$ by induction on n. This is trivial for $n = 1$ and 2. Assume that $n > 2$. By the induction hypothesis, applying the algorithm to the principal submatrix $A' = A[\{1, 2, \ldots, n-1\}]$ we obtain a Young tableau of $Y_{A'}$ of shape $\lambda_{n-1,p}$. We now apply step $(2)(i)$ of the algorithm by row-inserting into $Y_{A'}$ from right to left the sequence of integers

$$\underbrace{1, 1, \ldots, 1}_{a_{1n}}, \underbrace{2, 2, \ldots, 2}_{a_{2n}}, \ldots, \underbrace{n - 1, n - 1, \ldots, n - 1}_{a_{n-1,n}} \qquad (5.52)$$

to obtain a Young tableau Y^*. Successive insertion of the elements in (5.52) increases the size of row 1 by $\max\{a_{1n}, a_{2n}, \ldots a_{n-1,n}\} \leq p$. A similar conclusion holds when inserting the elements that are bumped from row 1 into row 2, and so forth. The resulting tableau has shape $(h_1, h_2, \ldots, h_{n-1})$ where $h_i \leq p(n - i)$ for $i = 1, 2, \ldots, n - 1$. The final step of the algorithm is to add boxes so that the shape is $\lambda_{n,p}$ and put n in each of the boxes added.

That $A \rightarrow Y_A$ is a bijection between $\mathcal{T}(R_n; p)$ and $\mathcal{Y}_{\lambda_{n,p}}(R_n)$ is a consequence of the invertibility of the row-insertion algorithm (the inverse was called the row-extraction algorithm), as developed in Section 4.1.   □

We now consider bounds for the numbers $|\mathcal{T}(R_n)|$. Let

$$\check{R}_n = (0, 1, 2, \ldots, n - 1)$$

be the nondecreasing score vector of a transitive tournament matrix of order n $(n \geq 1)$. We recall that if R_n is a nondecreasing, nonnegative integral vector, then by Landau's theorem, $\mathcal{T}(R_n)$ is nonempty if and only if $R_n \preceq \check{R}_n$. The components of the row sum vectors

$$\overline{R}_n = \begin{cases} ((n - 1)/2, (n - 1)/2, \ldots, (n - 1)/2) & (n \text{ odd}), \\ ((n - 2)/2, \ldots, \ldots, (n - 2)/2, n/2, \ldots, n/2) & (n \text{ even}) \end{cases}$$

of regular (n odd) and near-regular (n even) tournament matrices, being as equal as possible, satisfy $\overline{R}_n \preceq R_n, (n \geq 1)$, as is easily verified [8]. Thus for every nondecreasing R_n for which $\mathcal{T}(R_n)$ is nonempty we have

$$\overline{R}_n \preceq R_n \preceq \check{R}_n \quad (n \geq 1)$$

with equality holding only for $R_n = \overline{R}_n$ (for the left inequality) and $R_n = \check{R}_n$ (for the right inequality).

We also recall the score vector

$$\hat{R}_n = (1, 1, 2, \ldots, n - 3, n - 2, n - 2) \quad (n \geq 3)$$

and that for every strong, nondecreasing tournament score vector R_n,

$$\overline{R}_n \preceq R_n \preceq \hat{R}_n \quad (n \geq 3).$$

The number of matrices in $\mathcal{T}(\check{R}_n)$ equals 1. In the following theorem we count the number of tournament matrices in the class of irreducible tournament matrices $\mathcal{T}(\hat{R}_n)$ with the largest, with respect to majorization, score vector [8].

Theorem 5.8.4 *Let $n \geq 3$ be an integer. Then*

$$|\mathcal{T}(\hat{R}_n)| = 2^{n-2}.$$

Proof. By Theorem 5.5.2, the cardinality of $\mathcal{T}(\hat{R}_n)$ equals the number of balanced digraphs contained in the tournament $T(\hat{A}_n)$ where $\hat{A}_n$ is the matrix in $\mathcal{T}(\hat{A}_n)$ whose only 1 above the main diagonal is a 1 in position $(1, n)$ (see Section 5.7).

If D is a digraph with a vertex set $X = \{x_1, x_2, \ldots, x_n\}$, we let $\mathcal{B}(D)$ denote the set of all balanced digraphs with vertex set X whose edges are a subset of the edges of D. Note that the digraph with vertex set X and no edges belongs to $\mathcal{B}(D)$. Every directed cycle of $T(\hat{A}_n)$ uses the edge corresponding to the 1 in position $(1, n)$. Hence every balanced digraph contained in $T(\mathcal{A}_n)$ is a directed cycle. The number of such directed cycles of length k clearly equals

$$\binom{n-2}{k-2} \quad (k = 3, 4, \ldots, n)$$

and hence

$$|\mathcal{T}(\hat{R}_n)| = 1 + \sum_{k=3}^{n} \binom{n-2}{k-2} = \sum_{k=0}^{n-2} \binom{n-2}{k} = 2^{n-2}.$$

$\square$

Another way to prove Theorem 5.8.4 is to use the unique representation (see Section 5.7) of an arbitrary tournament matrix in $\mathcal{T}(\hat{R}_n)$ as

$$\hat{A}_{k_1} * \hat{A}_{k_2} * \cdots * \hat{A}_{k_p}.$$

We now turn to a theorem of Wan and Li [25] which shows that the number of matrices in a tournament class $\mathcal{T}(R_n)$ is a strictly monotone function with respect to the partial order of majorization.

Theorem 5.8.5 *Let $R_n = (r_1, r_2, \ldots, r_n)$ and $R'_n = (r'_1, r'_2, \ldots, r'_n)$ be nondecreasing, nonnegative integral vectors such that $R'_n \preceq R_n \preceq \hat{R}_n$. Assume that $R'_n \neq R_n$. Then*

$$|\mathcal{T}(R'_n)| > |\mathcal{T}(R_n)| \geq 1.$$

Proof. The conditions in the theorem imply that both $\mathcal{T}(R'_n)$ and $\mathcal{T}(R_n)$ are nonempty. By Corollary 1.7.5 it suffices to prove the theorem when there are distinct integers p and q with $p < q$ and an integer $d \geq 1$, such that[6]

$$r'_i = \begin{cases} r_i, & \text{if } i \neq p \text{ or } q, \\ r_i + d, & \text{if } i = p, \\ r_i - d, & \text{if } i = q. \end{cases}$$

There exists an integer $k \geq 1$ such that every matrix $A = [a_{ij}]$ in $\mathcal{T}(R_n)$ is of the form

$$\begin{bmatrix} A_1 & O & O & \dots & O \\ J & A_2 & O & \dots & O \\ J & J & A_3 & \dots & O \\ \vdots & \vdots & \vdots & \ddots & \vdots \\ J & J & J & \dots & A_k \end{bmatrix}$$

where $A_1, A_2, A_3, \dots, A_k$ are the irreducible components of A. We distinguish two cases.

Case 1: row p meets A_u and row q meets A_v where $u < v$. We have

$$r_q - r_p = (r'_q + d) - (r'_p - d) = (r'_q - r'_p) + 2d \geq 2d.$$

Hence there exist at least $2d$ integers j such that $a_{pj} = 0$ and $a_{qj} = 1$. For any choice of d of these integers, replacing the principal submatrix[7]

$$A[\{p, q, j\}] = \begin{bmatrix} 0 & 0 & 0 \\ 1 & 0 & 1 \\ 1 & 0 & 0 \end{bmatrix}$$

of A with

$$\begin{bmatrix} 0 & 0 & 1 \\ 1 & 0 & 0 \\ 0 & 1 & 0 \end{bmatrix}$$

or, in case $j = p$, replacing

$$A[\{p, q\}] = \begin{bmatrix} 0 & 0 \\ 1 & 0 \end{bmatrix}$$

by

$$\begin{bmatrix} 0 & 1 \\ 0 & 0 \end{bmatrix}$$

results in a matrix A' in $\mathcal{T}(R'_n)$. It follows that in this case, $|\mathcal{T}(R'_n)| \geq \binom{2d}{d}|\mathcal{T}(R_n)|$. Since $\binom{2d}{d} \geq 2$, we have $|\mathcal{T}(R'_n)| > \mathcal{T}(R_n)|$.

Case 2: rows p and q meet the same irreducible component of A. In this case there is no loss in generality in assuming that A is irreducible. The 2

---

[6]We could assume that $d = 1$, but for later referral we only assume that $d \geq 1$.
[7]Note that j may be less than p, between p and q, or greater than q.

by n submatrix $U = [u_{ij}]$ of A formed by rows p and q is a $(0, 1)$-matrix with row sum vector (r_p, r_q), column sum vector $E = (e_1, e_2, \ldots, e_n)$ where (i) $e_i = 0, 1$, or 2, (ii) $e_p + e_q = 1$, (iii) $\sum_{i=1}^{n} e_i = r_p + r_q$, and where (iv) $u_{1p} = 0$, (v) $u_{2q} = 0$, and (vi) $u_{1q} + u_{2p} = 1$. Let $\mathcal{E}$ be the collection of all vectors of size n satisfying the properties (i), (ii), and (iii) of E above, and let $\mathcal{U}(E)$ be the collection of all 2 by n matrices satisfying the properties (iv), (v), and (vi) of U above. By deleting rows and columns p and q of A we obtain a tournament matrix B in $\mathcal{T}(R_U)$ of order $n - 2$ where

$$R_U = (r_1, \ldots, r_{p-1}, r_{p+1}, \ldots, r_{q-1}, r_{q+1}, \ldots, r_n)$$

$$- (2 - e_1, \ldots, 2 - e_{p-1}, 2 - e_{p+1}, \ldots, 2 - e_{q-1}, 2 - e_{q+1}, \ldots, 2 - e_n).$$

Conversely, given a matrix U in $\mathcal{U}(E)$ and a matrix B in $\mathcal{T}(R_U)$, there is a unique tournament matrix A in $\mathcal{T}(R)$ such that $A[\{p, q\}, \cdot] = U$ and $A(\{p, q\}, \{p, q\}) = B$. As a result we have

$$|\mathcal{T}(R)| = \sum_{E \in \mathcal{E}} |\mathcal{T}(R_U)| \, |\mathcal{U}(E)|. \tag{5.53}$$

Let $g = g(E)$ be the number of components of E equal to 2. Then $\mathcal{U}(E)$ is nonempty if and only if $g \leq r_p - 1$, and if this holds,

$$|\mathcal{U}(E)| = \begin{cases} \binom{r_p + r_q - 2g - 1}{r_p - g - 1,} & \text{if } e_p = 1, \\ \binom{r_p + r_q - 2g - 1}{r_p - g,} & \text{if } e_q = 1. \end{cases} \tag{5.54}$$

For a matrix E in $\mathcal{E}$ we also let $\mathcal{U}'(E)$ be the collection of all 2 by n matrices defined as for $\mathcal{U}(E)$ with the exception that the row sum vector is $(r_p + 1, r_q - 1)$. We also have

$$|\mathcal{T}(R')| = \sum_{E \in \mathcal{E}} |\mathcal{T}(R_U)| \, |\mathcal{U}'(E)|, \tag{5.55}$$

and for $g = g(E) \leq r_p$,

$$|\mathcal{U}'(E)| = \begin{cases} \binom{r_p + r_q - 2g - 1}{r_p - g,} & \text{if } e_p = 1, \\ \binom{r_p + r_q - 2g - 1}{r_p - g + 1,} & \text{if } e_q = 1. \end{cases} \tag{5.56}$$

We now observe that

$$(r_p + r_q - 2g - 1) - 2(r_q - g) = r_p - r_q - 1 \geq 1$$

and hence

$$\binom{r_p + r_q - 2g - 1}{r_p - g} > \binom{r_p + r_q - 2g - 1}{r_p - g - 1}. \tag{5.57}$$

We also observe that

$$(r_p + r_q - 2g - 1) - 2(r_p - g + 1) \geq -1$$

and hence

$$\binom{r_p + r_q - 2g - 1}{r_p - g} \geq \binom{r_p + r_q - 2g - 1}{r_p - g}. \tag{5.58}$$

Since R is a strong row sum vector in this case, there exist a matrix $A = [a_{ij}]$ in $\mathcal{T}(R)$ such that $a_{pq} = 1$ and hence a vector E in $\mathcal{E}$ with $e_q = 1$. Using (5.53) to (5.58), we conclude that

$$|\mathcal{T}(R'_n)| > |\mathcal{T}(R_n)|,$$

and the proof is complete. $\square$

Theorems 5.8.5 and 5.8.4 have as corollary the following result of Gibson [16] that was conjectured by Brualdi and Li [8].

Corollary 5.8.6 *Let* $n \geq 3$ *be an integer and let* R_n *be a nondecreasing, strong row sum vector with* $R_n \neq \hat{R}_n$. *Then*

$$|\mathcal{T}(R_n)| > 2^{n-2}.$$

Proof. The corollary follows from Theorems 5.8.5 and 5.8.4 because R_n is strong and so $R_n \preceq \hat{R}_n$. $\square$

Corollary 5.8.7 *Let* R_n *be a nondecreasing, nonnegative integral vector such that* $\overline{R}_n \preceq R_n$. *Then*

$$|\mathcal{T}(\overline{R}_n)| \geq |\mathcal{T}(R_n)|$$

with equality if and only if $R_n = \overline{R}_n$. $\square$

By Corollary 5.8.7, the nonempty tournament class with the largest cardinality is the regular tournament class $\mathcal{T}(\overline{R}_n)$ (n odd) and the near-regular tournament class $\mathcal{T}(\overline{R}_n)$ (n even). It is thus of interest to find bounds for $|\mathcal{T}(\overline{R}_n)|$. The following bound [25] improves a bound in [8] and is a consequence of the proof (specifically Case 1) of Theorem 5.8.5.

Corollary 5.8.8 *Let* n *be a positive integer. Then*

$$|\mathcal{T}(\overline{R}_n)| \geq \begin{cases} \prod_{j=1}^{(n-1)/2} \binom{2j}{j}, & \text{if } n \text{ is odd,} \\[2em] \prod_{j=1}^{(n-2)/2} \binom{2j+1}{j}, & \text{if } n \text{ is even.} \end{cases}$$

Proof. We give the proof for n odd; the proof for n even follows in a similar way.

There is a sequence of distinct vectors $R_n^{(j)}$ $(j = 0, 1, \ldots, (n-1)/2)$ such that

$$\overline{R}_n = R_n^{(0)} \preceq R_n^{(1)} \preceq \cdots \preceq R_n^{(n-1)/2} = \check{R}_n,$$

and $R_n^{(j)}$ differs from $R_n^{(j-1)}$ in exactly two coordinates, the difference being j $(j = 1, 2, \ldots, (n-1)/2)$. We take the vector $R_n^{(j-1)}$ to be

$$(0, 1, \ldots, n - (n+1)/2 - j, (n-1)/2, \ldots, (n-1)/2, n - (n+1)/2 + j, \ldots, n-2, n-1)$$

for $j = 1, 2, \ldots, (n+1)/2$. In the proof of Theorem 5.8.5, Case 1 always applies. Hence

$$|\mathcal{T}(R_n^{(j)})| \geq \binom{2j}{j} |\mathcal{T}(R_n^{(j-1)})| \quad (j = 1, 2, \ldots, (n-1)/2),$$

and thus, since $|\mathcal{T}(\check{R}_n)| = 1$,

$$|\mathcal{T}(\overline{R}_n)| \geq \prod_{j=1}^{(n-1)/2} \binom{2j}{j}.$$

$\square$

The estimate in Corollary 5.8.8 for $|\mathcal{T}(\overline{R}_{10})|$ is $7\,056\,000$.

5.9 The Class $\mathcal{T}(R;2)$ of 2-Tournament Matrices

A tournament matrix $A = [a_{ij}]$ of order n represents the outcome of a round-robin tournament amongst n teams $p_1, p_2, \ldots, p_n$. Team p_i defeats team p_j if and only if $a_{ij} = 1$ $(i, j = 1, 2, \ldots, n; i \neq j)$. In the tournament $T(A)$, there is an edge from p_i to p_j if and only if p_i defeats p_j. The tournament $T(A)$ is obtained from the complete graph K_n by assigning an orientation to each edge. If p is a positive integer, then a p-tournament matrix represents the outcome of a round-robin tournament in which each team plays every other team p times. If $p = 2$, there is another interpretation possible for a p-tournament matrix.

Consider a round-robin tournament in which each team plays every other team exactly once and where ties are allowed. If we score 2 points for a win, 1 for a tie, and 0 for a loss, then we are led once again to 2-tournament matrices, that is, matrices $A = [a_{ij}]$ of order n satisfying

$$a_{ii} = 0 \quad (i = 1, 2, \ldots, n), \tag{5.59}$$
$$a_{ij} = 0, 1, \text{ or } 2 \quad (i, j = 1, 2, \ldots, n; i \neq j), \tag{5.60}$$

and
$$a_{ij} + a_{ji} = 2 \quad (1 \leq i < j \leq n). \tag{5.61}$$

The 2-tournament matrix interpretation of A is that each of the teams $p_1, p_2, \ldots, p_n$ plays two games with every other team, and a_{ij} records the number of wins of p_i over p_j $(1 \leq i, j \leq n; i \neq j)$. In our discussion below, we consider that each pair of distinct teams play one game and use the win–tie–loss interpretation.[8]

The general digraph GD(A) associated with a matrix A satisfying (5.59), (5.60), and (5.61) does not have any directed loops and between each pair of distinct vertices there are two edges, either two edges in the same direction (signifying a win) or one edge in each of the two directions (signifying a tie). If we suppress those pairs of edges in opposite directions[9] and delete one of two edges in the same direction, we are left with a digraph $\Gamma(A)$ which is an orientation of a graph of n vertices. Equivalently, let

$$A = A^{(1)} + 2A^{(2)}$$

where $A^{(1)}$ is a symmetric $(0,1)$-matrix with zero trace and $A^{(2)}$ is an *asymmetric* $(0,1)$-matrix with the sum of symmetrically opposite elements equal to 0 or 1. Then $\Gamma(A)$ is the digraph GD($A^{(2)}$). We call $\Gamma(A)$ the *oriented graph of the 2-tournament matrix A*. The matrix A can be recovered from $A^{(2)}$ (or from GD($A^{(2)}$)) by multiplying $A^{(2)}$ by 2 and then replacing each pair of symmetrically opposite 0's with 1's. Thus $A^{(2)}$, indicating as it does the positions of the 2's in A, affords an efficient representation of the outcomes of a round-robin tournament in which ties are possible.

Example. Let

$$A = \begin{bmatrix} 0 & 1 & 2 & 0 & 1 \\ 1 & 0 & 1 & 2 & 0 \\ 0 & 1 & 0 & 1 & 0 \\ 2 & 0 & 1 & 0 & 0 \\ 1 & 2 & 2 & 2 & 0 \end{bmatrix}.$$

Then $A = A^{(1)} + A^{(2)}$ where

$$A^{(1)} = \begin{bmatrix} 0 & 1 & 0 & 0 & 1 \\ 1 & 0 & 1 & 0 & 0 \\ 0 & 1 & 0 & 1 & 0 \\ 0 & 0 & 1 & 0 & 0 \\ 1 & 0 & 0 & 0 & 0 \end{bmatrix}, \text{ and } A^{(2)} = \begin{bmatrix} 0 & 0 & 1 & 0 & 0 \\ 0 & 0 & 0 & 1 & 0 \\ 0 & 0 & 0 & 0 & 0 \\ 1 & 0 & 0 & 0 & 0 \\ 0 & 1 & 1 & 1 & 0 \end{bmatrix}.$$

We recover A from $A^{(2)}$ by multiplying by 2 and replacing each pair of symmetrically opposite 0's with 1's.   □

---

[8] In the late 1980's I had the pleasure of discussing many of the issues in this section with T.S. Michael.

[9] These edges give redundant information since, knowing which games end in a win, we know the remaining games end in a tie.

We call a matrix $A = [a_{ij}]$ of order n satisfying (5.59), (5.60), and (5.61) a *2-tournament matrix*. According to our notation for general p, the collection of all 2-tournament matrices with a given row sum vector $R_n = (r_1, r_2, \ldots, r_n)$ is denoted by

$$\mathcal{T}(R_n; 2).$$

Let A be a 2-tournament matrix in $\mathcal{T}(R_n; 2)$. Then the column sum vector of A is given by $S_n = (s_1, s_2, \ldots, s_n)$ where $s_i = 2(n-1) - r_i$ $(i = 1, 2, \ldots, n)$. In terms of the oriented graph $\Gamma(A)$ associated as above with a 2-tournament matrix, the ith row sum r_i of A (the score of team p_i) equals $n - 1 + d_i^+ - d_i^-$ where d_i^+ is the outdegree and d_i^- is the indegree of the vertex i corresponding to team p_i.

The following theorem, the special case $p = 2$ of Theorem 2.2.4, characterizes those vectors R_n which are row sum vectors of 2-tournaments. It is stated without proof in Avery [4].

Theorem 5.9.1 *Let $R_n = (r_1, r_2, \ldots, r_n)$ be a nondecreasing, nonnegative integral vector. Then $\mathcal{T}(R_n; 2)$ is nonempty if and only if*

$$\sum_{i=1}^{k} r_i \geq k(k-1) \quad (k = 1, 2, \ldots, n), \quad \text{with equality for } k = n. \quad (5.62)$$

Proof. Let $A = [a_{ij}]$ be a matrix in $\mathcal{T}(R_n; 2)$. Then

$$\sum_{i=1}^{n} \sum_{j=1}^{n} a_{ij} = \sum_{1 \leq i < j \leq n} (a_{ij} + a_{ji}) = \sum_{1 \leq i < j \leq n} 2 = n(n-1).$$

Since $A[\{1, 2, \ldots, k\}]$ is a 2-tournament matrix $(k = 1, 2, \ldots, n - 1)$, (5.62) follows.

That (5.62) implies the existence of a 2-tournament with row sum vector R_n can be proved in a very similar way to the proof of the sufficiency part of Theorem 2.2.2 (Landau's theorem).   □

Let A be a 2-tournament matrix in $\mathcal{T}(R_n; 2)$, and let P be a permutation matrix of order 3. The addition of

$$P \begin{bmatrix} 0 & -1 & 1 \\ 1 & 0 & -1 \\ -1 & 1 & 0 \end{bmatrix} P^{\mathrm{T}}$$

to a principal submatrix of A of order 3, provided the result is a nonnegative matrix, gives a 2-tournament matrix B in $\mathcal{T}(R_n; 2)$. As with tournament matrices, we say that B is obtained from A by a Δ-*interchange*. Up to simultaneous row and column permutations, a Δ-interchange replaces a principal submatrix of A of order 3 with another as follows:

$$\begin{bmatrix} 0 & 2 & 0 \\ 0 & 0 & 2 \\ 2 & 0 & 0 \end{bmatrix} \text{ by } \begin{bmatrix} 0 & 1 & 1 \\ 1 & 0 & 1 \\ 1 & 1 & 0 \end{bmatrix}, \text{ or vice versa;} \quad (5.63)$$

or

$$\begin{bmatrix} 0 & 2 & 1 \\ 0 & 0 & 2 \\ 1 & 0 & 0 \end{bmatrix} \text{ by } \begin{bmatrix} 0 & 1 & 2 \\ 1 & 0 & 1 \\ 0 & 1 & 0 \end{bmatrix}, \text{ or vice versa.} \qquad (5.64)$$

The following theorem is given in [4] without proof, but it can be proved in a way very much similar to Theorem 5.5.3, and so we omit the proof.

Theorem 5.9.2 *Let A and B be two 2-tournament matrices in a nonempty class $\mathcal{T}(R_n; 2)$. Then B can be obtained from A by a sequence of Δ-interchanges.* □

The following lemma is from [4]. An oriented graph is *transitive* provided $a \to b$ is an edge whenever there is a vertex $c \neq a, b$ such that $a \to c$ and $c \to b$ are edges.

Lemma 5.9.3 *Let A be a 2-tournament matrix in $\mathcal{T}(R_n; 2)$ with the fewest number of 2's. Then the oriented graph $\Gamma(A)$ is transitive. In particular, there is a permutation matrix P such that all the 2's of PAP^{T} are below the main diagonal.*

Proof. We remark that to say the matrix A has the fewest number of 2's is to say that the oriented graph $\Gamma(A)$ has the fewest number of edges. If $\Gamma(A)$ is not transitive, then a Δ-interchange reduces the number of 2's in A by 3 or 1 (see (5.63) and (5.64)). □

If the oriented graph $\Gamma(A)$ is transitive, then we call A a *transitive 2-tournament matrix*. We now describe an algorithm [4] for constructing a 2-tournament matrix with a prescribed nondecreasing row sum vector $R = (r_1, r_2, \ldots, r_n)$ which is similar to Ryser's algorithm for constructing a tournament matrix with a prescribed nondecreasing row sum vector. The matrix constructed by this algorithm is a matrix with the fewest number of 2's for its row sum vector. Note that since R is nondecreasing and $r_1 + r_2 + \cdots + r_n = 2(n-1)$, then $r_n \geq n-1$. Let the column sum vector of a matrix in $\mathcal{T}(R; 2)$ be $S = (s_1, s_2, \ldots, s_n)$ where $s_i = 2(n-1) - r_i$ $(i = 1, 2, \ldots, n)$.

Avery's Algorithm for Constructing a Matrix in $\mathcal{T}(R; 2)$

(1) Begin with the zero matrix $A^{(n)} = O$ of order n.

(2) In the $n-1$ nondiagonal positions of column n of $A^{(n)}$, insert $s_n = 2(n-1) - r_n$ 1's in those positions corresponding to the s_n largest row sums, giving preference to the topmost positions in case of equal row sums. Then insert 1's in row n in the symmetrically opposite positions of these 1's, and 2's in the remaining positions of row n not on the main diagonal, thereby achieving the row sum r_n. The result

is a matrix $A^{(n-1)}$ of order n whose leading principal submatrix of order $n-1$ is a zero matrix, and whose nth row and column sums are r_n and s_n, respectively. Let $R^{(n-1)} = (r_1^{(n-1)}, r_2^{(n-1)}, \ldots, r_{n-1}^{(n-1)})$ and $S^{(n-1)} = (s_1^{(n-1)}, s_2^{(n-1)}, \ldots, s_{n-1}^{(n-1)})$ be the vectors obtained from R and S by subtracting, respectively, the row sum vector and column sum vector of $A^{(n-1)}$ and deleting the last component. We continue recursively.

(3) For $k = n-1, n-2, \ldots, 2$, do:

In the non-diagonal positions of column k of the principal submatrix $A^{(k)}[\{1, 2, \ldots, k\}] = O$ of $A^{(k)}$, insert $s_k^{(k)}$ 1's in those positions corresponding to the $s_k^{(k)}$ largest row sums, giving preference to the topmost positions in case of equal row sums. Then insert 1's in the symmetrically opposite positions of these 1's in row k of $A^{(k)}$, and 2's in the remaining positions of row k of $A^{(k)}$ not on the main diagonal, thereby achieving the row sum r_k. This results in a matrix $A^{(k-1)}$, and we let $R^{(k)} = (r_1^{(k-1)}, r_2^{(k-1)}, \ldots, r_{k-1}^{(k-1)})$ and $S^{(k)} = (s_1^{(k-1)}, s_2^{(k-1)}, \ldots, s_{k-1}^{(k-1)})$ be the vectors obtained from R and S by subtracting, respectively, the row sum vector and column sum vector of $A^{(k-1)}$ and deleting the last component.

(4) Output $\widetilde{A} = A^{(1)}$.

Theorem 5.9.4 *Let $R = (r_1, r_2, \ldots, r_n)$ be a nondecreasing, nonnegative integral vector satisfying (5.62). Then Avery's algorithm can be completed and the constructed matrix $\widetilde{A}$ is a transitive 2-tournament matrix with row sum vector R with the fewest number of 2's of all matrices in $\mathcal{T}(R;2)$.*

Proof. We first show that the algorithm can be completed to construct a matrix in $\mathcal{T}(R;2)$. The cases $n = 1$ and 2 are trivial. We let $n \geq 3$ and proceed by induction. Let A be a matrix in $\mathcal{T}(R;2)$ with the fewest number of 2's. Then by Lemma 5.9.3 the oriented graph $\Gamma(A)$ is transitive. Suppose there were an edge (p, n) in $\Gamma(A)$ from some vertex p to vertex n. Then for each i such that (n, i) is an edge, (p, i) is also an edge by transitivity. This implies that row p of A has a strictly larger sum than row n, contradicting the nondecreasing property of A. Thus there are no edges of the type (p, n) and hence no 2's in column n of A. We conclude that a matrix A in $\mathcal{T}(R;2)$ with the fewest number of 2's has no 2's in column n. We have $r_n = 2(n-1) - s_n$ where s_n (the column sum of A) is the number of 1's in row n of A and $n-1-s_n$ is the number of 2's.

Let X be a set of s_n vertices of $\Gamma(A)$ corresponding to the rows with largest sum, other than r_n. (In case of equal sums, X may not be uniquely determined.) Let $Y = \{1, 2, \ldots, n-1\} \setminus X$. Suppose there is an edge in $\Gamma(A)$ from vertex n to k of the vertices in X, say to the vertices $u_1, u_2, \ldots, u_k$ in X. Then there are exactly k vertices in Y, say $w_1, w_2, \ldots, w_k$, to which there is no edge in $\Gamma(A)$ from vertex n. Since $\Gamma(A)$ is transitive, there cannot be an edge from u_i to w_i $(i = 1, 2, \ldots, k)$. Moreover, if there were an edge from w_i to u_i, then the transitivity of $\Gamma(A)$ would imply that the row w_i had a larger sum than the row u_i, contradicting our choice of X. Thus there also is no edge from w_i to u_i $(i = 1, 2, \ldots, k)$. It follows that the principal submatrix of A of order 3 determined by rows u_i, w_i, and n is of the form

$$A[\{u_i, w_i, n\}] = \begin{bmatrix} 0 & 1 & 0 \\ 1 & 0 & 1 \\ 2 & 1 & 0 \end{bmatrix} \quad (i = 1, 2, \ldots, k).$$

Now k Δ-interchanges produce a matrix A' in $\mathcal{T}(R; 2)$ having no 2's in column n, for which there are edges in $\Gamma(A')$ from vertex n to each of the vertices in Y and none in X. The oriented graph $\Gamma(A')$ need not be transitive. The existence of the matrix A' implies that the vector $R^{(n-1)}$ is the row sum vector of a 2-tournament matrix. Hence by induction the algorithm can be completed to construct a matrix $\tilde{A}$ in $\mathcal{T}(R; 2)$.

We now show that $\Gamma(\tilde{A})$ is transitive by induction on n. This is clearly true for $n \leq 3$. Assume that $n \geq 4$. The induction hypothesis implies that $\Gamma(\tilde{A}[\{1, 2, \ldots, n - 1\}])$ is transitive, and hence any intransitive triple of $\Gamma(\tilde{A})$ includes vertex n. Let u, w, and n be an intransitive triple where $1 \leq u < w < n$. The algorithm implies that there does not exist an edge in $\Gamma(\tilde{A})$ from u to w, from u to n, or from w to n. This means that in $\Gamma(\tilde{A})$, there must be an edge from n to w and one from w to u, but no edge from n to u. The construction of $\tilde{A}$ implies that under these circumstances, with $u < w$, we have $r_u = r_w$ and $r_u^{(n-1)} = r_w - 1 = r_w^{(n-1)} - 1$. Since $\Gamma(\tilde{A}[\{1, 2, \ldots, n - 1\}])$ is transitive, the edge from w to u implies that the outdegrees and indegrees of w satisfy $d_w^+ \geq d_u^+ + 1$ and $d_w^- \leq d_u^- - 1$. This implies that $r_w^{(n-1)} \geq r_u^{(n-1)} + 2$ contradicting $r_u^{(n-1)} = r_w^{(n-1)} - 1$. Hence it follows by induction that $\Gamma(\tilde{A})$ is transitive.

Finally we show by induction on n that $\tilde{A}$ has the fewest number of 2's among all matrices in $\mathcal{T}(R; 2)$. The conclusion is easily checked for $n \leq 3$, and we assume that $n \geq 4$. By Lemma 5.9.3 we need only show that there is no transitive 2-tournament matrix in $\mathcal{T}(R; 2)$ with fewer 2's then $\tilde{A}$.

Let A be a 2-tournament matrix in $\mathcal{T}(R; 2)$. Consider the principal submatrices

$$\tilde{A}^* = \tilde{A}[\{1, 2, \ldots, n - 1\} \text{ and } A^* = A[\{1, 2, \ldots, n - 1\}]$$

of order $n - 1$ with row sum vectors $R^{(n-1)} = (r_1^{(n-1)}, r_2^{(n-1)}, \ldots, r_{n-1}^{(n-1)})$ and $R^* = (r_1^*, r_2^*, \ldots, r_{n-1}^*)$, respectively. Because both $\Gamma(\tilde{A})$ and $\Gamma(A)$ are transitive, vertex n in each has indegree equal to 0. Hence $r_i^{(n-1)} = r_i - 1$ for $2(n - 1) - r_n$ of the vertices of $\{1, 2, \ldots, n - 1\}$ and $r_i^{(n-1)} = r_i$ for the other vertices of $\{1, 2, \ldots, n-1\}$. A similar conclusion holds for the sums r_i^* $(i = 1, 2, \ldots, n - 1)$. We also know by the algorithm that if $r_i^{(n-1)} = r_i - 1$

and $r_j^{(n-1)} = r_j$, then $r_i \geq r_j$. Finally, we observe that if $R^{(n-1)} = R^*$ then, by the induction hypothesis, $\tilde{A}^*$, being the matrix constructed by the algorithm using $R^{(n-1)}$, contains no more 2's than A^* does.

We now assume that $R^{(n-1)} \neq R^*$. Thus there are vertices u and w different from n with $r_u > r_w$, $r_u^* = r_u = r_u^{(n-1)} + 1$, and $r_w^{(n-1)} = r_w = r_w^* + 1$. The induction assumption again implies that the matrix $\widetilde{A'}$ obtained by applying the algorithm with the vector R^* contains no more 2's than $\tilde{A}^*$. We distinguish two cases.

Case 1: there is an edge from u to w in $\Gamma(\widetilde{A'})$, that is, the (u,w)-entry of $\widetilde{A'}$ equals 2. In this case, we replace the (u,w)-entry of $\widetilde{A'}$ with a 1 and the (w,u)-entry (which equals 0) also with a 1 to form a matrix A_1. The sum of the elements in row u of A_1 is $r_u^* - 1 = r_u^{(n-1)}$, and the sum of the elements in row w is $r_w^* + 1 = r_w^{(n-1)}$. Thus the row sum vector R_1 of A_1 differs from $R^{(n-1)}$ in two fewer entries than does R^*. Moreover, A_1 has one fewer 2 than A^*, and by the induction assumption, the matrix $\widetilde{A_1}$ constructed by the algorithm with the row sum vector R_1 has no more 2's than A_1. Hence in this case, $\widetilde{A_1}$ has fewer 2's than $\widetilde{A'}$ and A^*.

Case 2: there does not exist an edge from u to w in $\Gamma(\widetilde{A'})$, that is, the (u,w)-entry and the (v,u)-entry of $\widetilde{A'}$ equal 2. Since $r_u^* > r_w^*$ and $\widetilde{A'}$ is transitive, there is a vertex p in $\Gamma(\widetilde{A'})$ such that either (i) (p,w) is an edge but (p,u) is not, or (ii) (u,p) is an edge but (w,p) is not. If (i) holds, then we change the (p,w)-entry of $\Gamma(\widetilde{A'})$ from 2 to 1 and the (p,u)-entry from 1 to 2. If (ii) holds then we change the (u,p)-entry of $\Gamma(\widetilde{A'})$ from 2 to 1 and the (w,p)-entry from 1 to 2. In both cases, row sum u is decreased from r_u^* to $r_u^* - 1 = r^{(n-1)}$, row sum w is increased from r_w^* to $r_w^* + 1 = r^{(n-1)}$, and the row sum p is unchanged. The resulting matrices A_2 have the same row sum vector R_1 as the matrix A_1 in Case 1 and the same number of 2's as $\widetilde{A'}$. By the inductive assumption, $\widetilde{A_1}$ has no more 2's than A_2. Therefore in this case, $\widetilde{A_1}$ has no more 2's than $\widetilde{A'}$ and A^*.

Together Cases 1 and 2 imply that if $R^* \neq R^{(n-1)}$, then there is a transitive 2-tournament matrix A_1 with row sum vector R_1 with no more 2's than A^* whose row sum vector differs from $R^{(n-1)}$ in two fewer entries than does R^*. If $R_1 \neq R^{(n-1)}$, we may repeat with R_1 in place of R^*. In this way we obtain a sequence of 2-transitive tournament matrices beginning with A^* and ending with $\tilde{A}^*$ each of which has no more 2's than its predecessor. Hence $\tilde{A}^*$ contains no more 2's than A^*, and hence $\widetilde{A}$ contains no more 2's than A. $\square$

Example. Let $n = 6$ and $R = (3,4,4,6,6,7)$. Applying Avery's algorithm we obtain the following transitive 2-tournament matrix in $\mathcal{T}(R;2)$ with the fewest number of 2's:

$$\tilde{A} = \begin{bmatrix} 0 & 1 & 1 & 0 & 1 & 0 \\ 1 & 0 & 1 & 1 & 0 & 1 \\ 1 & 1 & 0 & 1 & 1 & 0 \\ 2 & 1 & 1 & 0 & 1 & 1 \\ 1 & 2 & 1 & 1 & 0 & 1 \\ 2 & 1 & 2 & 1 & 1 & 0 \end{bmatrix}.$$

Another transitive 2-tournament matrix in $\mathcal{T}(R:2)$ also having the fewest number of 2's is

$$A = \begin{bmatrix} 0 & 1 & 1 & 0 & 0 & 1 \\ 1 & 0 & 1 & 1 & 1 & 0 \\ 1 & 1 & 0 & 1 & 1 & 0 \\ 2 & 1 & 1 & 0 & 1 & 1 \\ 2 & 1 & 1 & 1 & 0 & 1 \\ 1 & 2 & 2 & 1 & 1 & 0 \end{bmatrix}.$$

We have $\tilde{A} \ne A$ and, in addition, the oriented graphs $\Gamma(\tilde{A})$ and $\Gamma(A)$ are not isomorphic.   □

5.10   The Class $\mathcal{A}(R, *)_0$ of $(0,1)$-Matrices

In this section we consider a class of $(0,1)$-matrices in which only the row sum vector is prescribed. Let n be a positive integer and let $R = (r_1, r_2, \ldots, r_n)$ be a nonincreasing, nonnegative integral vector with $r_1 \le n - 1$. Let $\mathcal{A}(R, *)_0$ denote the class of all $(0,1)$-matrices of order n with zero trace and row sum vector R . Since

$$r_n \le r_{n-1} \le \cdots \le r_1 \le n - 1,$$

the class $\mathcal{A}(R, *)_0$ is nonempty; in fact, the matrix A of order n which has r_i 1's in any r_i nondiagonal positions in row i $(i = 1, 2, \ldots, n)$ is in $\mathcal{A}(R, *)_0$. The matrix A is the adjacency matrix of a digraph without loops in which the elements of R give the outdegrees of its vertices; the indegrees, that is, the column sums of A, are not specified. Achuthan, Ramachandra Rao, and Rao [1] investigated how close a matrix in $\mathcal{A}(R, *)_0$ can be to a tournament matrix in $\mathcal{T}(R)$. This problem can be viewed as a generalization of the question of existence of a tournament matrix with a prescribed row sum vector R.

Let $A = [a_{ij}]$ be a matrix in $\mathcal{A}(R, *)_0$. We define the 1-*symmetry* of A to be the number

$$s_1(A) = |\{(i,j) : a_{ij} = a_{ji} = 1, 1 \le i < j \le n\}|$$

of pairs of symmetric 1's in A. The 0-*symmetry* of A is the number

$$s_0(A) = |\{(i,j) : a_{ij} = a_{ji} = 0, 1 \le i < j \le n\}|$$

of pairs of symmetric 0's in A. We have

$$A = A_{\mathrm{s}} + A_{\mathrm{as}}$$

where A_{as} is the *asymmetric part* of A obtained by replacing each pair of symmetric 1's with 0's, and $A_{\mathrm{s}} = A - A_{\mathrm{as}}$ is the *symmetric part* of A. We have $A_{\mathrm{as}} = O$ if and only if $A \in \mathcal{A}(R)_0$.

We also define the *asymmetry* of A to be the number

$$\mathrm{as}(R) = |\{(i,j) : a_{ij} + a_{ji} = 1, 1 \leq i < j \leq n\}|$$

of symmetrical pairs of elements in which one is a 1 and the other is a 0. We have

$$s_1(A) + s_0(A) + \mathrm{as}(A) = \binom{n}{2}. \tag{5.65}$$

The 0-symmetry is determined by the 1-symmetry and row sum vector R by

$$s_0(A) = \binom{n}{2} - \sum_{i=1}^{n} r_i + s_1(A). \tag{5.66}$$

Also, $\mathrm{as}(A) = \binom{n}{2}$ if and only if A is a tournament matrix with row sum vector R.

Since the 0-symmetry is determined as in (5.66) by the 1-symmetry, and since the asymmetry is determined by the 1-symmetry from (5.65) and (5.66), we concentrate on the 1-symmetry. We let

$$\bar{s}_1(R) = \max\{s_1(A) : A \in \mathcal{A}_0(R, *)\}$$

be the *maximal 1-symmetry* of a matrix in $\mathcal{A}_0(R, *)$, and we let

$$\tilde{s}_1(R) = \min\{s_1(A) : A \in \mathcal{A}_0(R, *)\}$$

be the *minimal 1-symmetry* of a matrix in $\mathcal{A}_0(R, *)$. The following theorem is from [1].

Theorem 5.10.1 *Let n be a positive integer and let $R = (r_1, r_2, \ldots, r_n)$ be a nonincreasing, nonnegative integral vector with $r_1 \leq n - 1$. Let $A = [a_{ij}]$ be a matrix in $\mathcal{A}_0(R, *)$ for which $s_1(A) = \tilde{s}_1(R)$. Then there exists an integer t with $0 \leq t \leq n$ such that the following hold.*

(i) *There is a tournament matrix $B = [b_{ij}]$ of order t such that $B \leq A[\{1, 2, \ldots, t\}]$ (entrywise).*

(ii) $s_1(A) = s_1(A[\{1, 2, \ldots, t\}]) = \sum_{i=1}^{T} \sum_{j=1}^{T} a_{ij} - \binom{t}{2}.$

(iii) $A[\{1, 2, \ldots, t\}, \{t+1, t+2, \ldots, n\}] = J_{t, n-t}$ *and* $A[\{t+1, t+2, \ldots, n\}, \{1, 2, \ldots, t\}] = O_{n-t, t}.$

(iv) *We have $r_t \geq n - t$ if $t \neq 0$, and $r_{t+1} \leq n - t - 1$ if $t \neq n$.*

(v) $\tilde{s}_1(R) = \sum_{i=1}^{T} r_i - t(n - t) - \binom{t}{2}.$

Proof. Let

$$I = \{i : a_{ik} = a_{ki} = 1 \text{ for some } k \neq i\}.$$

Symmetrical pairs of 1's must have both their row index and column index in I and thus $s_1(A) = s_1(A[I])$. The set I is empty if and only if A has no symmetric pairs of 1's, that is, if and only if $A_s = O$. We first show:

(a) If i is in I and $j \neq i$, then at least one of a_{ij} and a_{ji} equals 1.

Suppose to the contrary that $a_{ij} = a_{ji} = 0$ for some $j \neq i$. Since i is in I, there exists a $k \neq i$ such that $a_{ik} = a_{ki} = 1$. By replacing a_{ik} with 0 and a_{ij} with 1 we obtain a matrix A' in $\mathcal{A}(R, *)_0$ with $s_1(A') < s_1(A)$, contradicting our choice of A.

Let

$$L = \{i : a_{ik} = a_{ki} = 0 \text{ for some } k \neq i\},$$

and let $K = \{1, 2, \ldots, n\} \setminus (I \cup L)$. We conclude that for i in K and $j \neq i$, we have $a_{ij} + a_{ji} \geq 1$. ¿From (a) we get:

(b) $I \cap L = \varnothing$, and for i in $I \cup K$ and $j \neq i$ at least one of a_{ij} and a_{ji} equals 1.

Let K' consist of all those i in K for which there exists j in L such that there is a path from vertex j to vertex i in the digraph $D(A_{as})$ of the asymmetric part of A, and let $K'' = K \setminus K'$. Then $A[L \cup K', K''] = O$, and from (b) we conclude:

(c) $A[K'', L \cup K'] = J$.

We now show that

(d) $A[I, L \cup K'] = J$.

Suppose to the contrary that there are an i in I and a j in $L \cup K'$ such that $a_{ij} = 0$. Then by (a) $a_{ji} = 1$. By the definition of K' there exists a path in $D(A_{as})$ from some vertex p in L to vertex j in $D(A_{as})$, where if j is in L we take $p = j$ and the path has length 0. Since $a_{ji} = 1$ and $a_{ij} = 0$, there is a path $p = u_1, u_2, \ldots, u_l, u_{l+1} = i$ from vertex p to vertex i in $D(A_{as})$. We may assume that i is the only vertex in I on this path, for otherwise we may replace i with the first vertex in I on this path. Since by (b), $I \cap L = \varnothing$, this path has length at least 1. Since i is in I, there is a vertex $q \neq 0$ in I such that $a_{iq} = a_{qi} = 1$. Since p is in L, there exists a vertex $s \neq p$ in K such that $a_{ps} = a_{sp} = 0$. Let A'' be the matrix obtained from A by changing $a_{iq}, a_{u_1, u_2}, \ldots, a_{u_l, u_{l+1}}$ from 1 to 0 and $a_{u_1, s}, a_{u_2, u_1}, \ldots, a_{u_l, u_{l-1}}$ from 0 to 1. Then A'' is in $\mathcal{A}(R)$ and $s(A'') = s(A) - 1$, contradicting our choice of A.

Let $I^* = I \cup K''$ and $I^{**} = L \cup K' = \{1, 2, \ldots, n\} \setminus I^*$. Also let $t = |I^*|$ so that $|I^{**}| = n - t$. It follows from (a), (b), (c), and (d) and the other stated properties of the sets I, L, K', K'' that assertions (i), (ii), and (iii) of the theorem are satisfied with $\{1, 2, \ldots, t\}$ replaced with I^* and with $\{t+1, t+2, \ldots, n\}$ replaced with I^{**}. This implies that

$$r_i \geq n - t \ (i \in I^*) \text{ and } r_j \leq n - t - 1 \ (j \in I^{**}).$$

Since R is nonincreasing, it follows that $I^* = \{1, 2, \ldots, t\}$ and $I^{**} = \{t+1, t+2, \ldots, n\}$. Since assertions (iv) and (v) of the theorem are consequences of (i), (ii), and (iii), the proof is complete. $\square$

From Theorem 5.10.1 we are able to obtain a formula for $\tilde{s}_1(R)$.

Theorem 5.10.2 *Let n be a positive integer and let $R = (r_1, r_2, \ldots, r_n)$ be a nonincreasing, nonnegative integral vector with $r_1 \leq n - 1$. let*

$$\beta(k) = \sum_{i=1}^{k} r_i - k(n - k) - \binom{k}{2} \quad (0 \leq k \leq n).$$

Then

$$\tilde{s}_1(R) = \max\{\beta(k) : 0 \leq k \leq n\}. \tag{5.67}$$

In fact,

$$\tilde{s}_1(R) = \max\{\beta(k) : k \in I\}, \tag{5.68}$$

where

$$I = \{k : 1 \leq k \leq n - 1, r_1 \geq n - k, \text{ and } r_{k+1} \leq n - k - 1\}.$$

Proof. Let A be a matrix in $\mathcal{A}(R, *)_0$ with $s_1(A) = \tilde{s}_1(R)$. Let k be an integer with $0 \leq k \leq n$, and let $A_k = A[\{1, 2, \ldots, k\}]$. We have

$$\sum_{i=1}^{k} \sum_{j=1}^{k} a_{ij} \geq \sum_{i=1}^{k} r_i - k(n - k),$$

and so

$$\tilde{s}_1(R) = s_1(A) \geq s_1(A_k) \geq \sum_{i=1}^{k} \sum_{j=1}^{k} a_{ij} - \binom{k}{2}$$

$$\geq \sum_{i=1}^{k} r_i - k(n - k)$$

$$= \beta(k).$$

Hence

$$\tilde{s}_1(R) \geq \max\{\beta(k) : 0 \leq k \leq n\} \geq 0.$$

From assertion (v) of Theorem 5.10.1, we now obtain (5.67). Equation (5.68) now follows from assertion (iv) of Theorem 5.10.1. $\square$

As a corollary we are able to obtain Landau's Theorem for the existence of a tournament matrix with a prescribed row sum vector, in fact a slight refinement of that theorem [1].

Corollary 5.10.3 *Let n be a positive integer and let $R' = (r'_1, r'_2, \ldots, r'_n)$ be a nondecreasing, nonnegative integral vector with $r'_n \leq n-1$. Then there exists a tournament matrix of order n with row sum vector R' if and only if*

$$\sum_{i=1}^{k} r'_i \geq \binom{k}{2} \quad (1 \leq k \leq n-1, r'_k \leq k-1, r'_{k+1} \geq k), \qquad (5.69)$$

and equality holds for $k = n$.

Proof. The necessity part of the theorem is straightforward. Assume that (5.69) holds, with equality when $k = n$. Let $R = (r_1, r_2, \ldots, r_n) = (r'_n, r'_{n-1}, \ldots, r'_1)$. Then $r_k = r'_{n+1-k}$ $(k = 1, 2, \ldots, n)$, and R is a nonincreasing vector with $r_1 \leq n-1$. Since $\sum_{i=1}^{n} r_i = \sum_{i=1}^{n} r'_i = \binom{n}{2}$, it suffices to show that $\tilde{s}_1(R) = 0$. Suppose to the contrary that $\tilde{s}_1(R) > 0$. Let t be the integer in Theorem 5.10.1. Then by (v) of that theorem, it follows that $1 \leq t \leq n-1$ and

$$\sum_{i=1}^{T} r_i > t(n-t) + \binom{t}{2}. \qquad (5.70)$$

Let $k = n - t$. Then $1 \leq k \leq n-1$, and by (iv) of Theorem 5.10.1, we have $r'_k = r_{n+1-k} = r_{t+1} \leq n-t-1 = k-1$ and $r'_{k+1} = r_{n-k} = r_t \geq n-t = k$. By (5.70) we get

$$\sum_{i=1}^{k} r'_i = \sum_{i=1}^{k} r_{n+1-i}$$

$$= \binom{n}{2} - \sum_{i=1}^{T} r_i$$

$$< \binom{n}{2} - t(n-t) - \binom{t}{2}$$

$$= \binom{k}{2},$$

a contradiction. This proves the corollary. $\square$

A formula for the maximal 1-symmetry $\bar{s}_1(R)$ of $\mathcal{A}(R, *)_0$ is also given in [1].

Theorem 5.10.4 *Let n be a positive integer and let $R = (r_1, r_2, \ldots, r_n)$ be a nonincreasing, nonnegative integral vector with $r_1 \leq n-1$. Then*

$$\bar{s}_1(R) = \left\lfloor \frac{1}{2} \min \left\{ \sum_{i=k+1}^{n} (r_i + \min\{k, r_i\}) + k(k-1) : k = 0, 1, \ldots, n \right\} \right\rfloor.$$

$$(5.71)$$

Proof. Let

$$\theta(R;k) = \sum_{i=1}^{k} r_k - k(k-1) - \sum_{i=k+1}^{n} \min\{k, r_k\} \quad (0 \le k \le n),$$

and let

$$\bar{\theta}(R) = \max\{\theta(R;k) : 0 \le k \le n\}.$$

Then (5.71) can be written as

$$\bar{s}_1(R) = \left\lfloor \frac{1}{2}\left\{\sum_{i=1}^{n} r_i - \bar{\theta}(R)\right\}\right\rfloor. \tag{5.72}$$

We first show that the quantity on the right side of (5.72) is an upper bound for $\bar{s}_1(R)$. If $\bar{\theta}(R) = 0$, this is trivial, and we now assume that $\bar{\theta}(R) > 0$. Let l be an integer with $0 \le l \le n$ such that $\bar{\theta}(R) = \theta(R;l)$, and let A be a matrix in $\mathcal{A}(R, *)_0$ with $s_1(A) = \bar{s}_1(R)$. Then the number of 1's in $A[\{1, 2, \ldots, l\}, \{l+1, l+2, \ldots, n\}]$ is at least $\sum_{i=1}^{l} r_i - l(l-1)$, while the number of 1's in $A[\{l+1, l+2, \ldots, n\}, \{1, 2, \ldots, l\}]$ is at most $\sum_{i=l+1}^{n} \min\{l, r_i\}$. It follows that

$$\mathrm{as}(A) \ge \sum_{i=1}^{l} r_i - l(l-1) - \sum_{i=l+1}^{n} \min\{l, r_i\} = \theta(R;l) = \bar{\theta}(R).$$

Since the number of 1's in A equals $2s_1(A) + \mathrm{as}(A) = \sum_{i=1}^{n} r_i$, we have

$$2\bar{s}_1(R) = 2s_1(A) \le \sum_{i=1}^{n} r_i - \theta(R;l).$$

Since $\bar{s}_1(R)$ is an integer, this proves the upper bound.

We now prove the existence of a matrix in $\mathcal{A}(R, *)_0$ with $s_1(A)$ equal to or greater than the quantity on the right side of (5.72). This will complete the proof of the theorem.

First suppose that $\bar{\theta}(R) = 0$ and $\sum_{i=1}^{n} r_i$ is an even integer. Then

$$\theta(R, k) = \sum_{i=1}^{k} r_k - k(k-1) - \sum_{i=k+1}^{n} \min\{k, r_k\} \le 0 \quad (k = 0, 1, \ldots, n).$$

By Theorem 1.5.5 there exist a symmetric $(0,1)$-matrix with zero trace and row sum vector R, and hence a matrix A in $\mathcal{A}(R, *)_0$ with $s_1(A) = (1/2)\sum_{i=1}^{n} r_i$.

Now assume that either $\bar{\theta}(R) > 0$ or $\sum_{i=1}^{n} r_i$ is an odd integer. Let k be the positive integer determined by

$$r_1 = r_2 = \cdots = r_k > r_{k+1}$$

where $k = n$ if $r_1 = r_2 = \cdots = r_n$. Let R' be obtained from R by replacing r_k with $r_k - 1$. Then R' is a nonincreasing vector of size n with terms not exceeding $n - 1$. In addition we have

$$\theta(R', j) = \begin{cases} \theta(R; j) - 1, & \text{if } j \geq k, \\ \theta(R; j), & \text{if } j < k \text{ and } r_k > j, \\ \theta(R; j) + 1, & \text{if } j < k \text{ and } r_k \leq j. \end{cases} \qquad (5.73)$$

We show that

$$0 \leq \bar{\theta}(R') \leq \bar{\theta}(R), \qquad (5.74)$$

and

$$\bar{\theta}(R') = \bar{\theta}(R) - 1 \text{ if } \bar{\theta}(R) > 0. \qquad (5.75)$$

Suppose to the contrary that (5.74) does not hold. Then there exists an integer j with $0 \leq j \leq n$ such that $\theta(R'; j) > \theta(R; j)$ and $\theta(R'; j) > 0$. It follows from (5.73) that $j < k$ and $r_k \leq j$. Thus

$$r_1 = r_2 = \cdots = r_k \leq j,$$

and

$$\theta(R; j) = jr_1 - j(j-1) - \sum_{i=j+1}^{n} r_i = (j-1)(r_1 - j) - \sum_{i=j+2}^{n} r_i \leq 0.$$

Now (5.73) implies that $\theta(R; j) = \theta(R'; j) - 1 \geq 0$, and hence equality holds above. This means that $r_1 = r_2 = \cdots = r_{j+1} = j$ and $r_{j+2} = r_{j+3} = \cdots = r_n = 0$. But then the matrix $(J_{j+1} - I_{j+1}) \oplus O_{n-j-1}$ is in $\mathcal{A}(R, *)_0$ and this implies that both $\bar{\theta}(R) = 0$ and $\sum_{i=1}^{n} r_i$ is even, contrary to our assumption. Thus (5.74) holds.

We now prove (5.75) holds. Assume that $\bar{\theta}(R) > 0$, and let l be any integer with $\theta(R; l) = \bar{\theta}(R)$. We show that $l \geq k$. From (5.73), we have $\theta(R'; l) = \theta(R; l) - 1 = \bar{\theta}(R) - 1$. Then (5.74) and (5.73), with the knowledge that in order to have $\theta(R; l) = \bar{\theta}(R)$ we must have $l \geq k$, implies that $\bar{\theta}(R') = \bar{\theta}(R) - 1$. So suppose to the contrary that $l < k$. Since $\theta(R; l) = \bar{\theta}(R) > 0$, we have

$$lr_1 \geq \sum_{i=1}^{l} r_i > l(l-1) + \sum_{i=l+1}^{n} r_i \min\{l, r_i\}. \qquad (5.76)$$

Thus $r_1 \geq l$. Hence $r_i \geq l$ $(i = 1, 2, \ldots, k))$, and since $l < k$, $r_1 = r_2 = \cdots = r_{l+1} \geq l$ $(i = 1, 2, \ldots, l)$. We now calculate that

$$
\begin{aligned}
\theta(R; l+1) - \theta(R; l) &= (l+1)r_1 - (l+1)l - \sum_{i=l+2}^{n} \min\{l+1, r_i\} \\
&\quad - lr_1 - l(l-1) + \sum_{i=l+1}^{n} \min\{l, r_i\} \\
&= r_1 - 2l + l - \sum_{i=l+2}^{n} (\min\{l+1, r_i\} - \min\{l, r_i\}) \\
&= r_1 - l - t, \qquad\qquad (5.77)
\end{aligned}
$$

where t is the number of integers i such that $l + 2 \leq i \leq n$ and $r_i \geq l + 1$. From (5.76) we get

$$
lr_1 > l(l-1) + l + lt = l(l+t).
$$

and hence $r_1 > l + t$. Hence by (5.77) we have $\theta(R; l+1) > \theta(R; l)$, contradicting $\theta(R; l) = \bar{\theta}(R)$. This contradiction proves that (5.75) holds.

We now replace R by R' and proceed recursively. After a finite number q of steps we obtain a nonincreasing, nonnegative integral vector $R^* = (r_1^*, r_2^*, \ldots, r_n^*)$ such that $\bar{\theta}(R^*) = 0$ and the sum of the elements of R^* is even. It follows from (5.74) and (5.75) that $q = \bar{\theta}(R)$ or $\bar{\theta}(R) - 1$, according as $\sum_{i=1}^{n} r_i - \bar{\theta}(R)$ is even or odd.

By Theorem 1.5.5 there exists a symmetric $(0, 1)$-matrix A^* with zero trace and row sum vector R^*. Replacing $r_i - r_i^*$ 0's in row i of A^* with 1's, $(i = 1, 2, \ldots, n)$, we obtain a $(0, 1)$-matrix A in $\mathcal{A}(R)$ with

$$
\begin{aligned}
s_1(A) \geq \frac{1}{2} \sum_{i=1}^{n} r_i^* &= \frac{1}{2} \left(\sum_{i=1}^{n} r_i - q \right) \\
&= \left\lfloor \frac{1}{2} \left\{ \sum_{i=1}^{n} r_i - \bar{\theta}(R) \right\} \right\rfloor.
\end{aligned}
$$

$\qquad\qquad\qquad\qquad\qquad\qquad\qquad\qquad\qquad\qquad\qquad\qquad\qquad\qquad \square$

References

[1] N. Achuthan, S.B. Rao, and A. Ramachandra Rao, The number of symmetric edges in a digraph with prescribed out-degrees, *Combinatorics and Applications*, Proceedings of the Seminar in Honor of Prof. S.S. Shrikhande on His 65th Birthday, ed. K.S. Vijayan and N.M. Singhi, Indian Statistical Institute, Delhi, 1982, 8–20.

[2] P. Acosta, A. Bassa, A. Chaiken *et al.*, On a conjecture of Brualdi and Shen on block transitive tournaments, *J. Graph Theory*, **44** (2003), 213–230.

[3] S. Ao and D. Hanson, Score vectors and tournaments with cyclic chromatic number 1 or 2, *Ars Combin.*, **49** (1998), 185–191.

[4] P. Avery, Score sequences of oriented graphs, *J. Graph Theory*, **15** (1991), 251–257.

[5] L.W. Beineke and J.W. Moon, On bipartite tournaments and scores, *The Theory and Application of Graphs* (Kalamazoo, Mich. 1980), Wiley, New York, 1981, 55–71.

[6] D.M. Bressoud, Colored tournaments and Weyl's denominator formula, *Europ. J. Combin.*, **8** (1987), 245–255.

[7] R.A. Brualdi and Q. Li, Upsets in round robin tournaments, *J. Combin. Theory, Ser. B*, **35** (1983), 62–77.

[8] R.A. Brualdi and Q. Li, The interchange graph of tournaments with the same score vector, *Progress in Graph Theory*, Academic Press, Toronto, 1984, 128–151.

[9] R.A. Brualdi and J. Shen, Landau's inequalities for tournament scores and a short proof of a theorem on transitive sub-tournaments, *J. Graph Theory*, **82** (2000), 244–254.

[10] D. de Caen and D.G. Hoffman, Impossibility of decomposing the complete graph on n points into $n-1$ isomorphic complete bipartite graphs, *SIAM J. Discrete Math.*, **2** (1989), 48–50.

[11] R.M. Calderbank and P. Hanlon, The extension to root systems of a theorem on tournaments, *J. Combin. Theory, Ser. A*, **41** (1986), 228–245.

[12] P. Camion, Chemins et circuits hamiltoniens des graphes complets, *C.R. Acad. Sci. Pari, Sér. A*, **259** (1959), 2151–2152.

[13] I. Charon and O. Hurdy, Links between the Slater index and the Ryser index of tournaments, *Graphs Combin.*, **19** (2003), 323–334.

[14] D.R. Fulkerson, Upsets in round robin tournaments, *Canad. J. Math.*, **17** (1965), 957–969.

[15] I. Gessel, Tournaments and Vandermonde's determinant, *J. Graph Theory*, **3** (1979), 305–307.

[16] P.M. Gibson, A bound for the number of tournaments with specified scores, *J. Combin. Theory, Ser. B*, **36** (1984), 240–243.

[17] P.M. Gibson, Disjoint 3-cycles in strong tournaments, unpublished manuscript.

[18] B. Guiduli, A. Gyárfás, S. Thomassé, and P. Weidl, 2-partition-transitive tournaments, *J. Combin. Theory, Ser. B*, **72** (1998), 181–196.

[19] R. Kannan, P. Tetali, and S. Vempala, Simple Markov chain algorithms for generating bipartite graphs and tournaments, *Random Struc. Algorithms*, **14** (1999), 293–308.

[20] H. G. Landau, On dominance relations and the structure of animal societies. III. The condition for a score structure, *Bull. Math. Biophys.*, **15** (1953), 143–148.

[21] L. McShine, Random sampling of labeled tournaments, *Electron. J. Combin.*, **7** (2000), # R8.

[22] J.W. Moon, An extension of Landau's theorem on tournaments, *Pacific J. Math.*, **13** (1963), 1343–1345.

[23] J.W. Moon, *Topics on Tournaments*, Holt, Rinehart, and Winston, New York, 1968.

[24] H.J. Ryser, Matrices of zeros and ones in combinatorial mathematics, *Recent Advances in Matrix Theory*, ed. Hans Schneider, U. Wisconsin Press, Madison, 1964, 103–124.

[25] H.-H. Wan and Q. Li, On the number of tournaments with prescribed score vector, *Discrete Math.*, **61** (1986), 213–219.

[26] A.-A.A. Yutsis, Tournaments and generalized Young tableau, *Mat. Zametki*, **27** (1980), 353–359.

6

Interchange Graphs

In this chapter we study the interchange graph $G(R, S)$ of a nonempty class $\mathcal{A}(R, S)$ of (0,1)-matrices with row sum vector R and column sum vector S, and investigate such graphical parameters as the diameter and connectivity. We also study the Δ-interchange graph of a nonempty class $\mathcal{T}(R)$ of tournament matrices with row sum vector R and show that it has a very special structure; in particular that it is a bipartite graph. In the final section we discuss how to generate uniformly at random a tournament matrix in a nonempty class $\mathcal{T}(R)$ and a matrix in a nonempty class $\mathcal{A}(R, S)$.

6.1 Diameter of Interchange Graphs $G(R, S)$

We assume throughout this section that $R = (r_1, r_2, \ldots, r_m)$ and $S = (s_1, s_2, \ldots, s_n)$ are nonnegative integral vectors for which the class $\mathcal{A}(R, S)$ is nonempty.

The vertex set of the *interchange graph* $G(R, S)$, as defined in Section 3.2, is the set $\mathcal{A}(R, S)$. Two matrices in $\mathcal{A}(R, S)$ are joined by an edge in $G(R, S)$ provided A differs from B by an interchange, equivalently, $A - B$ is an interchange matrix. By Theorem 3.2.3, given matrices A and B in $\mathcal{A}(R, S)$, a sequence of interchanges exists that transforms A into B, that is, there is a sequence of edges in $G(R, S)$ that connects A and B. Thus the interchange graph $G(R, S)$ is a connected graph. If P and Q are permutation matrices, then the interchange graphs $G(R, S)$ and $G(RP, SQ)$ are isomorphic. As a result, one may assume without loss of generality that R and S are nonincreasing.

Let G_1 and G_2 be two graphs with vertex sets V_1 and V_2, respectively. The *Cartesian product* $G_1 \times G_2$ has vertex set $V_1 \times V_2$, and there is an edge joining vertices (u_1, u_2) and (v_1, v_2) if and only if $u_1 = v_1$ and there is an edge joining u_2 and v_2 in G_2, or there is an edge joining u_1 and v_1 in G_1 and $u_2 = v_2$. The Cartesian product $G_1 \times G_2$ is a *trivial Cartesian product* provided G_1 or G_2 has a single vertex. The Cartesian product can be extended to more than two factors by induction.

Suppose that R and S are nonincreasing and that the class $\mathcal{A}(R,S)$ has an invariant position. As shown in Section 3.4 there exist integers k and l with $(k,l) \neq (0,n), (m,0)$ such that each matrix A in $\mathcal{A}(R,S)$ has the form

$$\begin{bmatrix} J_{k,l} & A_1 \\ A_2 & O_{m-k,n-l} \end{bmatrix}.$$

Let the row sum and column sum vectors of A_1 be R_1 and S_1, respectively. Also let R_2 and S_2 be, respectively, the row sum and column sum vectors of A_2. The matrices A in $\mathcal{A}(R,S)$ can be identified with the pairs (A_1, A_2) where A_1 is in $\mathcal{A}(R_1, S_1)$ and A_2 is in $\mathcal{A}(R_2, S_2)$. Since an interchange in A takes place either in A_1 or in A_2, it follows that there is an edge connecting matrices A and B in $G(R,S)$ if and only if for the corresponding pairs (A_1, A_2) and (B_1, B_2) either $A_1 = B_1$ and A_2 and B_2 are connected by an edge in $G(R_2, S_2)$, or A_1 and B_1 are connected by an edge in $G(R_1, S_1)$ and $A_2 = B_2$. Thus $G(R,S)$ is the *Cartesian product* $G(R_1, S_1) \times G(R_2, S_2)$ of $G(R_1, S_1)$ and $G(R_2, S_2)$.

In Section 3.2 we defined the *interchange distance* $\iota(A, B)$ between two matrices A and B in a class $\mathcal{A}(R, S)$ to be the smallest number of interchanges that transform A to B. Since each interchange corresponds to an edge in $G(R,S)$, Theorem 3.2.6 gives the following lemma. Recall that the *combinatorial distance* $d(A, B)$ between A and B is the number of nonzero elements of $A - B$, and the *balanced separation* $q(A, B)$ is the largest number of minimal balanced matrices in a minimal balanced decomposition of $A - B$.

Lemma 6.1.1 *Let A and B be matrices in a nonempty class $\mathcal{A}(R, S)$. Then the distance between A and B in the graph $G(R, S)$ is*

$$\iota(A, B) = \frac{d(A, B)}{2} - q(A, B).$$

$\square$

The *diameter* of a graph is the maximal distance between vertices of the graph. Since the graph $G(R, S)$ is connected, its diameter,

$$\mathrm{diam}(R, S) = \max\{\iota(A, B) : A, B \in \mathcal{A}(R, S)\},$$

is a finite number. From Lemma 6.1.1 we get the trivial bound $\mathrm{diam}(R, S) \leq (mn)/2 - 1$. It has been conjectured [1] that

$$\mathrm{diam}(R, S) \leq \frac{mn}{4}.$$

Qian [16] proved that

$$\mathrm{diam}(R, S)) \leq \frac{mn}{2} - \frac{m}{2} \ln \frac{n+2}{4}.$$

Brualdi and Shen [5] obtained a better (asymptotic) result for the diameter, namely,

$$\text{diam}(R, S)) \leq \frac{3 + \sqrt{17}}{16} \, mn \, (\approx 0.445mn).$$

In order to prove this inequality, we first obtain a result concerning the length of a shortest cycle in a balanced bipartite digraph. Recall that balanced bipartite digraphs correspond to $(0, 1, -1)$-matrices each of whose row and column sums equals 0 (balanced matrices). In a balanced digraph the indegree $\deg^-(v)$ of each vertex v equals its outdegree $\deg^+(v)$. We use the notation $[X, Y]$ to denote the set of all edges from the subset X of vertices to the subset Y of vertices of a digraph. In a balanced digraph with vertex set V, we have

$$|[X, V \setminus X]| = |[V \setminus X, X]|.$$

Lemma 6.1.2 *Let Γ be a balanced bipartite digraph with vertex bipartition $\{U, W\}$ where $|U| = m$ and $|W| = n$. Assume that Γ has more than*

$$\left(\sqrt{17} - 1\right) mn/4 \, (\approx 0.78mn) \tag{6.1}$$

edges. Then Γ contains a cycle of length 2 or 4.

    Proof. We prove the lemma by contradiction. Assume that Γ does not contain any 2-cycles or 4-cycles. Without loss of generality we assume that Γ is strongly connected. Let the number of edges of Γ be α. We shall prove that

$$m^2 n - \frac{m\alpha}{2} \geq 4 \sum_{w \in W} \left(\deg^-(w)\right)^2 \geq \frac{\alpha^2}{n}. \tag{6.2}$$

Upon dividing (6.2) by $m^2 n$ and applying the quadratic formula, we contradict (6.1).

    The inequality on the right in (6.2) is a consequence of the Cauchy–Schwarz inequality. We now use a counting argument to prove the inequality on the left.

    For each u in U and each positive integer i, let $\Gamma_i^+(u)$ and $\Gamma_i^-(u)$ denote, respectively, the set of all vertices that are at distance i *from* u and the set of all vertices that are at distance i *to* u. Since Γ is bipartite and has no 2-cycles, $\Gamma_1^+(u)$ and $\Gamma_1^-(u)$ are disjoint subsets of W. Since Γ is bipartite without 4-cycles, $\Gamma_2^+(u)$ and $\Gamma_2^-(u)$ are disjoint subsets of V_1, and $[\Gamma_2^+(u), \Gamma_1^-(u)] = [\Gamma_1^+(u), \Gamma_2^-(u)] = \oslash$. Thus

$$|[\Gamma_2^+(u), W \setminus \Gamma_1^-(u)]| = |[\Gamma_2^+(u), W \setminus \Gamma_1^-(u)]| + |[\Gamma_2^+(u), \Gamma_1^-(u)]|$$
$$= |[\Gamma_2^+(u), W]|. \tag{6.3}$$

Since Γ does not contain 2-cycles,

$$|\Gamma_2^+(u)|(n - |\Gamma_1^-(u)|) \geq |[\Gamma_2^+(u), W \setminus \Gamma_1^-(u)]| + |[V_2 \setminus \Gamma_1^-(u), \Gamma_2^+(u)]|. \tag{6.4}$$

Since $\Gamma_1^+(u)$ and $\Gamma_1^-(u)$ are disjoint subsets of W, we have

$$W \setminus \Gamma_1^-(u) \supseteq \Gamma_1^+(u),$$

and so

$$|[W \setminus \Gamma_1^-(u), \Gamma_2^+(u))]| \geq |[(\Gamma_1^+(u), \Gamma_2^+(u)]|. \tag{6.5}$$

Since Γ is balanced and $W \supseteq \Gamma_1^+(u)$, we also have

$$|[\Gamma_2^+(u), W]| = |[W, \Gamma_2^+(u)]| \geq |[\Gamma_1^+(u), \Gamma_2^+(u)]|. \tag{6.6}$$

Formulas (6.3) through (6.6) imply

$$|\Gamma_2^+(u)|(n - |\Gamma_1^-(u)|) \geq 2|[(\Gamma_1^+(u), \Gamma_2^+(u)]|. \tag{6.7}$$

A similar argument as above yields

$$|\Gamma_2^-(u)|(n - |\Gamma_1^+(u)|) \geq 2|[\Gamma_2^-(u), \Gamma_1^-(u)]|. \tag{6.8}$$

Since Γ is balanced, we have

$$|\Gamma_1^-(u)| = |\Gamma_1^+(u)|. \tag{6.9}$$

Since $\Gamma_2^+(u)$ and $\Gamma_2^-(u)$ are disjoint subsets of U, we have

$$m \geq |\Gamma_2^+(u)| + |\Gamma_2^-(u)|. \tag{6.10}$$

Adding inequalities (6.7) and (6.8) and using (6.9) and (6.10), we get

$$m(n - |\Gamma_1^+(u)|) \geq 2|[\Gamma_1^+(u), \Gamma_2^+(u)]| + 2|[\Gamma_2^-(u), \Gamma_1^-(u)]|. \tag{6.11}$$

Next we sum both sides of the inequality (6.11) over all u in U. On the left side we get

$$\sum_{u \in U} m\left(n - |\Gamma_1^+(u)|\right) = m^2 n - m \sum_{u \in U} \deg^+ u = m^2 n - \frac{m\alpha}{2}. \tag{6.12}$$

Both

$$\sum_{u \in U} |[\Gamma_1^+(u), \Gamma_2^+(u)]| \quad \text{and} \quad \sum_{u \in U} |[\Gamma_2^-(u), \Gamma_1^-(u)]|$$

equal the number of 2-paths in Γ with both initial vertex and terminal vertex in U. These 2-paths can also be counted by considering their middle vertices in W. Thus

$$\sum_{u \in U} |[\Gamma_1^+(u), \Gamma_2^+(u)]| = \sum_{u \in U} |[\Gamma_2^-(u), \Gamma_1^-(u)]| = \sum_{w \in W} \deg^+(w)\deg^-(w). \tag{6.13}$$

Since $\deg^+(v) = \deg^-(v)$ for all vertices v of Γ, the left inequality of (6.2) follows from (6.11), (6.12) and (6.13), and the proof of the lemma is complete. $\qquad\square$

Corollary 6.1.3 *Let C be a balanced $(0, 1, -1)$-matrix with more than*

$$\left(\sqrt{17} - 1\right) mn/4$$

nonzero entries. Then there exist a balanced matrix C' and an interchange matrix C_1 such that $C = C_1 + C'$. □

We now obtain a bound for the diameter of an interchange graph.

Theorem 6.1.4 *Let R and S be nonnegative integral vectors such that the class $\mathcal{A}(R, S)$ is nonempty. Then the diameter of the interchange graphs $G(R, S)$ satisfies*

$$\mathrm{diam}(R, S) \leq \frac{3 + \sqrt{17}}{16} mn \ (\approx 0.445mn).$$

Proof. We show that for any two matrices A and B in $\mathcal{A}(R, S)$

$$\iota(A, B) \leq \frac{3 + \sqrt{17}}{16} mn.$$

If $d(A, B) \leq (\sqrt{17} - 1)mn/4$, this inequality follows from Lemma 6.1.1. Now assume that $d(A, B) > (\sqrt{17} - 1)mn/4$. By Corollary 6.1.3, $C = A - B = C_1 + C'$ where C' is a balanced $(0, 1, -1)$-matrix and C_1 is an interchange matrix. If C_1 contains more than $(\sqrt{17} - 1)mn/4$ nonzero elements we may apply Corollary 6.1.3 to C'. Continuing recursively, we see that

$$A - B = C_1 + C_2 + \cdots + C_k + C^*,$$

where $C_1, C_2, \ldots, C_k$ are interchange matrices and C^* is a balanced $(0, 1, -1)$-matrix, and where

$$k = \left\lceil \frac{d(A, B) - \left(\sqrt{17} - 1\right) mn/4}{4} \right\rceil = \left\lceil \frac{4d(A, B) - \left(\sqrt{17} - 1\right) mn}{16} \right\rceil.$$

Thus $q(A, B) \geq k$. By Lemma 6.1.1,

$$i(A, B) \leq \frac{d(A, B)}{2} - \frac{4d(A, B) - \left(\sqrt{17} - 1\right) mn}{16} \leq \frac{3 + \sqrt{17}}{16} mn,$$

where the final inequality holds since $d(A, B) \leq mn$. □

Recently, Shen and Yuster [21] improved Corollary 6.1.3 and thus Theorem 6.1.4. They showed that more than $2mn/3$ edges guarantees the same conclusion of Corollary 6.1.3 and thus that $\mathrm{diam}(R, S) \leq 5mn/12$ ($\approx 0.417mn$).

Brualdi and Li [2] investigated interchange graphs of small diameter. We have $\mathrm{diam}(R, S) = 0$ if and only if $\mathcal{A}(R, S)$ contains a unique matrix, and this happens if and only if S is the conjugate R^* of R. A connected graph of diameter 1 is necessarily a complete graph of order $k \geq 2$.

Theorem 6.1.5 *Let R and S be nonnegative integral vectors of sizes m and n, respectively. Assume that the class $\mathcal{A}(R, S)$ is nonempty and that it has no invariant positions. Then the diameter of the interchange graph $G(R, S)$ equals 1 if and only if either*

$$m = 2 \leq n, \ R = (n-1, 1), \ \text{and } S = (1, 1, \ldots, 1), \ \text{or} \qquad (6.14)$$

$$n = 2 \leq m, \ R = (1, 1, \ldots, 1), \ \text{and } S = (m-1, 1). \qquad (6.15)$$

Proof. First assume that (6.14) holds. Then $\mathcal{A}(R, S)$ contains exactly n matrices, and they have the form

$$\begin{bmatrix} 1 & \cdots & 1 & 0 & 1 & \cdots & 1 \\ 0 & \cdots & 0 & 1 & 0 & \cdots & 0 \end{bmatrix}$$

where the unique 1 in row 2 is in any of the columns $j = 1, 2, \ldots, n$. Any matrix of this form can be transformed to any other by a single interchange, and hence $G(R, S)$ is the complete graph K_n. Hence $\text{diam}(R, S) = 1$. Similarly, if (6.15) holds then $G(R, S)$ is the complete graph K_m.

Now assume that $\text{diam}(R, S) = 1$. We assume that $m \leq n$ and prove that (6.14) holds. Since $\mathcal{A}(R, S)$ does not have invariant positions, we have $m \geq 2$. Let $A = [a_{ij}]$ be a matrix in $\mathcal{A}(R, S)$ and let p and k be integers with $1 \leq p \leq m$ and $1 \leq k \leq n$. Since $\mathcal{A}(R, S)$ does not have invariant positions, there exists a matrix $B = [b_{ij}]$ in $\mathcal{A}(R, S)$ such that $a_{pk} \neq b_{pk}$. Since $\text{diam}(R, S) = 1$, there exist integers q and l with $1 \leq q \leq m$ and $1 \leq l \leq n$ such that a $(p, q; k, l)$-interchange applied to A yields B. Suppose to the contrary that $m \geq 3$. We take $p = k = 1$ above, and choose integers u and v with $1 \leq u \leq m$, $u \neq 1, q$, and with $1 \leq v \leq n$, $v \neq 1, l$. There exist integers r and s such that a $(u, r; v, s)$-interchange can be applied to A to yield a matrix C in $\mathcal{A}(R, S)$. The matrices B and C differ in at least six positions, and hence their distance in $G(R, S)$ is at least 2. This contradiction shows that $m = 2$. Now since $\mathcal{A}(R, S)$ does not have any invariant positions, we have

$$S = (1, 1, \ldots, 1) \text{ and } R = (a, n - a) \text{ for some } a \text{ with } 1 \leq a \leq n - 1.$$

Suppose neither a nor $n - a$ equals 1. Then there are matrices A and B in $\mathcal{A}(R, S)$ of the form

$$A = \begin{bmatrix} 1 & 1 & 0 & 0 & X \\ 0 & 0 & 1 & 1 & Y \end{bmatrix} \text{ and } B = \begin{bmatrix} 0 & 0 & 1 & 1 & X \\ 1 & 1 & 0 & 0 & Y \end{bmatrix}.$$

Since the distance between A and B in $G(R, S)$ equals 2, we contradict the assumption that $\text{diam}(R, S) = 1$. Hence a or $n - a$ equals 1, completing the proof. $\qquad\square$

6.2   Connectivity of Interchange Graphs

The *vertex-connectivity* of a graph equals the minimal number of vertices whose removal, along with all incident edges, leaves a graph which is not connected, or a single vertex. Thus a nonconnected graph has vertex connectivity equal to 0, and a complete graph K_n has vertex-connectivity equal to $n-1$. A basic property of vertex-connectivity is the following.

Lemma 6.2.1 *Let G be a connected graph such that for any two vertices u and v whose distance in G equals 2, there exist three paths joining u and v, each pair of which have only the vertices u and v in common. Then the vertex-connectivity of G is at least 3.*

Proof. Let S be a set of vertices whose removal from G leaves a graph G' which is not connected. Assume to the contrary that $|S| \leq 2$. Since G is connected, $S \neq \emptyset$. First suppose that there are vertices u and v in different connected components of G' such that each of u and v is joined to a vertex w in S by an edge in G. Then the distance in G between u and v equals 2, but there do not exist three paths joining u and v having only u and v in common, a contradiction. It now follows that $|S| = 2$ and that G' has two connected components. Let $S = \{x, y\}$, then x and y are joined by an edge in G, x is joined by an edge in G only to vertices in one of the components of G', and y is joined by an edge in G only to vertices in the other component of G'. Let x be joined by an edge in G to a vertex u of G'. Then the distance in G between u and y equals 2 but there do not exist three paths between u and y which have only u and y in common, another contradiction. Thus $|S| \geq 3$. $\square$

In [1] it is proved that if a class $\mathcal{A}(R, S)$ contains at least three matrices, then the vertex-connectivity of its interchange graph $G(R, S)$ is at least equal to 2. Shao [19] proved that the vertex-connectivity of $G(R, S)$ is 3 or more, with a few exceptions, and we now turn to a proof of this theorem.

Let $\delta(R, S)$ denote the minimal degree of a vertex of a nonempty interchange graph $G(R, S)$. If $\delta(R, S) = 0$, then since $G(R, S)$ is connected, $G(R, S)$ has exactly one vertex and is a (complete) graph K_1 with one vertex. Now suppose that $\delta(R, S) = 1$. Let A be a matrix in $\mathcal{A}(R, S)$ such that A has degree 1 in $G(R, S)$. Then A has exactly one submatrix equal to

$$\begin{bmatrix} 1 & 0 \\ 0 & 1 \end{bmatrix} \quad \text{or} \quad \begin{bmatrix} 0 & 1 \\ 1 & 0 \end{bmatrix}.$$

Let B be the unique matrix obtainable from A by a single interchange. It is easy to verify that B also has degree 1 in $G(R, S)$. Since $G(R, S)$ is connected, this implies that $\mathcal{A}(R, S) = \{A, B\}$, and $G(R, S)$ is a complete graph K_2. We now assume that $\delta(R, S) \geq 2$.

We first characterize those interchange graphs with $\delta(R, S) = 2$. We shall make use of the bipartite digraph $\Gamma(\hat{A})$ from Chapter 3, where for a

$(0, 1)$-matrix A of size m by n, $\hat{A} = 2J_{m,n} - A$ is the $(1, -1)$-matrix obtained from A by replacing each 0 with a -1. Thus $\Gamma(\hat{A})$ is obtained from the complete bipartite graph $K_{m,n}$ by assigning a direction to each of its edges as specified by the 1's and -1's of $\hat{A}$. The degree of a matrix A in $G(R, S)$ equals the number of directed 4-cycles in $\Gamma(\hat{A})$.

Lemma 6.2.2 *Let A be a matrix in a nonempty class $\mathcal{A}(R, S)$ such that the degree of A in $G(R, S)$ equals 2. Then the digraph $\Gamma(\hat{A})$ has exactly two directed cycles, and they have length 4.*

Proof. Assume to the contrary that $\Gamma(\hat{A})$ contains a directed cycle of length greater than 4. Choose such a directed cycle $\gamma = (v_1, v_2, \ldots, v_t, v_1)$ with minimal length t. Then t is even and $t \geq 6$. First suppose that $t = 6$. Then the directed cycle γ has three chords yielding three directed cycles of length 4, a contradiction. Now suppose that $t \geq 8$. The minimality of γ implies that $(v_4, v_1), (v_5, v_2)$, and (v_6, v_3) are edges of $\Gamma(\hat{A})$, and this implies that A has three directed cycles of length 4, a contradiction. Thus every directed cycle of $\Gamma(\hat{A})$ has length 4, and the lemma follows. □

We can now characterize interchange graphs with minimal degree 2.

Theorem 6.2.3 *Let R and S be nonnegative integral vectors such that the class $\mathcal{A}(R, S)$ is nonempty. If the minimal degree $\delta(R, S)$ of the interchange graph $G(R, S)$ equals 2, then $G(R, S)$ is either a cycle of length 3 or a cycle of length 4. In particular $|\mathcal{A}(R, S)| = 3$ or 4.*

Proof. Assume that $\delta(R, S) = 2$ and let A be a matrix in $\mathcal{A}(R, S)$ with degree 2 in $G(R, S)$. By Lemma 6.2.2, $\Gamma(\hat{A})$ contains exactly two directed cycles γ_1 and γ_2, both of length 4. In a strong component of $\Gamma(\hat{A})$, each vertex and each edge belong to some directed cycle. Thus $\Gamma(\hat{A})$ has either one or two nontrivial strong components.

First suppose that $\Gamma(\hat{A})$ has two nontrivial strong components. Then it follows that each strong component is a directed cycle of length 4, and that $G(R, S)$ is a directed cycle of length 4.

Now suppose that $\Gamma(\hat{A})$ has one nontrivial strong component Γ', which necessarily contains both γ_1 and γ_2. Because $\Gamma(\hat{A})$ is obtained from $K_{m,n}$ by assigning a direction to each of its edges, it follows that Γ' is obtained from a complete bipartite graph $K_{a,b}$ by assigning a direction to each of its edges. Since each edge of Γ' is in at least one γ_1 and γ_2, we have $a, b \geq 2$ with $ab = 4, 6$, or 8. But $ab = 4$ implies that γ_1 and γ_2 are identical, and $ab = 8$ implies that $\{a, b\} = \{2, 4\}$ and $\Gamma(\hat{A})$ has four directed cycles of length 4. Thus the only possibility is $ab = 6$ and $\{a, b\} = \{2, 3\}$. In this case γ_1 and γ_2 have exactly two edges in common. Now it follows that $G(R, S)$ is a cycle with length equal to 3. □

Example. Let $R = (1, 2)$ and $S = (1, 1, 1)$. The matrix

$$A = \begin{bmatrix} 1 & 0 & 0 \\ 0 & 1 & 1 \end{bmatrix}$$

is in $\mathcal{A}(R, S)$ and $G(R, S)$ is a cycle of length 3. Now let $R = S = (3, 3, 1, 1)$. Then

$$A = \begin{bmatrix} 1 & 1 & 1 & 0 \\ 1 & 1 & 0 & 1 \\ 1 & 0 & 0 & 0 \\ 0 & 1 & 0 & 0 \end{bmatrix}$$

is in $\mathcal{A}(R, S)$ and $G(R, S)$ is a cycle of length 4. $\square$.

The following lemma will be useful in the proof of the theorem about vertex-connectivity, 3.

Lemma 6.2.4 *Let $\mathcal{A}(R, S)$ be a nonempty class containing matrices A and B such that the digraph obtained from $\Gamma(\hat{A})$ by removing those edges which belong to $\Gamma(\hat{B})$ is a directed cycle of length 6 or more. Then there are three paths joining A and B in $G(R, S)$ having only A and B in common.*

Proof. Let the directed cycle in the statement of the lemma, necessarily of even length, be $\gamma = (v_1, v_2, \ldots, v_{2k}, v_1)$ where $k \geq 3$. For each $i = 1, 2, 3$, the edges in $\Gamma(\hat{A})$ joining nonconsecutive vertices of γ divide γ into cycles, without regard to direction, of length 4, at least one of which is a directed cycle γ' in $\Gamma(\hat{A})$. By successively reversing the directions of these cycles, starting with γ' (i.e. applying interchanges), we reverse the direction of γ, and thus construct paths π_i $(i = 1, 2, 3)$ between A and B in $G(R, S)$. Each pair of these three paths, using as they do edges meeting v_1, v_2, and v_3, respectively, has no vertices other than A and B in common. $\square$

Theorem 6.2.5 *Let R and S be nonnegative integral vectors such that the class $\mathcal{A}(R, S)$ is nonempty. Assume that the minimal degree of $G(R, S)$ satisfies $\delta(R, S) \geq 3$. Then the vertex-connectivity of $G(R, S)$ is at least three.*

Proof. Let A and B be matrices in $\mathcal{A}(R, S)$. By Lemma 6.1.1, the distance $\iota(A, B)$ between A and B in $G(R, S)$ is given by

$$\iota(A, B) = \frac{d(A, B)}{2} - q(A, B),$$

where $d(A, B)$ is the number of nonzero elements of $A - B$ and $q(A, B)$ is the largest number of directed cycles partitioning the edges of the bipartite digraph $\Gamma(A - B)$. Let A and B be matrices in $\mathcal{A}(R, S)$ with $\iota(A, B) = 2$. By Lemma 6.2.1 it suffices to show that A and B are joined in $G(R, S)$ by three paths, each pair of which has only A and B in common. Since $\iota(A, B) = 2$, we have $q(A, B) \leq 2$. If $q(A, B) = 1$, then $\Gamma(A - B)$ is a directed cycle of length 6, and the result follows from Lemma 6.2.4.

We now assume that $q(A, B) = 2$. Thus the edges of $\Gamma(A - B)$ can be partitioned into two directed cycles γ_1 and γ_2, and since $\iota(A, B) = 2$,

γ_1 and γ_2 each have length 4. By reversing the edges first of γ_1 and then of γ_2, and then reversing the edges first of γ_2 and then of γ_1, we obtain paths (A, C_1, B) and (A, C_2, B) in $G(R, S)$ of length 2, where $C_1 \neq C_2$. Since $\delta(A, B) \geq 3$, there is a matrix $C_3 \neq C_1, C_2$ in $\mathcal{A}(R, S)$ joined to A by an edge in $G(R, S)$. We have $\iota(A_3, B) \leq \iota(A, B) + 1 \leq 3$, and hence $q(A_3, B) \leq 3$. We consider four cases.

Case 1: $q(C_3, B) = 1$. Then either C_3 is joined to B by an edge in $G(R, S)$, or Lemma 6.2.4 applies. In either case we obtain a path joining A and B that does not contain C_1 or C_2.

Case 2: $q(C_3, B) = 3$. Then three directed cycles γ_1', γ_2', and γ_3' of length 4 partition the edges of $\Gamma(B - C_3)$. By successively reversing the directions of the edges of these directed cycles in the orders $\gamma_1', \gamma_2', \gamma_3', \gamma_2', \gamma_3', \gamma_1'$, and $\gamma_3', \gamma_1', \gamma_2'$, we obtain three paths joining C_3 and B with only C_3 and B in common and so at least one of them does not contain C_1 or C_2.

Case 3: $q(C_3, B) = 2$ and $\iota(C_3, B) = 2$. Thus there are two directed cycles γ_1' and γ_2' of length 4 partitioning the edges of $\Gamma(B - C_3)$. Since $A \neq C_3$, $\{\gamma_1', \gamma_2'\} \neq \{\gamma_1, \gamma_2\}$. Without loss of generality, assume $\gamma_1' \notin \{\gamma_1, \gamma_2\}$. Then by reversing the directions of the edges of γ_2' followed by those of γ_1', we obtain a path between A and B which does not contain C_1 or C_2.

Case 4: $q(C_3, B) = 2$ and $\iota(C_3, B) = 3$. In this case there are directed cycles γ_1' and γ_2' of lengths 4 and 6, respectively, which partition the edges of $\Gamma(B - C_3)$. There are three edges α_1, α_2, and α_3 of $\Gamma(\hat{A})$ which form chords of γ_2', and one of them, say α_1, is not an edge of γ_2'. The following three paths between A and B have only A and B in common, and so at least one of them does not contain C_1 or C_2.

(i) Reverse first the directions of the edges of γ_1', and then those of γ_2' by using α_2 to reverse the directions of two directed cycles of length 4.

(ii) Reverse first the directions of the edges of γ_2' as in (i) and then the directions of the edges of γ_1'.

(iii) Reverse first the directions of the directed cycle of length 4 formed by the chord α_1 of γ_2', then the edges of γ_1', then the other directed cycle of length 4 formed by the chord α_1 (now in the opposite direction) of γ_2'.

Since one of these three paths does not contain C_1 or C_2, we have three paths between A and B, each pair of which has only A and B in common. This completes the proof of the theorem. $\square$

Theorems 6.2.3 and 6.2.5 immediately imply the following corollary.

Corollary 6.2.6 *If $\mathcal{A}(R, S)$ contains at least five matrices, then the vertex-connectivity of the interchange graph $G(R, S)$ is at least 3.* $\square$

Let G be a connected graph with at least one edge. The *edge-connectivity* of G equals the minimal number of edges whose removal from G leaves a graph which is not connected. By removing all the edges meeting a vertex v, we see that the edge-connectivity cannot exceed the minimal degree of a vertex of G. If there is a positive integer k such that for each pair of vertices v and w that are joined by an edge in G, there are k pairwise edge-disjoint paths joining v and w, then the edge-connectivity of G is at least k. The edge-connectivity of a connected graph with one vertex and no edges is defined to be ∞.

Chen, Guo, and Zhang [6] investigated the edge-connectivity of $G(R,S)$ and obtained the following result. Since the proof consists of analyzing many cases, we omit it.

Theorem 6.2.7 *Let R and S be nonnegative integral vectors such that the class $\mathcal{A}(R,S)$ is nonempty. Let A and B be matrices in $\mathcal{A}(R,S)$ which are joined by an edge in the interchange graph $G(R,S)$, and let their degrees in $G(R,S)$ be $\deg(A)$ and $\deg(B)$, respectively. Then there are $\min\{\deg(A),\deg(B)\}$ pairwise edge-disjoint paths with lengths not greater than 5 that join A and B in $G(R,S)$. In particular, the edge-connectivity of $G(R,S)$ equals its minimal degree.*

The connectivity of special interchange graphs $G(R,S)$ was investigated by Meng and Huang [13].

6.3 Other Properties of Interchange Graphs

In this section we discuss other properties of interchange graphs $G(R,S)$. Because of long and technical proofs, some results are stated without proof and the reader is referred to the original papers.

We begin with a result of Brualdi and Li [2] that characterizes those row sum vectors R and S for which $G(R,S)$ is bipartite. Since the Cartesian product of two graphs G_1 and G_2 is bipartite if and only if both G_1 and G_2 are bipartite, we assume that $\mathcal{A}(R,S)$ has no invariant positions. First we prove two lemmas.

Lemma 6.3.1 *Let $R = (r_1, r_2, \ldots, r_m)$ and $S = (s_1, s_2, \ldots, s_n)$ be nonincreasing, nonnegative integral vectors such that the class $\mathcal{A}(R,S)$ is nonempty. Assume that $n > 2$ and that $\mathcal{A}(R,S)$ does not have invariant positions. Let $\tilde{A}_{n-1}$ be the matrix obtained from the special matrix $\tilde{A}$ of $\mathcal{A}(R,S)$ by deleting its last column. Let R_{n-1} and S_{n-1} be, respectively, the row and column sum vectors of $\tilde{A}_{n-1}$. Then one of the following holds.*

(i) $\mathcal{A}(R_{n-1}, S_{n-1})$ *does not have invariant positions.*

(ii) $\mathcal{A}(R_{n-1}, S_{n-1})$ *has invariant 0-positions but does not have invariant 1-positions. Thus there exists an integer e with $1 \le e \le m - 1$ such* *that every matrix in $\mathcal{A}(R_{n-1}, S_{n-1})$ has the form*

$$\begin{bmatrix} A' \\ O_{m-e,n-1} \end{bmatrix}$$

where A' is in $\mathcal{A}(R', S')$ and $\mathcal{A}(R', S')$ does not have invariant positions.

(iii) *$\mathcal{A}(R_{n-1}, S_{n-1})$ has invariant 1-positions but does not have invariant 0-positions. Thus there exists an integer e with $1 \leq e \leq m - 1$ such that every matrix in $\mathcal{A}(R_{n-1}, S_{n-1})$ has the form*

$$\begin{bmatrix} J_{e,n-1} \\ A' \end{bmatrix}$$

where A' is in $\mathcal{A}(R', S')$ and $\mathcal{A}(R', S')$ does not have invariant positions.

Proof. To prove the lemma, we assume that $\mathcal{A}(R_{n-1}, S_{n-1})$ has invariant 1-positions and show that it cannot have any invariant 0-positions. There exist integers e and f with $1 \leq e \leq m - 1$ and $1 \leq f \leq n - 1$ such that every matrix in $\mathcal{A}(R_{n-1}, S_{n-1})$ has the form

$$\begin{bmatrix} J_{e,f} & A'' \\ A' & O \end{bmatrix},$$

where A' is in $\mathcal{A}(R', S')$ and $\mathcal{A}(R', S')$ does not have invariant 1-positions. Let $R' = (r'_{e+1}, \ldots, r'_m)$ where R' is nonincreasing and $r'_i < f$ ($i = e + 1, \ldots, m$). We first show that $f = n - 1$. Assume to the contrary that $f < n - 1$. Then

$$\tilde{A} = \begin{bmatrix} J_{e,f} & \tilde{A}'' & u \\ \tilde{A}' & O & v \end{bmatrix},$$

where $s_n \leq s_{n-1} \leq e$. Since $\mathcal{A}(R, S)$ does not have invariant positions, v is not a column of 0's. Thus there is a positive integer i such that $r_{e+i} = r'_{e+i} + 1 \leq r'_{e+1} \leq f$. Since $s_n \leq e$, u is not a column of 1's. The construction of $\tilde{A}$ now implies that $r_{e+i} = f$, that there exists an integer $t < e$ such that $r_{t+1} = \cdots = r_{e+i} = f$, that the first t elements of u equal 1, that the last $e - t$ elements of u equal 0, and that the last $e - t$ rows of $\tilde{A}''$ contain only 0's. But then $s_n \geq t + 1 > s_{n-1}$, a contradiction. We conclude that $f = n - 1$ and that every matrix in $\mathcal{A}(R_{n-1}, S_{n-1})$ has the form

$$\begin{bmatrix} J_{e,n-1} \\ A' \end{bmatrix},$$

where A' is in $\mathcal{A}(R', S')$ and $\mathcal{A}(R', S')$ does not have any invariant 1-positions.

We now show that $\mathcal{A}(R', S')$ does not have any invariant 0-positions. Suppose to the contrary that $\mathcal{A}(R', S')$ has an invariant 0-position. Since

$\mathcal{A}(R', S')$ does not have invariant 1-positions, it follows that R' or S' has at least one coordinate equal to 0. Thus $\tilde{A}$ has one of the forms

$$
\left[
\begin{array}{c|c}
J_{e,n-1} & \begin{matrix} 0 \\ \vdots \\ 0 \end{matrix} \\
\hline
& * \\
X & \vdots \\
& * \\
\hline
O & \begin{matrix} 1 \\ \vdots \\ 1 \end{matrix}
\end{array}
\right]
\quad \text{and} \quad
\left[
\begin{array}{c|c|c}
J_{e,n-1} & \multicolumn{2}{c}{\begin{matrix} 0 \\ \vdots \\ 0 \end{matrix}} \\
\hline
B & O & w
\end{array}
\right].
$$

Here, since $\mathcal{A}(R, S)$ and $\mathcal{A}(R', S')$ have no invariant 1-positions, w is not a column of 0's and no row of X contains only 1's. In the case of the first form, $r_1 = n - 1 \geq 2$ and $r_n = 1$, and column n contradicts the construction of $\tilde{A}$. In the case of the second form, $r_1 = n - 1$ and $r_i \leq n - 2$ for $i = e + 1, \ldots, m$, and again column n contradicts the construction of $\tilde{A}$. Therefore $\mathcal{A}(R', S')$ does not have invariant 0-positions, and the proof is complete. □

Lemma 6.3.2 *Let R and S be nonincreasing, nonnegative integral vectors such that $\mathcal{A}(R, S)$ is nonempty and does not have any invariant positions. Let p and q be integers with $1 \leq p < q \leq m$. Then there exists a matrix A in $\mathcal{A}(R, S)$ which has a $(p, q; k, l)$-interchange for some integers k and l.*

Proof. Since $\mathcal{A}(R, S)$ does not have any invariant positions, we have $m, n \geq 2$ and R and S are positive vectors. If $n = 2$, then $R = (1, 1, \ldots, 1)$ and $S = (a, n - a)$ for some integer a with $1 \leq a \leq n - 1$ and the lemma clearly holds. We now assume that $n \geq 3$ and proceed by induction on n. We consider the special matrix $\tilde{A}$ constructed by Ryser's algorithm in Section 3.1. Let $\tilde{A}_{n-1}$ be the matrix obtained from $\tilde{A}$ by deleting its last column. Then $\tilde{A}_{n-1}$ is the special matrix for a class $\mathcal{A}(R_{n-1}, S_{n-1})$. The class $\mathcal{A}(R_{n-1}, S_{n-1})$ satisfies (i), (ii), or (iii) of Lemma 6.3.1. If (i) holds, then the conclusion follows immediately from the inductive assumption. Assume that (ii) holds. Then there exists an integer e with $1 \leq e \leq m - 1$ such that

$$
\tilde{A} =
\left[
\begin{array}{c|c}
A' & \begin{matrix} u \\ 1 \end{matrix} \\
\hline
O_{m-e,n-1} & \begin{matrix} \vdots \\ 1 \end{matrix}
\end{array}
\right],
$$

where A' is in $\mathcal{A}(R', S')$, $\mathcal{A}(R', S')$ does not have any invariant positions, and $u = (u_1, \ldots, u_e)^{\mathrm{T}}$. If $1 \leq p < q \leq e$, the conclusion follows from the inductive assumption. Now assume that $1 \leq p \leq e < q \leq m$. If $u_p = 0$, then since row p of A' contain a 1, the conclusion holds. Suppose that $u_p = 1$. Then there exists an integer t such that $1 \le t \le e$ and $u_t = 0$. By the inductive assumption, there are integers i and j such that an $(e, t; i, j)$-interchange can be applied to A'. A $(k, q; j, n)$- or $(k, q; i, n)$-interchange can now be applied to $\tilde{A}$ to produce a matrix in $\mathcal{A}(R, S)$ for which the conclusion of the lemma holds. Finally, suppose that $e < p < q \le n$. We choose an integer t with $1 \le t \le e$ such that $u_t = 1$. Since row t of A' contains a 0, an interchange can be applied to $\tilde{A}$ to give a matrix for which the conclusion of the lemma holds.

Finally, assume that (iii) holds. Then there exists an integer e with $1 \le e \le m - 1$ such that

$$\tilde{A} = \left[\begin{array}{c|c} J_{e,n-1} & \begin{matrix} 0 \\ \vdots \\ 0 \end{matrix} \\ \hline A' & v \end{array} \right]$$

where A' is in $\mathcal{A}(R', S')$, $\mathcal{A}(R', S')$ does not have invariant positions, and $v = (v_{e+1}, \ldots, v_m)^{\mathrm{T}}$. An argument very similar to the above shows that either $\tilde{A}$ or a matrix obtained from $\tilde{A}$ by an interchange satisfies the conclusion of the lemma. □

Corollary 6.3.3 *Let $R = (r_1, r_2, \ldots, r_m)$ and S be nonnegative integral vectors such that the class $\mathcal{A}(R, S)$ is nonempty and does not have any invariant positions. Let p and q be integers such that $r_p > r_q$. Then there is a matrix A in $\mathcal{A}(R, S)$ with a 2 by 3 submatrix of the form*

$$A[\{p, q\}, \{i, j, k\}] = P \begin{bmatrix} 1 & 1 & 0 \\ 0 & 0 & 1 \end{bmatrix} Q, \tag{6.16}$$

for some permutation matrices P and Q. In particular, $G(R, S)$ contains a cycle of length 3.

    Proof. The conclusion follows from Lemma 6.3.2 and the fact that $r_p > r_q$. □

Theorem 6.3.4 *Let $R = (r_1, r_2, \ldots, r_m)$ and $S = (s_1, s_2, \ldots, s_n)$ be nonnegative integral vectors such that $\mathcal{A}(R, S)$ is nonempty and does not have any invariant positions. The following are equivalent:*

(i) *$G(R, S)$ is bipartite;*

(ii) *$G(R, S)$ does not have any cycles of length 3;*

(iii) *$m = n$, and either $R = S = (1, 1, \ldots, 1)$ or $R = S = (n - 1, n - 1, \ldots, n - 1)$.*

    Proof. That (i) implies (ii) is trivial. Assume that (ii) holds. It follows from Corollary 6.3.3 that $r_1 = r_2 = \cdots = r_m = c$ and $s_1 = s_2 = \cdots =$

$s_n = d$ for some integers c and d with $1 \leq c \leq n - 1$, $1 \leq d \leq m - 1$, and $cm = dn$. Let q be the GCD of m and n. Then $m = m'q$, $n = n'q$, and $cm' = dn'$ where m' and n' are relatively prime integers. There is an integer k such that $c = kn'$ and $d = km'$. Let C_q be the permutation matrix of order q corresponding to the permutation $(2, 3, \ldots, q, 1)$ of $\{1, 2, \ldots, q\}$, and consider the circulant matrix $B = I_q + C_q + \cdots + C_q^{q-1}$. We replace each 1 of B with $J_{m',n'}$ and each 0 with $O_{m',n'}$ and obtain a matrix A in $\mathcal{A}(R, S)$. If $1 < c < n - 1$, then in rows m' and $m' + 1$, A has a submatrix of the type given in (6.16), and hence $G(R, S)$ has a 3-cycle, contradicting (ii). A similar conclusion holds if $1 < d < m - 1$. Thus $c = 1$ or $n - 1$ and $d = 1$ or $m - 1$. If $c = d = 1$, or $c = n - 1$ and $d = m - 1$, then $m = n$ and (iii) holds. If $c = 1$ and $d = m - 1$, or $c = n - 1$ and $d = 1$, then $m = n = 2$ and (iii) again holds.

Now assume that (iii) holds. The case $R = S = (n - 1, n - 1, \ldots, n - 1)$ is complementary to the case $R = S = (1, 1, \ldots, 1)$, and we assume that the latter holds. Then $\mathcal{A}(R, S)$ consist of all the permutation matrices of order n with two permutation matrices joined by an edge in $G(R, S)$ if and only if the corresponding permutation matrices differ by a transposition. Then $G(R, S)$ is a bipartite graph with bipartition $\{U, W\}$ where U consists of all the permutation matrices of even parity and W consist of those of odd parity. Hence (i) holds and the proof is complete. $\qquad\square$

A graph G is called a *prime graph* provided whenever G is isomorphic to a Cartesian product $G_1 \times G_2$, G_1 or G_2 contains only one vertex, that is, G does not admit a nontrivial factorization into a Cartesian product of two graphs. A graph with a single vertex is prime. Sabidussi [18] (see also [8], [15], [23]) proved that every connected graph can be uniquely factored (up to isomorphism and order of the factors) into prime graphs.

Let $R = (r_1, r_2, \ldots, r_m)$ and $S = (s_1, s_2, \ldots, s_n)$ be nonincreasing, nonnegative integral vectors such that $\mathcal{A}(R, S)$ is nonempty and $\mathcal{A}(R, S)$ has an invariant position. As remarked in Section 6.1, $G(R, S)$ is the Cartesian product of two other graphs $G(R_1, S_1)$ and $G(R_2, S_2)$. This factorization may be trivial but cannot be if $1 \leq r_i \leq m - 1$ $(i = 1, 2, \ldots, m)$ and $1 \leq s_j \leq n - 1$ $(j = 1, 2, \ldots, n)$. Brualdi and Manber [4] characterized those classes without invariant positions in terms of its interchange graph as follows. We refer to [4] for the proof.

Theorem 6.3.5 *Let $R = (r_1, r_2, \ldots, r_m)$ and $S = (s_1, s_2, \ldots, s_n)$ be nonincreasing, nonnegative integral vectors such that $\mathcal{A}(R, S)$ is nonempty. Assume that $1 \leq r_i \leq n - 1$ $(i = 1, 2, \ldots, m)$ and $1 \leq s_j \leq m - 1$ $(j = 1, 2, \ldots, n)$. Then $G(R, S)$ is a prime graph if and only if $\mathcal{A}(R, S)$ does not have any invariant positions.*

Finally we remark that Zhang and Zhang [24] and Zhang [25] showed that $G(R, S)$ has a Hamilton cycle for certain special row and column sum vectors R and S. Li and Zhang [11] proved that $G(R, S)$ is edge-Hamiltonian (that is, for each edge there is a Hamilton cycle containing it) whenever $S = (1, 1, \ldots, 1)$; moreover, in this case, the connectivity and diameter of $G(R, S)$ were determined by Qian [17].

6.4   The Δ-Interchange Graph $G_\Delta(R)$

Let $R = (r_1, r_2, \ldots, r_n)$ be a nonnegative integral vector such that the class $\mathcal{T}(R)$ of tournament matrices with row sum vector R is nonempty. The Δ-*interchange graph* $G_\Delta(R)$, as defined in Section 5.5, has as vertex set the set $\mathcal{T}(R)$. Two matrices in $\mathcal{T}(R)$ are joined by an edge in $G_\Delta(R)$ provided A and B differ by a Δ-interchange. Identifying tournament matrices with tournaments, two tournaments with score vector R are joined by an edge in $G_\Delta(R)$ if and only if one can be obtained from the other by reversing the directions of the edges of a 3-cycle. Theorem 5.5.3 implies that $G_\Delta(R)$ is a connected graph. Since a tournament with score vector $R = (r_1, r_2, \ldots, r_n)$ has exactly

$$\alpha_R = \binom{n}{3} - \sum_{i=1}^{n} \binom{r_i}{2}$$

3-cycles, the graph $\mathcal{T}(R)$ is regular of degree α_R.

If A is a tournament matrix of order n with row sum vector R and P is a permutation matrix of order n, then $P^{\mathrm{T}} A P$ is a tournament matrix with row sum vector RP and $G_\Delta(R)$ is isomorphic to $G_\Delta(RP)$. Thus in investigating the Δ-interchange graph, we may assume without loss of generality that R is nondecreasing.

In Section 5.5 we defined the Δ-*interchange distance* $\iota_\Delta(A, B)$ between two tournament matrices A and B in $\mathcal{T}(R)$ to be the smallest number of Δ-interchanges that transform A into B. Since each Δ-interchange corresponds to an edge in $G_\Delta(R)$, we obtain the next lemma from Theorem 5.5.4. Recall that the *combinatorial (tournament) distance* $d_{\mathrm{t}}(A, B)$ equals the number of positions in which A has a 1 and B has a 0, and that $q(A, B)$ equals the cycle packing number of the balanced digraph $D((A - B)^+)$.

Lemma 6.4.1 *Let A and B be tournament matrices in a nonempty class $\mathcal{T}(R)$. Then the distance between A and B in the graph $G_\Delta(R)$ is*

$$\iota_\Delta(A, B) = d_{\mathrm{t}}(A, B) - 2q(A, B).$$

$\square$

Let $\mathrm{diam}(R)$ denote the *diameter of the Δ-interchange graph* $G_\Delta(R)$. It follows trivially from Lemma 6.4.1 that

$$\mathrm{diam}(R) \leq \binom{n}{2} - 2.$$

An elementary lower bound for the diameter of a class of irreducible tournament matrices is given in the next lemma.

Lemma 6.4.2 *Let R be a strong row sum vector of size $n \geq 3$. Then the diameter of the class $\mathcal{T}(R)$ satisfies*

$$\mathrm{diam}(R) \geq n - 2.$$

Proof. Let A be a tournament matrix in $\mathcal{T}(R)$ with associated tournament $T(A)$. Then $T(A)$ is strongly connected, and by Theorem 5.2.4, $T(A)$ has a directed Hamilton cycle. By reversing the directions of the edges of this cycle, we obtain a tournament $T(B)$ where B is also in $\mathcal{T}(R)$. We have $d_t(A, B) = n$ and $q(A, B) = 1$. Hence by Lemma 6.4.1, $\iota_\Delta(A, B) = n - 2$. $\qquad\square$

Example. Consider the strong row sum vector

$$\hat{R}_n = (1, 1, 2, 3, \ldots, n - 3, n - 2, n - 2) \quad (n \geq 3).$$

If A and B are in $\mathcal{T}(\hat{R}_n)$, then $d_t(A, B) \leq n$. If $A \neq B$, then since $q(A, B) \geq 1$, it follows from Lemma 6.4.1 that $\iota_\Delta(A, B) \leq n - 2$ and hence by Lemma 6.4.2 that $\mathrm{diam}(\hat{R}_n) = n - 2$. Recall that $\hat{A}_k$ is the tournament matrix of order k in $\mathcal{T}(\hat{R}_k)$ with exactly one 1 above the main diagonal and in the position $(1, k)$ $(k \geq 2)$. Then $\hat{A}_n$ and $B = \hat{A}_2 * \hat{A}_2 * \cdots * \hat{A}_2$ $(n - 1$ $\hat{A}_2$'s) are matrices in $\mathcal{T}(R)$ satisfying $d_t(\hat{A}_n, B) = n$ and $q(\hat{A}_n, B) = 1$. Hence $\iota_\Delta(\hat{A}_n, B) = n - 2$. There are other strong row sum vectors R of size n for which $\mathrm{diam}(R) = n - 2$. For example, let $R = (1, 2, 2, 2, 3)$. It is easy to verify that for any two matrices A and B in $\mathcal{T}(R)$, $d_t(A, B) \leq 7$. If $d_t(A, B) = 7$, then $q(A, B) \geq 2$ and hence $\iota_\Delta(A, B) \leq 7 - 2(2) = 3$. If $d_t(A, B) \leq 5$, then $q(A, B) \geq 1$, and again $\iota_\Delta(A, B) \leq 5 - 2(1) = 3$. Hence $\mathrm{diam}(R) \leq 3$ and by Lemma 6.4.2, $\mathrm{diam}(R) = 3$. $\qquad\square$

Brualdi and Li [3] have shown that a Δ-interchange graph $G(R)$ has a special structure determined by the minimal and maximal number of upsets possible in $\mathcal{T}(R)$.

Theorem 6.4.3 *Let R be a nonnegative integral vector such that $\mathcal{T}(R)$ is nonempty. Let $\tilde{v} = \tilde{v}(R)$ and $\bar{v} = \bar{v}(R)$. Then the vertices of the Δ-interchange graph $G_\Delta(R)$ can be partitioned into $\bar{v} - \tilde{v} + 1$ sets*

$$\mathcal{T}_{\tilde{v}}(R), \mathcal{T}_{\tilde{v}+1}(R), \ldots, \mathcal{T}_{\bar{v}}(R)$$

such that each edge of $G_\Delta(R)$ joins a vertex in $\mathcal{T}_k(R)$ and a vertex in $\mathcal{T}_{k+1}(R)$ for some k with $\tilde{v} \leq k \leq \bar{v} - 1$.

Proof. By Lemma 5.6.1 a Δ-interchange applied to a tournament matrix A either increases or decreases the number $v(A)$ of upsets of A by 1. Thus the only edges in $G_\Delta(R)$ join matrices in $\mathcal{T}(R)$ whose numbers of upsets differ by 1. The theorem now follows. $\qquad\square$

A graph G is called *k-partite* provided its vertex set can be partitioned into k parts so that each edge joins two vertices in different parts. The interchange graph $G_\Delta(R)$ is a $(\bar{v}(R) - \tilde{v}(R) + 1)$-partite graph, but a very special kind in which the parts can be ordered so that edges only join vertices in consecutive parts. By grouping together the matrices in $\mathcal{T}(R)$ with an even number of upsets and those with an odd number of upsets, we see that $\mathcal{T}(R)$ is a bipartite graph.

Corollary 6.4.4 *The Δ-interchange graph of a nonempty class $\mathcal{T}(R)$ of tournament matrices is a bipartite graph.* $\square$

Theorem 6.4.3 enables us to give a lower bound for the diameter of *regular* and *near-regular* tournament matrices, that is, those with row sum vector

$$((n-1)/2, (n-1)/2, \ldots, (n-1)/2) \quad (n \text{ odd})$$

and

$$((n-2)/2, \ldots, (n-2)/2, n/2, \ldots, n/2) \quad (n \text{ even}),$$

respectively.

Corollary 6.4.5 *If n is odd and $R_n = ((n-1)/2, (n-1)/2, \ldots, (n-1)/2)$, then*

$$\mathrm{diam}(R_n) \geq \frac{(n-1)^2}{4}.$$

If n is even and $R_n = ((n-2)/2, \ldots, (n-2)/2, n/2, \ldots, n/2)$, then

$$\mathrm{diam}(R_n) \geq \frac{n(n-2)}{4}.$$

Proof. By Theorems 5.7.5 and 5.7.9

$$\tilde{v}(R_n) = \binom{\lfloor \frac{n+1}{2} \rfloor}{2} \quad \text{and} \quad \bar{v}(R_n) = \binom{n}{2} - \binom{\lfloor \frac{n+2}{2} \rfloor}{2}.$$

Hence it follows from Theorem 6.4.3 that

$$\mathrm{diam}(R_n) \geq \binom{n}{2} - \binom{\lfloor \frac{n+2}{2} \rfloor}{2} - \binom{\lfloor \frac{n+1}{2} \rfloor}{2}$$

and the corollary follows. $\square$

It has been conjectured [3] that the maximum diameter of an interchange graph $G_\Delta(R_n)$, where R_n is a row sum vector of size n, equals

$$\frac{(n-1)^2}{4} \quad (n \text{ odd}) \quad \text{and} \quad \frac{n(n-2)}{4} \quad (n \text{ even}).$$

A tournament matrix in $\mathcal{T}_{\tilde{v}}$ is joined by an edge in $G_\Delta(R)$ only to tournament matrices with $\tilde{v} + 1$ upsets. A tournament matrix in $\mathcal{T}_{\bar{v}}$ is joined by an edge only to tournaments with $\bar{v} - 1$ upsets. If $\tilde{v} < k < \bar{v}$, then a tournament matrix with k upsets may be joined by an edge both to tournaments with $k - 1$ upsets and to tournaments with $k + 1$ upsets. The fact that both possibilities need not occur shows that the structure of Δ-interchange graphs can be very complicated. Before giving an example we characterize tournament matrices for which no interchange increases, respectively, decreases the number of upsets [3].

Let A be a tournament matrix with nondecreasing row sum vector and with corresponding tournament $T(A)$. Let the vertex set of $T(A)$ be $\{x_1, x_2, \ldots, x_n\}$. Those edges of $T(A)$ from a vertex x_i to a vertex x_j with $1 \leq i < j \leq n$ are called *upset edges*; those from x_j to x_i are called *nonupset edges*. The *upset digraph* of A is the digraph $T_u(A)$ with the same vertices as $T(A)$ whose edges are the upset edges of $T(A)$. The *nonupset digraph* of A is the digraph $T_{nu}(A)$ whose edges are the nonupset edges of $T(A)$.

Lemma 6.4.6 *Let A be a tournament matrix. Then there does not exist a Δ-interchange which decreases the number of upsets of A if and only if the upset digraph $T_u(A)$ is transitive. Similarly, there does not exist a Δ-interchange which increases the number of upsets of A if and only if the nonupset digraph $T_{nu}(A)$ is transitive.*

Proof. First assume that $T_u(A)$ is not transitive. Then there are vertices x_i, x_j, and x_k of $T(A)$ where $1 \leq i < j < k \leq n$ such that there are upset edges from x_i to x_j and x_j to x_k but a nonupset edge from x_k to x_i. Then $x_i \to x_j \to x_k \to x_i$ is a 3-cycle, and a Δ-interchange decreases the number of upsets of A.

Now assume that $T_u(A)$ is transitive. Then any 3-cycle in $T(A)$ contains only one upset edge. Hence a Δ-interchange cannot decrease the number of upsets in A. The assertion concerning the nonupset digraph is proved in a similar way. $\square$

Example. Let $R = (2, 2, 2, 2, 2)$. The formula for the minimal number of upsets in Theorem 5.6.3 gives $\tilde{v}(R) = 3$. A tournament matrix in $\mathcal{T}(R)$ with three upsets is

$$A = \begin{bmatrix} 0 & 0 & 0 & 1 & 1 \\ 1 & 0 & 0 & 0 & 1 \\ 1 & 1 & 0 & 0 & 0 \\ 0 & 1 & 1 & 0 & 0 \\ 0 & 0 & 1 & 1 & 0 \end{bmatrix}.$$

Since rows $1, 2$, and 3 can each contain at most two upset edges, and row 4 can contain at most one, we get that $\bar{v}(R) \leq 7$. Since A^{T} is in $\mathcal{T}(R)$ and has seven upsets, we have $\bar{v}(R) = 7$. Thus $G_\Delta(R)$ is a 5-partite graph. Now consider the tournament matrix

$$B = \begin{bmatrix} 0 & 0 & 1 & 0 & 1 \\ 1 & 0 & 0 & 1 & 0 \\ 0 & 1 & 0 & 0 & 1 \\ 1 & 0 & 1 & 0 & 0 \\ 0 & 1 & 0 & 1 & 0 \end{bmatrix}$$

in $\mathcal{T}(R)$ with four upsets. The tournament matrix B^{T} is in $\mathcal{T}(R)$ and has six upsets. Since $T_{\mathrm{u}}(B)$ is transitive and $T_{\mathrm{nu}}(B^{\mathrm{T}})$ is not transitive, we conclude that B is a vertex in $\mathcal{T}_4(R)$ that is joined by an edge only to vertices in $\mathcal{T}_5(R)$ and B^{T} is a vertex in $\mathcal{T}_6(R)$ that is joined by an edge only to vertices in $\mathcal{T}_5(R)$.   □

We now consider the vertex-connectivity of a Δ-interchange graph. Shao [20] showed that with just a few exceptions, the vertex-connectivity is at least 3.[1] The proof is similar to the proof of Theorem 6.2.5 that the vertex-connectivity of an interchange graph of minimal degree at least 3 is also at least 3, and so we shall be more concise. An interchange graph $G_\Delta(R)$ is regular, and we let $\delta(R)$ denote its degree of regularity.

If $\delta(R) = 1$, then $G_\Delta(R)$ is a complete graph K_2.

Theorem 6.4.7 *Let R be a nonnegative integral vector such that $\mathcal{T}(R)$ is nonempty and $\delta(R) = 2$. Then $G_\Delta(R)$ is a cycle of length 4.*

Proof. Let T_1 and T_2 be tournaments in $\mathcal{T}(R)$ whose distance in $G_\Delta(R)$ equals 2. Then T_2 can be obtained from T_1 by reversing the directions of the edges of two 3-cycles with no edges in common or one 4-cycle. In either case we easily find two paths (T_1, T', T_2) and (T_1, T'', T_2) of length 2 where $T' \neq T'''$. Putting these two paths together we get a cycle of length 4. Since $G_\Delta(R)$ is regular of degree 2, $G_\Delta(R)$ is a cycle of length 4.   □

Example. Let $R = (1, 1, 1)$. Then

$$A = \begin{bmatrix} 0 & 1 & 0 \\ 0 & 0 & 1 \\ 1 & 0 & 0 \end{bmatrix}$$

is in $\mathcal{T}(R)$ and $G_\Delta(R)$ is a complete graph K_2. Now let $R = (2, 2, 1, 1)$. Then

$$A = \begin{bmatrix} 0 & 0 & 1 & 1 \\ 1 & 0 & 0 & 1 \\ 0 & 1 & 0 & 0 \\ 0 & 0 & 1 & 0 \end{bmatrix}$$

is in $\mathcal{T}(R)$ and $G_\Delta(R)$ is a cycle of length 4.   □

Theorem 6.4.8 *Let R be a nonnegative integral vector such that $\mathcal{T}(R)$ is nonempty and $\delta(R) \geq 3$. Then the vertex-connectivity of $G_\Delta(R)$ is at least 3.*

Proof. By Lemma 6.2.1 it suffices to show that for any two tournaments T and T' in $\mathcal{T}(R)$ whose distance $\iota_\Delta(T, T')$ in $G_\Delta(R)$ equals 2, there exist three paths between T and T', each pair of which has only T and T' in

---

[1] We are indebted to Jiayu Shao for providing us with a translated outline of the paper [20].

common. As in the proof of Theorem 6.4.7, we can find two paths $\gamma_1 = (T, T_1, T')$ and (T, T_2, T') of length 2 where $T_1 \neq T_2$. Since $d \geq 3$, there exists a tournament matrix $T_3 \neq T_1, T_2$ such that there is an edge between T and T_3 in $G_\Delta(R)$. We have $\iota_\Delta(T', T_3) \leq 3$. Since $G_\Delta(R)$ is bipartite, $\iota_\Delta(T', T_3) \neq 2$ and it is easy to see that $\iota_\Delta(T', T_3) \neq 1$. Hence $\iota_\Delta(T', T_3) = 3$. Thus

$$3 = \iota_\Delta(T', T'') \geq q(T', T_3),$$

where recall that $q = q(T', T_3)$ equals the cycle packing number of the balanced digraph Γ^* obtained from T' by deleting the edges that belong to T_3. There are three cases to consider.

Case 1: $q = 1$. In this case Γ^* is a directed cycle of length 5.

Case 2: $q = 2$. In this case the edges of Γ^* are partitioned into a directed cycle of length 3 and one of length 4.

Case 3: $q = 3$. In this case the edges of Γ^* are partitioned into three directed cycles of length 3.

In each case we can find a path $\gamma_3 = (T_3, T_4, T_5, T')$ of length 3 from T_3 to T' in $G_\Delta(R)$ with $\{T_4, T_5\} \cap \{T_1, T_2\} = \varnothing$. Thus γ_1, γ_2, and $\gamma_3 = (T, T_3, T_4, T_5, T')$ are three paths between T and T', each pair of which has only T and T' in common. The theorem now follows. $\qquad\square$

6.5   Random Generation of Matrices in $\mathcal{A}(R,S)$ and $\mathcal{T}(R)$

Given a nonnegative integral row sum vector $R = (r_1, r_2, \ldots, r_n)$ for which the class $\mathcal{T}(R)$ is nonempty, we first consider the problem of generating a tournament matrix in $\mathcal{T}(R)$ uniformly at random, equivalently, a labeled tournament with score vector R uniformly at random. This problem was considered by Kannan, Tetali, and Vempala [10] and solved in the special case of "near-regularity." The problem was completely solved by McShine [12]. In discussing the problem and solution, we shall need to assume that the reader is familiar with certain facts about Markov chains [9, 22]. Our discussion is very dependent on [12] to which we refer for the details of the proof.

We define a Markov chain $\mathcal{M}_n(R)$ which is a random walk on the Δ-interchange graph $G_\Delta(R)$. In doing so we shall blur the distinction between tournament matrices and tournaments. The *state space* of $\mathcal{M}_n(R)$ is the set $\mathcal{T}(R)$. Transition from one state to another is defined as follows.

If the current state is A_t (a matrix in $\mathcal{T}(R)$) then

(i) Choose a 3-cycle (Δ-interchange) uniformly at random in A_t.

(ii) Choose r equal to 0 or 1 uniformly at random.

(iii) If $r = 0$, set $A_{t+1} = A_t$.

(iv) If $r = 1$, then reverse the edges of the 3-cycle (perform the Δ-interchange) to obtain a new matrix A_{t+1} in $\mathcal{T}(R_n)$.

Since $G_\Delta(R)$ is connected, the Markov chain $\mathcal{M}_n(R)$ is irreducible. The graph $G_\Delta(R)$ is regular of degree

$$\alpha_R = \binom{n}{2} - \sum_{i=1}^{n} \binom{r_i}{2}.$$

Thus in step (i) of the algorithm we choose each 3-cycle with probability $1/\alpha_R$. Since the self-loop probabilities equal $1/2 \neq 0$ ($r = 0$ or 1 with equal probability), the Markov chain is also aperiodic and has a unique stationary distribution π.

Let $P = [p_{ij}]$ be the transition matrix of $\mathcal{M}(R_n)$. The matrix P is an irreducible, primitive, symmetric matrix of order $|\mathcal{T}(R_n)|$, and its rows and columns are indexed by the matrices in $\mathcal{T}(R_n)$. For A and B in $\mathcal{T}(R)$, we denote the (A, B)-entry of P by $P(A, B)$. The elements on the main diagonal of P are all equal to $1/2$. In addition, in each row and column, P has α_R nonzero elements off the main diagonal and each of these nonzero elements equals $1/(2\alpha_R)$. In particular, P is a doubly stochastic matrix.[2] The stationary distribution π is an eigenvector of P for its eigenvalue 1 and is the vector, indexed by the tournament matrices in $\mathcal{T}(R_n)$, with all elements equal to $1/|\mathcal{T}(R_n)|$. The entries of π are denoted by $\pi(A)$ for A in $\mathcal{T}(R_n)$. Since P is irreducible,

$$\lim_{k \to \infty} P^k = \Pi \tag{6.17}$$

where each row (and since P is symmetric, each column as well) of the matrix Π is the vector π.

To measure the rate of convergence of (6.17) of a Markov chain with transition matrix P and stationary vector π, we use the *total variation distance* between P^T ($t = 1, 2, 3, \ldots$) and π, given by

$$||P^\mathrm{T}, \pi||_\mathrm{tvd} = \max_{A \in \mathcal{T}(R_n)} \left\{ \frac{1}{2} \sum_{B \in \mathcal{T}(R_n)} |P^\mathrm{T}(A, B) - \pi(B)| \right\}.$$

For $\epsilon > 0$, the *mixing time* of $\mathcal{M}_n(R_n)$ is defined to be

$$\tau(\mathcal{M}_n(R_n), \epsilon) = \min\{t : ||P^{t'}, \pi||_\mathrm{tvd} \leq \epsilon \text{ for all } t' \geq t\}.$$

The following theorem is due to McShine [12].

---

[2] A nonnegative matrix with all row and column sums equal to 1. Doubly stochastic matrices are studied in Chapter 9.

Theorem 6.5.1 *The mixing time of the Markov chain $\mathcal{M}_n(R_n)$ satisfies*

$$\tau(\mathcal{M}_n(R_n), \epsilon) \leq C \, \alpha_{R_n} (\ln n + \ln \epsilon^{-1}) \quad (0 < \epsilon \leq 1),$$

where C is an absolute constant.

Because the bound in Theorem 6.5.1 is itself bounded by a polynomial in n and $\ln \epsilon^{-1}$, the Markov chain $\mathcal{M}_n(R_n)$ (more properly the family of Markov chains $(\mathcal{M}_n(R_n) : n = 1, 2, 3, \ldots)$ is called *rapidly mixing* [22].

We now briefly consider the random generation of a matrix in a nonempty class $\mathcal{A}(R,S)$ where R and S are positive integral vectors of sizes m and n, respectively [7, 22, 14]. We define a Markov chain $\mathcal{M}(R,S)$ which is a random walk on the interchange graph $G(R,S)$. The *state space* of $\mathcal{M}(R,S)$ is the set $\mathcal{A}(R,S)$. Transition from one state to another proceeds as follows.

If the current state is the matrix A_t in $\mathcal{A}(R,S)$, then

(i) Choose a pair $\{i,j\}$ of distinct rows uniformly at random and choose a pair $\{k,l\}$ of distinct columns uniformly at random.

(ii) If the matrix $A[\{i,j\},\{k,l\}]$ of order 2 is one of

$$\begin{bmatrix} 1 & 0 \\ 0 & 1 \end{bmatrix} \text{ and } \begin{bmatrix} 0 & 1 \\ 1 & 0 \end{bmatrix},$$

then perform the corresponding interchange and set A_{t+1} equal to the resulting matrix.

(iii) Else, set $A_{t+1} = A_t$.

Since the interchange graph $G(R,S)$ is connected, this Markov chain is also irreducible, but the degrees of the vertices of the chain are not all equal as matrices can contain different numbers of interchanges.[3] If m and n are at least 2 and not both equal to 2, then since R and S are positive, the chain is easily seen to be aperiodic with positive self-loop probabilities, and has a unique stationary distribution π. The transition matrix P of $\mathcal{M}(R,S)$ is an irreducible, primitive, symmetric doubly stochastic matrix and

$$\lim_{k \to \infty} P^k = \Pi$$

where Π is the constant vector each of whose entries equals $1/|\mathcal{A}(R,S)|$. Thus the limiting distribution is the equiprobable distribution. Discussion on the rate of convergence of this Markov chain and methods for speeding up convergence can be found in [7].

---

[3] If the algorithm were modified so that one chooses an interchange at random from the available interchanges for A_t (thus A_{t+1} would never equal A_t), then the limiting distribution would not necessarily be constant. In this case, the result would be biased in favor of matrices that contain more interchanges (have larger degree in the interchange graph $G(R,S)$) [7].

References

[1] R.A. Brualdi, Matrices of zeros and ones with fixed row and column sum vectors, *Linear Algebra Appl.*, **33** (1980), 159–231.

[2] R.A. Brualdi and Q. Li, Small diameter interchange graphs of classes of matrices of zeros and ones, *Linear Algebra Appl.*, **46** (1982), 177–194.

[3] R.A. Brualdi and Q. Li, The interchange graph of tournaments with the same score vector, *Progress in Graph Theory*, Academic Press, Toronto, 1984, 128–151.

[4] R.A. Brualdi and R. Manber, Prime interchange graphs of classes of matrices of zeros and ones, *J. Combin. Theory Ser. B*, **35** (1983), 156–170.

[5] R.A. Brualdi and J. Shen, Disjoint cycles in Eulerian digraphs and the diameter of interchange graphs, *J. Combin. Theory, Ser. B*, **85** (2002), 189–196.

[6] R.S. Chen, X.F. Guo, and F.J. Zhang, The edge connectivity of interchange graphs of classes of matrices of zeros and ones, *J. Xinjiang Univ. Natur. Sci.*, **5** (1) (1988), 17–25.

[7] G.W. Cobb and Y.-P. Chen, An application of Markov chain Monte Carlo to community ecology, *Amer. Math. Monthly*, **110** (2003), 265–288.

[8] W. Imrich, Über das schwache kartesische Produkt von Graphen, *J. Combin. Theory*, **11** (1971), 1–16.

[9] M. Jerrum, Mathematical foundations of the Markov chain Monte Carlo method, *Probabilistic Methods for Algorithmic Discrete Mathematics*, Algorithms and Combinatorics, 16, ed. M. Habib, C. McDiarmid, J. Ramírez-Alfonsín, and B. Reed, Springer, Berlin, 1998, 116–165.

[10] R. Kannan, P. Tetali, and S. Vempala, Simple Markov chain algorithms for generating bipartite graphs and tournaments, *Random Struc. Algorithms*, **14** (1999), 293–308.

[11] X.L. Li and F.J. Zhang, Hamiltonicity of a type of interchange graphs, Second Twente Workshop on Graphs and Combinatorial Optimization (Enschede, 1991), *Discrete Appl. Math.*, **51** (1994), 107–111.

[12] L. McShine, Random sampling of labeled tournaments, *Electron. J. Combin.*, **7** (2000), # R8.

[13] J. Meng and Q. Huang, Cayley graphs and interchange graphs, *J. Xinjiang Univ.*, **9** (1) (1992), 5–10.

[14] I. Miklós and J. Podani, Randomization of presence–absence matrices: comments and new algorithms, *Ecology*, **85** (2004), 86–92.

[15] D.J. Miller, Weak Cartesian product of graphs, *Colloq. Math.*, **21** (1970), 55–74.

[16] J.G. Qian, On the upper bound of the diameter of interchange graphs, *Discrete Math.*, **195** (1999), 277–285.

[17] J.G. Qian, Some properties of a class of interchange graphs, *Discrete Math.*, **195** (1999), 277–285.

[18] G. Sabidussi, Graph multiplication, *Math. Z.*, **72** (1950), 446–457.

[19] J.Y. Shao, The connectivity of interchange graph of class $\mathcal{A}(R,S)$ of (0,1)-matrices, *Acta Math. Appl. Sinica*, **2** (4) (1985), 304–308.

[20] J.Y. Shao, The connectivity of the interchange graph of the class of tournaments with a fixed score vector (in Chinese), *Tonji Daxue Xuebao*, 15 (1987), 239–242.

[21] J. Shen and R. Yuster, A note on the number of edges guaranteeing a C_4 in Eulerian bipartite graphs, *Electron. J. Combin.*, **9** (2001), #N6.

[22] A.J. Sinclair, *Algorithms for Random Generation and Counting: A Markov Chain Approach*, Birkhäuser, Boston, 1993.

[23] V.G. Vizing, The Cartesian product of graphs (in Russian), *Vyčisl. Sistemy*, **9** (1963) 30–43.

[24] F. Zhang and Y. Zhang, A type of (0, 1)-polyhedra, *J. Xinjiang Univ.*, **7** (4) (1990), 1–4.

[25] H. Zhang, Hamiltonicity of interchange graphs of a type of matrices of zeros and ones, *J. Xinjiang Univ.*, **9** (3) (1992), 5–10.

7

Classes of Symmetric Integral Matrices

We study in this chapter classes of symmetric matrices with a prescribed row sum vector R and with various restrictions on their elements. These include symmetric $(0, 1)$-matrices (possibly with trace equal to 0) and symmetric, nonnegative integral matrices (possibly with trace equal to 0 and with a uniform upper bound on their elements). A symmetric, nonnegative integral matrix of order n corresponds to a general graph with n vertices; if the matrix has zero trace, then it corresponds to a multigraph. We investigate certain graph-theoretical parameters of our classes.

7.1 Symmetric Interchanges

Let n be a positive integer and let $R = (r_1, r_2, \ldots, r_n)$ be a nonnegative integral vector. We define three types of interchanges which replace a symmetric, nonnegative integral matrix $A = [a_{ij}]$ of order n having row sum vector R with another matrix of the same type. Consider the pairs of symmetric $(0, \pm 1)$-matrices

$$\begin{bmatrix} 1 & -1 \\ -1 & 1 \end{bmatrix} \text{ and } \begin{bmatrix} -1 & 1 \\ 1 & -1 \end{bmatrix}, \tag{7.1}$$

$$\begin{bmatrix} 0 & 1 & -1 \\ 1 & -1 & 0 \\ -1 & 0 & 1 \end{bmatrix} \text{ and } \begin{bmatrix} 0 & -1 & 1 \\ -1 & 1 & 0 \\ 1 & 0 & -1 \end{bmatrix}, \tag{7.2}$$

$$\begin{bmatrix} 0 & -1 & 1 & 0 \\ -1 & 0 & 0 & 1 \\ 1 & 0 & 0 & -1 \\ 0 & 1 & -1 & 0 \end{bmatrix} \text{ and } \begin{bmatrix} 0 & 1 & -1 & 0 \\ 1 & 0 & 0 & -1 \\ -1 & 0 & 0 & 1 \\ 0 & -1 & 1 & 0 \end{bmatrix}. \tag{7.3}$$

Let P be a permutation matrix of order n. A *symmetric interchange matrix* of order n is a matrix of one of the three types:

(i) $P^{\mathrm{T}}(B \oplus O_{n-2})P$ $(n \geq 2)$, where B is one of the matrices in (7.1),

(ii) $P^{\mathrm{T}}(B \oplus O_{n-3})P$ $(n \geq 3)$, where B is one of the matrices in (7.2), and

(iii) $P^{\mathrm{T}}(B \oplus O_{n-4})P$ $(n \geq 4)$, where B is one of the matrices in (7.3).

Thus a symmetric interchange matrix has one of the matrices in (7.1), (7.2), and (7.3) as a principal submatrix and has zeros elsewhere. Note that only the symmetric interchange matrices of type (iii) have all 0's on the main diagonal.

Let A_1 and A_2 be nonnegative integral matrices of order n with A_2 symmetric, and suppose that $A_1 = A_2 + C$ where C is a symmetric interchange matrix. Then A_1 is symmetric, A_1 and A_2 have the same row sum vector, and A_1 is obtained from A_2 by adding (i) one of the matrices in (7.1) to a principal submatrix $A_2[\{p,q\}]$ of order 2 of A_2, (ii) one of the matrices in (7.2) to a principal submatrix[1] $A[\{p,q,k\}]$ of order 3 of A_2, or (iii) one of the matrices in (7.3) to a principal submatrix[2] $A_2[\{p,q,k,l\}]$ of order 4 of A_2. In the last case, if A_1 has zero trace, so does the matrix A_2. We say that A_1 is obtained from A_2 by a *symmetric interchange of type* (i) *with indices* $\{p,q\}$, *of type* (ii) *with indices* $\{p,q,k\}$, *or of type* (iii) *with indices* $\{p,q,k,l\}$, respectively.

In graphical terms, a symmetric interchange of type (i) replaces loops at two distinct vertices a and b with an edge $\{a,b\}$ joining them, or vice versa; a symmetric interchange of type (ii) replaces an edge $\{a,b\}$ joining two distinct vertices a and b and a loop at a third vertex c with an edge $\{a,c\}$ joining a and c and a loop at vertex b, or vice versa; a symmetric interchange of type (iii) replaces a pair of disjoint edges $\{a,b\}$ and $\{c,d\}$ on distinct vertices a,b,c,d with the pair $\{a,c\}$ and $\{b,d\}$ of disjoint edges.

Let $R = (r_1, r_2, \ldots, r_n)$ be a nonnegative integral vector of size n. We consider the previously defined classes:

(i) the set $\mathcal{A}(R)$ of all symmetric $(0,1)$-matrices with row sum vector R; and

(ii) the subset $\mathcal{A}(R)_0$ of $\mathcal{A}(R)$ consisting of the set of all symmetric $(0,1)$-matrices with row sum vector R and with zero trace.

Let A_1 and A_2 be matrices in $\mathcal{A}(R)$. Then the matrix $A_1 - A_2$ is a symmetric, balanced $(0, \pm 1)$-matrix. If A_1 and A_2 are in $\mathcal{A}(R)_0$, then the matrix $A_1 - A_2$ has zero trace. This motivates our study of symmetric, balanced $(0, \pm 1)$-matrices.

---

[1]Here p, q, k may not be in increasing order, and so this is not strictly speaking a principal submatrix, but a simultaneous row and column permutation of a principal submatrix of order 3.

[2]Similarly, p, q, k, l may not be in increasing order, and so this is, in general, a simultaneous row and column permutation of a principal submatrix of order 4.

Let $C = [c_{ij}]$ be a symmetric, balanced $(0, \pm1)$-matrix of order n. Assume that $C \neq O$. We consider the *signed graph* $G^{\pm} = G^{\pm}(C)$ with vertex set $V = \{v_1, v_2, \ldots, v_n\}$ in which there is an edge joining v_i and v_j of sign c_{ij} provided $c_{ij} \neq 0$ $(i, j = 1, 2, \ldots, n)$. A closed trail[3] of $G^{\pm}$ is called *balanced* provided the signs of its edges alternate between $+1$ and -1. The balanced closed trail γ is called *minimal* provided it does not contain a shorter, balanced closed trail, that is, a balanced closed trail cannot be obtained by deleting one or more of the signed edges of γ. Since C is a balanced matrix, $G^{\pm}$ has a balanced closed trail and hence a minimal balanced closed trail γ. Deleting the signed edges of γ from $G^{\pm}$ leaves a balanced signed graph. Hence it follows that there are balanced closed trails $\gamma_1, \gamma_2, \ldots, \gamma_q$ whose signed edges partition the signed edges of $G^{\pm}$. We may view each balanced closed trail γ_i as a signed graph with the same vertex set V as $G^{\pm}$. Thus γ_i is the signed graph corresponding to a symmetric, balanced $(0, \pm1)$-matrix C_i of order n $(i = 1, 2, \ldots, q)$. With this understanding,

$$C = C_1 + C_2 + \cdots + C_q, \tag{7.4}$$

where for $i \neq j$, the set of positions of the nonzero elements of C_i is disjoint from the set of positions of the nonzero elements of C_j. A decomposition (7.4) of C into minimal, symmetric balanced matrices is called a *minimal symmetric, balanced decomposition* of C.

Example. Consider the symmetric $(0, \pm1)$-matrix

$$C = \begin{bmatrix} 1 & 1 & -1 & -1 & 0 \\ 1 & 0 & -1 & 0 & 0 \\ -1 & -1 & 0 & 1 & 1 \\ -1 & 0 & 1 & 1 & -1 \\ 0 & 0 & 1 & -1 & 0 \end{bmatrix}.$$

Then C is balanced, and a minimal symmetric, balanced decomposition of C is $C = C_1 + C_2$ where

$$C_1 = \begin{bmatrix} 1 & 0 & 0 & -1 & 0 \\ 0 & 0 & 0 & 0 & 0 \\ 0 & 0 & 0 & 0 & 0 \\ -1 & 0 & 0 & 1 & 0 \\ 0 & 0 & 0 & 0 & 0 \end{bmatrix} \quad \text{and} \quad C_2 = \begin{bmatrix} 0 & 1 & -1 & 0 & 0 \\ 1 & 0 & -1 & 0 & 0 \\ -1 & -1 & 0 & 1 & 1 \\ 0 & 0 & 1 & 0 & -1 \\ 0 & 0 & 1 & -1 & 0 \end{bmatrix}.$$

The signed graph $G^{\pm}(C_1)$ consists of two loops of sign $+1$ joined by an edge of sign -1. The signed graph $G^{\pm}(C_2)$ consists of two cycles of length 3, one with two edges of sign $+1$ and one of sign -1 and the other with two edges of sign -1 and one of sign $+1$; the two cycles are joined at the vertex where two edges of the same sign meet. $\quad\square$

---

[3]A closed trail in a graph is a sequence of distinct edges of the form $\{x_1, x_2\}, \{x_2, x_3\}, \ldots, \{x_{m-1}, x_m\}, \{x_m, x_1\}$.

7.2   Algorithms for Symmetric Matrices

Let $R = (r_1, r_2, \ldots, r_n)$ be a nonnegative integral vector. In addition to the sets $\mathcal{A}(R)$ and $\mathcal{A}(R)_0$ we shall also consider the following classes each of which has been previously defined.

(iii) The set $\mathcal{N}(R)$ of nonnegative symmetric matrices with row sum vector R. (Here R need not be an integral vector.)

(iv) The subset $\mathcal{N}(R)_0$ of $\mathcal{N}(R)$ of those matrices with zero trace.

(v) The subset $\mathcal{Z}^+(R)$ of $\mathcal{N}(R)$ of those matrices with integral elements.

(vi) The subset $\mathcal{Z}^+(R)_0$ of $\mathcal{Z}^+(R)$ of those matrices with zero trace.

(vii) The subset $\mathcal{Z}^+(R; p)$ of $\mathcal{Z}^+(R)$ of those matrices each of whose elements does not exceed a given positive integer p.

(viii) The subset $\mathcal{Z}^+(R; p)_0$ of $\mathcal{Z}^+(R; p)$ of those matrices with zero trace.

When $p = 1$, we have $\mathcal{Z}^+(R; 1) = \mathcal{A}(R)$ and $\mathcal{Z}^+(R; 1)_0 = \mathcal{A}(R)_0$ as defined in Section 7.1.

The diagonal matrix with main diagonal equal to R belongs to $\mathcal{N}(R)$ and $\mathcal{Z}^+(R)$. Havel [9] and Hakimi [7] independently gave an algorithm for the construction of a graph (with no loops) with degree sequence R, that is, for a matrix in $\mathcal{A}(R)_0$. This algorithm was generalized by Chungphaisan [3] for the construction of a multigraph (so no loops) with degree sequence R in which the multiplicity of each edge is at most p, that is, for a nonnegative integral matrix in $\mathcal{Z}^+(R; p)_0$. In case $p = 2$, one particularly revealing way to carry out Chungphaisan's algorithm is described in [1]. We generalize these algorithms to include graphs or multigraphs with loops, that is, matrices in $\mathcal{A}(R)$ and $\mathcal{Z}^+(R; p)$.[4]

Assume without loss of generality that R is nonincreasing. We first give algorithms to construct matrices in $\mathcal{N}(R)_0$ and $\mathcal{Z}^+(R)_0$. By Theorem 2.3.1 the class $\mathcal{N}(R)_0$ is nonempty if and only if

$$r_1 \leq r_2 + r_3 + \cdots + r_n. \tag{7.5}$$

If (7.5) does not hold and we let $a = r_1 - (r_2 + r_3 + \cdots + r_n)$, then there exists a matrix in $\mathcal{N}(R)$ whose only nonzero off-diagonal element is a in position $(1, 1)$. If R is also an integral vector, then [17] $\mathcal{Z}^+(R)_0$ is nonempty if and only if (7.5) holds and

$$r_1 + r_2 + \cdots + r_n \text{ is an even integer.} \tag{7.6}$$

---

[4]In computing the degrees of vertices of a graph, it is usual to consider that a loop adds 2 to the degree of the vertex to which it is incident; see e.g. [6]. This is quite natural when viewing a diagram of a graph and, in addition, this retains the property that an edge adds 2 to the sum of the degrees of the vertices of a graph. From the point of view of a matrix, the elements on the main diagonal contribute to row and column sums in the same way as nondiagonal elements.

These conditions are clearly necessary for the class $\mathcal{Z}^+(R)_0$ to be nonempty; in Theorem 7.2.2, we prove the validity of an algorithm to construct a matrix in $\mathcal{Z}^+(R)_0$ when (7.5) and (7.6) hold. If (7.5) does not hold, then again defining $a = r_1 - (r_2 + r_3 + \cdots + r_n)$, we see that by replacing r_1 with $r_1 - a$, both (7.5) and (7.6) now hold, and hence [6] there exists a matrix in $\mathcal{Z}^+(R)$ whose only nonzero off-diagonal element is a in position $(1,1)$.

Owens and Trent [17], Owens [16], and Kleitman [10] have given algorithms to determine a matrix $A = [a_{ij}]$ in $\mathcal{Z}^+(R)$ which is closest to a $(0,1)$-matrix, that is, for which

$$\sum_{i,j} \max\{0, a_{ij} - 1\}$$

is minimal.

In Section 4.4 we extended the notion of an interchange to the class $\mathcal{Z}^+(R, S)$ of nonnegative integral matrices with row sum vector R and column sum vector S. We may further extend this to the class $\mathcal{N}(R, S)$ of nonnegative matrices with row sum vector R and column sum vector S as follows. Let $A = [a_{ij}]$ be a matrix in $\mathcal{N}(R, S)$, and let

$$A[\{p, q\}, \{k, l\}] = \begin{bmatrix} a_{pk} & a_{pl} \\ a_{qk} & a_{ql} \end{bmatrix}$$

be a submatrix of A of order 2. Let $b = \min\{a_{pk}, a_{ql}\} > 0$. Then we say that the matrix B obtained from A by replacing the submatrix $A[\{p, q\}, \{k, l\}]$ with

$$\begin{bmatrix} a_{pk} - b & a_{pl} + b \\ a_{qk} + b & a_{ql} - b \end{bmatrix}$$

is obtained from A by a $(p, q; k, l)$-*interchange*. Similar language is used for the matrix C obtained from A by replacing $A[\{p, q\}; \{k, l\}]$ with the matrix

$$A[\{p, q\}, \{q, l\}] = \begin{bmatrix} a_{pk} + c & a_{pl} - c \\ a_{qk} - c & a_{ql} + c \end{bmatrix},$$

where $c = \min\{a_{pl}, a_{qk}\} > 0$. Matrices B and C also belong to $\mathcal{N}(R, S)$. By adapting earlier arguments, we see that *any two matrices in $\mathcal{N}(R, S)$ can be gotten from one another by a sequence of interchanges.*

Algorithm I: Constructing a Matrix in $\mathcal{N}(R)_0$

(1) Construct a matrix A in $\mathcal{N}(R, R)$ as described in Section 2.1.

(2) Use (at most $n - 1$) $(p, q; p, q)$-interchanges with $p \neq q$ to obtain a matrix $B = [b_{ij}]$ in $\mathcal{N}(R, R)$ with at most one nonzero element on the main diagonal.

(3) If all elements on the main diagonal of B equal 0, then output the matrix $A^* = (B + B^{\mathrm{T}})/2$ in $\mathcal{N}(R)_0$.

(4) Otherwise, let p be the unique integer such that $b_{pp} > 0$. Use (at most $(n-1)(n-2)$) $(p,q;p,l)$-interchanges with $q \neq l$ to obtain a matrix C with all elements on the main diagonal equal to 0, and then output the matrix $A^* = (C + C^T)/2$.

Theorem 7.2.1 *Let $R = (r_1, r_2, \ldots, r_n)$ be a nonincreasing, nonnegative vector satisfying* (7.5). *Then Algorithm* I *above can be carried out to give a matrix in* $\mathcal{N}(R)_0$.

Proof. That we can use at most $n-1$ $(p,q;p,q)$-interchanges to obtain a matrix B with at most one nonzero on the main diagonal is clear. It is also clear that if there are only 0's on the main diagonal of B, then $A^* = (B + B^T)/2$ is in $\mathcal{N}(R)_0$. Suppose that in step (4), we are unable to obtain a 0 in position (p,p). Then simultaneously permuting rows and columns to bring the element in position (p,p) to position $(1,1)$, we obtain a matrix of the form

$$\begin{bmatrix} x & \alpha \\ \beta & O_{n-1} \end{bmatrix}$$

where $x > 0$ and $\beta = \alpha^T$, since R is both the row and column sum vector. Hence we obtain

$$r_p = r_1 + \cdots + r_{p-1} + r_{p+1} + \cdots + r_n + x$$

contradicting (7.5). Thus step (4) can be carried out to obtain the matrix C and the theorem follows.   □

Even if the vector R is an integral vector, Algorithm I above may not produce an integral matrix (some elements may be fractions with denominator equal to 2). The following algorithm will always produce a matrix in a nonempty class $\mathcal{Z}^+(R)_0$ if R is integral. If R is not integral, it offers an alternative construction for a matrix in $\mathcal{N}(R)_0$ by ignoring the integrality conditions.

Algorithm II: Constructing a Matrix in $\mathcal{Z}^+(R)_0$

(1) Given a nonnegative integral vector $R_n = R = (r_1, r_2, \ldots, r_n)$. While $n \geq 2$, choose nonnegative integers $x_1, x_2, \ldots, x_{n-1}$ such that

$$x_1 + x_2 + \cdots + x_{n-1} = r_n,$$

and

$$x_i \geq \frac{2(r_i + r_n) - \sum_{j=1}^n r_j}{2} \quad (i = 1, 2, \ldots, n-1).$$

(i) Put $a_{ii} = 0$ $(i = 1, 2, \ldots, n)$.

(ii) Put $a_{in} = a_{ni} = x_i$ $(i = 1, 2, \ldots, n-1)$.

(iii) Replace n with $n - 1$ and R_n with

$$R_{n-1} = (r_1 - x_1, r_2 - x_2, \ldots, r_{n-1} - x_{n-1}).$$

(2) Output the matrix $A = [a_{ij}]$ of order n.

Theorem 7.2.2 *Let $n \geq 2$ be an integer, and let $R = (r_1, r_2, \ldots, r_n)$ be a nonincreasing, nonnegative integral vector satisfying (7.5) and (7.6). Then Algorithm II above can be carried out to give a matrix in $\mathcal{Z}^+(R)_0$.*

Proof. If $n = 2$, then since $R = (r_1, r_2)$ satisfies (7.5) and is nonincreasing, $r_1 = r_2$. Algorithm II produces the matrix

$$\begin{bmatrix} 0 & r_1 \\ r_1 & 0 \end{bmatrix}$$

in $\mathcal{Z}^+(R)_0$. Now suppose that $n \geq 3$. Since $\sum_{j=1}^{n} r_i$ is an even integer by (7.6),

$$\frac{2(r_i + r_n) - \sum_{i=1}^{n} r_i}{2} \quad \text{is an integer} \quad (i = 1, 2, \ldots, n - 1). \tag{7.7}$$

Since (7.5) holds and R is nonincreasing,

$$2r_i \leq \sum_{j=1}^{n} r_j \quad (i = 1, 2, \ldots, n).$$

From this it follows that

$$r_i \geq \frac{2(r_i + r_n) - \sum_{j=1}^{n} r_j}{2} \quad (i = 1, 2, \ldots, n - 1). \tag{7.8}$$

We next show that the sum of the lower bounds given for the x_i in Algorithm II is also a lower bound for r_n. Suppose to the contrary that we have

$$\sum_{i=1}^{n-1} \left(2(r_i + r_n) - \sum_{i=1}^{n} r_i \right) > 2r_n.$$

Then

$$\sum_{i=1}^{n} 2r_i - 2r_n + (n - 1)2r_n - (n - 1)\sum_{i=1}^{n} r_i > 2r_n$$

and hence

$$-\sum_{i=1}^{n} r_i + 2r_n > 0.$$

Using the nonincreasing property of R, we see that this contradicts (7.5). Hence

$$\sum_{i=1}^{n-1} \left(\frac{2(r_i + r_n) - \sum_{i=1}^{n} r_i}{2} \right) \leq r_n. \tag{7.9}$$

Finally we note that the lower bound given for x_i is equivalent to

$$2(r_i - x_i) \le \sum_{j=1}^{n-1}(r_j - x_j).\tag{7.10}$$

Formulas (7.7) to (7.10) imply that the recursive construction in the algorithm can be carried out to produce a matrix in $\mathcal{Z}^+(R)_0$. □

Let $R = (r_1, r_2, \ldots, r_n)$ be a nonincreasing, nonnegative integral vector and let p be a positive integer. According to Theorem 1.5.2, the class $\mathcal{Z}^+(R; p)$ is nonempty if and only if

$$pkl \ge \sum_{j=1}^{l} r_j - \sum_{i=k+1}^{n} r_i \quad (1 \le l \le k \le n).\tag{7.11}$$

We now describe an algorithm for constructing a matrix in $\mathcal{Z}^+(R; p)$, that is, a symmetric, nonnegative integral matrix with row sum vector R in which all the elements are bounded above by p.

Algorithm III: Constructing a Matrix in $\mathcal{Z}^+(R; p)$

(1) Given a nonnegative integral vector

$$R_n = R = (r_1, r_2, \ldots, r_n).$$

Let $\widetilde{A} = [a_{ij}]$ be the zero matrix of order n and set $k = n$.

(2) While $k \ge 1$, do:

    (i) Set $u = r_k$. While $u > 0$, do:

        (a) Determine the smallest integer t with $1 \le t \le k$ such that $a_{tk} < p$ and $r_t > 0$, and the largest integer i with $t \le i \le k$ such that $r_t = r_{t+1} = \cdots = r_i$.

        (b) Add 1 to a_{ik} and, if $i \ne k$, add 1 to a_{ki}; subtract 1 from r_i.

        (c) Subtract 1 from u.

    (ii) Subtract 1 from k, and let $R_k = (r_1, r_2, \ldots, r_k)$.

(3) Output the matrix $\widetilde{A}$.

Example. Let $R = (7, 7, 4, 3)$ and $p = 2$. The matrix $\widetilde{A}$ produced by the algorithm III is

$$\begin{bmatrix} 2 & 2 & 2 & 1 \\ 2 & 2 & 1 & 2 \\ 2 & 1 & 1 & 0 \\ 1 & 2 & 0 & 0 \end{bmatrix}.$$

□

Theorem 7.2.3 *Let $n \geq 1$ and $p \geq 1$ be integers, and let $R = (r_1, r_2, \ldots, r_n)$ be a nonincreasing, nonnegative integral vector satisfying (7.11). Then Algorithm III above can be carried out to give a matrix $\widetilde{A}$ in $\mathcal{Z}^+(R; p)$.*

Proof. The case $n = 1$ is trivial, and so we assume that $n \geq 2$. Since (7.11) holds, $\mathcal{Z}^+(R; p)$ is nonempty. By induction it suffices to show that, when $k = n$, step (2)(i) of the algorithm can be carried out to construct column n of $\widetilde{A}$ with elements at most p and summing to r_n, and that there is a matrix in $\mathcal{Z}^+(R; p)$ whose column n equals column n of $\widetilde{A}$. The first assertion is equivalent to

$$\sum_{i=1}^{n} \min\{p, r_i\} \geq r_n. \tag{7.12}$$

If $p \geq r_n$, then $\min\{p, r_n\} = r_n$ so that (7.12) holds. Now assume that $p < r_n$. Then $p < r_n \leq r_{n-1} \leq \cdots \leq r_1$ and $\min\{p, r_i\} = p$ for $i = 1, 2, \ldots, n$. Hence in this case (7.12) is equivalent to $pn \geq r_n$. By taking $l = 1$ and $k = n$ in (7.11) we obtain $pn \geq r_1 \geq r_n$.

We now turn to establishing the second assertion. Let $A = [a_{ij}]$ be any matrix in $\mathcal{Z}^+(R; p)$. We show that there exists a sequence of symmetric interchanges which transforms A into a matrix whose column n equals column n of $\widetilde{A}$. It follows from step (2) of the algorithm that the elements $b_1, b_2, \ldots, b_n$ in column n of $\widetilde{A}$ are chosen to be as large as possible, with favor given to the bottommost positions in case of equal row sums, subject to the conditions that (i) $b_i \leq \min\{p, r_i\}$ $(i = 1, 2, \ldots, n)$ and (ii) $r_1 - b_1 \geq r_2 - b_2 \geq \cdots \geq r_n - b_n \geq 0$. Since R is nonincreasing, $b_i \geq b_{i+1} - 1$ with equality only when $r_i = r_{i+1}$ $(i = 1, 2, \ldots, n - n - 1)$.

Case 1: there are integers k and l with $k < l$ such that $r_k = r_l$ and $a_{kn} > a_{ln}$. First suppose that $l = n$. If $a_{kk} < p$, then an interchange of type (i) produces a matrix $B = [b_{ij}]$ in $\mathcal{Z}^+(R; p)$ with $b_{kn} = a_{kn} - 1$ and $b_{nn} = a_{nn} + 1$. Assume that $a_{kk} = p$. Then there exists an integer $t \neq k, n$ such that $a_{kt} < a_{nt}$. If $a_{tt} \neq 0$, then a symmetric interchange of type (i) with indices $\{t, k\}$ followed by one with indices $\{k, n\}$ produces a matrix $B = [b_{ij}]$ in $\mathcal{Z}^+(R; p)$, with the intermediary matrix also in $\mathcal{Z}^+(R; p)$, satisfying $b_{kn} = a_{kn} - 1$ and $b_{nn} = a_{nn} + 1$. We now suppose that $a_{tt} = 0$. If $r_t = r_k$, then a symmetric interchange of type (i) with indices $\{t, n\}$ produces a matrix $B = [b_{ij}]$ in $\mathcal{Z}^+(R; p)$ satisfying $b_{tn} = a_{tn} - 1$ and $b_{nn} = a_{nn} + 1$ (so k is replaced with t). Assume that $r_t > r_k$. Then there exists $s \neq t, k, n$ such that $a_{ts} > a_{ks}$. Then a symmetric interchange of type (ii) with indices $\{s, t, k\}$ followed by one of type (i) with indices $\{k, n\}$ produces a matrix $B = [b_{ij}]$ in $\mathcal{Z}^+(R; p)$, with the intermediary matrix also in $\mathcal{Z}^+(R; p)$, such that $b_{kn} = a_{kn} - 1$ and $b_{nn} = a_{nn} + 1$.

Now suppose that $l < n$. There exists a $t \neq n$ such that $a_{kt} < a_{lt}$. If we can choose t so that $t \neq k, l$, then a symmetric interchange of type (iii) produces a matrix $B = [b_{ij}]$ in $\mathcal{Z}^+(R; p)$ with $b_{kn} = a_{kn} - 1$ and

$b_{ln} = a_{ln} + 1$. Thus we assume that the only possible values of t are k and l; in particular, this implies that $a_{kk} < p$. We assume that $t = k$, the case $t = l$ being handled in a similar way. Then a symmetric interchange of type (ii) with indices $\{t, k, l\}$ produces a matrix $B = [b_{ij}]$ with $b_{kn} = a_{kn} - 1$ and $b_{ln} = a_{ln} + 1$.

Case 2: there are integers k and l with $k < l$ such that $r_k > r_l$, $r_k - a_{kn} > r_l - a_{ln} + 1$, $a_{kn} < p$, and $a_{ln} > 0$. First suppose that $l = n$. If $a_{kk} > 0$, then a symmetric interchange of type (i) with indices $\{k, n\}$ gives a matrix $B = [b_{ij}]$ in $\mathcal{Z}^+(R; p)$ with $b_{kn} = a_{kn} + 1$ and $b_{nn} = a_{nn} - 1$. Thus we may assume that $a_{kk} = 0$. Since $r_k - a_{kn} > r_l - a_{ln}$, there exists an integer $t \neq k, n$ such that $a_{kt} > a_{nt}$. If $a_{tt} < p$, then a symmetric interchange of type (i) with indices $\{k, t\}$ followed by one with indices $\{k, n\}$ gives a matrix B with the same properties, with the intermediary matrix also in $\mathcal{Z}^+(R; p)$. Thus we may also assume that $a_{tt} = p$. Now a symmetric interchange of type (ii) with indices $\{t, k, n\}$ gives a matrix $B = [b_{ij}]$ in $\mathcal{Z}^+(R; p)$ with $b_{tn} = a_{tn} + 1$ and $b_{nn} = a_{nn} - 1$. Thus we have succeeded in decreasing a_{nn}. This reduces this case to $l \neq n$ which we now assume.

There exists an integer $t \neq n$ such that $a_{kt} > a_{lt}$. If $t \neq k, l$, then a symmetric interchange of type (iii) gives a matrix $B = [b_{ij}]$ in $\mathcal{Z}^+(R; p)$ with $b_{kn} = a_{kn} + 1$ and $b_{ln} = a_{ln} - 1$. Now assume that $t = k$. Then $a_{kk} > a_{lk}$. If $a_{ll} < p$, then a symmetric interchange with indices $\{k, l, n\}$ gives such a matrix B. Suppose that $a_{ll} = p$. A symmetric interchange of type (i) with indices $\{k, l\}$ returns us to the above circumstances with $a_{ll} < p$ unless $a_{kk} = 1$ (and so $a_{lk} = 0$). Since $p \geq a_{kn}$ and $a_{ln} \geq 1 = a_{kk}$, and since $r_k - a_{kn} > r_l - a_{ln}$, a symmetric interchange of type (iii) now produces the matrix B. Finally, assume that $t = l$, so that $a_{kl} > a_{ll}$. If $a_{kk} > 0$, then a symmetric interchange of type (ii) with indices $\{k, l, n\}$ gives a matrix $B = [b_{ij}]$ in $\mathcal{Z}^+(R; p)$ with $b_{kn} = a_{kn} + 1$ and $b_{ln} = a_{ln} - 1$. So we assume that $a_{kk} = 0$. Since $r_k - a_{kn} > r_l - a_{ln}$, we now see that there is a symmetric interchange of type (iii) that produces the matrix B. The theorem now follows by induction. $\qquad\square$

In the proof of Theorem 7.2.3, we showed that any matrix in $\mathcal{Z}^+(R; p)$ can be transformed into the matrix $\widetilde{A}$ by a sequence of symmetric interchanges with all intermediary matrices in $\mathcal{Z}^+(R; p)$. This immediately implies the following corollary.

Corollary 7.2.4 *Let A and B be matrices in $\mathcal{Z}^+(R; p)$. Then there is a sequence of symmetric interchanges which transforms A into B where all intermediary matrices belong to $\mathcal{Z}^+(R; p)$.* $\qquad\square$

Havel [9] and Hakimi [7, 8] gave an algorithm to construct a (simple) graph with a given degree sequence, that is, a matrix in the class $\mathcal{A}(R)_0$ (see also [11]. This algorithm was generalized by Chungphaisan [3] to construct a multigraph with degree sequence R in which there is a uniform bound p

on the multiplicities of the edges, that is, a matrix in $\mathcal{Z}^+(R;p)_0$. This algorithm is similar to Algorithm III.

Algorithm IV: Constructing a Matrix in $\mathcal{Z}^+(R;p)_0$

(1) Given a nonincreasing, nonnegative integral vector

$$R_n = R = (r_1, r_2, \ldots, r_n).$$

Let $\widetilde{A} = [a_{ij}]$ be the zero matrix of order n and set $k = n$.

(2) While $k \geq 1$, do:

    (i) Set $u = r_k$. While $u > 0$, do:

        (a) Determine the smallest integer t with $1 \leq t < k$ such that $a_{tk} < p$ and $r_t > 0$, and the largest integer i with $t \leq i \leq k$ such that $r_t = r_{t+1} = \cdots = r_i$.

        (b) Add 1 to a_{ik} and, if $i \neq k$, add 1 to a_{ki}; subtract 1 from r_i.

        (c) Subtract 1 from u.

    (ii) Subtract 1 from k, and let $R_k = (r_1, r_2, \ldots, r_k)$.

(3) Output the matrix $\widetilde{A}$.

Let $R = (r_1, r_2, \ldots, r_n)$ be a nonincreasing, nonnegative integral vector and let p be a positive integer. According to Theorem 1.5.4, the class $\mathcal{Z}^+(R;p)_0$ is nonempty if and only if

$$\sum_{i=1}^{k} r_i \leq pk(k-1) + \sum_{i=k+1}^{n} \min\{r_i, pk\} \quad (k = 1, 2, \ldots, n). \qquad (7.13)$$

The proofs of the following theorem and corollary are very similar to the proofs of Theorem 7.2.3 and Corollary 7.2.4; consequently we omit them.

Theorem 7.2.5 *Let $n \geq 1$ and $p \geq 1$ be integers, and let $R = (r_1, r_2, \ldots, r_n)$ be a nonincreasing, nonnegative integral vector satisfying (7.13). Then Algorithm IV above can be carried out to give a matrix $\widetilde{A}$ in $\mathcal{Z}^+(R;p)_0$.* $\qquad\square$

The following result is from [3], [5].

Corollary 7.2.6 *Let A and B be matrices in $\mathcal{Z}^+(R;p)_0$. Then there is a sequence of symmetric interchanges of type (iii) which transforms A into B where all intermediary matrices belong to $\mathcal{Z}^+(R;p)_0$.* $\qquad\square$

7.3   The Class $\mathcal{A}(R)_0$

Let $R = (r_1, r_2, \ldots, r_n)$ be a nonnegative integral vector where we assume that

$$\tau = r_1 + r_2 + \cdots + r_n = 2e \qquad (7.14)$$

is an even integer. In this section, based largely on Michael [14, 15], we investigate in more detail the class $\mathcal{A}(R)_0$ of symmetric $(0, 1)$-matrices with row sum vector R and zero trace, equivalently, the class of graphs (so no loops are permitted) with e edges and with degree sequence equal to R. By Corollary 7.2.6, any two matrices in $\mathcal{A}(R)_0$ can be transformed into each other by a sequence of symmetric interchanges of type (iii) where all intermediary matrices are in $\mathcal{A}(R)_0$. In analogy to the interchange distance for the class $\mathcal{A}(R, S)$ investigated in Section 3.2, the *symmetric interchange distance* $\iota_{\mathrm{s}}(A, B)$ between matrices A and B in $\mathcal{A}(R)_0$ is defined to be the minimal number of symmetric interchanges that transforms A into B. However, the symmetric interchange distance is much more complicated; in fact, it has been shown by Will [22] that computing the symmetric interchange distance is an NP-complete problem.

The *structure matrix* associated with R is the matrix $T_0 = T_0(R) = [t_{kl}]$ where

$$t_{kl} = kl - \min\{k, l\} + \sum_{i=k+1}^{n} r_i - \sum_{j=1}^{l} r_j \quad (k, l = 0, 1, \ldots, n). \qquad (7.15)$$

If A is a matrix in $\mathcal{A}(R)_0$ partitioned as

$$\begin{bmatrix} A_{11} & A_{12} \\ A_{21} & A_{22} \end{bmatrix}$$

where A_{11} is a k by l matrix with $0 \leq k, l \leq n$, then t_{kl} equals the number of nondiagonal 0's in A_{11} plus the number of 1's in A_{22}. Thus the structure matrix $T_0(R)$ is nonnegative if the class $\mathcal{A}(R)_0$ is nonempty.

As with the structure matrix associated with a pair of nonnegative integral vectors, the elements of the structure matrix $T_0(R)$ satisfy a number of elementary properties which we collect in the next lemma. The straightforward verifications are omitted.

Lemma 7.3.1 *Let* $R = (r_1, r_2, \ldots, r_n)$ *be a nonnegative integral vector with* $r_1 + r_2 + \cdots + r_n$ *even, and let* $T_0(R) = [t_{ij}]$ *be its associated structure matrix of order* $n + 1$. *Then we have the following.*

(i) $T_0(R)$ *is a symmetric matrix.*

(ii) *The elements in row* i *of* $T_0(R)$ *satisfy the recurrence relation*

$$t_{ij} - t_{i,j-1} = \begin{cases} i - 1 - r_j, & \text{if } i \leq j, \\ i - r_j, & \text{if } i \geq j \end{cases} \quad (i, j = 1, 2, \ldots, n).$$

(iii) *The elements in row 0 and column 0 of $T_0(R)$ satisfy $t_{0n} = t_{n0} = 0$, and $t_{0k} = t_{k0} = \sum_{i=k+1}^{n} r_i$ $(k = 0, 1, \ldots, n-1)$.*

(iv) *The elements of $T_0(R)$ satisfy the recurrence relation*

$$t_{ij} = t_{i-1,j} + t_{i,j-1} - t_{i-1,j-1} + \begin{cases} 1 \ \text{if} \ i \neq j, \\ 0 \ \text{if} \ i = j \end{cases} \quad (i, j = 1, 2, \ldots, n).$$

(v) *The diagonal elements of $T_0(R)$ satisfy the recurrence relation*

$$t_{ii} = t_{i-1,i-1} + 2(i - 1 - r_i) \quad (i = 1, 2, \ldots, n),$$

and are even integers.

(vi) *If R is nonincreasing, then the elements in row i form a log-convex sequence, that is,*

$$t_{i,j-1} + t_{i,j+1} \geq 2t_{ij} \quad (i = 0, 1, \ldots, n; j = 1, 2, \ldots, n),$$

and the elements on the main diagonal form a strictly log-convex sequence, that is,

$$t_{i+1,i+1} + t_{i-1,i-1} > 2t_{ii} \quad (i = 1, 2, \ldots, n - 1).$$

$\square$

Since $T_0(R)$ is a symmetric matrix, the equalities (7.15) are equivalent to the equalities

$$t_{kl} = k(l - 1) + \sum_{i=k+1}^{n} r_i - \sum_{j=1}^{l} r_j \quad (0 \leq k \leq l \leq n). \tag{7.16}$$

From the elements in row and column 0 as given in (iii) in Lemma 7.3.1 and the recurrence relation (iv), the elements of $T_0(R)$ are easily computed.

Example. Let $R = (4, 4, 3, 3, 2)$. Then

$$T_0(R) = \begin{bmatrix} 16 & 12 & 8 & 5 & 2 & 0 \\ 12 & 8 & 6 & 5 & 4 & 4 \\ 8 & 6 & 4 & 5 & 6 & 8 \\ 5 & 5 & 5 & 6 & 9 & 13 \\ 2 & 4 & 6 & 9 & 12 & 18 \\ 0 & 4 & 8 & 13 & 18 & 24 \end{bmatrix}.$$

We note that (7.15) implies that the structure matrix $T_0(R)$ can be factored as

$$T_0(R) = E_{n+1} \begin{bmatrix} \tau & -r_1 & -r_2 & \cdots & -r_n \\ -r_1 & & & & \\ -r_2 & & J_n - I_n & & \\ \vdots & & & & \\ -r_n & & & & \end{bmatrix} E_{n+1}^{\mathrm{T}}$$

where E_{n+1} denotes the matrix of order $n+1$ with 1's on and below the main diagonal and 0's elsewhere. Since E_{n+1} and $J_n - I_n$ ($n \geq 2$) are nonsingular matrices, this factorization implies that when $n \geq 2$, the rank of T_0, over any field of characteristic 0, is either n or $n + 1$. More specifically, except for the trivial cases where R equals $(0, 0, \ldots, 0)$ or $(n-1, n-1, \ldots, n-1)$, the rank of T_0 equals $n + 1$ whenever $\mathcal{A}(R)_0$ is nonempty.

Let G be a graph in which there may be loops.[5] A *matching* M in G is a collection of edges no two of which have a vertex in common. A *perfect matching* is a matching M which spans the vertex set of G, that is, one with the property that each vertex of G belongs to some edge of M. Let the cardinality of a matching M be k, and suppose that M has k_0 loops and k_1 nonloop edges where $k = k_0 + k_1$. The *span of the matching* M is the set of vertices of the edges in M and has cardinality equal to $k_0 + 2k_1$. If A is the adjacency matrix of G, then the matching M corresponds to a symmetric, subpermutation matrix $P \leq A$ with $k_0 + 2k_1$ 1's. We refer to P as a *matching of the matrix* A of cardinality k and *span size* $k_0 + 2k_1$. The matrix P has term rank $\rho(P)$ equal to its span size $k_0 + 2k_1$. The *matching number* $\mu(A)$ equals the largest span size of a matching of A.[6]

Example. Consider the symmetric matrix

$$A = \begin{bmatrix} 0 & 1 & 1 \\ 1 & 0 & 1 \\ 1 & 1 & 0 \end{bmatrix}.$$

Then $\mu(A) = 2$ and $\rho(A) = 3$. If we replace the 0 in position $(1, 1)$ with a 1, then the matching number equals 3.   □

Now let R be a nonnegative integral vector for which the class $\mathcal{A}(R)_0$ is nonempty. Each matrix A in $\mathcal{A}(R)_0$ has zero trace and is the adjacency matrix of a graph without loops. For such a matrix A the cardinality of a matching equals half its span size. We let

$$\overline{\mu}_0 = \overline{\mu}_0(R) = \max\{\mu(A) : A \in \mathcal{A}(R)_0\}$$

denote the *maximal matching number* of a matrix in $\mathcal{A}(R)_0$, and

$$\widetilde{\mu}_0 = \widetilde{\mu}_0(R) = \min\{\mu(A) : A \in \mathcal{A}(R)_0\}$$

denote the *minimal matching number* of a matrix in $\mathcal{A}(R)_0$. The numbers $\overline{\mu}_0(R)$ and $\widetilde{\mu}_0(R)$ are analogs for $\mathcal{A}(R)_0$ of the maximal and minimal term ranks, $\overline{\rho}(R, S)$ and $\widetilde{\rho}(R, S)$, respectively, for a nonempty class $\mathcal{A}(R, S)$.

---

[5]Since we are considering $\mathcal{A}(R)_0$, the graphs in this section do not have any loops. In the next section we consider $\mathcal{A}(R)$ whose associated graphs may have loops.

[6]Our use of matching number for graphs with no loops is twice that as used in [14]. The reason for our definition is that in the case of graphs with loops, a nonloop edge spans two vertices while a loop edge spans only one.

Since a symmetric interchange of type (iii) in Section 7.1 changes the matching number of a matrix by 0 or 2, it follows from Corollary 7.2.6 that for each even integer m with $\widetilde{\mu}_0(R) \le m \le \overline{\mu}_0(R)$, there exists a matrix in $\mathcal{A}(R)_0$ with matching number m.

We now turn to obtaining a formula for the maximal matching number $\overline{\mu}_0(R)$ of a nonempty class $\mathcal{A}(R)_0$ which is quite analogous to the formula for the maximal term rank $\overline{\rho}(R, S)$ of a nonempty class $\mathcal{A}(R, S)$. According to Corollary 1.5.3, if R is nonincreasing, the class $\mathcal{A}(R)_0$ is nonempty if and only if the structure matrix $T_0(R)$ is a nonnegative matrix. As observed by Michael [15], the nonemptiness of $\mathcal{A}(R)_0$ is equivalent to the nonnegativity of $T_0(R)$ under an assumption on R weaker than the nonincreasing property.[7] It is this fact that is used in the derivation we give of a formula for $\overline{\mu}_0(R)$.

Recall that a sequence $U = (u_1, u_2, \ldots, u_n)$ of integers is *nearly nonincreasing* provided

$$u_i \ge u_j - 1 \quad (1 \le i < j \le n).$$

Theorem 7.3.2 *Let $R = (r_1, r_2, \ldots, r_n)$ be a nearly nonincreasing, nonnegative integral vector such that $r_1 + r_2 + \cdots + r_n$ is an even integer. Then $\mathcal{A}(R)_0$ is nonempty if and only if the structure matrix $T_0(R)$ is nonnegative.*

Proof. We have already observed that $T_0(R)$ is nonnegative if $\mathcal{A}(R)_0$ is nonempty. Now assume that $T_0(R)$ is nonnegative. We prove that there exists a matrix in $\mathcal{A}(R)_0$ by induction on the nonnegative integer $e = (r_1 + r_2 + \cdots + r_n)/2$. If $e = 0$, then R is a zero vector and $\mathcal{A}(R)_0$ contains a zero matrix.

We now assume that $e > 0$ and, without loss of generality, that $r_i > 0$ $(i = 1, 2, \ldots, n)$. Let

$$q = \min\{i : r_i \ge r_k \text{ for } k = 1, 2, \ldots, n\}.$$

First suppose that $q = n$. Then $R = (r, \ldots, r, r + 1)$ for some integer $r \le n - 2$. Since $r_1 + r_2 + \cdots + r_n$ is even, n and r are both odd. It is now straightforward to construct a matrix in $\mathcal{A}(R)_0$ and to check that $T(R)$ is nonnegative.

It remains to consider the case where $q < n$. We have $r_1 = \cdots = r_{q-1} = r_q - 1$, and we let $\hat{R} = (\hat{r}_1, \hat{r}_2, \ldots, \hat{r}_n)$ be the vector obtained from R by subtracting 1 from each of r_q and r_n. We have $\hat{r}_1 + \hat{r}_2 + \cdots + \hat{r}_n = 2(e - 1)$. We show that the structure matrix $\hat{T}_0 = [\hat{t}_{ij}]$ corresponding to $\hat{R}$ is nonnegative. Simple computation shows that

$$\hat{t}_{ij} = \begin{cases} t_{ij} - 2, & \text{if } 0 \le i, j < q, \\ t_{ij} - 1, & \text{if } 0 \le i < q \le j < n \text{ or } 0 \le j < q \le i < n, \end{cases}$$

---

[7] As shown in Theorem 2.1.4, this is also true for $\mathcal{A}(R, S)$ and its structure matrix $T(R, S)$.

and $\hat{t}_{ij} \geq t_{ij}$, otherwise. Because T_0 is symmetric and nonnegative, it suffices to prove that

$$t_{ij} \geq 2 \; (0 \leq j \leq i < q) \text{ and } t_{ij} \geq 1 \; (0 \leq j < q \leq i < n). \qquad (7.17)$$

If $j \leq i$ and $j < q$, then we calculate that

$$t_{ij} = ij - j + \sum_{k=i+1}^{n} r_k - \sum_{k=1}^{j} r_k = j(i - r_q) + \sum_{k=i+1}^{n} r_k,$$

and thus, since $r_q \geq 1$ and $r_n \geq 1$, the inequalities (7.17) hold when $r_q \leq i$. We now assume that $r_q > i$. If $j < q \leq i$, then

$$t_{ij} = t_{iq} + (q - j)(r_q - i) + 1$$

and the second inequalities in (7.17) hold. Finally, we assume that $j \leq i < q$. Then using (ii) of Lemma 7.3.1, we have

$$t_{ij} = t_{i,i+1} + (i - j + 1)(r_q - i).$$

Suppose to the contrary that the inequality $t_{ij} < 2$. Then we must have $i = j$, $t_{i,i+1} = 0$, and $r_q = i$. The recurrence relation (ii) of Lemma 7.3.1 now implies that $t_{ii} = 1$, contradicting (v) of that lemma. This contradiction completes the verification of (7.17) and establishes the nonnegativity of $\hat{T}_0$.

By induction there is a matrix $\hat{A} = [\hat{a}_{ij}]$ in the class $\mathcal{A}(\hat{R})_0$. If $\hat{a}_{qn} = 0$, then the matrix A obtained from $\hat{A}$ by replacing each of $\hat{a}_{qn}$ and $\hat{a}_{nq}$ with a 1 is in $\mathcal{A}(R)_0$. Suppose that $\hat{a}_{qn} = 1$. The inequality $0 \leq t_{qn} = q((n-1) - (r_q - 1)) - 1$ implies that $\hat{r}_q = r_q - 1 < n - 1$. Thus $\hat{a}_{qk} = 0$ for some index $k \neq q$. Since $r_k \geq r_n - 1$, there is an index $h \neq n$ such that $\hat{a}_{qh} = 1$ and $\hat{a}_{nh} = 0$. The matrix obtained from $\hat{A}$ by applying a symmetric interchange of type (iii) with indices $\{q, h, k, n\}$ is in $\mathcal{A}(\hat{R})_0$, and the elements in position (q, n) and (n, q) equal 0. If we replace each of these 0's with a 1, we obtain a matrix in $\mathcal{A}_0(R)_0$.   □

The following lemma is the symmetric analog of Theorem 3.5.2 [13, 11, 12]. It can be proved in a very similar way using symmetric interchanges of type (iii) in place of interchanges.

Lemma 7.3.3 *Let $R = (r_1, r_2, \ldots, r_n)$ and $R' = (r'_1, r'_2, \ldots, r'_n)$ be nonnegative integral vectors where $r'_i = r_i$ or $r_i - 1$ $(i = 1, 2, \ldots, n)$. Assume that each of the classes $\mathcal{A}(R)_0$ and $\mathcal{A}(R')_0$ is nonempty. Then there exist matrices A in $\mathcal{A}(R)_0$ and A' in $\mathcal{A}(R')_0$ such that $A' \leq A$ (entrywise).*   □

The next lemma enables us to apply Theorem 7.3.2 to obtain a formula for $\overline{\mu}(R)$. In the proof, the matrix E_{ij} with $i \neq j$ denotes the symmetric matrix of order n with 1's in positions (i, j) and (j, i) and 0's elsewhere.

Lemma 7.3.4 *Let* $R = (r_1, r_2, \ldots, r_n)$ *be a nonincreasing, nonnegative integral vector. Suppose there is a matrix* $A = [a_{ij}]$ *in* $\mathcal{A}(R)_0$ *with a matching* $P = [p_{ij}]$ *of span size* q. *Then there exists a matrix* B *in* $\mathcal{A}(R)_0$ *with a matching of span size* q *whose row sum vector is of the form* $(1, \ldots, 1, 0, \ldots, 0)$.

Proof. Suppose that there is an index k such that row $k+1$ of P contains a 1 but row k does not. Then $p_{k+1,l} = 1$ and $p_{kl} = 0$ for some index $l \neq k$. If $a_{kl} = 1$, then we replace P with the matching $P - E_{k+1,l} + E_{kl}$ of A of span size q. We now assume that $a_{kl} = 0$. Since R is nonincreasing, there is an index $j \neq k$ such that $a_{kj} = 1$ and $a_{k+1,j} = 0$. If $p_{k+1,j} = 1$, we replace P with the matching $P - E_{k+1,j} + E_{kj}$ of A of span size q. Now assume also that $p_{k+1,j} = 0$. We then replace A with the matrix $A' = A - E_{kl} + E_{k+1,j} - E_{kj} - E_{k+1,j}$ in $\mathcal{A}(R)_0$ and P with the matching $P + E_{kl} - E_{k+1,l}$ of A' of span size q. In all cases row k of P now contains a 1 and row $k+1$ does not. Iteration of this argument completes the proof. $\square$

Theorem 7.3.5 *Let* $R = (r_1, r_2, \ldots, r_n)$ *be a nonincreasing, nonnegative vector for which the class* $\mathcal{A}(R)_0$ *is nonempty, and let* $T_0(R) = [t_{ij}]$ *be the structure matrix corresponding to* R. *Then the maximal matching number satisfies*

$$\bar{\mu}_0(R) = 2 \min \left\{ \left\lfloor \frac{t_{kl} + k + l}{2} \right\rfloor : k, l = 0, 1, \ldots, n \right\}. \tag{7.18}$$

Proof. Let q be an integer with $0 \leq q \leq n/2$. By Lemma 7.3.4 there is a matrix in $\mathcal{A}(R)_0$ with a matching of span size q if and only if there is a matrix in $\mathcal{A}(R)_0$ with a matching of span size q whose row sum vector $R_1 = (1, \ldots, 1, 0, \ldots, 0)$ (q 1's). It follows from Lemma 7.3.3 that the latter holds if and only if the class $\mathcal{A}(R')_0$ is nonempty where $R' = R - R_1 = (r'_1, r'_2, \ldots, r'_n)$. The vector R' is nearly nonincreasing and the sum of its elements is even. By Theorem 7.3.2, $\mathcal{A}(R')_0$ is nonempty if and only if its structure matrix $T_0(R') = [t'_{ij}]$ is nonnegative.

For $k, l = 0, 1, \ldots, n$ we have

$$t'_{kl} = kl - \min\{k, l\} + \sum_{i=k+1}^{n} r'_i - \sum_{j=1}^{l} r'_j$$

$$= kl - \min\{k, l\} + \sum_{i=k+1}^{n} (r_i - 1) - \max\{0, 2q - k\}$$

$$- \sum_{j=1}^{l} (r_j - 1) + \min\{l, 2q\}$$

$$= kl - \min\{k, l\} + \sum_{i=k+1}^{n} r_i - \sum_{j=1}^{l} r_j + \min\{k, 2q\} - 2q + \min\{l, 2q\}$$

$$= t_{kl} - 2q + \min\{k, 2q\} + \min\{l, 2q\}.$$

Thus $\mathcal{A}(R')_0$ is nonempty if and only if

$$2q \le t_{kl} + \min\{k, 2q\} + \min\{l, 2q\} \quad (k, l = 0, 1, \dots, n). \qquad (7.19)$$

If $k \ge 2q$ or $l \ge 2q$, then (7.19) holds. If $k, l < 2q$, then (7.19) holds if and only if

$$2q \le t_{kl} + k + l.$$

The theorem now follows. $\qquad\qquad\qquad\qquad\qquad\qquad\qquad\qquad\qquad\qquad\square$

In our study of the term rank of matrices in a class $\mathcal{A}(R, S)$, we established the existence of a special matrix in $\mathcal{A}(R, S)$ with maximal term rank $\overline{\rho}(R, S)$. In the following theorem we show the existence of a special matrix in $\mathcal{A}(R)_0$ whose matching number equals the maximal matching number for $\mathcal{A}(R)_0$.

Theorem 7.3.6 *Let R be a nonincreasing, nonnegative integral vector such that the class $\mathcal{A}(R)_0$ is nonempty. Let $\overline{\mu}_0 = \overline{\mu}_0(R)$. Then there exists a matrix $A = [a_{ij}]$ in $\mathcal{A}(R)_0$ with $\mu(A) = \overline{\mu}_0$ with a matching $P = [p_{ij}]$ of span size $\overline{\mu}_0$ satisfying $p_{1,\overline{\mu}_0} = p_{2,\overline{\mu}_0 - 1} = \cdots = p_{\overline{\mu}_0, 1} = 1$.*

Proof. In fact we show how, starting from a matrix $A = [a_{ij}]$ in $\mathcal{A}(R)_0$ and a matching $Q = [q_{ij}]$ of A of span size l (necessarily an even integer), to obtain a matrix in $\mathcal{A}(R)_0$ with a matching of span size l with 1's in positions $(1, l), (2, l - 1), \dots, (l, 1)$. By Lemma 7.3.4 we may assume that the span of Q is $\{1, 2, \dots, l\}$. Suppose that Q has a 0 in a position $(k, l - k + 1)$ (and thus also in position $(l - k + 1, k)$), and choose k to be the minimal such integer. There exist integers k' and $(l - k + 1)'$ such that $q_{kk'} = q_{k'k} = q_{l-k+1,(l-k+1)'} = q_{(l-k+1)',l-k+1} = 1$. The minimality of k implies that k' and $(l - k + 1)'$ are strictly between k and $l - k + 1$. If either $a_{k,l-k+1} = a_{k',(l-k+1)'} = 0$ or $a_{k,l-k+1} = a_{k',(l-k+1)'} = 1$, then applying at most one symmetric interchange of type (iii) to A, we obtain a matrix in $\mathcal{A}(R)_0$ with a matching $P' = [p'_{ij}]$ with span $\{1, 2, \dots, l\}$ such that $p'_{k,l-k+1} = p'_{k',(l-k+1)'} = 1$. Suppose that $a_{k,l-k+1} = 1$ and $a_{k',(l-k+1)'} = 0$. Because $r_{k'} \ge r_{l-k+1}$, there exists an index j such that $a_{k'j} = 1$ and $a_{j,l-k+1} = 0$. Applying a symmetric interchange of type (iii) to A we obtain a matrix in $\mathcal{A}(R)_0$ having P' as a matching. A similar argument is available in case $a_{k,l-k+1} = 0$ and $a_{k',(l-k+1)'} = 1$. In each case we obtain a matrix in $\mathcal{A}(R)_0$ with a matching of span $\{1, 2, \dots, l\}$ with more 1's in the positions $\{(1, l), (2, l-1), \dots, (l, 1)\}$. Iteration of this process completes the proof. $\qquad\qquad\qquad\qquad\qquad\qquad\qquad\qquad\qquad\qquad\square$

We conclude this section by considering another natural parameter for a nonempty class $\mathcal{A}(R)_0$. Let A be a $(0, 1)$-matrix with zero trace. We let

$$\omega_0(A) = \max\{k : J_k - I_k \text{ is a principal submatrix of } A\}.$$

Since A is the adjacency matrix of a graph G, $\omega_0(A)$ is the largest size of a complete subgraph (*clique*) of G. As a result we call $\omega_0(A)$ the *clique number* of A. We let

$$\overline{\omega}_0(R) = \max\{\omega_0(A) : A \in \mathcal{A}(R)_0\}$$

denote the *maximal clique number* of a matrix in $\mathcal{A}(R)_0$ and

$$\widetilde{\omega}_0(R) = \min\{\omega_0(A) : A \in \mathcal{A}(R)_0\}$$

denote the minimal clique number of a matrix in $\mathcal{A}(R)_0$. These clique numbers were investigated by Punnim [19] in the context of graphs. Since a symmetric interchange of type (iii) in Section 7.1 changes the clique number by at most 1, it follows from Corollary 7.2.6 that for each integer k with $\widetilde{\omega}_0 \leq k \leq \overline{\omega}_0$, there exists a matrix in $\mathcal{A}(R)_0$ with clique number equal to k.

Complementary to $\omega(A)$ is $\zeta_0(A)$, the largest order of a principal zero submatrix of A. It follows that $\omega_0(A) = \zeta_0(J_n - I_n - A)$. The parameters $\overline{\zeta}_0(R)$ and $\widetilde{\omega}_0(R)$ are defined in an analogous way.

Let $\Delta(R)$ denote the maximum entry of the nonnegative integral vector R.

Lemma 7.3.7 *Let A be a matrix in the class $\mathcal{A}(R)_0$. Then*

$$\zeta_0(A) \geq \frac{n}{\Delta(R) + 1},$$

and hence,

$$\omega_0(A) \geq \frac{n}{n - \Delta(R)}.$$

Proof. We first show that A has a principal zero submatrix of order at least $n/(\Delta + 1)$. If $\Delta = 0$, this is obviously true. We assume that $\Delta \geq 1$ and proceed by induction on Δ. Let A_1 be a principal zero submatrix of A of order $\zeta(A)$, and let A_2 be the principal submatrix of A of order $n - \zeta(A)$ obtained by deleting the rows and columns of A_1, and let R_2 be the row sum vector of $_2$. We have that $\Delta(R_2) \leq \Delta(R) - 1$, by the maximality property of $\omega_0(A)$. Hence by induction

$$\zeta_0(R) \geq \zeta_0(R_2) \geq \frac{n - \zeta_0(A)}{\Delta(R_2) + 1} \geq \frac{n - \zeta_0(A)}{\Delta(R)},$$

and it follows that $\zeta_0(R) \geq n/(\Delta(R) + 1)$. The second inequality in the lemma follows by applying the first inequality to the matrix $J_n - I_n - A$. $\square$

We now assume that R is the vector $(k, k, \ldots, k)$ of size n with $0 \leq k \leq n-1$, and consider the subclass $\mathcal{A}^*(n, k)_0$ of $\mathcal{A}(n, k)$ consisting of symmetric $(0, 1)$-matrices of order n with zero trace having exactly k 1's in each row and column. Matrices in $\mathcal{A}^*(n, k)_0$ correspond to graphs with n vertices each of degree k. It follows from Theorem 1.5.4 that the class $\mathcal{A}^*(n, k)_0$ is nonempty if and only if kn is an even integer (that is, k is even if n is odd). For the class $\mathcal{A}^*(n, k)_0$, we write $\overline{\omega}_0(n, k), \widetilde{\omega}_0(n, k), \overline{\zeta}_0(n, k)$, and $\widetilde{\zeta}_0(n, k)$ in place of $\overline{\omega}_0(R), \widetilde{\omega}_0(R), \overline{\zeta}_0(R)$, and $\widetilde{\omega}_0(R)$, respectively.

The following is an immediate corollary of Lemma 7.3.7.

Corollary 7.3.8 *Let k be an integer with $0 \le k \le n - 1$ such that nk is an even integer. Then*

$$\widetilde{\zeta}_0(n, k) \ge \frac{n}{k + 1} \quad \text{and} \quad \widetilde{\omega}_0(n, k) \ge \frac{n}{n - k}.$$

$\square$

We now obtain a formula for $\overline{\omega}_0(n, k)$.

Theorem 7.3.9 *Let k be a nonnegative integer with $0 \le k \le n - 1$ such that nk is an even integer. Then*

$$\overline{\omega}_0(n, k) = \begin{cases} k + 1, & \text{if } n = k + 1, \\ \lfloor \frac{n}{2} \rfloor, & \text{if } k + 2 \le n \le 2k + 1, \\ k + 1, & \text{if } 2k + 2 \le n. \end{cases}$$

Proof. If $n = k + 1$, then $J_n - I_n = J_{k+1} - I_{k+1} \in \mathcal{A}^*(n, k)_0$ and hence $\omega_0(n, k) = k$. Now assume that $n \ge 2k + 2$, and let $m = n - (k + 1) \ge k + 1$. If k is odd, then n, and hence m, is an even integer. There exists a matrix $B \in \mathcal{A}^*(m, k)_0$, and the matrix $(J_{k+1} - I_{k+1}) \oplus B \in \mathcal{A}^*(n, k)$ satisfies $\omega_0(A) = k + 1$. Hence $\overline{\omega}_0(n, k) = K + 1$. Finally, assume that $k + 2 \le n \le 2k + 1$. Let A be a matrix in $\mathcal{A}^*(n, k)$ with $\omega_0(A) = t$. Thus A has a principal submatrix equal to $J_t - I_t$ and the number p of 1's of A above its main diagonal equals $kn/2$ and satisfies

$$\frac{kn}{2} \le \binom{t}{2} + \binom{n - t}{2} + (t)(k - t + 1).$$

Hence $(n - 2t)(n - k - 1) \ge 0$. If $t > n/2$, then this implies that $n - 1 \le k$, a contradiction. Hence $t \le \lfloor n/2 \rfloor$. It remains to show that, in this case where $k + 2 \le n \le 2k + 1$, there exists a matrix $A \in \mathcal{A}^*(n, k)_0$ with $\omega_0(A) = \lfloor n/2 \rfloor$. We show how to construct such a matrix when n is even.

Let $n = 2m$. Thus $k + 2 \le 2m \le 2k + 1$. We describe the matrix in terms of its graph. First we take two complete graphs on disjoint sets of m vertices each. We then construct a bipartite graph between U and W with each vertex incident with $k - m + 1 \ge 1$ edges. The adjacency matrix of this graph belongs to $\mathcal{A}^*(2m, k)_0$ and satisfies $\omega_0(A) = m$. A slight variation of this construction can be used when n is odd. $\square$

The numbers $\widetilde{\omega}(n, k)_0$ are also determined in [19]. The values of these numbers depend on special relations that may exist between n and k, and we refer to [19]. In [18], the minimal and maximal values of the chromatic number of the graphs whose adjacency matrices are in $\mathcal{A}^*(n, k)_0$ are investigated and are determined in some general cases.

7.4 The Class $\mathcal{A}(R)$

Let $R = (r_1, r_2, \ldots, r_n)$ be a nonnegative integral vector. In this section we study the class $\mathcal{A}(R)$ of symmetric $(0, 1)$-matrices with row sum vector equal to R. We obtain analogs for this class of results for $\mathcal{A}(R)_0$ in Section 7.3. We first remark that by Corollary 7.2.4 any two matrices in $\mathcal{A}(R)$ can be transformed into each other by a sequence of symmetric interchanges of types (i), (ii), and (iii) with all intermediary matrices in $\mathcal{A}(R)$. Without loss of generality we assume that R is nonincreasing.

In our study of $\mathcal{A}(R)$, the structure matrix associated with R is now taken to be the structure matrix

$$T = T(R) = T(R, R) = [t_{ij}]$$

of order $n + 1$ applied in Chapter 3 in our study of the classes $\mathcal{A}(R, S)$. The motivation for this is that, according to Lemma 1.5.1, the class $\mathcal{A}(R)$ is nonempty if and only if the class $\mathcal{A}(R, R)$ is nonempty. Since the column sum vector equals the row sum vector R, the elements of the structure matrix are

$$t_{kl} = kl - \sum_{j=1}^{l} r_j + \sum_{i=k+1}^{n} r_i \quad (k, l = 0, 1, \ldots, n).$$

In addition, the structure matrix T is now a symmetric matrix. It follows from Corollary 1.5.3 that $\mathcal{A}(R)$ is nonempty if and only if the structure matrix $T(R)$ is nonnegative.

Let R be a nonnegative integral vector for which the class $\mathcal{A}(R)$ is nonempty. We let

$$\overline{\mu} = \overline{\mu}(R) = \max\{\mu(A) : A \in \mathcal{A}(R)\}$$

denote the *maximal matching number* of a matrix in $\mathcal{A}(R)$, and

$$\widetilde{\mu} = \widetilde{\mu}(R) = \min\{\mu(A) : A \in \mathcal{A}(R)\}$$

denote the *minimal matching number* of a matrix in $\mathcal{A}(R)$.

The next two lemmas are analogs of Lemmas 7.3.3 and 7.3.4 for the class $\mathcal{A}(R)$. We omit the very similar proofs.

Lemma 7.4.1 *Let $R = (r_1, r_2, \ldots, r_n)$ and $R' = (r'_1, r'_2, \ldots, r'_n)$ be nonnegative integral vectors where $r'_i = r_i$ or $r_i - 1$ $(i = 1, 2, \ldots, n)$. If both of the classes $\mathcal{A}(R)$ and $\mathcal{A}(R')$ are nonempty, then there exist a matrix A in $\mathcal{A}(R)$ and a matrix A' in $\mathcal{A}(R')$ such that $A' \leq A$ (entrywise).* $\square$

Lemma 7.4.2 *Let $R = (r_1, r_2, \ldots, r_n)$ be a nonincreasing, nonnegative integral vector. Suppose there is a matrix $A = [a_{ij}]$ in $\mathcal{A}(R)$ with a matching $P = [p_{ij}]$ of span size q. Then there exists a matrix B in $\mathcal{A}(R)$ with a matching of span size q whose row sum vector is of the form $(1, \ldots, 1, 0, \ldots, 0)$.* $\square$

We also have the following analog of Theorem 7.3.2.

Theorem 7.4.3 *Let $R = (r_1, r_2, \ldots, r_n)$ be a nearly nonincreasing, nonnegative integral vector. Then $\mathcal{A}(R)$ is nonempty if and only if the structure matrix $T(R)$ is nonnegative.* $\square$

Lemmas 7.4.1 and 7.4.2 and Theorem 7.4.3 now yield the following formula for the maximal matching number of a matrix in a nonempty class $\mathcal{A}(R)$.

Theorem 7.4.4 *Let $R = (r_1, r_2, \ldots, r_n)$ be a nonincreasing nonnegative vector for which the class $\mathcal{A}(R)$ is nonempty, and let $T(R) = [t_{ij}]$ be the structure matrix corresponding to R. Then the maximal matching number satisfies*

$$\overline{\mu}(R) = \min\left\{ t_{kl} + k + l : k, l = 0, 1, \ldots, n \right\}. \tag{7.20}$$

$\square$

Since the formula for $\overline{\mu}(R)$ given in (7.20) is the same as the formula for the maximal term rank in $\mathcal{A}(R, R)$ given by Theorem 3.5.4, we obtain the following.

Corollary 7.4.5 *Let R be a nonincreasing, nonnegative integral vector such that the class $\mathcal{A}(R, R)$, and hence the class $\mathcal{A}(R)$, is nonempty. Then*

$$\overline{\rho}(R, R) = \overline{\mu}(R).$$

$\square$

We observe that if there is a matrix in $\mathcal{A}(R)$ with a matching $P = [p_{ij}]$ of span size t, then there is a matrix in $\mathcal{A}(R)$ with a matching Q of span size t in which Q has no 1's on its main diagonal, if t is even, and exactly one 1 on its main diagonal, if t is odd. Suppose that there are distinct integers k and l such that $p_{kk} = p_{ll} = 1$. If $a_{kl} = a_{lk} = 1$, then we may replace p_{kl} and p_{lk} with 1's and replace p_{kk} and p_{ll} with 0's, and obtain another matching of A of span size t. If $a_{kl} = a_{lk} = 0$, then we may apply a symmetric interchange with indices $\{k, l\}$ and replace A with a matrix A' in $\mathcal{A}(R)$ and P with a matching P' of A' of span size t. In both cases we reduce the number of 1's of the matching on the main diagonal by 2. In fact, a proof similar to the proof of Theorem 7.3.6 establishes the existence of a special matrix in $\mathcal{A}(R)$ with maximal matching number $\overline{\mu}(R)$.

Theorem 7.4.6 *Let R be a nonincreasing, nonnegative integral vector such that the class $\mathcal{A}(R)$ is nonempty. Let $\overline{\mu} = \overline{\mu}(R)$ be the maximal matching number of $\mathcal{A}(R)$. Then there exists a matrix $A = [a_{ij}]$ in $\mathcal{A}(R)$ with $\mu(A) = \overline{\mu}$ with a matching $P = [p_{ij}]$ of span size $\overline{\mu}$ satisfying $p_{1,\overline{\mu}} = p_{2,\overline{\mu}-1} = \cdots = p_{\overline{\mu},1} = 1$.* $\square$

Analogous to the parameter $\omega_0(A)$ defined in the preceding section for matrices with zero trace, for an arbitrary square $(0, 1)$-matrix A we let

$$\omega(A) = \max\{k : J_k \text{ is a principal submatrix of } A\}.$$

Regarding A as the adjacency matrix of a graph G where loops are allowed, we see that $\omega(A)$ is the largest size of a complete subgraph of G, where complete now includes the property that there is a loop at each vertex (a *full-clique*). We call $\omega(A)$ the *full-clique number* of A. We let

$$\overline{\omega}(R) = \max\{\omega(A) : A \in \mathcal{A}(R)\}$$

denote the *maximal full-clique number* of a matrix in $\mathcal{A}(R)_0$ and

$$\widetilde{\omega}_0(R) = \min\{\omega(A) : A \in \mathcal{A}(R)\}$$

denote the *minimal full-clique number* of a matrix in $\mathcal{A}(R)$. Since symmetric interchanges of types (i), (ii), and (iii) in Section 7.1 change the full-clique number by at most 1, it follows from Corollary 7.2.6 that for each integer k with $\widetilde{\omega} \leq k \leq \overline{\omega}$, there exists a matrix in $\mathcal{A}(R)$ with clique number equal to k.

Let k and n be integers with $0 \leq k \leq n, n > 0$. Consider the class $\mathcal{A}^*(n, k)$ of $\mathcal{A}(n, k)$ consisting of symmetric $(0, 1)$-matrices of order n having exactly k 1's in each row and column. Since we are allowed to have 1's on the main diagonal, the class $\mathcal{A}^*(n, k)$ is always nonempty. We use $\overline{\omega}(n, k)$ and $\widetilde{\omega}(n, k)$ in place of $\overline{\omega}(R)$ and $\widetilde{\omega}(R)$, respectively.

The formula for $\overline{\omega}(n, k)$ given in the next theorem is the analog of Theorem 7.3.9 and can be proved in a similar way.

Theorem 7.4.7 *Let k be a nonnegative integer with $0 \leq k \leq n, n > 0$. Then*

$$\overline{\omega}(n, k) = \begin{cases} k, & \text{if } n = k, \\ \lfloor \frac{n}{2} \rfloor, & \text{if } k + 1 \leq n \leq 2k - 1, \\ k + 1, & \text{if } 2k \leq n. \end{cases}$$

$\square$

Related material to that covered in this chapter can be found in [4], [20], and [21].

References

[1] R.A. Brualdi and T.S. Michael, The class of 2-multigraphs with a prescribed degree sequence, *Linear Multilin. Alg.*, **24** (1989), 81–102.

[2] W.K. Chen, *Applied Graph Theory*, North-Holland, Amsterdam, 1976.

[3] V. Chungphaisan, Conditions for sequences to be r-graphic, *Discrete Math.*, **7** (1974), 31–39.

[4] R.B. Eggleton and D.A. Holton, The graph of type $(0, \infty, \infty)$ realizations of a graphic sequence, *Combinatorial Mathematics VI* (Proc. Sixth Austral. Conf., Univ. New England, Armidale, 1978), Lecture Notes in Math. 748, Springer, Berlin, 1978, 41–54.

[5] D.R. Fulkerson, A.J. Hoffman, and M.H. McAndrew, Some properties of graphs with multiple edges, *Canad. J. Math.*, **17** (1965), 166–177.

[6] A.J. Goldman and R.H. Byrd, Minimum-loop realization of degree sequences, *J. Res. Nat. Bur. Stands.*, **87** (1982), 75–78.

[7] S.L. Hakimi, On realizability of a set of integers as degrees of the vertices of a linear graph I, *J. Soc. Indust. Appl. Math.*, **10** (1962), 496–506.

[8] S.L. Hakimi, On realizability of a set of integers as degrees of the vertices of a linear graph II: uniqueness, *J. Soc. Indust. Appl. Math.*, **11** (1963), 135–147.

[9] V. Havel, A remark on the existence of finite graphs (in Czech), *Časopis Pěst. Math.*, **80** (1955), 477–485.

[10] D.J. Kleitman, Minimal number of multiple edges in realization of an incidence sequence without loops, *SIAM J. Appl. Math.*, **18** (1970), 25–28.

[11] D.J. Kleitman and D.L. Wang, Algorithms for constructing graphs and digraphs with given valencies and factors, *Discrete Math.*, **6** (1973), 79–88.

[12] S. Kundu, The k-factor conjecture is true, *Discrete Math.*, **6** (1973), 367–376.

[13] L. Lovász, Valencies of graphs with 1-factors, *Period. Math. Hung.*, **5** (1974), 149-151.

[14] T.S. Michael, The structure matrix and a generalization of Ryser's maximum term rank formula, *Linear Algebra Appl.*, **145** (1991), 21–31.

[15] T.S. Michael, The structure matrix of the class of r-multigraphs with a prescribed degree sequence, *Linear Algebra Appl.*, **183** (1993), 155–177.

[16] A.B. Owens, On determining the minimum number of multiple edges for an incidence sequence, *SIAM J. Appl. Math.*, **18** (1970), 238–240.

[17] A.B. Owens and H.M. Trent, On determining minimal singularities for the realization of an incidence sequence, *SIAM J. Appl. Math.*, **15** (1967), 406–418.

[18] N. Punnim, Degree sequences and chromatic numbers of graphs, *Graphs Combin.*, **18** (2002), 597–603.

[19] N. Punnim, The clique numbers of regular graphs, *Graphs Combin.*, **18** (2002), 781–785.

[20] R. Taylor, Constrained switching in graphs, *Combinatorial Mathematics, VII (Geelong, 1980)*, Lecture Notes in Math., 884, Springer, Berlin–New York, 1981, 314–336.

[21] R. Taylor, Switchings constrained to 2-connectivity in simple graphs, *SIAM J. Algebraic Discrete Methods*, **3** (1982), 114–121.

[22] T. Will, Switching distance between graphs with the same degrees, *SIAM J. Discrete Math.*, **12** (1999), 298–306.

8

Convex Polytopes of Matrices

In this chapter we study continuous analogs of the matrix classes investigated in earlier chapters, namely classes of nonnegative matrices with prescribed row and column sums. Such classes form convex polytopes in a Euclidean space. Extreme points, and more generally facets, of these polytopes can be characterized. In addition, many of the combinatorial parameters investigated earlier make sense for these classes, and the extreme values of these parameters are studied. In the next chapter, we study more thoroughly the special case of the convex polytope of doubly stochastic matrices (all row and column sums equal to 1).

8.1 Transportation Polytopes

Let $R = (r_1, r_2, \ldots, r_m)$ and $S = (s_1, s_2, \ldots, s_n)$ be positive real vectors. Recall from Chapter 2 that $\mathcal{N}(R, S)$ denotes the class of all nonnegative matrices with row sum vector R and column sum vector S, and that $\mathcal{N}(R, S)$ is nonempty if and only if

$$r_1 + r_2 + \cdots + r_m = s_1 + s_2 + \cdots + s_n.$$

In the natural way the class $\mathcal{N}(R, S)$ can be regarded as a subset of Euclidean space $\Re^{mn}$. With this understanding, since $cA + (1 - c)B$ is in $\mathcal{N}(R, S)$ whenever A and B are in $\mathcal{N}(R, S)$ and $0 \le c \le 1$, $\mathcal{N}(R, S)$ is a *convex polytope*, that is, a bounded subset of $\Re^{mn}$ which is the intersection of a finite number of closed half-spaces. Each matrix $A = [a_{ij}]$ in $\mathcal{N}(R, S)$ is the solution of the *transportation problem* in which it is desired to transport an infinitely divisible material from m [18] sources $X_1, X_2, \ldots, X_m$ to n destinations $Y_1, Y_2, \ldots, Y_n$ where r_i is the supply at source X_i and s_j is the demand at destination Y_j. It is for this reason that $\mathcal{N}(R, S)$ is sometimes called a *transportation polytope*.

The *dimension* of a subset X of Euclidean space is the dimension of the smallest linear subspace one of whose translates contains X.

Theorem 8.1.1 *If $\mathcal{N}(R,S)$ is nonempty, then* $\dim \mathcal{N}(R,S) = (m-1)(n-1)$.

Proof. The matrix of coefficients of the system of equations

$$\sum_{j=1}^{n} x_{ij} = 0 \quad (i = 1, 2, \ldots, m),$$

$$\sum_{i=1}^{m} x_{ij} = 0 \quad (j = 1, 2, \ldots, n)$$

has rank equal to $m+n-1$ (since the sum of the first m equations equals the sum of the last n equations), and hence its null space $\mathcal{W}_{m,n}$ has dimension $mn - m - n + 1 = (m-1)(n-1)$. The m by n matrix $A_{R,S} = [a_{ij}]$, where

$$a_{ij} = \frac{r_i s_j}{\sum_{i=1}^{m} r_i} \quad (1 \le i \le m, 1 \le j \le n),$$

is a positive matrix in $\mathcal{N}(R,S)$, from which the conclusion follows.   □

It follows from the proof of Theorem 8.1.1 that $\mathcal{N}(R,S)$ is contained in the translation of the subspace $\mathcal{W}_{m,n}$ by $A_{R,S}$ and hence is contained in the affine space

$$A_{R,S} + \mathcal{W}_{m,n} = \{A_{R,S} + X : X \in \mathcal{W}_{m,n}\}$$

of dimension $mn - m - n + 1$. The faces of $\mathcal{N}(R,S)$ are the intersections of $\mathcal{N}(R,S)$ with affine subspaces of $\Re^{mn}$ (or affine subspaces of $\mathcal{W}_{m,n}$). The 0-dimensional faces of $\mathcal{N}(R,S)$ are the *extreme points* (*vertices*) of $\mathcal{N}(R,S)$; the $(mn - m - n)$-dimensional faces are its *facets*.

Recall that the pattern of a matrix A is the subset $\mathcal{P}(A)$ of its positions which contain a nonzero element, and this subset is often identified with the corresponding $(0,1)$-matrix. Also recall that the bipartite graph associated with an m by n matrix $A = [a_{ij}]$ is the graph $\mathrm{BG}(A)$ with bipartition $U = \{u_1, u_2, \ldots, u_m\}$ and $W = \{w_1, w_2, \ldots, w_n\}$ whose edges are those pairs $\{u_i, w_j\}$ for which $a_{ij} \neq 0$.

Theorem 8.1.2 *Let A be a matrix in $\mathcal{N}(R,S)$, Then the following are equivalent.*

(i) *A is an extreme point of $\mathcal{N}(R,S)$.*

(ii) *The bipartite graph $\mathrm{BG}(A)$ has no cycles, that is, the connected components of $\mathrm{BG}(A)$ are trees.*

(iii) *There is no other matrix in $\mathcal{N}(R, S)$ whose pattern is a subset of the pattern of A, equivalently, whose bipartite graph is contained in that of A.*

Proof. First assume $BG(A)$ has a cycle. Then there is a balanced matrix[1] C whose bipartite graph is contained in $BG(A)$. For ϵ a small positive number $A \pm \epsilon C$ is also in $\mathcal{N}(R, S)$ and

$$A = \frac{1}{2}(A + \epsilon C) + \frac{1}{2}(A - \epsilon C).$$

Here $A \pm \epsilon C$ are distinct matrices in $\mathcal{N}(R, S)$ whose patterns are contained in $\mathcal{P}(A)$, and thus A is not an extreme point. Therefore (i) implies (ii). A similar argument shows that (iii) implies (i): if B is a matrix in $\mathcal{N}(R, S)$ with pattern contained in the pattern of A, replace C with $A - B$ in the preceding argument. Now assume that (ii) holds. Since $BG(A)$ does not have any cycles, it has a pendant edge and the corresponding matrix element is uniquely determined. Now iteratively choosing pendant edges it is easy to see that there is no other matrix in $\mathcal{N}(R, S)$ whose pattern is contained in that of A. Thus (iii) holds and from (iii) we also see that (i) holds. $\square$

Since a graph with k vertices and no cycles has at most $k - 1$ edges, the following corollary is immediate.

Corollary 8.1.3 *The cardinality of the pattern of an extreme point of $\mathcal{N}(R, S)$ is at most $m + n - 1$.* $\square$

¿From the equivalence of (i) and (ii) in Theorem 8.1.2 we obtain the following constructive characterization of extreme points.

Corollary 8.1.4 *The extreme points of the nonempty polytope $\mathcal{N}(R, \dot{S})$ are exactly those matrices that result by carrying out the recursive construction below in all possible ways.*

(i) *Let $R = (r_1, r_2, \ldots, r_m)$ and $S = (s_1, s_2, \ldots, s_n)$ where $r_1 + r_2 + \cdots + r_m = s_1 + s_2 + \cdots + s_n$. Begin with $\widehat{R} = R$ and $\widehat{S} = S$, $K = \{1, 2, \ldots, m\}$ and $L = \{1, 2, \ldots, n\}$, and $A = [a_{ij}]$ equal to the zero matrix of size m by n.*

(ii) *Choose $i \in K$ and $j \in L$. Replace a_{ij} with $\min\{r_i, s_j\}$. If $r_i = \min\{r_i, s_j\}$, then delete i from K; if $s_j = \min\{r_i, s_j\}$, then delete j from L. Replace r_i in $\widehat{R}$ and s_j in $\widehat{S}$ with $r_i - \min\{r_i, s_j\}$ and $s_j - \min\{r_i, s_j\}$, respectively.*

(iii) *Repeat step (ii) until K and L are empty (and $\widehat{R}$ and $\widehat{S}$ are zero vectors).*

---

[1] See Section 3.2.

If R and S are integral vectors, then each extreme point of $\mathcal{N}(R,S)$ is an integral matrix. □

The above results can be found in many places, including [10], [18], [22], [23].

Recall that an m by n matrix A is *decomposable* provided there exist permutation matrices P and Q such that $PAQ = A_1 \oplus A_2$ where A_1 and A_2 each have at least one row or column, and A is *indecomposable* otherwise. The matrix A is indecomposable if and only if the bipartite graph $BG(A)$ is connected. If all rows and columns of A are nonzero, and A is decomposable, the matrices A_1 and A_2 must be nonvacuous. An indecomposable matrix does not have a row or column of all 0's.

In [23] the following results are established concerning the number of extreme points of the polytopes $\mathcal{N}(R,S)$; some of these results are due to Demuth [11] (see also [12]). Let $R = (r_1, r_2, \ldots, r_m)$ and $S = (s_1, s_2, \ldots, s_n)$ be positive vectors such that the polytope $\mathcal{N}(R,S)$ is nonempty. Then $\mathcal{N}(R,S)$ is *degenerate* provided there exist nonempty sets I and J with $I \subset \{1, 2, \ldots, m\}$ and $J \subset \{1, 2, \ldots, n\}$ such that

$$\sum_{i \in I} r_i = \sum_{j \in J} s_j.$$

The polytope $\mathcal{N}(R,S)$ is *nondegenerate* if it is not degenerate. The polytope $\mathcal{N}(R,S)$ is a nondegenerate polytope if and only if for every matrix $A \in \mathcal{N}(R,S)$ the bipartite graph of A is connected, equivalently, if and only if the bipartite graph of each extreme point has $m + n - 1$ (the maximum number) edges, that is, is a tree.

Let $e(R,S)$ be the number of extreme points of $\mathcal{N}(R,S)$. Let

$$\tilde{e}_{mn} = \min\{e(R,S) : \mathcal{N}(R,S) \neq \oslash\}$$

be the *minimum number of extreme points of a nonempty polytope* $\mathcal{N}(R,S)$ where R and S are of sizes m and n, respectively. Let

$$\bar{e}_{mn} = \max\{e(R,S) : \mathcal{N}(R,S) \neq \oslash\}$$

be the *maximum number of extreme points of a nonempty polytope* $\mathcal{N}(R,S)$. Let $\tilde{e}_{mn}^{\mathrm{nd}}(R,S)$ and $\bar{e}_{mn}^{\mathrm{nd}}(R,S)$ be the corresponding extremal values when restricted to nondegenerate polytopes $\mathcal{N}(R,S)$.

Theorem 8.1.5 $\bar{e}_{mn} = \bar{e}_{mn}^{\mathrm{nd}}$.

The numbers $\bar{e}_{mn}$ have been computed for some values of m and n but not in general. Let R and S be vectors of size m and n, respectively, of the form

$$R = (n, n, \ldots, n) \text{ and } S = (m, m, \ldots, m),$$

and let $\mathcal{N}_{m,n} = \mathcal{N}(R,S)$. Then $\mathcal{N}_{m,n}$ is nonempty, and Klee and Witzgall [23] conjectured, and Bolker [3] proved, that if m and n are relatively prime, then the number of vertices of $\mathcal{N}_{mn}$ is the maximum $\bar{e}_{mn}$. A more general result can be found in [34].

Pak [30] considered the polytope $\mathcal{N}_{m,m+1}$ of m by $m+1$ nonnegative matrices with all row sums equal to $m+1$ and all column sums equal to m and gave a recurrence relation for the number of faces of each possible dimension. Since m and $m+1$ are relatively prime they have the maximum number of vertices for transportation polytopes of their size.

In contrast to $\bar{e}_{mn}$, the numbers $\tilde{e}_{mn}$ have been completely determined [34].

Theorem 8.1.6 *If* $m \leq n$, *then*

$$\tilde{e}_{mn} = \frac{n!}{(n-m+1)!} \quad and \quad \tilde{e}_{mn}^{\mathrm{nd}} = n^{m-1}.$$

Example. Let m and n be positive integers with $m \leq n$. Let $R = (1, \ldots, 1, n-m+1)$ and $S = (1, \ldots, 1)$ be vectors of size m and n, respectively. Then $\mathcal{N}(R, S)$ is nonempty, and its extreme points are those m by n $(0, 1)$-matrices with $n-m+1$ 1's in the last row, whose deletion along with the columns containing those 1's leaves a permutation matrix of order $m-1$. A simple counts gives

$$e(R, S) = \binom{n}{n-m+1}(m-1)! = \frac{n!}{(n-m+1)!}.$$

Hence $e(R, S) = \tilde{e}_{mn}$.

Now let $R = (1, \ldots, 1, mn-m+1)$ and $S = (m, m, \ldots, m)$. Since the elements of R and S are integers, it follows from Corollary 8.1.4 that each extreme point of $\mathcal{N}(R, S)$ is an integral matrix. In an extreme point each of the first $m-1$ rows contains exactly one 1, whose placement uniquely determines the remaining elements in the last row, each of which must be positive. The number of positive elements in the extreme point equals $m-1+n$. Hence the bipartite graph of every extreme point is a tree, and $\mathcal{N}(R, S)$ is nondegenerate. Therefore the number of extreme points equals $n^{m-1} = \tilde{e}_{mn}^{\mathrm{nd}}$. $\qquad\square$

We now characterize the patterns of matrices in $\mathcal{N}(R, S)$ [4].

Theorem 8.1.7 *Let* $R = (r_1, r_2, \ldots, r_m)$ *and* $S = (s_1, s_2, \ldots, s_n)$ *be positive vectors with* $\sum_{i=1}^{m} r_i = \sum_{j=1}^{n} s_j$. *Let* W *be an* m *by* n $(0, 1)$-matrix. *There exists a matrix in* $\mathcal{N}(R, S)$ *with pattern equal to* W *if and only if the following* (R, S)-support condition is satisfied.

For each K *with* $\oslash \subset K \subset \{1, 2, \ldots, m\}$ *and each* L *with* $\oslash \subset L \subset \{1, 2, \ldots, n\}$ *such that* $W[K, L] = O$,

$$\sum_{l \in L} s_l \geq \sum_{k \in K} r_k \tag{8.1}$$

with equality if and only if $W(K, L] = O$.

Proof. Let A be a matrix with pattern equal to $W = [w_{ij}]$. Let K and L be as prescribed in the (R, S)-support condition. If $W(K, L] = O$, then clearly equality holds in (8.1). If $W(K, L] \neq O$, then strict inequality holds in (8.1). Hence the (R, S)-support condition is satisfied.

Now assume that the (R, S)-support condition holds. Let the row and column sum vectors of W be $(a_1, a_2, \ldots, a_m)$ and $(b_1, b_2, \ldots, b_n)$, respectively. Let ϵ be a small positive number. If A is a matrix in $\mathcal{N}(R, S)$ with pattern equal to W, then the matrix $A_\epsilon = A - \epsilon W$ is a nonnegative matrix with row sum vector $R_\epsilon = (r_1 - \epsilon a_1, r_2 - \epsilon a_2, \ldots, r_m - \epsilon a_m)$ and column sum vector $S_\epsilon = (s_1 - \epsilon b_1, s_2 - \epsilon b_2, \ldots, s_n - \epsilon b_n)$. Moreover, the pattern of A_ϵ is contained in the pattern W. Thus it suffices to show that the inequalities (8.1) imply that there is a matrix A^* in $\mathcal{N}(R_\epsilon, S_\epsilon)$ with pattern contained in the pattern W; taking $A = A^* + \epsilon W$ will then complete the proof. Applying Corollary 1.4.2 with $c_{ij} = 0$ if $w_{ij} = 0$ and $c_{ij} = \infty$ if $w_{ij} = 1$, we see that such a matrix A_ϵ exists if and only if, whenever $W[K, L) = O$,

$$\sum_{l \in L}(s_l - \epsilon b_l) \geq \sum_{k \in K}(r_k - \epsilon a_k),$$

equivalently

$$\sum_{l \in L} s_l \geq \sum_{k \in K} r_k + \epsilon n_{K,L}, \qquad (8.2)$$

where $n_{K,L}$ equals the number of 1's in $W(K, L]$. It follows from (8.1) that the inequalities (8.2) hold for ϵ small enough.   □

We remark that the patterns of real matrices (negative elements permitted) whose row and column sum vectors are prescribed real vectors R and S, respectively, are characterized in [21]. If in Theorem 8.1.7 we assume that W is indecomposable, then (8.1) reduces to

$$\sum_{l \in L} s_l > \sum_{k \in K} r_k, \quad \text{whenever } W[K, L) = O.$$

Let W be an m by n $(0, 1)$-matrix, and let $\mathcal{N}_W(R, S)$ denote the *class of all nonnegative matrices in $\mathcal{N}(R, S)$ with pattern equal to the $(0, 1)$-matrix W*. In the next theorem we obtain the analog for $\mathcal{N}_W(R, S)$ of the interchange theorem for $\mathcal{A}(R, S)$. Recall that a minimal balanced matrix is a $(0, \pm 1)$-matrix whose associated bipartite digraph, apart from possible isolated vertices, is a directed cycle. If C is a minimal balanced matrix and θ is a positive number, then $A \pm \theta C$ are matrices obtained from A by a *cycle interchange*.

Theorem 8.1.8 *Let A and B be two matrices in $\mathcal{N}_W(R, S)$. Then there exists a sequence of matrices $A_1 = A, A_2, \ldots, A_k = B$ in $\mathcal{N}_W(R, S)$ such that A_{i+1} is obtained from A_i by a cycle interchange for $i = 1, 2, \ldots, k - 1$.*

Proof. The row and column sums of the matrix $A - B$ are equal to zero. Hence there is a minimal balanced matrix C such that the pattern of C is contained in that of $A - B$ with the positions of the 1's (respectively, -1's) of C a subset of the positions of the positive elements (respectively, negative elements) of $A - B$. Hence there is a positive number θ equal to the smallest absolute value of the nonzero elements of C such that $A + \theta C$ or $A - \theta C$ is a matrix in $\mathcal{N}_W(R, S)$ agreeing with B in at least one more position than A does. Arguing by induction, we complete the proof.   □

The class $\mathcal{N}_W(R, S)$ is not in general closed and so is not in general a convex polytope. The closure of this set in $\Re^{mn}$ is $\mathcal{N}_{\leq W}(R, S)$, the *convex polytope of all nonnegative matrices with row sum vector R and column sum vector S with pattern contained in W.*[2] One can also consider $\mathcal{N}_{\leq W}(\leq R, \leq S)$, the *convex polytope of all nonnegative matrices with row and column sum vectors bounded componentwise from above by R and S, respectively, and pattern contained in W.* In case W is the m by n matrix J_{mn} of all 1's, we write $\mathcal{N}(\leq R, \leq S)$ instead of $\mathcal{N}_{\leq J_{mn}}(\leq R, \leq S)$.

We first verify that the closure of $\mathcal{N}_W(R, S)$ equals $\mathcal{N}_{\leq W}(R, S)$. The polytopes $\mathcal{N}_{\leq W}(R, S)$ are the *faces* of $\mathcal{N}(R, S)$. By restricting W so that it satisfies the (R, S)-support condition in Theorem 8.1.7, we obtain a bijection between the nonempty faces of $\mathcal{N}(R, S)$ and the m by n $(0, 1)$-matrices satisfying the (R, S)-support condition. The faces of $\mathcal{N}(R, S)$ are considered in more detail in Section 8.4.

Theorem 8.1.9 *The closure of $\mathcal{N}_W(R, S)$ in $\Re^{mn}$ is $\mathcal{N}_{\leq W}(R, S)$.*

Proof. That the closure is contained in $\mathcal{N}_{\leq W}(R, S)$ is clear. Let A be a matrix in $\mathcal{N}_{\leq W}(R, S)$. Then A is in $\mathcal{N}_{W'}(R, S)$ for some pattern W' with $W' \leq W$. Thus the theorem asserts that the closure of $\mathcal{N}_W(R, S)$ equals

$$\bigcup_{W'} \{ \mathcal{N}_{W'}(R, S) \neq \oslash : W' \leq W \}.$$

Let W' be properly contained in W. It suffices to show that $\mathcal{N}_{W'}(R, S)$ is contained in the closure of $\mathcal{N}_{W''}(R, S)$ for some subset W'' of W. Without loss of generality we assume that W is an indecomposable pattern.

Let A be a matrix in $\mathcal{N}_{W'}(R, S)$ and let B be a matrix in $\mathcal{N}_W(R, S)$. Then $B - A$ is a matrix all of whose line sums equal zero, and its pattern is contained in W. For ϵ sufficiently small, $A + \epsilon(B - A)$ is a matrix in $\mathcal{N}_{W''}(R, S)$ for some subset W'' of W. Hence A is in the closure of $\mathcal{N}_{W''}(R, S)$.   □

The following analog of Theorem 8.1.2 holds.

Theorem 8.1.10 *Let A be a matrix in $\mathcal{N}_{\leq W}(R, S)$. Then the following are equivalent.*

---

[2]Recall that we are identifying the pattern with a $(0,1)$-matrix.

(i) *A is an extreme point of $\mathcal{N}_{\leq W}(R, S)$.*

(ii) *The bipartite graph $BG(A)$ does not have any cycles, that is, the connected components of $BG(A)$ are trees.*

(iii) *There is no other matrix in $\mathcal{N}_{\leq W}(R, S)$ whose pattern is a subset of the pattern of A, that is, whose bipartite graph is contained in that of A.*   □

We now consider the convex polytopes $\mathcal{N}_{\leq W}(\leq R, \leq S)$. Note that $\mathcal{N}_{\leq W}(\leq R, \leq S)$ always contains a zero matrix and so is never empty. In order for a matrix to be an extreme point of this convex polytope it is necessary, but no longer sufficient, that the associated bipartite graph not have any cycles. Let A be a matrix in $\mathcal{N}_{\leq W}(\leq R, \leq S)$. The ith row sum of A is a *deficient row sum* provided it is strictly less than r_i $(1 \leq i \leq m)$; we also refer to a *deficient row*. A *deficient column sum* (respectively, *deficient column*) is defined in a similar way.

Theorem 8.1.11 *Let A be a matrix in the convex polytope $\mathcal{N}_{\leq W}(\leq R, \leq S)$. Then A is an extreme point of $\mathcal{N}_{\leq W}(\leq R, \leq S)$ if and only if the connected components of the bipartite graph $BG(A)$ are trees (so do not have any cycles) at most one of whose vertices corresponds to a deficient row or column.*

Proof. As already remarked the bipartite graph of an extreme point does not contain any cycles and hence its connected components are trees. Let $A = [a_{ij}]$ be a matrix in $\mathcal{N}_{\leq W}(\leq R, \leq S)$ the connected components of whose bipartite graph are trees. If no row or column sum of A is deficient, then A is an extreme point of $\mathcal{N}_W(R, S)$ and hence of $\mathcal{N}_{\leq W}(\leq R, \leq S)$. Now assume that some connected component of $BG(A)$ has at least two deficient row or column sums. We argue the case in which there are both a deficient row sum and a deficient column sum; the other cases of two deficient row sums or two deficient column sums can be argued in a similar way. Without loss of generality we assume that $BG(A)$ is a tree.

Let row sum $r_i' < r_i$ and column sum $c_j' < c_j$ be deficient. Since a tree is a connected graph, there exist positive elements

$$a_{ij_1}, a_{i_2,j_1}, a_{i_2,j_2}, a_{i_3,j_2}, \ldots, a_{i_k,j_{k-1}}, a_{i_k,j} \tag{8.3}$$

where $i, i_2, i_3, \ldots, i_k$ are distinct and $j_1, j_2, \ldots, j_{k-1}, j$ are also distinct. Let ϵ be a positive number less than each of the positive elements (8.3) and less than the positive numbers $r_i - r_i'$ and $s_j - s_j'$. Let P_ϵ be the matrix obtained from A by replacing the elements (8.3) with $\epsilon, -\epsilon, \epsilon, -\epsilon, \ldots, \epsilon$ respectively, and all other elements with 0. Then $A + P_\epsilon$ and $A - P_\epsilon$ are distinct matrices in $\mathcal{N}_{\leq W}(\leq R, \leq S)$ and

$$A = \frac{1}{2}(A + P_\epsilon) + \frac{1}{2}(A - P_\epsilon).$$

Hence A is not an extreme point.

Now assume that each connected component of $\mathrm{BG}(A)$ is a tree with at most one deficient row or column sum. If A is not an extreme point, then A is the convex combination of distinct matrices A_1 and A_2 in $\mathcal{N}_{\leq W}(\leq R, \leq S)$ whose patterns are subsets of the pattern of A and whose only possible deficient row and column sums are those deficient in A. Since each connected component of $\mathrm{BG}(A)$ is a tree, it follows that there is no other matrix in $\mathcal{N}_{\leq W}(\leq R, \leq S)$ whose pattern is a subset of the pattern of A and whose only possible deficient row and column sums are those deficient in A. Hence A is an extreme point of $\mathcal{N}_{\leq W}(\leq R, \leq S)$.   □

Example. Let $R = (2, 5, 3, 2, 6)$ and $S = (3, 4, 2, 4, 3, 2)$. Then

$$\begin{bmatrix} 1 & 0 & 0 & 0 & 0 & 0 \\ 2 & 3 & 0 & 0 & 0 & 0 \\ 0 & 1 & 2 & 0 & 0 & 0 \\ 0 & 0 & 0 & 2 & 0 & 0 \\ 0 & 0 & 0 & 1 & 3 & 2 \end{bmatrix}$$

is an extreme point of $\mathcal{N}_{\leq J_{5,6}}(\leq R, \leq S)$. The bipartite graph of this matrix consists of two trees, one of which has a deficient row sum and the other of which has a deficient column sum.   □

We now show how to get the extreme points of $\mathcal{N}(\leq R, \leq S)$ from the extreme points of $\mathcal{N}(R, S)$ [6].

Theorem 8.1.12 *The extreme points of $\mathcal{N}(\leq R, \leq S)$ are exactly those matrices obtained as follows.*

  (i) *Choose an extreme point A of $\mathcal{N}(R, S)$.*

 (ii) *In each tree which is a connected component of $\mathrm{BG}(A)$, delete a (possibly empty) set of edges of a subtree.*

(iii) *Let A' be the matrix obtained from A by replacing with zero those elements corresponding to the edges deleted in* (ii).

Proof. Let A' be a matrix obtained by the procedure described in the theorem. Since a subtree of each connected component (a tree) of $\mathrm{BG}(A)$ is deleted in (ii), the components of the bipartite graph of A' are trees with at most one deficient row or column sum. Hence by Theorem 8.1.11, A' is an extreme point of $\mathcal{N}(\leq R, \leq S)$. (This direction does not use the assumption that $W = J_{m,n}$ as implied by our notation.)

Now assume that $A' = [a'_{ij}]$ is an extreme point of $\mathcal{N}(\leq R, \leq S)$. By Theorem 8.1.11 each connected component of $\mathrm{BG}(A')$ is a tree with at most one deficient row or column sum. Let $I = \{i_1, i_2, \ldots, i_k\}$ and $J = \{j_1, j_2, \ldots, j_l\}$ be the indices of the deficient row and column sums, respectively. Note that since $\sum_{i=1}^{m} r_i = \sum_{j=1}^{n} s_j$, $I \neq \varnothing$ if and only if $J \neq \varnothing$. If $I = J = \varnothing$, there is nothing to prove. Thus we assume

$I \neq \oslash$ and $J \neq \varnothing$, and thus $A'[I, J] = O$. Let $R^* = (r_{i_1}, r_{i_2}, \ldots, r_{i_k})$ and $S^* = (s_{j_1}, s_{j_2}, \ldots, s_{j_l})$. Then the sum of the elements of R^* equals the sum of the elements of S^*. Hence $\mathcal{N}(R^*, S^*)$ is a nonempty, convex polytope. Let F be an extreme point of this polytope and let A be obtained from A' by replacing the zero matrix $A[I, J]$ with F. The connected components of the bipartite graph $BG(A)$ are trees and A belongs to $\mathcal{N}(R, S)$. Hence A is an extreme point of $\mathcal{N}(R, S)$, and the theorem follows. $\square$

Example. Theorem 8.1.12 does not hold in general for the convex polytopes $\mathcal{N}_{\leq W}(\leq R, \leq S)$. Let

$$W = \begin{bmatrix} 1 & 1 & 1 & 1 \\ 1 & 1 & 0 & 1 \\ 1 & 1 & 1 & 1 \\ 1 & 1 & 1 & 1 \end{bmatrix},$$

and let $R = (2, 2, 1, 2)$ and $S = (2, 1, 2, 2)$. Then

$$A' = \begin{bmatrix} 1 & 1 & 0 & 0 \\ 1 & 0 & 0 & 0 \\ 0 & 0 & 0 & 1 \\ 0 & 0 & 1 & 1 \end{bmatrix}$$

is an extreme point of $\mathcal{N}_{\leq W}(\leq R, \leq S)$ which cannot be obtained as specified in Theorem 8.1.12. $\square$

Let $R = (r_1, r_2, \ldots, r_m)$ and $S = (s_1, s_2, \ldots, s_n)$ be positive vectors with, as before,

$$\tau = r_1 + r_2 + \cdots + r_m = s_1 + s_2 + \cdots + s_n.$$

In analogy with the polytopes $\mathcal{N}(\leq R, \leq S)$ and $\mathcal{N}_{\leq W}(\leq R, \leq S)$ we now consider the (unbounded) convex polyhedra:

(i) $\mathcal{N}(\geq R, \geq S)$ consisting of *all nonnegative matrices whose row and column sum vectors are bounded below componentwise by R and S, respectively;*

(ii) $\mathcal{N}_{\leq W}(\geq R, \geq S)$ consisting of *all nonnegative matrices in $\mathcal{N}(\geq R, \geq S)$ with pattern contained in W;* and

(iii) $\mathcal{N}^+(R, S)$ consisting of *all nonnegative matrices which can be expressed in the form $A + N$ where $A \in \mathcal{N}(R, S)$ and N is a nonnegative matrix.*

We have

$$\mathcal{N}(R, S) \subseteq \mathcal{N}^+(R, S) \subseteq \mathcal{N}(\geq R, \geq S).$$

While it is obvious that $\mathcal{N}(R,S) \neq \mathcal{N}^+(R,S)$, it is not so obvious that, in general, $\mathcal{N}^+(R,S) \neq \mathcal{N}(\geq R, \geq S)$. A characterization of matrices in $\mathcal{N}^+(R,S)$ is given in Theorem 8.1.13 below. The following example illustrates the fact that not every matrix in $\mathcal{N}(\geq R, \geq S)$ can be written in the form $A + N$ for some $A \in \mathcal{N}(R,S)$ and some nonnegative matrix N.

Example. Let $R = S = (3,1,1)$. Then the matrix

$$C = \begin{bmatrix} 0 & 1 & 2 \\ 1 & 0 & 0 \\ 2 & 0 & 0 \end{bmatrix}$$

belongs to $\mathcal{N}(\geq R, \geq S)$. Since there does not exist a matrix in $\mathcal{N}(R,S)$ with zeros in those positions in which C has zeros, we conclude that C is not in $\mathcal{N}^+(R,S)$. $\quad\square$

We now obtain necessary and sufficient conditions for a matrix to belong to $\mathcal{N}^+(R,S)$ [9, 27, 20].

Theorem 8.1.13 *Let $R = (r_1, r_2, \ldots, r_n)$ and $S = (s_1, s_2, \ldots, s_n)$ be positive vectors with $r_1 + r_2 + \cdots + r_m = s_1 + s_2 + \cdots + s_n$. Let $C = [c_{ij}]$ be a nonnegative matrix. Then $C \in \mathcal{N}^+(R,S)$ if and only if*

$$\sum_{i \in I, j \in J} c_{ij} \geq \sum_{i \in I} r_i - \sum_{j \notin J} s_j \ (I \subseteq \{1, 2, \ldots, m\}, J \subseteq \{1, 2, \ldots, n\}).$$

Proof. The theorem is an immediate consequence of Corollary 1.4.2 in Section 1.4. (Note the paragraph preceding this corollary which explains that the result remains valid when the words "nonnegative integral" are replaced with "nonnegative.") $\quad\square$

The following corollaries are also from [27].

Corollary 8.1.14 *A matrix is an extreme point of $\mathcal{N}^+(R,S)$ if and only if it is an extreme point of $\mathcal{N}(R,S)$.*

Proof. Let A be an extreme point of $\mathcal{N}(R,S)$. If $A = cA_1 + (1-c)A_2$ where $0 \leq c \leq 1$ and where $A_1, A_2 \in \mathcal{N}^+(R,S)$, then necessarily $A_1, A_2 \in \mathcal{N}(R,S)$. It follows that A is an extreme point of $\mathcal{N}^+(R,S)$. Now suppose that B is an extreme point of $\mathcal{N}^+(R,S)$ and suppose that $B \notin \mathcal{N}(R,S)$. Then there exists a nonnegative matrix $X \neq O$ such that $B = A + X$ for some $A \in \mathcal{N}(R,S)$. We may write $X = (1/2)X_1 + (1/2)X_2$ for some nonnegative matrices $X_1, X_2 \neq O$. Then

$$B = \frac{1}{2}(B + X_1) + \frac{1}{2}(B + X_2),$$

and we conclude that B is not an extreme point of $\mathcal{N}^+(R,S)$. $\quad\square$

Corollary 8.1.15 *The extreme points of $\mathcal{N}(\geq R, \geq S)$ which are contained in $\mathcal{N}(R, S)$ are precisely the extreme points of $\mathcal{N}(R, S)$.*

Proof. This corollary is a simple consequence of the fact that a matrix $A = [a_{ij}]$ in $\mathcal{N}(\geq R, \geq S)$ belongs to $\mathcal{N}(R, S)$ if and only if $\sum_{i,j} a_{ij} = r_1 + r_2 + \cdots + r_m$. $\square$

Example. Let $R = S = (3, 1, 1)$ and consider again the matrix

$$C = \begin{bmatrix} 0 & 1 & 2 \\ 1 & 0 & 0 \\ 2 & 0 & 0 \end{bmatrix}$$

in $\mathcal{N}(\geq R, \geq S)$. Suppose that $C = cC_1 + (1 - c)C_2$ where $0 < c < 1$ and $C_1, C_2 \in \mathcal{N}(\geq R, \geq S)$. Then both C_1 and C_2 have zeros in those positions in which C has zeros. Since the second row and column sums of C_1 and C_2 must be at least 1 and since the second row and column sums of C equal 1, the elements of both C_1 and C_2 in positions (1,2) and (2,1) must be 1. Continuing like this we see that the elements in both C_1 and C_2 in positions $(1, 3)$ and $(3, 1)$ equal 2. Hence $C = C_1 = C_2$ and we conclude that C is an extreme point of $\mathcal{N}(\geq R, \geq S)$. Since $C \notin \mathcal{N}(R, S)$, C is not an extreme point of $\mathcal{N}(R, S)$. $\square$

Let A be a matrix in $\mathcal{N}_{\leq W}(\geq R, \geq S)$. In order for a matrix to be an extreme point of this convex polyhedron it is necessary, but no longer sufficient, that the associated bipartite graph not have any cycles. The ith row sum of A is an *excessive row sum* (or the ith row is an *excessive row*) provided it is strictly greater than r_i $(1 \leq i \leq m)$. An *excessive column sum* (or *excessive column*) is defined in a similar way. Analogous to Theorem 8.1.11 we have the following characterization of extreme points of $\mathcal{N}_{\leq W}(\geq R, \geq S)$.

Theorem 8.1.16 *Let A be a matrix in the convex polyhedron $\mathcal{N}_{\leq W}(\geq R, \geq S)$. Then A is an extreme point of $\mathcal{N}_{\leq W}(\geq R, \geq S)$ if and only if the connected components of the bipartite graph $BG(A)$ are trees (so do not have any cycles) at most one of whose vertices corresponds to an excessive row or column.* $\square$

8.2   Symmetric Transportation Polytopes

Let $R = (r_1, r_2, \ldots, r_n)$ be a positive real vector. Recall that $\mathcal{N}(R)$ denotes the class of all symmetric nonnegative matrices with row and column sum vector equal to R and that this class is always nonempty. We may regard $\mathcal{N}(R)$ as a subset of Euclidean space $\Re^{n(n+1)/2}$, and with this understanding, $\mathcal{N}(R)$ is a convex polytope. This polytope is investigated in [6] and [14]. The dimension of $\mathcal{N}(R)$ equals $n(n - 1)/2$. This follows from the fact that the row sum constraints are n linearly independent equations in $n(n+1)/2$ variables.

Recall from Chapter 1 that the pattern of a symmetric matrix $A = [a_{ij}]$ of order n is represented by a graph $G(A)$ with vertex set $V = \{v_1, v_2, \ldots, v_n\}$ and an edge joining v_i and v_j if and only if $a_{ij} \neq 0$ $(i, j = 1, 2, \ldots, n)$. Nonzero elements on the main diagonal of the matrix correspond to loops in the graph.

We now turn to the analog of Theorem 8.1.2 for symmetric transportation polytopes. A connected graph is a *near-tree* provided it consists of a cycle with a (possibly trivial) tree rooted at each vertex;[3] these trees are to be pairwise vertex-disjoint. The result of contracting the cycle of a near-tree to a single vertex is a tree. An *odd near-tree* is a near-tree whose cycle has odd length. A graph obtained from a tree by inserting a loop at one of its vertices is an odd near-tree.

The following theorem is from [6], although the equivalence of (i) and (ii) below is given in [8] in a different form. Also, in [26], an equivalent theorem is proved using somewhat different language.

Theorem 8.2.1 *Let A be a matrix in $\mathcal{N}(R)$. Then the following are equivalent.*

(i) *A is an extreme point of $\mathcal{N}(R)$.*

(ii) *The connected components of the graph $G(A)$ are either trees or odd near-trees.*

(iii) *There is no other matrix in $\mathcal{N}(R)$ whose pattern is a subset of the pattern of A, equivalently, whose graph is contained in that of A.*

Proof. First assume that $G(A)$ has a cycle of even length, or two odd length cycles connected by a path. Then there is a nonzero symmetric balanced $(0, 1, -1)$-matrix C whose graph is contained in $G(A)$. Hence for ϵ small enough $A + \epsilon C$ and $A - \epsilon C$ are distinct matrices in $\mathcal{N}(R)$, have the same graph as A, and satisfy

$$A = \frac{1}{2}(A + \epsilon C) + \frac{1}{2}(A - \epsilon C).$$

Thus (i) implies (ii). If (iii) holds, then clearly A is an extreme point, and so (iii) implies (i). Now assume that (ii) holds. Let $B = [b_{ij}]$ be a matrix in $\mathcal{N}(R)$ whose graph is contained in $G(A)$. At a position corresponding to a pendant edge of $G(A)$, B must agree with A. By recursively considering pendant edges, we conclude that B agrees with A at all positions except possibly those corresponding to the edges of the odd cycle of each odd near-tree of $G(A)$. Consider a symmetric matrix $C = [c_{ij}]$ of odd order k with row sum vector $R' = (r'_1, r'_2, \ldots, r'_k)$ whose graph is a cycle of odd length k.

---

[3]In other language, a near-tree is a connected graph with cyclotomic number equal to 1.

The matrix of coefficients for the resulting system of linear equations can be taken to be the matrix of order k of the form

$$\begin{bmatrix} 1 & 0 & 0 & \ldots & 0 & 1 \\ 1 & 1 & 0 & \ldots & 0 & 0 \\ 0 & 1 & 1 & \ldots & 0 & 0 \\ \vdots & \vdots & \vdots & \ddots & \vdots & \vdots \\ 0 & 0 & 0 & \ldots & 1 & 0 \\ 0 & 0 & 0 & \ldots & 1 & 1 \end{bmatrix}.$$

Since k is odd, this matrix is nonsingular, and hence the c_{ij} are uniquely determined. We conclude that no other matrix in $\mathcal{N}(R)$ has a pattern which is contained in A, and therefore (iii) holds.   □

The number of edges of a near-tree of order k is k. Hence the cardinality of the pattern of a symmetric matrix whose graph is a near-tree is $2k-1$ or $2k$, depending on whether the cycle of the near-tree has length equal to or greater than 1. Thus, since a tree of order k has $k-1$ edges and the corresponding pattern has cardinality $2(k-1)$, Theorem 8.2.1 implies the following.

Corollary 8.2.2 *The cardinality of the pattern of an extreme point of* $\mathcal{N}(R)$ *is at most* $2n$.   □

The number of extreme points of symmetric transportation polytopes is investigated in [26].

In Theorem 8.2.1 we have characterized the patterns of the extreme points of $\mathcal{N}(R)$. We now characterize the patterns of matrices in $\mathcal{N}(R)$. We note that this is the same as characterizing matrices in $\mathcal{N}(R,R)$ with a symmetric pattern. This is because $A \in \mathcal{N}(R,R)$ implies that $(A+A^{\mathrm{T}})/2 \in \mathcal{N}(R)$; if A has a symmetric pattern, then $(A+A^{\mathrm{T}})/2$ has the same pattern as A. Thus we may make use of Theorem 8.1.7 in describing the patterns of matrices in $\mathcal{N}(R)$.

Theorem 8.2.3 *Let* $R = (r_1, r_2, \ldots, r_n)$ *be a positive vector. Let* W *be a symmetric* $(0,1)$-*matrix of order* n. *There exists a matrix in* $\mathcal{N}(R)$ *with pattern equal to* W *if and only if the following symmetric* R-*support condition is satisfied.*

For all partitions I, J, H *of* $\{1, 2, \ldots, n\}$ *such that* $W[J \cup H, H] = O$,

$$\sum_{i \in I} r_i \geq \sum_{h \in H} r_h.$$

with equality if and only if $W[I, I \cup J] = O$.

Proof. Symbolically, the symmetric R-support condition in the theorem asserts that whenever the rows and columns of W can be simultaneously permuted to give

$$\begin{bmatrix} W[I,I] & W[I,J] & W[I,H] \\ W[J,I] & W[J,J] & O \\ W[H,I] & O & O \end{bmatrix},$$

where, by symmetry,

$$W[J, I] = W[I, J]^{\mathrm{T}} \text{ and } W[H, I] = W[I, H]^{\mathrm{T}},$$

then $\sum_{i \in I} r_i \geq \sum_{h \in H} r_h$ with equality if and only if $W[I, I]$ and $W[I, J]$ are zero matrices.

The symmetric R-support condition is easily seen to be a necessary condition for $\mathcal{N}(R)$ to contain a matrix with pattern equal to W. Now assume that the symmetric R-support condition holds. As already noted, we need only establish the existence of a matrix in $\mathcal{N}(R, R)$ with pattern equal to W. By Theorem 8.1.7 such a matrix exists if and only if

$$\sum_{l \in L} r_l \geq \sum_{k \in K} r_k$$

whenever $\oslash \subset K, L \subseteq \{1, 2, \ldots, n\}$ and $W[K, L] = O$, with equality if and only if $W(K, L) = O$.

Consider such a K and L with $W[K, L] = O$. Then I, J, H are a partition of $\{1, 2, \ldots, n\}$ where

$$I = L \setminus K, \ J = K \setminus L, \text{ and } H = \{1, 2, \ldots, n\} \setminus (I \cup J).$$

We first verify that $W[J \cup H, J] = O$. Since $J \subset K$ and $J \cap L = \emptyset$, we have $W[J, J] = O$. Next we observe that $H = (\{1, 2, \ldots, n\} \setminus (K \cup L)) \cup (K \cap L)$. Then $W[K \cap L, K \setminus L] = O$, since $K \cap L \subseteq K$ and $(K \setminus L) \cap L = \emptyset$. Using the assumption that W is symmetric, we have $W[\{1, 2, \ldots, n\} \setminus (K \cup L), J]^{\mathrm{T}} = H[J, \{1, 2, \ldots, n\} \setminus (K \cup L)] = O$, since $J \subseteq K$ and $(\{1, 2, \ldots, n\} \setminus (K \cup L)) \cap L = \emptyset$. We next observe that $\sum_{l \in L} r_l \geq \sum_{k \in K} r_k$ if and only if $\sum_{i \in I} r_i \geq \sum_{j \in J} r_j$ with equality holding in the first inequality if and only if it holds in the second. A verification similar to the one just completed additionally shows that if $W(K, L) = O$, then $W[I, I \cup H] = O$.

In the other direction, let I, J, H be a partition of $\{1, 2, \ldots, n\}$ with $W[J \cup H, J] = O$. Define $K = J \cup H$ and $L = I \cup H$. Then $W[K, L] = O$. Moreover, $\sum_{i \in I} r_i \geq \sum_{j \in J} r_j$ if and only if $\sum_{l \in L} r_l \geq \sum_{k \in K} r_k$ with equality in one if and only if equality in the other. We also have $W[I, I \cup H] = O$ if and only if $W(K, L) = O$. The theorem now follows. $\square$

We remark that the patterns of real, symmetric matrices whose row sum vector is a prescribed vector R are characterized in [14].

As a special case of Theorem 8.2.3 we obtain a criterion for the existence of a matrix in $\mathcal{N}(R)$ with a prescribed main diagonal.

Corollary 8.2.4 *Let $d_1, d_2, \ldots, d_n$ be nonnegative numbers with $d_i \leq r_i$ for $i = 1, 2, \ldots, n$. There exists a symmetric, nonnegative matrix with row sum vector $R = (r_1, r_2, \ldots, r_n)$ and main diagonal $(d_1, d_2, \ldots, d_n)$ if and only if*

$$\sum_{i=1}^{n} r_i - \sum_{i=1}^{n} d_i \geq 2 \max\{r_j - d_j : 1 \leq j \leq n\}. \tag{8.4}$$

*In particular, there exists a symmetric, nonnegative matrix with row sum
vector R and zero trace, if and only if*

$$\sum_{i=1}^{n} r_i \geq 2 \max\{r_j : 1 \leq j \leq n\}. \tag{8.5}$$

Proof. Let $A = [a_{ij}]$ be a matrix as specified in the theorem. Then for
each $k = 1, 2, \ldots, n$,

$$\sum_{i=1}^{n} r_i - \sum_{i=1}^{n} d_i = \sum_{i=1}^{n} \sum_{j=1}^{n} a_{ij} - \sum_{i=1}^{n} a_{ii}$$

$$\geq \sum_{i \neq j} a_{ij}$$

$$\geq \sum_{\{i : i \neq k\}} (a_{ik} + a_{ki})$$

$$= 2(r_k - d_k).$$

Hence (8.4) holds.

Now assume that (8.4) holds. Let $R' = (r'_1, r'_2, \ldots, r'_n)$ where $r'_i = r_i - d_i$
for $i = 1, 2, \ldots, n$. Then R' is nonnegative and there is a matrix as specified
in the theorem if and only if there is a matrix in $\mathcal{N}(R')$ with 0's on its main
diagonal. First assume that $r'_i > 0$ for $i = 1, 2, \ldots, n$. Let $W = J_n - I_n$.
According to Theorem 8.2.3 there is a matrix in $\mathcal{N}(R')$ with pattern equal
to W provided for each $k = 1, 2, \ldots, n$,

$$\sum_{i \neq k} r'_i \geq r_k,$$

equivalently,

$$\sum_{i=1}^{n} r'_i \geq 2r'_k.$$

This latter inequality is equivalent to (8.4), and hence the specified matrix
exists. If some of the r'_i are 0, then the desired matrix has certain zero
rows and columns. After deleting these rows and columns, we may apply
the above argument.   □

For the special case of trace equal to zero in Corollary 8.2.4, see also
Theorem 2.3.1.

Let W be a symmetric $(0,1)$-matrix of order n, and let $\mathcal{N}_W(R)$ denote the class of all symmetric, nonnegative matrices in $\mathcal{N}(R)$ with pattern
equal to W. The class $\mathcal{N}_W(R)$ is not in general closed and so is not a
convex polytope. The closure of this polytope in $\Re^{n(n+1)/2}$ is the convex
polytope $\mathcal{N}_{\leq W}(R)$ of all symmetric, nonnegative matrices of order n with
row sum vector R and with pattern contained in W; this fact will be verified in the next theorem. In addition to the polytope $\mathcal{N}_{\leq W}(R)$ we shall

also consider the convex polytope $\mathcal{N}_{\leq W}(\leq R)$ of all symmetric, nonnegative matrices with row sum vector bounded componentwise by R and with pattern contained in W. If W is the all 1's matrix J_n of order n, then we write $\mathcal{N}(\leq R)$ instead of $\mathcal{N}_{\leq J_n}(\leq R)$. The polytopes $\mathcal{N}_{\leq W}(R)$ are the facets of $\mathcal{N}(R)$. There is a bijection between the facets of $\mathcal{N}(R)$ and the $(0,1)$-matrices of order n which satisfy the symmetric R-support condition of Theorem 8.2.3.

Recall that a symmetric matrix A of order n is *symmetrically decomposable* provided there exists a permutation matrix P such that $PAP^{\mathrm{T}} = A_1 \oplus A_2$ where A_1 and A_2 are each matrices of order at least 1; if A is not symmetrically decomposable then A is *symmetrically indecomposable*. The matrix A is symmetrically indecomposable if and only if its graph $G(A)$ is connected.

Theorem 8.2.5 *The closure of $\mathcal{N}_W(R)$ in $\Re^{n(n+1)/2}$ is $\mathcal{N}_{\leq W}(R)$.*

Proof. The theorem asserts that the closure of $\mathcal{N}_W(R)$ equals

$$\bigcup \{\mathcal{N}_{W'}(R) \neq \oslash : W' \leq W\}.$$

The closure of $\mathcal{N}_W(R)$ is clearly contained in $\mathcal{N}_{\leq W}(R)$. Thus we need to prove that every matrix in $\mathcal{N}_{\leq W}(R)$ is in the closure of $\mathcal{N}_W(R)$. Without loss of generality we assume that W is symmetrically indecomposable, that is, the graph $G(W)$ is connected.

Let W' be a proper subset of W such that $\mathcal{N}_{W'}(R)$ is nonempty. It suffices to show that $\mathcal{N}_{W'}(R)$ is contained in the closure of $\mathcal{N}_{W''}(R)$ for some symmetric subset W'' of W. Let A be a matrix in $\mathcal{N}_{W'}(R)$ and let B be a matrix in $\mathcal{N}_W(R)$. Then $B - A$ is a symmetric matrix with all row sums equal to zero, and the pattern of $B - A$ is contained in W. For ϵ sufficiently small, $A + \epsilon(B - A)$ is a matrix in $\mathcal{N}_{W''}(R)$ for some subset W'' of W. Hence A is in the closure of $\mathcal{N}_{W''}(R)$. □

The following analog of Theorem 8.2.1 holds.

Theorem 8.2.6 *Let A be a matrix in $\mathcal{N}_{\leq W}(R)$, Then the following are equivalent.*

(i) *A is an extreme point of $\mathcal{N}_{\leq W}(R)$.*

(ii) *The connected components of the graph $G(A)$ are either trees or odd near-trees.*

(iii) *There is no other matrix in $\mathcal{N}_{\leq W}(R)$ whose pattern is a subset of the pattern of A, equivalently, whose graph is contained in that of A.*

□

We now consider the convex polytopes $\mathcal{N}_{\leq W}(\leq R)$. The connected components of the graphs of the extreme points of these polytopes are necessarily trees or odd near-trees. As in the nonsymmetric case, the ith row of a matrix in $\mathcal{N}_{\leq W}(\leq R)$ is a *deficient row* provided its sum is strictly less than r_i.

Theorem 8.2.7 *Let A be a matrix in $\mathcal{N}_{\leq W}(\leq R)$. Then A is an extreme point of $\mathcal{N}_{\leq W}(\leq R)$ if and only if the connected components of the graph $G(A)$ are trees at most one of whose vertices corresponds to a deficient row, or odd near-trees none of whose vertices corresponds to a deficient row.*

Proof. Consider a connected component of $G(A)$ which is a tree with at least two of its vertices corresponding to deficient rows. As in the proof of Theorem 8.1.11, a path between two vertices corresponding to deficient rows gives a symmetric matrix P_ϵ such that $A + P_\epsilon$ and $A - P_\epsilon$ are distinct matrices in $\mathcal{N}_{\leq W}(\leq R)$ for ϵ small enough. Since

$$A = \frac{1}{2}(A + P_\epsilon) + \frac{1}{2}(A - P_\epsilon),$$

A is not an extreme point.

Now consider a connected component which is an odd near-tree having at least one vertex corresponding to a deficient row. First assume this vertex is a vertex of the cycle $(v_{i_1}, v_{i_2}, \ldots, v_{i_k}, v_{i_1})$ of odd length k of the near-tree. Let P_ϵ be the symmetric matrix which has ϵ in the (i_1, i_2) and (i_2, i_1) positions, $-\epsilon$ in the (i_2, i_3) and (i_3, i_2) positions, $\ldots$, ϵ in the (i_k, i_1) and (i_1, i_k) positions, and 0's elsewhere. Then for ϵ small enough $A + P_\epsilon$ and $A - P_\epsilon$ are distinct matrices in $\mathcal{N}_{\leq W}(\leq R)$ with

$$A = \frac{1}{2}(A + P_\epsilon) + \frac{1}{2}(A - P_\epsilon).$$

Hence A is not an extreme point. Now assume that the vertex corresponding to a deficient row is a vertex w of one of the trees rooted at the odd length cycle as designated above, but not the root itself. There is a path from this vertex w to a vertex, say vertex v_{i_1}, of the cycle. Starting at vertex v_{i_1}, we alternate $\pm\epsilon$ at positions corresponding to the edges of the cycle until we return to vertex v_{i_1} and then alternate $\mp 2\epsilon$ at the positions corresponding to the edges of the path from v_{i_1} to w. We obtain a matrix Q_ϵ such that for ϵ small enough, $A + Q_\epsilon$ and $A - Q_\epsilon$ are distinct matrices in $\mathcal{N}_{\leq W}(\leq R)$ with

$$A = \frac{1}{2}(A + Q_\epsilon) + \frac{1}{2}(A - Q_\epsilon).$$

Hence A is not an extreme point.

Now assume that the connected components of the graph $G(A)$ are trees at most one of whose vertices corresponds to a deficient row, or odd near-trees none of whose vertices corresponds to a deficient row. That such a

matrix A is an extreme point of $\mathcal{N}_{\leq W}(A)$ follows from previous arguments.
A component which is an odd near-tree with no vertices corresponding to
deficient rows was shown to be uniquely determined by its pattern in the
proof of Theorem 8.2.1. The proof that a component which is a tree with
at most one vertex corresponding to a deficient row is uniquely determined
by its pattern is similar to that given in the nonsymmetric case in the proof
of Theorem 8.1.11.   □

Finally, we obtain the analog of Theorem 8.1.12 for the polytope $\mathcal{N}(\leq R)$.

Theorem 8.2.8 *The extreme points of $\mathcal{N}(\leq R)$ are exactly those matrices
obtained as follows.*

(i) *Choose an extreme point A of $\mathcal{N}(R)$.*

(ii) *In each tree which is a connected component of $G(A)$, delete a (possibly empty) set of edges of a subtree.*

(iii) *In each odd near-tree which is a connected component of $G(A)$, delete
all or none of its edges, unless the cycle of the near-tree is a loop, in
which case just the loop may be deleted.*

(iv) *Let A' be the symmetric matrix obtained from A by replacing with 0
those elements corresponding to the edges deleted in (ii) and (iii).*

Proof. It is easily checked that a matrix A' obtained by the procedure
described in the theorem satisfies the conditions for an extreme point of
$\mathcal{N}(\leq R)$ as given in Theorem 8.2.7. Conversely, let the matrix A' be obtained from an extreme point of $\mathcal{N}(R)$ according to the procedure in the
theorem. Consider a component of $G(A)$ which is a tree T with exactly one
vertex corresponding to a deficient row, say vertex v_i corresponding to row
i with row sum $r_i' < r_i$. Then we attach a loop to vertex v_i and replace
the 0 in position (i,i) of A' with $r_i - r_i'$. We do this for each such tree
T. If in $G(A')$ there is a nonempty set of isolated vertices $v_{i_1}, v_{i_2}, \ldots, v_{i_k}$,
then with $K = \{i_1, i_2, \ldots, i_k\}$ the principal submatrix $A'[K]$ of A' equals
O. We then replace this zero matrix by any extreme point of $\mathcal{N}(R^*)$ where
$R^* = (r_{i_1}, r_{i_2}, \ldots, r_{i_k})$. Let A be the matrix resulting from all these replacements. Then by Theorem 8.2.1 A is an extreme point of $\mathcal{N}(R)$, and
A' can be obtained from A as prescribed in the theorem.   □

We conclude this section with an example illustrating Theorem 8.2.8.

Example. Let $R = (3, 1, 1, 1, 3, 1, 1)$ and let A be the matrix in $\mathcal{N}(R)$

defined by

$$A = \begin{bmatrix} 1 & 1 & 1 & 0 & 0 & 0 & 0 \\ 1 & 0 & 0 & 0 & 0 & 0 & 0 \\ 1 & 0 & 0 & 0 & 0 & 0 & 0 \\ 0 & 0 & 0 & 0 & 1 & 0 & 0 \\ 0 & 0 & 0 & 1 & 0 & 1 & 1 \\ 0 & 0 & 0 & 0 & 1 & 0 & 0 \\ 0 & 0 & 0 & 0 & 1 & 0 & 0 \end{bmatrix}.$$

Then A is an extreme point of $\mathcal{N}(R)$ whose graph has two components, an odd near-tree with a cycle of length 1 and a tree. If we delete the edge of the cycle of length 1 of the near-tree and a subtree of the tree consisting of a path of length 2, we obtain the following extreme point of $\mathcal{N}(\leq R)$:

$$A' = \begin{bmatrix} 0 & 1 & 1 & 0 & 0 & 0 & 0 \\ 1 & 0 & 0 & 0 & 0 & 0 & 0 \\ 1 & 0 & 0 & 0 & 0 & 0 & 0 \\ 0 & 0 & 0 & 0 & 0 & 0 & 0 \\ 0 & 0 & 0 & 0 & 0 & 1 & 0 \\ 0 & 0 & 0 & 0 & 1 & 0 & 0 \\ 0 & 0 & 0 & 0 & 0 & 0 & 0 \end{bmatrix}.$$

$\square$

In the next section we investigate the term rank of matrices in $\mathcal{N}(R, S)$ and its symmetric analog for matrices in $\mathcal{N}(R)$.

8.3   Term Rank and Permanent

Recall that the *term rank* $\rho(A)$ of a nonnegative matrix A equals the maximal cardinality of a set of positive elements of A with no two of these positive elements from the same row or column of A. By Theorem 1.2.1 we know that $\rho(A)$ is equal to the minimal number of rows and columns of A which contain all the positive elements of A.

Let $R = (r_1, r_2, \ldots, r_m)$ and $S = (s_1, s_2, \ldots, s_n)$ again be positive vectors with

$$r_1 + r_2 + \cdots + r_m = s_1 + s_2 + \cdots + s_n.$$

Thus $\mathcal{N}(R, S)$ is nonempty, and we define

$$\tilde{\rho}(\mathcal{N}(R, S)) = \min\{\rho(A) : A \in \mathcal{N}(R, S)\}$$

to be the *minimal term rank* of matrices in $\mathcal{N}(R, S)$ and

$$\bar{\rho}(\mathcal{N}(R, S)) = \max\{\rho(A) : A \in \mathcal{N}(R, S)\}$$

to be the *maximal term rank* of matrices in $\mathcal{N}(R, S)$. Since R and S are positive vectors, $\mathcal{N}(R, S)$ contains a positive matrix and hence

$$\bar{\rho}(\mathcal{N}(R, S)) = \min\{m, n\}.$$

Jurkat and Ryser [22] obtained a formula for the minimal term rank. An alternative proof was given in [4] using Theorem 8.1.7. It is this proof that we give here. In the theorem we assume without loss of generality that R is nondecreasing and that S is nonincreasing.

Theorem 8.3.1 *Let $R = (r_1, r_2, \ldots, r_m)$ and $S = (s_1, s_2, \ldots, s_n)$ be positive vectors such that $\mathcal{N}(R, S) \neq \varnothing$. Assume that $r_1 \leq r_2 \leq \cdots \leq r_m$ and that $s_1 \geq s_2 \geq \cdots \geq s_n$. Then*

$$\tilde{\rho}(\mathcal{N}(R, S)) = \min\{m - e + f\} \tag{8.6}$$

where the minimum is taken over all pairs e, f such that

$$s_1 + \cdots + s_f \geq r_1 + \cdots + r_e \ (0 \leq e \leq m, \ 0 \leq f \leq n). \tag{8.7}$$

Proof. Consider an e and f for which equality holds in (8.7). Let

$$W = \begin{bmatrix} J_{e,f} & O \\ X & J_{m-e,n-f} \end{bmatrix}$$

where $X = J$ or $X = O$ depending on whether or not strict inequality occurs in the first inequality in (8.7). It follows from Theorem 8.1.7 that there exists a matrix A in $\mathcal{N}(R, S)$ with pattern equal to W. Since the positive elements of A are contained in $m - e$ rows and f columns, $\rho(A) \leq m - e + f$. Hence $\tilde{\rho}(\mathcal{N}(R, S)) \leq \min\{m - e + f\}$ where the minimum is taken over all pairs e, f satisfying (8.7).

Now let A be an arbitrary matrix in $\mathcal{N}(R, S)$ with rows $i_{e+1}, i_{e+2}, \ldots, i_m$ and columns $j_1, j_2, \cdots, j_f$ containing all the positive elements of A.[4] Let $\{i_1, i_2, \ldots, i_e\} = \{1, 2, \ldots, m\} \setminus \{i_{e+1}, i_{e+2}, \ldots, i_m\}$. There exist permutation matrices P and Q such that

$$PAQ = \begin{bmatrix} A_1 & O \\ A_{21} & A_2 \end{bmatrix}$$

where A_1 is the e by f matrix $A[\{i_1, i_2, \ldots, i_e\}, \{j_1, j_2, \ldots, j_f\}]$. Clearly,

$$s_{j_1} + \cdots + s_{j_f} \geq r_{i_1} + \cdots + r_{i_e},$$

and hence by the monotonicity assumptions $s_1 + \cdots + s_f \geq r_1 + \cdots + r_e$. Therefore, $\rho(A)$ is at least as large as the minimum in (8.6) and hence (8.6) holds. □

We now assume that $m = n$ and investigate the permanent of matrices in a class $\mathcal{N}(R, S)$. We assume that $R = (r_1, r_2, \ldots, r_n)$ and $S = (s_1, s_2, \ldots, s_n)$ are positive, and without loss of generality we now assume that R and S are nondecreasing. Let

$$\widetilde{\mathrm{per}}(\mathcal{N}(R, S)) = \min\{\mathrm{per} \ A : A \in \mathcal{N}(R, S)\}$$

---

[4]If either $e = m$ or $f = 0$, then the corresponding sequences are taken to be empty.

and

$$\overline{\text{per}}(\mathcal{N}(R,S)) = \max\{\text{per } A : A \in \mathcal{N}(R,S)\}$$

denote the *minimal* and *maximal permanents*, respectively, of matrices in $\mathcal{N}(R,S)$

If $\tilde{\rho}(\mathcal{N}(R,S)) < n$, then $\widetilde{\text{per}}(\mathcal{N}(R,S)) = 0$. Thus we now assume that $\tilde{\rho}(\mathcal{N}(R,S)) = n$. Then every matrix in $\mathcal{N}(R,S)$ has a positive permanent. Since $\mathcal{N}(R,S)$ is a compact set and the permanent is a continuous function, $\widetilde{\text{per}}(\mathcal{N}(R,S)) > 0$. It is a very difficult problem to compute $\widetilde{\text{per}}(\mathcal{N}(R,S))$ even in special cases.[5] We now turn to the maximal permanent of matrices in $\mathcal{N}(R,S)$ and its evaluation by Jurkat and Ryser [22]. Since there is a positive matrix in $\mathcal{N}(R,S)$ and $\mathcal{N}(R,S)$ is compact, this maximum is attained and is positive.

Theorem 8.3.2 *Let $R = (r_1, r_2, \ldots, r_n)$ and $S = (s_1, s_2, \ldots, s_n)$ be positive vectors with $r_1 \leq r_2 \leq \cdots \leq r_n$ and $s_1 \leq s_2 \leq \cdots \leq s_n$. Then*

$$\overline{\text{per}}(\mathcal{N}(R,S)) = \prod_{i=1}^{n} \min\{r_i, s_i\}. \tag{8.8}$$

Proof. We may use our recursive construction for matrices in $\mathcal{N}(R,S)$ (see Section 2.1) and construct a matrix $A = [a_{ij}]$ in $\mathcal{N}(R,S)$ with $a_{ii} = \min\{r_i, s_i\}$ for $i = 1, 2, \ldots, n$. By the Laplace expansion for the permanent, per $A = \prod_{i=1}^{n} \min\{r_i, s_i\}$. It remains to show that the permanent of each matrix in $\mathcal{N}(R,S)$ does not exceed $\prod_{i=1}^{n} \min\{r_i, s_i\}$

We first note that a straightforward induction [22] establishes that

$$\prod_{i=1}^{n} \min\{r_i, s_i\} \leq \prod_{i=1}^{n} \min\{r_i, s_{j_i}\}$$

holds for each permutation $j_1, j_2, \ldots, j_n$ of $\{1, 2, \ldots, n\}$. We proceed by induction and note that the case $n = 1$ is trivial. Assume that $n \geq 1$ and for definiteness, that $r_1 \leq s_1$. Let $A = [a_{ij}]$ be a matrix in $\mathcal{N}(R,S)$. Then

$$\text{per } A = \sum_{j=1}^{n} a_{1j} \text{ per } A(1,j)$$

where $A(1,j)$ is the matrix obtained from A by striking out row 1 and column j. Let $r_2(j), r_3(j), \ldots, r_n(j)$ and $s_1(j), \ldots, s_{j-1}(j), s_{j+1}(j) \ldots, s_n(j)$ be, respectively, the row and column sums of $A(1,j)$. Let $i_2, i_3, \ldots, i_n$ be a permutation of $\{2, 3, \ldots, n\}$ for which

$$r_{i_1}(j) \leq r_{i_2}(j) \leq \cdots \leq r_{i_{n-1}}(j).$$

---

[5]For instance, if R and S are vectors of size n consisting of all 1's, then $\mathcal{N}(R,S)$ is the polytope Ω_n of doubly stochastic matrices of order n. That the minimal permanent of matrices in Ω_n equals $n!/n^n$ was the van der Waerden conjecture, now theorem; see Chapter 9.

Let $k_1, k_2, \ldots, k_{n-1}$ be a permutation of $\{1, \ldots, j-1, j+1, \ldots, n\}$ such that

$$s_{k_1}(j) \leq s_{k_2}(j) \leq \cdots \leq s_{k_{n-1}}(j).$$

Then for each j we have

$$r_{i_t}(j) \leq r_{i_t} \text{ and } s_{k_t}(j) \leq s_{k_t} \ (t = 1, 2, \ldots, n-1).$$

Hence by the inductive hypothesis,

$$\text{per } A(1, j) \leq \prod_{t=1}^{n-1} \min\{r_{i_t}, s_{k_t}\} \ (j = 1, 2, \ldots, n).$$

By calculation we obtain

$$
\begin{aligned}
\text{per } A &= \sum_{j=1}^{n} a_{1j} \, \text{per } A(1, j) \\
&\leq \left(\sum_{j=1}^{n} a_{1j} \right) \prod_{t=1}^{n} \min\{r_{i_t}, s_{k_t}\} \\
&= r_1 \prod_{t=1}^{n} \min\{r_{i_t}, s_{k_t}\} \\
&\leq r_1 \prod_{i=2}^{j} \min\{r_i, s_{i-1}\} \prod_{i=j+1}^{n} \min\{r_i, s_i\} \\
&\leq r_1 \prod_{i=2}^{j} \min\{r_i, s_i\} \prod_{i=j+1}^{n} \min\{r_i, s_i\} \\
&= r_1 \prod_{i=2}^{n} \min\{r_i, s_i\}.
\end{aligned}
$$

Since $r_1 \leq s_1$, the proof is complete. $\square$

We once again turn our attention to the class $\mathcal{N}(R)$ of symmetric non-negative matrices with row sum vector R. Let A be a symmetric matrix. The *symmetric term rank* $\rho_s(A)$ is the maximal cardinality of a symmetric set of nonzero elements of A with no two of these elements from the same row or column. We have $\rho_s(A) \leq \rho(A)$ with strict inequality possible.

Example. Consider the symmetric matrix

$$
\begin{bmatrix}
1 & 0 & 0 & 0 \\
0 & 0 & 1 & 1 \\
0 & 1 & 0 & 1 \\
0 & 1 & 1 & 0
\end{bmatrix}.
$$

Then $\rho_s(A) = 3$ (for instance, take the 1's in the symmetric set of positions $\{(1,1),(2,3),(3,2)\}$) but $\rho(A) = 4$.   □

In the graph $G(A)$ a symmetric set of nonzero positions with no two from the same row or column corresponds to a collection of pairwise vertex-disjoint loops and edges.

Let $R = (r_1, r_2, \ldots, r_n)$ be a positive vector. Then $\mathcal{N}(R)$ is nonempty, and we define

$$\tilde{\rho}_s(\mathcal{N}(R)) = \min\{\rho_s(A) : A \in \mathcal{N}(R)\}$$

to be the *minimal symmetric term rank* of matrices in $\mathcal{N}(R)$ and

$$\bar{\rho}_s(\mathcal{N}(R)) = \max\{\rho_s(A) : A \in \mathcal{N}(R)\}$$

to be the *maximal symmetric term rank* of matrices in $\mathcal{N}(R)$. Since R is a positive vector, $\mathcal{N}(R)$ contains a positive matrix (and a diagonal matrix with only positive elements on the main diagonal), and hence

$$\bar{\rho}_s(\mathcal{N}(R)) = n.$$

A formula for the minimal symmetric term rank of matrices in $\mathcal{N}(R)$ is given in [6]. To obtain this formula we need to use a variation of the Tutte–Berge theorem on matchings in graphs as quoted in [6] and proved in [7]. Let G be a graph that may have loops. Recall that a *matching* in G is a set of pairwise vertex-disjoint edges of G. The *span of a matching* is the set of vertices of its edges. A matching of k edges none of which is a loop has $2k$ vertices in its span. A matching of k edges of which k_1 are loops has $k_1 + 2(k - k_1) = 2k - k_1$ vertices in its span. Let $m(G)$ denote the *maximal number of vertices of a matching* not containing any loops. If loops are allowed, then we use $m_q(G)$ to denote the *maximal number of vertices in a matching*. If A is a symmetric matrix and G is the graph of A, then $\rho_s(A) = m_q(G)$.

Let G be a graph with a vertex set V of size n. Let X be a subset of V and let $G(V \setminus X)$ be the subgraph of G induced on $V \setminus X$. Thus $G(V \setminus X)$ is obtained from G by removing all vertices of X and all edges at least one of whose vertices belongs to X. Let $C_G(X)$ denote the set of connected components of $G(V \setminus X)$ which have an odd number of vertices, and let $C_G^*(X)$ denote the set of connected components of $G(V \setminus X)$ which have an odd number of vertices and no loops. Let $c_G(X) = |C_G(X)|$, and let $c_G^*(X) = |C_G^*(X)|$. We then have the following theorem, a proof of which is given in [7] and can be obtained by straightforward modification of any of the several proofs available for the Tutte–Berge theorem such as those given in [1], [5], [33].

Theorem 8.3.3 *Let G be a graph with vertex set V. Then*

$$m(G) = \min\{n - c_G(X) + |X| : X \subseteq V\}, \tag{8.9}$$

and

$$m_q(G) = \min\{n - c_G^*(X) + |X| : X \subseteq V\}. \tag{8.10}$$

We now obtain a formula for the minimal symmetric term rank over the class $\mathcal{N}(R)$ of symmetric nonnegative matrices with row sum vector R.

Theorem 8.3.4 *Let $R = (r_1, r_2, \ldots, r_n)$ be a positive nonincreasing vector. Then*

$$\tilde{\rho}(\mathcal{N}(R)) = \min \left\{ e + f - \left\lfloor \frac{f - e}{3} \right\rfloor \right\} \tag{8.11}$$

where the minimum is taken over all pairs of integers e, f with $0 \leq e \leq f \leq n$ and $r_1 + \cdots + r_e \geq r_{f+1} + \cdots + r_n$.

Proof. Let

$$\alpha = e^* + f^* - \left\lfloor \frac{f^* - e^*}{3} \right\rfloor$$

be the minimum in (8.11) where $0 \leq e^* \leq f^* \leq n$ and $r_1 + \cdots + r_{e^*} \geq r_{f^*+1} + \cdots + r_n$. We choose e^* largest and then f^* smallest subject to these conditions. We show first that $\rho_s(A) \geq \alpha$ for every matrix A in $\mathcal{N}(R)$. Then we construct a matrix in $\mathcal{N}(R)$ whose symmetric term rank equals α.

Let A be a matrix in $\mathcal{N}(R)$. Applying Theorem 8.3.3 and using the correspondence between graphs and matrices, we see that there exists a permutation matrix P such that PAP^{T} has the form

$$\left[\begin{array}{c|ccc|c} A_0 & \multicolumn{3}{c|}{A_1} & A_2 \\ \hline & X_1 & O & \cdots & O & \\ & O & X_2 & \cdots & O & \\ A_1^{\mathrm{T}} & \vdots & \vdots & \ddots & \vdots & O \\ & O & O & \cdots & X_t & \\ \hline A_2^{\mathrm{T}} & \multicolumn{3}{c|}{O} & O \end{array} \right], \tag{8.12}$$

where $A_0 = A[\{i_1, i_2, \ldots, i_{e'}\}]$ is a matrix of order e' and the zero matrix $O = A[\{i_{f'+1}, \ldots, i_n\}]$ in the lower right corner is of order $n - f'$ (thus the sum of the orders of $X_1, X_2, \ldots, X_t$ equals $f' - e'$), and $X_1, X_2, \ldots, X_t$ are square matrices of either even order, odd order with at least one nonzero diagonal element, or odd order at least 3 with only zeros on the main diagonal. Moreover,

$$\rho_s(A) = n - ((n - f') + a) + e' = e' + f' - a,$$

where a equals the number of matrices X_i of odd order at least 3 with only zeros on the main diagonal. We have $a \leq \lfloor (f' - e')/3 \rfloor$. It follows from (8.12) that

$$r_{i_1} + \cdots + r_{i_{e'}} \geq r_{i_{f'+1}} + \cdots + r_{i_n},$$

and hence by the monotonicity of R that

$$r_1 + \cdots + r_{e'} \geq r_{f'+1} + \cdots + r_n.$$

Hence

$$\rho_s(A) \geq e' + f' - \left\lfloor \frac{f' - e'}{3} \right\rfloor \geq \alpha.$$

We now obtain a matrix in $\mathcal{N}(R)$ with symmetric term rank equal to α. If $r_1 \geq r_2 + \cdots + r_n$, then with $e = f = 1$, we see that $\alpha = 2$. In this case the matrix

$$\begin{bmatrix} c & r_2 & \cdots & r_n \\ \hline r_2 & & & \\ \vdots & & O & \\ r_n & & & \end{bmatrix}, \quad c = r_1 - (r_2 + \cdots + r_n), \quad (8.13)$$

is a matrix in $\mathcal{N}(R)$ with symmetric term rank equal to α. We now assume that $r_1 < r_2 + \cdots + r_n$ and thus that $r_k < \sum_{i \neq k} r_i$ for each $k = 1, 2, \ldots, n$. Let

$$B = \begin{bmatrix} B_0 & & & B_{01} & & & J \\ \hline & B_1 & O & \cdots & O & O & \\ & O & B_2 & \cdots & O & O & \\ B_{01}^{\mathrm{T}} & \vdots & \vdots & \ddots & \vdots & \vdots & O \\ & O & O & \cdots & B_k & O & \\ & O & O & \cdots & O & B_{k+1} & \\ \hline J^{\mathrm{T}} & & & O & & & O \end{bmatrix}, \quad (8.14)$$

where B_0 is of order e^*, B_{01} is an e^* by $f^* - e^*$ matrix, and J is an e^* by $n - f^*$ matrix of all 1's. The matrices B_0 and B_{01} in (8.14) are either zero matrices or all 1's matrices depending on whether or not equality holds in the inequality $r_1 + \cdots + r_{e^*} \geq r_{f^*+1} + \cdots + r_n$. The integer k is defined by $k = \lfloor (f^* - e^*)/3 \rfloor$. The matrices $B_1, \ldots, B_k$ are matrices of order 3 equal to

$$\begin{bmatrix} 0 & 1 & 1 \\ 1 & 0 & 1 \\ 1 & 1 & 0 \end{bmatrix}. \quad (8.15)$$

The matrix B_{k+1} is a (possibly vacuous) identity matrix of order $(f^* - e^*) - 3k$.

Suppose for the three rows $p, p+1, p+2$ of B corresponding to some B_i with $1 \leq i \leq k$ (so in particular, $e^* + 1 \leq p \leq f^* - 2$), we have $r_p \geq r_{p+1} + r_{p+2}$. By monotonicity we then have

$$r_{e^*+1} \geq r_p \geq r_{p+1} + r_{p+2} \geq r_{f^*-1} + r_{f^*}.$$

Hence

$$r_1 + \cdots + r_{e^*} + r_{e^*+1} \geq r_{f^*-1} + r_{f^*} + r_{f^*+1} + \cdots + r_n.$$

We calculate that

$$(e^* + 1) + (f^* - 2) - \left\lfloor \frac{(f^* - 2) - (e^* + 1)}{3} \right\rfloor = e^* + 1 + f^* - 2 - \left\lfloor \frac{f^* - e^*}{3} \right\rfloor + 1$$

$$= e^* + f^* - \left\lfloor \frac{f^* - e^*}{3} \right\rfloor = \alpha.$$

By our choice of e^* and f^*, it now follows that $r_p < r_{p+1} + r_{p+2}$ for the three rows $p, p+1, p+2$ corresponding to each B_i with $1 \le i \le k$.

It now is straightforward to check that B satisfies the symmetric R-support condition of Theorem 8.2.3, and hence there exists a matrix A in $\mathcal{N}(R)$ with pattern equal to B. Let G be the graph of B and let X be the set of e^* vertices of G corresponding to the first e^* rows and columns. Then by Theorem 8.3.3

$$\rho_s(A) \le n - c_G^*(X) + |X|$$

$$= n - \left(\left\lfloor \frac{f^* - e^*}{3} \right\rfloor + n - f^* \right) + e^*$$

$$= e^* + f^* + \left\lfloor \frac{f^* - e^*}{3} \right\rfloor$$

$$= \alpha.$$

Thus $\rho_s(A) = \alpha$ and the proof of the theorem is complete. $\qquad\square$

As an immediate corollary we can characterize those classes $\mathcal{N}(R)$ for which every matrix in the class satisfies $\rho_s(A) = n$.

Corollary 8.3.5 *Let $R = (r_1, r_2, \ldots, r_n)$ be a positive nonincreasing vector. Then $\rho_s(A) = n$ for every matrix $A \in \mathcal{N}(R)$ if and only if*

$$e + f - \left\lfloor \frac{f - e}{3} \right\rfloor \ge n$$

for all integers e and f such that $0 \le e \le f \le n$ and $r_1 + \cdots + r_e \ge r_{f+1} + \cdots + r_n$. $\qquad\square$

We now consider the subpolytope $\mathcal{N}(R)_0$ of $\mathcal{N}(R)$ of symmetric, nonnegative matrices with row sum vector R and with trace equal to zero. As usual we assume that $R = (r_1, r_2, \ldots, r_n)$ is nonincreasing. The polytope $\mathcal{N}(R)_0$ is the closure of $\mathcal{N}_{\le J-I}(R)$, and by Corollary 8.2.4 is nonempty if and only if

$$\sum_{i=1}^{n} r_i \ge 2r_1.$$

A characterization of its extreme points is contained in Theorem 8.2.6. Let

$$\tilde{\rho}_s(\mathcal{N}(R)_0) = \min\{\rho_s(A) : A \in \mathcal{N}(R)_0\}$$

and

$$\bar{\rho}_s(\mathcal{N}(R)_0) = \max\{\rho_s(A) : A \in \mathcal{N}(R)_0\}$$

be, respectively, the *minimal symmetric term rank* and *maximal symmetric term rank* of matrices in $\mathcal{N}(R)_0$.

Theorem 8.3.6 *Let $R = (r_1, r_2, \ldots, r_n)$ be a positive nonincreasing vector with $r_1 + \cdots + r_n \geq 2r_1$. Then*

$$\tilde{\rho}_s(\mathcal{N}(R)_0) = \min\left\{e + f - \frac{f - e}{3}\right\} \tag{8.16}$$

where the minimum is taken over all pairs of integers e, f with $0 \leq e \leq f \leq n$ such that $f - e$ is divisible by 3 and $r_1 + \cdots + r_e \geq r_{f+1} + \cdots + r_n$.

Proof. The proof is similar to the proof of Theorem 8.3.4 and we shall use some of its notation. The hypotheses and Corollary 8.2.4 imply that $\mathcal{N}(R)_0$ is nonempty. Let

$$\alpha^0 = e^* + f^* - \left\lfloor \frac{f^* - e^*}{3} \right\rfloor$$

be the minimum in (8.16) where $0 \leq e^* \leq f^* \leq n$, $f^* - e^*$ is divisible by 3, and $r_1 + \cdots + r_{e^*} \geq r_{f^*+1} + \cdots + r_n$. We choose e^* largest and then f^* smallest subject to these conditions.

Let A be a matrix in $\mathcal{N}(R)_0$. There is a permutation matrix P such that PAP^T has the form (8.12) where the matrices X_i are matrices of odd order $m_i \geq 3$ ($i = 1, 2, \ldots, t$) and where $\rho_s(A) = e' + f' - t$. We have

$$r_{i_1} + \cdots + r_{i_{e'}} \geq r_{i_{f'+1}} + \cdots + r_{i_n},$$

and hence by the monotonicity of R that

$$r_1 + \cdots + r_{e'} \geq r_{f'+1} + \cdots + r_n.$$

Let $2k = (m_1 - 3) + (m_2 - 3) + \cdots + (m_t - 3)$. With $e = e' + k$ and $f = f' - k$, we now have that $e \leq f$, $f - e = m_1 + \cdots + m_t - 2k = 3t$, and from the monotonicity of R,

$$r_1 + r_2 + \cdots + r_e \geq r_{f+1} + r_{f+2} + \cdots + r_n.$$

Moreover,

$$\alpha^0 \leq e + f - \frac{f - e}{3}$$
$$= e' + f' - 2k - \frac{f' - e' - 2k}{3}$$
$$= e' + f' - t = \rho_s(A).$$

We now show how to obtain a matrix A' in $\mathcal{N}(R)_0$ with symmetric term rank at most equal to α^0, and hence equal to α^0. If $r_1 \geq r_2 + \cdots + r_n$, then since $r_1 + \cdots + r_n \geq 2r_1$ by hypothesis, $r_1 = r_2 + \cdots + r_n$. The matrix A' in (8.13) with $c = 0$ is in $\mathcal{N}(R)_0$ and has symmetric term rank equal to α^0. Now assume that $r_1 < r_2 + \cdots + r_n$. We now consider the matrix B in (8.14) where B_{k+1} is now vacuous and $B_1, B_2, \ldots, B_t$ are each given by (8.15). The matrices B_0 and B_{01} equal $J - I$ and J, respectively, if $r_1 + \cdots + r_{e^*} > r_{f^*+1} + \cdots + r_n$, and equal O otherwise. As in the proof of Theorem 8.3.4 there is a matrix A in $\mathcal{N}(R)_0$ with $\rho_s(A) \leq \alpha^0$. Hence $\tilde{\rho}_s(\mathcal{N}(R)_0) = \alpha^0$. $\qquad\square$

As a corollary we can characterize those classes $\mathcal{N}(R)_0$ for which every matrix has symmetric term rank equal to n.

Corollary 8.3.7 *Let $R = (r_1, r_2, \ldots, r_n)$ be a positive nonincreasing vector with $r_1 + \cdots + r_n \geq 2r_1$. Every matrix A in $\mathcal{N}(R)_0$ satisfies $\rho_s(A) = n$ if and only if $e + f - (f - e)/3 \geq n$ for all e and f such that $0 \leq e \leq f \leq n$, and $f - e$ is divisible by 3 and $r_1 + \cdots + r_e \geq r_{f+1} + \cdots + r_n$.* $\qquad\square$

A formula for $\bar{\rho}_s(\mathcal{N}(R)_0)$ has not been determined.

8.4 Faces of Transportation Polytopes

Throughout this section we assume that $R = (r_1, r_2, \ldots, r_m)$ and $S = (s_1, s_2, \ldots, s_n)$ are positive vectors for which

$$r_1 + r_2 + \cdots + r_m = s_1 + s_2 + \cdots + s_n,$$

with this common value denoted by τ. The polytope $\mathcal{N}(R, S)$ is nonempty and has dimension $d = (m-1)(n-1)$. In Theorem 8.1.2 we characterized the extreme points of $\mathcal{N}(R, S)$ as those matrices in $\mathcal{N}(R, S)$ whose associated bipartite graphs have no cycles. The extreme points of a polytope are its faces of dimension 0. In this section we consider general faces of $\mathcal{N}(R, S)$. The polytope $\mathcal{N}(R, S)$ is defined by the linear constraints

$$\sum_{j=1}^{n} x_{ij} = r_i \ (i = 1, 2, \ldots, m), \ \sum_{i=1}^{m} x_{ij} = s_j \ (j = 1, 2, \ldots, n), \qquad (8.17)$$

and

$$x_{ij} \geq 0 \ (i = 1, 2, \ldots, m; j = 1, 2, \ldots, n). \qquad (8.18)$$

The faces are obtained by choosing $H \subseteq \{1, 2, \ldots, m\} \times \{1, 2, \ldots, n\}$ and setting $x_{ij} = 0$ for each $(i, j) \in H$. Thus each face of $\mathcal{N}(R, S)$ is of the form[6]

$$\mathcal{N}(R, S)_H = \{A = [a_{ij}] : A \in \mathcal{N}(R, S), \ a_{ij} = 0 \text{ for } (i, j) \in H\}.$$

---

[6]Let $W = [w_{ij}]$ be the m by n $(0, 1)$-matrix such that $w_{ij} = 0$ if and only if $(i, j) \in H$. Then $\mathcal{N}(R, S)_H = \mathcal{N}_{\leq W}(R, S)$. To emphasize here the dependence of a face on those positions which are prescribed to be zero, we use the notation $\mathcal{N}(R, S)_H$ in this section.

For some choices of H we may have $\mathcal{N}(R,S)_H = \varnothing$. We may also have sets $H_1 \neq H_2$ for which $\mathcal{N}(R,S)_{H_1} = \mathcal{N}(R,S)_{H_2}$.

Example. Let $m = n = 3$ and let $R = S = (1,1,1)$. If we choose $H = \{(1,2),(1,3),(2,2),(2,3)\}$, then $\mathcal{N}(R,S)_H$ is the set of all nonnegative matrices of order 3 with all row and column sums equal to 1 and with 0's as indicated in

$$\begin{bmatrix} x & 0 & 0 \\ y & 0 & 0 \\ z & u & v \end{bmatrix}.$$

This set is clearly empty. Now let

$$H = \{(1,2),(1,3)\} \text{ and } K = \{(1,2),(1,3),(2,1),(3,1)\}.$$

Then $\mathcal{N}(R,S)_H$ and $\mathcal{N}(R,S)_K$ consist of all nonnegative matrices of order 3 with all row and column sums equal to 1 and with 0's as indicated, respectively, in

$$\begin{bmatrix} x & 0 & 0 \\ y & w & p \\ z & u & v \end{bmatrix}, \text{ and } \begin{bmatrix} x & 0 & 0 \\ 0 & w & p \\ 0 & u & v \end{bmatrix}.$$

Since all row and column sums equal 1, a matrix in $\mathcal{N}(R,S)_H$ must have $y = z = 0$. Hence $\mathcal{N}(R,S)_H = \mathcal{N}(R,S)_K$. □

The *facets* of a polytope are the faces of dimension one less that that of the polytope. Since the dimension of the transportation polytope $\mathcal{N}(R,S)$ is $d = (m-1)(n-1)$, the facets of $\mathcal{N}(R,S)$ are its faces of dimension $d-1 = (m-1)(n-1)-1$. The following characterization of facets is from Klee and Witzgall [23].

Theorem 8.4.1 *Each facet of a nonempty transportation polytope $\mathcal{N}(R,S)$ is of the form $\mathcal{N}(R,S)_{H_{ij}}$ where $H_{ij} = \{(i,j)\}$ with $1 \leq i \leq m$ and $1 \leq j \leq n$. The face $\mathcal{N}(R,S)_{H_{ij}}$ is*

(i) *empty if $r_i + s_j > \tau$;*

(ii) *an extreme point if $r_i + s_j = \tau$; and*

(iii) *a facet if $r_i + s_j < \tau$.*

Proof. If $r_i + s_j > \tau$, then there clearly cannot exist a matrix in $\mathcal{N}(R,S)_{H_{ij}}$ and hence $\mathcal{N}(R,S)_{H_{ij}}$ is empty. Now suppose that $r_i + s_j = \tau$. Without loss of generality assume that $i = j = 1$. Then $r_1 + s_1 = \tau$ implies that $r_1 = s_2 + \cdots + s_n$ and $s_1 = r_2 + \cdots + r_m$. There is a unique matrix in $\mathcal{N}(R,S)_{H_{ij}}$, namely,

$$\left[\begin{array}{c|ccc} 0 & s_2 & \cdots & s_n \\ \hline r_2 & & & \\ \vdots & & O & \\ r_m & & & \end{array}\right].$$

Hence $\mathcal{N}(R,S)_{H_{ij}}$ is a face of dimension 0, that is, an extreme point.

Now assume that $r_i + s_j < \tau$, and again without loss of generality assume that $i = j = 1$. Then $r_1 + s_1 < \tau$ implies that $r_1 < s_2 + \cdots + s_n$ and $s_1 < r_2 + \cdots + r_m$. Hence there exist positive numbers $c_2, \ldots, c_n$ with $c_j < s_j$ for $j = 2, \ldots, n$ and $c_2 + \cdots + c_n = r_1$, and there exist positive numbers $d_2, \ldots, d_m$ with $d_i < r_i$ for $i = 2, \ldots, m$ and $d_2 + \cdots + d_m = s_1$. The vectors $C = (s_2 - c_2, \ldots, s_n - c_n)$ and $D = (r_2 - d_2, \ldots, r_m - d_m)$ are positive vectors with $(r_2 - d_2) + \cdots + (r_m - d_m) = (s_2 - c_2) + \cdots + (s_n - c_n)$. Hence $\mathcal{N}(C, D)$ is nonempty and indeed contains a positive matrix. Therefore $\mathcal{N}(R,S)_{H_{1,1}}$ contains a matrix whose only zero is in position $(1, 1)$.

The matrix of coefficients of the system of equations

$$\sum_{j=2}^{n} x_{1j} = 0,$$

$$\sum_{j=1}^{n} x_{ij} = 0 \ (i = 2, \ldots, m),$$

$$\sum_{i=2}^{m} x_{i1} = 0,$$

$$\sum_{i=1}^{m} x_{ij} = 0 \ (j = 2, \ldots, n)$$

has rank equal to $m + n - 1$ (since the sum of the first m equations equals the sum of the last n equations), and hence its null space has dimension

$$(mn - 1) - m - n + 1 = mn - m - n = (m - 1)(n - 1) - 1.$$

Since $\mathcal{N}(R,S)_{H_{1,1}}$ contains a positive matrix it now follows that the dimension of $\mathcal{N}(R,S)_{H_{11}}$ equals $(m - 1)(n - 1) - 1$ and $\mathcal{N}(R,S)_{H_{1,1}}$ is a facet of $\mathcal{N}(R,S)$ $\square$

The possible number of facets of transportation polytopes is given in the next theorem [23]. In the theorem we assume without loss of generality that $m \leq n$.

Theorem 8.4.2 *Let m and n be positive integers with $m \leq n$, and let $R = (r_1, r_2, \ldots, r_m)$ and $S = (s_1, s_2, \ldots, s_n)$ be positive vectors such that $\mathcal{N}(R, S) \neq \oslash$. Let $f(R, S)$ be the number of facets of $\mathcal{N}(R, S)$. Then*

$$f(R, S) = 1 \qquad \text{if } m = 1;$$
$$f(R, S) = 2 \qquad \text{if } m = n = 2;$$
$$(m - 1)n \leq f(R, S) \leq mn \qquad \text{if } m \geq 2 \text{ and } n \geq 3.$$

Moreover, for $m > 1$ and $n > 2$, each integer p with $(m - 1)n \leq p \leq mn$ is the number of facets of some transportation polytope $\mathcal{N}(R, S)$.

Proof. If $m = 1$, then the dimension of $\mathcal{N}(R, S)$ is 0 and its only facet is the empty facet. If $m = n = 2$, then the dimension of $\mathcal{N}(R, S)$ is 1 and hence $\mathcal{N}(R, S)$ has two facets, namely its two extreme points.

Now assume that $m \geq 2$ and $n \geq 3$ so that the dimension of $\mathcal{N}(R, S)$ is at least 2. It follows from Theorem 8.4.1 that $f(R, S) \leq mn$.

Let $\tau = r_1 + \cdots + r_m = s_1 + \cdots + s_n$, and let i and j be integers with $1 \leq i \leq m$ and $1 \leq j \leq n$. By Theorem 8.4.1, $\mathcal{N}(R, S)_{H_{ij}}$ is a facet of $\mathcal{N}(R, S)$ if and only if $r_i < \tau - s_j$. Let k and l be integers with $k \neq i$ and $l \neq j$ satisfying $1 \leq k \leq m$ and $1 \leq l \leq n$. Suppose that neither $\mathcal{N}(R, S)_{H_{ij}}$ nor $\mathcal{N}(R, S)_{H_{kl}}$ is a facet. Then we have

$$r_i \geq \tau - s_j > s_l \geq \tau - r_k \geq r_i,$$

where the strict inequality is a consequence of our assumption that $n \geq 3$. This contradiction implies that if neither $\mathcal{N}(R, S)_{H_{ij}}$ nor $\mathcal{N}(R, S)_{H_{kl}}$ is a facet, then either $k = i$ or $l = j$. This in turn implies that the number of faces of $\mathcal{N}(R, S)_{H_{ij}}$ that are not facets is at most $\max\{m, n\} = n$. Hence $f(R, S) \geq mn - n$.

We now show that, under the assumptions $m \geq 2$ and $n \geq 3$, for each integer k with $0 \leq k \leq n$, there exist positive vectors R and S such that exactly k of $\mathcal{N}(R, S)_{H_{ij}}$ are not facets, that is, $r_i + s_j \geq \tau$ for exactly k pairs $(i, j) \in \{1, 2, \ldots, m\} \times \{1, 2, \ldots, n\}$. First suppose that $k = 0$. Let $r_i = n$ for $1 \leq i \leq m$ and let $s_j = m$ for $1 \leq j \leq n$. Then $\tau = mn$ and $r_i + s_j = n + m < mn = \tau$, the strict inequality being a consequence of our assumptions that $m \geq 2$ and $n \geq 3$. Now suppose that $1 \leq k \leq n$. Let $r_2, \ldots, r_m$ be arbitrary positive numbers and let $\tau' = r_2 + \cdots + r_m$. Let $s_1, s_2, \ldots, s_n$ be positive numbers satisfying

$$s_j > \tau' \ (1 \leq j \leq k) \text{ and } s_j < \tau' \ (k < j \leq n).$$

Let $\tau = s_1 + s_2 + \cdots + s_n$ and let $r_1 = \tau - \tau'$. Then $r_1 > 0$, and $\tau = r_1 + r_2 + \cdots + r_m$. We have $r_1 + s_j \geq \tau$ if and only if $s_j \geq \tau - r_1 = \tau'$; hence exactly k of $\mathcal{N}(R, S)_{H_{1j}}$ are not faces. Since $k \geq 1$, it follows from what we have already proved above that, if $\mathcal{N}(R, S)_{H_{ij}}$ with $i \neq 1$ is not a facet, then $k = 1$ and $j = 1$. Suppose there were such a pair (i, j). Then $r_i + s_1 \geq \tau$ and hence $r_i \geq \tau - s_1 > s_n$, a contradiction since $i \neq 1$. Hence the number of facets of $\mathcal{N}(R, S)$ equals $mn - k$. □

Let $W = [w_{ij}]$ be an m by n $(0, 1)$-matrix. The polytope $\mathcal{N}_{\leq W}(R, S)$ is a (possibly empty) face of $\mathcal{N}(R, S)$. We call such polytopes $\mathcal{N}_{\leq W}(R, S)$ *restricted transportation polytopes.*[7] Gibson [19] investigated the number of facets a restricted transportation polytope can have in terms of its dimension. A polytope of dimension 0 (a point) has one facet (the empty facet);

---

[7] In terms of the transportation problem formulated at the beginning of this section, $\mathcal{N}_{\leq W}(R, S)$ represents solutions of a transportation problem in which goods cannot be transported between certain source–demand pairs.

a polytope of dimension 1 (a line segment) has two facets (its two vertices). Let $d > 1$ be an integer.

Let $\tilde{f}(d)$ equal the *minimum number of facets of a restricted transportation polytope of dimension d* and let $\bar{f}(d)$ be the *maximum number of facets of a restricted transportation polytope of dimension d*. These extremal values are taken over all restricted transportation polytopes without regard for row and column sum vectors R and S, and their sizes. Since a polytope of dimension d has at least $d + 1$ facets, we have $\tilde{f}(d) \geq d + 1$.

For given R and S of sizes m and n, respectively, there is a bijective correspondence between the nonempty faces of $\mathcal{N}(R, S)$ and the m by n $(0, 1)$-matrices W satisfying the (R, S)-support condition. Thus every restricted transportation problem is of the form $\mathcal{N}_{\leq W}(R, S)$ for some R and S and some matrix W satisfying the (R, S)-support condition. In what follows we shall distinguish between those W which are indecomposable and those which are decomposable. In the case of indecomposable W, equality never occurs in the (R, S)-support condition. If W is decomposable, the polytope $\mathcal{N}_{\leq W}(R, S)$ is the orthogonal sum of two "smaller" restricted transportation polytopes and its dimension is the sum of the dimensions of these two polytopes.

In the next lemma we collect a number of important properties that are useful in evaluating $\bar{f}(d)$.

Lemma 8.4.3 *Let $R = (r_1, r_2, \ldots, r_m)$ and $S = (s_1, s_2, \ldots, s_n)$ be positive vectors such that $r_1 + r_2 + \cdots + r_m = s_1 + s_2 + \cdots + s_n$. Let $W = [w_{ij}]$ be an m by n $(0, 1)$-matrix satisfying the (R, S)-support condition.*

(i) *If W is indecomposable then $\dim \mathcal{N}_{\leq W}(R, S) = \sigma(W) - m - n + 1$.*

(ii) *Let p and q be integers such that $w_{pq} = 1$. Let W' be the matrix obtained from W by replacing w_{pq} with 0. Then there exists at most one m by n $(0, 1)$-matrix $W^* \leq W'$ satisfying the (R, S)-support condition such that $\mathcal{N}_{\leq W^*}(R, S)$ is a facet of $\mathcal{N}_{\leq W}(R, S)$.*

(iii) *Let U be an m by n $(0, 1)$-matrix satisfying the (R, S)-support condition. Then $\mathcal{N}_{\leq U}(R, S)$ is a facet of $\mathcal{N}_{\leq W}(R, S)$ if and only if one of the following holds.*

   (a) *U is indecomposable and is obtained from W by replacing $w_{pq} = 1$ with 0 for some p and q.*

   (b) *U is decomposable and there exist permutation matrices P and Q and indecomposable matrices $U_1, U_2, \ldots, U_k$ such that*

$$PUQ = U_1 \oplus U_2 \oplus \ldots \oplus U_k \qquad (8.19)$$

and

$$PWQ = \begin{bmatrix} U_1 & O & \cdots & O & E_1 \\ E_2 & U_2 & \cdots & O & O \\ \vdots & \vdots & \ddots & \vdots & \vdots \\ O & O & \cdots & U_{k-1} & O \\ O & O & \cdots & E_k & U_k \end{bmatrix} \qquad (8.20)$$

where $E_1, E_2, \ldots, E_k$ are $(0,1)$-matrices having exactly one 1.

Proof. (i) Since W is indecomposable, the graph $\mathrm{BG}(W)$ is a connected bipartite graph with $m + n$ vertices. Let T be a spanning tree of this graph. The tree T has $m + n - 1$ edges. By recursively considering pendant edges of T and removing them, we see that the elements of a matrix A in $\mathcal{N}_{\leq W}(R, S)$ corresponding to the edges of T are uniquely determined by the elements corresponding to the edges of $\mathrm{BG}(W)$ that are not edges of T. Since the total number of edges of $\mathrm{BG}_{\leq W}(R, S)$ equals $\sigma(W)$, it follows that $\dim \mathcal{N}_{\leq W}(R, S) \leq \sigma(W) - (m + n - 1)$. Since W satisfies the (R, S)-support condition, there exists a matrix $A = [a_{ij}]$ in $\mathcal{N}_W(R, S)$. Again let T be a spanning tree of $\mathrm{BG}(W)$. Let E be the set of positions of W, of cardinality $\sigma(W) - (m + n - 1)$, corresponding to the edges of $\mathrm{BG}(W)$ that are not edges of T. Each such edge belongs to a cycle $\gamma_{(p,q)}$ of $\mathrm{BG}(W)$ whose other edges are edges of T. For each $(p, q) \in E$, let $C_{(p,q)}$ be a balanced $(0, 1, -1)$-matrix such that $\mathrm{BG}(C_{(p,q)})$ is the cycle $\gamma_{(p,q)}$, apart from isolated vertices. Then for all $\epsilon_{(p,q)}$ small enough, with $(p, q) \in E$,

$$A + \sum_{(p,q) \in E} \epsilon_{(p,q)} C_{(p,q)} \in \mathcal{N}_{\leq W}(R, S).$$

Hence $\dim \mathcal{N}_{\leq W}(R, S) \geq \sigma(W) - (m + n - 1)$, and (i) now follows.

(ii) Let W^* be obtained from W' by replacing with 0 all those 1's for which no matrix in $\mathcal{N}_{\leq W'}(R, S)$ has a positive entry in the corresponding position. Then W^* satisfies the (R, S)-support condition and $\mathcal{N}_{W^*}(R, S)$ is a proper face of $\mathcal{N}_W(R, S)$ and is contained in every face of $\mathcal{N}_{\leq W'}(R, S)$. Thus $\mathcal{N}_{\leq W^*}(R, S)$ is the only possible facet of $\mathcal{N}_{\leq W}(R, S)$ of the type described in (ii).

(iii) If U satisfies (a) or (b), then it follows from the dimension formula given in (i) that $\mathcal{N}_{\leq U}(R, S)$ is a facet of $\mathcal{N}_{\leq W}(R, S)$. Now suppose that $\mathcal{N}_{\leq U}(R, S)$ is a facet of $\mathcal{N}_{\leq W}(R, S)$. If U is indecomposable, then since U satisfies the (R, S)-support condition,

$$(\sigma(W) - m - n + 1) - 1 = \dim \mathcal{N}_{\leq U}(R, S) = \sigma(U) - m - n + 1$$

and hence $\sigma(U) = \sigma(W) - 1$. Hence (a) holds in this case. Now suppose that U is decomposable. Then there exist permutation matrices P and Q

such that (8.19) holds where $U_1, U_2, \ldots, U_k$ are indecomposable. Let

$$PWQ = \begin{bmatrix} U_1 & W_{12} & \cdots & W_{1,k-1} & W_{1k} \\ W_{21} & U_2 & \cdots & W_{2,k-1} & W_{2k} \\ \vdots & \vdots & \ddots & \vdots & \vdots \\ W_{k-1,1} & W_{k-1,2} & \cdots & U_{k-1} & W_{k-1,k} \\ W_{k1} & W_{k2} & \cdots & W_{k,k-1} & U_k \end{bmatrix}. \tag{8.21}$$

Since $\mathcal{N}_{\leq U}(R, S)$ is a facet of $\mathcal{N}_{\leq W}(R, S)$ it follows from the dimension formula in (i) that $\sigma(W) - \sigma(U) = k$. Consider the digraph G with vertices $\{1, 2, \ldots, k\}$ where (i, j) is an arc if and only if $W_{ij} \neq O$. This digraph has k arcs. Since W and U satisfy the (R, S)-support condition, it easily follows that G is a directed cycle of length k. Hence there exists a permutation matrix M such that $M^{\mathrm{T}}(PWQ)M$ has the form (8.21) where the blocks in the main diagonal form a permutation $U_{i_1}, U_{i_2}, \ldots, U_{i_k}$ of $U_1, U_2, \ldots, U_k$. Since $M^{\mathrm{T}}UM = U_{i_1} \oplus U_{i_2} \oplus \cdots \oplus U_{i_k}$, (iii) now follows. $\quad\square$

The following lemma will also be used in our evaluation of $\bar{f}(d)$.

Lemma 8.4.4 *Let $R = (r_1, r_2, \ldots, r_m)$ and $S = (s_1, s_2, \ldots, s_n)$ be positive vectors such that $r_1 + r_2 + \cdots + r_m = s_1 + s_2 + \cdots + s_n$. Let $W = [w_{ij}]$ be an m by n indecomposable $(0, 1)$-matrix satisfying the (R, S)-support condition. Let $k > 2$ be an integer, and let*

$$\gamma : x_1 \to x_2 \to \cdots \to x_{k-1} \to x_k \tag{8.22}$$

be a path of length k in the bipartite graph $BG(W)$ where each of the vertices $x_2, \ldots, x_{k-1}$ has degree 2. Then there are at most two m by n $(0, 1)$-matrices U such that (i) $x_1 \to x_2 \to \cdots \to x_{k-1} \to x_k$ is not a path of $BG(U)$, (ii) U satisfies the (R, S)-support condition, and (iii) $\mathcal{N}_{\leq U}(R, S)$ is a facet of $\mathcal{N}_{\leq W}(R, S)$.

Proof. Let U be a matrix satisfying (i), (ii), and (iii) of the lemma. Since W satisfies the (R, S)-support condition, $\mathcal{N}_{\leq U}(R, S)$ is a proper subface of $\mathcal{N}_{\leq W}(R, S)$. Hence there is at most one facet of $\mathcal{N}_{\leq W}(R, S)$ corresponding to the removal of an edge of the path (8.22). Let $A = [a_{ij}]$ be a matrix in $\mathcal{N}_U(R, S)$. Let y_i be the element in A in the position corresponding to the arc $x_i \to x_{i+1}$ of γ $(i = 1, 2, \ldots, k-1)$. Then every matrix B in $\mathcal{N}_U(R, S)$ has the same elements as A in these positions. Define

$$h(U) = (y_1, y_2, \ldots, y_{k-1}).$$

Since γ is not a path in $BG(U)$, $h(U)$ contains at least one 0. From this it follows that for two matrices $U \neq U'$ satisfying (i), (ii) and (iii), $h(U) \neq h(U')$.

Let $c_t = r_i$ or s_j according to whether vertex x_t corresponds to row i or column j $(t = 1, 2, \ldots, k-1)$. First consider the case in which $x_1 \neq x_k$.

Then $g(U)$ is in the convex polytope $\mathcal{P}$ given by

$$\{(z_1, z_2, \ldots, z_{k-1}) : z_1 \leq c_1, z_{k-1} \leq c_{k-1}, z_{p-1} + z_p = c_p, 2 \leq p \leq k-1\}.$$

The polytope $\mathcal{P}$ has dimension at most equal to 1, since each vector in it is determined uniquely by the value of any z_i. Moreover, each vector $g(U)$, having a zero component, must be an extreme point of $\mathcal{P}$. It follows that there are at most two U's satisfying (i), (ii), and (iii). A similar proof works in case $x_1 = x_k$. $\square$

Theorem 8.4.5 *Let $d > 1$ be an integer. Then*

$$d + 1 \leq \tilde{f}(d) \leq \bar{f}(d) \leq 6(d-1).$$

Proof. That $\tilde{f}(d) \geq d+1$ is clear since a d-dimensional polytope always has at least $d + 1$ extreme points. We now show that every d-dimensional face $\mathcal{N}_{\leq W}(R, S)$ of $\mathcal{N}(R, S)$, where W satisfies the (R, S)-support condition, has at most $6(d-1)$ facets.

If W contains a row or column, say row m, with exactly one 1, then we may delete row m and reduce the appropriate column sum by r_m thereby obtaining a d-dimensional face of a transportation polytope congruent to $\mathcal{N}_{\leq W}(R, S)$. Thus we may assume that each row and each column of W contain at least two 1's. Without loss of generality we assume that the rows and columns of W with exactly two 1's come first. Thus

$$W = \begin{bmatrix} W_{11} & W_{12} \\ W_{21} & W_{22} \end{bmatrix}$$

where W_{11} is a p by q matrix, the rows of W containing exactly two 1's are rows $1, 2, \ldots, p$, and the columns of W containing exactly two 1's are columns $1, 2, \ldots, q$. Here p and q may be 0.

We first assume that W is indecomposable. Since $d > 1$ it follows from (i) of Lemma 8.4.3 that $p < m$ or $q < n$. Let

$$\mathcal{F}_i = \{\mathcal{N}_{\leq U_i}(R, S) : i = 1, 2, \ldots, t\} \tag{8.23}$$

be the distinct facets of $\mathcal{N}_{\leq W}(R, S)$ where the U_i satisfy the (R, S)-support condition. Then there exist sets

$$M_i \subseteq \{(i, j) : 1 \leq i \leq m, 1 \leq j \leq n\} \ (i = 1, 2, \ldots, t)$$

such that $w_{kl} = 1$ for all (k, l) in M_i and U_i is obtained from W by replacing these 1's with 0's. By (ii) of Lemma 8.4.3, the sets $M_1, M_2, \ldots, M_t$ are pairwise disjoint.

Let

$$W^* = \{(i, j) : a_{ij} = 1, i > p \text{ or } j > q\}.$$

We define a map g from the facets (8.23) to W^* as follows.

(a) First suppose that U_i is a facet obtained as in Lemma 8.4.4 from a path γ of the form (8.22) with $k > 2$, where we assume that γ is maximal relative to the stated conditions. Since $d > 1$, each of the two vertices x_1 and x_k has degree at least 3. Let a_{st} and a_{uv} be the elements of A corresponding to the edges $x_1 \to x_2$ and $x_{k-1} \to x_k$ of γ. We then define

$$g(U_i) = \{(s,t),(u,v)\}.$$

(b) Otherwise, we define

$$g(U_i) = \{(s,t)\}$$

where (s,t) is any element of M_i.

We now show that

$$t = |g^{-1}(W^*)| \leq |W^*|. \tag{8.24}$$

Suppose $g(U_i) \cap g(U_j) \neq \varnothing$ for some $i \neq j$. Since the sets $M_1, M_2, \ldots, M_t$ are pairwise disjoint, it follows that there exist distinct (s,t) and (u,v) in W^* such that

$$g(U_i) = g(U_j) = \{(s,t),(u,v)\}.$$

It also follows from Lemma 8.4.4 that there is no other U_k with $k \neq i,j$ such that $g(U_k) = \{(s,t),(u,v)\}$. Hence (8.24) holds.

We now conclude that

$$t \leq |W^*| = \sigma(W_{12}) + \sigma(W_{21}) + \sigma(W_{22}).$$

We have

$$\sigma(W) = 2p + \sigma(W_{21}) + \sigma(W_{22}) = 2q + \sigma(W_{12}) + \sigma(W_{22}).$$

Hence from the dimension formula given in (i) of Lemma 8.4.3, we have

$$\begin{aligned}
6(d-1) &= 3(\sigma(W) - 2m) + 3(\sigma(A) - n) \\
&= \sigma(W_{21}) + \sigma(W_{22}) + 2(\sigma(W_{21}) + \sigma(W_{22}) - 3(m-p)) \\
&\quad + \sigma(W_{12}) + \sigma(W_{22}) + 2(\sigma(W_{12}) + \sigma(W_{22}) - 3(n-q)).
\end{aligned}$$

Since rows $p+1, p+2, \ldots, m$ and columns $q+1, q+2, \ldots, n$ each contain at least 3 1's we now obtain

$$6(d-1) \geq \sigma(W_{12}) + \sigma(W_{21}) + 2\sigma(A_{22}) \geq t.$$

This proves the assertion in the case that W is indecomposable.

Now assume that W is decomposable. Without loss of generality we assume that

$$W = W_1 \oplus \cdots \oplus W_k, R = (R_1, \ldots, R_k), S = (S_1, S_2, \ldots, S_k)$$

where W_i is an indecomposable matrix satisfying the (R_i, S_i)-support condition for $i = 1, \ldots, k$. We may also assume that for some positive integer $h \leq k$, $d_i = \dim \mathcal{N}_{\leq W_i}(R_i, S_i) \geq 1$ $(1 \leq i \leq h)$ and $d_i = \dim \mathcal{N}_{\leq W_i}(R_i, S_i) = 0$ $(h+1 \leq i \leq k)$. Let t be the number of facets of $\mathcal{N}_W(R, S)$, and let t_i be the number of facets of $\mathcal{N}_{\leq W_i}(R_i, S_i)$. Then $t = t_1 + \cdots + t_h$. If $d_i = 1$ for $i = 1, \ldots, h$, then

$$t = t_1 + \cdots + t_h = 2(d_1 + \cdots + d_h) = 2d < 6(d-1).$$

Now suppose that $d_l > 1$ for some l. Then, using the indecomposable case, we have

$$t = t_l + \sum_{i=1(i \neq l)}^{h} t_i \leq 6(d_l - 1) + \sum_{i=1(i \neq l)}^{h} 6d_i = 6(d-1).$$

The proof of the theorem is now complete. $\square$

Gibson [19] has shown that the bounds in the theorem cannot be improved, indeed that for any integer f with $d+1 \leq f \leq 6(d-1)$ there is a face with dimension d of some transportation polytope which has f facets.

The *graph of a polytope*[8] $\mathcal{P}$ is the graph $G(\mathcal{P})$ whose vertices are the extreme points[9] of $\mathcal{P}$ and whose edges are pairs of vertices which determine one-dimensional faces of $\mathcal{P}$. Aside from its intrinsic interest, one reason for studying $G(\mathcal{P})$ is its connection with linear programming. In the simplex method for solving linear programs, one moves from one vertex of $G(\mathcal{P})$ to another along its edges, that is, one traverses a path of $G(\mathcal{P})$. In this regard, of particular interest is the diameter of the graph of a polytope, which gives a bound on the maximum number of iterations needed for solving, using the simplex method, a linear program on a polytope. Now let $R = (r_1, r_2, \ldots, r_m)$ and $S = (s_1, s_2, \ldots, s_n)$ be positive vectors with $r_1 + r_2 + \cdots + r_m = s_1 + s_2 + \cdots + s_n$. The polytope $\mathcal{N}(R, S)$ is then nonempty and, by Theorem 8.1.2, the bipartite graphs of its extreme points have connected components that are trees. We now identify the edges of the graph $G(\mathcal{N}(R, S))$ in terms of the bipartite graphs of its vertices [34]. If G_1 and G_2 are graphs with the same vertex set V, then the *graph union* $G_1 \sqcup G_2$ is the graph with vertex set V whose edges are the union of the edges of G_1 and G_2.

Theorem 8.4.6 *Let A and B be two distinct vertices of the $\mathcal{N}(R, S)$. Then A and B are joined by an edge in the graph $G(\mathcal{N}(R, S))$ if and only if $BG(A) \sqcup BG(B)$ contains exactly one cycle.*

Proof. Let W_A and W_B be the patterns of A and B, respectively, and let W be the union of W_A and W_B. Then the smallest face of $\mathcal{N}(R, S)$

---

[8]Also called the *vertex–edge graph of a polytope*.
[9]It is for this reason that we refer to the extreme points of a polytope as its vertices in this paragraph.

containing both A and B is $\mathcal{N}_{\leq W}(R,S)$. Thus A and B are joined by an edge in $G(\mathcal{N}(R,S))$ if and only if $\dim \mathcal{N}_{\leq W}(R,S) = 1$. Since $(A+B)/2$ is in $\mathcal{N}_W(R,S)$, W satisfies the (R,S)-support condition, and hence from Lemma 8.4.3 we see that $\dim \mathcal{N}_{\leq W}(R,S) = 1$ if and only if $\mathrm{BG}(W)$ has exactly one cycle. The lemma now follows. $\qquad\square$

In [28] the numbers of extreme points of transportation polytopes $\mathcal{N}(R,R)$ with the same row and column sum vector is investigated. In [31] the numbers of extreme points of transportation-like polytopes are investigated.

In [15] (nondegenerate) transportation polytopes with a fixed number of faces having the maximum and minimum numbers of vertices are characterized; the maximum and minimum diameters are also investigated. These and other results can be found in the book [34]. In addition, in [16] (nondegenerate) transportation polytopes with the maximal number of faces of a specified dimension are studied. In [25] transportation polytopes with a minimum number of faces of a given dimension k are investigated, and a formula is derived for that number.

References

[1] I. Anderson, Perfect matchings of a graph, *J. Combin. Theory, Ser. B*, **10** (1971), 183–186.

[2] M.L. Balinski, On two special classes of transportation polytopes, *Math. Programming Study*, **1** (1974), 43–58.

[3] E.D. Bolker, Transportation polytopes, *J. Combin. Theory*, **13** (1972), 251–262.

[4] R.A. Brualdi, Convex sets of non-negative matrices, *Canad. J. Math.*, **20** (1968), 144–157.

[5] R.A. Brualdi, Matchings in arbitrary graphs, *Proc. Cambridge Philos. Soc.*, **69** (1971), 401–407.

[6] R.A. Brualdi, Combinatorial properties of symmetric non-negative matrices, *Teorie combinatorie, Atti dei Convegni Lincei, 17*, ed. B. Segre, Accademia Nazionale dei Lincei, Rome, 1976, 99–120.

[7] R.A. Brualdi and H.J. Ryser, *Combinatorial Matrix Theory*, Cambridge U. Press, Cambridge, 1991.

[8] G. Converse and M. Katz, Symmetric matrices with given row sums, *J. Combin. Theory*, **18** (1975), 171–176.

[9] A.B. Cruse, A proof of Fulkerson's characterization of permutation matrices, *Linear Algebra Appl.*, **12** (1975), 21–28.

[10] G.B. Dantzig, *Linear Programming and Extensions*, Princeton U. Press, Princeton, 1962.

[11] O. Demuth, A remark on the transportation problem (in Czech, with German, and Russian summaries), *Časopis Pěst. Math.*, **86** (1961), 103–110.

[12] A. Doig, The minimum number of basic feasible solutions to a transport problem, *Operational Res. Quart.*, **14** (1963), 387–391.

[13] J. Dubois, Polytopes de transport symétriques, *Discrete Appl. Math.*, **4** (1973), 1–27.

[14] E.E. Eischen, C.R. Johnson, K. Lange, and D.P. Stanford, Patterns, linesums, and symmetry, *Linear Algebra Appl.*, **357** (2002), 273–289.

[15] V.A. Emeličev, M.K. Kravcov, and A.P. Kračkovskiĭ, On some classes of transportation polytopes, *Soviet Math. Dokl.*, **19** (1978), 889–893.

[16] V.A. Emeličev, M.K. Kravcov, and A.P. Kračkovskiĭ, Transportation polytopes with maximal number of *k*-faces, *Soviet Math. Dokl.*, **31** (1985), 504–508.

[17] L.R. Ford, Jr. and D.R. Fulkerson, *Flows in Networks*, Princeton U. Press, Princeton, 1962.

[18] D.R. Fulkerson, Hitchcock transportation problem, *RAND Corporation Report*, P-890 (1956).

[19] P.M. Gibson, Facets of faces of transportation polytopes, Congressus Numerantium No. XVII (1976), *Utilitas Math.*, Winnipeg, Manitoba, 323–333.

[20] D.J. Houck and A.O. Pittenger, Analytic proofs of a network feasibility theorem and a theorem of Fulkerson, *Linear Algebra Appl.*, **24** (1979), 151–156.

[21] C.R. Johnson and D.P. Stanford, Patterns that allow given row and column sums, *Linear Algebra Appl.*, **327** (2001), 53–60.

[22] W.B. Jurkat and H.J. Ryser, Term ranks and permanents of non-negative matrices, *J. Algebra*, **5** (1967), 342–357.

[23] V. Klee and C. Witzgall, Facets and vertices of transportation polytopes, *Mathematics of Decision Sciences, Part I*, ed. G.B. Dantzig and A. Veinott, American Mathematical Society, Providence, 1967, 257–282.

[24] M.K. Kravtsov, A proof of the Dubois hypothesis on the maximum number of vertices of a symmetric transportation polytope (in Russian), *Dokl. Akad. Nauk BSSR*, **33** (1989), 9–12.

[25] M.K. Kravtsov, Transportation polytopes with a minimal number of *k*-faces, *Discrete Math. Appl.*, **3** (1993), 115–125.

[26] M. Lewin, On the extreme points of the polytope of symmetric matrices with given row sums, *J. Combin. Theory, Ser. A*, **23** (1977), 223–232.

[27] C.-K. Li, On certain convex matrix sets, *Discrete Math.*, **79** (1990), 323–326.

[28] R. Loewy, D.R. Shier, and C.R. Johnson, Perron eigenvectors and the symmetric transportation polytope, *Linear Algebra Appl.*, **150** (1991), 139–155.

[29] J.A. Oviedo, Adjacent extreme points of a transportation polytope, *Optimization*, **37** (1996), 279–287.

[30] I. Pak, On the number of faces of certain transportation polytopes, *Europ. J. Combin.*, **21** (2000), 689–694.

[31] H. Perfect and L. Mirsky, Extreme points of certain convex polytopes, *Monatsh. Math.*, **68** (1964), 143–149.

[32] W. Watkins and R. Merris, Convex sets of doubly-stochastic matrices, *J. Combin. Theory, Ser. A.*, **16** (1974), 129–120.

[33] D.B. West, *Introduction to Graph Theory*, 2nd ed., Prentice-Hall, Upper Saddle River, 2001.

[34] V.A. Yemelichev, M.M. Kovalev, and M.K. Kravtsov, *Polytopes, Graphs and Optimisation*, Cambridge U. Press, Cambridge, 1984.

9

Doubly Stochastic Matrices

The polytope Ω_n of doubly stochastic matrices of order n was introduced in Section 1.7 where its connection with majorization of vectors was developed. In this chapter we investigate in detail this special transportation polytope and some subpolytopes of Ω_n that arise naturally. We study special properties of the faces and graph of Ω_n. Each instance of vector majorization gives rise to a polytope of doubly stochastic matrices which we also investigate. Doubly stochastic matrices arise in applications, in particular in the optimal assignment problem.

Each doubly stochastic matrix has a positive permanent. Since Ω_n is a compact set, the permanent achieves a minimum on Ω_n and we establish the van der Waerden conjecture for this minimum. Also, we determine which nonnegative matrices can be rescaled to a doubly stochastic matrix. We conclude this chapter with some related results.

9.1 Random Functions

One motivation for the study of doubly stochastic matrices, namely their connection with the majorization order, was given in Chapter 1. We begin this chapter by discussing a second motivation.

Let n be a positive integer, and let $\mathcal{F}_n$ denote the set of all n^n functions

$$f : \{1, 2, \ldots, n\} \longrightarrow \{1, 2, \ldots, n\}.$$

A random function in $\mathcal{F}_n$ can be defined by specifing a probability distribution on the functions in $\mathcal{F}_n$. One way to obtain such a probability distribution is as follows. Let

$$a_{i1}, a_{i2}, \ldots, a_{in} \quad (i = 1, 2, \ldots, n)$$

be n-tuples of nonnegative numbers with

$$a_{i1} + a_{i2} + \cdots + a_{in} = 1 \quad (i = 1, 2, \ldots, n),$$

that is, probability distributions on $\{1, 2, \ldots, n\}$. We choose the values $f(i)$ of our function independently according to these distributions:

$$f(i) = j \text{ with probability } a_{ij} \quad (i, j = 1, 2, \ldots, n).$$

If $j_1, j_2, \ldots, j_n$ is an n-tuple of integers taken from $\{1, 2, \ldots, n\}$, then the probability that the function f satisfies $f(i) = j_i$ for $i = 1, 2, \ldots, n$ equals $a_{1j_1} a_{2j_2} \ldots a_{nj_n}$. Since

$$1 = \prod_{i=1}^{n} (a_{i1} + a_{i2} + \cdots + a_{in}) = \sum_{j_1=1}^{n} \sum_{j_2=1}^{n} \cdots \sum_{j_n=1}^{n} a_{1j_1} a_{2j_2} \ldots a_{nj_n},$$

we have a probability distribution on $\mathcal{F}_n$. Let E_j be the expected number of times that j occurs as a value of a random function $f \in \mathcal{F}_n$. Then

$$E_j = 1 \cdot a_{1j} + 1 \cdot a_{2j} + \cdots + 1 \cdot a_{nj} \quad (j = 1, 2, \ldots, n).$$

Thus the matrix $A = [a_{ij}]$ of order n, defining a random function f in $\mathcal{F}_n$, is a nonnegative matrix with row sums $1, 1, \ldots, 1$ and column sums $E_1, E_2, \ldots, E_n$. We may regard A as a *probabilistic analog of a function* from $\{1, 2, \ldots, n\}$ to $\{1, 2, \ldots, n\}$.

The probability that a random function in $\mathcal{F}_n$ is a bijection equals the permanent of A as given by

$$\text{per } A = \sum_{(j_1, j_2, \ldots, j_n)} a_{1j_1} a_{2j_2} \ldots a_{nj_n},$$

where the sum is over all permutations $(j_1, j_2, \ldots, j_n)$ of $\{1, 2, \ldots, n\}$. This permanent may be zero since, for instance, A may have a column of all zeros. If the expected number E_j of times that j occurs as a value of f equals 1 for each j, that is, if the column sums of A are all equal to 1, then A may be regarded as a *probabilistic analog of a bijection* from $\{1, 2, \ldots, n\}$ to $\{1, 2, \ldots, n\}$. Thus doubly stochastic matrices are probabilistic analogs of functions which are bijections. It follows from Theorem 1.7.1 that the permanent of a doubly stochastic matrix is positive. Thus if a function in $\mathcal{F}_n$ is chosen at random with distribution defined by a doubly stochastic matrix, then the probability that the function is a bijection is positive.

In this chapter we investigate doubly stochastic matrices in detail.

9.2 Basic Properties

Let $R = S = (1, 1, \ldots, 1)$ be vectors of all 1's of size n. A matrix A belongs to the transportation polytope $\mathcal{N}(R, S)$ if and only if all its elements are nonnegative and all its row and column sums equal 1. This polytope was denoted by Ω_n in Section 1.7 and is the polytope of *doubly stochastic matrices*, called the *assignment polytope*[1] or *Birkhoff polytope* of order n. By Theorem 8.1.1 the dimension of Ω_n equals $(n-1)^2$, and by Theorem 1.7.1 the extreme points of Ω_n are the $n!$ permutation matrices of order n. That the extreme points of Ω_n are permutation matrices also follows from Theorem 8.1.2: By that theorem the connected components of the bipartite graph of an extreme point of Ω_n are trees. The fact that the row and column sums are all equal implies immediately that these trees must be single edges and thus the matrix is a permutation matrix. The polytope Ω_n has $n!$ extreme points attaining the minimum $\tilde{e}(m, n)$ in Theorem 8.1.5.

Recall that a square matrix A of order n is *fully indecomposable* provided either $n = 1$ and A is not a zero matrix, or $n \geq 2$ and there do not exist permutation matrices P and Q such that

$$PAQ = \begin{bmatrix} X_1 & X_{12} \\ O & X_2 \end{bmatrix} \tag{9.1}$$

where X_1 and X_2 are square matrices of order at least 1; equivalently, for $n \geq 2$, A is fully indecomposable provided it does not have a zero submatrix of size k by $n - k$ for any k with $0 < k < n$. The square matrix A is *partly decomposable* provided it is not fully indecomposable. The fully indecomposable matrix A is *nearly decomposable* provided each matrix obtained from A by replacing a nonzero element with zero is not fully indecomposable. As remarked following Theorem 1.3.6, a nearly decomposable matrix of order $n \geq 3$ has at most $3(n - 1)$ nonzero elements.

Now assume that A is doubly stochastic and (9.1) holds. Then the property that all row and column sums of A equal 1 implies that $X_{12} = O$.[2] By iterating this argument on X_1 and X_2 we see that there are permutation matrices P' and Q' such that $P'AQ' = A_1 \oplus A_2 \oplus \cdots \oplus A_t$ where $A_1, A_2, \ldots, A_t$ are fully indecomposable matrices. The matrices $A_1, A_2, \ldots, A_t$ are the *fully indecomposable components* of A. Thus in the terminology of Section 1.3 (see Theorem 1.3.5) A has *total support*. A fully indecomposable component of order 1 is a *trivial component*. That a doubly stochastic matrix A has total support also follows from the fact that the extreme points of Ω_n are permutation matrices; A is then a convex combination of permutation matrices and so every nonzero element of A belongs to a nonzero diagonal [117].

From the above discussion we conclude that, apart from row and column permutations, the pattern of a doubly stochastic matrix is a direct sum of fully indecomposable $(0, 1)$-matrices and that for $R = S = (1, 1, \ldots, 1)$, a $(0, 1)$-matrix W satisfies the (R, S)-support condition if and only if it has total support. Applying Lemma 8.4.3 we conclude that there is a bijection

---

[1]See Section 9.12.

[2]By taking B equal to the pattern of A, this also is a special case of the (R, S)-support condition of Theorem 8.2.3.

between the nonempty faces of Ω_n and the $(0,1)$-matrices of order n with total support.

Let W be a $(0,1)$-matrix of order n with total support. Then

$$\Omega_n(W) = \{A \in \Omega_n : A \leq W \ (\text{entrywise})\}$$

denotes the face of Ω_n which has zeros wherever W has zeros and possibly elsewhere.[3] Now assume that W is, in addition, fully indecomposable. Then W is certainly indecomposable. Hence by Lemma 8.4.3 the dimension of $\Omega_n(W)$ equals $\sigma(W) - 2n + 1$, where $\sigma(W)$ equals the number of 1's in W. If W is, apart from row and column permutations, a direct sum of k fully indecomposable matrices, then $\dim \Omega_n(W) = \sigma(W) - 2n + k$.

We collect some of the above discussion in the following theorem.

Theorem 9.2.1 *Let n be a positive integer. Then the patterns of doubly stochastic matrices of order n are precisely the $(0,1)$-matrices of order n with total support. There is a bijection between the nonempty faces of the polytope Ω_n of doubly stochastic matrices and the $(0,1)$-matrices W of order n with total support. This bijection is given by*

$$W \longleftrightarrow \Omega_n(W) = \{A \in \Omega_n : A \leq W\};$$

moreover, $\dim \Omega_n(W) = \sigma(W) - 2n + k$ where k is the number of fully indecomposable components of W.   □

The extreme points of the face $\Omega_n(W)$ are the permutation matrices P with $P \leq W$. Hence the number of extreme points of the face $\Omega_n(W)$ equals the permanent of W. Since a convex polytope of dimension t has at least $t + 1$ extreme points, and has exactly $t + 1$ extreme points if and only if it is a simplex, we get the following inequality for the permanent of a $(0,1)$-matrix with total support, originally proved by Minc [97]. The geometric observation is from [27].

Corollary 9.2.2 *Let W be a $(0,1)$-matrix of order n with total support and let k be the number of fully indecomposable components of W. Then*

$$\text{per } W \geq \sigma(A) - 2n + k + 1,$$

with equality if and only if the face $\Omega_n(W)$ of Ω_n is a simplex.   □

If $A = [a_{ij}]$ is a nonnegative matrix of order n with total support, then we can use A to construct a doubly stochastic matrix with the same pattern as A. Let $B = [b_{ij}]$ be the matrix of order n defined by

$$b_{ij} = \frac{a_{ij}\text{per } A(i,j)}{\text{per } A} \quad (i, j = 1, 2, \ldots, n).$$

---

[3]In Chapter 8 this face was denoted by $\mathcal{N}_{\leq W}(R, S)$, $R = S = (1, 1, \ldots, 1)$.

Since A has total support, $a_{ij} \neq 0$ implies that per $A(i,j) \neq 0$ and hence B has the same pattern as A. It follows from the Laplace expansion for permanents that B is doubly stochastic.

Since $\dim \Omega_n = (n-1)^2$, the classical theorem of Carathéodory implies that a matrix $A \in \Omega_n$ can be written as a convex combination of at most $\dim \Omega_n + 1 = n^2 - 2n + 2$ permutation matrices. More generally, if W is the pattern of A and W has k fully indecomposable components, then A can be written as a convex combination of at most $\dim \Omega_n(W) + 1 = \sigma(W) - 2n + k + 1$ permutation matrices. We now describe an algorithm, called *Birkhoff's algorithm*.[4] Given a doubly stochastic matrix A, the algorithm extracts a permutation matrix in a representation of A as a convex combination of permutation matrices and reduces the dimension of the smallest face containing the resulting matrix by at least 1.

It is convenient to formulate the algorithm for *quasi doubly stochastic matrices*, that is, nonnegative matrices A all of whose row and column sums have the same positive value $\tau(A)$. By dividing the matrix by this common value, we get a doubly stochastic matrix. In particular, a $(0,1)$-matrix is the pattern of a quasi doubly stochastic matrix if and only if it is the pattern of a doubly stochastic matrix. When referring to the face of Ω_n containing a quasi doubly stochastic matrix we always mean the face containing the normalized doubly stochastic matrix.

The Birkhoff Algorithm for Quasi Doubly Stochastic Matrices

Begin with a quasi doubly stochastic matrix $A = [a_{ij}]$ of order n with pattern W and with row and column sums equal to $\tau(A)$.

(1) Set A_1 equal to A and W_1 equal to W.

(2) While $A_i \neq O$:

    (a) Determine a permutation matrix $P_i = [p_{ij}]$ such that $P_i \leq W_i$.

    (b) Choose a smallest positive element $c_i = a_{rs}$ of A_i in a position (r,s) corresponding to a 1 in P_i.

    (c) Set $A_{i+1} = (A_i - c_i P_1)$, a quasi doubly stochastic matrix with $\tau(A_{i+1}) = \tau(A_i) - c_i$, and let W_{i+1} be the pattern of A_{i+1}.

(3) Output $A = c_1 P_1 + c_2 P_2 + \cdots$

The following theorem is proved in [76], [111], [18]. The proof we give is from [18]. See also [84].

Theorem 9.2.3 *Let A be a doubly stochastic matrix of order n, and let W be the pattern of A. Let k be the number of fully indecomposable components of W. The Birkhoff algorithm expresses A as a convex combination of at most $\sigma(W) - 2n + k + 1$ permutation matrices.*

---

[4] This algorithm is implicit in Birkhoff's original proof that the extreme points of Ω_n are permutation matrices.

Proof. Each application of step (2) of the Birkhoff algorithm decreases the dimension of the smallest face containing the quasi doubly stochastic matrix. This is because for each i, the matrix W_i has total support and A_i is contained in the face $\Omega_n(W_i)$ and, by definition of W_i, in no smaller face. On the other hand, the matrix A_{i+1} is contained in $\Omega_n(W_{i+1})$. Since W_{i+1} has at least one more zero than W_i, $\Omega_n(W_{i+1})$ is a proper subface of $\Omega_n(W_i)$ and hence $\dim \Omega_n(W_{i+1}) < \dim \Omega_n(W_i)$. Thus the algorithm clearly terminates and gives A as a convex combination of at most $\sigma(W) - 2n + k + 1$ permutation matrices.   □

A representation of a doubly stochastic matrix A obtained by applying the Birkhoff algorithm is called a *Birkhoff representation* of A.[5] There may be many Birkhoff representations of a doubly stochastic matrix and the number of permutation matrices is not necessarily the same in all such representations. In [11] a construction is given for doubly stochastic matrices with the property that every Birkhoff representation requires the maximum number of permutation matrices (see Theorem 9.2.3).

Example. Let

$$A = \frac{1}{6} \begin{bmatrix} 1 & 4 & 0 & 1 \\ 2 & 1 & 3 & 0 \\ 2 & 1 & 1 & 2 \\ 1 & 0 & 2 & 3 \end{bmatrix}.$$

Then

$$A = \frac{1}{6}P_1 + \frac{1}{6}P_2 + \frac{1}{3} + \frac{1}{3}P_4 = \frac{1}{6}Q_1 + \frac{1}{6}Q_2 + \frac{1}{6}Q_3 + \frac{1}{6}Q_4 + \frac{1}{3}Q_5$$

are Birkhoff representations of A where

$$P_1 = \begin{bmatrix} 0 & 0 & 0 & 1 \\ 0 & 0 & 1 & 0 \\ 0 & 1 & 0 & 0 \\ 1 & 0 & 0 & 0 \end{bmatrix}, \quad P_2 = \begin{bmatrix} 1 & 0 & 0 & 0 \\ 0 & 1 & 0 & 0 \\ 0 & 0 & 1 & 0 \\ 0 & 0 & 0 & 1 \end{bmatrix},$$

$$P_3 = \begin{bmatrix} 0 & 1 & 0 & 0 \\ 1 & 0 & 0 & 0 \\ 0 & 0 & 0 & 1 \\ 0 & 0 & 1 & 0 \end{bmatrix}, \quad P_4 = \begin{bmatrix} 0 & 1 & 0 & 0 \\ 0 & 0 & 1 & 0 \\ 1 & 0 & 0 & 0 \\ 0 & 0 & 0 & 1 \end{bmatrix},$$

---

[5]A different way to carry out the algorithm is to interchange the order of steps (2)(a) and (2)(b) above, that is, choose first a smallest positive entry of A_i and then a permutation matrix $P_i \leq W_i$ containing the corresponding 1 in the pattern. This was the original procedure of Birkhoff. Doing (2)(a) before (2)(b) is less restrictive (since we choose a smallest [positive] entry in the positions corresponding to the 1's in P_i, not the smallest positive entry in A_i), but still produces a convex representation using no more than the number of permutation matrices guaranteed by Carathéodory's theorem.

and

$$Q_1 = \begin{bmatrix} 0 & 0 & 0 & 1 \\ 1 & 0 & 0 & 0 \\ 0 & 1 & 0 & 0 \\ 0 & 0 & 1 & 0 \end{bmatrix}, \quad Q_2 = \begin{bmatrix} 1 & 0 & 0 & 0 \\ 0 & 1 & 0 & 0 \\ 0 & 0 & 0 & 1 \\ 0 & 0 & 1 & 0 \end{bmatrix}, \quad Q_3 = \begin{bmatrix} 0 & 1 & 0 & 0 \\ 1 & 0 & 0 & 0 \\ 0 & 0 & 1 & 0 \\ 0 & 0 & 0 & 1 \end{bmatrix},$$

$$Q_4 = \begin{bmatrix} 0 & 1 & 0 & 0 \\ 0 & 0 & 1 & 0 \\ 0 & 0 & 0 & 1 \\ 1 & 0 & 0 & 0 \end{bmatrix}, \quad Q_5 = \begin{bmatrix} 0 & 1 & 0 & 0 \\ 0 & 0 & 1 & 0 \\ 1 & 0 & 0 & 0 \\ 0 & 0 & 0 & 1 \end{bmatrix}.$$

$\square$

Suppose that A is an integral, quasi doubly stochastic matrix with all row and column sums equal to q. Applying the Birkhoff algorithm to A we obtain a representation of A as a nonnegative, *integer* linear combination of permutation matrices.

Merris and Watkins [96] show that if m is a positive integer, the convex set of doubly stochastic matrices each of whose entries does not exceed $1/m$ is the convex hull of those doubly stochastic matrices of order n having m entries in each row and column equal to $1/m$.

9.3 Faces of the Assignment Polytope

Let n be a positive integer. By Theorem 9.2.1 the nonempty faces of Ω_n are of the form

$$\Omega_n(W) = \{A \in \Omega_n : A \leq W\}$$

where W is a $(0,1)$-matrix of order n with total support. We have $\Omega_n(J_n) = \Omega_n$. If $n \geq 3$ then Ω_n has n^2 facets and these are in one-to-one correspondence with the $(0,1)$-matrices of order n with exactly one 0. The faces of Ω_n are systematically investigated in [27] (see also [28]).

If W' is a $(0,1)$-matrix with total support with $W' \leq W$ and $W' \neq W$, then $\Omega_n(W')$ is a proper subface of $\Omega_n(W)$. Let W be a $(0,1)$-matrix of order n with total support and suppose that W has a trivial fully indecomposable component. Further, suppose this trivial component lies in row i and column j. Then each matrix in $\Omega_n(W)$ has a 1 in row i and column j and 0's elsewhere in row i and column j. If W^* is the matrix of order $n-1$ obtained from W by striking out the row i and column j, then $\Omega_n(W)$ and $\Omega_{n-1}(W^*)$ have the same essential combinatorial and geometric structure. Thus trivial fully indecomposable components can be omitted from our discussions.

We first consider faces of Ω_n which have a small number of vertices for their dimensions. The 0-dimensional faces (extreme points) of $\Omega_n(W)$ are the permutation matrices P with $P \leq W$, and thus are in one-to-one correspondence with certain permutations of $\{1, 2, \ldots, n\}$.

Now consider one-dimensional faces $\Omega_n(W)$. Since a fully indecomposable $(0,1)$-matrix W of order $m > 1$ contains at least two 1's in each row and column, by Theorem 9.2.1 $\Omega_n(W)$ has dimension at least $2m - 2m + 1 = 1$. It follows that W has exactly one nontrivial fully indecomposable component and this component has exactly two 1's in each row and column. Thus there exist permutation matrices P and Q such that

$$
PWQ = \left[\begin{array}{c|c}
I_{n-m} & O \\
\hline
O & \begin{array}{cccccc}
1 & 1 & 0 & \cdots & 0 & 0 \\
0 & 1 & 1 & \cdots & 0 & 0 \\
0 & 0 & 1 & \cdots & 0 & 0 \\
\vdots & \vdots & \vdots & \ddots & \vdots & \vdots \\
0 & 0 & 0 & \cdots & 1 & 1 \\
1 & 0 & 0 & \cdots & 0 & 1
\end{array}
\end{array}\right] . \tag{9.2}
$$

The extreme points and one-dimensional faces of Ω_n form the *graph* $G(\Omega_n)$ *of the polytope* Ω_n. The vertex set of $G(\Omega_n)$ can be identified with the set S_n of permutations of $\{1, 2, \ldots, n\}$. Two permutations σ and π are joined by an edge in this graph if and only if π has exactly one nontrivial cycle with respect to σ, that is, $\sigma^{-1}\pi$ is a permutation in S_n whose cycle decomposition contains exactly one cycle of length greater than 1.

By Corollary 9.2.2, if A is a $(0,1)$-matrix of order n with total support and W has k fully indecomposable components, then per $W \geq \sigma(W) - 2n + k + 1$ with equality if and only if $\Omega_n(W)$ is a simplex. If each fully indecomposable component of W is trivial, then W is a permutation matrix and $\Omega_n(W)$ is a simplex of dimension 0. Now suppose that W has at least two nontrivial fully indecomposable components. Since each such component contains at least two permutation matrices, there are two permutation matrices P and Q in $\Omega_n(W)$ that are not joined by an edge in $G(\Omega_n)$, and hence $\Omega_n(W)$ is not a simplex. We thus have the following corollary.

Corollary 9.3.1 *Let W be a $(0,1)$-matrix of order $n > 1$ with total support. Then $\Omega_n(W)$ is a simplex if and only if W has exactly one nontrivial fully indecomposable component W' and the permanent of W' equals $\sigma(W') - 2m + 2$ where W' is of order m.* □

A simplex of dimension 2 is a triangle. Besides triangles, the only other faces of dimension 2 of Ω_n are rectangles. Let W be a $(0,1)$-matrix of order n with total support such that $\dim \Omega_n(W) = 2$. First assume that W has exactly one nontrivial fully indecomposable component, and without loss of generality let W be fully indecomposable. Then by Theorem 9.2.1, $\sigma(W) = 2n + 1$. Hence one row sum of W equals 3 and the others equal 2. Such a matrix has permanent equal to 3 with each 1 in the row with sum 3 part of exactly one permutation matrix P with $P \leq W$ (otherwise the matrix is not fully indecomposable). Hence $\Omega_n(W)$ has dimension 2 and has three extreme points, and $\Omega_n(W)$ is a triangle. Now assume that W

has at least two (and hence exactly two) nontrivial fully indecomposable components. Since the dimension of $\Omega_n(W)$ equals 2, it follows that after row and column permutations, $W = W_1 \oplus W_2 \oplus I$ where W_1 and W_2 are fully indecomposable $(0,1)$-matrices with exactly two 1's in each row and column. It is now simple to check that $\Omega_n(W)$ has four extreme points forming a rectangle.

Example. Consider the polytope Ω_3. This polytope is contained in $\Re^9$, has dimension 4, and has six vertices. For every pair of distinct permutations σ and π of $\{1,2,3\}$, $\sigma^{-1}\pi$ has exactly one cycle of length greater than 1. From our discussion above it follows that each pair of distinct permutation matrices of order 3 determine a face of Ω_3 of dimension 1. Hence the graph $G(\Omega_3)$ is the complete graph K_6. Notice that while every pair of extreme points of Ω_3 determines an edge of Ω_n, Ω_3 is not a simplex. The polytope Ω_3 is an example of a polytope which is 2-neighborly but not a simplex. $\quad\square$

A polytope $\mathcal{P}$ is a *2-neighborly polytope* provided every pair of distinct extreme points determine an edge (a one-dimensional face). A simplex is a 2-neighborly polytope, but as illustrated in the preceding example, there are 2-neighborly polytopes that are not simplexes (but only in dimension 4 or higher).

Let W be a nonnegative integral matrix of order n with total support. Then W is a *2-neighborly matrix* provided for each pair of nonempty proper subsets I and J of $\{1, 2, \ldots, n\}$ of the same cardinality, either the permanent of the submatrix $W[I, J]$ does not exceed 1 or that of its complementary submatrix does not:

$$\text{per } W[I, J] \leq 1 \text{ or per } W(I, J) \leq 1.$$

Lemma 9.3.2 *Let W be a $(0,1)$-matrix of order n with total support. Then W is a 2-neighborly matrix if and only if $\Omega_n(W)$ is a 2-neighborly polytope.*

Proof. First assume that the matrix W is not 2-neighborly, and let I and J be nonempty proper subsets of $\{1, 2, \ldots, n\}$ of the same cardinality such that per $W[I, J] \geq 2$ and per $W(I, J) \geq 2$. Without loss of generality, we assume that $I = J = \{1, 2, \ldots, k\}$ for some integer k with $1 \leq k \leq n-1$. Let P_1 and P_2 be distinct permutation matrices with $P_1, P_2 \leq W[I, J]$, and let Q_1 and Q_2 be distinct permutation matrices with $Q_1, Q_2 \leq W(I, J)$. Then $P = P_1 \oplus Q_1$ and $Q_2 \oplus Q_2$ are permutation matrices with $P, Q \leq W$ such that P and Q do not determine an edge of $\Omega_n(W)$. Hence $\Omega_n(W)$ is not 2-neighborly.

Conversely, now assume that $\Omega_n(W)$ is not a 2-neighborly polytope. Then there exist two distinct permutation matrices P and Q such that P and Q do not determine an edge of $\Omega_n(W)$. Without loss of generality assume that $P = I_n$. Then the permutation π corresponding to Q has at least two nontrivial cycles implying there exist $I = J = \{1, 2, \ldots, k\}$ for some k with $1 \leq k < n$ such that per $W[I, J] \geq 2$ and per $W(I, J) \geq 2$. Hence the matrix W is not 2-neighborly. $\quad\square$

Before continuing our discussion of faces of Ω_n we introduce a general construction for matrices which is useful in inductive proofs involving faces of the assignment polytope.

Let $A = [a_{ij}]$ be an m by n real matrix, and let $u^1, u^2, \ldots, u^m$ and $v^1, v^2, \ldots, v^n$ be, respectively, the row and column vectors of A. If u^k contains exactly two nonzero elements, then A is *contractible on row* k. Similarly, A is *contractible on column* l provided v^l contains exactly two nonzero elements. Suppose that A is contractible on column l, and a_{pl} and a_{ql} are its two nonzero elements where $p < q$. Then the $m-1$ by $n-1$ matrix A' obtained from A by replacing row q of A with $a_{ql}u^p + a_{pl}u^q$ and then deleting row p and column l is called the *contraction of A on column l* (with respect to rows p and q).[6] The *contraction of A on row k* is defined in a similar way. A real matrix B is a *contraction* of matrix A provided either $B = A$ or B can be obtained from A by a sequence of row and column contractions.

Example. Consider the matrix

$$A = \begin{bmatrix} 1 & 4 & 2 \\ 2 & 3 & 0 \\ 0 & 1 & 5 \end{bmatrix}.$$

Then a contraction on column 1 followed by a contraction on row 1 (or column 1) of the resulting matrix results in a matrix of order 1:

$$\begin{bmatrix} 1 & 4 & 2 \\ 2 & 3 & 0 \\ 0 & 1 & 5 \end{bmatrix} \rightarrow \begin{bmatrix} 11 & 4 \\ 1 & 5 \end{bmatrix} \rightarrow \begin{bmatrix} 59 \end{bmatrix}.$$

Notice that 59 is the permanent of A.   □

The following lemma is useful in characterizing 2-neighborly matrices.

Lemma 9.3.3 *Let A be a fully indecomposable, nonnegative integral matrix of order $n \geq 2$ which can be contracted to a $(0,1)$-matrix B. Then A is a $(0,1)$-matrix.*

Proof. We may assume that B is obtained from A by a single contraction on column 1 with respect to rows 1 and 2. Let

$$A = \begin{bmatrix} a & u \\ b & v \\ 0 & C \end{bmatrix} \quad \text{and } B = \begin{bmatrix} av + bu \\ C \end{bmatrix}.$$

Since A is fully indecomposable, neither u nor v can be a zero vector. Since B is a $(0,1)$-matrix, it follows that $a = b = 1$, u and v are $(0,1)$-vectors, and C is a $(0,1)$-matrix. Hence A is a $(0,1)$-matrix.   □

---

[6]If A is the vertex–edge incidence matrix of a graph, then contraction on a column corresponds to the graph operation of edge-contraction.

We are primarily interested in contractions of $(0, 1)$-matrices. Since a contraction of a $(0, 1)$-matrix is, in general, only a nonnegative integral matrix, we begin with nonnegative integral matrices.

Theorem 9.3.4 *Let $A = [a_{ij}]$ be a nonnegative integral matrix of order $n > 1$, and let B be a contraction of A. Then the following hold.*

(i) per A = per B.

(ii) *If A is a 2-neighborly matrix, then B is a 2-neighborly matrix.*

(iii) *Suppose that B is obtained from A by a single contraction on column k with respect to rows p and q, where rows p and q of A each contain at least two nonzero elements. Then A is fully indecomposable if and only if B is fully indecomposable.*

(iv) *Suppose that B is obtained from A by a single contraction on column k and the two nonzero elements of column k are 1's. If B is a 2-neighborly matrix, then A is a 2-neighborly matrix.*

Proof. To prove each of the assertions, we may assume that B is obtained from A by a contraction on column 1 with respect to rows 1 and 2. Thus

$$A = \begin{bmatrix} a & u \\ b & v \\ 0 & C \end{bmatrix} \text{ and } B = \begin{bmatrix} av + bu \\ C \end{bmatrix}, \tag{9.3}$$

where $a \neq 0$, $b \neq 0$, and C is an $n - 2$ by $n - 1$ matrix.

The multilinearity of the permanent function implies that

$$\text{per } B = a \text{ per } \begin{bmatrix} v \\ C \end{bmatrix} + b \text{ per } \begin{bmatrix} u \\ C \end{bmatrix} = \text{per } A,$$

proving (i).

To prove (ii) we assume that A is a 2-neighborly matrix. Let I and J be sets of the same cardinality such that $\emptyset \neq I, J \subset \{1, 2, \ldots, n - 1\}$. One of the matrices $B_1 = B[I, J]$ and $B_2 = B(I, J)$ is a submatrix of C, and we may assume that B_2 is a submatrix of C. There exist sets K, L of the same cardinality with $\emptyset \subset K, L \subset \{1, 2, \ldots, n\}$ such that

$$A_1 = A[K, L] = \begin{bmatrix} a & u' \\ b & v' \\ 0 & C' \end{bmatrix},$$

where B_1 is a contraction of A_1 on column 1 with respect to rows 1 and 2. By (i) per A_1 = per B_1. Moreover, $A(K, L) = B_2$. Since A is a 2-neighborly matrix, it now follows that per $B_1 \leq 1$ or per $B_2 \leq 1$. Hence B is a 2-neighborly matrix.

For (iii), we suppose that B is obtained from A by a single contraction on column k with respect to rows p and q, where rows p and q of A each contain at least two nonzero elements. First assume that B is partly decomposable. Then B has an r by s zero submatrix $O_{r,s}$ where r and s are positive integers with $r + s = n - 1$. If $O_{r,s}$ is also a submatrix of C then clearly A has an r by $s + 1$ zero submatrix and $r + s + 1 = n$, implying that A is partly decomposable. Now assume that $O_{r,s}$ is not a submatrix of C. Since a and b are positive, and u and v are nonnegative vectors, we conclude that A has an $r + 1$ by s zero submatrix with $r + 1 + s = n$, and again A is partly decomposable. For the converse, we now assume that A is partly decomposable. There are positive integers p and q with $r + s = n$ such that A contains a p by q zero submatrix $O_{p,q}$. If $O_{p,q}$ is contained in the last $n - 2$ rows of A, then $q \geq 2$ and B contains a p by $q - 1$ zero submatrix with $p + q - 1 = n - 1$, implying B is partly decomposable. Otherwise, since a and b are positive, $O_{p,q}$ is contained in the last $n - 1$ columns of A. The assumption that rows p and q of A contain at least two nonzero elements implies that $O_{p,q}$ is neither u nor v, and $p \geq 2$. Thus in contracting A to get B, $O_{p,q}$ becomes a $p - 1$ by q zero matrix with $p - 1 + q = n - 1$, and B is partly decomposable.

To prove (iv), we now assume that $a = b = 1$ so that

$$B = \begin{bmatrix} u + v \\ C \end{bmatrix}.$$

Assume that B is a 2-neighborly matrix. Let I and J be sets of equal cardinality such that $\oslash \subset I, J \subset \{1, 2, \ldots, n\}$. Let $A_1 = A[I, J]$ and $A_2 = A(I, J)$. As a first case, assume that A_1 or A_2 is a submatrix of C. Then as in the proof of (ii), we see that there exist complementary submatrices $B_1 = B[K, L]$ and $B_1 = B(K, L)$ of B such that per $A_1 = $ per B_1 and per $A_2 = $ per B_2. Since B is 2-neighborly either per $A_1 \leq 1$ or per $A_2 \leq 1$. Thus in this case, A is a 2-neighborly matrix. The remaining case is that neither A_1 nor A_2 is a submatrix of C. First assume that A_1 or A_2 has order 1. Without loss of generality we may assume also that $A_1 = [a_{1j}]$ where $2 \leq j \leq n$. Then per $A_2 = $ per B_2 where B_2 is the submatrix of B which is complementary to its submatrix $[a_{1j} + a_{2j}]$ of order 1. Since the elements of B are nonnegative integers, it now follows from the assumption that B is 2-neighborly that either per $A_1 = a_{1j} \leq 1$ or per $A_2 \leq 1$. Now assume that the orders of A_1 and A_2 are both at least 2. Our supposition that neither A_1 nor A_2 is a submatrix of C now allows us to assume that $1 \in I \cap J$ and $2 \notin I$. It now follows that there are complementary submatrices B_1 and B_2 of B such that B_1 is a contraction of A and $A_2 \leq B_2$ (entrywise). Hence per $A_1 = $ per B_1 and per $A_2 \leq $ per B_2. Since B is a 2-neighborly matrix, either per $A_1 \leq 1$ or per $A_2 \leq 1$. Thus in this case A is also a 2-neighborly matrix. This completes the proof of (iv).   □

We observe that if A and B have the form (9.3), so that B is a contraction of A on column 1 with respect to rows 1 and 2, and if $a = b = 1$, then the sums of the elements of A and B satisfy $\sigma(B) = \sigma(A) - 2$, and the row sum vector of B is obtained from that of A by replacing its first two components with their sum minus 2, and the column sum vector of B is obtained from that of A by deleting its first component.

Since a 2-neighborly $(0,1)$-matrix can have at most one nontrivial fully indecomposable component, we assume that the matrix in the next theorem, characterizing 2-neighborly matrices, is fully indecomposable.

Theorem 9.3.5 *Let A be a fully indecomposable, nonnegative integral matrix of order $n \geq 2$. Then A is a 2-neighborly matrix if and only if A satisfies one of the two properties:*

(i) *The all 1's matrix J_3 of order 3 is a contraction of A.*

(ii) *There exist an integer p with $0 \leq p \leq n-1$ and permutation matrices P and Q such that*

$$PAQ = \begin{bmatrix} A_3 & A_1 \\ A_2 & O \end{bmatrix}, \qquad (9.4)$$

where A_3 is an $n-p$ by $p+1$ nonnegative integral matrix, and A_1 and A_2^T are $(0,1)$-matrices with exactly two 1's in each column. Such a matrix A satisfies per $A = \sigma(A_3)$ where $\sigma(A_3)$ is the sum of the elements of A_3.

Proof. First assume that A satisfies (i). Since J_3 is a 2-neighborly $(0,1)$-matrix, and the fully indecomposable matrix A can be contracted to J_3, it follows from Lemma 9.3.3 that A is a $(0,1)$-matrix and then from (iv) of Theorem 9.3.4 that A is a 2-neighborly matrix. Now assume that A satisfies (ii). We prove by induction on n that A is a 2-neighborly matrix. First suppose that $n = 2$. Then in (ii), $p = 1$ or 2 and the zero matrix O is vacuous. The matrix in (9.4) is of the form

$$\begin{bmatrix} a & b \\ 1 & 1 \end{bmatrix} \quad \text{or} \quad \begin{bmatrix} a & 1 \\ b & 1 \end{bmatrix}.$$

Thus A is 2-neighborly when $n = 2$. Now suppose that $n > 2$, and without loss of generality that A itself has the form given in (9.4). First consider the case in which $0 \leq p \leq n-2$ so that A_1 is not vacuous. The matrix A' obtained from A by contraction on column n (with respect to the rows containing its two 1's) has the form

$$\begin{bmatrix} A_3' & A_1' \\ A_2 & O \end{bmatrix}.$$

By (iii) of Theorem 9.3.4 A' is a fully indecomposable, nonnegative integral matrix of order $n-1$, and A_3' is of size $(n-1)-p$ by $p+1$ with $0 \leq p \leq n-2$ and $\sigma(A') = \sigma(A)$. If $p = n-2$, then A_1' is vacuous and it follows that A' satisfies the corresponding property (ii) with n replaced with $n-1$. Now assume that $0 \leq p \leq n-3$ so that A_1' is not vacuous. Then all the column sums of A_1' equal 2. Since A' is fully indecomposable, A_1' is a $(0,1)$-matrix with exactly two 1's in each column. Hence by induction A' is 2-neighborly.

Now by (iv) of Theorem 9.3.4, A is 2-neighborly. Finally consider the case in which $p = n - 1$. Then a similar argument using a contraction on row n shows that A is 2-neighborly. Thus when A satisfies (i) or (ii), A is 2-neighborly. Moreover, it follows from Theorem 9.3.4 that per $A = \sigma(A_3)$ since, in our induction, $\sigma(A_3') = \sigma(A_3)$.

We now assume that $A = [a_{ij}]$ is 2-neighborly and show by induction on n that A satisfies (i) or (ii). If $n = 2$, then $A = J_2$ and (ii) holds with $p = 1$. Now let $n = 3$. If $A = J_3$, then (i) holds. Otherwise, A has at least one 0 which we may assume to be a_{33}. If A has the form given in (9.4) with $p = 1$, then we are done. Otherwise row 3 or column 3 of A has an element greater than 1. Without loss of generality, we may assume that $a_{13} > 1$. The assumption that A is 2-neighborly implies that per $A[\{2,3\},\{1,2\}] = 1$. The full indecomposability assumption on A implies that after row and column permutations

$$A = \begin{bmatrix} b & c & d \\ 1 & 0 & e \\ f & 1 & 0 \end{bmatrix},$$

where $cef \neq 0$; the 2-neighborly assumption on A implies that at most one of c, e, and f is greater than 1. Since A does not have the form specified in (ii) with $p = 1$, either $c > 1$ or $e > 1$, and $f = 1$. Suppose $c > 1$. Then $e = 1$ and A satisfies (ii) with $p = 2$. Finally suppose that $e > 1$ and thus $c = 1$. Then the 2-neighborly assumption implies that $b = 0$, and it follows that A satisfies (ii) with $p = 0$. This completes the argument for $n = 3$.

We now assume that $n > 3$. We first use the inductive assumption to show that A has a row or column with two 1's and all other elements equal to 0. Assume to the contrary that no row or column of A has this property. A nearly decomposable matrix can be obtained from A by replacing certain nonzero elements with 0's, and a nearly decomposable matrix of order at least 2 contains both a row and a column with exactly two nonzero elements. Thus the sets U, respectively V, consisting of all fully indecomposable $(0,1)$-matrices X such that $X \leq A$ and X has exactly two 1's in row i, respectively column i, are nonempty. For each $i \in U$, let m_i be the number of positive elements in row i of A, and for each $i \in V$, let n_i be the number of positive elements in column i of A. Let

$$s = \min\{\{m_i : i \in U\} \cup \{n_i : i \in V\}\}. \tag{9.5}$$

Without loss of generality we assume that $s = n_k$ for some $k \in V$ and then take $k = 1$. There exists a fully indecomposable matrix A' obtained from A by replacing certain positive elements in column k with 0's such that column k of A' has positive elements only in two positions, which we can take to be rows 1 and 2. Since $A' \leq A$, A' is also a 2-neighborly matrix. Let

$$A' = \begin{bmatrix} a & \alpha \\ b & \beta \\ 0 & C \end{bmatrix},$$

and let

$$B = \begin{bmatrix} a\beta + b\alpha \\ C \end{bmatrix}$$

be the matrix obtained from A' by contraction on column 1. By (ii) of Theorem 9.3.4, B is a 2-neighborly matrix. Hence by the inductive assumption B satisfies (i) or (ii). First assume that B satisfies (i), that is, B can be contracted to J_3. If $n = 4$, then $B = J_3$ and B is equal to one of the matrices

$$\begin{bmatrix} 1 & 1 & 0 & 0 \\ 1 & 0 & 1 & 1 \\ 0 & 1 & 1 & 1 \\ 0 & 1 & 1 & 1 \end{bmatrix}, \quad \begin{bmatrix} 1 & 0 & 0 & 1 \\ 1 & 1 & 0 & 0 \\ 0 & 1 & 1 & 1 \\ 0 & 1 & 1 & 1 \end{bmatrix}.$$

It can be verified that replacing with a positive integer any 0 in column 1 of either of these two matrices results in a matrix that is not 2-neighborly. Hence $A = A'$ and this completes the verification in the induction for $n = 4$ in the case that B can be contracted to J_3. We now continue with this case assuming that $n > 4$. Since B is a fully indecomposable matrix of order $n-1$ that can be contracted to J_3, the sum of the entries of B equals $2(n-1)+3$. Since $n-1 > 3$, this implies that B is a (0,1)-matrix and that B, and hence A, has a column with two 1's and $n-2$ 0's. Therefore if B can be contracted to J_3, A has a column with two 1's and $n-2$ 0's.

Now assume that B satisfies the conditions of (ii). Thus there are permutation matrices G and H such that

$$GBH = \begin{bmatrix} B_3 & B_1 \\ B_2 & O \end{bmatrix}$$

where B_3 is an $n-1-p$ by $p+1$ matrix and $0 \le p \le n-2$. If $p \ne n-2$ then B, and hence A, has a column with two elements equal to 1, and the other elements equal to 0. Now assume that $p = n-2$. Then

$$GB = \begin{bmatrix} B_3 \\ B_2 \end{bmatrix}$$

where B_2 is an $n-2$ by $n-1$ (0,1)-matrix with exactly two 1's in each row (in this case we may assume, as shown, that H is the identity matrix). If the first row of B corresponds to a row of B_2, then the first row of B contains two 1's with all other elements equal to 0. Now assume that the first row of B corresponds to the (unique) row of B_3. Thus $B_3 = a\beta + b\alpha$ and $B_2 = C$. We now show that the integer s defined in (9.5) satisfies $s \le 3$. There exists a nearly decomposable (0,1)-matrix F with $F \le A$. The number of 1's in a nearly decomposable matrix of order $n \ge 3$ is at most $3(n-1)$. Since $n > 4$, F has at least three rows with exactly two 1's. Hence there is a $t \ge 3$ such that row t of F contains exactly two 1's. Then t is in the set U, and s as defined in (9.5) satisfies $s \le 3$. Since $n > 4$, there is an integer $i \ge 3$ such that $a_{i1} = 0$ and row i of A has two elements equal to 1 and the other elements equal to 0.

In summary, with $n > 4$ we have shown at this point that A has either a row or a column with two elements equal to 1 and the other elements equal to 0. Without loss of generality we may now assume that A and the matrix B obtained from A by contraction on column 1 with respect to rows 1 and 2 satisfy

$$A = \begin{bmatrix} 1 & \alpha \\ 1 & \beta \\ 0 & C \end{bmatrix} \text{ and } B = \begin{bmatrix} \alpha + \beta \\ C \end{bmatrix}.$$

Since B is a 2-neighborly matrix, the inductive assumption implies that either J_3 is a contraction of B and therefore of A, or B satisfies the conditions in (ii) in the statement of the theorem. Assume that B satisfies (ii). Then there exist an integer q with $0 \le q \le n - 2$ and permutation matrices P' and Q' such that

$$B' = P'BQ' = \begin{bmatrix} B_3' & B_2' \\ B_1' & O \end{bmatrix},$$

where B_3' is an $n - 1 - q$ by $q + 1$ nonnegative integral matrix, and B_1' and $B_2'^{\mathrm{T}}$ are $(0,1)$-matrices with exactly two 1's in each column. If the first row of B corresponds to one of the first $n - 1 - q$ rows of B', then A satisfies (ii) with $p = q$ (A_3 is an $n - q$ by $q + 1$ matrix). If the first row of B corresponds to one of the last rows of B', then A satisfies (ii) with $p = q + 1$ (A_3 is an $n - q + 1$ by $q + 2$ matrix). The induction is complete, and so is the proof of the theorem. $\square$

We now use Theorem 9.3.5 to characterize the simplicial faces of Ω_n. It follows from Corollary 9.3.1 that there is no loss in generality in assuming A is fully indecomposable in the next theorem.

Theorem 9.3.6 *Let A be a fully indecomposable $(0, 1)$-matrix of order $n > 2$. Then the face $\Omega_n(A)$ of Ω_n is a simplex if and only if there exist an integer p with $1 \le p \le n - 2$ and permutation matrices P and Q such that*

$$PAQ = \begin{bmatrix} A_3 & A_1 \\ A_2 & O \end{bmatrix}, \tag{9.6}$$

where A_3 is an $n - p$ by $p + 1$ $(0,1)$-matrix, and A_1 and A_2^{T} are $(0,1)$-matrices with exactly two 1's in each column.

Proof. First assume that $\Omega_n(A)$ is a simplex. Then $\Omega_n(A)$ is 2-neighborly and hence satisfies (i) or (ii) of Theorem 9.3.5. Suppose that A satisfies (i) so that A can be contracted to J_3. Then it follows from Theorems 9.2.1 and 9.3.4 that per $A - \dim \Omega_n(A) = $ per $J_3 - \dim \Omega_3(J_3) = 2$. Hence $\Omega_n(A)$ is not a simplex. Thus A satisfies (ii) of Theorem 9.3.5. We now show that p can be chosen in the range $1 \le p \le n - 2$. Suppose A satisfies (ii) with $p = 0$. Then in (9.6) we have that A_3 is n by 1, and A_1 is n by $n - 1$ with exactly two 1's in each column (A_2 and O are empty).

Thus some row of A_1, say row n of A_1, contains at most one 1. Since A is fully indecomposable, this row contains exactly one 1 and A has a 1 in the $(n, 1)$-position, Hence A satisfies (ii) with $p = 1$. A similar argument shows that if A satisfies (ii) with $p = n - 1$ then it also satisfies (ii) with $p = n - 2$. Thus A satisfies (9.6).

Now assume that A satisfies (9.6). It follows from Theorem 9.3.5 that per $A = \sigma(A_3) = \dim \Omega_n(A) + 1$, and therefore $\Omega_n(A)$ is a simplex. $\quad\square$

We now identify the maximal simplexes among the faces of Ω_n.

Corollary 9.3.7 *Let A be a $(0, 1)$-matrix of order $n > 2$ with total support such that the face $\Omega_n(A)$ of Ω_n is a simplex. Then there does not exist a face of Ω_n which is a simplex and properly contains $\Omega_n(A)$ if and only if A is fully indecomposable and there exist an integer p with $1 \le p \le n - 2$ and permutation matrices P and Q such that*

$$PAQ = \begin{bmatrix} J_{n-p,p+1} & A_1 \\ A_2 & O \end{bmatrix}, \tag{9.7}$$

where A_1 and A_2^{T} are $(0, 1)$-matrices with exactly two 1's in each column.

Proof. A fully indecomposable matrix A as specified in (9.7) has the property that replacing any 0 with a 1 gives a matrix B with at most $n - 2$ rows and columns with exactly two 1's which therefore cannot satisfy the conditions of Theorem 9.3.6. Hence $\Omega_n(B)$ is not a simplex.

For the converse, first assume that A is not fully indecomposable. Then using Theorem 9.3.6 one readily shows that there is a $(0, 1)$-matrix B of order n such that $\Omega_n(B)$ is a simplex properly containing $\Omega_n(A)$. Now assume that A is fully indecomposable. Since $\Omega_n(A)$ is a simplex, by Theorem 9.3.6 A satisfies (9.6) with $1 \le p \le n - 2$. If $A_3 \ne J_{n-p,p+1}$, then replacing any 0 in A_3 with 1 gives a matrix B which, by Theorem 9.3.6 again, has the property that $\Omega_n(B)$ is a simplex. Hence $\Omega_n(A)$ is not a maximal simplicial face of Ω_n. $\quad\square$

It follows from Corollary 9.3.7 that the maximal dimension of a simplicial face of Ω_n equals $\lfloor (n + 1)/2 \rfloor \lceil (n + 1)/2 \rceil - 1$ for all $n > 2$.

By Theorem 9.3.6, if A is a fully indecomposable matrix of order $n > 2$, then $\Omega_n(A)$ is a simplex if and only if A has the form given in (9.6) for some permutation matrices P and Q and some integer p with $1 \le p \le n - 2$. To complete a characterization of the simplicial faces of Ω_n we must determine when a matrix of this form is fully indecomposable. We do that in the next theorem under the more general conditions given in (9.4).

Recall that the *vertex–edge incidence matrix* of a graph with k vertices $v_1, v_2, \ldots, v_k$ and m edges $e_1, e_2, \ldots, e_m$ is the k by m $(0,1)$-matrix with a 1 in row i and column j exactly when v_i is a vertex of edge e_j. Such a matrix has exactly two 1's in each column and no two columns are identical.

Theorem 9.3.8 *Let p and n be integers with $n \geq 2$ and $0 \leq p \leq n - 1$. Let A be a nonnegative integral matrix of order n of the form*

$$PAQ = \begin{bmatrix} A_3 & A_1 \\ A_2 & O \end{bmatrix}, \tag{9.8}$$

where A_3 is an $n-p$ by $p+1$ matrix and where A_1 and A_2^T are $(0,1)$-matrices with exactly two 1's in each column. Then A is fully indecomposable if and only if

(i) *A_1 and A_2^T are vertex–edge incidence matrices of trees, and*

(ii) *A has at least positive two elements in each row and column.*

Proof. We first assume that A is fully indecomposable. Then (ii) clearly holds, and we show that A_1 is the vertex–edge incidence matrix of a tree. A similar argument will show that A_2^T is also the vertex–edge incidence matrix of a tree, thus verifying (i). If $p = n - 1$, then A_1 is vacuous so we assume that $p \leq n - 2$. The matrix A_1 is an $n - p$ by $n - p - 1$ $(0, 1)$-matrix with exactly two 1's in each column. Since A is fully indecomposable A_1 cannot have two identical columns. Thus A_1 is the vertex–edge incidence matrix of some graph G with $n - p$ vertices and $n - p - 1$ edges. To show that G is a tree we need only show that G is a connected graph. Suppose to the contrary that G has at least two connected components. Then this implies that there exist permutation matrices U and V such that $U A_1 V = B_1 \oplus B_2$ where B_1 is an r by s matrix with $0 < r < n - p$. Thus A contains both an $n - r$ by s zero matrix with $n - r + s \geq n$ if $s \geq r$, and a $p + r$ by $n - p - 1 - s$ zero matrix with $p + r + (n - p - 1 - s) = n + r - 1 - s \geq n$ if $s \leq r - 1$. This implies that A is partly decomposable in contradiction to our assumption.

Now assume that A_1 and A_2^T are vertex–edge incidence matrices of trees and A has at least two positive elements in each row and column. We prove that A is fully indecomposable by induction on n. This is readily checked if $n = 2$ or 3. We now assume that $n \geq 4$ so that either A_1 has at least two columns or A_2 has at least two rows. It suffices to consider the case in which A_1 has at least two columns. Since A_1 is the vertex–edge incidence matrix of a tree G, G has a pendant edge (an edge one of whose vertices meets no other edge). We may assume that the last column of A_2 corresponds to this edge and that its two vertices correspond to rows 1 and 2 of A_2. Thus column n of A contains exactly two 1's and these 1's are in rows 1 and 2. Let B be the matrix obtained from A by a contraction on column n with respect to rows 1 and 2. Then B has the form

$$B = \begin{bmatrix} A_3' & A_1' \\ A_2 & O \end{bmatrix}.$$

The matrix A_2', being a contraction on the last column (corresponding to a pendant edge of G) of A_2 with respect to its first two rows, is also the vertex–edge incidence matrix of a tree. In particular, A_1' has at least two 1's in each column, and hence the last $(n-1) - (p+1)$ columns of B each contain at least two 1's. The matrix A_2, being the transpose of the vertex–edge incidence matrix of a tree, has at least one 1 in each column. Suppose some column k of A_2 contains exactly one 1. Then our hypotheses imply that column k of A_3, and hence column k of A_3', contains at least one positive element. Therefore each of the first $p+1$ columns of B contains at least two positive elements. Since A has at least two positive elements in each row, it is clear that each row of B, with the possible exception of row 1, contains at least two positive elements. But since row 1 or row 2 of A_1 corresponds to a pendant edge of the graph G, either row 1 or row 2 of A_3 contains a positive element, and hence row 1 of B contains at least two positive elements. We apply the inductive assumption to B and conclude that B is fully indecomposable. Now by Theorem 9.3.4, A is fully indecomposable, and the proof of the theorem is complete. $\qquad\square$

We now characterize fully indecomposable $(0,1)$-matrices A of order n which can be contracted to J_3. It follows from Theorems 9.3.4 and 9.3.8 that these are the matrices for which $\Omega_n(A)$ is a 2-neighborly polytope but not a simplex.

Theorem 9.3.9 *Let A be a fully indecomposable, nonnegative integral matrix of order $n \geq 3$. Then A can be contracted to J_3 if and only if A satisfies each of the following properties:*

(i) *per $A = 6$;*

(ii) *the sum of the elements of A equals $2n + 3$;*

(iii) *there exist permutation matrices P and Q such that if $B = PAQ$, then per $B[\{1, 2, \ldots, n-3\}] > 0$ (for $n \geq 3$), and each of the last three row and column sums of B equals 3.*

Proof. First assume that A can be contracted to J_3. We prove that properties (i), (ii), and (iii) are satisfied by induction on n. If $n = 3$, then $A = J_3$ and these properties hold. Now assume that $n \geq 4$. There exist matrices $A_0 = A, A_1, \ldots, A_{n-4}, A_{n-3} = J_3$ such that A_i is a contraction of A_{i-1} for $i = 1, 2, \ldots, n-3$. Since J_3 is a $(0,1)$-matrix, A is also a $(0,1)$-matrix, and (i) of Theorem 9.3.4 and the discussion following the proof of that theorem imply that properties (i) and (ii) hold. We may assume that

$$A = \begin{bmatrix} 1 & \alpha \\ 1 & \beta \\ 0 & C \end{bmatrix} \quad \text{and} \quad A_1 = \begin{bmatrix} \alpha + \beta \\ C \end{bmatrix}.$$

Since A_1 can be contracted to J_3, we apply the inductive assumption to conclude that there are permutation matrices P_1 and Q_1 such that

$$P_1 A_1 Q_1 = B = \begin{bmatrix} B_{11} & B_{12} \\ B_{21} & B_{22} \end{bmatrix}$$

where B_{22} is a square matrix of order 3, per $B_{11} > 0$, and each of the last three row and column sums of B equals 3. Since A_1 can be contracted to J_3, the sum $\sigma(\alpha + \beta) = \sigma(\alpha) + \sigma(\beta)$ of the elements in $\alpha + \beta$ equals 2 or 3. First suppose this sum is 2, so that α and β each contain exactly one 1. Then there exist permutation matrices P and Q such that

$$PAQ = C = \begin{bmatrix} C_{11} & C_{12} \\ C_{21} & C_{22} \end{bmatrix}$$

where C_{22} is a matrix of order 3, each of the last three row and column sums of C equals 3, and B_{11} is a contraction of C_{11}. Since per $B_{11} > 0$, it follows from (i) of Theorem 9.3.4 that per $C_{11} > 0$. Thus A satisfies (iii) in this case. If $\sigma(\alpha) + \sigma(\beta) = 3$ and if, without loss of generality, $\sigma(\alpha) = 1$ and $\sigma(\beta) = 2$, then a similar argument shows that A satisfies (iii) in this case as well.

For the converse, we prove by induction on n that the assumptions (i), (ii), and (iii) together imply that $A = [a_{ij}]$ can be contracted to J_3 (implying, in particular, that A is a (0,1)-matrix). First let $n = 3$. Then each row and column sum of A equals 3. If $A \neq J_3$, then A contains a 0. We may assume, since A is fully indecomposable, that $a_{11} = 0$ and then that $a_{12} = a_{21} = 1$ and $a_{13} = a_{31} = 2$. It then follows that A is one of the two matrices

$$\begin{bmatrix} 0 & 1 & 2 \\ 1 & 1 & 1 \\ 2 & 1 & 0 \end{bmatrix} \quad \text{and} \quad \begin{bmatrix} 0 & 1 & 2 \\ 1 & 2 & 0 \\ 2 & 0 & 1 \end{bmatrix}.$$

Since neither of these matrices has permanent equal to 6, we contradict (iii). Hence $A = J_3$. Now let $n \geq 4$. Then (ii) and (iii) imply that there exist integers r and s with $1 \leq r < s \leq n$ such that the matrix $B = [b_{ij}]$ satisfies $b_{r1} = b_{s1} = 1$ and $b_{i1} = 0$ for $i \neq r, s$. Let C be the matrix obtained from B by a contraction on column 1 with respect to rows r and s. Applying Theorem 9.3.4 we see that C is a fully indecomposable, nonnegative integral matrix satisfying per $C = 6$, $\sigma(C) = 2(n-1)+3$, and each of the last three column sums of C equals 3. Let $B_1 = B[\{1, 2, \ldots, n-3\}]$ and, when $n \geq 5$, let $C_1 = C[\{1, 2, \ldots, n-4\}]$. Because per $B_1 > 0$, the first column of B_1 contains a 1 and hence $r \leq n-3$, implying that the last three row sums of C are also equal to 3. If $n = 4$, then C satisfies (iii), and so (i), (ii), and (iii). Now let $n \geq 5$. Assume first that $s > n - 3$. Then b_{r1} is the only positive element in the first column of B_1, and so $0 < $ per $B_1 = b_{r1}$ per $B_1(r, 1)$ and per $B_1(r, 1) > 0$. Since $C_1 = B_1(r, 1)$ when $s > n - 3$, we conclude that C satisfies (i), (ii), and (iii). Now assume that $s \leq n - 3$. Then C_1 is a contraction of B_1, and hence by (i) of Theorem 9.3.4, per $C_1 = $ per $B_1 > 0$. Thus in this case C also satisfies (i), (ii), and (iii). Applying the inductive assumption, we conclude that B and hence A can be contracted to J_3. $\square$

Theorems 9.3.5 and 9.3.9 show that the faces of Ω_n which are 2-neighborly polytopes but not simplexes are four-dimensional polytopes with six vertices. In [27], the following theorem is proved which characterizes faces of

Ω_n whose number of extreme points exceeds its dimension by 2. We refer the interested reader to [27] for a proof.

Theorem 9.3.10 *Let A be a $(0,1)$-matrix of order $n \geq 3$ with total support. Then for some d, $\Omega_n(A)$ has dimension d and exactly $d+2$ extreme points if and only if one of the following holds:*

(i) *$d = 2$ and $\Omega_n(A)$ is a rectangle;*

(ii) *$d \geq 3$, and A has exactly one nontrivial fully indecomposable component B with B satisfying one of*

  (a) *B can be contracted to K_3 (so $\Omega_n(A)$ is a four-dimensional 2-neighborly polytope with six vertices),*

  (b) *there exist positive integers r, s and t with $r \geq 2$, $s \geq 2$, and $r + s + t = n + 1$, and there exist permutation matrices P and Q such that*

$$PAQ = \begin{bmatrix} A_3 & A_1 & A_4 \\ A_2 & O & O \\ A_5 & O & A_6 \end{bmatrix}$$

  where A_3 is an r by s matrix, A_6 is a t by t matrix with exactly $2t - 1$ 1's, A_1 and A_2^T have exactly two 1's in each column, and A_4 and A_5^T contain exactly two 1's and these 1's occur in their first column.

Thus far we have primarily considered faces of Ω_n with a small number of extreme points for their dimension, the smallest being $k+1$ extreme points for dimension k. We now look at the opposite circumstance and consider faces of dimension k with a large number of extreme points. Since the nonempty faces of Ω_n are in bijective correspondence with $(0,1)$-matrices A of order n with total support, the maximal number of extreme points a face of Ω_n of dimension k can have equals the maximal permanent of such matrices A with $\dim \Omega_n(A) = k$. We give without proof a theorem of Foregger [63] for this maximum (see also [39]).

Theorem 9.3.11 *Let A be a fully indecomposable $(0,1)$-matrix of order $n > 1$. Then*

$$\operatorname{per} A \leq 2^{\sigma(A)-2n} + 1. \tag{9.9}$$

Equality holds if and only if there exist permutation matrices P and Q and an integer $r \geq 2$ such that

$$PAQ = \begin{bmatrix} A_1 & O & \cdots & O & E_r \\ E_1 & A_2 & \cdots & O & O \\ \vdots & \vdots & \ddots & \vdots & \vdots \\ O & O & \cdots & A_{r-1} & O \\ O & O & \cdots & E_{r-1} & A_r \end{bmatrix} \tag{9.10}$$

where

(i) E_i *contains exactly one* 1 $(1 \leq i \leq r)$;

(ii) $r - k + 1$ *of the* A_i *equal the identity matrix* I_1 *of order* 1;

(iii) *the other* $k - 1$ *of the* A_i *are fully indecomposable matrices of order at least* 2 *with exactly two* 1*'s in each row and column.*

If A is a fully indecomposable $(0, 1)$-matrix of order n, then $\dim \Omega_n(A) = \sigma(A) - 2n + 1$. Thus the inequality (9.9) can be restated as

$$\text{per } A \leq 2^{\dim \Omega_n(A)-1} + 1, \tag{9.11}$$

and so the maximal number of extreme points of a face of Ω_n of dimension k corresponding to a fully indecomposable matrix is $2^{k-1} + 1$. If A is not fully indecomposable, the number of extreme points can be larger. We include the proof of this result [27] as given in [39].

Theorem 9.3.12 *Let* n *be a positive integer. The maximal number of extreme points of a face of* Ω_n *of dimension* k *equals* 2^k, *with equality if and only if the face is a* k-*dimensional parallelepiped. The face* $\Omega_n(A)$ *corresponding to a* $(0, 1)$-*matrix* A *of order* n *with total support is a* k-*dimensional parallelepiped for some* k *if and only if there exist permutation matrices* P *and* Q *such that*

$$PAQ = A_1 \oplus A_2 \oplus \cdots \oplus A_k \oplus I_p$$

where $A_1, A_2, \ldots, A_k$ *are fully indecomposable* $(0, 1)$-*matrices with exactly two* 1*'s in each row and column and* I_p *is an identity matrix of order* p *for some* $p \geq 0$.

Proof. We may assume that $A = A_1 \oplus A_2 \oplus \cdots \oplus A_t$ where A_i is a fully indecomposable matrix of order $n_i \geq 1$ and $t \geq 1$. Then

$$\text{per } A = \prod_{i=1}^{t} \text{per } A_i.$$

We may assume that $n_i \geq 2$ $(i = 1, 2, \ldots, t)$, thereby eliminating the possibility of I_p with $p \geq 1$ as given in the statement of the theorem. Suppose for some i we have $\sigma(A_i) = 2n_i$. Then since A_i is fully indecomposable, A_i has exactly two 1's in each row and column and per $A_i = 2$. If all $n_i = 2$, then $k = t$ and we have per $A = 2^t = 2^k$. If not, then it easily follows that we may now assume that $\sigma(A_i) - 2n_i \geq 1$ for $i = 1, 2, \ldots, t$. Applying Theorem 9.3.11 we get

$$\text{per } A = \prod_{i=1}^{t} \text{per } A_i \leq \prod_{i=1}^{t} (2^{\sigma(A_i)-2n_i} + 1)$$

$$\leq 2^{\left(\sum_{i=1}^{t} \sigma(A_i)-2n_i\right)+t-1} + 1$$

$$= 2^{\sigma(A)-2n+t-1} + 1$$

$$< 2^{\sigma(A)-2n+t} = 2^k.$$

The theorem now follows. $\square$

The k-dimensional faces of Ω_n with at least $2^{k-1}+1$ extreme points are characterized in [27]. In particular, the number of extreme points of such faces is of the form $2^{k-1}+2^m$ for $m=0,1,\ldots,k-1$. In addition, faces which are pyramids over some base are also characterized.

To conclude this section we briefly discuss some additional affine and combinatorial properties of Ω_n derived in [29]. Let $\mathcal{P}$ and $\mathcal{Q}$ be two polytopes. A map $h:\mathcal{P}\to\mathcal{Q}$ is an *affine map* provided

$$h(cx+(1-c)y)=ch(x)+(1-c)h(y)\quad(x,y\in\mathcal{P},0\le c\le 1).$$

The polytopes $\mathcal{P}$ and $\mathcal{Q}$ are *affinely equivalent* provided there exists a bijective affine map $h:\mathcal{P}\to\mathcal{Q}$. The polytopes $\mathcal{P}$ and $\mathcal{Q}$ are *combinatorially equivalent* provided the partially ordered set[7] of faces of $\mathcal{P}$ is isomorphic to the partially ordered set of faces of $\mathcal{Q}$. In general, combinatorial equivalence is a weaker property than affine equivalence.[8]

Let A and B be two $(0,1)$-matrices of orders m and n, respectively. Let $A_1,A_2,\ldots,A_r$ be the nontrivial fully indecomposable components of A, and let $B_1,B_2,\ldots,B_s$ be those of B. It is not difficult to show [29] that the faces $\Omega_m(A)$ and $\Omega_n(B)$ are affinely equivalent (respectively, combinatorially equivalent) if and only if $r=s$ and for some permutation τ of $\{1,2,\ldots,r\}$, $\Omega_m(A_i)$ and $\Omega_n(B_i)$ are affinely equivalent (respectively, combinatorially equivalent) for $i=1,2,\ldots,r$.

The $(0,1)$-matrices A and B are *contraction equivalent* provided either $A=B$ or there exist $(0,1)$-matrices $A_0,A_1,\ldots,A_p$ such that $A_0=A$, $A_p=B$, and for each $i=1,2,\ldots,t$, A_i is a contraction of A_{i-1} or A_{i-1} is a contraction of A_i.

Theorem 9.3.13 *Let A and B be contraction equivalent $(0,1)$-matrices of orders m and n, respectively, with total support. Then $\Omega_m(A)$ and $\Omega_n(B)$ are affinely equivalent.*

Proof. It suffices to show that A and B are affinely equivalent when B is a contraction of A. Moreover, we may assume that B is obtained from A by a contraction on column 1 with respect to rows 1 and 2. Thus

$$A\begin{bmatrix}1 & \alpha\\1 & \beta\\0 & C\end{bmatrix}\quad\text{and}\quad B=\begin{bmatrix}\alpha+\beta\\C\end{bmatrix}$$

where, since B is a $(0,1)$-matrix, $\alpha\beta^{\mathrm{T}}=0$. Let $h:\Omega_n(A)\to\Omega_{n-1}(B)$ be defined as follows:

$$\text{For } X=\begin{bmatrix}x_1 & \mu_1\\x_2 & \mu_2\\0 & X_1\end{bmatrix}\in\Omega_n(A),\ h(X)=\begin{bmatrix}\mu_1+\mu_2\\X_1\end{bmatrix}.\qquad(9.12)$$

---

[7]This partially ordered set is a finite lattice.
[8]It was conjectured in [29] however that if two faces of doubly stochastic polytopes are combinatorially equivalent, then they are affinely equivalent.

Then h is an affine map with $h(X) \in \Omega_{n-1}B)$. We show that h is a bijection implying that $\Omega_n(A)$ and $\Omega_{n-1}(B)$ are affinely equivalent.

First suppose that $h(X) = h(Y)$ where X is as in (9.12) and

$$
Y = \begin{bmatrix} y_1 & \nu_1 \\ y_2 & \nu_2 \\ 0 & Y_1 \end{bmatrix} \in \Omega_n(A).
$$

Then $X_1 = Y_1$, and from $\alpha\beta^{\mathrm{T}} = 0$, it follows that $\mu_1 = \nu_1$ and $\mu_2 = \nu_2$. Since X and Y are doubly stochastic, we have $x_1 = y_1$ and $x_2 = y_2$ as well. Hence $X = Y$ and h is injective.

Now U be a matrix in $\Omega_{n-1}(B)$. Then U can be expressed in the form

$$
W = \begin{bmatrix} \xi_1 + \xi_2 \\ U_1 \end{bmatrix}
$$

where $\xi_1 \leq \alpha$, $\xi_2 \leq \beta$, and $U_1 \leq C$. Let $c = 1 - \sigma(\xi_1)$ and $d = 1 - \sigma(\xi_2)$. Then the matrix

$$
V = \begin{bmatrix} c & \xi_1 \\ d & \xi_2 \\ 0 & U_1 \end{bmatrix}
$$

is in $\Omega_n(A)$ and satisfies $h(V) = U$. Thus h is surjective, and hence bijective. $\square$

Applying Theorems 9.3.4, 9.3.8, and 9.3.13 we immediately obtain that up to affine equivalence, Ω_3 is the only 2-neighborly, nonsimplex polytope that can be a face of any Ω_n.

Theorem 9.3.14 *Let n be a positive integer. Then a face of Ω_n which is 2-neighborly polytope but not a simplex is affinely equivalent to Ω_3.* $\square$

While nearly decomposable matrices have special combinatorial properties not shared by all fully indecomposable matrices, from the point of view of affine equivalence there are no differences. This is made precise in the following theorem from [29].

Theorem 9.3.15 *Let A be a fully indecomposable $(0,1)$-matrix of order n with $\dim \Omega_n(A) > 1$. Then A is contraction equivalent to a nearly decomposable matrix B. Thus $\Omega_n(A)$ is affinely equivalent to $\Omega_m(B)$ where m is the order of B.*

Finally we note that in [28] it is proved that a face of Ω_n of dimension $d \geq 3$ has at most $3(d-1)$ facets. In [106] it is proved that $(0,1)$-polytopes (polytopes all of whose extreme points have values 0 and 1) satisfy the Hirsch conjecture which links the dimension d, number f of facets, and diameter[9] δ by the inequality $\delta \leq f - d$. Since Ω_n is a $(0,1)$-polytope, Ω_n satisfies the Hirsch conjecture.

---

[9]The diameter of the graph of Ω_n.

9.4   Graph of the Assignment Polytope

Let n be a positive integer. As defined in the preceding section, the graph of
the polytope Ω_n is the graph $G(\Omega_n)$ whose vertices are the extreme points
of Ω_n (so the permutation matrices of order n) and whose edges are the
pairs $\{P, Q\}$ of permutation matrices P and Q such that P and Q are the
extreme points of a one-dimensional face of Ω_n. Thus the *vertex set* $V(\Omega_n)$
of $G(\Omega_n)$ can be taken as the set S_n of $n!$ permutations of $\{1, 2, \ldots, n\}$.
The *edge set* $E_n = E(\Omega_n)$ can be taken as the pairs $\{\sigma, \pi\}$ of permutations
in S_n, such that the permutation $\sigma^{-1}\pi$, equivalently $\pi^{-1}\sigma$, has exactly one
permutation cycle of length greater than 1. Thus the graph of Ω_n has a very
simple realization independent of Ω_n. The graph $G(\Omega_n)$ is the *Cayley graph*
$G(S_n, C_n)$ where C_n is the set of permutations in the symmetric group S_n
that have exactly one nontrivial cycle defined as follows: the vertices of
$G(S_n, C)$ are the permutations in S_n, and for each such permutation σ, σ
is joined by an edge to all permutations $\sigma\tau$ where $\tau \in C_n$.

For $n = 1, 2$, and 3, $G(\Omega_n) = K_n$, the complete graph with n vertices.
If A is a $(0, 1)$-matrix of order n with total support, then the graph of
the face $\Omega_n(A)$ is the graph $G(\Omega_n(A))$ induced by $G(\Omega_n)$ on the set of
permutation matrices P with $P \le A$.

In the next theorem we collect some elementary facts about the graph
$G(\Omega_n)$ which appear in [6] and [28]. First we recall some standard graph
concepts needed for this section. The *diameter* of a graph is the maximal
distance between vertices, and the *complement of a graph* G is the graph
$\overline{G}$ with the same set of vertices as G with two vertices joined by an edge
in $\overline{G}$ if and only if they are not joined by an edge in G. A *Hamilton cycle*
in a graph is a cycle that visits each vertex exactly once. If G_1 and G_2
are graphs with vertex sets V_1 and V_2, respectively, then recall that their
Cartesian product $G \times G_2$ is the graph with vertex set $V_1 \times V_2$ and with an
edge joining (v_1, v_2) and (u_1, u_2) if and only if $v_1 = u_1$ and there is an edge
in G_2 joining v_2 and u_2, or $v_2 = u_2$ and there is an edge in G_1 joining v_1
and u_1. Cartesian product, relative to graph isomorphism, is a commutative
and associative binary operation. In particular, for graphs $G_1, G_2, \ldots, G_k$,
the k-fold Cartesian product $G_1 \times G_2 \times \cdots \times G_k$ is unambiguous.

Theorem 9.4.1 *Let n be a positive integer. The following properties are
satisfied by the graph $G(\Omega_n)$.*

(i) *The number of vertices is $n!$*

(ii) *$G(\Omega_n)$ is vertex-symmetric with the degree of each vertex equal to*

$$\sum_{k=2}^{n} \binom{n}{k} (k-1)!$$

(iii) *$G(\Omega_n)$ is a connected graph whose diameter equals 1 for $n = 1, 2$, and
3 and equals 2 for $n \ge 4$.*

(iv) *For $n \geq 5$, the complement of $G(\Omega_n)$ is a connected graph and its diameter also equals 2.*

(v) *$G(\Omega_n)$ has a Hamilton cycle.*

(vi) *If A is a $(0,1)$-matrix of order n with total support whose fully indecomposable components are the matrices $A_1, A_2, \ldots, A_k$ $(k \geq 2)$ of orders $n_1, n_2, \ldots, n_k$, respectively, then the graph $G(\Omega_n(A))$ of the face $\Omega_n(A)$ is isomorphic to the Cartesian product*

$$G(\Omega_{n_1}(A_1)) \times G(\Omega_{n_2}(A_2)) \times \cdots \times G(\Omega_{n_k}(A_k)).$$

In particular, $G(\Omega_n)$ is an induced subgraph of $G(\Omega_{n+1})$.

(vi) *Let A and B be contraction equivalent $(0,1)$-matrices of orders m and n, respectively, with total support. Then $G(\Omega_m(A))$ and $G(\Omega_n(B))$ are isomorphic graphs.*

Proof. There are $n!$ permutations in S_n and the number of permutations with exactly one cycle of length greater than 1 equals $\sum_{k=2}^{n} \binom{n}{k} (k-1)!$ If π is a permutation in S_n then the map $f_\pi : S_n \to S_n$ defined by $f_\pi(\sigma) = \sigma\pi$ gives an automorphism of $G(\Omega_n)$. Thus (i) and (ii) hold.

Since $G(\Omega_n) = K_n$ for $n = 1, 2$, and 3, the diameter equals 1 in these cases. Now let $n \geq 4$. Then not every pair of distinct vertices of $G(\Omega_n)$ is joined by an edge. Hence the diameter is at least 2. Since $G(\Omega_n)$ is vertex-symmetric it suffices to show that the identity permutation ι and a permutation σ with at least two cycles of length greater than 1 are joined by a path of length 2. This is equivalent to showing that such a permutation σ can be expressed as a product of two permutations, each with exactly one cycle of length greater than 1. Let the permutation cycles of length greater than 1 of σ be $(i_1, i_2, \ldots, i_{p_1}), (j_1, j_2, \ldots, j_{p_2}), \ldots, (k_1, k_2, \ldots, k_{p_m})$ where $p_i \geq 1$ for $i = 1, 2, \ldots, m$. Then one verifies that $\sigma = \tau \cdot \pi$ where τ and π are the permutations in S_n with exactly one cycle of length greater than 1 given by

$$\tau = (i_1, i_2, \ldots, i_{p_1}, j_1, j_2, \ldots, j_{p_2}, \ldots, k_1, k_2, \ldots, k_{p_m}), \text{ and}$$

$$\pi(k_1, \ldots, j_1, i_1), \text{ respectively.}$$

This proves (iii).[10]

For $n \leq 4$ the complementary graph of $G(\Omega_n)$ is not connected; in fact, $G(\Omega_4)$ is isomorphic to six pairwise vertex-disjoint copies of the complete

---

[10]Geometrically, as described in [28], σ and ι determine a face of Ω_n of dimension m which is a rectangular parallelepiped. Then one constructs a pyramid over this parallelepiped of dimension $m + 1$ whose top vertex is the middle vertex in a path of length 2 from σ to ι in the graph $G(\Omega_n)$. The details are to let $A = [a_{ij}]$ be the $(0,1)$-matrix of order n where $a_{ij} = 1$ if and only if $j = i$ or $\sigma(i) = j$. Without loss of generality one may assume that $A = A_1 \oplus A_2 \oplus \cdots \oplus A_r$ where the A_i are fully indecomposable

graph K_4 [28].[11] Now let $n \geq 5$. Let τ and σ be two permutations that are not joined by an edge in $\overline{G(\Omega_n)}$. We need to show they are joined by a path of length 2 in the complementary graph. Without loss of generality we may assume that τ is the identity permutation and thus that σ has exactly one cycle of length greater than 1. We then want to express σ as the product of two permutations in S_n neither of which has exactly one cycle of length greater than 1. There is no loss in generality in assuming that the unique cycle of length greater than 1 of σ is $(1, 2, \ldots, k)$. We define permutations π and β_k in S_n by their cycle decompositions:

$$\pi = (1,3) \cdot (4,5),$$
$$\beta_2 = (4,5) \cdot (1,3,2), \beta_3 = (4,5) \cdot (2,3), \beta_4 = (1,4,5) \cdot (2,3),$$
$$\beta_5 = (1,4) \cdot (2,3), \text{ and } \beta_k = (1,4,6,7,\ldots,k) \cdot (2,3) \, (k \geq 6).$$

It is now straightforward to verify that $\sigma = \pi \cdot \beta_k$ for all $k \geq 2$, completing the proof of (iv).

The *reflected Gray code* (see e.g. [22]) implies that $G(\Omega_n)$ has a Hamilton cycle, indeed that the permutations in S_n can be ordered $\sigma_1, \sigma_2, \ldots, \sigma_{n!}$ so that each of $\sigma_i \cdot \sigma_{i+1}^{-1}$ $(i = 1, 2, \ldots, n! - 1)$ and $\sigma_{n!} \sigma_1^{-1}$ has exactly one cycle of length greater than or equal to 2 and this cycle is of length exactly 2. This proves (v). The proof of (vi) is an elementary verification. Assertion (vi) is a direct consequence of Theorem 9.3.13, since affine equivalent polytopes have isomorphic graphs.[12]   □

We now consider the graph $G(\Omega_n(A))$ of a face of Ω_n where A is a (0,1)-matrix of order n with total support. By (vi) of Theorem 9.4.1 this graph is isomorphic to the Cartesian product of the graphs of the faces corresponding to its fully indecomposable components. Recall that a graph G is *prime (with respect to the Cartesian product)* provided it cannot be expressed as the Cartesian product in a nontrivial way, more precisely, provided it has at least two vertices and whenever G is isomorphic to the Cartesian product $G_1 \times G_2$ of two graphs G_1 and G_2, then either G_1 or

---

matrices of which $r - m$ are identity matrices of order 1. Then let

$$B = \begin{bmatrix} A_1 & O & \cdots & O & E_1 \\ E_2 & A_2 & \cdots & O & O \\ \vdots & \vdots & \ddots & \vdots & \vdots \\ O & O & \cdots & A_{r-1} & O \\ O & O & \cdots & E_r & A_r \end{bmatrix}$$

where each E_i contains exactly one 1. Then $\dim \Omega_n(B) = \dim \Omega_n(A) + 1$, and indeed $\Omega_n(B)$ is a pyramid over the parallelepiped $\Omega_n(A)$. In this way we can find many paths from σ to ι each of which takes place in a face of Ω_n of dimension only 1 more than that of the smallest face of Ω_n containing the permutation matrices determined by σ and ι. One of these paths has length 2.

[11]This is because the graph of the subgroup of order 4 of the permutation group S_4 equal to $\{\iota, (1,2) \cdot (3,4), (1,3) \cdot (2,4), (1,4) \cdot (2,3)\}$ is totally disconnected, and thus so is the graph of each of its cosets. Since each vertex of $G(\Omega_4)$ has degree 20, it follows that the complement of $G(\Omega_4)$ consists of six vertex-disjoint copies of K_4.

[12]A direct proof that $G(\Omega_n(A))$ and $G(\Omega_n(B))$ are isomorphic is given in [28].

G_2 has exactly one vertex (and so G is isomorphic to either G_1 or G_2). A theorem of Sabidussi [123] asserts that each graph G with at least two vertices can be uniquely expressed as a Cartesian product of prime graphs: if G is isomorphic to $G_1 \times G_2 \times \cdots \times G_p$ and to $H_1 \times H_2 \times \cdots \times H_q$ where the G_i and the H_j are prime graphs, then $p = q$ and there exists a permutation $(j_1, j_2, \ldots, j_p)$ of $\{1, 2, \ldots, p\}$ such that G_i is isomorphic to H_{j_i} for $i = 1, 2, \ldots, p$. Using this result, Brualdi and Gibson [28] proved the following theorem relating primeness of graphs to fully indecomposable components.[13]

Theorem 9.4.2 *Let A be a $(0,1)$-matrix of order n with total support.*

(i) *$G(\Omega_n(A))$ is a prime graph if and only if A has exactly one nontrivial fully indecomposable component.*

(ii) *If there are prime graphs $G_1, G_2, \ldots, G_m$ such that $G(\Omega_n(A))$ is isomorphic to $G_1 \times G_2 \times \cdots \times G_m$, then A has exactly m nontrivial fully indecomposable components $A_1, A_2, \ldots, A_m$ and there exists a permutation $(j_1, j_2, \ldots, j_m)$ of $\{1, 2, \ldots, m\}$ such that $G(\Omega_n(A_i))$ is isomorphic to G_{j_i} for $i = 1, 2, \ldots, m$.*

Example. We may have $G(\Omega_n(A))$ isomorphic to $G(\Omega_n(B))$ even though $\dim \Omega_n(A) \neq \dim \Omega_n(B)$. Let

$$
A = \begin{bmatrix} 1 & 1 & 0 & 0 \\ 1 & 0 & 1 & 1 \\ 0 & 1 & 1 & 1 \\ 0 & 1 & 1 & 1 \end{bmatrix} \text{ and } B = \begin{bmatrix} 1 & 1 & 1 & 1 \\ 1 & 1 & 1 & 1 \\ 0 & 1 & 1 & 0 \\ 0 & 0 & 1 & 1 \end{bmatrix}.
$$

Then $\dim \Omega_4(A) = 4$ and $\dim \Omega_4(B) = 5$. Since $\operatorname{per} A = \operatorname{per} B = 6$, $\Omega_4(A)$ and $\Omega_4(B)$ each have six vertices, and their graphs are isomorphic to the complete graph K_6 on six vertices. In fact, $\Omega_4(A)$ is a 2-neighborly polytope affinely equivalent to Ω_3; $\Omega_4(B)$ is a five-dimensional simplex. □

A *Hamilton path* in a graph is a path which visits each vertex exactly once. A graph is *Hamilton-connected* provided for each pair u, v of distinct vertices there is a Hamilton path joining u and v. A Hamilton-connected graph with at least three vertices has a Hamilton cycle and, in fact, each edge belongs to a Hamilton cycle. Bipartite graphs with at least three vertices are never Hamilton-connected; if the two sets U and V in the bipartition of the vertices have the same cardinality, then two vertices in U, or two vertices in V, are not joined by a Hamilton path, while if $|U| > |V|$, then for each vertex $u \in U$ and each vertex $v \in V$, there is no Hamilton path joining u and v. We now generalize (v) of Theorem 9.4.1 by showing that the graph of every face of Ω_n of dimension at least 2 has a Hamilton cycle, and indeed is Hamilton-connected except when the face is affinely a

---

[13]In Theorem 6.3.5 we applied Sabidussi's theorem to the interchange graph of a class $\mathcal{A}(R, S)$.

rectangular parallelepiped of some dimension, equivalently when the graph is a *d-hypercube* $K_2^{(d)} = K_2 \times K_2 \times \cdots \times K_2$ (*d* copies of the complete graph K_2 of order 2) for some $d \geq 0$. (Here $K_2^{(0)}$ is the trivial graph with one vertex and no edges.) If $d \geq 2$, then $K_2^{(d)}$ has a Hamilton cycle but is not Hamilton-connected.

Theorem 9.4.3 *Let $G_1, G_2, \ldots, G_k$ be Hamilton-connected graphs. If G_j has at least three vertices for at least one j, then $G = G_1 \times G_2 \times \cdots \times G_k$ is Hamilton-connected. If G_i has at most two vertices for all $i = 1, 2, \ldots, k$, then G is a d-hypercube for some $d \geq 0$, and is not Hamilton-connected if $d \geq 2$.*

Proof. If $k = 1$, the theorem is obvious. Assume that $k \geq 2$. If each G_i has at most two vertices, then $G_i = K_1$ or K_2 for each i and hence G is a *d*-hypercube for some *d*.

Now assume that some G_j has at least three vertices. It suffices to prove that G is Hamilton-connected when $k = 2$, for then we may argue by induction for general k. Thus we assume that G_1 is a Hamilton-connected graph with at least three vertices and that G_2 is a Hamilton-connected graph. Let $G = G_1 \times G_2$. If G_2 has only one vertex, then G is isomorphic to G_1 and so is Hamilton-connected.

Case 1: G_2 has exactly two vertices v and w. Let α and β be distinct vertices of G. First suppose that $\alpha = (u, v)$ and $\beta = (u', v)$ where $u \neq u'$. There exists a Hamilton path $u = u_1, u_2, \ldots, u_m = u'$ in G_1 from u to u'. In addition, there exists a Hamilton path $u = x_1, x_2, \ldots, x_m = u_2$ in G_1 from u to u_2. Since G_2 is Hamilton-connected, there is an edge in G_2 joining v and w. Then

$$(u, v) = (u_1, v), (u_1, w) = (x_1, w), (x_2, w), \ldots, (x_m, w) = (u_2, w),$$
$$(u_2, v), (u_3, v), \ldots, (u_m, v) = (u', v)$$

is a Hamilton path in G from α to β.

Now suppose that $\alpha = (u, v)$ and $\beta = (u', w)$. Since G_1 has at least three vertices, there exists a vertex u'' of G_1 different from u and u'. There exist Hamilton paths

$$u = u_1, u_2, \ldots, u_m = u'' \text{ and } u'' = x_1, x_2, \ldots, x_m = u'$$

in G from u to u'' and from u'' to u', respectively. Then

$$(u, v) = (u_1, v), (u_2, v), \ldots, (u_m, v) = (u'', v), (u'', w) = (x_1, w),$$

$$(x_2, w), (x_3, w), \ldots, (x_m, w) = (u', w)$$

is a Hamilton path in G from α to β.

Case 2: G_2 has at least three vertices. Let $\alpha = (u, v)$ and $\beta = (u', v')$ be distinct vertices of G. Since in this case both G_1 and G_2 have at least three vertices, there is no loss in generality in assuming that $v \neq v'$. There is a Hamilton path $v = v_1, v_2, \ldots, v_n = v'$ in G_2 from v to v'. Since G_1 has at least three vertices, there exists a sequence $u_1, u_2, \ldots, u_n$ of vertices of G_1 with $u_1 = u$, $u_n = u'$ and $u_i \neq u_{i+1}$ for $i = 1, 2, \ldots, n-1$. There exist Hamilton paths

$$u_i = x_{i1}, x_{i2}, \ldots, x_{im} = u_{i+1} \quad (i = 1, 2, \ldots, n-1)$$

in G_1. Then

$$(u, v) = (x_{11}, v_1), (x_{12}, v_1), \ldots, (x_{1m}, v_1) = (u_2, v_1) = (x_{21}, v_1),$$

$$(x_{22}, v_2), \ldots, (x_{2m}, v_2) = (u_3, v_2) = (x_{31}, v_2),$$

$$\ldots, (x_{n-1,1}, v_n), (x_{n-1,2}, v_n), \ldots, (x_{n-1,m}, v_n) = (u_n, v_n) = (u', v')$$

is a Hamilton path in G from α to β.

Hence G is Hamilton-connected and the theorem follows.   □

In establishing the Hamilton properties of faces of Ω_n, we shall make use of the following three lemmas. The first assertion in the next lemma is a consequence of the fact that $K_2^{(m)}$ is a connected bipartite graph; the second assertion can be proved by induction using the fact that $K_2^{(m+1)}$ is isomorphic to $K_2^{(m)} \times K_2$.

Lemma 9.4.4 *For each integer $m \geq 1$, the hypercube graph $K_2^{(m)}$ can be uniquely bipartitioned into two sets X and Y so that each edge joins a vertex in X to a vertex in Y. For each pair of vertices x and y with $x \in X$ and $y \in Y$, there exists a Hamilton path joining x and y.*   □

Lemma 9.4.5 *Let A be a fully indecomposable $(0,1)$-matrix of order $n \geq 2$. Let r be an integer with $1 \leq r \leq n$. Let s and t be distinct integers such that row r contains 1's in columns s and t. Finally, let $P = [p_{ij}]$ be a permutation matrix with $P \leq A$ such that $p_{rs} = 1$. Then there exists a permutation matrix $Q = [q_{ij}]$ with $Q \leq A$ and $q_{rt} = 1$ such that P and Q are joined by an edge in $G(\Omega_n(A))$.*

Proof. Without loss of generality we may assume that $r = s = 1$ and that $P = I_n$. Since A is fully indecomposable, there exists a permutation matrix $Q = [q_{ij}]$ with $Q \leq A$ and $q_{1t} = 1$. The permutation σ of $\{1, 2, \ldots, n\}$ corresponding to Q can be written as a product of disjoint cycles. Since $I_n \leq A$, there is a permutation matrix $R = [r_{ij}]$ with $R \leq A$ such that $r_{1t} = 1$ and the permutation of $\{1, 2, \ldots, n\}$ corresponding to R has exactly one cycle of length greater than 1. Hence I_n and R are joined by an edge in the graph $G(\Omega_n(A))$.   □

Lemma 9.4.6 *Let A be a nearly decomposable $(0,1)$-matrix and assume that $\dim \Omega_n(A) \geq 2$. Then there exists a fully indecomposable, but not nearly decomposable, $(0,1)$-matrix B such that B is contraction equivalent to A.*

Proof. By Theorem 1.3.6, we may assume that

$$
A = \begin{bmatrix}
\begin{array}{cccccc|c}
1 & 0 & 0 & \cdots & 0 & 0 & \\
1 & 1 & 0 & \cdots & 0 & 0 & \\
0 & 1 & 1 & \cdots & 0 & 0 & \\
\vdots & \vdots & \vdots & \ddots & \vdots & \vdots & F_1 \\
0 & 0 & 0 & \cdots & 1 & 0 & \\
0 & 0 & 0 & \cdots & 1 & 1 & \\
\hline
\multicolumn{6}{c|}{F_2} & A_1
\end{array}
\end{bmatrix}
$$

where A_1 is a nearly decomposable matrix of order m for some integer m with $1 \leq m < n$. The matrix F_1 contains exactly one 1 and this 1 is in its first row and last column; the matrix F_2 contains exactly one 1 and this 1 is in its last row and last column. Since $\dim \Omega_n(A) \geq 2$, $m \neq 1$, and it follows from Theorem 1.3.6 that $m \geq 3$ and the element in the last row and last column of A_1 is a 0. The matrix obtained from A_1 by replacing the 0 in its last row and column satisfies the conclusions of the lemma.   □

Theorem 9.4.7 *Let $A = [a_{ij}]$ be a $(0,1)$-matrix of order n with total support. Then $G(\Omega_n(A))$ is Hamilton-connected unless $\Omega_n(A)$ is a rectangular parallelepiped of dimension $m \geq 2$, equivalently, unless $G(\Omega_n(A))$ is a hypercube graph $K_n^{(m)}$ for some $m \geq 2$. The graph $G(\Omega_n(A))$ has a Hamilton cycle provided that $\dim(\Omega_n(A) \geq 2$.*

Proof. Using Theorem 9.4.3 we need only prove that if $\Omega_n(A)$ is not a rectangular parallelepiped of dimension $m \geq 2$, then $G(\Omega_n(A))$ is Hamilton-connected. Since the only planar faces of Ω_n are triangular or rectangular, we assume that $m \geq 3$ and proceed by induction in m.

Assume first that A has $t \geq 2$ nontrivial fully indecomposable components. Since $\Omega_n(A)$ is not a rectangular parallelepiped, it follows from the inductive assumption, (vi) of Theorem 9.4.1, and Theorem 9.4.3 that $G(\Omega_n(A))$ is Hamilton-connected. Thus we may now assume that $t = 1$ and indeed that A itself is fully indecomposable. By Lemma 9.4.6, we may assume that A is not nearly decomposable. Thus there exists an element $a_{rs} = 1$ of A such that the matrix B obtained from A by replacing this 1 with a 0 is fully indecomposable. We may assume that $r = s = 1$ and that $I_n \leq A$. Then $\Omega_n(B)$ is a facet of $\Omega_n(A)$, and by the induction assumption $G(\Omega_n(B))$ is Hamilton-connected. Let C be the matrix obtained from A by replacing with 0 all elements in row 1 and column 1 except for $a_{11} = 1$.

Since A is fully indecomposable, $\Omega_n(C)$ is a nonempty face of $\Omega_n(A)$ with $\dim \Omega_n(C) < \dim \Omega_n(A)$.[14] The set of extreme points of $\Omega_n(A)$ is the disjoint union of the sets of extreme points of $\Omega_n(B)$ (those permutation matrices $P = [p_{ij}] \leq A$ with $p_{11} = 0$) and $\Omega_n(C)$ (those with $p_{11} = 1$).

If $\Omega_n(C)$ has only one extreme point, then this extreme point is I_n and geometrically, $\Omega_n(A)$ must a pyramid with base $\Omega_n(B)$ and apex I_n.[15] Thus in $G(\Omega_n(A))$, I_n is joined by an edge to each vertex of $G(\Omega_n(B))$. The Hamilton-connectedness of $G(\Omega_n(B))$ now implies that $G(\Omega_n(A))$ is Hamilton-connected.

We now assume that $\Omega_n(C)$ has at least two extreme points. The inductive assumption implies that $G(\Omega_n(C))$ is either Hamilton-connected or a hypercube graph $K_2^{(m)}$ for some $m \geq 2$. Let P and Q be two vertices of $G(\Omega_n(A))$. The remainder of the proof is broken up into three cases.

Case 1: P is a vertex of $G(\Omega_n(B))$ and Q is a vertex of $G(\Omega_n(C))$. Making use of Lemma 9.4.4 we see that $G(\Omega_n(C))$ has a Hamilton path $Q = R_1, R_2, \ldots, R_l$ beginning at its vertex Q. Since B is fully indecomposable, row 1 of A contains at least three 1's. By Lemma 9.4.6, R_q is joined by an edge in $G(\Omega_n(A))$ to a vertex S of $G(\Omega_n(B))$ with $S \neq P$. Since $G(\Omega_n(B))$ is Hamilton-connected there is a Hamilton path $S = S_1, S_2, \ldots, S_h = P$ in $G(\Omega_n(B))$. Thus

$$Q = R_1, R_2, \ldots, R_l, S = S_1, S_2, \ldots, S_h = P$$

is a Hamilton path in $G(\Omega_n(A)$ joining Q and P.

Case 2: P and Q are both vertices of $G(\Omega_n(B))$. Since $G(\Omega_n(B))$ is Hamilton-connected, there is a Hamilton path $P = S_1, S_2, \ldots, S_h = Q$ in $G(\Omega_n(B))$ which joins P and Q. Lemma 9.4.5 implies that each S_i is joined by an edge of $G(\Omega_n(A))$ to some vertex of $G(\Omega_n(C))$, and each vertex of $G(\Omega_n(C))$ is joined by an edge to at least one of the S_j. From this we conclude that there is an integer i with $1 \leq i < h$ such that S_i is joined by an edge to U and S_{i+1} is joined by an edge to W where U and W are distinct vertices of $G(\Omega_n(C))$. If $G(\Omega_n(C))$ is a hypercube, U and W can be chosen so that they are in different parts of the unique bipartition of the vertices of $G(\Omega_n(C))$. Now applying the inductive assumption and using Lemma 9.4.4, we obtain a Hamilton path $U = R_1, R_2, \ldots, R_l = W$ in $G(\Omega_n(C))$. Then

$$P = S_1, \ldots, S_i, U = R_1, R_2, \ldots, R_l = W, S_{i+1}, \ldots, S_h = Q$$

is a Hamilton path in $G(\Omega_n(A))$ joining P and Q.

---

[14]But note that C is not fully indecomposable and indeed may not even have total support.

[15]Arguing combinatorially, one can show that after simultaneous row and column permutations involving the last $n-1$ rows and columns of C, the matrix of C becomes a triangular matrix with $I_n \leq C$. This implies that for each permutation matrix $P = [p_{ij}] \leq A$ with $P \neq I_n$, $p_{11} = 0$ and the corresponding permutation of $\{1, 2, \ldots, n\}$ has exactly one nontrivial cycle.

Case 3: P and Q are both vertices of $G(\Omega_n(C))$. First consider the subcase where there is a Hamilton path $P = R_1, R_2, \ldots, R_l = Q$ in $G(\Omega_n(A))$ joining P and Q. Lemma 9.4.5 implies that each of R_1 and R_2 is joined by edges of $G(\Omega_n(A))$ to at least two vertices of $G(\Omega_n(B))$. Hence there exist distinct vertices X and Y of $G(\Omega_n(B))$ such that R_1 is joined to X by an edge and R_2 is joined by an edge to Y. By the inductive assumption, there is a Hamilton path in $G(\Omega_n(B))$ joining X and Y, and we easily obtain now a Hamilton path in $G(\Omega_n(A))$ joining P and Q.

Now consider the subcase in which there does not exist a Hamilton path in $G(\Omega_n(C))$ joining P and Q. The inductive assumption and Lemma 9.4.4 imply that $G(\Omega_n(C))$ is a hypercube graph $K_2^{(m)}$ for some $m \geq 2$, and P and Q are in the same part of the unique bipartition of the vertices of $K^{(m)}$. Lemma 9.4.4 implies that $G(\Omega_n(C))$ has a Hamilton cycle $R_1, R_2, \ldots, R_l, R_1$ where $P = R_1$ and $Q = R_j$ for some $j \neq 1, 2, l$. Lemma 9.4.5 implies as before that there are distinct vertices X and Y of $G(\Omega_n(B))$ such that R_{j-1} is joined by an edge to X and R_l is joined by an edge to Y. By the inductive assumption, there is a Hamilton path $X = S_1, S_2, \ldots, S_h = Y$ in $G(\Omega_n(B))$ joining X and Y. Now

$$P = R_1, R_2, \ldots, R_{j-1}, X = S_1, S_2, \ldots, S_h = Y, R_q, R_{q-1}, \ldots, R_j = Q$$

is a Hamilton path joining P and Q.

The theorem now follows by induction. $\square$

We remark that in a series of two papers Nadeff and Pulleyblank [107, 108] have greatly generalized Theorem 9.4.7 as follows.

Theorem 9.4.8 *The graph of a polyhedron each of whose extreme points is $(0,1)$-valued either is Hamilton-connected or is a hypercube graph.*

Now we consider the connectivity of the graph of the polytope Ω_n. Recall that the *connectivity* of a graph G is the minimal number $\kappa(G)$ of vertices of G whose removal (along with incident edges) leaves a disconnected or trivial graph. The connectivity of the complete graph of order n satisfies $\kappa(K_n) = n - 1$ $(n \geq 1)$; the connectivity of all other graphs of order n is strictly less than $n - 1$. The connectivity of an arbitrary graph is bounded above by the minimal degree of its vertices as can be seen by the removal of all the vertices adjacent to a specified vertex. By (ii) of Theorem 9.4.1 each vertex of the graph $G(\Omega_n)$ has degree

$$d_n = \sum_{k=2}^{n} \binom{n}{k} (k-1)!$$

Each vertex of the complementary graph $\overline{G(\Omega_n)}$ has degree

$$\bar{d}_n = n! - 1 - d_n.$$

Balinski and Russakoff [6] conjectured and Imrich [73] proved that the connectivity of $G(\Omega_n)$ equals d_n. Imrich also determined the connectivity of the complementary graph $\overline{G(\Omega_n)}$. We shall not give a self-contained proof of these results, but shall rely on some results of Watkins [144] on connectivity of vertex-symmetric graphs.[16]

Let G be a connected graph with vertex set V, and assume that G is not a complete graph. A set X of vertices is a *separating set* for G provided the graph $G \setminus X$ obtained from G by removal of the vertices in X along with all edges incident with X is a disconnected graph. The set X is a *minimal separating set* provided $\kappa(G) = |X|$. Let X be a minimal separating set of G. Then the graph $G \setminus X$ is a disconnected graph, the vertex sets of whose connected components are called *parts* (with respect to X). The smallest number of vertices in a part taken over all minimal separating sets of G is denoted by $p(G)$. An *atomic part* is a part S with $|S| = p(G)$. The following theorem is proved in [144].

Theorem 9.4.9 *Let G be a non-complete, connected graph. Then the following properties hold.*

(i) *$p(G) \geq 2$ if and only if $\kappa(G)$ is strictly less than the degree of each vertex of G.*

(ii) *Two atomic parts are either identical or disjoint.*

(iii) *Let G be vertex-symmetric, and let the constant degree of its vertices be $\delta(G)$. If $\kappa(G) < \delta(G)$, then $\kappa(G) = mp(G)$ for some integer $m \geq 2$.*

Theorem 9.4.10 *Let n be a positive integer. Then*

$$\kappa(G(\Omega_n)) = d_n.$$

The connectivity of the complement $\overline{G(\Omega_n)}$ satisfies

$$\kappa(\overline{G(\Omega_n)}) = n! - 1 - d_n \quad (n \geq 5).$$

Proof. In this proof we view the graph $G(\Omega_n)$ as the Cayley graph $G(S_n, C_n)$ where C_n is the set of permutations in S_n with exactly one nontrivial cycle. The set C_n is naturally partitioned as $C_n = \bigcup_{i=2}^{n} C_n^{(i)}$ where $C_n^{(i)}$ is the set of permutations in C_n whose nontrivial cycle has length i $(i = 2, 3, \ldots, n)$. The sets C_i are closed under inversion and conjugation. Let $\langle C_i \rangle$ denote the subgroup of S_n generated by C_i.

We first show that

$$d_n \leq 2|\langle C_i \rangle| \quad (i = 2, 3, \ldots, n). \tag{9.13}$$

We do this by showing that the alternating subgroup A_n of order $n!/2$ of S_n is contained in $\langle C_i \rangle$ for each i. Let

$$\sigma = (x_1, x_2, \ldots, x_k)(y_1, y_2, \ldots, y_l) \ldots (u_1, u_2, \ldots, u_m)$$

---

[16]Graphs for which, given vertices u and w, there is an automorphism taking u to w.

be a permutation in C_i given in terms of its cycle decomposition. Then the permutation

$$\tau = (u_m, \ldots, u_2, u_1) \ldots (y_l, \ldots, y_2, y_1)(x_1, x_2, x_k \ldots, x_4, x_3)$$

is also in C_i, and $\sigma\tau = (x_1, x_2)(x_1, x_3)$. It follows that $\langle C_i \rangle$ contains every permutation of S_n of the form $(a, b)(a, c)$. Hence $\langle C_i \rangle$ also contains every permutation $((a, b)(ac))((c, a)(c, d)) = (a, b)(c, d)$. Since each permutation in A_n is a product of an even number of transpositions, $A_n \subseteq \langle C_i \rangle$.

We now assume that $\kappa(G(\Omega_n)) < d_n$ and obtain a contradiction. Since $G(\Omega_n)$ is vertex-symmetric, (iii) of Theorem 9.4.9 implies that $\kappa(G(\Omega_n)) = mp(G)$ for some integer $m \geq 2$. Let X be an atomic part of $G(\Omega_n)$. By (i) of Theorem 9.4.9, $|X| = p(G) \geq 2$ and hence there exists an edge joining two vertices in X, say vertices α and $\alpha\sigma$ where σ is a permutation in C_k for some k. We next show that X contains all permutations of the form $\alpha\gamma$ where $\gamma \in C_k$. Let γ be an arbitrary permutation in C_k. There is an automorphism ϕ of the group S_n such that $\phi(\sigma) = \tau$. Let $\psi = \alpha\phi\alpha^{-1}$. Then $\psi(X)$ is also an atomic part of $G(\Omega_n))$, and we have

$$\psi(\alpha) = \alpha\phi\alpha^{-1}(\alpha) = \alpha\phi(\iota) = \alpha,$$
$$\psi(\alpha\sigma) = \alpha\phi\alpha^{-1}(\alpha\sigma) = \alpha\phi(\sigma) = \alpha\tau.$$

Hence, the atomic parts X and $\psi(X)$ have the vertex α in common and so by (ii) of Theorem 9.4.9, $X = \psi(X)$ and X contains $\alpha\tau$. Now a simple induction allows us to conclude that X contains all permutations of the form $\alpha\tau_1\tau_2 \ldots \tau_s$ with $\tau_j \in C_k$ for each j. Hence X contains all permutations of the form $\alpha\gamma$ with $\gamma \in \langle C_k \rangle$.

Now using (9.13), we then obtain

$$2p(G) \leq \kappa(G(\Omega_n)) < d_n \leq 2|C_k| \leq 2|X| = 2p(G),$$

a contradiction. Thus $\kappa(G(\Omega_n)) = d_n$.

Each vertex of the complementary graph $\overline{G(\Omega_n)}$ has degree equal to

$$\kappa(\overline{G(\Omega_n)}) = n! - 1 - d_n \ (n \geq 5).$$

The graph $\overline{G(\Omega_n)}$ is also a Cayley graph $G(S_n, D_n)$ where $D_n = S_n \setminus (C_n \cup \{\iota\})$. The set D_n is closed under conjugation since C_n is, and for $n \geq 5$, D_n contains an odd permutation. A similar argument now shows that the connectivity of $\overline{G(\Omega_n)}$ equals this number provided $n \geq 5$. $\quad\square$

We note that $\overline{G(\Omega_n)}$ is not connected for $2 \leq n \leq 4$; for $n = 2$ and 3, $\overline{G(\Omega_n)}$ has no edges, while for $n = 4$, $\overline{G(\Omega_n)}$ is the pairwise disjoint union of six copies of K_4.

The final parameter we discuss for $G(\Omega_n)$ is the chromatic index [19]. If G is any graph, the *chromatic index* of G is the smallest number $\chi'(G)$ of colors which can be used to color the edges of G so that the edges of the same color form a matching of G, that is, each pair of edges of the same color is vertex-disjoint. A theorem of Vizing asserts that if the maximal degree of a vertex of G is δ, then $\chi'(G) = \delta$ or $\delta + 1$. If G is a regular graph of degree δ, then $\chi'(G) = \delta$ if and only if the edges of G can be partitioned into d_n perfect matchings. We show that the chromatic index of $G(\Omega_n)$ equals d_n, that is, the edges of $G(\Omega_n)$ can be partitioned into perfect matchings.

The following lemma contains standard, elementary results from graph theory. If a graph with n vertices has a perfect matching, then n is even and each perfect matching contains $n/2$ edges.

Lemma 9.4.11 *Let n be a positive integer.*

(i) *If n is even, the edges of K_n can be partitioned into $n-1$ perfect matchings.*

(ii) *If n is odd, the edges of K_n can be partitioned into n matchings each of which contains $(n-1)/2$ edges.*

(iii) *If G is a regular bipartite graph, then the edges of G can be partitioned into perfect matchings.*

Our proof for the chromatic index of $G(\Omega_n)$ is based on a series of lemmas.

Let H be a subgroup of index t of S_n, and let $H = H_1, H_2, \ldots, H_t$ be a listing of the right cosets of S_n relative to H. Let $G(\Omega_n)[H_i]$ denote the subgraph of $G(\Omega_n)$ induced on the vertices in H_i $(1 \le i \le t)$. Let $G(\Omega_n)[H_i, H_j]$ denote the bipartite subgraph of $G(\Omega_n)$ whose vertex set is $H_i \cup H_j$ and whose edge set consists of all the edges of $G(\Omega_n)$ which join a vertex in H_i to a vertex in H_j $(1 \le i < j \le t)$.

Lemma 9.4.12 *The graphs $G(\Omega_n)[H_i]$ $(i = 1, 2, \ldots, t)$ are isomorphic.*

Proof. Let σ_i be a coset representative for H_i so that $H_i = H\sigma_i$ $(i = 1, 2, \ldots, t)$. The mapping $f_i : H \to H_i$ defined by $f(\tau) = \tau\sigma_i$ is easily seen to be a graph isomorphism from $G(\Omega_n)[H]$ to $G(\Omega_n)[H_i]$. This conclusion holds even if $i = 1$ and σ_1 is any coset representative of $H_1 = H$, and thus gives an automorphism of $G(\Omega_n)[H]$.   □

Lemma 9.4.13 *Each of the graphs $G(\Omega_n)[H_i, H_j]$ with $i \neq j$ is a regular graph.*

Proof. We first show that the vertices in H_i have the same degree in $G(\Omega_n)[H_i, H_j]$. Let σ and τ be two permutations in H_i, so that $\tau\sigma^{-1} \in H$. Let ρ be a permutation in H_j. The mapping $f : H_j \to H_j$ defined by $f(\rho) = \tau\sigma^{-1}\rho$ is a bijection, and since $\tau^{-1}(\tau\sigma^{-1}\rho) = \sigma^{-1}\rho$, there is an edge joining σ and ρ if and only if there is one joining τ and $\tau\sigma^{-1}\rho$. Hence

σ and τ have the same degree, and thus so do all vertices in H_i. Similarly, all vertices in H_j have the same degree. Since $|H_i| = |H_j|$, we conclude that $G(\Omega_n)[H_i, H_j]$ is a regular graph for $i \neq j$. □

The proof for the chromatic index of $G(\Omega_n)$ will be by induction on n. The collection of permutations $\sigma \in S_n$ such that $\sigma(n) = n$ form a subgroup of S_n of index n isomorphic to S_{n-1}. In what follows we identify this subgroup with S_{n-1}. Let

$$S_{n-1}^n = S_{n-1}, S_{n-1}^{n-1} = S_n(n-1, n), \ldots, S_{n-1}^1 = S_{n-1}(1, n)$$

be a listing of the right cosets of S_n with respect to its subgroup S_{n-1}.

Lemma 9.4.14 *Each of the bipartite graphs $G(\Omega_n)[S_{n-1}^i, S_{n-1}^j]$ with $i \neq j$ is a regular bipartite graph of degree m_{n-1} where*

$$m_{n-1} = \frac{d_n - d_{n-1}}{n-1} = \sum_{k=0}^{n-2} \binom{n-2}{k} k!$$

Proof. Lemma 9.4.13 implies that the graphs in the statement of the lemma are regular. First consider the graph $G(\Omega_n)[S_{n-1}, S_{n-1}^i]$ with $1 \leq i \leq n-1$. Let σ be a permutation in S_{n-1}. Then there is an edge joining the identity ι and $\sigma(i, n)$ if and only if $\sigma(i, n)$ has exactly one nontrivial cycle. Since $\sigma(n) = n$, this is so if and only if $\sigma = \iota$ or σ has exactly one nontrivial cycle and $\sigma(i) \neq i$. Therefore the degree of ι, and so the degree of regularity of $G(\Omega_n)[S_{n-1}^n, S_{n-1}^i]$, equals 1 plus the number of permutations in S_{n-1} with exactly one nontrivial cycle where this cycle is of the form $(i, i_1, \ldots, i_k)$ with $1 \leq k \leq n-2$. Hence the degree of regularity of $G(\Omega_n)[S_{n-1}^n, S_{n-1}^i]$ equals

$$\sum_{k=0}^{n-2} \binom{n-2}{k} k!$$

Now consider one of the graphs $G(\Omega_n)[S_{n-1}^i, S_{n-1}^j]$ with $1 \leq i, j \leq n-1$ and $i \neq j$. Let σ be a permutation in S_{n-1}. There is an edge joining $(i, n) \in S_{n-1}^i$ and $\sigma(j, n) \in S_{n-1}^j$ if and only if $(i, n)\sigma(j, n)$ has exactly one nontrivial cycle. Allowing σ to vary over all permutations in S_{n-1} we conclude that all of the graphs $G(\Omega_n)[S_{n-1}^i, S_{n-1}^j]$ with $1 \leq i \neq j \leq n-1$ have the same degree of regularity, denoted by m_{n-1}. The degree of regularity of each of the graphs $G(\Omega_n)[S_{n-1}^i]$ is d_{n-1}. Hence, using our first calculation above, we get that

$$d_n = d_{n-1} + \sum_{k=0}^{n-2} \binom{n-2}{k} k! + (n-2)m_{n-1}.$$

Simple calculation gives

$$d_n - d_{n-1} = (n-1) \sum_{k=0}^{n-2} \binom{n-2}{k} k!$$

Putting these two equations together, we get

$$m_{n-1} = \sum_{k=0}^{n-2} \binom{n-2}{k} k!,$$

and the lemma follows.   □

The final lemma we need is the following.

Lemma 9.4.15 *For* $n \geq 3$, $d_n \geq m_n$.

Proof. We have

$$d_n = \sum_{k=2}^{n} \binom{n}{k} (k-1)! \text{ and } m_n = \sum_{k=1}^{n} \binom{n-1}{k-1} (k-1)!$$

We also have

$$\binom{n}{k}(k-1)! \geq \binom{n-1}{k-1}(k-1)! \quad (2 \leq k \leq n),$$

with equality if and only if $k = n$. Since the first term in the summation for m_n equals 1 and since all terms in both summations are integers, we have that $d_n \geq m_n$ for $n \geq 3$.   □

Theorem 9.4.16 *The chromatic index of the graph* $G(\Omega_n)$ *satisfies*

$$\chi'(G(\Omega_n)) = d_n \quad (n \geq 2).$$

Proof. We show by induction on n that the edges of $G(\Omega_n)$ can be partitioned into perfect matchings. Since $G(\Omega_2) = K_2$ and $G(\Omega_3) = K_6$, the conclusion holds in these cases. Let $n \geq 4$. By Lemma 9.4.12 the graphs $G(\Omega_n)[S_{n-1}^i]$ $(i = 1, 2, \ldots, n)$ are all isomorphic to $G(\Omega_{n-1})$, and hence by the inductive assumption, their edges can be partitioned into perfect matchings. By Lemma 9.4.14 the graphs $G(\Omega_n)[S_{n-1}^i, S_{n-1}^j]$ with $i \neq j$ are all regular bipartite graphs of degree m_{n-1} and hence by Lemma 9.4.11 their edges can be partitioned into perfect matchings.

First suppose that n is even. Using (i) of Lemma 9.4.11 we see that the union of the edges of the $\binom{n}{2}$ graphs $G(\Omega_n)[S_{n-1}^i, S_{n-1}^j]$ can be partitioned into perfect matchings. It now follows that the edges of $G(\Omega_n)$ can be partitioned into perfect matchings.

Now suppose that n is odd. By (ii) of Lemma 9.4.11, the edges of K_n can be partitioned into matchings each of which contains $(n-1)/2$ edges. Using this we conclude that the union of the edges of the $\binom{n}{2}$ graphs

$G(\Omega_n)[S_{n-1}^i, S_{n-1}^j]$ can be partitioned into nm_{n-1} matchings each containing $(n-1)!(n-1)/2$ edges. Each of these matchings together with a perfect matching of each of the graphs $G(\Omega_n)[S_{n-1}^i]$ gives a perfect matching of $G(\Omega_n)$. By Lemma 9.4.15, $d_{n-1} - m_{n-1} \geq 0$. Hence the nm_{n-1} matchings mentioned above and m_{n-1} of the d_{n-1} perfect matchings of each of the graphs $G(\Omega_n)[S_{n-1}^i]$ can be paired to give nm_{n-1} perfect matchings of $G(\Omega_n)$. The remaining $d_{n-1} - m_{n-1}$ perfect matchings of each of the $G(\Omega_n)[S_{n-1}^i]$ combine to give $d_{n-1} - m_{n-1}$ perfect matchings of $G(\Omega_n)$. The result is a partition of the edges of $G(\Omega_n)$ into perfect matchings. The theorem now follows.   □

9.5   Majorization Polytopes

Let $X = (x_1, x_2, \ldots, x_n)$ and $Y = (y_1, y_2, \ldots, y_n)$ be two n-vectors of real numbers, and let

$$X' = (x_{[1]}, x_{[2]}, \ldots, x_{[n]}) \text{ and } Y' = (y_{[1]}, y_{[2]}, \ldots, y_{[n]})$$

be obtained from X and Y, respectively, by rearranging components in nonincreasing order. Recall from Chapter 1 that X is majorized by Y, denoted $X \preceq Y$, provided that the partial sums satisfy

$$\sum_{i=1}^k x_{[i]} \leq \sum_{i=1}^k y_{[i]} \quad (1 \leq k \leq n)$$

with equality for $k = n$. From Theorem 1.7.6 we get that $X \preceq Y$ if and only if there exists a doubly stochastic matrix A such that $X = YA$. Let X be majorized by Y and let

$$\Omega_n(X \preceq Y) = \{A : A \in \Omega_n, X = YA\}.$$

Then $\Omega_n(X \preceq Y)$ is a nonempty, convex polytope contained in Ω_n, called a *majorization polytope* (for the majorization $X \preceq Y$). In this section we investigate majorization polytopes. Let P and Q be permutation matrices. Then

$$\Omega_n(XP \preceq YQ) = Q^{-1} \cdot \Omega_n(X \preceq Y) \cdot P,$$

and we henceforth assume, without loss of generality, that the components of X and of Y are in nonincreasing order.

As proved in Lemma 1.7.4, if X and Y differ in t components, then there is a doubly stochastic matrix in $\Omega_n(X \preceq Y)$ equal to a product of at most $t-1$ elementary doubly stochastic matrices. There are many theorems showing the existence of other special matrices in majorization polytopes. Their proofs usually involve some technical calculations with inequalities, and so we shall simply summarize the conclusions and refer to the literature for their proofs.

Let $U = [u_{ij}]$ be a complex, unitary matrix (or a real, orthogonal matrix) of order n. Then $UU^* = U^*U = I_n$ implies that the matrix $\widehat{U} = [|u_{ij}|^2]$ obtained from U by replacing each entry by the square of its absolute value is a doubly stochastic matrix of order n. A doubly stochastic matrix which arises in this way from a unitary matrix is called a *ortho-stochastic matrix*. The matrix

$$\frac{1}{2}\begin{bmatrix} 1 & 1 & 0 \\ 1 & 0 & 1 \\ 0 & 1 & 1 \end{bmatrix}$$

is a doubly stochastic matrix which is not an orthostochastic matrix. A matrix $A = [a_{ij}]$ of order n is *uniformly tapered* provided

$$a_{11} \geq a_{12} \geq \cdots \geq a_{1n},$$
$$a_{nn} \geq a_{n-1,n} \geq \cdots \geq a_{1n},$$

and

$$a_{i,j} + a_{i+1,j+1} \geq a_{i,j+1} + a_{i+1,j} \ (2 \leq i,j \leq n).$$

The doubly stochastic matrix

$$\frac{1}{4}\begin{bmatrix} 3 & 1 & 0 \\ 1 & 2 & 1 \\ 0 & 1 & 3 \end{bmatrix}$$

is uniformly tapered. A matrix $A = [a_{ij}]$ of order n is a *Hessenberg matrix* provided $a_{ij} = 0$ whenever $j \geq i + 2$. The matrix

$$\frac{1}{8}\begin{bmatrix} 4 & 4 & 0 & 0 \\ 2 & 2 & 4 & 0 \\ 1 & 1 & 2 & 4 \\ 1 & 1 & 2 & 4 \end{bmatrix}$$

is a doubly stochastic Hessenberg matrix.

Theorem 9.5.1 *Let $X = (x_1, x_2, \ldots, x_n)$ and $Y = (y_1, y_2, \ldots, y_n)$ be non-increasing, real vectors with $X \preceq Y$. Then $\Omega_n(X \preceq Y)$ contains a matrix of each of the following types:*

(i) *a product of at most $n - 1$ elementary doubly stochastic matrices;*

(ii) *an orthostochastic matrix* (A. Horn [71]);

(iii) *a uniformly tapered, doubly stochastic matrix* (Hoffman [69]);

(iv) *a positive semidefinite symmetric, doubly stochastic matrix* (Chao and Wong [43]);

(v) *a matrix of the form PHQ where P and Q are permutation matrices and H is a doubly stochastic Hessenberg matrix* (Brualdi and Hwang [33]).

We remark that in [109], the orthostochastic matrices of order 3 are given a parametrization. In [122] uniformly tapered, doubly stochastic matrices are investigated.

In addition to the general existence results in Theorem 9.5.1, there are results which characterize those majorizations for which the corresponding majorization polytope contains other special matrices. Again we shall be content to state these results and refer to the literature for their proofs. Let $X = (x_1, x_2, \ldots, x_n)$ and $Y = (y_1, y_2, \ldots, y_n)$ be nonincreasing vectors with $X \preceq Y$. Then $x_1 + x_2 + \cdots + x_k \leq y_1 + y_2 + \cdots + y_k$ for $k = 1, 2, \ldots, n$ with equality for $k = n$. The majorization $X \preceq Y$ has a *coincidence at k* provided $k < n$ and $x_1 + x_2 + \cdots + x_k = y_1 + y_2 + \cdots + y_k$. If $X \preceq Y$ has a coincidence at k where $k < n$ and $y_k > y_{k+1}$, then $X \preceq Y$ is *decomposable at k* and is called *k-decomposable*. The notion of a coincidence of a majorization leads to conformal decompositions

$$X = X^{(1)} \oplus X^{(2)} \oplus \cdots \oplus X^{(r)} \text{ and } Y = Y^{(1)} \oplus Y^{(2)} \oplus \cdots \oplus Y^{(r)}$$

where $X^{(i)} \preceq Y^{(i)}$ and $X^{(i)} \preceq Y^{(i)}$ has no coincidences. The majorizations $X^{(i)} \preceq Y^{(i)}$ are called the *c-components* of $X \preceq Y$ $(i = 1, 2, \ldots, r)$. Similarly, the notion of decomposability of a majorization leads to conformal decompositions

$$X = \hat{X}^{(1)} \oplus \hat{X}^{(2)} \oplus \cdots \oplus \hat{X}^{(s)} \text{ and } Y = \hat{Y}^{(1)} \oplus \hat{Y}^{(2)} \oplus \cdots \oplus \hat{Y}^{(s)}$$

where $\hat{X}^{(i)} \preceq \hat{Y}^{(i)}$ and $\hat{X}^{(i)} \preceq \hat{Y}^{(i)}$ is not k-decomposable for any k. The majorizations $\hat{X}^{(i)} \preceq \hat{Y}^{(i)}$ are called the *d-components* of $X \preceq Y$ $(i = 1, 2, \ldots, s)$. The collection of c-components of $X \preceq Y$ is the collection of c-components obtained from each of its d-components.

Theorem 9.5.2 *Let $X = (x_1, x_2, \ldots, x_n)$ and $Y = (y_1, y_2, \ldots, y_n)$ be nonincreasing, real vectors with $X \preceq Y$. Then $\Omega_n(X \preceq Y)$ contains*

(i) *a positive, doubly stochastic matrix if and only if either X is a constant vector, or $X \preceq Y$ does not have a coincidence at k for any $k = 1, 2, \ldots, n-1$* [23],

(ii) *a fully indecomposable, doubly stochastic matrix if and only $X \preceq Y$ is not k-decomposable for any $k = 1, 2, \ldots, n-1$* [23],

(iii) *a positive definite symmetric, doubly stochastic matrix if and only if, where $X^{(i)} \preceq Y^{(i)}$ are the c-components of $X \preceq Y$ $(i = 1, 2, \ldots, r)$, $X^{(i)}$ is not a constant vector of size 2 or more for each $i = 1, 2, \ldots, r$* [34],

(iv) *a unique doubly stochastic matrix if and only if*

    (a) *the components of X are all different, and*

    (b) *X and Y can be conformally partitioned as $X_{(1)} \oplus X_{(2)} \oplus \cdots \oplus X_{(p)}$ and $Y_{(1)} \oplus Y_{(2)} \oplus \cdots \oplus Y_{(p)}$, respectively, where $X_{(i)}$ and $Y_{(i)}$ have size 1 or 2 and where $X_{(i)} \preceq Y_{(i)}$ for $i = 1, 2, \ldots, p$ [23].*

In [23] a formula is given for the dimension of a majorization polytope which implies assertion (iv) in Theorem 9.5.2. The extreme points of a majorization $\Omega_3(X \preceq Y)$ where X and Y are vectors of size 3 are described in [44].

We prove only one theorem in this section, namely a theorem of Levow [81], in answer to a question of Mirsky [105], characterizing those majorizations whose majorization polytopes contain a nonsingular matrix. We shall begin by outlining a slightly different proof of Lemma 1.7.4 which is convenient for the theorem that follows. Recall that $T_n^{(i,j)}$ is the permutation matrix of order n corresponding to the permutation that interchanges i and j and leaves every other element fixed, and that a matrix of the form $E_n(i, j; c) = cI_n + (1-c)T_n^{(i,j)}$ $(0 \le c \le 1)$ is an elementary doubly stochastic matrix. This matrix is nonsingular if and only if $c \ne 1/2$.

Lemma 9.5.3 *Let $X = (x_1, x_2, \ldots, x_n)$ and $Y = (y_1, y_2, \ldots, y_n)$ be non-increasing vectors with $X \preceq Y$. Then there is a doubly stochastic matrix A with $X = YA$ such that A is a product of at most $n - 1$ elementary doubly stochastic matrices.*

    Proof. We prove the lemma by induction on n, with the cases $n = 1$ and 2 easily verified. Assume that $n \ge 3$. If $X \preceq Y$ has a coincidence at k, then it follows by induction that $X = AY$ where A is a doubly stochastic matrix equal to the product of at most $(k - 1) + (n - k - 1) = n - 2$ elementary doubly stochastic matrices. Otherwise, let $\delta_k = \sum_{i=1}^{k}(y_i - x_i)$ for $k = 1, 2, \ldots, n - 1$. Since there is no coincidence, $\delta_k > 0$ for each k, and we let $\delta = \min\{\delta_k : 1 \le k \le n - 1\}$. Choose t so that $\delta_t = \delta$, and define

$$X' = (y_1 + \delta, y_2, \ldots, y_{n-1}, y_n - \delta).$$

Then $X \preceq X' \preceq Y$, and the majorization $X \preceq X'$ has a coincidence at t. Hence as above, there is a doubly stochastic matrix A such that $X = X'A$ where A is a product of at most $n-2$ elementary doubly stochastic matrices. In addition, $X' = YE_n(1, n; c)$ where

$$c = \frac{y_1 - y_n + \delta}{y_1 - y_n + 2\delta} \ge \frac{1}{2}.$$

(We note for later use that $c = 1/2$ if and only if $y_1 = y_n$, that is, Y is a constant vector.) Thus $X = YE_n(i, j; c)A$, where $E_n(i, j; c)A$ is a product of at most $n - 1$ elementary doubly stochastic matrices. $\quad\square$

    Again let $X = (x_1, x_2, \ldots, x_n)$ and $Y = (y_1, y_2, \ldots, y_n)$ be nonincreasing vectors. If for some integer k with $1 \le k \le n-1$, there are doubly stochastic matrices $A_1 \in \Omega_k$ and $A_2 \in \Omega_{n-k}$ with $X = (A_1 \oplus A_2)Y$, then X is majorized by Y and $X \preceq Y$ has a coincidence at k. Conversely, if $X \preceq Y$ has a coincidence at k, there exist $A_1 \in \Omega_k$ and $A_2 \in \Omega_{n-k}$ such that $X = (A_1 \oplus A_2)Y$ but not every $A \in \Omega_n(X \preceq Y)$ need be such a direct sum. The next lemma shows that this is the case if $X \preceq Y$ is k-decomposable.

Lemma 9.5.4 *Let $X = (x_1, x_2, \ldots, x_n)$ and $Y = (y_1, y_2, \ldots, y_n)$ be nonincreasing vectors with $X \preceq Y$ and let $A = [a_{ij}]$ be a matrix in $\Omega_n(X \preceq Y)$. Assume that $X \preceq Y$ is k-decomposable. Then there exist matrices $A_1 \in \Omega_k$ and $A_2 \in \Omega_{n-k}$ such that $A = A_1 \oplus A_2$.*

Proof. Using the facts $A \in \Omega_n$, $y_j \geq y_k$ for $j = 1, 2, \ldots, k$ and, since $X \preceq Y$ is k-decomposable, that $y_j - y_k < 0$ for $j = k+1, k+2, \ldots, n$, we calculate that

$$\sum_{i=1}^{k}(x_i - y_k) = \sum_{i=1}^{k}\sum_{j=1}^{n} a_{ij}(y_j - y_k)$$

$$= \sum_{j=1}^{k}(y_j - y_k)\sum_{i=1}^{k} a_{ij} + \sum_{j=k+1}^{n}(y_j - y_k)\sum_{i=1}^{k} a_{ij}$$

$$\leq \sum_{j=1}^{k}(y_j - y_k)\sum_{i=1}^{k} a_{ij}$$

$$\leq \sum_{j=1}^{k}(y_j - y_k)\sum_{i=1}^{n} a_{ij}$$

$$\leq \sum_{j=1}^{k}(y_j - y_k)$$

$$= \sum_{j=1}^{k}(x_j - y_k).$$

Hence we have equality throughout and, since $y_j - y_k < 0$ for $j > k$, this implies that $\sum_{i=1}^{k} a_{ij} = 0$ for $j = k+1, k+2, \ldots, n$. Since A is a nonnegative matrix we conclude that $a_{ij} = 0$ for $i = 1, 2, \ldots, k$ and $j = k+1, k+2, \ldots, n$. Thus A is partly decomposable, and since A is doubly stochastic, $A = A[\{1, 2, \ldots, k\}] \oplus A[\{k+1, k+2, \ldots, n\}]$ where the two matrices in the direct sum are doubly stochastic. $\square$

In the next lemma we show the existence of a nonsingular matrix in $\Omega_n(X \preceq Y)$ in case $X \preceq Y$ has no coincidences and X is not a scalar vector. We henceforth assume that $n \geq 2$ as the case $n = 1$ is trivial ($I_1 \in \Omega_1(X \preceq Y)$).

Lemma 9.5.5 *Let $n \geq 2$ and assume that $X = (x_1, x_2, \ldots, x_n)$ and $Y = (y_1, y_2, \ldots, y_n)$ be nonincreasing vectors with $X \preceq Y$. Assume also that*

$X \preceq Y$ has no coincidences. If X is not a scalar vector, then there is a nonsingular matrix A in $\Omega_n(X \preceq Y)$. If X is a scalar vector, then there exists a nonsingular doubly stochastic matrix B such that

$$B\left(\frac{1}{2}I_n + \frac{1}{2}T_n^{(1,n)}\right)$$

is in $\Omega_n(X \preceq Y)$.

Proof. We prove the lemma by induction on n. If $n = 2$, then we may take A to be a nonsingular elementary doubly stochastic matrix (see the proof of Lemma 9.5.3), and B to be the identity matrix I_2. Now assume that $n \geq 3$. We consider five cases which together exhaust all possibilities.

Case 1: $x_1 > x_2 > \cdots > x_n$. In this case the vector $X' = (x_1', x_2', \ldots, x_n')$ constructed in the proof of Lemma 9.5.3 satisfies $x_1' > x_2' > \cdots > x_n'$ and $X' \preceq Y$ has a coincidence. By induction there are a nonsingular doubly stochastic matrix B such that $X' = YB$ and, as in the proof of Lemma 9.5.3, a nonsingular elementary doubly stochastic matrix C such that $X = X'C$. Thus BC is a nonsingular matrix in $\Omega_n(X \preceq Y)$.

Case 2: for some i and j,

$$x_1 \geq \cdots \geq x_{i-1} > x_i = x_{i+1} = \cdots = x_j > x_{j+1} \geq \cdots \geq x_n.$$

Then for $\epsilon > 0$ sufficently small and

$$X' = (x_1, \ldots, x_{i-1}, x_i + \epsilon, x_{i+1}, \ldots, x_j, \ldots, x_n - \epsilon)$$

we have X' is nonincreasing, $X \preceq X' \preceq Y$, and $X' \preceq Y$ has no coincidences. We have $X = X'B$ where B is a nonsingular, elementary doubly stochastic matrix. Either X' falls into Case 1, or it falls into Case 2 with one less pair of equal consecutive elements and we may repeat with $X' \preceq Y$ replacing $X \preceq Y$. In this way we find a nonsingular doubly stochastic matrix A with $X = YA$.

Case 3: for some i, $y_1 = y_2 \geq \cdots \geq y_i > y_{i+1} \geq \cdots \geq y_n$ or $y_1 \geq \cdots \geq y_i > y_{i+1} \geq \cdots \geq y_{n-1} = y_n$. An argument similar to that in Case 2 using the vector $Y' = (y_1 + \epsilon, y_2, \ldots, y_{n-1}, y_n - \epsilon)$ can be used in this case.

Case 4: $x_1 = x_2 = \cdots = x_n$ and $y_1 - x_1 \neq x_n - y_n$. As in the proof of Lemma 9.5.3 we let $\delta = \min\{\delta_k : 1 \leq k \leq n-1\}$ where $\delta_k = \sum_{i=1}^{k}(y_i - x_i)$. Let $X' = (y_1 + \delta, y_2, \ldots, y_{n-1}, y_n - \delta)$. Then $X \preceq X' \preceq Y$, and $X = X'E_n(1, n; 1/2)$. We show that the c-components $X'^{(i)} \preceq Y^{(i)}$ of $X' \preceq Y$ either are of size 1 or have the property that the first component of $X'^{(i)}$ is strictly greater than its last component. Hence our previous cases and the inductive assumption will imply that $X' = YB$ for some nonsingular, doubly stochastic matrix B.

Since X is a constant vector with $X \preceq Y$, there exists an integer t with $1 < t < n$ such that $y_t \geq x_1 \geq y_{t+1}$. Since Y is a nonincreasing vector and $\delta_k = \delta_{k-1} + (y_k - x_k)$, we have that $\delta_k \geq \delta_{k+1}$ for $1 \leq k < t$ and $\delta_k \leq \delta_{k+1}$ for $t \geq k < n$. Thus $\delta = \delta_t$. Since $y_1 - x_1 \neq x_1 - y_n$, we have $\delta_1 \neq \delta_{n-1}$. Thus $X' \preceq Y$, and it is easy to check that its c-components consist of one or more components of length 1 at one end and one component without coincidences associated with a segment of the form $(x_1, \ldots, x_1, x_1 - \delta)$ or one of the form $(x_1 + \delta, x_1, \ldots, x_1)$.

Case 5: $x_1 = x_2 = \cdots = x_n$ and $y_1 - x_1 = x_1 - y_n$. If $y_1 = y_n$, then $x_1 = y_1$ contradicting the assumption that $X \preceq Y$ has no coincidences. Thus $x_1 > x_n$. Let k be the smallest integer such that $y_1 = \cdots = y_{k-1} > y_k \geq y_{k+1} \geq \cdots \geq y_n$. There is an $\epsilon > 0$ such that $Y' = (y_1, \ldots, y_{k-1} - \epsilon, y_k + \epsilon, \ldots, y_n)$ is nonincreasing and $X \preceq Y' \preceq Y$, and $X \preceq Y'$ has no coincidences. In this way we can reduce this case to Case 4.   □

We now characterize those indecomposable majorizations $X \preceq Y$ whose majorization polytopes contain a nonsingular matrix.

Theorem 9.5.6 *Let* $x = (x_1, x_2, \ldots, x_n)$ *and* $Y = (y_1, y_2, \ldots, y_n)$ *be nonincreasing vectors such that* $X \preceq Y$. *Assume that* $X \preceq Y$ *is not k-decomposable for any* $k = 1, 2, \ldots, n-1$. *Then* $\Omega_n(X \preceq Y)$ *contains a nonsingular matrix if and only if for at least one of the c-components* $X^{(i)} \preceq Y^{(i)}$ *of* $X \preceq Y$, $X^{(i)}$ *either is of size 1 or is not a constant vector.*

Proof. Let

$$X = X^{(1)} \oplus X^{(2)} \oplus \cdots \oplus Y^{(r)} \text{ and } Y = Y^{(1)} \oplus Y^{(2)} \oplus \cdots \oplus Y^{(r)}$$

be the conformal decompositions of X and Y where $X^{(i)} \preceq Y^{(i)}$ and $X^{(i)} \preceq Y^{(i)}$ has no coincidences $(i = 1, 2, \ldots, r)$. Let $X^{(i)} = (x_1^i, \ldots, x_{n_i}^i)$ and $Y^{(i)} = (y_1^i, \ldots, y_{n_i}^i)$ for $i = 1, 2, \ldots, r$. We first show that if $n_i \geq 2$ and $x_i^i = \cdots = x_{n_i}^i$ for $i = 1, 2, \ldots, r$, then every matrix $A = [a_{ij}] \in \Omega_n(X \preceq Y)$ is singular. We do this by constructing a nonzero vector $u = (u_1, u_2, \ldots, u_n)$ such that $u^T A = 0$.

Let $m_0 = 0$ and $m_j = n_1 + \cdots + n_j$ $(j = 1, 2, \ldots, r)$. Since each $X^{(i)} \preceq Y^{(i)}$ has no coincidences and $n_i \geq 2$, we have $x_{m_i + 1} \neq y_{m_i + 1}$ $(i = 0, 1, \ldots, r - 1)$, and $x_{m_i - 1} \neq y_{m_i - 1}$ $(i = 1, 2, \ldots, r)$. Let $u_i = y_i - x_i$ for $i = 1, 2, \ldots, n_1 (= m_1)$, and let

$$u_i = \frac{x_{m_j}}{y_{m_j+1} - x_{m_j+1}}(y_i - x_j) \ (i = m_j + 1, \ldots, m_{j+1}; j = 2, 3, \ldots, r - 1).$$

Let y_{k_i} be the first component of Y which equals y_{m_i-1} $(i = 2, \ldots, r)$, and set $k_1 = 1$. Let y_{l_i} be the last component of Y which is equal to y_{m_i} $(i = 1, 2, \ldots, r)$. Let $m_0 = 0$. Let s be an integer with $m_{i-1} + 1 \leq s \leq m_i$.

We show that $u^T A = 0$. We have

$$\sum_{t=1}^{n} u_t a_{ts} = \sum_{t=k_i}^{l_i} u_t a_{ts}$$

$$= u_{m_i-1} \sum_{t=k_i}^{m_i-1} a_{ts} + \frac{u_{m_i-1}}{y_{m_i-1+1} - x_{m_i-1}} \sum_{t=m_i-1+1}^{m_i} a_{ts}(y_t - x_t)$$

$$+ u_{m_i-1} \frac{y_{m_i} - x_{m_i}}{y_{m_i-1+1} - x_{m_i-1+1}} \sum_{t=m_i+1}^{l_i} a_{ts}$$

$$= \frac{u_{m_i-1}}{y_{m_i-1+1} - x_{m_i-1+1}} \sum_{t=k_i}^{l_i} (y_t - x_t) a_{ts}$$

$$= \frac{u_{m_i-1}}{y_{m_i-1+1} - x_{m_i-1+1}} \sum_{t=1}^{n} (y_t - x_t) a_{ts}$$

$$= 0.$$

Therefore $uA = 0$

Now assume that for at least one of the c-components $X^{(i)} \preceq Y^{(i)}$ of $X \preceq Y$, $X^{(i)}$ is either of size 1 or not a constant vector. It follows from Lemma 9.5.5 that

$$X^{(i)} = Y^{(i)} B_{i,1} B_{i,2} \quad (i = 1, 2, \ldots, r)$$

where $B_{i,1}$ is nonsingular and $B_{i,2}$ is either an identity matrix or the singular, elementary doubly stochastic matrix $E_{n_i}(1, n_i; 1/2)$. Let $B_i = B_{i,1} B_{i,2}$ for $i = 1, 2, \ldots, r$, and let $B = B_1 \oplus B_2 \oplus \cdots \oplus B_r$; we have $X = YB$. By Lemma 9.5.5 again, at least one of the matrices $B_{i,2}$ is an identity matrix. If all are, then B is a nonsingular matrix and we are done. So we assume that at least one of the matrices $B_{i,2}$ does not equal I. There exists a k such that $B_{k,2} \neq I$ and $B_{k+1,2} = I$, or a k such that $B_{k,2} = I$ and $B_{k+1,2} \neq I$. We argue the first case only, since a similar argument will take care of the second case. The matrix B_k has rank $n_k - 1$, and its leading submatrix B'_k of order $n_k - 1$ is nonsingular. Let B'_{k+1} be the leading submatrix of order $n_{k+1} - 1$ of B_{k+1}.

The elements on the main diagonals of the matrices $B_1, B_2, \ldots, B_r$ are strictly positive. Let α be the last diagonal element of B_k and let β be the first diagonal element of B_{k+1}. Let $\epsilon = \min\{\alpha/2, \beta/2\}$. Let E_k be the matrix obtained from $B_k \oplus B_{k+1}$ by replacing its principal submatrix of order 2 formed by rows and columns n_k and n_{k+1} with

$$\begin{bmatrix} \alpha - \epsilon & \epsilon \\ \epsilon & \beta - \epsilon \end{bmatrix}.$$

Then E_k is a doubly stochastic matrix and

$$\begin{bmatrix} X^{(k)} \\ X^{(k+1)} \end{bmatrix} = \begin{bmatrix} Y^{(k)} \\ Y^{(k+1)} \end{bmatrix} E_k.$$

Moreover,

$$\det E_k = -\epsilon^2 \det B_k' \det B_{k+1}' + \det E_k[\{1, 2, \dots, n_k\}]$$
$$\times \det E_k[\{n_k + 1, n_k + 2, \dots, n_k + n_{k+1}\}]$$
$$= \epsilon \det B_k' \det B_{k+1}' \neq 0.$$

Thus E_k is a nonsingular matrix, the elements on its main diagonal are positive, and

$$X = Y(B_1 \oplus \cdots \oplus B_{k-1} \oplus E_k \oplus B_{k+2} \oplus \cdots \oplus B_r).$$

Repeated application of this construction results in a nonsingular doubly stochastic matrix A with $X = YA$. □

Finally, we drop the indecomposability assumption in order to obtain the following result. Recall the d-components $\hat{X}^{(i)} \preceq \hat{Y}^{(i)}$ $(i = 1, 2, \dots, s)$ of a majorization $X \preceq Y$.

Theorem 9.5.7 *Let $X = (x_1, x_2, \dots, x_n)$ and $Y = (y_1, y_2, \dots, y_n)$ be non-increasing vectors such that $X \preceq Y$. Then $\Omega_n(X \preceq Y)$ contains a nonsingular matrix if and only if each d-component $\hat{X}^{(i)} \preceq \hat{Y}^{(i)}$ has the property that for at least one of its c-components $\hat{X}_{(i,j)} \preceq \hat{Y}_{(i,j)}$, $\hat{X}_{(i,j)}$ either is of size 1 or is not a constant vector.*

Proof. This theorem is an immediate consequence of Theorem 9.5.6 and Lemma 9.5.4. □

It is interesting to compare Theorem 9.5.7 with (iii) of Theorem 9.5.2 which requires that $X \preceq Y$ contain a positive definite (in particular, non-singular matrix). We conclude this section with an example [34].

Example. Let $X = (1, 1, 0)$ and $Y = (2, 0, 0)$. Then $X \preceq Y$ and the matrix

$$A = \frac{1}{4} \begin{bmatrix} 2 & 2 & 0 \\ 2 & 1 & 0 \\ 0 & 1 & 3 \end{bmatrix}$$

is a nonsingular matrix with $X = YA$. It is easy to check, without using (iii) of Theorem 9.5.2, that there is no positive definite symmetric, doubly stochastic matrix B with $X = YB$. A symmetric matrix $B \in \Omega_3(X \preceq Y)$ would have to be of the form

$$B = \begin{bmatrix} 1/2 & 1/2 & 0 \\ 1/2 & b_{22} & b_{23} \\ 0 & b_{32} & b_{33} \end{bmatrix}.$$

If B were positive definite, then $b_{22} > 1/2$, contradicting the double stochasticity of B.

9.6   A Special Subpolytope of Ω_n

Mirsky [105] raised the question of determining the *convex hull $\dot{\Omega}_n$ of the permutation matrices P of order n with $P \neq I_n$*. Thus $\dot{\Omega}_n$ is a subpolytope of Ω_n, and a doubly stochastic matrix A belongs to $\dot{\Omega}_n$ provided there is a representation of A as a convex sum of permutation matrices which does not use the identity matrix. The extreme points of $\dot{\Omega}_n$ are the $n! - 1$ non-identity permutation matrices of order n. Provided $n \geq 3$, the dimension of $\dot{\Omega}_n$ is $(n-1)^2$, the value of the dimension of Ω_n. A satisfactory answer to Mirsky's question would be a set of linear inequalities that are necessary and sufficient for a doubly stochastic matrix to belong to $\dot{\Omega}_n$; if this set of inequalities were minimal, then they would describe the facets of $\dot{\Omega}_n$.

Example. Every doubly stochastic matrix of order n with a zero on its main diagonal belongs to $\dot{\Omega}_n$. For $n \geq 3$, the doubly stochastic matrix $(1/n)J_n$ each of whose elements equals $1/n$ is in $\dot{\Omega}_n$. The matrix

$$\frac{1}{2} \begin{bmatrix} 1 & 1 & 0 \\ 0 & 1 & 1 \\ 1 & 0 & 1 \end{bmatrix}$$

is a doubly stochastic matrix of order 3 not belonging to $\dot{\Omega}_3$. A similar example works for any $n \geq 3$.   □

Cruse [48] characterized the polytope $\dot{\Omega}_n$ by finding an infinite collection of linear inequalties whose satisfaction is necessary and sufficient for a doubly stochastic matrix to belong to $\dot{\Omega}_n$. These inequalities, when viewed as points in $\Re^{n^2}$, form a convex polytope. Thus if the extreme points of this polytope were known, they would give a finite collection of facet-defining contraints for $\dot{\Omega}_n$. We follow the proof of Cruse's theorem as presented by Brualdi and Hwang [32].

Recall from Chapter 2 that a generalized tournament matrix of order n is a nonnegative matrix $T = [t_{ij}]$ satisfying $t_{ii} = 0$ for $i = 1, 2, \ldots, n$ and $t_{ij} + t_{ji} = 1$ for all i and j with $1 \leq i < j \leq n$. We also recall from Chapter 5 that a tournament matrix $A = [a_{ij}]$ is transitive if and only if

$$2 \geq a_{ij} + a_{jk} + a_{ki} \geq 1 \quad (i, j, k \text{ distinct}),$$

with the left inequality a consequence of the fact that the right inequality holds for all distinct i, j, and k. A generalized transitive tournament matrix is defined to be a generalized tournament matrix which satisfies these inequalities. Thus a *generalized transitive tournament matrix* is a nonnegative matrix $A = [a_{ij}]$ of order n satisfying each of the following constraints:

(i) $a_{ii} = 0 \quad (i = 1, 2, \ldots, n)$;

(ii) $a_{ij} + a_{ji} = 1 \quad (i \neq j)$;

(iii) $a_{ij} + a_{jk} + a_{ki} \geq 1$ $(i, j, k$ distinct$)$ and so $2 \geq a_{ij} + a_{jk} + a_{ki} \geq 1$ $(i, j, k$ distinct$)$.

It follows by induction that (ii) and (iii) imply

(iv) $a_{i_1 i_2} + a_{i_2 i_3} + \cdots + a_{i_{t-1} i_t} + a_{i_t i_1} \geq 1$ $(i_1, i_2, \ldots, i_t$ distinct; $t \geq 3)$.

For instance, if i, j, k, l are distinct, then

$$a_{ij} + a_{jk} + a_{ki} \geq 1 \text{ and } a_{ik} + a_{kl} + a_{li} \geq 1.$$

Since $a_{ik} + a_{ki} = 1$, adding we get $a_{ij} + a_{jk} + a_{kl} + a_{li} \geq 1$.

The set of all generalized tournament matrices of order n forms a convex polytope $\mathcal{T}_n^g$ whose extreme points are the tournament matrices of order n. The set of all generalized transitive tournament matrices of order n also forms a convex polytope $\mathcal{T}_n^{gt}$. If $n \leq 5$, the transitive tournament matrices are the only extreme points of $\mathcal{T}_n^{gt}$. For $n \geq 6$, while the transitive tournament matrices of order n are extreme points of $\mathcal{T}_n^{gt}$, there are other extreme points (which therefore have at least one element not equal to 0 or 1). An example of an extreme point of $\mathcal{T}_6^{gt}$ which is not a transitive tournament matrix is the matrix [48, 68]

$$\frac{1}{2}\begin{bmatrix} 0 & 1 & 2 & 2 & 1 & 1 \\ 1 & 0 & 2 & 2 & 2 & 2 \\ 0 & 0 & 0 & 1 & 1 & 0 \\ 0 & 0 & 1 & 0 & 0 & 1 \\ 1 & 0 & 1 & 2 & 0 & 1 \\ 1 & 0 & 2 & 1 & 1 & 0 \end{bmatrix}.$$

A transitive tournament matrix T of order n corresponds to a linear ordering of the set $\{1, 2, \ldots, n\}$; there exists a permutation matrix P such that PTP^T is a triangular matrix with 1's in all positions below the main diagonal. The polytope $\dot{\mathcal{T}}_n^{gt}$ is defined to be the convex hull of the transitive tournament matrices and is also known as the *linear ordering polytope* [120]. We have $\dot{\mathcal{T}}_n^{gt} = \mathcal{T}_n^{gt}$ if and only if $n \leq 5$ [55].

For m by n matrices $X = [x_{ij}]$ and $Y = [y_{ij}]$, let their *inner product* $X \circ Y$ be

$$X \circ Y = \sum_{i=1}^{m} \sum_{j=1}^{n} x_{ij} y_{ij}.$$

Let P be a permutation matrix of order n different from I_n, corresponding therefore to a permutation of $\{1, 2, \ldots, n\}$ with at least one cycle of length greater than 1. If T is a generalized transitive tournament matrix, then it follows from (ii) and (iv) above that $P \circ T \geq 1$. Hence if A is a doubly stochastic matrix in $\dot{\Omega}_n$, we have

$$A \circ T \geq 1 \quad (A \in \dot{\Omega}_n, T \in \mathcal{T}_n^{gt}).$$

We shall show that these inequalities characterize $\dot{\Omega}_n$ as a subset of Ω_n. But first we obtain a characterization of $\dot{\Omega}_n$ of another sort.

A *cycle matrix* is a $(0, 1)$-matrix A of order n whose associated digraph is a directed cycle of length at least 2, possibly with some isolated vertices.

Thus a cycle matrix corresponds to a nontrivial cycle of length at least 2 of a permutation of $\{1, 2, \ldots, n\}$. For example,

$$\begin{bmatrix} 0 & 0 & 1 & 0 & 0 \\ 0 & 0 & 0 & 0 & 0 \\ 0 & 0 & 0 & 0 & 1 \\ 0 & 0 & 0 & 0 & 0 \\ 1 & 0 & 0 & 0 & 0 \end{bmatrix}$$

is a cycle matrix corresponding to the permutation cycle $(1, 3, 5, 1)$. A *loop matrix* is a $(0, 1)$-matrix of order n with exactly one 1, and this 1 is on the main diagonal. The digraph of a loop matrix is a directed cycle of length 1 (a loop) and isolated vertices. A square matrix is called *sum-symmetric* provided its row sum vector equals its column sum vector. Cycle and loop matrices are sum-symmetric. The set Θ_n of nonnegative, sum-symmetric matrices of order n forms a convex cone in $\Re^{n \times n}$:

(i) $A \in \Theta_n$ and $c \geq 0$ imply $cA \in \Theta_n$;

(ii) $A, B \in \Theta_n$ imply $A + B \in \Theta_n$.

The next lemma [2] shows that the extreme rays of Θ_n are those rays determined by cycle and loop matrices.

Lemma 9.6.1 *A matrix A of order n is in Θ_n if and only if A can be expressed as*

$$A = c_1 Z_1 + c_2 Z_2 + \cdots + c_k Z_k$$

where $Z_1, Z_2, \ldots, Z_k$ are loop or cycle matrices and $c_1, c_2, \ldots, c_k$ are positive numbers.

Proof. Every matrix satisfying the condition in the lemma is sum-symmetric. The proof of the converse is by induction on the number of positive elements of A. If $A = O$, then we may use an empty sum. Suppose $A \neq O$. The associated digraph of A will have a directed cycle (length 1 or more) provided A has an irreducible component of order at least 2 or a nonzero component of order 1. If this is not the case, then there exists a permutation matrix P such that PAP^{T} is a strictly triangular matrix (zeros on and below its main diagonal). Since A is sum-symmetric, so is PAP^{T}. But this easily implies that PAP^{T}, and hence A, is a zero matrix. Hence there are a cycle or loop matrix Z_1 and a positive number c_1 such that $A - c_1 Z_1$ is a nonnegative sum-symmetric matrix with at least one more zero than A. The lemma now follows by induction. $\qquad\square$

Let $\mathcal{C}_n$ denote the convex hull of the cycle matrices of order n. We now give a characterization of matrices in $\dot{\Omega}_n$.

Theorem 9.6.2 *Let A be a doubly stochastic matrix of order n. Then A is in $\dot{\Omega}_n$ if and only if A has a decomposition of the form*

$$A = C + N \text{ where } C \in \mathcal{C}_n \text{ and } N \text{ is a nonnegative matrix.}$$

Proof. First assume that $A \in \dot{\Omega}_n$. Then there exist nonidentity permutation matrices $P_1, P_2, \ldots, P_k$ and positive numbers $a_1, a_2, \ldots, a_k$ summing to 1 such that

$$A = a_1 P_1 + a_2 P_2 + \cdots + a_k P_k. \tag{9.14}$$

Since $P_i \neq I_n$ there exist at least one cycle matrix C_i and a $(0,1)$-matrix N_i such that $P_i = C_i + N_i$ for $i = 1, 2, \ldots, k$. Then

$$A = \sum_{i=1}^{k} a_i C_i + \sum_{i=1}^{k} a_i N_i$$

where the first summation gives a matrix in $\mathcal{C}_n$ and the second summation gives a nonnegative matrix.

Now assume that A is a doubly stochastic matrix with $A = C + N$ where $C \in \mathcal{C}_n$ and N is nonnegative. The matrix C has a decomposition of the form $C = \sum_{i=1}^{T} d_i C_i$ where the C_i are cycle matrices and the d_i are positive numbers summing to 1. If N has a zero on its main diagonal then so does A; hence, since A is doubly stochastic, A is a convex combination of permutation matrices, none of which can equal I_n. Next suppose that N is a diagonal matrix. Let P_i be the unique permutation matrix of order n which agrees with C_i off the main diagonal $(i = 1, 2, \ldots, t)$. Then $A' = \sum_{i=1}^{T} d_i P_i$ is a doubly stochastic matrix in $\dot{\Omega}_n$ which agrees with A off the main diagonal. Since A and A' are both doubly stochastic, then $A' = A$ and we conclude $A \in \dot{\Omega}_n$.

Finally assume that N is not a diagonal matrix and has no zeros on its main diagonal. Since A is a doubly stochastic matrix and C is a sum-symmetric matrix, N is a sum-symmetric matrix. Lemma 9.6.1 implies that N is a positive linear combination of cycle and loop matrices. Combining this with (9.14), we conclude that

$$A = \sum_{i=1}^{l} a_i' C_i' + N'$$

where the a_i' are positive numbers, the C_i' are cycle matrices, and $N' = [n_{ij}']$ is a diagonal matrix each of whose diagonal elements is positive. Since N is not a diagonal matrix, the number $\omega = \sum_{i=1}^{l} a_i'$ is strictly greater than 1. There exists a nonnegative, diagonal matrix $N'' = [n_{ij}'']$ such that the matrix $A' = [a_{ij}']$ defined by

$$A' = \sum_{i=1}^{l} \frac{a_i'}{\omega} C_i' + N''$$

is a doubly stochastic matrix. By what we have already proved, A' is a convex combination of nonidentity permutation matrices. Since $\omega > 1$, it follows that $n_{ii}' \leq n_{ii}''$ $(i = 1, 2, \ldots, n)$ with strict inequality for at least one i. Let

$$\delta = \min \left\{ \frac{n_{ii}'}{n_{ii}''} : i = 1, 2, \ldots, n \right\}.$$

Then $\delta < 1$, and the matrix

$$A'' = \frac{1}{1-\delta}(A - \delta A')$$

is a doubly stochastic matrix with a least one 0 on its main diagonal. Again by what we have already proved, A'' is a convex combination of nonidentity permutation matrices. Since

$$A = \delta A' + (1 - \delta)A'',$$

A is also a convex combination of nonidentity permutation matrices. □

Let $\mathcal{P}$ be a polyhedron of m by n nonnegative matrices.[17] Thus $\mathcal{P}$ is the intersection of a finite number of half-spaces in $\Re^{m \times n}$. The polyhedron $\mathcal{P}$ is of *blocking type* provided $A \in \mathcal{P}$ and $X \geq A$ imply that $X \in \mathcal{P}$. The polyhedron $\mathcal{P}$ is of blocking type if and only if there exist a finite number of nonnegative matrices $U_1, U_2, \ldots, U_l$ such that

$$\mathcal{P} = \{X \in \Re^{m \times n} : U_i \circ X \geq 1, i = 1, 2, \ldots, l\}. \tag{9.15}$$

Also the polyhedron $\mathcal{P}$ has blocking type if and only if there exist nonnegative matrices $M_1, M_2, \ldots, M_k$ such that

$$\mathcal{P} = \{X + N : X \text{ is in the convex hull of } M_1, M_2, \ldots, M_k; N \geq O\}. \tag{9.16}$$

The polyhedron $\mathcal{P}$ given by (9.16) is denoted by

$$\langle M_1, M_2, \ldots, M_k \rangle$$

and is sometimes called the *dominant* of the convex hull of $M_1, M_2, \ldots, M_k$. If $\mathcal{K}$ is a polytope whose extreme points $M_1, M_2, \ldots, M_k$ are not known, then we use the notation

$$\langle \mathcal{K} \rangle$$

instead of $\langle M_1, M_2, \ldots, M_k \rangle$.

Let $\mathcal{P}$ be a polyhedron of blocking type. The *blocker* of $\mathcal{P}$ is the polyhedron

$$B(\mathcal{P}) = \{Y \in \Re^{m \times n} : Y \geq O, Y \circ X \geq 1 \text{ for all } X \in \mathcal{P}\}. \tag{9.17}$$

The following characterizations are due to Fulkerson (see [125]).

Lemma 9.6.3 *Let $\mathcal{P}$ be a polyhedron of m by n matrices of blocking type. Then the following hold.*

---

[17]A more general discussion, not restricted to matrices, can be found in [125], although the proofs are the same.

(i) $B(\mathcal{P})$ is of blocking type.

(ii) $B(B(\mathcal{P})) = \mathcal{P}$.

(iii) An m by n matrix X belongs to $\mathcal{P}$ if and only if $Y \circ X \geq 1$ for all $Y \in B(\mathcal{P})$.

(iv) $\mathcal{P} = \langle M_1, M_2, \ldots, M_k \rangle$, if and only if

$$B(\mathcal{P}) = \{Z \in \Re^{m \times n} : Z \geq O, Z \circ M_i \geq 1, i = 1, 2, \ldots, m\}.$$

Proof. Since $\mathcal{P}$ is of blocking type there exist nonnegative matrices $M_1, M_2, \ldots, M_k$ such that $\mathcal{P} = \langle M_1, M_2, \ldots, M_k \rangle$. It follows from the definition of blocker in (9.17) that

$$B(\mathcal{P}) = \{Y \in \Re^{m \times n} : Y \geq O, Y \circ M_i \geq 1 \text{ for all } i = 1, 2, \ldots, k\};$$

hence $B(\mathcal{P})$ is of blocking type, and both (i) and the forward direction of (iv) hold. For (ii) we clearly have $\mathcal{P} \subseteq B(\mathcal{P})$. Suppose there exists a matrix Y in $B(B(\mathcal{P})) \setminus P$. Let $\mathcal{P}$ be given as in (9.15). Then there exists one of the nonnegative matrices U_j such that $U_j \circ Y < 1$. Since $X \circ U_j \geq 1$ for all $X \in \mathcal{P}$, U_j is in $B(\mathcal{P})$ contradicting $Y \in B(B(\mathcal{P}))$. This proves (ii). The assertion (iii) now follows from (ii) as $X \in \mathcal{P}$ if and only if $X \in B(B(\mathcal{P}))$, if and only if $X \geq O$ and $Z \circ X \geq 1$ for all $Z \in B(\mathcal{P})$. It remains to verify the backward direction of (iv). Assume that

$$B(\mathcal{P}) = \{Z \in \Re^{m \times n}, Z \geq O, Z \circ M_i \geq 1, i = 1, 2, \ldots, m\},$$

and let $\mathcal{Q} = \langle M_1, M_2, \ldots, M_k \rangle$. Therefore $B(\mathcal{Q}) = B(\mathcal{P})$, and hence by (ii), $\mathcal{P} = \mathcal{Q}$. $\qquad\square$

Let
$$\mathcal{C}_n^* = \langle C : C \text{ a cycle matrix of order } n \rangle = \langle C \in \mathcal{C}_n \rangle,$$

and let
$$\mathcal{T}_n^{\text{gt}*} = \langle \mathcal{T}_n^{\text{gt}} \rangle.$$

Both $\mathcal{C}_n^*$ and $\mathcal{T}_n^{\text{gt}*}$ are polyhedra of blocking type. We now show that each is the blocker of the other. By (ii) of Lemma 9.6.3 we need only show one of them is the blocker of the other.

Theorem 9.6.4 Let n be a positive integer. Then $B(\mathcal{C}_n^*) = \mathcal{T}_n^{\text{gt}*}$, equivalently, a nonnegative matrix A of order n has a decomposition of the form $A = T + N$ where T is a generalized transitive tournament matrix and N is a nonnegative matrix if and only if $C \circ A \geq 1$ for every cycle matrix C of order n.

Proof. Since $T \circ C \geq 1$ for every generalized transitive tournament matrix T and every cycle matrix C, it follows that if $A = T + N$ where T is a generalized transitive tournament matrix and N is a nonnegative matrix, then $A \circ C \geq 1$.

Now assume that $A \circ C \geq 1$ for every cycle matrix C of order n. Choose $B = [b_{ij}] \leq A$ such that $B \circ C \geq 1$ for every cycle matrix C and the sum of the elements of B is as small as possible. We prove that B is a generalized transitive tournament matrix by showing that $b_{pq} + b_{qp} = 1$ whenever $p \neq q$. Suppose to the contrary that there exist distinct p and q with $b_{pq} + b_{qp} \neq 1$ so that $b_{pq} + b_{qp} > 1$. The minimality condition on B implies that $n \geq 3$ and that $b_{pq} > 0$ and $b_{qp} > 0$. The minimality condition on B also implies that there exist cycle matrices C_1 and C_2 such that the element of C_1 in position (p, q) equals 1 and the element in position (q, p) of C_2 equals 1, and $B \circ C_1 = B \circ C_2 = 1$. There exists a cycle matrix C whose set of nonzero positions is a subset of the union of the nonzero positions of C_1 and C_2 but does not contain (p, q) or (q, p). We then have

$$B \circ C \leq B \circ C_1 + B \circ C_2 - (b_{pq} + b_{qp}) < 1 + 1 - 1 = 1,$$

contradicting our assumption. We conclude that B is a generalized transitive tournament matrix and $A = B + (A - B)$ where $A - B$ is nonnegative. $\quad\square$

Corollary 9.6.5 *Let n be a positive integer. Then $B(\mathcal{T}_n^{gt*}) = \mathcal{C}_n^*$, equivalently, a nonnegative matrix A of order n has a decomposition of the form $A = C + N$ where C is a convex combination of cycle matrices and N is a nonnegative matrix if and only if $T \circ A \geq 1$ for every generalized transitive tournament matrix T of order n.* $\quad\square$

From Theorem 9.6.2 and Corollary 9.6.5 we immediately obtain the following characterization of matrices in $\dot{\Omega}_n$, answering the question of Mirsky.

Theorem 9.6.6 *A doubly stochastic matrix A of order n is a convex combination of the nonidentity permutation matrices of order n if and only if $T \circ A \geq 1$ for every generalized transitive tournament matrix of order n.* $\quad\square$

As a corollary we get a simple characterization of the symmetric matrices in $\dot{\Omega}_n$.

Corollary 9.6.7 *Let $A = [a_{ij}]$ be a symmetric, doubly stochastic matrix of order n. Then A is a convex combination of the nonidentity permutation matrices of order n if and only if*

$$a_{11} + a_{22} + \cdots + a_{nn} \leq n - 2.$$

Proof. Let $T = [t_{ij}]$ be a generalized transitive tournament matrix of order n. Then

$$
\begin{aligned}
T \circ A &= \sum_{1 \le i < j \le n} (a_{ij}t_{ij} + a_{ji}t_{ji}) \\
&= \sum_{1 \le i < j \le n} a_{ij}(t_{ij} + t_{ji}) \\
&= \sum_{1 \le i < j \le n} a_{ij} \\
&= \left(n - \sum_{i=1}^{n} a_{ii} \right) \Big/ 2.
\end{aligned}
$$

The corollary now follows from Theorem 9.6.6. □

We have previously noted that $\mathcal{T}_n^{\text{gt}}$ is the convex hull of the transitive tournament matrices if $n \le 5$, from which it follows that a doubly stochastic matrix of order $n \le 5$ is a convex combination of nonidentity permutation matrices if and only if $T \circ A \ge 1$ for all transitive tournament matrices of order n. In [32] it was shown that this conclusion also holds for $n = 6$ even though $\mathcal{T}_6^{\text{gt}}$ is not the convex hull of the transitive tournament matrices of order 6. More information about the extreme points of the polytope of generalized transitive tournament matrices can be found in [12], [13], [14], [15], [112], [113], [68], [120], [136].

9.7 The Even Subpolytope of Ω_n

We now consider another subpolytope of the doubly stochastic polytope Ω_n about which there is less information known. An *even permutation matrix* of order n is a permutation matrix corresponding to an even permutation of $\{1, 2, \ldots, n\}$. There are $n!/2$ matrices in the set $\mathcal{E}_n$ of even permutation matrices of order n. A doubly stochastic matrix which can be expressed as a convex combination of even permutation matrices is called an *even doubly stochastic matrix*. The set of even doubly stochastic matrices of order n forms a convex polytope Ω_n^{ev}. Hoffman (see [104]) proposed the problem of determining criteria for deciding whether or not a doubly stochastic matrix is even.

The convex polytope Ω_n^{ev} is investigated in [36]. Its extreme points are the matrices in $\mathcal{E}_n$. If $n = 1$ or 2, the identity matrix is the only even doubly stochastic matrix. The even doubly stochastic matrices of order 3 are the matrices

$$
\begin{bmatrix}
a & b & c \\
c & a & b \\
b & c & a
\end{bmatrix}
$$

where a, b, and c are nonnegative numbers summing to 1. Thus Ω_n^{ev} is a

Doubly Stochastic Matrices

regular simplex of dimension 2 in $\Re^9$. We show that the dimension of Ω_n^{ev} equals the dimension of Ω_n when $n \geq 4$.

Theorem 9.7.1 *Let $n \geq 4$ be an integer. Then* $\dim \Omega_n^{\mathrm{ev}} = (n-1)^2$.

Proof. Since $\dim \Omega_n = (n-1)^2$, it suffices to show that there exist even permutation matrices $P_0 = I_n, P_1, P_2, \ldots, P_{(n-1)^2}$ such that the $(n-1)^2$ matrices $P_i - P_0$ with $1 \leq i \leq (n-1)^2$ are linearly independent. We do this by construction of $P_1, P_2, \ldots, P_{(n-1)^2}$. If $n = 4$, there are nine even permutations of $\{1, 2, 3, 4\}$ that do not fix 1, and the corresponding permutation matrices work in this case.

Now let $n \geq 5$ and consider the $(0, 1)$-matrix B of order n which has a total of $2n - 1$ 1's, occupying the positions on the main diagonal and the subdiagonal below it. The only permutation matrix Q with $Q \leq B$ is $P_0 = I_n$. We now show that it is possible to order the (positions of the) 0's of A in such a way that replacing the 0's by 1's in succession in that order produces matrices $A_0 = A, A_1, A_2, \ldots, A_{(n-1)^2} = J_n$ with the property that there are even permutation matrices $P_1, P_2, \ldots, P_{(n-1)^2}$ such that $P_i \leq A_i$ but $P_i \not\leq A_{i-1}$ $(1 \leq i \leq (n-1)^2)$.

First replace with 1's the 0's in positions (1,3), (2,4), and (3,5). Each replacement creates a new (even) permutation matrix with a cycle of length 3 and two fixed points. Next replace with a 1 the 0 in position (1,5) creating a new (even) permutation matrix with a cycle of length 5. Now replace with a 1 the 0 in position (5,1) creating a new (even) permutation matrix with a cycle of length 3 and two fixed points. Observe that the matrix A_5 constructed up to this point has 1's in positions (1,5) and (5,1) corresponding to a transposition. If $n = 5$ or 6, it is not difficult to complete the argument. For instance, if $n = 5$, we can continue to replace 0's with 1's in positions in the order

$$(2,3), (3,4), (4,2), (1,2), (4,5), (3,1), (5,3), (1,4), (4,1), (2,5), (5,2).$$

Now assume that $n \geq 7$. We next replace with a 1 the 0 in position $(n-1, n)$ giving 1's in both positions $(n-1, n)$ and $(n, n-1)$, and thus creating a transposition disjoint from the one corresponding to (1,5) and (5,1). Let A' be the leading principal submatrix of order $n-2$ of the matrix constructed at this point. We order the positions of its 0's subject only to the condition that the positions above its main diagonal come first. Using the fact that there are 1's now in positions $(n-1, n)$ and $(n, 1)$, it is easy to see that replacing with 1's these 0's in the order specified creates each time a new even permutation matrix. Next we do a similar construction using the lower principal submatrix of order $n - 2$ and taking advantage of the 1's in positions (1,2) and (2,1). At this point we have a matrix $A_{(n-1)^2-7}$ with exactly seven 0's, and it is not difficult to complete the proof. $\qquad\square$

Mirsky [104] characterized the main diagonals of even doubly stochastic matrices. Before proving this theorem we record three lemmas. Let Γ_n be the set of all (0,1)-vectors of size n which do not have exactly one or two 1's. The proof of the first lemma is straightforward.

Lemma 9.7.2 *The set Γ_n equals the set of diagonal vectors of the even permutation matrices of order n.* □

Lemma 9.7.3 *Let $u, u^1, u^2, \ldots, u^m$ be real vectors of size n. Then u is in the convex hull of $u^1, u^2, \ldots, u^m$ if and only if for every real vector x of size n,*

$$u \circ x \leq \max\{u^k \circ x : 1 \leq k \leq m\}.$$

Proof. This is a standard result in the elementary theory of convex sets. □

Lemma 9.7.4 *Let $n \geq 3$ be an integer, and let $f = (f_1, f_2, \ldots, f_n)$ be a real vector satisfying*

$$1 \geq f_1 \geq f_2 \geq \cdots \geq f_n \geq 0, \text{ and} \tag{9.18}$$
$$n - 3 + 2f_n - f_1 - \cdots - f_{n-1} \geq 0. \tag{9.19}$$

Then f is in the convex hull of the set Γ_n.

Proof. We first show that if $g = (g_1, g_2, \ldots, g_n)$ satisfies $g_1 \geq g_2 \geq \cdots \geq g_n$, then

$$\sum_{k=1}^{n} f_k g_k \leq \sum_{k=1}^{r} g_k \tag{9.20}$$

for some r with $0 \leq r \leq n$ and $r \neq n-2, r \neq n-1$. Let p be the number of nonnegative g_i. Then (9.20) holds with $r = p$, and so we now assume that $p = n-2$ or $n-1$. In both of these cases, $g_{n-2} \geq 0$.

First suppose that $g_{n-2} + g_{n-1} + g_n \geq 0$. Then using the monotonicity assumption on g and (9.18) and (9.19), we obtain

$$\sum_{k=1}^{n} g_k - \sum_{k=1}^{n} f_k g_k = \sum_{k=1}^{n} (1 - f_k) g_k. \tag{9.21}$$

Letting α be the common value in (9.21), we obtain

$$\alpha \geq \sum_{k=1}^{n-2} (1 - f_k) g_{n-2} + (1 - f_{n-1}) g_{n-1} + (1 - f_n)(-g_{n-2} - g_{n-1})$$
$$= (n - 3 + f_n - f_1 - \cdots - f_{n-2}) g_{n-2} - (f_{n-1} - f_n) g_{n-1}$$
$$\geq (n - 3 + f_n - f_1 - \cdots - f_{n-2}) g_{n-2} - (f_{n-1} - f_n) g_{n-2}$$
$$= (n - 3 + 2f_n - f_1 - \cdots - f_{n-1}) g_{n-2} \geq 0.$$

Hence (9.20) holds with $r = n$.

Now suppose that $g_{n-2} + g_{n-1} + g_n < 0$. Then letting

$$\beta = \sum_{k=1}^{n-3} g_k - \sum_{k=1}^{n} f_k g_k,$$

we obtain

$$\beta = \sum_{k=1}^{n-3}(1 - f_k)g_k - f_{n-2}g_{n-2} - f_{n-1}g_{n-1} - f_n g_n$$

$$\geq \sum_{k=1}^{n-3}(1 - f_k)g_{n-2} - f_{n-2}g_{n-2} - f_{n-1}g_{n-1} + f_n(g_{n-2} + g_{n-1})$$

$$= (n - 3 + f_n - f_1 - \cdots - f_{n-2})g_{n-2} - (f_{n-1} - f_n)g_{n-1}$$

$$\geq (n - 3 + 2f_n - f_1 - \cdots - f_{n-1})g_{n-2} \geq 0.$$

Hence (9.20) holds with $r = n - 3$.

Now let u be any vector and let v be the nonincreasing vector obtained by rearranging the components of u. Then since f is nonincreasing, $u \circ f \leq v \circ f$ and hence by (9.20),

$$u \circ f \leq v \circ f \leq \max\{v \circ x : x \in \Gamma_n\} = \max\{u \circ x : x \in \Gamma_n\}.$$

Applying Lemma 9.7.3 we conclude that f belongs to the convex hull of Γ_n.   □

Theorem 9.7.5 *Let n be a positive integer, and let $d_1, d_2, \ldots, d_n$ be n real numbers. Then there exists an even doubly stochastic matrix whose main diagonal elements are $d_1, d_2, \ldots, d_n$ if and only if*

(i) $0 \leq d_i \leq 1$ *for $i = 1, 2, \ldots, n$, and*

(ii) $\sum_{k=1}^{n} d_k - 3\min\{d_i : 1 \leq i \leq n\} \leq n - 3$.

Proof. The theorem holds trivially when $n = 1$ or 2 and so we assume that $n \geq 3$. First suppose that $d_1, d_2, \ldots, d_n$ are the diagonal elements of an even doubly stochastic matrix A of order n. Then by Lemma 9.7.2, the vector $(d_1, d_2, \ldots, d_n)$ is in the convex hull of Γ_n. Since every vector in Γ_n obviously satisfies (i) and (ii), so does d.

Now suppose that the vector $d = (d_1, d_2, \ldots, d_n)$ satisfies (i) and (ii). Let f be the monotone nonincreasing vector obtained by rearranging the components of d. Then f satisfies (9.18) and (9.19) of Lemma 9.7.4 and so f is in the convex hull of Γ_n:

$$f = \sum_{u \in \Gamma_n} c_u u \quad \left(c_u \geq 0, \ \sum_{u \in \Gamma_n} c_u = 1 \right).$$

By Lemma 9.7.2 each vector $u \in \Gamma_n$ is the diagonal vector of an even permutation matrix P_u. Thus

$$A = \sum_{u \in \Gamma_n} c_u P_u$$

is an even doubly stochastic matrix whose diagonal vector is f. There is a permutation matrix Q such that QAQ^T has diagonal vector d. Since QPQ^T is an even permutation matrix whenever P is, QAQ^T is an even doubly stochastic matrix whose diagonal vector is d.   □

Corollary 9.7.6 *Let $A = [a_{ij}]$ be an even doubly stochastic matrix. Then for every even permutation π of $\{1, 2, \ldots, n\}$ and every integer k with $1 \leq k \leq n$,*

$$\sum_{i=1}^{n} a_{i\pi(i)} - 3a_{k\pi(k)} \leq n - 3. \tag{9.22}$$

Proof. Since the product of two even permutation matrices is an even permutation, if P is an even permutation corresponding to the even permutation $\pi = (\pi(1), \pi(2), \ldots, \pi(n))$, then PA is an even doubly stochastic matrix whose diagonal elements are $a_{1\pi(1)}, a_{2\pi(2)}, \ldots, a_{n\pi(n)}$. We now apply Theorem 9.7.5 to complete the proof.   □

Example. In responding to a query of Mirsky, von Below [9] showed that the inequalities (9.22) in Corollary 9.7.6 are not sufficient for a doubly stochastic matrix to be even if $n \geq 4$. The doubly stochastic matrix

$$\frac{1}{6} \begin{bmatrix} 1 & 3 & 2 & 0 \\ 3 & 2 & 0 & 1 \\ 0 & 1 & 3 & 2 \\ 2 & 0 & 1 & 3 \end{bmatrix}$$

satisfies (9.22) but is not even. For $n \geq 5$, the doubly stochastic matrix of order n given by

$$B_n = \frac{1}{2} \begin{bmatrix} 0 & 1 & 1 & 0 & 0 & \cdots & 0 & 0 \\ 1 & 1 & 0 & 0 & 0 & \cdots & 0 & 0 \\ 0 & 0 & 1 & 1 & 0 & \cdots & 0 & 0 \\ 0 & 0 & 0 & 1 & 1 & \cdots & 0 & 0 \\ 0 & 0 & 0 & 0 & 1 & \cdots & 0 & 0 \\ \vdots & \vdots & \vdots & \vdots & \vdots & \ddots & \vdots & \vdots \\ 0 & 0 & 0 & 0 & 0 & \cdots & 1 & 1 \\ 1 & 0 & 0 & 0 & 0 & \cdots & 0 & 1 \end{bmatrix}$$

satisfies (9.22) but is not even. In fact B_n can be expressed as a convex combination of permutation matrices in only one way, namely as $(1/2)(P + Q)$

where P is the permutation matrix corresponding to the transposition which interchanges 1 and 2, and Q is the permutation matrix $2B_n - P$. If $n = 1, 2$, or 3, then the inequalities (9.22) are sufficient for a doubly stochastic matrix to be even. For $n = 2$, (9.22) implies $A = I_2$. For $n = 3$, (9.22) implies

$$
A = \begin{bmatrix} a & b & c \\ c & a & b \\ b & c & a \end{bmatrix},
$$

an even doubly stochastic matrix.   □

In [36], additional necessary conditions for a doubly stochastic matrix to be even were obtained, and we discuss these now. If $A = [a_{ij}]$ is a doubly stochastic matrix, then we let $\mathcal{E}_n(A)$ denote the *set of all even permutations* π of $\{1, 2, \ldots, n\}$ such that $a_{i\pi(i)} \neq 0$ for $i = 1, 2, \ldots, n$. These even permutations have the property that the positions of the 1's of the corresponding permutation matrix are occupied in A by nonzero elements, and thus they are the even permutation matrices in the nonzero pattern of A.

Theorem 9.7.7 *Let $A = [a_{ij}]$ be an even doubly stochastic matrix of order n. Then the following conditions hold.*

(i) $\sum_{\pi \in \mathcal{E}_n(A)} \min\{a_{i\pi(i)} : 1 \leq i \leq n\} \geq 1$.

(ii) $\sum_{\pi \in \mathcal{E}_n(A), \pi(k)=l} \min\{a_{i\pi(i)} : i \neq k\} \geq a_{kl} \quad (k, l = 1, 2, \ldots, n)$.

(iii) *If $a_{i_1 j_1}, a_{i_2 j_2}, \ldots, a_{i_t j_t}$ are positive elements of A such that for each permutation $\pi \in \mathcal{E}_n(A)$, $\pi(i_p) = j_p$ for at most one $p = 1, 2, \ldots, t$, then*

$$
a_{i_1 j_1} + a_{i_2 j_2} + \cdots + a_{i_t j_t} \leq 1.
$$

(iv) *Let σ be any odd permutation of $\{1, 2, \ldots, n\}$ and let k be an integer such that for all $\pi \in \mathcal{E}_n(A)$, $\sigma(i) = \pi(i)$ for at most k integers $i = 1, 2, \ldots, n$. Then*

$$
\sum_{i=1}^{n} a_{i\sigma(i)} \leq k.
$$

(In particular, the integer $k = n - 2$ satisfies the stated assumption.)

Proof. Since A is an even doubly stochastic matrix, A can be written as

$$
A = \sum_{\pi \in \mathcal{E}_n(A)} c_\pi P_\pi, \tag{9.23}
$$

where the c_π are nonnegative numbers summing to 1 and P_π is the permutation matrix corresponding to π.

Let π be a permutation in $\mathcal{E}_n(A)$. Then $a_{i\pi(i)} \geq c_\pi$ for each $i = 1, 2, \ldots, n$, and hence

$$
\min\{a_{i\pi(i)} : 1 \leq i \leq n\} \geq c_\pi.
$$

Thus

$$\sum_{\pi \in \mathcal{E}_n(A)} \min\{a_{i\pi(i)} : 1 \leq i \leq n\} \geq \sum_{\pi \in \mathcal{E}_n(A)} c_\pi \geq 1$$

and (i) holds.

Now let k and l be integers with $1 \leq k, l \leq n$. It follows from (9.23) that for each $\pi \in \mathcal{E}_n(A)$ with $\pi(k) = l$,

$$c_\pi \leq \min\{a_{i\pi(i)} : i \neq k\}, \text{ and}$$

$$a_{kl} = \sum_{\pi \in \mathcal{E}_n(A), \pi(k)=l} c_\pi.$$

Using these two relations we get (ii).

Let $a_{i_1 j_1}, a_{i_2 j_2}, \ldots, a_{i_t j_t}$ be a set of positive elements of A such that for each permutation $\pi \in \mathcal{E}_n(A)$, $\pi(i_p) = j_p$ for at most one $p = 1, 2, \ldots, t$. Each even permutation $\pi \in \mathcal{E}_n(A)$ has 1's in at most one of the positions $(i_1, j_1), (i_2, j_2), \ldots, (i_t, j_t)$. Let $\mathcal{E}_n(A)(i_p, j_p)$ be the set of permutations $\pi \in \mathcal{E}_n(A)$ such that $\pi(i_p) = j_p$ $(i = 1, 2, \ldots, p)$. Then the sets $\mathcal{E}_n(A)(i_p, j_p)$ with $1 \leq p \leq t$ are pairwise disjoint, and hence

$$\sum_{p=1}^{T} a_{i_p j_p} = \sum_{p=1}^{T} \sum_{\pi \in \mathcal{E}_n(A)(i_p, j_p)} c_\pi$$

$$\leq \sum_{\pi \in \mathcal{E}_n(A)} c_\pi = 1,$$

proving (iii).

Finally we prove (iv). Let $P_\pi = [p_{ij}]$ be a permutation matrix corresponding to a permutation π in $\mathcal{E}_n(A)$. Then the assumptions in (iv) imply that $\sum_{i=1}^{n} p_{i\sigma(i)} \leq k$, and hence

$$\sum_{i=1}^{n} a_{i\sigma(i)} \leq \sum_{\pi \in \mathcal{E}_n(A)} k c_\pi = k,$$

proving (iv). $\qquad\qquad\qquad\qquad\qquad\qquad\qquad\qquad\qquad\qquad\qquad\quad$ $\square$

Example. The following matrices from [36] illustrate some independence of conditions (i)–(iv) in Theorem 9.7.7 and (9.22) in Corollary 9.7.6. As already pointed out, the matrix

$$B_5 = \frac{1}{2} \begin{bmatrix} 0 & 1 & 1 & 0 & 0 \\ 1 & 1 & 0 & 0 & 0 \\ 0 & 0 & 1 & 1 & 0 \\ 0 & 0 & 0 & 1 & 1 \\ 1 & 0 & 0 & 0 & 1 \end{bmatrix}$$

satisfies (9.22) but it does not satisfy (i). We conclude that $B_5 \notin \Omega_5^{\mathrm{ev}}$ (as previously indicated), and (9.22) does not imply (i). The matrix

$$\frac{1}{6} \begin{bmatrix} 3 & 3 & 0 & 0 & 0 \\ 3 & 3 & 0 & 0 & 0 \\ 0 & 0 & 2 & 4 & 0 \\ 0 & 0 & 2 & 0 & 4 \\ 0 & 0 & 2 & 2 & 2 \end{bmatrix}$$

satisfies (9.22) and (i) but does not satisfy (ii) when $k = l = 1$. Thus this matrix is not in Ω_5^{ev}, and (9.22) and (i) do not imply (ii). The matrix

$$A = \frac{1}{3} \begin{bmatrix} 1 & 1 & 1 & 0 & 0 \\ 1 & 1 & 0 & 0 & 1 \\ 0 & 1 & 1 & 1 & 0 \\ 0 & 0 & 1 & 1 & 1 \\ 1 & 0 & 0 & 1 & 1 \end{bmatrix}$$

satisfies (9.22), (i), and (ii). Consider the set of positions

$$X = \{(1,1),(2,5),(3,2),(4,3),(5,4)\}.$$

Each of the positions in X is occupied by a 1 in exactly one of the five permutations in $E_5(A)$. The sum of the elements of A in the positions in X is 5/3. Hence $A \notin \Omega_5^{\mathrm{ev}}$, and (9.22), (i), and (ii) together do not imply (iii). The matrix A also does not satisfy (iv) for the odd permutation σ where $\sigma(1) = 1, \sigma(2) = 5, \sigma(3) = 2, \sigma(4) = 3, \sigma(5) = 4$.   □

Example. Consider the (triangular) face $\Omega_4(A)$ of Ω_4 of dimension 2 where

$$A = \begin{bmatrix} 1 & 1 & 0 & 0 \\ 1 & 0 & 1 & 0 \\ 1 & 0 & 0 & 1 \\ 0 & 1 & 1 & 1 \end{bmatrix}.$$

The three vertices of this face are even permutation matrices and hence $\Omega_4(A) \subseteq \Omega_4^{\mathrm{ev}}$. It follows that $\Omega_4(A)$ is a face of Ω_5^{ev}. In general, some faces of Ω_n^{ev} can be obtained in this way by identifying fully indecomposable, $(0,1)$-matrices A of order n with the property that if P is a permutation matrix with $P \leq A$, then P is even. For example, let $n = k_1 + k_2 + \cdots + k_t$ where the k_i are odd integers greater than 1. Let P_k be a permutation matrix of order k corresponding to a permutation cycle of length k. Then the matrix

$$A = (I_{k_1} + P_{k_1}) \oplus (I_{k_2} + P_{k_2}) \oplus \cdots \oplus (I_{k_t} + P_{k_t})$$

has the property that all the extreme points of $\Omega_n(A)$ are even permutation matrices, and hence $\Omega_n(A)$ is a t-dimensional face of Ω_n^{ev}. It seems difficult to characterize all the faces of Ω_n^{ev}. The one-dimensional faces, and thus the graph, can be characterized. □

We now consider the graph $G(\Omega_n^{\text{ev}})$ whose vertices are the even permutations σ of $\{1, 2, \ldots, n\}$ (or permutation matrices P_σ of order n).

Theorem 9.7.8 *Let σ and π be even permutations of $\{1, 2, \ldots, n\}$, and let the cycle decomposition of the permutation $\sigma^{-1}\pi$ contain p cycles of odd length greater than 1 and q cycles of even length. Then there is an edge in $G(\Omega_n^{\text{ev}})$ joining σ and π if and only if either $p = 1$ and $q = 0$, or $p = 0$ and $q = 2$.*

Proof. Since σ and π are even permutations, $\sigma^{-1}\pi$ is also an even permutation and hence q is an even integer. It suffices to prove the theorem when $\sigma = \iota$, the identity permutation. Thus we need to show that $\{cI_n + (1-c)P_\pi : 0 \leq c \leq 1\}$ is a one-dimensional face of Ω_n^{ev} if and only if π has either $p = 1$ cycle of odd length greater than 1 and $q = 0$ cycles of even length, or $p = 0$ cycles of odd length greater than 1 and $q = 2$ cycles of even length.

Let $X_c = cI_n + (1-c)P_\pi$, with $0 < c < 1$, be a convex combination of I_n and P_π. If $p = 1$ and $q = 0$, or $p = 0$ and $q - 2$, then $\mathcal{E}_n(X_c) = \{\iota, \pi\}$. Thus in this case it follows that I_n and P_π are the extreme points of a one-dimensional face. To complete the proof we show that in all other cases, I_n and P_π are not the extreme points of a one-dimensional face.

First suppose that $p = 2$. Then there exist distinct permutations τ_1 and τ_2 in $\mathcal{E}_n(X_c)$ which have exactly one cycle of length greater than 1 and this cycle has odd length. We then have $X_{1/2} = (1/2)(P_{\tau_1} + P_{\tau_2})$ implying that I_n and P_π do not determine a one-dimensional face. The case $p > 2$ is similarly argued.

Now suppose that $p = 0$ and $q = 4$. In this case there exist distinct permutations τ_1 and τ_2 in $\mathcal{E}_n(X_c)$ which have exactly two cycles of length greater than 1 and these cycles have even length. An argument similar to the preceding one completes the proof. The case $p = 0$ and $q \geq 6$ is similarly argued.

Finally suppose that $p = 1$ and $q = 2$. In this case there exist distinct permutations τ_1 and τ_2 such that τ_1 has one cycle of length greater than 1 and it has odd length, and τ_2 has two cycles of length greater than 1 and they have even length. Arguing as in the first case, and in a similar way for $p = 1$ and $q \geq 4$, completes the proof. □

Next we consider the diameter of $G(\Omega_n^{\text{ev}})$. For $n = 1$ and 2, this graph has only one vertex, while for $n = 3$, it is the complete graph K_3.

Theorem 9.7.9 *The diameter of $G(\Omega_n^{\text{ev}})$ equals 2 for $n \geq 4$.*

Proof. It is enough to show that for each even permutation $\pi \neq \iota_n$ of $\{1, 2, \ldots, n\}$, there is a path of length 2 from ι_n to π. There is no loss in generality in assuming that π has no fixed points and that for some k, π has k cycles which cyclically permute $\{1, 2, \ldots, n_1\}$, $\{n_1+1, n_1+2, \ldots, n_1+n_2\}$, $\ldots$, $\{n_1 + n_2 + \cdots, n_{k-1}+1, \ldots, n\}$ where $n = n_1 + n_2 + \cdots + n_k$ and, since π is an even permutation, $k \equiv n \bmod 2$. Let σ be the permutation which permutes $\{1, 2, \ldots, n\}$ in a cycle of length n. Then $\pi^{-1}\sigma$ has exactly one cycle of length greater than 1 and this cycle has length k. Hence if n, and so k, is even, it follows from Theorem 9.7.8, that ι is joined to σ by an edge and σ is joined to π. Now assume that n, and so k, is odd. We then let σ be the permutation fixing 1 and permuting $\{2, 3, \ldots, n\}$ in a cycle of length $n - 1$. The permutation $\pi^{-1}\sigma$ has exactly one cycle of length greater than 1, and in this case its length is $k + 1$. Hence both σ and $\pi^{-1}\sigma$ are even permutations, and ι is joined by an edge to σ and σ is joined by an edge to π. In each case we have a path of length 2 from ι to π.   □

Example. The obvious analog of Birkhoff's algorithm for Ω_n does not work for Ω_n^{ev}. For example, let

$$A = \frac{1}{3} \begin{bmatrix} 1 & 1 & 0 & 0 & 1 \\ 1 & 1 & 1 & 0 & 0 \\ 0 & 1 & 1 & 1 & 0 \\ 0 & 0 & 1 & 1 & 1 \\ 1 & 0 & 0 & 1 & 1 \end{bmatrix}.$$

Then the $(0,1)$-matrix $3A$ is the sum of three even permutation matrices and hence A is an even doubly stochastic matrix. Let P be the even permutation matrix

$$\begin{bmatrix} 0 & 1 & 0 & 0 & 0 \\ 1 & 0 & 0 & 0 & 0 \\ 0 & 0 & 0 & 1 & 0 \\ 0 & 0 & 1 & 0 & 0 \\ 0 & 0 & 0 & 0 & 1 \end{bmatrix}$$

satisfying $(1/3)P \le A$. Then $(3/2)(A-P)$ is a doubly stochastic matrix but not an even doubly stochastic matrix. Hence A is not a convex combination of even permutation matrices including P. Another illuminating example is the matrix

$$B = \frac{1}{3} \begin{bmatrix} 1 & 1 & 1 & 0 \\ 1 & 1 & 0 & 1 \\ 0 & 1 & 1 & 1 \\ 1 & 0 & 1 & 1 \end{bmatrix},$$

an even doubly stochastic matrix with $|\mathcal{E}_4(B)| = 6$. The $(0,1)$-matrix $3B$ is not the sum of three even permutation matrices, but

$$\frac{1}{2} \sum_{P \in \mathcal{E}_4(B)} P = 3B$$

and

$$B = \frac{1}{6} \sum_{P \in \mathcal{E}_4(B)} P.$$

                                                                                    □

In [36] a special generalized matrix function, called the *even permanent*, was defined on matrices $X = [x_{ij}]$ of order n by

$$\text{evenper}(X) = \sum_{\pi \in A_n} x_{1\pi(1)} x_{2\pi(2)} \cdots x_{n\pi(n)}$$

where the summation is over all even permutations of order n. It was conjectured that for $X \in \Omega_n^{\text{ev}}$,

$$\text{evenper}(X) \geq \frac{n!}{2n^n}$$

with equality if and only if X is the even doubly stochastic matrix J_n each of whose elements equals $1/n$. The even permanent is investigated in [118], [119]. It is also conjectured in [36] that the number of facets of the polytope Ω_n^{ev} is not bounded by a polynomial function of n, that is, Ω_n^{ev} cannot be described as the solution set of polynomially many linear inequalities. This conjecture has been independently verified in [70] and [50]. In [70] a large family of $n!\,(n-1)!/2$ distinct facets is constructed for $n \geq 6$. This family of facets is based on the construction of one facet of Ω_n^{ev}; specifically, it is shown that the set of permutation matrices of order n corresponding to the permutations of $\{1, 2, \ldots, n\}$ which are equal to the product of the transposition interchanging 1 and 2 and a cycle

$$i_1 \to i_2 \to \cdots \to i_k \to i_1 \quad (i_1 < i_2 < \cdots < i_k, k \text{ even})$$

is the set of extreme points of a facet of Ω_n^{ev}. In [50] it is proved that for $n \geq 4$, none of the inequalities in (ii) of Theorem 9.7.5 that are satisfied by every even doubly stochastic matrix determines a facet of Ω_n^{ev}. In addition, an explicit construction is given of a family of $(1/2)n(n-1)\,n!$ linear inequalities, each of which determines a distinct facet of Ω_n^{ev} for $n \geq 5$; a complete description of Ω_4^{ev} is also given.

We now give an elementary linear characterization of another subpolytope of Ω_n. Let $\mathcal{L}_n$ denote the set of all permutation matrices of order n corresponding to permutations of $\{1, 2, \ldots, n\}$ which are transpositions. Thus a permutation matrix $P = [p_{ij}]$ of order n belongs to $\mathcal{L}_n$ if and only if there exist distinct integers k and l with $p_{kl} = p_{lk} = 1$ and $p_{ii} = 1$ for $i \neq k, l$. We denote this permutation matrix by $T_n^{k,l}$. Let Ω_n^{tr} denote the convex hull of $\mathcal{L}_n$. Since each $T_n^{k,l}$ is symmetric, Ω_n^{tr} consists of symmetric matrices. Also each matrix in Ω_n^{tr} has trace equal to $n-2$, since this is true for the matrices in $\mathcal{L}_n$. The following theorem [75] characterizes Ω_n^{tr}.

Theorem 9.7.10 *Let* $n \geq 2$ *be an integer. The polytope* Ω_n^{tr} *consists of all symmetric doubly stochastic matrices with trace equal to* $n-2$.

Proof. In view of our observations above, we need only show that every symmetric doubly stochastic matrix $A = [a_{ij}]$ of order n with trace equal to $n - 2$ is in Ω_n^{tr}. We have

$$\sum_{i=1}^{n} \sum_{j=i+1}^{n} a_{ij} = \frac{1}{n} \sum_{i \neq j} a_{ij} = \frac{1}{n}(n - \mathrm{trace}(A)).$$

Let

$$B = [b_{kl}] = A - \sum_{i=1}^{n} \sum_{j=i+1}^{n} a_{ij} T_n^{i,j}.$$

Then B is a diagonal matrix and

$$b_{kk} = a_{kk} - \sum \{a_{ij} : i, j \neq k, j > i\}$$

$$= x_{kk} - \sum_{i<k} \left(\sum_{j>i} (a_{ij} - a_{pk}) \right) - \sum_{i>k} \sum_{j>i} a_{ij}$$

$$= x_{kk} + \sum_{i<k} a_{ik} - \sum_{i \neq k} \sum_{j>i} a_{ij}$$

$$= \sum_{i<k} a_{ik} - \left(\sum_{i=1}^{n} \sum_{j=i+1}^{n} a_{ij} - \sum_{j=k+1}^{n} a_{kj} \right)$$

$$= \left(\sum_{i \leq k} a_{ik} + \sum_{j>k} a_{kj} \right) - \sum_{i=1}^{n} \sum_{j=p+1}^{n} x_{ij}$$

$$= \left(\sum_{i \leq k} a_{ik} + \sum_{j>k} a_{jk} \right) - \sum_{i=1}^{n} \sum_{j=p+1}^{n} x_{ij}$$

$$= 1 - \frac{1}{2}(n - \mathrm{trace}(A))$$

$$= \frac{1}{2}(\mathrm{trace}(A) - (n - 2)) = 0.$$

Thus the diagonal matrix B has a zero main diagonal and $B = O$. Hence $A = \sum_{i=1}^{n} \sum_{j=i+1}^{n} a_{ij} T_n^{i,j}$ implying that $A \in \Omega_n^{\mathrm{tr}}$. □

We remark that essentially the same argument shows that the convex hull of $\mathcal{L}_n \cup \{I_n\}$ consists of the symmetric doubly stochastic matrices of order n with trace at least $n - 2$.

To conclude this section, we consider certain subpolytopes of Ω_n whose rows and columns are constrained by majorization.

Let $\mathcal{S}^n$ denote the set of stochastic vectors in $\Re^n$, that is, nonnegative vectors $d = (d_1, d_2, \ldots, d_n)$ such that $\sum_{i=1}^{n} d_i = 1$. Let

$$\Omega_n(d) = \{A \in \Omega_n : \text{each line of } A \text{ is majorized by } d\}.$$

Assume, without loss of generality, that d is nonincreasing. Then a stochastic vector $x = (x_1, x_2, \ldots, x_n)$ satisfies $x \preceq d$ if and only if

$$\sum_{i \in K} x_i \leq \sum_{i=1}^{|K|} d_i \quad (K \subseteq \{1, 2, \ldots, n\}).$$

It follows that $\Omega_n(d)$ is a bounded set defined by a finite number of linear constraints, and hence $\Omega_n(d)$ is a subpolytope of Ω_n. The polytopes $\Omega_n(d)$ were introduced and investigated in [24] and are called *majorization-constrained assignment polytopes* or *majorization-constrained Birkhoff polytopes*. The vector $(1/n, 1/n, \ldots, 1/n)$ in $\mathcal{S}^n$ is majorized by each vector $d \in \mathcal{S}^n$, and hence $(1/n)J_n$ is in $\Omega_n(d)$ for each $d \in \mathcal{S}^n$. The vector $e_1 = (1, 0, \ldots, 0)$ in $\mathcal{S}^n$ majorizes each vector in $\mathcal{S}^n$ implying that $\Omega_n(e_1) = \Omega_n$. Some additional elementary properties of the polytopes $\Omega_n(d)$ are collected in the next theorem.

Theorem 9.7.11 *Let* $d = (d_1, d_2, \ldots, d_n)$ *and* $d' = (d'_1, d'_2, \ldots, d'_n)$ *be nonincreasing vectors in* $\mathcal{S}_n$. *Then:*

(i) $d \preceq d'$ *if and only if* $\Omega_n(d) \subseteq \Omega_n(d')$; *moreover,* $\Omega_n(d) = \Omega_n(d')$ *if and only if* $d = d'$;

(ii) *if* $A \in \Omega_n(d)$ *and* $U_1, U_2 \in \Omega_n$, *then* $U_1 A U_2 \in \Omega_n(d)$;

(iii) $\dim \Omega_n(d) = 0$ *if* d *is the vector* $(1/n, 1/n, \ldots, 1/n)$, *and* $\dim \Omega_n(d) = (n-1)^2$ *otherwise;*

(iv) $\Omega_n(d)$ *contains a nonsingular matrix if and only if* d *does not equal the vector* $(1/n, 1/n, \ldots, 1/n)$.

    Proof. (i) The circulant matrix $\mathrm{Circ}(d_1, d_2, \ldots, d_n)$ belongs to $\Omega_n(d)$ and $\mathrm{Circ}(d'_1, d'_2, \ldots, d'_n)$ belongs to $\Omega_n(d')$. Thus if $\Omega_n(d) \subseteq \Omega_n(d')$, then $d \preceq d'$. The converse implication is obvious. (ii) This assertion follows from Lemma 1.7.3. (iii) If $d = (1/n, 1/n, \ldots, 1/n)$, then $\Omega_n(d)$ contains only $(1/n)J_n$, and hence its dimension equals 0. Now assume that $d \neq (1/n, 1/n, \ldots, 1/n)$ so that $d_1 > d_n$. Let $x = (1/n, 1/n, \ldots, 1/n)$. Then $\sum_{i=1}^{k} x_i < \sum_{i=1}^{k} d_i$ for $1 \leq k \leq n-1$. It follows that there is a sufficiently small positive ϵ such that replacing any submatrix $(1/n)J_n[\{i, j\}, \{p, q\}]$ of $(1/n)J_n$ of order 2 by

$$\begin{bmatrix} 1/n \pm \epsilon & 1/n \mp \epsilon \\ 1/n \mp \epsilon & 1/n \pm \epsilon \end{bmatrix}$$

results in a matrix also in $\Omega_n(d)$. This implies that $\dim \Omega_n(d) = \dim \Omega_n = (n-1)^2$. (iv) If $d = (1/n, 1/n, \ldots, 1/n)$, then $\Omega_n(d)$ contains only $(1/n)J_n$, a singular matrix. If $d \neq (1/n, 1/n, \ldots, 1/n)$, then there is a small positive number ϵ such that the matrix $(1 - \epsilon)(1/n)J_n + \epsilon I_n$ is in $\Omega_n(d)$ and is nonsingular. $\qquad\square$

We now turn to obtaining a combinatorial characterization of the extreme points of a majorization-constrained polytope $\Omega_n(d)$ as given in [24]. Let $x = (x_1, x_2, \ldots, x_n)$ and $d = (d_1, d_2, \ldots, d_n)$ be nonincreasing vectors in $\mathcal{S}^n$ such that $x \preceq d$. Let

$$x = x^{(1)} \oplus x^{(2)} \oplus \cdots \oplus x^{(r)} \text{ and } d = d^{(1)} \oplus d^{(2)} \oplus \cdots \oplus d^{(r)}$$

be the conformal decompositions of x and d, as defined in Section 9.5. Thus $x^{(i)} \preceq d^{(i)}$ and $x^{(i)} \preceq d^{(i)}$ has no coincidences $(i = 1, 2, \ldots, r)$. We call the partition of the indices of x (and d) as determined by this conformal partition, the *d-partition* of x.

Lemma 9.7.12 *Let x, y, and z be nonincreasing vectors in $\mathcal{S}^n$ each of which is majorized by the vector $d \in \mathcal{S}^n$. Suppose that $x = cy + (1 - c)z$ where $0 < c < 1$. Then the d-partition of y and of z is a refinement of the d-partition of x.*

Proof. Let $S_1, S_2, \ldots, S_r$ be the d-partition of x where $S_i = \{s_1 + \cdots + s_{i-1} + 1, \ldots, s_i\}$ $(1 \le i \le r)$, $s_0 = 0$ and $s_r = n$. Since $x = cy + (1 - c)z$ and x and y are both majorized by d, we have that

$$\sum_{j=1}^{s_1} d_j = \sum_{j=1}^{s_1} x_j \le \max\left\{ \sum_{j=1}^{s_1} y_j, \sum_{j=1}^{s_1} z_j \right\} \le \sum_{j=1}^{s_1} d_j.$$

Since $0 < c < 1$, we conclude that

$$\sum_{j=1}^{s_1} y_j = \sum_{j=1}^{s_1} z_j = \sum_{j=1}^{s_1} d_j.$$

Proceeding by induction, we have

$$\sum_{j=s_{i-1}+1}^{s_i} y_j = \sum_{j=s_{i-1}+1}^{s_i} z_j = \sum_{j=s_{i-1}+1}^{s_i} d_j \quad (i = 1, 2, \ldots, r).$$

The lemma now follows. $\square$

Consider the positions $V = \{(i, j) : 1 \le i, j \le n\}$ of a matrix of order n. Let

$$V_i = \{(i, k) : 1 \le k \le n\} \text{ and } W_i = \{(k, i) : 1 \le k \le n\} \quad (1 \le i \le n)$$

be the positions in row i and column i, respectively. Assume that there is a bi-family $\mathcal{P}$ of ordered partitions

$$\mathcal{R}_i = (R_{i1}, R_{i2}, \ldots, R_{ip_i}) \text{ and } \mathcal{C}_i = (C_{1i}, C_{2i}, \ldots, C_{q_i i}) \quad (1 \le i \le n)$$

of the V_i and W_i, respectively. Let

$$r_{ij} = |R_{ij}| \quad (1 \le i \le n; 1 \le j \le p_i)$$

and
$$s_{ij} = |C_{ij}| \quad (1 \le j \le n; 1 \le i \le q_j).$$

On each of the sets R_{ij} we consider a complete graph $K_{r_{ij}}$ whose edges are *red edges*. On each of the sets C_{ij} we consider a complete graph $K_{c_{ij}}$ whose edges are *blue edges*. The *colored graph* $G(\mathcal{P})$ has vertex set equal to V and edge set equal to the union of the edge sets of all the complete red-graphs $K_{r_{ij}}$ and complete blue-graphs $K_{c_{ij}}$. This graph is independent of the actual ordering of the parts in each partition of $\mathcal{P}$.

Now let $A = [a_{ij}]$ be a matrix in $\Omega_n(d)$. Since each line of A is majorized by d, there are a d-partition $\mathcal{R}_i = (R_{i1}, R_{i2}, \ldots, R_{ip_i})$ of the set V_i ($1 \le i \le n$) and a d-partition $\mathcal{C}_i = (C_{1i}, C_{2i}, \ldots, C_{q_i,i})$ of the set W_i, $1 \le i \le n$). These d-partitions result by applying Lemma 9.7.12 to the nonincreasing rearrangements of the rows and columns of A. Thus A and d induce a bi-family $\mathcal{P}_A^d$ of partitions of V as described above and a colored graph $G(\mathcal{P}_A^d)$. It follows that if (i, j) is a vertex of at least one edge of $G(\mathcal{P}_A^d)$, then $a_{ij} > 0$.

Theorem 9.7.13 *The matrix $A = [a_{ij}]$ in $\Omega_n(d)$ is an extreme point of $\Omega_n(d)$ if and only if there does not exist a cycle in the colored graph $G(\mathcal{P}_A^d)$ whose edges are alternately red and blue.*

Proof. We first assume that there is an alternating red–blue cycle γ in $G(\mathcal{P}_A^d)$. This implies the existence of a $(0, \pm 1)$-matrix $H = [h_{ij}] \ne O$ of order n such that the sum of the entries of H in each R_{ij} and that in each C_{ij} equal 0. Since $a_{ij} > 0$ for those (i, j) with $h_{ij} \ne 0$, there is a sufficiently small positive number ϵ such that the matrices $A \pm \epsilon D$ are in $\Omega_n(d)$. We have
$$A = \frac{1}{2}(A + \epsilon H) + \frac{1}{2}(A - \epsilon H),$$
showing that A is not an extreme point of $G(\mathcal{P}_A^d)$.

Now assume that A is not an extreme point of $G(\mathcal{P}_A^d)$. Thus there exist matrices B and C in $\Omega_n(d)$ with $A \ne B, C$ such that
$$A = \frac{1}{2}B + \frac{1}{2}C.$$

By Lemma 9.7.12, the d-partitions of row i (respectively, column i) of B and C are refinements of the d-partition of row i (respectively, column i) of A ($i = 1, 2, \ldots, n$). Since $A \ne B$ and A and B are doubly stochastic, there exists (k, l) such that $a_{kl} > b_{kl}$. Since the d-partition of row k of B is a refinement of the d-partition of row k of A, there exists $p \ne l$ such that $a_{kp} < b_{kp}$, and (k, l) and (k, p) belong to the same part of the d-partition of row k of A. Since the d-partition of column p of B is a refinement of the d-partition of column p of A, there exists q such that $a_{qp} > b_{qp}$ where (k, p) and (q, p) belong to the same part of the d-partition of column p of A. Continuing in this fashion, we ultimately obtain an alternating red–blue cycle of $G(\mathcal{P}_A^d)$. $\square$

In [24] further properties of the polytopes $\Omega_n(d)$ are obtained; in particular, the case where d has only two nonzeros is further investigated. In addition, doubly stochastic matrices whose diagonals are constrained by majorization are considered.

9.8   Doubly Substochastic and Superstochastic Matrices

A doubly stochastic matrix is a nonnegative matrix each of whose row and column sums equals 1. These row and column constraints force a doubly stochastic matrix to be square. A *doubly substochastic matrix* is a nonnegative matrix each of whose row and column sums is at most 1. Such a matrix need not be square. In investigating doubly substochastic matrices, however, it can usually be assumed that the matrices are square. This is because one can attach zero rows or zero columns to an m by n doubly substochastic matrix in order to obtain a doubly stochastic matrix of order $\max\{m, n\}$. Any submatrix of a doubly stochastic matrix is doubly substochastic.

Example. The matrix

$$\begin{bmatrix} 0.2 & 0.4 & 0.1 \\ 0.3 & 0.5 & 0.2 \end{bmatrix}$$

is a doubly substochastic matrix, as is the matrix

$$\begin{bmatrix} 0.2 & 0.4 & 0.1 \\ 0.3 & 0.5 & 0.2 \\ 0 & 0 & 0 \end{bmatrix}.$$

□

The product of two doubly stochastic matrices is also doubly substochastic. Decreasing the elements of a doubly stochastic matrix always results in a square, doubly substochastic matrix. An important property of doubly stochastic matrices is the converse [110].

Lemma 9.8.1 *Let A be a doubly substochastic matrix of order n. Then there exist a doubly stochastic matrix B of order n with $A \le B$.*

Proof. Suppose that $A = [a_{ij}]$ is not already doubly stochastic, and let the row and column sum vectors of A be $R = (r_1, r_2, \ldots, r_n)$ and $S = (s_1, s_2, \ldots, s_n)$. Then there exist i and j with $1 \le i, j \le n$ such that $r_i < 1$ and $s_j < 1$. Replace a_{ij} with $a_{ij} + \min\{1 - r_i, 1 - s_j\}$. Then either the sum of row i is now 1 or the sum of column j is. Continuing like this we eventually get all row and column sums equal to 1.   □

Corollary 9.8.2 *Let A be a square, nonnegative matrix. Then A is doubly substochastic if and only if there exists a doubly stochastic matrix B with $A \le B$.*   □

It is natural, by analogy with the definition of a doubly substochastic matrix, to define a doubly superstochastic matrix to be a nonnegative matrix each of whose row and column sums is at least 1. But such a definition would not be restrictive enough as it leads to a class of matrices which do not have many interesting combinatorial properties. For instance, the matrix

$$B = \begin{bmatrix} 1 & 0 & 0 \\ 1 & 0 & 0 \\ 0 & 1 & 1 \end{bmatrix}$$

would be a doubly superstochastic matrix under such a definition. The matrix B has the property that there is no doubly stochastic matrix A with $A \le B$, in fact no doubly superstochastic matrix A with $A \ne B$ and $A \le B$. We mimic the characterization of doubly substochastic matrices given in Corollary 9.8.2 and define a *doubly superstochastic matrix* to be a matrix B for which there exists a doubly stochastic matrix A with $A \le B$. Such a matrix is necessarily nonnegative. The product of two doubly superstochastic matrices is doubly superstochastic.

We let $\Omega_{m,n}^{\mathrm{sub}}$ denote the set of all m by n substochastic matrices, with this notation shortened to Ω_n^{sub} if $m = n$. The polytopes $\Omega_{m,n}^{\mathrm{sub}}$ are special cases of the polytopes $\mathcal{N}(\le R, \le S)$ of nonnegative matrices with row and column sum vectors bounded above componentwise by R and S, respectively, as discussed in Chapter 8. We also let Ω_n^{sup} denote the set of all doubly superstochastic matrices of order n. These are the polyhedra $\mathcal{N}^+(R, S)$ with $R = S = (1, 1, \ldots, 1)$, investigated in Chapter 8. We first investigate $\Omega_{m,n}^{\mathrm{sub}}$.

A *subpermutation matrix* is a $(0, 1)$-matrix with at most one 1 in each row and column. Its rank[18] equals the number of its 1's. The square subpermutation matrices are obtained from permutation matrices by replacing some (possibly none) of their 1's with 0's. A zero matrix is a subpermutation matrix. We have the following characterization of $\Omega_{m,n}^{\mathrm{sub}}$ [103]. We give a direct proof of this theorem although it follows as a special case of Theorem 8.1.2.

Theorem 9.8.3 *The set $\Omega_{m,n}^{\mathrm{sub}}$ is a convex polytope whose extreme points are the m by n subpermutation matrices.*

Proof. If A and B are substochastic, so is $cA + (1 - c)B$ for $0 \le c \le 1$. Thus $\Omega_{m,n}^{\mathrm{sub}}$ is convex and, since it is defined by a finite number of linear inequalities, $\Omega_{m,n}^{\mathrm{sub}}$ is a convex polytope. The subpermutation matrices are clearly extreme points. We prove that they are the only extreme points by induction on $m + n$. If $m + n = 2$, this is obvious. Suppose $m + n \ge 3$, and let $A = [a_{ij}]$ be an extreme point of $\Omega_{m,n}^{\mathrm{sub}}$. If all row and columns of A contain at least two positive elements, then the bipartite graph $\mathrm{BG}(A)$ representing the nonzero pattern of A has a cycle, and as in the proof of Theorem 1.7.1, A is not an extreme point. Thus some row or column of

---

[18]Either term rank or ordinary linear algebra rank, since they are equal in this instance.

A has at most one positive element. Without loss of generality assume that column 1 of A contains at most one positive elements. If column 1 contains no positive elements, then the proof is completed by induction. Now assume that column 1 contains exactly one positive element $d > 0$, which we can assume to be in row 1. Thus

$$
A = \left[\begin{array}{c|c} \begin{matrix} d \\ 0 \\ \vdots \\ 0 \end{matrix} & A' \end{array} \right]
$$

where $A' \in \Omega^{\mathrm{sub}}_{m,n-1}$. Suppose A' were not an extreme point. Then by induction, $A' = \sum_{i=1}^{t} c_i P_i$ where the P_i are subpermutation matrices and the c_i are nonnegative numbers summing to 1. Because $d > 0$, not all of the P_i can have a 1 in their first row. Suppose $P_1, \ldots, P_k$ do, and $P_{k+1}, \ldots, P_t$ do not, where $k < t$. Then $\sum_{i=1}^{k} c_i \leq 1 - d$ and hence $\sum_{i=k+1}^{t} c_i \geq d$. Choose l maximal so that $\sum_{i=k+1}^{l} c_i \leq d$; here l may be k and this summation empty. Let P_i' be obtained from P_i by bordering it with an initial column of all 0's, and let P_i'' be obtained from P_i by bordering it with a unit column whose 1 is in its first position. Then

$$
A = \sum_{i=1}^{k} c_i P_i' + \sum_{i=k+1}^{l} c_i P_i'' + \left(d - \sum_{i=k+1}^{l} c_i \right) P_{l+1}''
$$
$$
+ \left(c_{l+1} - \left(d - \sum_{i=k+1}^{l} c_i \right) \right) P_{l+1}' + \sum_{i=l+2}^{t} c_i P_i'.
$$

We have

$$
c_{l+1} - \left(d - \sum_{i=k+1}^{l} c_i \right) \geq 0
$$

and, by choice of l,

$$
\sum_{i=k+1}^{l+1} c_i > d.
$$

This represents A as a nontrivial convex combination of matrices in $\Omega^{\mathrm{sub}}_{m,n}$, contradicting the assumption that A is an extreme point. We conclude that A' is an extreme point of $\Omega^{\mathrm{sub}}_{m,n-1}$. Thus A' is a (0,1)-matrix with at most one 1 in each of its rows and columns. Since the first row sum of A' is strictly less than 1, the first row of A' is a zero row. Let P be the subpermutation matrix of size $m - 1$ by $n - 1$ obtained by deleting row 1 of A'. If $d < 1$, then

$$
A = d \left[\begin{array}{c|c} 1 & 0 \ldots 0 \\ \hline \begin{matrix} 0 \\ 0 \\ \vdots \\ 0 \end{matrix} & P \end{array} \right] + (1-d) \left[\begin{array}{c|c} 0 & 0 \ldots 0 \\ \hline \begin{matrix} 0 \\ 0 \\ \vdots \\ 0 \end{matrix} & P \end{array} \right],
$$

contradicting again that A is an extreme point. Thus A is a subpermutation matrix. □

Let $\Omega_{m,n}^{\mathrm{sub}_k}$ be the *convex hull of the subpermutation matrices of a fixed rank* k. Then $\Omega_{m,n}^{\mathrm{sub}_k}$ is a subpolytope of $\Omega_{m,n}^{\mathrm{sub}}$. The following theorem of Mendelsohn and Dulmage [91] characterizes this convex hull by linear constraints.[19]

Theorem 9.8.4 *Let m, n, and k be positive integers with $k \leq m \leq n$. Let A be an m by n nonnegative matrix. Then A is in $\Omega_{m,n}^{\mathrm{sub}_k}$ if and only if the sum of the elements of A equals k, and each row or column sum of A is at most equal to 1.*

Proof. Without loss of generality we assume that $m = n$. The conditions in the theorem are clearly necessary for A to belong to $\Omega_{m,n}^{\mathrm{sub}_k}$, and we turn to their sufficiency.

Let the row and sum vectors of A be $R = (r_1, r_2, \ldots, r_n)$ and $S = (s_1, s_2, \ldots, s_n)$, respectively. Then $r_i, s_i \leq 1$ for $i = 1, 2, \ldots, n$, and we have $\sum_{i=1}^{n} r_i = \sum_{i=1}^{n} s_i = k$. Since

$$\sum_{i=1}^{n} (1 - r_i) = n - k = \sum_{i=1}^{n} (1 - s_i),$$

there exists a nonnegative matrix B with nonnegative row sum vector $R' = (1 - r_1, 1 - r_2, \ldots, 1 - r_n)$ and column sum vector $(1, 1, \ldots, 1)$ ($n - k$ 1's), and a nonnegative matrix C with row sum vector $(1, 1, \ldots, 1)$ ($n - k$ 1's) and column sum vector $S = (1 - s_1, 1 - s_2, \ldots, 1 - s_n)$. The matrix

$$A' = \begin{bmatrix} A & B \\ C & O \end{bmatrix}$$

is a doubly stochastic matrix of order $2n - k$. Hence A' can be represented in the form

$$A' = c_1 P_1' + c_2 P_2' + \cdots + c_l P_l' \tag{9.24}$$

where the c_i are positive numbers summing to 1 and the P_i' are permutation matrices of order $2n - k$. Each permutation matrix P_i' contains $n - k$ 1's in positions of the submatrix B of A' and $n - k$ 1's in positions of the submatrix C. Hence each permutation matrix P_i' contains exactly $2n - k - (n - k) - (n - k) = k$ 1's in positions from the submatrix A of A'. For $i = 1, 2, \ldots, l$, let $P_i = P_i'[\{1, 2, \ldots, n\}]$ be the leading submatrix of order n of P_i'. Then from (9.24) we get

$$A = c_1 P_1 + c_2 P_2 + \cdots + c_l P_l$$

where the P_i are subpermutation matrices of rank k. The theorem now follows. □

---

[19] In [91] one of the constraints given is not needed.

For a doubly substochastic matrix $A = [a_{ij}]$ of order n, let Ω_n^A denote the *set of all doubly stochastic matrices* $X = [x_{ij}]$ *of order n such that* $x_{ij} \geq a_{ij}$ *for all* $i, j = 1, 2, \ldots, n$. By Lemma 9.8.1, Ω_n^A is nonempty. The matrices in Ω_n^A are the *doubly stochastic extensions* of A. If A is doubly stochastic, then $\Omega_n^A = \{A\}$. In general, Ω_n^A is a convex subpolytope of Ω_n. For $n > 1$, we have $\Omega_n^A = \Omega_n$ if and only if $A = O$.

The following theorem [56, 53] shows that the extreme points of Ω_n^A can be characterized in the same way as the extreme points of Ω_n. Let $A = [a_{ij}]$ be a doubly substochastic matrix of order n. If $B = [b_{ij}]$ is in Ω_n^A, then $\mathrm{BG}(B - A)$ is the bipartite graph whose edges correspond to the elements b_{ij} of B with $b_{ij} > a_{ij}$.

Theorem 9.8.5 *Let A be a doubly stochastic matrix. A matrix $B \in \Omega_n^A$ is an extreme point of Ω_n^A if and only if the bipartite graph $\mathrm{BG}(B - A))$ has no cycles.*

Proof. The proof is very similar to proofs already given. If $\mathrm{BG}(B - A)$ has a cycle, then this cycle can be used to define two matrices B_1 and B_2 in Ω_n^A with $B_1 \neq B_2$ such that $B = (1/2)(B_1 + B_2)$. If $\mathrm{BG}(B - A)$ does not have any cycles, then knowing that B belongs to Ω_n^A, the elements of B are uniquely determined by those of A. It follows that B is an extreme point of $\mathrm{BG}(B - A)$. $\qquad \square$

The definition for a matrix A of order n to be doubly superstochastic is equivalent to the existence of a doubly stochastic matrix B and a nonnegative matrix X such that $A = B + X$. The set Ω_n^{sup} of doubly superstochastic matrices of order n is clearly an unbounded convex set. The following corollary is a special case of Theorem 8.1.13 and Corollary 8.1.14.

Corollary 9.8.6 *A nonnegative matrix $A = [a_{ij}]$ of order n is doubly superstochastic if and only if*

$$\sum_{i \in I, j \in J} a_{ij} \geq |I| + |J| - n \quad (I, J \subseteq \{1, 2, \ldots, n\}).$$

The extreme points of Ω_n^{sup} are precisely the permutation matrices of order n. $\qquad \square$

Doubly substochastic and superstochastic matrices arise in weakened forms of majorization and we discuss this now [90, 10, 80]. Let $X = (x_1, x_2, \ldots, x_n)$ and $Y = (y_1, y_2, \ldots, y_n)$ be two n-vectors of real numbers which are assumed to be nonincreasing:

$$x_1 \geq x_2 \geq \cdots \geq x_n, \quad y_1 \geq y_2 \geq \cdots \geq y_n.$$

Then X is *weakly majorized from below* by Y, denoted $X \preceq_{\mathrm{w}} Y$, provided that the partial sums satisfy

$$\sum_{i=1}^{k} x_i \leq \sum_{i=1}^{k} y_i \quad (1 \leq k \leq n),$$

Unlike majorization, equality need not occur for $k = n$. X is *weakly majorized from above* by Y, denoted $X \preceq^w Y$, provided that the partial sums satisfy

$$\sum_{i=k+1}^{n} x_i \geq \sum_{i=k+1}^{n} y_i \quad (0 \leq k \leq n-1).$$

We have that $X \preceq Y$ implies that $X \preceq_w Y$ and $X \preceq^w Y$. In addition, if either $X \preceq_w Y$ or $X \preceq^w Y$, and $\sum_{i=1}^{n} x_i = \sum_{i=1}^{n} y_i$, then $X \preceq Y$.

Theorem 9.8.7 *Let $X = (x_1, x_2, \ldots, x_n)$ and $Y = (y_1, y_2, \ldots, y_n)$ be non-increasing, nonnegative vectors. Then $X \preceq_w Y$ if and only if $X = YP$ for some doubly substochastic matrix P. Also, $X \preceq^w Y$ if and only if $X = YQ$ for some doubly superstochastic matrix Q.*

Proof. First we note the assumption of the nonnegativity of X and Y, an assumption that is not needed for ordinary majorization.

Suppose first that $X = YP$ where P is a doubly substochastic matrix. There exists a doubly stochastic matrix A such that $P \leq A$. Let $Z = (z_1, z_2, \ldots, z_n) = YA$. Then $Z \preceq Y$ and $x_i \leq z_i$ for $i = 1, 2, \ldots, n$, and hence $X \preceq_w Y$. Now suppose that $X \preceq_w Y$. If $X = 0$, then with $P = O$, $X = PY$. Assume that $X \neq 0$, and let

$$\delta = \sum_{i=1}^{n} y_i - \sum_{i=1}^{n} x_i.$$

Let ϵ be the smallest positive component of X, and let t be the smallest integer such that $\epsilon t \geq \delta$. The vector X' obtained from X by including t new components equal to δ/t and the vector Y' obtained from Y by including t new components equal to 0 satisfy $X' \preceq Y'$. Thus there exists a doubly stochastic matrix A such that $X' = Y'A$. The doubly substochastic submatrix $A[\{1, 2, \ldots, n\}]$ of A satisfies $X = A[\{1, 2, \ldots, n\}]Y$.

Now suppose that $X = YP$ where P is a doubly superstochastic matrix. Let $P = A + B$ where A is doubly stochastic and B is a nonnegative matrix. If $Z = (z_1, z_2, \ldots, z_n) = YA$, then $Z \preceq Y$ where $z_i \leq x_i$ for $i = 1, 2, \ldots, n$. It easily follows now that $X \preceq^w Y$. Now assume that $X \preceq^w Y$. Then there exists a vector $Z = (z_1, z_2, \ldots, z_n)$ such that that $Z \preceq Y$ and $x_i \geq z_i$ for $i = 1, 2, \ldots, n$. There exists a diagonal matrix D with diagonal elements $d_i = x_i/z_i \geq 1$ such that $X = ZD$. There also exists a doubly stochastic matrix A such that $Z = YA$. Then $X = YAD$ where AD is a doubly superstochastic matrix. $\qquad\square$

9.9 Symmetric Assignment Polytope

We now consider symmetric doubly stochastic matrices, and denote by Ξ_n the *set of all symmetric doubly stochastic matrices of order n*. We have that Ξ_n is a subpolytope of Ω_n. The polytope Ξ_n is a special instance of a symmetric transportation polytope, and as a result its extreme points are characterized by Theorem 8.2.1. Since, in the case of doubly stochastic matrices, all row and columns equal 1, some simplification occurs in the description of the extreme points [77].

Theorem 9.9.1 *Let n be a positive integer. A matrix A of order n is an extreme point of Ξ_n if and only if there is a permutation matrix P such that $P^{\mathrm{T}} A P$ is a direct sum of matrices each of which is of one of the types:*

(i) I_1 (*identity matrix of order 1*);

(ii) $\begin{bmatrix} 0 & 1 \\ 1 & 0 \end{bmatrix}$;

(iii) *a matrix of odd order at least 3 of the form*

$$\begin{bmatrix} 0 & \frac{1}{2} & 0 & 0 & \cdots & 0 & \frac{1}{2} \\ \frac{1}{2} & 0 & \frac{1}{2} & 0 & \cdots & 0 & 0 \\ 0 & \frac{1}{2} & 0 & \frac{1}{2} & \cdots & 0 & 0 \\ 0 & 0 & \frac{1}{2} & 0 & \cdots & 0 & 0 \\ \vdots & \vdots & \vdots & \vdots & \ddots & \vdots & \vdots \\ 0 & 0 & 0 & 0 & \cdots & 0 & \frac{1}{2} \\ \frac{1}{2} & 0 & 0 & 0 & \cdots & \frac{1}{2} & 0 \end{bmatrix},$$

that is, a matrix $(P + P^{\mathrm{T}})/2$ where P is a full cycle permutation matrix of odd order at least 3.

Proof. According to Theorem 8.2.1, the extreme points of Ξ_n are those matrices in Ξ_n the connected components of whose associated bipartite graphs are trees or odd near-trees. The doubly stochastic property implies that the components which are trees are single edges (type (ii) above), and the components that are odd near-trees are odd cycles (type (i) above for cycles of length 1 and type (iii) above for odd cycles of length 3 or more). □

From Theorem 8.2.8 we obtain the extreme points of the *convex set Ξ_n^{sub} of symmetric, doubly substochastic matrices of order n* [78].

Corollary 9.9.2 *Let n be a positive integer. A matrix A of order n is an extreme point of Ξ_n^{sub} if and only if there is a permutation matrix P such that $P^{\mathrm{T}} A P$ is a direct sum of matrices each of which is either the zero matrix O_1 of order 1 or one of the matrices (i), (ii), and (iii) given in Theorem 9.9.1.* □

Using a theorem of Edmonds [57] (for an alternative proof, see [5]) Cruse [47] characterized by linear constraints the convex hull of the integral extreme points of Ξ_n, that is, the convex hull of the symmetric permutation matrices of order n. We shall also make use of Edmonds' theorem without proof. This theorem is usually formulated as a characterization by linear constraints of the matching polytope of a graph (the convex hull of its matchings), but we shall formulate it in terms of matrices.

Let n be a positive integer, and let Υ_n be the convex hull of the symmetric subpermutation matrices of order n with zero trace. Let W be a symmetric $(0,1)$-matrix of order n with zero trace, and let $\Upsilon_n\langle W \rangle$ denote the subset of Υ_n consisting of the convex hull of those symmetric subpermutation matrices P with $P \leq W$ (elementwise inequality).

Theorem 9.9.3 *Let $W = [w_{ij}]$ be a symmetric $(0,1)$-matrix with zero trace. Then a symmetric matrix $A = [a_{ij}]$ of order n belongs to $\Upsilon_n\langle W \rangle$ if and only if the following linear constraints are satisfied:*

(i) $0 \leq a_{ij} \leq w_{ij}$ $(i, j = 1, 2, \ldots, n)$;

(ii) $\sum_{j=1}^{n} a_{ij} \leq 1$ $(i = 1, 2, \ldots, n)$;

(iii) $\sum_{i=1}^{n} a_{ij} \leq 1$ $(j = 1, 2, \ldots, n)$;

(iv) *for each $K \subseteq \{1, 2, \ldots, n\}$ of odd cardinality $2k + 1$ with $k \geq 1$,*

$$\sum_{i,j \in K} a_{ij} \leq 2k.$$

Note that in the presence of (ii) or (iii), condition (i) is equivalent to $a_{ij} \geq 0$ with equality if $w_{ij} = 0$ $(i, j = 1, 2, \ldots, n)$. By taking W to be the symmetric matrix $J_n - I_n$ whose only 0's are the 0's on the main diagonal, we see that Theorem 9.9.3 characterizes the convex hull of the symmetric subpermutation matrices of order n with zero trace.

We use Theorem 9.9.3 to characterize the convex hull of the symmetric permutation matrices, sometimes called the *symmetric assignment polytope* and denoted by Σ_n.

Theorem 9.9.4 *A symmetric matrix $A = [a_{ij}]$ of order n belongs to Σ_n if and only if the following linear constraints are satisfied:*

(i*) $a_{ij} \geq 0$ $(i, j = 1, 2, \ldots, n)$;

(ii*) $\sum_{j=1}^{n} a_{ij} = 1$ $(i = 1, 2, \ldots, n)$;

(iii*) $\sum_{i=1}^{n} a_{ij} = 1$ $(j = 1, 2, \ldots, n)$;

(iv*) *for each $K \subseteq \{1, 2, \ldots, n\}$ of odd cardinality $2k + 1$ with $k \geq 1$,*

$$\sum_{i,j \in K, i \neq j} a_{ij} \leq 2k.$$

Proof. Let Σ_n^* denote the convex polytope consisting of those symmetric matrices satisfying the constraints (i*)–(iv*). (Note that constraint (iv*) rules out matrices in Ξ_n which have a direct summand of type (iii) in Theorem 9.9.1.) These constraints are satisfied by a matrix in Σ_n as they are satisfied by all symmetric permutation matrices. Thus $\Sigma_n \subseteq \Sigma_n^*$. To show that $\Sigma_n^* \subseteq \Sigma_n$, it suffices to show that the extreme points of Σ_n^* are integral. This is so because constraints (i*)–(iv*) then imply that each extreme point of Σ_n^* is a symmetric permutation matrix.

Let $A = [a_{ij}]$ be a matrix in Σ_n^*. Let $W = [w_{ij}]$ be the symmetric $(0,1)$-matrix of order $2n$ with zero trace defined by

$$W = \left[\begin{array}{c|c} J_n - I_n & I_n \\ \hline I_n & O_n \end{array} \right].$$

We use A to define a matrix $B = [b_{ij}]$ in $\Upsilon_{2n}\langle W \rangle$; put $b_{ij} = a_{ij}$ if $1 \leq i \neq j \leq n$, and put $b_{i,n+i} = b_{n+i,i} = a_{ii}$ for $i = 1, 2, \ldots, n$. All other b_{ij} are set equal to 0. Note that $b_{ij} \neq 0$ only if $w_{ij} \neq 0$. The matrix B is a symmetric nonnegative matrix, and we now show that B belongs to $\Upsilon_{2n}\langle W \rangle$.

Since A is doubly stochastic, it also follows that the first n row and column sums of B equal 1, and the last n row and column sums are at most 1. It remains to verify that (iv) of Theorem 9.9.3 holds.

Let $K \subseteq \{1, 2, \ldots, n\}$ be of odd cardinality $2k + 1$ with $k \geq 1$. Write $K = K_1 \cup (n + K_2)$ where $K_1, K_2 \subseteq \{1, 2, \ldots, n\}$. If $K_2 = \varnothing$, then $|K_1| = 2k + 1$ and by (iv*)

$$\sum_{i,j \in K} b_{ij} = \sum_{i,j \in K_1} b_{ij} = \sum_{i,j \in K_1, i \neq j} a_{ij} \leq 2k.$$

Now suppose that $K_2 \neq \varnothing$ so that $|K_1| \leq 2k$. We distinguish two possibilities.

Case 1: $K_1 \cap K_2 = \varnothing$. In this case,

$$\sum_{i,j \in K} b_{ij} = \sum_{i,j \in K_1, i \neq j} a_{ij}$$
$$\leq \sum_{i \in K_1} \sum_{j=1}^{n} a_{ij} = |K_1| \leq 2k.$$

Case 2: $K_1 \cap K_2 \neq \varnothing$, say $p \in K_1 \cap K_2$ so that $p \in K_1$ and $n + p \in K_2$. Let $K' = K \setminus \{p, n+p\}$ so that $|K| = 2(k-1) + 1$. We then have

$$\sum_{i,j \in K} b_{ij} = \sum_{j \in K} b_{pj} + \sum_{i \in K} b_{ip} + \sum_{i,j \in K'} b_{ij}.$$

We now use induction on k. If $k = 1$, then $\sum_{i,j \in K'} b_{ij} = 0$, and so

$$\sum_{i,j \in K} b_{ij} = \sum_{j \in K} b_{pj} + \sum_{i \in K} b_{ip} \leq 1 + 1 = 2 = 2k.$$

Now assume that $k > 1$. By induction $\sum_{i,j \in K'} b_{ij} \leq 2(k-1)$, and hence

$$\sum_{i,j \in K} b_{ij} = \sum_{j \in K} b_{pj} + \sum_{i \in K} b_{ip} + \sum_{i,j \in K'} b_{ij} \leq 1 + 1 + 2(k-1) = 2k.$$

Thus we now know that B belongs to $\Upsilon_{2n}\langle W \rangle$. Now assume that A contains at least one nonintegral element. Then B also contains a non-integral element, and hence by Theorem 9.9.3, B is not an extreme point of $\Upsilon_{2n}\langle W \rangle$. Hence there exist distinct matrices $B_1 = [b_{ij}^{(1)}]$ and $B_2 = [b_{ij}^{(2)}]$ in $\Upsilon_{2n}\langle W \rangle$ such that $B = (1/2)(B_1 + B_2)$. We use B_1 and B_2 to construct distinct matrices $A_1, A_2 \in \Sigma_n$ such that $A = (1/2)(A_1 + A_2)$, implying that A is not an extremal matrix of Σ_n. The matrices A_1 and A_2 are obtained from B_1 and B_2 in the obvious way, namely, for $i = 1$ and 2,

$$A_i = B_i[\{1, 2, \ldots, n\}] + B_i[\{1, 2, \ldots, n\}, \{n+1, n+2, \ldots, 2n\}].$$

The only nonobvious property to check is that A_1 and A_2 belong to Σ_n. That $A_1 = [a_{ij}^{(1)}]$ and $A_2 = [a_{ij}^{(2)}]$ are symmetric nonnegative matrices follows since B_1 and B_2 are. That the row and column sums equal 1 follows from the facts that these row and columns sums are at most 1 (since that is true for B_1 and B_2) and that the row and column sums of A are 1. Now let K be a subset of $\{1, 2, \ldots, n\}$ of odd cardinality $2k+1$ where $k \geq 1$. Then for $q = 1, 2$,

$$\sum_{i,j \in K, i \neq j} a_{ij}^{(q)} = \sum_{i,j \in K} b_{ij}^{(q)} \leq 2k,$$

and hence A_1 and A_2 are in Σ_n.

It now follows that the extreme matrices of Σ_n have only integral elements, and hence $\Sigma_n \subseteq \Sigma_n^*$, and this completes the proof that $\Sigma_n = \Sigma_n^*$. $\square$

9.10   Doubly Stochastic Automorphisms

Let G be a graph with vertex set $\{1, 2, \ldots, n\}$ which may have a loop at each vertex. Recall that the adjacency matrix of G is the $(0,1)$-matrix $A = [a_{ij}]$ of order n in which $a_{ij} = 1$ if and only if there is an edge joining vertex i to vertex j; the loops of G correspond to 1's on the main diagonal of A. An *automorphism* of the graph G is a permutation σ of the vertex set $\{1, 2, \ldots, n\}$ such that there is an edge joining vertices i and j if and only if there is an edge joining vertices $\sigma(i)$ and $\sigma(j)$. The automorphism σ corresponds to a permutation matrix P of order n such that $PAP^{\mathrm{T}} = A$, equivalently, $PA = AP$. Let

$$\mathcal{P}(A) = \{P : PA = AP, P \text{ a permutation matrix}\}$$

be the *set of all automorphisms of G viewed as permutation matrices* and thus as points in $\Re^{n^2}$. A permutation matrix is a doubly stochastic matrix, and Tinhofer [137] relaxed the notion of an isomorphism by replacing P with an arbitrary doubly stochastic matrix. A *doubly stochastic automorphism* of G (or of its adjacency matrix A) is a doubly stochastic matrix X of order n such that $XA = AX$. Let

$$\Omega_n[A] = \{X : XA = AX, X \in \Omega_n\}$$

be the set of doubly stochastic automorphisms of G. Then $\Omega_n[A]$ is a convex subpolytope of Ω_n, since it is obtained from Ω_n by imposing additional linear constraints. Let the *convex hull of* $\mathcal{P}(A)$ be denoted by convhull($\mathcal{P}(A)$). Then convhull($\mathcal{P}(A)$) satisfies

$$\text{convhull}(\mathcal{P}(A)) \subseteq \Omega_n(A).$$

The graph G is called *compact* provided every doubly stochastic automorphism is a convex combination of automorphisms, that is,

$$\text{convhull}(\mathcal{P}(A)) = \Omega_n[A].$$

Let G be the graph of order n with n loops, whose adjacency matrix is I_n. Then $\mathcal{P}(I_n)$ is the set of all permutation matrices of order n and $\Omega_n(I_n) = \Omega_n$. Thus Birkhoff's theorem asserts that convhull($\mathcal{P}(I_n)$) $= \Omega_n(I_n)$.

Example. Let G be the graph of order 7 consisting of a cycle of length 3 and a disjoint cycle of length 4. The adjacency matrix of G is the matrix

$$A = \begin{bmatrix} 0 & 1 & 1 & 0 & 0 & 0 & 0 \\ 1 & 0 & 1 & 0 & 0 & 0 & 0 \\ 1 & 1 & 0 & 0 & 0 & 0 & 0 \\ 0 & 0 & 0 & 0 & 1 & 0 & 1 \\ 0 & 0 & 0 & 1 & 0 & 1 & 0 \\ 0 & 0 & 0 & 0 & 1 & 0 & 1 \\ 0 & 0 & 0 & 1 & 0 & 1 & 0 \end{bmatrix}.$$

Then, since all line sums of A equal 2, $(1/7)J_7$ is in $\Omega_7[A]$, but $(1/7)J_7 \notin$ convhull($\mathcal{P}(A)$), since there is no automorphism of G which takes vertex 1 to vertex 4. Thus convhull($\mathcal{P}(A)$) $\neq \Omega_7[A]$. More generally, assume that the adjacency matrix A of G is regular in the sense that all row and column sums are equal. Then $(1/n)J_n \in \Omega_n[A]$, and thus a necessary condition for compactness is that G be *vertex-transitive*, that is, given any two vertices there is an automorphism of G taking one to the other.

Now let G be the complete graph K_n of order n with adjacency matrix $J_n - I_n$. It is easy to check that G is a compact graph. Similarly, it follows that the graph K_n^* obtained from K_n by including a loop at each vertex is a compact graph, since its adjacency matrix is $A = J_n$. $\qquad\square$

Assume the graph G with adjacency matrix A is compact. Then the graph with adjacency matrix $J_n - A$ is also compact. If G has no loops so that the trace of A equals 0, then the graphs with adjacency matrices $A+I_n$ and $J_n - I_n - A$ are readily seen to be compact as well. We also note that if G is compact and H is a graph isomorphic to G, then H is also compact. In fact, the adjacency matrix of H is QAQ^T for some permutation matrix Q, and $\mathcal{P}(B) = Q \cdot \mathcal{P}(A) \cdot Q^T$; thus

$$\text{convhull}(\mathcal{P}(B)) = Q \cdot \text{convhull}(\mathcal{P}(A)) \cdot Q^T$$

and

$$\Omega_n(B) = Q \cdot \Omega_n[A] \cdot Q^T.$$

Tinhofer [139] proved that the vertex-disjoint union of isomorphic and connected compact graphs is also compact. The following lemma will be used in the proof of this result.

Lemma 9.10.1 *Let G be a graph with adjacency matrix A of order n, and let $X = [x_{ij}] \in \Omega_n[A]$. Let $R = (r_1, r_2, \ldots, r_n)$ be the row and column sum vector of A. If $x_{kl} > 0$, then $r_k = r_l$ $(k, l = 1, 2, \ldots, n)$.*

Proof. Without loss of generality we may assume that $r_1 \geq r_2 \geq \cdots \geq r_n$. Let 1^n be the row vector of all 1's of size n. Since $X \in \Omega_n[A]$, we have

$$R = 1^n A = (1^n X)A = 1^n(XA) = 1^n(AX) = (1^n A)X = RX. \qquad (9.25)$$

Suppose that $r_1 = \cdots = r_m > r_{m+1}$. Then from (9.25) we get that for $1 \leq p \leq m$,

$$r_1 = r_p = \sum_{i=1}^n r_i x_{ip} = r_1 \left(\sum_{i=1}^m x_{ip} \right) + \sum_{i=m+1}^n r_i x_{ip}$$

implying that $x_{ip} = 0$ for $m + 1 \leq i \leq n$ and $1 \leq p \leq m$ and, since X is doubly stochastic, $x_{pj} = 0$ for $1 \leq p \leq m$ and $m + 1 \leq j \leq n$. Continuing inductively we complete the proof of the lemma. $\qquad \square$

Theorem 9.10.2 *Let G be a connected, compact graph of order n with adjacency matrix $A = [a_{ij}]$, and let H be a graph isomorphic to G with adjacency matrix $B = [b_{ij}]$. Then the graph $G \oplus H$ with adjacency matrix $A \oplus B$ is compact.*

Proof. Let

$$X = [x_{ij}] = \begin{bmatrix} X_1 & X_{12} \\ X_{21} & X_2 \end{bmatrix}$$

be a doubly stochastic matrix of order $2n$ where each of the four blocks of X is square of order n. Then $X \in \Omega_n[A \oplus B]$ if and only if

$$X_1 A = AX_1, \ X_{12}B = AX_{12}, \ X_{21}A = BX_{21}, \ \text{and} \ X_2 B = BX_2. \quad (9.26)$$

Assume that $X \in \Omega_n[A \oplus B]$. By Lemma 9.10.1, if $x_{ij} > 0$ then the ith and jth row sums of $A \oplus B$ are equal. Using this fact and the equation $X_1 A = AX_1$ we obtain that the column sum vectors $c_{X_1} = (c_1, c_2, \ldots, c_n)$ of X_1 and $c_A = (d_1, d_2, \ldots, d_n)$ of A satisfy

$$\sum_{i=1}^{n} c_i a_{ij} = d_j c_j \quad (1 \leq j \leq n). \tag{9.27}$$

Let A' be the matrix obtained from A by multiplying the elements of column j by d_j^{-1} for $j = 1, 2, \ldots, n$. Rewriting (9.27), we get

$$c_{X_1} A' = c_{X_1}.$$

The matrix A' is a column-stochastic matrix[20] and since G is a connected graph, A is irreducible. From the Perron–Frobenius theory of nonnegative matrices (see e.g. [8]), A' has a unique (up to scalar multiples) left eigenvector corresponding to its eigenvalue 1. Since the vector 1^n of all 1's is such a left eigenvector, it follows that c_{X_1} is a constant vector, that is, X_1 has constant column sums. Similar calculations show that each of the matrices X_1, X_{12}, X_{21}, and X_2 has both constant row and constant column sums. Since each of these matrices is square, the constant row sum of each is identical to its constant column sum. Since X is doubly stochastic, these constant sums are $a, 1 - a, 1 - a, a$ for X_1, X_{12}, X_{21}, and X_2, respectively.

Since G and H are isomorphic there is a permutation matrix Q such that $B = QAQ^{\mathrm{T}}$. Then

$$\frac{1}{a} X_1, \quad \frac{1}{1-a} X_{12} Q, \quad \frac{1}{1-a} Q^{\mathrm{T}} X_{21}, \quad \text{and} \quad \frac{1}{a} Q^{\mathrm{T}} X_2 Q$$

are in $\Omega(A)$. Since the graph of A is compact, it now follows from (9.26) that

$$\frac{1}{a} X_1 = \sum_{P \in \mathcal{P}(A)} \sigma_P P \quad \text{and} \quad \frac{1}{a} X_2 = \sum_{P \in \mathcal{P}(A)} \sigma'_P Q P Q^{\mathrm{T}}$$

where the σ_P and σ'_P satisfy

$$0 \leq \sigma_P, \sigma'_P \leq 1, \quad \text{and} \quad \sum_{P \in \mathcal{P}(A)} \sigma_P = \sum_{P \in \mathcal{P}(A)} \sigma'_P = a.$$

It is now easily seen that the matrix $\sigma X_1 \oplus \sigma X_2$ is in the convex hull of the set of matrices

$$\left\{ \begin{bmatrix} S & O \\ O & T \end{bmatrix} : S \in \mathcal{P}(A), T \in \mathcal{P}(B) \right\} \subseteq \mathcal{P}(A \oplus B).$$

Similarly, the matrix

$$\begin{bmatrix} O & (1-a)X_{12} \\ (1-a)X_{21} & O \end{bmatrix}$$

---

[20] A nonnegative matrix with each column sum equal to 1.

is in the convex hull of $\mathcal{P}(A \oplus B)$. Since $a + (1 - a) = 1$, the matrix

$$X = \frac{1}{a} \begin{bmatrix} aX_1 & O \\ O & \sigma X_2 \end{bmatrix} + \frac{1}{1-a} \begin{bmatrix} O & (1-a)X_{12} \\ (1-a)X_{21} & O \end{bmatrix}$$

is in $\mathcal{P}(A \oplus B)$. Hence $G \oplus H$ is compact. $\qquad \square$

We now show that cycles are compact graphs [137].

Theorem 9.10.3 *Let G be a graph which is a cycle of length n with adjacency matrix*

$$A = [a_{ij}] = \begin{bmatrix} 0 & 1 & 0 & 0 & \cdots & 0 & 1 \\ 1 & 0 & 1 & 0 & \cdots & 0 & 0 \\ 0 & 1 & 0 & 1 & \cdots & 0 & 0 \\ 0 & 0 & 1 & 0 & \cdots & 0 & 0 \\ \vdots & \vdots & \vdots & \vdots & \ddots & \vdots & \vdots \\ 0 & 0 & 0 & 0 & \cdots & 0 & 1 \\ 1 & 0 & 0 & 0 & \cdots & 1 & 0 \end{bmatrix}.$$

Then the graph G is compact.

Proof. In this proof we take all indices modulo n. Thus $a_{ij} = 1$ if and only $j \equiv i + 1$ or $i - 1 \bmod n$. Assume that $X = [x_{ij}] \in \Omega_n[A]$. Then

$$x_{i+1,j} + x_{i-1,j} = x_{i,j-1} + x_{i,j+1} \quad (1 \le i \le j).$$

This implies that

$$x_{i+1,j-i} - x_{i,j-i-1} = x_{i,j-i+1} - x_{i-1,j-i} = \cdots = x_{1j} - x_{n,j-1}, \quad (9.28)$$

and

$$x_{i+1,j+i} - x_{i,j+i+1} = x_{i,j+i-1} - x_{i-1,j+i} = \cdots = x_{1j} - x_{n,j+1}. \quad (9.29)$$

For j fixed the expressions of the right sides of (9.28) and (9.29) are constant.

First assume that $x_{1j} - x_{n,j-1} > 0$. Then $x_{i+1,j-i} > x_{i,j-i-1} \ge 0$ for $1 \le i \le n$. The matrix X has positive elements in the positions corresponding to the automorphism $P = [p_{ij}]$ of G given by $i \to j - i + 1$ (this automorphism comes from a reflection when the vertices of G are regarded as the vertices of a regular n-gon) where $p_{ik} = 1$ if and only if $k \equiv j - i + 1 \bmod n$. An analogous result holds when $x_{1j} - x_{n,j-1} < 0$. Now assume that $x_{1j} - x_{n,j+1} > 0$. Then $x_{i+1,j+i} > x_{i,j+i+1} \ge 0$, and X has positive elements in the positions corresponding to the automorphism $P = [p_{ij}]$ of G given by $i \to j + i - 1$ (this automorphism comes from a rotation of a regular n-gon) where $p_{ik} = 1$ if and only if $k \equiv j + i - 1 \bmod n$. Again an analogous result holds when $x_{1j} - x_{n,j+1} < 0$.

In each of the four cases, let $\lambda = \min\{x_{ik} : p_{ik} = 1\}$. If $\lambda = 1$, then $X = P$. Assume that $\lambda < 1$, and define

$$Y = \frac{X - \lambda P}{1 - \lambda}.$$

Then $Y \in \Omega_n$ and has at least one fewer nonzero element than X. Arguing by induction, we conclude that Y and hence X are in convhull$(\mathcal{P}(A))$.

To complete the proof, we now assume that $x_{1j} = x_{n,j-1} = x_{n,j+1}$ for $1 \le j \le n$. Equations (9.28) and (9.29) now imply that

$$x_{i+1,j} = x_{i,j+1} = x_{i-1,j} = x_{i,j-1} \quad (1 \le i, j \le n).$$

From this we get that there are numbers α and β such that

$$x_{ij} = \begin{cases} \alpha & \text{if } i - j \equiv 0 \bmod 2, \\ \beta & \text{if } i - j \equiv 1 \bmod 2. \end{cases}$$

Since X is doubly stochastic, $\alpha\beta \ne 0$. Now it is easy to see that X is a convex combination of automorphisms of G (corresponding to reflections) and hence $X \in$ convhull$(\mathcal{P}(A))$. $\square$

An immediate corollary of Theorems 9.10.2 and 9.10.3 is the following result [21] which reduces to Birkhoff's theorem when $k = 1$.

Corollary 9.10.4 *Let k be a positive integer, and let G be a graph which is the vertex-disjoint union of cycles of length k. Then G is a compact graph.*
                                                                              $\square$

In [137], [138], it is also shown that the adjacency matrices A of trees of order n, and a class of graphs called strong tree-cographs, also satisfy $\Omega_n[A] = $ convhull$(\mathcal{P}(A))$. In [66] graphs for which the adjacency matrix A satisfies $\Omega_n[A] = \{I_n\}$ are characterized. In [124] circulant graphs which are compact are investigated. In [146] other compact graphs are identified.

Various algorithmic aspects of compact graphs are investigated in [137], [139]. In particular, in [137] a complete set of invariants is given in order for two graphs G and H with adjacency matrices A and B, respectively, to be *doubly stochastic isomorphic*, that is, in order that there exist a doubly stochastic matrix X such that $XA = AX$. It is also shown that there is a polynomial algorithm to decide whether or not two compact graphs are isomorphic.

A property for graphs more restrictive than compactness was introduced in [21]. Let G be a graph with adjacency matrix A of order n. The set $\mathcal{N}(A)$ of all nonnegative matrices X of order n such that $XA = AX$ is a cone pointed at the zero matrix O. This cone contains the cone

$$\text{cone}(\mathcal{P}(A)) = \left\{\sum_P c_P P : c_P \ge 0\right\}$$

generated by the automorphisms of G. We have

$$\text{cone}(\mathcal{P}(A)) \subseteq \mathcal{N}(A).$$

A graph is *supercompact* provided we have the equality

$$\text{cone}(\mathcal{P}(A)) = \mathcal{N}(A),$$

that is, provided the extreme rays of the cone $\mathcal{N}(A)$ are the rays from O toward a matrix $P \in \mathcal{P}(A)$. Note that if $X \in \text{cone}(\mathcal{P}(A))$ and $X \neq O$, then X has constant row and column sums $c \neq 0$ and hence $(1/c)X \in \text{convhull}(\mathcal{P}(A))$. We recall that a graph is *regular of degree k* provided all its vertices have the same degree k, equivalently, the row and column sums of its adjacency matrix are all equal to k.

Lemma 9.10.5 *Let G be a graph of order n with adjacency matrix A. If G is supercompact, then G is compact and G is a regular.*

Proof. Assume G is supercompact so that $\text{cone}(\mathcal{P}(A)) = \mathcal{N}(A)$. Since clearly $A \in \mathcal{N}(A)$, $A \in \mathcal{P}(A)$ and hence A has constant row and column sums. Now let $X \in \Omega_n[A]$. Then $X \in \mathcal{N}(A)$ and hence $X \in \text{cone}(\mathcal{P}(A))$. Since X is doubly stochastic, we have $X \in \text{convhull}(\mathcal{P}(A))$. Thus $\Omega_n[A] \subseteq \text{convhull}(\mathcal{P}(A))$ and hence G is compact.   □

Since trees are compact graphs and a tree is regular if and only if its order is 2, trees of order $n \geq 3$ are compact but not supercompact.

Example. Let G be the graph of order 4 consisting of two vertex-disjoint edges. The adjacency matrix of G is

$$A = \begin{bmatrix} 0 & 0 & 1 & 0 \\ 0 & 0 & 0 & 1 \\ 1 & 0 & 0 & 0 \\ 0 & 1 & 0 & 0 \end{bmatrix}.$$

The matrix

$$X = \begin{bmatrix} 1 & 0 & 0 & 0 \\ 0 & 0 & 0 & 0 \\ 0 & 0 & 1 & 0 \\ 0 & 0 & 0 & 0 \end{bmatrix}$$

is in $\mathcal{N}(A)$. Since X does not have constant row and column sums, $X \notin \text{cone}(\mathcal{P}(A))$, and hence G is not supercompact. The graph G however is compact. The compactness of G can be determined by direct calculation, or by using the compactness of complete graphs (K_2 in this case) and Theorem 9.10.2. Thus compact regular graphs need not be supercompact, and the converse of Lemma 9.10.5 does not hold.   □

We now show that graphs which are cycles are supercompact [21].

Theorem 9.10.6 *Let G be a graph which is a cycle of length n. Then G is supercompact.*

Proof. Let A be the adjacency matrix of G. In Theorem 9.10.3 it was shown that cycles are compact graphs. Let $X = [x_{ij}]$ be a nonzero matrix in $\mathcal{N}(A)$. Following the proof of Theorem 9.10.3 we see that there is a automorphism $P = [p_{ij}]$ of G such that $x_{ij} > 0$ whenever $p_{ij} = 1$ (the double stochasticity of X was not used in the proof to obtain this conclusion). Let

$$\epsilon = \min\{x_{ij} : p_{ij} = 1, 1 \le i, j \le n\}.$$

Then $\epsilon > 0$ and the matrix $X - \epsilon P \in \mathcal{N}(A)$, and $X - \epsilon P$ has at least one more zero than X. Arguing inductively, we conclude that $X \in \text{cone}(\mathcal{P}(A))$, and thus $\mathcal{N}(A) = \text{cone}(\mathcal{P}(A))$. □

Other constructions of compact graphs are given in [21], [139]. In particular, it is shown in [21] that complete bipartite graphs $K_{m,m}$, as well as graphs obtained from $K_{m,m}$ by removing m pairwise vertex-disjoint edges (a perfect matching), are compact. Finally we mention that in [60] the notion of compactness is extended to finite permutation groups in general.

9.11   Diagonal Equivalence

In this section we show that every nonnegative square matrix with total support can be scaled to a doubly stochastic matrix. More generally, we show that every m by n nonnegative matrix which has the pattern of some matrix in a transportation polytope $\mathcal{N}(R, S)$ can be scaled to a matrix in $\mathcal{N}(R, S)$. In other words, since scaling does not alter the pattern of a matrix, if the obvious necessary condition for scaling holds then scaling is possible.

Let A be an m by n matrix. A *scaling* of A is a matrix $D_1 A D_2$ where D_1 and D_2 are nonsingular, diagonal matrices of orders m and n, respectively. Given a nonnegative matrix A we seek diagonal matrices D_1 and D_2 with positive main diagonals such that $D_1 A D_2 \in \mathcal{N}(R, S)$ for prescribed positive row and column sum vectors $R = (r_1, r_2, \ldots, r_m)$ and $S = (s_1, s_2, \ldots, s_n)$.

Recall that the patterns of doubly stochastic matrices are the $(0, 1)$-matrices with total support, and that the patterns of matrices in $\mathcal{N}(R, S)$ are those m by n $(0, 1)$-matrices W which satisfy the (R, S)-support condition:

For each K with $\oslash \subset K \subset \{1, 2, \ldots, m\}$ and each L with $\oslash \subset L \subset \{1, 2, \ldots, n\}$ such that $W[K, L] = O$,

$$\sum_{l \in L} s_l \ge \sum_{k \in K} r_k \tag{9.30}$$

with equality and only if $W(K, L] = O$. Recall also that an m by n matrix W having no zero rows or columns is decomposable provided for K and L as above, $W[K, L]$ and $W(K, L]$ are both zero matrices; otherwise W

is indecomposable. If W is decomposable and satisfies the (R, S)-support condition, then for K and L with $W[K, L] = O$ and $W(K, L) = O$ we have $\sum_{l \in L} s_l = \sum_{k \in K} r_k$.

We show that if the nonnegative matrix A satisfies the (R, S)-support condition, then it can be scaled to a matrix in $\mathcal{N}(R, S)$. An important special case occurs when $R = S = (1, 1, \ldots, 1)$ and the scaling is to a doubly stochastic matrix.

A brief history of this question is the following. Sinkhorn [127] in 1964 showed that a positive square matrix can be scaled to a doubly stochastic matrix. His proof was based on an iterative procedure which alternately scaled the row sums to 1 and then the column sums to 1. He showed that this procedure converged to a doubly stochastic matrix. Earlier, in 1940, Deming and Stephen [52] had already proposed this iterative procedure but did not give a convergence proof. Independently, Sinkhorn and Knopp [132] and Brualdi, Parter, and Schneider [37] showed that a square, nonnegative matrix can be scaled to a doubly stochastic matrix exactly when it has total support; the proof in [132] uses the aforementioned iterative procedure. Other proofs have been given in [54] and [85]. In addition, Parlett and Landis [115] proposed three other methods for iteratively scaling a matrix to a doubly stochastic matrix. Menon [92] proved a nonnegative matrix can be scaled to a matrix in $\mathcal{N}(R, S)$ provided it has the same zero–nonzero pattern as some matrix in $\mathcal{N}(R, S)$, a question resolved in [1] (see Theorem 8.1.7). Menon and Schneider [93] gave another proof of this result. Sinkhorn [128] used the technique of Djoković [54] and London [85] to also give another proof. In [129], Sinkhorn showed that the iterative scheme of alternately scaling row and column sums to R and S works in this more general situation as well. Another proof is given in [91]. Here we follow the proof in [129].

The following inequality plays an important role in this proof. We adopt the convention that $0^0 = 1$.

Lemma 9.11.1 *Let* $x_1, x_2, \ldots, x_n$ *and* $\lambda_1, \lambda_2, \ldots, \lambda_n$ *be nonnegative real numbers. Let* $\lambda = \lambda_1 + \lambda_2 + \cdots + \lambda_n$. *Then*

$$\left(\sum_{i=1}^{k} \lambda_k x_k \right)^{\lambda} \geq \lambda^{\lambda} \prod_{k=1}^{n} x_k^{\lambda_k}.$$

Proof. The stated inequality is equivalent to

$$\log \left(\sum_{i=1}^{k} \lambda_k x_k \right)^{\lambda} \geq \log \left(\lambda^{\lambda} \prod_{k=1}^{n} x_k^{\lambda_k} \right), \text{ and so to}$$

$$\lambda \log \left(\sum_{k=1}^{n} \lambda_k x_k \right) \geq \lambda \log \lambda + \sum_{k=1}^{n} \lambda_k \log x_k, \text{ or to}$$

$$\log \left(\sum_{k=1}^{n} \frac{\lambda_k}{\lambda} x_k \right) \geq \sum_{k=1}^{n} \frac{\lambda_k}{\lambda} \log x_k.$$

Hence the inequality is a consequence of the concavity of the logarithm function.   □

Theorem 9.11.2 *Let A be an m by n nonnegative matrix with no zero rows or columns. Let $R = (r_1, r_2, \ldots, r_m)$ and $S = (s_1, s_2, \ldots, s_n)$ be positive vectors with $r_1 + r_2 + \cdots + r_m = s_1 + s_2 + \cdots + s_n$. Then there exist a matrix $B \in \mathcal{N}(R, S)$ and positive diagonal matrices D_1 and D_2 such that $D_1 A D_2 = B$ if and only if the pattern of A satisfies the (R, S)-support condition. If A is indecomposable, then the matrix B in $\mathcal{N}(R, S)$ is unique, and the matrices D_1 and D_2 are unique up to reciprocal scalar factors.*

Proof. The (R, S)-support condition is obviously necessary for there to exist positive diagonal matrices D_1 and D_2 with $D_1 A D_2 \in \mathcal{N}(R, S)$. Now assume that A satisfies the (R, S)-support condition. We may assume that A is indecomposable, and thus for all K and L with $\oslash \subset K \subset \{1, 2, \ldots, m\}$ and $\oslash \subset L \subset \{1, 2, \ldots, n\}$ such that $A[K, L] = O$, we have $A(K, L] \neq O$ (thus $\sum_{l \in L} s_l > \sum_{k \in K} r_k$). Otherwise we could use induction to complete the proof.

We define a function ϕ on the positive real n-tuples space $\Re_+^n$ by

$$\phi(x) = \frac{\prod_{i=1}^m \left(\sum_{j=1}^n a_{ij} x_j \right)^{r_i}}{\prod_{j=1}^n x_j^{s_j}}.$$

Since $\sum_{i=1}^m r_i = \sum_{j=1}^n s_j$, we have $\phi(\lambda x) = \phi(x)$ for all positive λ. Thus the minimum value of ϕ on $\Re_+^n$, if it exists, is assumed on

$$W = \{x \in \Re_+^n : ||x||^2 = x_1^2 + x_2^2 + \cdots + x_n^2 = 1\}.$$

We show that ϕ assumes a minimum on $\Re_+^n$ by showing that as x approaches the boundary of R_+^n through points in W, then $\phi(x)$ approaches ∞. Let

$$L = \{j : x_j \to 0, \text{ as } x \text{ approaches the boundary}\},$$

and let

$$K = \{i : a_{ij} = 0 \text{ for all } j \notin L\}.$$

We have $\oslash \subset L \subset \{1, 2, \ldots, n\}$. The positivity of the s_j implies that $K \neq \{1, 2, \ldots, m\}$. If $K = \oslash$, then the definition of ϕ implies that $\phi(x) \to \infty$. Thus we may assume that $K \neq \oslash$. We now write $\phi(x) = \phi_1(x) \phi_2(x)$ where

$$\phi_1(x) = \frac{\prod_{i \in K} \left(\sum_{j=1}^n a_{ij} x_j \right)^{r_i}}{\prod_{j \in L} x_j^{s_j}} = \frac{\prod_{i \in K} \left(\sum_{j \in L} a_{ij} x_j \right)^{r_i}}{\prod_{j \in L} x_j^{s_j}},$$

and

$$\phi_2(x) = \frac{\prod_{i \notin K} \left(\sum_{j=1}^{n} a_{ij} x_j\right)^{r_i}}{\prod_{j \notin L} x_j^{s_j}}.$$

Since $\phi_2(x)$ has a positive limit, we need only consider $\phi_1(x)$.

Let $B = [b_{ij}]$ be any matrix in $\mathcal{N}(R, S)$. Then $B[K, L] = O$ and so $\sum_{j \in L} b_{ij} = r_j$ for each $i \in K$. Also by our assumption $B(K, L] \neq O$. Hence by Lemma 9.11.1 we get

$$\left(\sum_{j \in L} a_{ij} x_j\right)^{r_i} \geq r_i^{r_i} \frac{\prod_{j \in L} a_{ij}^{b_{ij}}}{\prod_{j \in L} b_{ij}^{b_{ij}}} \prod_{j \in L} x_j^{b_{ij}} = \theta_i \prod_{j \in L} x_j^{b_{ij}} \quad (i \in K),$$

where θ_i is defined implicity in this expression. It follows that

$$\phi_1(x) \geq \frac{\prod_{i \in K} \theta_i}{\prod_{j \in L} x_j^{s_j - \sum_{i \in K} b_{ij}}}.$$

For each $j \in L$, we have $\sum_{i \in K} b_{ij} \leq s_j$, and since $B(K, L] \neq O$, this sum is positive for at least one $j_0 \in L$. We conclude now that $\phi_1(x) \to \infty$, and finally that $\phi(x) \to \infty$, as x approaches the boundary of R_+^n through points in W. Thus $\phi(x)$ achieves a minimum on R_+^n at a point $\bar{x} = (\bar{x}_1, \bar{x}_2, \ldots, \bar{x}_n) \in R_+^n$. At such a minimum point $\bar{x}$, we have

$$\frac{\partial \log \phi(x)}{\partial x_k} = 0 \quad (k = 1, 2, \ldots, n).$$

Therefore

$$\sum_{i=1}^{m} r_i \left(\frac{a_{ik}}{\sum_{j=1}^{n} a_{ij} \bar{x}_k}\right) - \frac{s_k}{\bar{x}_k} = 0 \quad (k = 1, 2, \ldots, n).$$

We now put

$$\bar{y}_i = \frac{r_i}{\sum_{j=1}^{n} a_{ij} \bar{x}_j} \quad (i = 1, 2, \ldots, m),$$

and then define D_1 and D_2 to be the positive diagonal matrices with diagonal elements $\bar{y}_1, \bar{y}_2, \ldots, \bar{y}_m$ and $\bar{x}_1, \bar{x}_2, \ldots, \bar{x}_n$, respectively. We then have that $D_1 A D_2 \in \mathcal{N}(R, S)$.

We now turn to the uniqueness properties of B, D_1, and D_2. Suppose that D_3 and D_4 are positive diagonal matrices such that the matrix $C = D_3 A D_4$ is in $\mathcal{N}(R, S)$. For $i = 1, 3$, let the diagonal elements of D_i be $d_1^i, d_2^i, \ldots, d_m^i$, and for $i = 2, 4$, let the diagonal elements of D_i be $d_1^i, d_2^i, \ldots, d_n^i$. Let $p_k = d_k^3/d_k^1$ $(k = 1, 2, \ldots, m)$ and let $q_l = d_l^4/d_l^2$ $(l = 1, 2, \ldots, n)$. Let $p_u = \min\{p_i : i = 1, 2, \ldots, m\}$ and let $p_v = \max\{p_i : i = 1, 2, \ldots, m\}$, and let $q_e = \min\{q_l : l = 1, 2, \ldots, n\}$ and $q_f = \max\{q_l : l =$

$1, 2, \ldots, n\}$. Then

$$p_u = r_u \left(\sum_{j=1}^{n} d_u^1 a_{uj} q_j d_j^2 \right)^{-1}$$

$$\geq \frac{r_u}{q_f} \left(\sum_{j=1}^{n} p_u a_{uj} d_j^2 \right)^{-1}$$

$$= \frac{1}{q_f}.$$

Thus

$$p_u q_f \geq 1. \tag{9.31}$$

Similarly we have

$$q_e p_v \geq 1. \tag{9.32}$$

Combining inequalities (9.31) and (9.32), we conclude that $q_l = q_f$ ($l = 1, 2, \ldots, n$) and $p_i = p_u$ ($i = 1, 2, \ldots, m$), and then that $q_f = p_u^{-1}$. Hence $D_3 = p_e D_1$ and $D_4 = (p_e)^{-1} D_3$, and $D_1 A D_2 = D_3 A D_4$. □

The following corollary is an immediate consequence of Theorem 9.11.2.

Corollary 9.11.3 *Let A be a nonnegative matrix of order n without any zero rows or columns. Then there exist a doubly stochastic matrix B and positive diagonal matrices D_1 and D_2 such that $D_1 A D_2 = B$ if and only if A has total support. If A is a fully indecomposable matrix, then the doubly stochastic matrix B is unique, and the matrices D_1 and D_2 are unique up to reciprocal scalar factors.* □

We now let $R = (r_1, r_2, \ldots, r_n)$ be a positive real vector and let A be a symmetric, nonnegative matrix of order n, and we consider the possibility of symmetrically scaling A to a matrix in the symmetric transportation polytope $\mathcal{N}(R)$. Thus we seek a positive diagonal matrix D such that DAD has row sum vector R, and thus also column sum vector R. We recall from Chapter 8 that the patterns of matrices in $\mathcal{N}(R)$ are those $(0,1)$-matrices W which satisfy the symmetric R-support condition:
For all partitions I, J, H of $\{1, 2, \ldots, n\}$ such that $W[J \cup H, H] = O$,

$$\sum_{i \in I} r_i \geq \sum_{h \in H} r_h, \tag{9.33}$$

with equality if and only if $W[I, I \cup J] = O$.
The following lemma and theorem are proved in [16].

Lemma 9.11.4 *Let A be a decomposable but irreducible, symmetric non-negative matrix of order n. Then there exists a permutation matrix P such that*

$$PAP^{\mathrm{T}} = \begin{bmatrix} & & & & & & B_1 \\ & O & & & & \cdot \cdot & \\ & & & & B_k & & \\ & & & B_{k+1} & & & \\ & & B_k^{\mathrm{T}} & & & & \\ & & & & & O & \\ & \cdot \cdot & & & & & \\ B_1^{\mathrm{T}} & & & & & & \end{bmatrix},$$

where $B_1, \ldots, B_k, B_{k+1}$ are square indecomposable matrices, with B_{k+1} a possibly vacuous symmetric matrix.

Proof. We first show that there exists a permutation matrix P such that PAP^{T} has the form

$$\begin{bmatrix} O & O & B_1 \\ O & A_1 & O \\ B_1^{\mathrm{T}} & O & O \end{bmatrix}, \tag{9.34}$$

where B_1 is a nonvacuous indecomposable square matrix and A_1 is an irreducible symmetric matrix (possibly vacuous). Since A is decomposable, there exist partitions α_1, α_2 and β_1, β_2 of $\{1, 2, \ldots, n\}$ such that $A[\alpha_1, \beta_2]$ and $A[\alpha_2, \beta_1]$ are zero matrices. Let $\alpha = \beta_1 \setminus \alpha_1, \gamma = \alpha_1 \setminus \beta_1$, and $\beta = \{1, 2, \ldots, n\} \setminus (\alpha \cup \gamma)$. If $\beta_1 \subseteq \alpha_1$, then we have $A[\beta_1, \beta_2] = O$ and, by symmetry, $A[\beta_2, \beta_1] = O$, contradicting our assumption that A is irreducible. Thus $\beta_1 \not\subseteq \alpha_1$, and similarly $\alpha_1 \not\subseteq \beta_1$. Thus α and γ are nonempty disjoint subsets of $\{1, 2, \ldots, n\}$. Since $\gamma \subseteq \alpha_1$ and $\gamma \cap \beta_1 = \varnothing$ (so $\gamma \subseteq \beta_1$) we have that $A[\gamma, \gamma] = O$. The matrix $A[\beta, \gamma]$ is made up of $A[\alpha_1 \cap \beta_1, \gamma]$, which equals O since $\gamma \subset \beta_2$, and $A[\{1, 2, \ldots, n\} \setminus (\alpha_1 \cup \beta_1), \gamma]$, which also equals O, since $\{1, 2, \ldots, n\} \setminus (\alpha_1 \cup \beta_1) \subseteq \beta_2$ and $\gamma \subseteq \alpha_1$. Hence $A[\beta, \gamma] = O$, and by symmetry $A[\gamma, \beta] = O$. Since $A[\alpha_1, \beta_2]$ and $A[\alpha_2, \beta_1]$ are zero matrices, we have

$$\sum_{i \in \alpha_1} r_i = \sum_{j \in \beta_1} r_j \text{ and so } \sum_{i \in \alpha} r_i = \sum_{j \in \gamma} r_j.$$

The nonnegativity of A now implies that $A[\alpha, \alpha], A[\alpha, \beta]$, and $A[\beta, \alpha]$ are all zero matrices. Thus there exists a permutation matrix P such that

$$PAP^{\mathrm{T}} = \begin{bmatrix} O & O & A[\alpha, \gamma] \\ O & A[\beta, \beta] & O \\ A[\alpha, \gamma]^{\mathrm{T}} & O & O \end{bmatrix}.$$

The irreducibility of A now implies that $A[\alpha, \gamma]$ is not decomposable. If $A[\beta, \beta]$ is not vacuous, then $A[\beta, \beta]$ is irreducible for otherwise A would be reducible. If $A[\beta, \beta]$ is decomposable, we may repeat the above argument with A replaced with $A[\beta, \beta]$. We may now argue by induction to complete the proof. $\qquad\qquad\qquad\qquad\qquad\qquad\qquad\qquad\qquad\qquad\qquad\qquad\qquad$ $\square$

Theorem 9.11.5 *Let A be a symmetric nonnegative matrix of order n with no zero rows, and let $R = (r_1, r_2, \ldots, r_n)$ be a positive real vector. Then there exist a matrix $B \in \mathcal{N}(R)$ and a positive diagonal matrix D such that $DAD = B$ if and only if A satisfies the symmetric R-support condition. If A is fully indecomposable, then the matrix B in $\mathcal{N}(R)$ and the positive diagonal matrix D are unique.*

Proof. The symmetric (R, S)-support condition is clearly necessary for there to exist a positive diagonal matrix D such that DAD is in $\mathcal{N}(R)$. We now assume that A satisfies the symmetric (R, S)-support condition. As shown in Theorem 8.2.3 the symmetric R-support condition is a necessary and sufficient condition for there to exist a matrix in $\mathcal{N}(R, R)$ with the same pattern as the symmetric matrix A and it is equivalent to the (R, R)-support condition when the matrix A is symmetric.

First suppose that A is indecomposable. Then by Theorem 9.11.2 there exist positive diagonal matrices D_1 and D_2, unique up to reciprocal scalar factors, and a matrix $B \in \mathcal{N}(R, R)$ such that $D_1 A D_2 = B$. Since A is symmetric, we have $(D_1 A D_2)^{\mathrm{T}} = D_2 A D_1$ is also a matrix in $\mathcal{N}(R, R)$. Thus by Theorem 9.11.2, there exists a nonzero real number c such that $D_2 = c D_1$. Let $D = \sqrt{c} D_1$. Then $DAD = B$. Since A is symmetric, B is symmetric and hence $B \in \mathcal{N}(R)$. The uniqueness assertion is a consequence of the uniqueness assertion of Theorem 9.11.2 using the fact that the diagonal elements of the diagonal matrix D are to be positive.

Suppose the matrix A is reducible. Since A is symmetric, there is a permutation matrix P such that $PAP^{\mathrm{T}} = A_1 \oplus A_2$ where A_1 and A_2 are nonvacuous (necessarily symmetric) matrices, and we may complete the proof by induction.

We now assume that A is irreducible but decomposable. We may assume that A has the form as given in Lemma 9.11.4. Implicit in this form is a partition of $\{1, 2, \ldots, n\}$ into sets $\pi_1, \ldots, \pi_k, \pi_{k+1}, \rho_k, \ldots, \rho_1$ where the set π_{k+1} may be vacuous, and thus a decomposition of R into subvectors $R_{\pi_1}, \ldots, R_{\pi_k}, R_{\pi_{k+1}}, R_{\rho_k}, \ldots, R_{\rho_1}$. It is straightforward to verify that B_i satisfies the (R_{π_i}, R_{ρ_i})-support condition for $i = 1, 2, \ldots, k$, and that B_{k+1}, if nonvacuous, satisfies the symmetric $R_{\pi_{k+1}}$-support condition. By Theorem 9.11.2 there exist positive diagonal matrices D_i and E_i such that $D_i B_i E_i \in \mathcal{N}(R_{\pi_i}, R_{\rho_i})$ for $i = 1, 2, \ldots, k$. Since B_{k+1} is indecomposable, then as we have already seen, there exists a positive diagonal matrix D_{k+1} such that $D_{k+1} B_{k+1} D_{k+1} \in \mathcal{N}(R_{\pi_{k+1}})$. Let D be the positive diagonal matrix with blocks $D_1, \ldots, D_k, D_{k+1}, E_k, \ldots, E_1$. Then $DAD \in \mathcal{N}(R)$. $\square$

The following corollary [49] is an immediate consequence of Theorem 9.11.5.

Corollary 9.11.6 *Let A be a nonnegative symmetric matrix of order n with no zero rows or columns. Then there exist a symmetric doubly stochastic matrix B and a positive diagonal matrix D such that $DAD = B$ if and only if A has total support. If A is a fully indecomposable matrix, then the doubly stochastic matrix B and diagonal matrix D are unique.* □

9.12   Applications of Doubly Stochastic Matrices

In this section we discuss briefly two applications in which doubly stochastic matrices play an important role, namely the optimal assignment problem and satellite-switched, time-division multiple-access systems (SS/TDMA systems). The former naturally leads into consideration of nonnegative matrices each of whose positive diagonals have the same sum.

We begin with the optimal assignment problem.

Let $B = [b_{ij}]$ be an m by n real matrix where we assume that $m \le n$. If $P = [p_{ij}]$ is a real matrix of the same size as B, then the inner product of P and B is

$$P \circ B = \sum_{i=1}^{m} \sum_{j=1}^{n} p_{ij} b_{ij}.$$

The *optimal assignment problem* asks for the determination of $h(B)$ where

$$h(B) = \max\{P \circ B : P \text{ an } m \text{ by } n \text{ subpermutation matrix of rank } m\}.$$

(If we replace B by $-B$, the optimal assignment problem is equivalent to the determination of the minimal value of $P \circ B$.) If B is a $(0,1)$-matrix, then the optimal assignment problem is equivalent to the determination of the term rank of B [38].

If $m < n$, then we may border B with $n-m$ rows of 0's without changing the solution of the problem. Thus we may assume that $m = n$ without loss of generality. In this case, since Ω_n is the convex hull of the permutation matrices of order n and since $P \circ B$ is a linear function of the elements of P, we have

$$h(B) = \max\{S \circ B : S \in \Omega_n\}.$$

Thus the optimal assignment problem can be solved by methods of linear programming [126, Chapter 17]. Another way to describe the optimal assignment problem is to use the trace function of a matrix. The discussion that follows gives some insight into the structure of the optimal solutions of the assignment problem.

Let $\sigma = (i_1, i_2, \ldots, i_n)$ be a permutation of order $\{1, 2, \ldots, n\}$, and let $Q_\sigma = [q_{ij}]$ be the permutation matrix of order n in which $q_{1i_1} = q_{2i_2} = \cdots = q_{ni_n} = 1$. The *diagonal of B corresponding to Q_σ* (or to σ) is the ordered set $(b_{1i_1}, b_{2i_2}, \ldots, b_{ni_n})$ of elements of B. This diagonal is a *positive diagonal*, respectively, a *nonzero diagonal*, of B provided each of its elements is positive, respectively, nonzero. The *diagonal sum* of B corresponding to Q_σ (or σ) is

$$Q_\sigma \circ B = b_{1i_1} + b_{2i_2} + \cdots + b_{ni_n} = \mathrm{trace}(Q_\sigma^{\mathrm{T}} B) = \mathrm{trace}(BQ_\sigma^{\mathrm{T}}).$$

It follows that

$$h(B) = \max\{\mathrm{trace}(P^{\mathrm{T}} B) : P \text{ a permutation matrix of order } n\}.$$

A *maximal-sum diagonal* of B is a diagonal whose corresponding diagonal sum equals $h(B)$.

The following theorem is due to Balasubramanian [3].

Theorem 9.12.1 *Let $A \in \Omega_n$ and let B be a real matrix of order n. Then $\mathrm{trace}(AB) = h(AB) = h(B)$ if and only if for every permutation matrix Q corresponding to a positive diagonal of A, $\mathrm{trace}(Q^{\mathrm{T}} B) = \mathrm{trace}(B)$.*

Proof. First assume that $\mathrm{trace}(AB) = h(AB) = h(B)$. Since A is doubly stochastic, there exist permutation matrices $P_1, P_2, \ldots, P_k$ and positive numbers $c_1, c_2, \ldots, c_k$ with $c_1 + c_2 + \cdots + c_k = 1$, such that A is the convex combination

$$A = c_1 P_1^{\mathrm{T}} + c_2 P_2^{\mathrm{T}} + \cdots + c_k P_k^{\mathrm{T}}. \tag{9.35}$$

Then

$$h(AB) = \mathrm{trace}(AB) = \sum_{i=1}^{k} c_i \mathrm{trace}(P_i^{\mathrm{T}} B) \le \sum_{i=1}^{k} c_i h(B) = h(B).$$

Since $h(AB) = h(B)$, we have $\mathrm{trace}(P_i^{\mathrm{T}} B) = h(B)$ for $i = 1, 2, \ldots, k$. Now it follows from Birkhoff's theorem (more precisely, its proof) that if the diagonal of A corresponding to a permutation matrix Q is positive, then we may choose a convex combination (9.35) equal to A where $P_1 = Q$. Hence $\mathrm{trace}(Q^{\mathrm{T}} B) = h(B)$ for each permutation matrix Q corresponding to a positive diagonal of A.

Now assume that $\mathrm{trace}(Q^{\mathrm{T}} B) = \mathrm{trace}(B)$ for every permutation matrix Q corresponding to a positive diagonal of A. Taking a convex combination of permutation matrices equal to A as in (9.35), we have $\mathrm{trace}(P_i^{\mathrm{T}} B) = h(B)$ for $i = 1, 2, \ldots, k$ and hence

$$\mathrm{trace}(AB) = \sum_{i=1}^{k} c_i \mathrm{trace}(P_i^{\mathrm{T}} B) = \sum_{i=1}^{k} c_i h(B) = h(B).$$

We also have

$$\mathrm{trace}(AB) \le h(AB)$$
$$= \max_{P} \{\mathrm{trace}(P^{\mathrm{T}} AB)\}$$

$$= \max_P \left\{ \sum_{i=1}^{k} c_i \text{trace}(P^T P_i^T B) \right\}$$

$$\leq \sum_{i=1}^{k} c_i h(B) = h(B),$$

where the maxima are taken over all permutation matrices P of order n. Therefore

$$h(B) = \text{trace}(AB) \leq h(AB) \leq h(B),$$

and hence

$$h(AB) = \text{trace}(AB) = h(B).$$

$\square$

The following important corollary is an immediate consequence of Theorem 9.12.1. In less formal terms it asserts that a diagonal of B formed solely out of positions occupied by maximal-sum diagonals of B is itself a maximal-sum diagonal. A different proof is given in [38] based on some results in [121].

Corollary 9.12.2 *Let $B = [b_{ij}]$ be a real matrix of order n, and assume that $P_1, P_2, \ldots, P_k$ are permutation matrices such that $P \circ B = h(B)$. Let $D = [d_{ij}]$ be the $(0,1)$-matrix where $d_{ij} = 1$ if and only if at least one of $P_1, P_2, \ldots, P_k$ has a 1 in position (i,j) $(i, j = 1, 2, \ldots, n)$. Then if P is any permutation matrix with $P \leq D$, then $P \circ B = h(B)$.*

Proof. We note that the matrix D has total support. Let $A = (1/k)(P_1^T + P_2^T + \cdots + P_k^T) \in \Omega_n$. We have $\text{trace}(P_i^T B) = h(P_i^T B) = h(B)$ for $i = 1, 2, \ldots, k$ and hence $\text{trace}(AB) = h(AB) = h(B)$. Theorem 9.12.1 now implies that if P is any permutation matrix with $P \leq D$, then $P^T \circ B = h(B)$ $\square$

In investigating $h(B)$ and the maximal-sum diagonals of B, there is no loss in generality in assuming that B is a positive matrix. This is because we may choose a sufficiently large positive number c such that the matrix $cJ_n + B$ is positive. We have $h(cJ_n + B) = nc + h(B)$ and, in addition, for each permutation σ of $\{1, 2, \ldots, n\}$ the diagonal of B corresponding to σ is a maximal-sum diagonal if and only if the diagonal of $cJ_n + B$ corresponding to σ is a maximal-sum diagonal.

Let $B = [b_{ij}]$ be a positive matrix of order n and let $M = [m_{ij}]$ be the nonnegative matrix obtained from B by defining, for $i, j = 1, 2, \ldots, n$,

$$m_{ij} = \begin{cases} b_{ij}, & \text{if } b_{ij} \text{ belongs to a maximal-sum diagonal of } B, \\ 0, & \text{otherwise.} \end{cases}$$

The matrix M has total support, and it follows from Corollary 9.12.2 that $h(B) = h(M)$ and that the maximal-sum diagonals of B are in one-to-one correspondence with the positive diagonals of M. In particular, all of the positive diagonals of M have the same sum. We say that such a nonnegative matrix M has *positive diagonals of constant sum*. One way to obtain a matrix of total support with positive diagonals of constant sum is the following.

Let $R = (r_1, r_2, \ldots, r_n)$ and $S = (s_1, s_2, \ldots, s_n)$ be two n-tuples of positive numbers. Let $U = [u_{ij}]$ be a $(0,1)$-matrix of order n with total support. Let $A = [a_{ij}]$ be the matrix obtained from U by defining, for $i, j = 1, 2, \ldots, n$,

$$a_{ij} = \begin{cases} r_i + s_j, & \text{if } u_{ij} = 1, \\ 0, & \text{if } u_{ij} = 0. \end{cases}$$

Then the positive diagonals of A all have the sum

$$\sum_{i=1}^{n} r_i + \sum_{j=1}^{n} s_j.$$

Sinkhorn and Knopp [133] proved the converse.[21] Another proof and more general results can be found in [59].

Theorem 9.12.3 *Let $A = [a_{ij}]$ be a real matrix of total support for which the nonzero diagonals have constant sum. Then there exist n-tuples $R = (r_1, r_2, \ldots, r_n)$ and $S = (s_1, s_2, \ldots, s_n)$ such that $a_{ij} = r_i + s_j$ for all i and j satisfying $a_{ij} \neq 0$.*

Proof. It suffices to prove the theorem under the assumption that A is fully indecomposable. If $n = 1$, the conclusion is obvious. Let $n \geq 2$. The matrix A has at least $2n$ nonzero elements. If A has exactly $2n$ nonzero elements, then we may assume that

$$A = \begin{bmatrix} a_1 & b_1 & 0 & \cdots & 0 \\ 0 & a_2 & b_2 & \cdots & 0 \\ 0 & 0 & a_3 & \cdots & 0 \\ \vdots & \vdots & \vdots & \ddots & \vdots \\ b_n & 0 & 0 & \cdots & a_n \end{bmatrix}$$

---

[21] Sinkhorn and Knopp proved that if the positive diagonal products of a nonnegative matrix A of order n with total support are all equal and without loss of generality assumed to be 1, then there exist diagonal matrices D and E with positive main diagonals such that $A = DUE$ where U is a $(0, 1)$-matrix (necessarily the $(0, 1)$-matrix obtained from A by replacing each positive element with a 1). Let A' be the matrix obtained from A by replacing each positive element a_{ij} (all of which can be assumed to be greater than 1 without loss of generality) with its logarithm $a'_{ij} = \ln a_{ij}$. Then all the positive diagonals of A have the same product if and only if all the nonzero diagonals of A' have the same sum. Let D' and E' be the diagonal matrices obtained from D and E by replacing the positive diagonal elements d_i and e_i on the main diagonals of D and E, respectively, with $d'_i = \ln d_i$ and $e'_i = \ln e_i$. Then $d'_i + a'_{ij} + e'_j = 0$, equivalently $a'_{ij} = (-d'_i) + (-e'_j)$, whenever $a'_{ij} \neq 0$. This whole argument is reversible using the exponential function.

where $(a_1, a_2, \ldots, a_n)$ and $(b_1, b_2, \ldots, b_n)$ are the only two nonzero diagonals, and $a_1 + a_2 + \cdots + a_n = b_1 + b_2 + \cdots + b_n$. In this case we let

$$r_k = \sum_{i=1}^{k} a_i - \sum_{i=1}^{k-1} b_i \quad \text{and} \quad s_k = \sum_{i=1}^{k-1} b_i - \sum_{i=1}^{k-1} a_i \quad (k = 1, 2, \ldots, n).$$

We have $r_k + s_k = a_k$ and $r_k + s_{k+1} = b_k$ (subscripts taken modulo n when appropriate). We now proceed by induction on the number of nonzero elements and assume that A has at least $2n + 1$ nonzero elements. Let the constant sum of the nonzero diagonals of A be c.

Case 1: there is a nonzero element a_{pq} of A such that replacing a_{pq} by 0 gives a fully indecomposable matrix $B = [b_{ij}]$. Without loss of generality we assume that a_{pq} is a_{11}. Each nonzero diagonal of B sums to c and hence by induction there exist $R = (r_1, r_2, \ldots, r_n)$ and $S = (s_1, s_2, \ldots, s_n)$ such that $b_{ij} = r_i + s_j$ whenever $b_{ij} > 0$. Since A is fully indecomposable there is a nonzero diagonal of A which contains a_{11} and without loss of generality assume this nonzero diagonal comes from the main diagonal of A. Hence $a_{11} = c - (a_{22} + \cdots + a_{nn})$. Since $\sum_{i=1}^{n}(r_i + s_i) = c$, it follows that $a_{11} = r_1 + s_1$. Therefore $a_{ij} = r_i + s_j$ whenever $a_{ij} \neq 0$.

Case 2: if Case 1 does not hold, then A is nearly decomposable. We may assume that A has the form given in Theorem 1.3.6:

$$A = \begin{bmatrix} a_1 & 0 & 0 & \cdots & 0 & 0 & \\ b_1 & a_2 & 0 & \cdots & 0 & 0 & \\ 0 & b_2 & a_3 & \cdots & 0 & 0 & \\ \vdots & \vdots & \vdots & \ddots & \vdots & \vdots & F_1 \\ 0 & 0 & 0 & \cdots & a_{n-m-1} & 0 & \\ 0 & 0 & 0 & \cdots & b_{n-m-1} & a_{n-m} & \\ \hline & & F_2 & & & & A_1 \end{bmatrix} \qquad (9.36)$$

where A_1 is a nearly decomposable matrix of order $m \geq 1$, the matrix F_1 contains exactly one nonzero element f_1 and it is in its first row and last column, the matrix F_2 contains exactly one nonzero element f_2 and it is in its last row and last column, and if $m \neq 1$, then $m \geq 3$ and the element in the last row and last column of A_1 is a 0.

If $m = 1$, then A has exactly $2n$ nonzero elements, a situation already treated above. Hence we may assume that $m \geq 3$ and the element in the last row and last column of A_1 equals 0. All the nonzero diagonals of A_1 have sum equal to $c - \sum_{i=1}^{n-m} a_i$. There is a nonzero diagonal of A containing the elements $f_1, f_2, b_1, b_2, \ldots, b_{n-m-1}$ and this implies that the matrix $A_1' = [a_{ij}']$ obtained from A_1 by replacing the 0 in its last row and last column with $d = f_1 + f_2 + b_1 + b_2 + \cdots + b_{n-m-1}$ has the property that each of its nonzero diagonals has the same sum. By induction there exist vectors $R' = (r_1', r_2', \ldots, r_m')$ and $S' = (s_1, s_2, \ldots, s_m)$ such that $a_{ij}' = r_i' + s_j'$ whenever $a_{ij}' \neq 0$. To complete the proof we argue as we did when A contained exactly $2n$ nonzero elements (apply a similar argument to the matrix of order $n - m + 1$ obtained from A in (9.36) by "shrinking" A_1 to the matrix of order 1 whose unique element is d). $\qquad\qquad\square$

Referring to Theorem 9.12.3, let $B = [b_{ij}]$ be the matrix of order n where $b_{ij} = r_i + s_j$ for $i, j = 1, 2, \ldots, n$. The sum of the elements in each diagonal of B equals the constant sum of the nonzero diagonals of A. Since A is fully indecomposable, each submatrix $A(i, j)$ of A of order $n - 1$, complementary to the element a_{ij} of A, contains a nonzero diagonal. It follows that each b_{ij} is uniquely determined by a nonzero diagonal of $A(i, j)$. Hence the matrix B is uniquely determined by A. The n-tuples $R = (r_1, r_2, \ldots, r_n)$ and $S = (s_1, s_2, \ldots, s_n)$ are not uniquely determined since we may replace R with $R' = (r_1 - t, r_2 - t, \ldots, r_n - t)$ and S with $S = (s_1 + t, s_2 + t, \ldots, s_n + t)$ for any real number t. In fact, the full indecomposability of A implies that any vectors $R' = (r_1', r_2', \ldots, r_n')$ and $S' = (s_1', s_2', \ldots, s_n')$ satisfying $a_{ij} = r_i' + s_j'$ whenever $a_{ij} \neq 0$ are obtained from R and S in this way.

The multiplicative version of Theorem 9.12.3 is contained in the next corollary.

Corollary 9.12.4 *Let $A = [a_{ij}]$ be a nonnegative matrix of total support for which the positive diagonals have constant product. Then there exist positive n-tuples $R = (r_1, r_2, \ldots, r_n)$ and $S = (s_1, s_2, \ldots, s_n)$ such that $a_{ij} = r_i s_j$ for all i and j for which $a_{ij} > 0$. If D_1 and D_2 are the diagonal matrices whose main diagonals are $(1/r_1, 1/r_2, \ldots, 1/r_n)$ and $(1/s_1, 1/s_2, \ldots, 1/s_n)$, respectively, then $D_1 A D_2$ is a $(0,1)$-matrix B. The diagonal matrices D_1 and D_2 are unique up to scalar multiples.*

Proof. The existence of the diagonal matrices D_1 and D_2 follows from Theorem 9.12.3. There is no loss in generality in assuming that A is fully indecomposable, and then that the main diagonal of A is a positive diagonal. Suppose that E_1 and E_2 are diagonal matrices with positive diagonal elements such that $E_1 A E_2 = B$. Then $(D_1 E_1^{-1}) A (E_2^{-1} D_2) = A$. Let $D = D_1 E_1^{-1}$ and $E = E_2^{-1} D_2$. Then $DAE = A$. Since the main diagonal contains only positive elements, it follows that $E = D^{-1}$. We now show $D = cI_n$ for some nonzero number c. If not, then without loss of generality, we may assume that $d_1 = \cdots = d_k \neq d_{k+1}, \ldots, d_n$ where $(d_1, d_2, \ldots, d_n)$ is the main diagonal of D and k is an integer with $1 \leq k \leq n - 1$. Then $DAD^{-1} = A$ implies that $A[\{1, 2, \ldots, k\}, \{k+1, k+2, \ldots, n\}] = O$, contradicting the full indecomposability of A. Hence $D = cI_n$, and $D_1 E_1^{-1} = cI_n$ so that $D_1 = cE_1$ and $D_2 = c^{-1} E_2$. $\qquad\qquad\square$

For doubly stochastic matrices we have the following.

Corollary 9.12.5 *Distinct doubly stochastic matrices of order n do not have proportional diagonal products.*

Proof. Let A and B be doubly stochastic matrices of order n whose corresponding diagonal products are proportional:

$$\prod_{i=1}^{n} a_{i\sigma(i)} = d \prod_{i=1}^{n} b_{i\sigma(i)}, \quad \text{for all permutations } \sigma \text{ of } \{1, 2, \ldots, n\} \ .$$

Here d is some positive number. Since doubly stochastic matrices have total support, $a_{ij} > 0$ if and only if $b_{ij} > 0$. Let $C = [c_{ij}]$ be the matrix of order n defined by

$$c_{ij} = \begin{cases} \frac{a_{ij}}{b_{ij}} & \text{if } b_{ij} > 0, \\ \\ 0 & \text{if } b_{ij} = 0. \end{cases}$$

Then all the positive diagonal products of C have value 1. By Corollary 9.12.4 there are diagonal matrices D and E with positive main diagonals such that DCE is a (0,1)-matrix. This implies that $DAE = B$. Since A and B are both doubly stochastic, it now is a consequence of the uniqueness assertion of Corollary 9.11.6 that $A = B$. $\quad\square$

We now turn to a different problem concerning diagonal sums of doubly stochastic matrices. Let A be a matrix of order n. Two diagonals of A are *disjoint diagonals* provided they do not contain an element from the same position of A. A diagonal of A is disjoint from a permutation matrix P provided it does not contain an element from a position occupied by a 1 in P, equivalently, provided it is disjoint from the diagonal of A corresponding to P. Similarly, two permutation matrices are disjoint provided they do not contain a 1 in the same position. More generally, if X is a set of positions of A, then a diagonal of A (respectively, a permutation matrix of order n) is disjoint from X provided it does not contain an element from (respectively, a 1 in) a position of X. The following theorem arose from a conjecture of Wang [142] which was proved independently by Sinkorn [130] and Balasubramanian [4]. The more general theorem below is from Achilles [1].

For a matrix A, $\zeta(A)$ denotes the set of positions of A containing zeros.

Theorem 9.12.6 *Let A and B be doubly stochastic matrices of order n. Let $X \subseteq \zeta(A)$ and $Y \subseteq \zeta(B)$. Assume that all diagonals of A disjoint from X have the same sum d_A, and that all diagonals of B disjoint from Y have the same sum d_B. If $X \subseteq Y$, then $d_A \leq d_B$. If $X = Y$, then $A = B$.*

Proof. First assume that $X \subseteq Y$. Let $\mathcal{P}(A)$ denote the set of all permutation matrices P such that P is disjoint from X, and let $\mathcal{P}(B)$ be defined in a similar way. Then $P \circ A = d_A$ for $P \in \mathcal{P}(A)$ and $P \circ B = d_B$ for $P \in \mathcal{P}(B)$. It follows from Birkhoff's theorem that

$$A = \sum_{P \in \mathcal{P}(A)} e_P P$$

where $0 \leq e_P \leq 1$ for $P \in \mathcal{P}(A)$ and $\sum_{P \in \mathcal{P}(A)} e_P = 1$, and that

$$B = \sum_{P \in \mathcal{P}(B)} f_P P$$

where $0 \leq f_P \leq 1$ for $P \in \mathcal{P}(B)$ and $\sum_{P \in \mathcal{P}(B)} f_P = 1$. Then

$$A \circ A = A \circ \left(\sum_{P \in \mathcal{P}(A)} e_P P \right)$$

$$= \sum_{P \in \mathcal{P}(A)} e_P (A \circ P)$$

$$= \sum_{P \in \mathcal{P}(A)} e_P d_A = d_A.$$

Similarly, $B \circ B = d_B$.

Since $X \subseteq Y$, we have $A \circ P = d_A$ for $P \in \mathcal{P}(B)$, and hence

$$A \circ B = A \circ \left(\sum_{P \in \mathcal{P}(B)} f_P P \right)$$

$$= \sum_{P \in \mathcal{P}(B)} f_P d_A = d_A.$$

Using the Cauchy–Schwarz inequality, we obtain

$$0 \leq (A - B) \circ (A - B) = A \circ A - 2A \circ B + B \circ B$$

$$= d_A - 2d_A + d_B$$

$$= d_B - d_A. \tag{9.37}$$

Hence $d_A \leq d_B$. If $X = Y$, then we also get that $d_B \leq d_A$ and hence $d_A = d_B$; now from (9.37) we get that $A = B$. □

The following corollary is from [130], [4].

Corollary 9.12.7 *Let A be a doubly stochastic matrix of order n. Let m be an integer with $1 \leq m < n$, and let $P_1, P_2, \ldots, P_m$ be pairwise disjoint permutation matrices of order n corresponding to identically zero diagonals of A. If every diagonal of A disjoint from $P_1, P_2, \ldots, P_m$ has the same sum, then all elements of A off these m identically zero diagonals equal $1/(n-m)$.*

Proof. Let X be the set of positions of A occupied by the m identically zero diagonals. Let $B = [b_{ij}]$ be the doubly stochastic matrix of order n defined by

$$b_{ij} = \begin{cases} 0 & \text{if } (i,j) \in X, \\ \frac{1}{n-m} & \text{otherwise.} \end{cases}$$

Then $\zeta(B) = X$ and $d_B = n/(n-m)$. Taking $Y = X$ in Theorem 9.12.6, we conclude that $A = B$. □

We remark here that in [121], [38], [143] a class of matrices with monomial elements of the form x^k, intimately related to the optimal assignment problem, is investigated. A normalized form, motivated by the optimal assignment problem, is defined and it is shown that each class contains only finitely many normalized forms.

We now turn to an interesting application of doubly stochastic matrices to SS/TDMA systems as related in [21] (see also [41]). Consider a communication satellite equipped with a certain number k of transponders which connect *source stations* to *destination stations*. The satellite is able to process on board so that in parallel the transponders can be switched to connect upbeams from any k source stations to any k destination stations. Each source station has a certain amount of information, called *traffic*, that it needs to transmit to each destination station. The problem is to devise an efficient and conflict-free scheme for transmitting all the specified traffic. A conflict-free scheme is one in which at any time each transponder links at most one source station to a destination station. Traffic between source and destination stations is measured in terms of a unit of channel capacity called a *slot time*. A traffic scheme's efficiency is measured in terms of the total number of slot times it takes to transmit all the specified traffic from source to destination stations, and the time necessary to reconfigure the system to link different source stations with different destination stations.

Let there be m source stations $X_1, X_2, \ldots, X_m$ and n destination stations $Y_1, Y_2, \ldots, Y_n$. Traffic requirements can then be specified by an m by n nonnegative integral matrix $T = [t_{ij}]$ where t_{ij} equals the amount of traffic to be transmitted from X_i to Y_j ($i = 1, 2, \ldots, m; j = 1, 2, \ldots, n$). At any given time each of the k transponders connects a source station to a destination station or is idle. Such a *switching configuration* can be specified by an m by n subpermutation matrix P of rank at most k.[22] The amount of traffic sent in a switching configuration is specified by a nonnegative integral matrix with at most k positive elements and with no two positive elements in the same row or column. This matrix is of the form DP where D is a diagonal matrix of order m. We call such a matrix a *k-switching matrix*.

Example. Suppose that $m = 4, n = 5$, and $k = 3$, and let the traffic matrix be

$$T = \begin{bmatrix} 3 & 2 & 4 & 2 & 0 \\ 0 & 5 & 0 & 3 & 0 \\ 0 & 0 & 3 & 4 & 2 \\ 2 & 0 & 2 & 0 & 6 \end{bmatrix}.$$

---

[22] This subpermutation matrix specifies which source stations are linked by a transponder to which destination stations, but does not specify how the transponders are assigned.

The traffic between source and destination stations can be accommodated by decomposing T into a sum of 3-switching matrices:

$$T = \begin{bmatrix} 0 & 0 & 4 & 0 & 0 \\ 0 & 5 & 0 & 0 & 0 \\ 0 & 0 & 0 & 0 & 0 \\ 0 & 0 & 0 & 0 & 6 \end{bmatrix} + \begin{bmatrix} 3 & 0 & 0 & 0 & 0 \\ 0 & 0 & 0 & 3 & 0 \\ 0 & 0 & 3 & 0 & 0 \\ 0 & 0 & 0 & 0 & 0 \end{bmatrix}$$

$$+ \begin{bmatrix} 0 & 2 & 0 & 0 & 0 \\ 0 & 0 & 0 & 0 & 0 \\ 0 & 0 & 0 & 4 & 0 \\ 2 & 0 & 0 & 0 & 0 \end{bmatrix} + \begin{bmatrix} 0 & 0 & 0 & 2 & 0 \\ 0 & 0 & 0 & 0 & 0 \\ 0 & 0 & 0 & 0 & 2 \\ 0 & 0 & 2 & 0 & 0 \end{bmatrix}. \qquad (9.38)$$

Here we have four switching configurations each with transmission time equal to its maximal element. The total transmission time is the sum of the maximal elements in each of the switching configurations, and so is $15 = 6 + 3 + 4 + 2$.

An alternative way to accommodate the traffic in this case is to use the decomposition into five 3-switching matrices:

$$T = \begin{bmatrix} 3 & 0 & 0 & 0 & 0 \\ 0 & 3 & 0 & 0 & 0 \\ 0 & 0 & 3 & 0 & 0 \\ 0 & 0 & 0 & 0 & 0 \end{bmatrix} + \begin{bmatrix} 0 & 0 & 0 & 2 & 0 \\ 0 & 2 & 0 & 0 & 0 \\ 0 & 0 & 0 & 0 & 0 \\ 2 & 0 & 0 & 0 & 0 \end{bmatrix}$$

$$+ \begin{bmatrix} 0 & 0 & 4 & 0 & 0 \\ 0 & 0 & 0 & 0 & 0 \\ 0 & 0 & 0 & 4 & 0 \\ 0 & 0 & 0 & 0 & 4 \end{bmatrix} + \begin{bmatrix} 0 & 2 & 0 & 0 & 0 \\ 0 & 0 & 0 & 2 & 0 \\ 0 & 0 & 0 & 0 & 0 \\ 0 & 0 & 0 & 0 & 2 \end{bmatrix}$$

$$+ \begin{bmatrix} 0 & 0 & 0 & 0 & 0 \\ 0 & 0 & 0 & 1 & 0 \\ 0 & 0 & 0 & 0 & 2 \\ 0 & 0 & 2 & 0 & 0 \end{bmatrix}. \qquad (9.39)$$

With this decomposition, the total transmission time is $13 = 3+2+4+2+2$.

□

The *duration* of a particular traffic scheme can be measured by $T_k + st_k$ where T_k is the total transmission time, s is the number of k-switching matrices, and t_k is the time needed to configure or reconfigure the system to match certain of the source stations, transponders, and destination stations. The two traffic schemes in the above example have durations $15 + 4t_3$ and $13+5t_3$, respectively. Which is more efficient depends on the reconfiguration time t_3. In general, when t_k is small, the duration can be approximated by the smallest transmission time T_k^* taken over all traffic schemes with k-switching matrices that can accommodate the traffic; when t_k is large, it can be approximated by $T_k^{**} + s_k$ where s_k is the smallest number of

k-switching matrices needed to accommodate the traffic, and T_k^{**} is the smallest transmission time taken over all traffic schemes with s_k switching matrices. In what follows we determine T_k^* and s_k.

Theorem 9.12.8 *Let $T = [t_{ij}]$ be an m by n nonnegative integral matrix, and let $k \leq \min\{m, n\}$ be a positive integer. Let M be the maximal row and column sum of T, and let $\sigma(T)$ be the sum of all the elements of T. The minimal transmission time T_k^* for T equals*

$$\min \left\{ M, \left\lceil \frac{\sigma(T)}{k} \right\rceil \right\}. \tag{9.40}$$

Proof. The quantity q in (9.40) is easily seen to be a lower bound for T_k^*, and so we need only prove there is a traffic scheme with transmission time q. By attaching zero rows or zero columns we may assume that T is a square matrix of order n with $k \leq n$. We have $\sigma(T) \leq qk$. If $\sigma(T) < qk$, then there exist i and j such that the ith row sum and the jth column sum of T are strictly less than M. Increasing the element of T in position (i, j) and proceeding inductively, we arrive at a nonnegative integral matrix T' with maximal row and column sum no larger than q and $\sigma(T') = qk$.

The matrix $T'' = (1/q)T'$ is a doubly substochastic matrix of order n. By Theorem 9.8.4, T'' is a convex combination of subpermutation matrices of rank k, and each such subpermutation contains a 1 in every row or column of T'' that sums to 1. Since T' is an integral matrix, there is a subpermutation matrix P of rank k such that $P \leq T'$ and $T' - P$ has maximal row and column sums no larger than $q - 1$. We have $\sigma(T' - P) = (q - 1)k$, and proceeding inductively we are able to write T' as the sum of q subpermutation matrices of rank at most k. It now follows that T can be written as a sum of subpermutation matrices of rank at most k. This gives a traffic scheme with total transmission time q. $\qquad\square$

In a similar way one proves the following theorem.

Theorem 9.12.9 *Let $T = [t_{ij}]$ be an m by n nonnegative integral matrix, let K be the maximal number of positive elements in a row or column of T, and let $\kappa(T)$ be the total number of positive elements of T. The minimal number s_k of k-switching matrices needed to accommodate the traffic in T equals*

$$\min \left\{ K, \left\lceil \frac{\kappa(T)}{k} \right\rceil \right\}. \tag{9.41}$$

$\qquad\square$

We remark that the determination of the numbers T_k^{**} is apparently a computationally difficult problem, since the problem in general is NP-complete [67]. A generalization of SS/TDMA systems is given in [82].

9.13 Permanent of Doubly Stochastic Matrices

We recall that the permanent of a matrix $A = [a_{ij}]$ of order n is defined by

$$\text{per}(A) = \sum a_{1i_1} a_{2i_2} \cdots a_{ni_n}$$

where the summation extends over all permutations of $\{1, 2, \ldots, n\}$. Since a doubly stochastic matrix is a convex combination of permutations matrices, its permanent is positive. Since Ω_n is a polytope, it is compact and hence

$$\min\{\text{per}(A) : A \in \Omega_n\}$$

exists and is a positive number. In 1926 van der Waerden [141] posed the problem of determining this minimum and *conjectured* that

$$\min\{\text{per}(A) : A \in \Omega_n\} = \frac{n!}{n^n}$$

and that the only doubly stochastic matrix with this minimum permanent is the matrix $(1/n)J_n$ of order n each of whose elements equals $1/n$. This conjecture motivated considerable interest in the permanent resulting in many partial results. In 1981 it was independently resolved by Egoryčev [58] and Falikman [61]. The interested reader can consult the book [102] for further historical information and additional references.

In this section we give a proof of the van der Waerden conjecture based on the proofs of Egoryčev and Falikman and as simplified in [100], [102] (see also [8]).

A doubly stochastic matrix A of order n is a *minimizing matrix* (with respect to the permanent) provided $\text{per}(A) \leq \text{per}(B)$ for all $B \in \Omega_n$. We first derive some properties of minimizing matrices.

The two lemmas are due to Marcus and Newman [88].

Lemma 9.13.1 *Let A be a minimizing matrix in Ω_n. Then A is fully indecomposable.*

Proof. Suppose to the contrary $A = [a_{ij}]$ is partly decomposable. Since a doubly stochastic matrix remains doubly stochastic after row and column permutations, and has total support, we assume without loss of generality that $A = [a_{ij}] = B \oplus C$ where $B = [b_{ij}] \in \Omega_k$ for some k with $1 \leq k \leq n-1$, $C = [c_{ij}] \in \Omega_{n-k}$, and $b_{kk} > 0$ and $c_{11} > 0$. The matrices B and C also have total support and hence $\text{per}(B(k,k)) > 0$ and $\text{per}(C(1,1)) > 0$. For ϵ a small positive number, the matrix $A(\epsilon)$ obtained from A by replacing b_{kk} and c_{11} by $b_{kk} - \epsilon$ and $c_{11} - \epsilon$, respectively, and $a_{k,k+1}$ and $a_{k+1,k}$ by ϵ is a doubly stochastic matrix.

A straightforward calculation establishes that

$$\text{per}(A(\epsilon)) = \text{per}(A) - \epsilon(\text{per}(A(k,k)) + \text{per}(A(k+1,k+1))) + \epsilon^2 f(A)$$

where $f(A)$ is some function of A. Since B and C have total support, $\mathrm{per}(A(k,k)) + \mathrm{per}(A(k+1,k+1)) > 0$. Hence we may choose ϵ small enough so that $\mathrm{per}(A(\epsilon)) < \mathrm{per}(A)$, a contradiction. We conclude that A is fully indecomposable. $\qquad\qquad\qquad\qquad\qquad\qquad\qquad\qquad\qquad\qquad\square$

Lemma 9.13.2 *Let $A = [a_{ij}]$ be a minimizing matrix in Ω_n. If $a_{ij} > 0$, then $\mathrm{per}(A(i,j)) = \mathrm{per}(A)$.*

Proof. We apply the method of Lagrange multipliers to the smallest face containing A, that is, to the face $\Omega_n(F)$ where $F = [f_{ij}]$ is the $(0,1)$-matrix of order n obtained from A by replacing each positive element with 1. Let $W = \{(i,j) : f_{ij} = 1\}$. The matrix A is in the interior of this face and the permanent function achieves an absolute minimum on $\Omega_n(F)$ at A. A matrix X belongs to $\Omega_n(F)$ if and only if the following constraints are satisfied:

$$x_{ij} \geq 0, \quad \text{for all } i,j = 1,2,\ldots,n, \text{ with equality if } (i,j) \notin W,$$

$$\sum_{j=1}^{n} x_{ij} = 1 \quad (i = 1,2,\ldots,n),$$

$$\sum_{i=1}^{n} x_{ij} = 1 \quad (j = 1,2,\ldots,n).$$

For $X \in \Omega_n(F)$, let

$$g(x_{ij} : (i,j) \in W) = \mathrm{per}(X) + \sum_{i=1}^{n} \lambda_i \left(1 - \sum_{j=1}^{n} x_{ij}\right) + \sum_{j=1}^{n} \mu_j \left(1 - \sum_{i=1}^{n} x_{ij}\right).$$

We have

$$\frac{\partial g(x_{ij} : (i,j) \in W)}{\partial x_{ij}} = \mathrm{per}(X(i,j)) - \lambda_i - \mu_j.$$

Hence, since the permanent has a minimum at A,

$$\mathrm{per}(A(i,j)) = \lambda_i + \mu_j \quad ((i,j) \in W). \tag{9.42}$$

Expanding the permanent along rows and columns of A, we get the two equations

$$\mathrm{per}(A) = \sum_{j=1}^{n} a_{ij}(\lambda_i + \mu_j) = \lambda_i + \sum_{j=1}^{n} a_{ij}\mu_j \quad (1 \leq i \leq n), \tag{9.43}$$

$$\mathrm{per}(A) = \sum_{i=1}^{n} a_{ij}(\lambda_i + \mu_j) = \mu_j + \sum_{i=1}^{n} a_{ij}\lambda_i \quad (1 \leq j \leq n). \tag{9.44}$$

With $e = (1, 1, \ldots, 1)$ (the row vector of all 1's of size n), $\lambda = (\lambda_1, \lambda_2, \ldots, \lambda_n)$, and $\mu = (\mu_1, \mu_2, \ldots, \mu_n)$, we get

$$\text{per}(A)e = \lambda + \mu A^{\text{T}}, \tag{9.45}$$
$$\text{per}(A)e = \lambda A + \mu. \tag{9.46}$$

Multiplying (9.45) on the right by A and (9.46) by A^{T}, we get

$$\text{per}(A)e = \lambda A + \mu A^{\text{T}} A \text{ and } \text{per}(A)e = \lambda A A^{\text{T}} + \mu A^{\text{T}}.$$

Subtracting from (9.46) and (9.45), respectively, we get

$$\mu A^{\text{T}} A = \mu \text{ and } \lambda A A^{\text{T}} = \lambda. \tag{9.47}$$

Since AA^{T} and $A^{\text{T}}A$ are also doubly stochastic, we also have

$$e A^{\text{T}} A = e \text{ and } e A^{\text{T}} A = e. \tag{9.48}$$

The matrices $A^{\text{T}}A$ and AA^{T}, being the product of two fully indecomposable, doubly stochastic matrices, are fully indecomposable, doubly stochastic matrices.[23] Since AA^{T} and $A^{\text{T}}A$ are fully indecomposable, they are irreducible, and hence by the Perron–Frobenius theory of nonnegative matrices, 1 is a simple eigenvalue of each of them. Hence from (9.47) and (9.48) we get that λ and μ are multiplies of e:

$$\lambda = ce \text{ and } \mu = de.$$

Returning to (9.42), we now get

$$\text{per}(A(i,j)) = c + d \quad ((i,j) \in W).$$

Now from (9.44) or (9.45) we get, as A is doubly stochastic,

$$\text{per}(A) = c + d = \text{per}(A(i,j)) \quad ((i,j) \in W).$$

$$\square$$

The next lemma is due to London [84] who showed that if A is a minimizing matrix, then all the permanents $\text{per}\, A(k, l)$ are at least as large as the permanent of A. The proof given here is due to Minc [98].

Lemma 9.13.3 *Let $A = [a_{ij}]$ be a minimizing matrix in Ω_n. Then*

$$\text{per}(A(k,l)) \geq \text{per}(A) \quad (k, l = 1, 2, \ldots, n).$$

---

[23]For a proof of the full indecomposability of AA^{T}, write A^{T} as a convex combination of permutation matrices and note that, for some $\epsilon > 0$ and permutation matrix P, $AA^{\text{T}} \geq \epsilon AP$.

Proof. By symmetry it suffices to show that $\operatorname{per}(A(1,1)) \geq \operatorname{per}(A)$. By Lemma 9.13.1, A is a fully indecomposable matrix. It follows from Theorem 1.3.3 that $\operatorname{per}(A(1,1)) > 0$. Without loss of generality we may assume that $a_{22}a_{33}\ldots a_{nn} > 0$. Consider the function

$$f(t) = \operatorname{per}(tI_n + (1-t)A) \quad (0 \leq t \leq 1).$$

Since A is a minimizing matrix, we have $f'(0) \geq 0$ (derivative from the right). An easy calculation shows that

$$f'(0) = \sum_{i=1}^{n} \operatorname{per}(A(i,i)) - n\operatorname{per}(A) \geq 0. \tag{9.49}$$

Since $a_{ii} > 0$ for $i = 2, \ldots, n$, Lemma 9.13.2 implies that $\operatorname{per}(A(i,i)) = \operatorname{per}(A)$ for $i = 2, \ldots, n$. Hence (9.49) now gives that $\operatorname{per}(A(1,1)) \geq \operatorname{per}(A)$. $\qquad\square$

The following lemma is also from [88].

Lemma 9.13.4 *Let $A = [a_{ij}]$ be a matrix of order n, and let k and l be integers with $1 \leq k < l \leq n$. Suppose that*

$$\operatorname{per}(A(i,k)) = \operatorname{per}(A(i,l)) \quad (i = 1, 2, \ldots, n). \tag{9.50}$$

Then the matrix B obtained from A by replacing columns k and l by their average satisfies $\operatorname{per}(B) = \operatorname{per}(A)$.

Proof. Without loss of generality we assume that $k = 1$ and $l = 2$. Write A in terms of its columns as $A = [\alpha_1, \alpha_2, \ldots, \alpha_n]$. Using the multilinearity of the permanent, we obtain

$$\begin{aligned}
\operatorname{per}(B) &= \operatorname{per}(A[(\alpha_1+\alpha_2)/2, (\alpha_1+\alpha_2)/2, \alpha_3, \ldots, \alpha_n]) \\
&= \operatorname{per}(A[\alpha_1, \alpha_1, \alpha_3, \ldots, \alpha_n])/4 + \operatorname{per}(A[\alpha_1, \alpha_2, \alpha_3, \ldots, \alpha_n])/4 \\
&\quad + \operatorname{per}(A[\alpha_2, \alpha_1, \alpha_3, \ldots, \alpha_n])/4 + \operatorname{per}(A[\alpha_2, \alpha_2, \alpha_3, \ldots, \alpha_n])/4.
\end{aligned}$$

Expanding each of the four permanents above by column 1 or 2 and using (9.50), we obtain that $\operatorname{per}(B) = \operatorname{per}(A)$. $\qquad\square$

We now prove a lemma of Falikman [61].

Lemma 9.13.5 *Let n be an integer with $n \geq 2$, and let*

$$A = [a_{ij}] = [\alpha_1, \alpha_2, \ldots, \alpha_{n-1}]$$

be a positive n by $n-1$ matrix with columns $\alpha_1, \alpha_2, \ldots, \alpha_{n-1}$. Let $\beta = (b_1, b_2, \ldots, b_n)^{\mathrm{T}}$ be a real column vector. Suppose that

$$\operatorname{per}([\alpha_1, \ldots, \alpha_{n-2}, \alpha_{n-1}, \beta]) = 0.$$

Then

$$\operatorname{per}([\alpha_1, \ldots, \alpha_{n-2}, \beta, \beta]) \leq 0$$

with equality if and only if $\beta = 0$.

Proof. We consider the bilinear form

$$f(x, y) = \text{per}([\alpha_1, \ldots, \alpha_{n-2}, x, y])$$

where $x = (x_1, x_2, \ldots, x_n)^{\text{T}}$ and $y = (y_1, y_2, \ldots, y_n)^{\text{T}}$ are real column vectors, and prove the lemma by induction on n. Our hypothesis is that $f(\alpha_{n-1}, \beta) = 0$, and we need to show that $f(\beta, \beta) \le 0$ with equality if and only if $\beta = 0$.

First assume that $n = 2$. Then $f(\alpha_1, \beta) = a_{11}b_2 + a_{21}b_1 = 0$ implies that $b_2 = -a_{21}b_1/a_{11}$. Thus $f(\beta, \beta) = 2b_1b_2 = -2a_{21}b_1^2/a_{11}$. The positivity of A now implies that this last quantity is nonpositive and equals 0 if and only if $b_1 = 0$. But by above $b_2 = 0$ if and only if $b_1 = 0$, that is, if and only if $\beta = 0$.

We now assume that $n > 2$ and proceed by induction. If x is a multiple of the unit column vector e_n with a 1 in position n, then clearly $f(x, e_n) = 0$. We next suppose that x is not a multiple of e_n and that $f(x, e_n) = 0$, and prove that $f(x, x) < 0$. For a column vector u of size n, let u^* denote the column vector of size $n-1$ obtained from u by deleting its last element. Then, using the Laplace expansion for the permanent along row n and the assumption that $f(x, e_n) = 0$, we see that

$$f(x, x) = \sum_{j=1}^{n-2} a_{nj} f_j(x^*, x^*),$$

where

$$f_j(x^*, y^*) = \text{per}[\alpha_i^*, \ldots, \alpha_{j-1}^*, \alpha_{j+1}^*, \ldots, \alpha_{n-2}^*, x^*, y^*].$$

Since $f(x, e_n) = 0$, it follows that $f_j(x^*, a_j^*) = 0$ for $j = 1, 2, \ldots, n-2$. Since A is a positive matrix, we also have $f_j(a_j^*, a_j^*) > 0$, for $j = 1, 2, \ldots, n-2$. Since x is not a multiple of e_n, x^* is not a zero vector and the induction hypothesis implies that $f_j(x^*, x^*) < 0$ for $j = 1, 2, \ldots, n-2$ and hence that $f(x, x) \le 0$.

Now assume that

$$f(\alpha_{n-1}, \beta) = 0. \tag{9.51}$$

Since A is a positive matrix $f(\alpha_{n-1}, e_n) > 0$. Let

$$\epsilon = -\frac{f(\beta, e_n)}{f(\alpha_{n-1}, e_n)}.$$

Then we see that

$$f(\beta + \epsilon \alpha_{n-1}, e_n) = 0$$

and, by what we have just proved,

$$f(\beta + \epsilon \alpha_{n-1}, \beta + \epsilon \alpha_{n-1}) \le 0;$$

equivalently, because of (9.51),

$$f(\beta, \beta) + \epsilon^2 f(\alpha_{n-1} \alpha_{n-1}) \le 0. \tag{9.52}$$

Since A is a positive matrix, (9.52) implies that $f(\beta, \beta) \leq 0$. If $f(\beta, \beta) = 0$, then (9.52) implies that $\epsilon = 0$, that is, $f(\beta, e_n) = 0$. As proved above, this implies that $\beta = ce_n$ for some scalar c. We now calculate that

$$0 = f(\alpha_{n-1}, \beta) = f(\alpha_{n-1}, ce_n) = cf(\alpha_{n-1}, e_n),$$

which gives $c = 0$ and hence $\beta = 0$.   □

We now come to a crucial inequality in proving the van der Waerden conjecture. Egoryčev obtained this inequality from an inequality known as *Alexandrov's inequality for mixed discriminants*. As in [102] we obtain it from Lemma 9.13.5.

Lemma 9.13.6 *Let* $A = [\alpha_1, \alpha_2, \ldots, \alpha_n]$ *be a matrix of order* n *whose first* $n - 1$ *column vectors* $\alpha_1, \alpha_2, \ldots, \alpha_{n-1}$ *are positive vectors. Then*

$$\operatorname{per}(A)^2 \geq \operatorname{per}([\alpha_1, \ldots, \alpha_{n-2}, \alpha_{n-1}, \alpha_{n-1}])\operatorname{per}([\alpha_1, \ldots, \alpha_{n-2}, \alpha_n, \alpha_n]).$$
$$(9.53)$$

Equality holds in (9.53) *if and only if* α_{n-1} *and* α_n *are linearly dependent vectors.*

Proof. We use the bilinear form

$$f(x, y) = \operatorname{per}([\alpha_1, \ldots, \alpha_{n-2}, x, y])$$

defined in the proof of Lemma 9.13.5. Let

$$\epsilon = \frac{f(\alpha_{n-1}, \alpha_n)}{f(\alpha_{n-1}, \alpha_{n-1})} = \frac{\operatorname{per}(A)}{f(\alpha_{n-1}, \alpha_{n-1})}.$$

Let $\beta = \alpha_n - \epsilon \alpha_{n-1}$. Then

$$f(\alpha_{n-1}, \beta) = f(\alpha_{n-1}, \alpha_n) - \epsilon f(\alpha_{n-1}, \alpha_{n-1}) = 0,$$

and hence by Lemma 9.13.5 we get

$$\begin{aligned}
0 \geq f(\beta, \beta) &= f(\alpha_n, \beta) - \epsilon f(\alpha_{n-1}, \beta) \\
&= f(\alpha_n, \beta) \\
&= f(\alpha_n, \alpha_n) - \epsilon f(\alpha_n, \alpha_{n-1}) \\
&= f(\alpha_n, \alpha_n) - \frac{f(\alpha_{n-1}, \alpha_n)^2}{f(\alpha_{n-1}, \alpha_{n-1})}.
\end{aligned}$$

Hence

$$f(\alpha_{n-1}, \alpha_n)^2 \geq f(\alpha_{n-1}, \alpha_{n-1})f(\alpha_n, \alpha_n)$$

proving (9.53). By Lemma 9.13.5, equality holds if and only if $\beta = 0$. If $\beta = 0$, then α_{n-1} and α_n are linearly dependent. If α_{n-1} and α_n are linearly dependent, then clearly equality holds in (9.53).   □

We remark for later use that a continuity argument shows that the inequality (9.53) holds if $\alpha_1, \alpha_2, \ldots, \alpha_{n-1}$ are only assumed to be nonnegative, but the condition for equality is no longer true.

We now obtain Egoryčev's crucial result that a matrix with smallest permanent in Ω_n has the property that all the permanents of its submatrices of order $n-1$ are equal (and so equal the permanent of the matrix itself).

Lemma 9.13.7 *Let $A = [a_{ij}]$ be a permanent minimizing matrix in Ω_n. Then*
$$\mathrm{per}(A(i,j)) = \mathrm{per}(A) \quad (i,j = 1, 2, \ldots, n).$$

Proof. The lemma holds for $n = 1$ and we assume that $n \geq 2$. The matrix A is fully indecomposable by Lemma 9.13.1, and by Lemma 9.13.3 satisfies
$$\mathrm{per}(A(i,j)) \geq \mathrm{per}(A) > 0 \quad (i,j = 1, 2, \ldots, n). \tag{9.54}$$
Suppose to the contrary that for some k and l, $\mathrm{per}(A(k,l)) > \mathrm{per}(A)$. Since A is fully indecomposable, there exists an integer $s \neq l$ such that $a_{ks} > 0$. We have
$$a_{il}\mathrm{per}((A(i,s)) \geq a_{il}\mathrm{per}(A) \quad (i = 1, 2, \ldots, n),$$
$$a_{is}\mathrm{per}(A(i,l)) \geq a_{is}\mathrm{per}(A) \quad (i = 1, 2, \ldots, n).$$
We also have
$$a_{ks}\mathrm{per}(A(k,l)) > a_{ks}\mathrm{per}(A).$$
But then by Lemma 9.13.6 (see also the remark following its proof), we have
$$\mathrm{per}(A)^2 \geq \left(\sum_{i=1}^{n} a_{is}\mathrm{per}(A(i,l))\right)\left(\sum_{i=1}^{n} a_{il}\mathrm{per}(A(i,s))\right)$$
$$> \left(\sum_{i=1}^{n} a_{is}\mathrm{per}(A)\right)\left(\sum_{i=1}^{n} a_{il}\mathrm{per}(A)\right)$$
$$= \mathrm{per}(A)^2,$$
a contradiction. $\qquad\qquad\square$

Our final lemma [88] is a consequence of the multilinearity of the permanent and we omit the simple proof.

Lemma 9.13.8 *Let $A = [\alpha_1, \alpha_2, \ldots, \alpha_n]$ be a matrix of order n with column vectors $\alpha_1, \alpha_2, \ldots, \alpha_n$, and let k and l be integers with $1 \leq k < l \leq n$ such that*
$$\mathrm{per}(A(i,k)) = \mathrm{per}((A(i,l)) \quad (i = 1, 2, \ldots, n).$$
Let B be the matrix obtained from A by replacing columns α_k and α_l by their average $(\alpha_k + \alpha_l)/2$. Then $\mathrm{per}(B) = \mathrm{per}(A)$. $\qquad\square$

We now obtain the solution of the van der Waerden conjecture.

Theorem 9.13.9 *Let A be a doubly stochastic matrix of order n. Then*

$$\text{per}(A) \geq \frac{n!}{n^n} \text{ with equality if and only if } A = (1/n)J_n.$$

Proof. Since the permanent of $(1/n)J_n$ equals $n!/n^n$ we need only show that a minimizing matrix A in Ω_n must equal $(1/n)J_n$. We may assume that $n \geq 2$. By Lemma 9.13.1 A is fully indecomposable and by Lemma 9.13.7

$$\text{per}(A(i,j)) = \text{per}(A) \quad (i,j = 1, 2, \ldots, n).$$

Let p be an integer with $1 \leq p \leq n$. We apply the averaging process in Lemma 9.13.8 to columns other than column p of A and, since A has at least two positive elements in each row, obtain another minimizing doubly stochastic matrix $B = [b_{ij}]$ each of whose columns, except possibly column p, is positive. By Lemmas 9.13.6 and 9.13.7 we have that for each integer $q \neq p$ with $1 \leq q \leq n$,

$$\text{per}(B)^2 = \left(\sum_{k=1}^{n} b_{kp}\text{per}(B(k,q)) \right) \left(\sum_{k=1}^{n} b_{kq}\text{per}(B(k,p)) \right).$$

Lemma 9.13.6 now implies that columns p and q of B are multiples of one another and, since B is doubly stochastic, are actually equal. Since this conclusion holds for all p and q with $p \neq q$, $B = (1/n)J_n$. Since column p of B equals column p of A, and since p is any integer between 1 and n, we now conclude that $A = (1/n)J_n$. □

Tverberg [140] conjectured and Friedland [64] proved an inequality for the permanent more general than that given in Theorem 9.13.9. We state this without proof. For a matrix A of order n and an integer k with $1 \leq k \leq n$, let $\text{per}_k(A)$ denote the sum of the permanents of all the submatrices of A of order k. Thus $\text{per}_1(A)$ is the sum of the elements of A, and $\text{per}_n(A) = \text{per}(A)$.

Theorem 9.13.10 *Let A be a doubly stochastic matrix of order n, and let k be an integer with $1 \leq k \leq n$. Then*

$$\text{per}_k(A) \geq \binom{n}{k}^2 \frac{k!}{n^k} \text{ with equality if and only if } A = (1/n)J_n.$$

Lemma 9.13.3 asserts that a minimizing matrix A in Ω_n necessarily satisfies $\text{per}(A(i,j)) \geq \text{per}(A)$ for all i and j with $1 \leq i, j \leq n$. Brualdi and Foregger [25] conjectured that besides the matrix $(1/n)J_n$, there is, up to row and column permutations, only one doubly stochastic matrix which satisfies all these inequalities, namely the matrix $A = (1/2)(I_n + P_n)$ where P_n is the permutation matrix with 1's in positions $(1,2), \ldots,$

$(n-1, n), (n, 1)$. Bapat [7] proved this conjecture using some of the results discovered by Egoryčev in his proof of the van der Waerden conjecture.

There has been considerable work on determining the minimal permanent on certain faces $\Omega_n(A)$ of the polytope Ω_n ($n \geq 3$). It is not our intention to discuss this work in any detail here, but we mention two quite different faces for which the minimum permanent has been obtained. The first of these is the facets $\Omega_n(A)$ where A is a $(0, 1)$-matrix of order n with exactly one 0. The union of these facets constitutes the boundary of Ω_n. Knopp and Sinkhorn [79] showed that the minimum value of the permanent on the boundary of Ω_n equals

$$(n-2)! \left(\frac{n-2}{(n-1)^2} \right)^{n-2} \qquad (n \geq 3).$$

For $n > 3$, a matrix A on the boundary of Ω_n has permanent equal to this minimum value if and only if there are permutation matrices P and Q such that

$$PAQ = \begin{bmatrix} 0 & 1/n-1 & \cdots & 1/n-1 \\ 1/n-1 & & & \\ \vdots & & \frac{n-2}{(n-1)^2} J_{n-1} & \\ 1/n-1 & & & \end{bmatrix}. \qquad (9.55)$$

If $n = 3$, the minimum value is also achieved at all matrices of the form

$$\frac{1}{2} \begin{bmatrix} 0 & 1 & 1 \\ 1 & t & 1-t \\ 1 & 1-t & t \end{bmatrix} \qquad (0 \leq t \leq 1).$$

It is interesting to note that the matrices A satisfying (9.55) are those matrices on the boundary that are closest to $(1/n)J_n$ in the Euclidean norm.

In contrast, the other face $\Omega_n(A)$ we mention has many 0's [20]. Let $A_n = [a_{ij}]$ be the $(0, 1)$-matrix of order n defined by

$$a_{ij} = \begin{cases} 0, & \text{if } i + j \leq n - 1, \\ 1, & \text{otherwise.} \end{cases}$$

For example, if $n = 5$, then

$$A_5 = \begin{bmatrix} 0 & 0 & 0 & 1 & 1 \\ 0 & 0 & 1 & 1 & 1 \\ 0 & 1 & 1 & 1 & 1 \\ 1 & 1 & 1 & 1 & 1 \\ 1 & 1 & 1 & 1 & 1 \end{bmatrix}.$$

The dimension of $\Omega_n(A_n)$ is roughly half the dimension of Ω_n. The minimum value of the permanent on Ω_n equals $1/2^{n-1}$ and is achieved at all the matrices $H = [h_{ij}]$ satisfying $h_{ij} = 1/2$ if $i + j = n$, or $i = 1$ and

$j = n$, or $i = n$ and $j = 1$, and h_{ij} is unspecified, otherwise. Thus the minimum is achieved at a large subpolytope $\mathcal{P}_n$ of $\Omega_n(A_n)$.[24] Let B_n be the (0,1)-matrix of order n with 1's on the main diagonal, superdiagonal, and subdiagonal, and 0's elsehere. The subface $\Omega_n(B_n)$ of $\Omega_n(A_n)$ consists of the *tridiagonal* doubly stochastic matrices of order n. In [51] $\Omega_n(B_n)$, in particular the structure of the faces of $\Omega_n(B_n)$, is investigated.

An unresolved conjecture related to the van der Waerden conjecture is a conjecture due to Dittert (reported in [101]). Let Ψ_n be the convex polyhedron of all nonnegative matrices of order n, the sum of whose elements equals n. Clearly, $\Omega_n \subseteq \Psi_n$. Let $A \in \Psi_n$ and let the row sum vector and column sum vector of A be $(r_1, r_2, \ldots, r_n)$ and $(s_1, s_2, \ldots, s_n)$, respectively. Define a function v on Ψ_n by

$$v(A) = \prod_{i=1}^{n} r_i + \prod_{i=1}^{n} s_j - \operatorname{per}(A).$$

If A is doubly stochastic, then $v(A) = 2 - \operatorname{per}(A)$. If $A = (1/n)J_n$, then $v(J_n) = 2 - n!/n^n$. The *Dittert conjecture* asserts that

$$v(A) \leq 2 - \frac{n!}{n^n} \quad (A \in \Psi_n),$$

with equality if and only if $A = (1/n)J_n$. It follows from Theorem 9.13.9 that the Dittert conjecture holds for doubly stochastic matrices. Some results concerning this conjecture can be found in [131], [72]. An update on the status of certain open problems concerning the permanent can be found in [45].

We briefly discuss another generalization of Ω_n and the permanent function as given in Brualdi, Hartfiel, and Hwang [31]. Now we assume that $R = (r_1, r_2, \ldots, r_m)$ and $S = (s_1, s_2, \ldots, s_n)$ are positive integral vectors with

$$r_1 + r_2 + \cdots + r_m = s_1 + s_2 + \cdots + s_n. \tag{9.56}$$

We also assume that the class $\mathcal{A}(R, S)$ of m by n (0,1)-matrices with row sum vector R and column sum vector S is nonempty. We regard a matrix $A = [a_{kl}]$ as an (R, S)-*assignment* of m individuals $p_1, p_2, \ldots, p_m$ to n jobs $j_1, j_2, \ldots, j_n$, where the individual p_k is assigned to work on the r_k jobs j_l for which $a_{kl} = 1$. We think of r_k as the capacity (number of jobs) of p_k ($k = 1, 2, \ldots, m$), and s_l as the requirement (number of individuals) of job j_l. In an (R, S)-assignment, each individual works to capacity and each job has its requirement satisfied.

Let $W = [w_{kl}]$ be an m by n (0, 1)-matrix where $w_{kl} = 1$ if and only if p_k is qualified for j_l. Then we define an (R, S)-*assignment function* by

$$p_{R,S}(W) = |\{A \in \mathcal{A}(R, S) : A \leq W\}|,$$

---

[24] In fact, this subpolytope has dimension equal to the dimension of $\Omega_{n-1}(A_{n-1})$ and consequently the ratio of the dimension of $\mathcal{P}_n$ to that of $\Omega_n(A_n)$ goes to zero as n increases.

the number of (R, S)-assignments subject to the qualification constraints imposed by W. If $m = n$ and $R = S = (1, 1, \ldots, 1)$, then $p_{R,S}(W)$ is the permanent per(W) of W. Thus (R, S)-assignment functions extend the permanent function on $(0, 1)$-matrices to more-general row and column sum vectors.

We now generalize the assignment polytope to more-general row and column sums. As above, let $R = (r_1, r_2, \ldots, r_m)$ and $S = (s_1, s_2, \ldots, s_n)$ be positive integral vectors satisfying (9.56). Define $\Omega_{R,S}$ to be the set of all m by n matrices $X = [x_{ij}]$ satisfying

$$
\left.
\begin{aligned}
0 \leq x_{ij} \leq 1 \quad & (i = 1, 2, \ldots, m; j = 1, 2, \ldots, n), \\[2mm]
\textstyle\sum_{j=1}^{n} x_{ij} = r_i \quad & (i = 1, 2, \ldots, m), \\[2mm]
\textstyle\sum_{i=1}^{m} x_{ij} = s_j \quad & (i = 1, 2, \ldots, n).
\end{aligned}
\right\}
\qquad (9.57)
$$

We call $\Omega_{R,S}$ the (R, S)-*assignment polytope*. When R and S equal the vector of all 1's of size n, $\Omega_{R,S}$ is the assignment polytope Ω_n.

Each matrix in $\mathcal{A}(R, S)$ is in $\Omega_{R,S}$ and, being a $(0, 1)$-matrix, is an extreme point of $\Omega_{R,S}$. We now prove that $\mathcal{A}(R, S)$ is precisely the set of extreme points of $\Omega_{R,S}$. We could appeal to the integrality theorem for network flows[25] but we give the simple direct proof.

Theorem 9.13.11 *Let* $R = (r_1, r_2, \ldots, r_m)$ *and* $S = (s_1, s_2, \ldots, s_n)$ *be positive integral vectors satisfying* (9.56). *Then* $\Omega_{R,S} \neq \oslash$ *if and only if* $\mathcal{A}(R, S) \neq \oslash$. *In fact,* $\Omega_{R,S}$ *is the convex hull of the matrices in* $\mathcal{A}(R, S)$.

Proof. Let $X = [x_{ij}]$ be a matrix in $\Omega_{R,S}$. If X is an integral matrix, then $X \in \mathcal{A}(R, S)$. Assume that X has at least one nonintegral element, that is, an element which is strictly between 0 and 1. Using a familiar argument, we see that there are nonintegral elements occupying positions corresponding to a cycle in the bipartite graph of X. Let C be the $(0, 1, -1)$-matrix obtained by alternating 1's and -1's in these positions. Then for ϵ a small positive number, $X \pm \epsilon C \in \Omega_{R,S}$ and

$$
X = \frac{1}{2}(X + \epsilon C) + \frac{1}{2}(X - \epsilon C),
$$

and X is not an extreme point of $\Omega_{R,S}$. $\square$

Assuming that $\mathcal{A}(R, S)$ has no invariant 1's, the dimension of the polytope $\Omega_{R,S}$ equals $(m-1)(n-1)$, and hence $\Omega_{R,S}$ has at least $(m-1)(n-1)+1$ extreme points.[26]

---

[25]See e.g. Chapter 6 of [39] where integrality is implicit in the definitions.
[26]So $|\mathcal{A}(R, S)| \geq (m - 1)(n - 1) + 1$ if R and S are positive integral vectors such that the class $\mathcal{A}(R, S)$ is nonempty and has no invariant 1's.

We extend the assignment function to a matrix in a nonempty polytope $\Omega_{R,S}$ by defining[27]

$$p_{R,S}(X) = \sum_{A \in \mathcal{A}(R,S)} \prod_{\{(i,j):a_{ij}=1\}} x_{ij} \quad (X = [x_{ij}] \in \Omega_{R,S}).$$

In [31] the problems of determining

$$M_{R,S} = \max\{p_{R,S}(X) : X \in \Omega_{R,S}\}$$

and

$$m_{R,S} = \min\{p_{R,S}(X) : X \in \Omega_{R,S}\}$$

were raised.

For Ω_n, where R and S equal the vector of all 1's of size n, we have $M_{R,S} = 1$ and by the now proved van der Waerden conjecture, $m_{R,S} = n!/n^n$. The matrix $(1/n)J_n$, on which the permanent achieves its minimum on Ω_n, is the barycenter

$$\frac{1}{n!} \sum_P P$$

of Ω_n, where the summation extends over all permutation matrices of order n. The matrices in Ω_n achieving the maximum permanent 1 are precisely the permutation matrices of order n. The *barycenter* of $\Omega_{R,S}$ is given by

$$J_{R,S} = \frac{1}{|\mathcal{A}(R,S)|} \sum_{A \in \mathcal{A}(R,S)} A.$$

Example. Let $R = S = (2,1,1)$. Then $|\mathcal{A}(R,S)| = 5$ and it can be easily verified that

$$J_{R,S} = \frac{1}{5} \begin{bmatrix} 4 & 3 & 3 \\ 3 & 1 & 1 \\ 3 & 1 & 1 \end{bmatrix},$$

where

$$p_{R,S}(J_{R,S}) = \frac{9}{25}.$$

The matrix

$$X_t = \begin{bmatrix} 1 & 1/2 & 1/2 \\ 1/2 & t & 1/2 - t \\ 1/2 & 1/2 - t & t \end{bmatrix} \quad (0 \le t \le 1/2)$$

belongs to $\Omega_{R,S}$, and

$$p_{R,S}(X_t) = \frac{5}{16} \quad (0 \le t \le 1/2).$$

---

[27]We may apply this definition to any m by n matrix. It is also true [31] that such an assignment function can be evaluated as a permanent of a larger matrix.

In fact, a tedious calculation shows that

$$m_{R,S} = \frac{5}{16}$$

and that the matrices X_t are precisely the matrices in $\Omega_{R,S}$ achieving $m_{R,S}$.

$\square$

It follows from the above example that, in general, $m_{R,S} \ne p_{R,S}(J_{R,S})$. However, the following theorem is proved in [31] using the solution of the van der Waerden conjecture; we omit the proof.

Theorem 9.13.12 *Let $R = (1, 1, \ldots, 1)$ be the vector of all 1's of size m and let $S = (s_1, s_2, \ldots, s_n)$ be a positive vector where $s_1 + s_2 + \cdots + s_n = m$. Then*

$$p_{R,S}(X) \ge \left(\prod_{k=1}^{n} \frac{s_k^{s_k}}{s_k!} \right) \frac{m!}{m^m} \quad (X \in \Omega_{R,S}),$$

with equality if and only if $X = J_{R,S}$.

The following theorem [31] shows that for $\Omega_{R,S}$ as in Theorem 9.13.12, we also have $M_{R,S} = 1$.

Theorem 9.13.13 *Let R and S be as in Theorem 9.13.12. Then*

$$p_{R,S}(X) \le 1 \quad (X \in \Omega_{R,S}),$$

with equality if and only if $X \in \mathcal{A}(R, S)$.

It was conjectured in [31] that, in general, $M_{R,S} = 1$ and that the matrices X in $\Omega_{R,S}$ with $p_{R,S}(X) = 1$ are precisely the matrices in $\mathcal{A}(R, S)$. This conjecture was disproved by Gibson [65] with the following example.

Example. Let $R = S = (n - 1, n - 1, \ldots, n - 1)$ be vectors of size $n \ge 2$. Then $\mathcal{A}(R, S)$ consists of those matrices $J_n - P$ where P is a permutation matrix of order n. The barycenter $J_{R,S}$ is the matrix of order n each of whose elements equals $(n - 1)/n$. Hence

$$p_{R,S}(J_{R,S}) = n! \left(\frac{n-1}{n} \right)^{n(n-1)}.$$

We have that $p_{R,S}(J_{R,S}) > 1$ for $n \ge 5$, and indeed that

$$\lim_{n \to \infty} p_{R,S}(J_{R,S}) = \infty.$$

$\square$

Because of the apparent difficulty in determining $m_{R,S}$ and $M_{R,S}$ in general, Gibson [65] has raised the question of determining

$$\min\{m_{R,S} : \mathcal{A}(R, S) \ne \oslash\}$$

and

$$\max\{M_{R,S} : \mathcal{A}(R, S) \ne \oslash\},$$

where the minimum and maximum are taken over all positive integral vectors R and S of size m and n, respectively, for which $\mathcal{A}(R, S) \ne \oslash$.

9.14   Additional Related Results

In this final section we collect a few additional results on or related to doubly stochastic matrices that have not been reported in the other sections, and include some additional references.

We begin with a spectral characterization of stochastic matrices. A *stochastic matrix* A of order n may be either *row-stochastic* (all row sums equal to 1) or *column-stochastic* (all column sums equal to 1). A stochastic matrix has an eigenvalue equal to 1:

$$Ae = e, \quad \text{if } A \text{ is row-stochastic,}$$
$$A^{\mathrm{T}}e = e, \quad \text{if } A \text{ is column-stochastic,}$$

where e is the column n-vector of all 1's. A doubly stochastic matrix of order n satisfies both of these equations. If P and Q are permutation matrices and A is a stochastic matrix, then PAQ is also a stochastic matrix and hence 1 is also an eigenvalue of PAQ. Brualdi and Wielandt [40] used this observation to obtain a spectral characterization of stochastic matrices. This characterization applies if the elements of A come from an arbitrary field where, for instance, row sums equal 1 means that row sums equal the multiplicative identity 1 of the field.

Theorem 9.14.1 *Let A be a matrix of order n over a field F. Then 1 is an eigenvalue of PA for all permutation matrices P of order n if and only if A is stochastic.*

Proof. We have already observed that 1 is an eigenvalue of all PA if A is stochastic and P is a permutation matrix. Now assume that 1 is an eigenvalue of PA for every permutation matrix P of order n. Let X^P be the eigenspace of PA for the eigenvalue 1, that is,

$$X^P = \{x : PAx = x, x \in F^n\}.$$

Then X^P has dimension at least 1. By replacing A by QA for some permutation matrix Q, we may assume that X^I has the smallest dimension of these subspaces.

Let $x^I = (x_1^I, x_2^I, \ldots, x_n^I)^{\mathrm{T}}$ be a nonzero vector in X^I. If $x_1^I = x_2^I = \cdots = x_n^I$, then A is row-stochastic. We now assume that x^I is not a constant vector and prove that A is column-stochastic. If necessary, we may replace A by RAR^{T} for some permutation matrix R in order to get an integer k such that $x_i^I \neq x_j^I$ for all i and j with $1 \leq i \leq k < j \leq n$. Such a replacement preserves the fact that X^I has minimal dimension. If $k > 1$, then for $2 \leq i \leq k$, let P_i be the permutation matrix corresponding to the transposition which interchanges i and n. For $k + 1 \leq j \leq n$, let Q_j be the permutation matrix which interchanges 1 and j. The minimality of the dimension of X^I implies that there are vectors $x^i = (x_1^i, x_2^i, \ldots, x_n^i) \in X^{P_i}$

and vectors $y^j = (y_1^j, y_2^j, \dots, y_n^j) \in X^{Q_j}$ such that $x_i^i \neq x_n^i$ $(i = 2, \dots, k)$ and $y_1^j \neq y_j^j$ $(j = k+1, \dots, n)$.

We now show that the n vectors $x^I, x^2, \dots, x^k, y^{k+1}, \dots, y^n$ are linearly independent. Suppose that

$$bx^I + \sum_{i=2}^{k} c_i x^i + \sum_{j=k+1}^{n} d_j y^j = 0, \tag{9.58}$$

where the first summation is vacuous if $k = 1$. Applying A to this equation and using the fact that $P_i^{-1} = P_i$ and $Q_j^{-1} = Q_j$, we get

$$bx^I + \sum_{i=2}^{k} c_i P_i x^i + \sum_{j=k+1}^{n} d_j Q_j y^j = 0. \tag{9.59}$$

Subtracting (9.59) from (9.58) we get

$$\sum_{i=2}^{k} c_i (x^i - P_i x^i) + \sum_{j=k+1}^{n} d_j (y^j - Q_j y^j) = 0. \tag{9.60}$$

Let p be an integer with $2 \leq p \leq k$. Since the P^i with $i \neq p$ and the Q^j do not change the pth coordinates of their corresponding vectors, (9.60) implies that

$$c_p (x_p^p - x_p^n) = 0$$

and therefore

$$c_p = 0 \quad (p = 2, \dots, k). \tag{9.61}$$

Now let q be an integer with $k + 1 \leq q \leq n - 1$. Then the P_i and the Q_j with $j \neq q$ do not change the qth coordinates of their corresponding vectors, and (9.60) now implies that

$$d_q = 0 \quad (q = k+1, \dots, n-1). \tag{9.62}$$

Using (9.61) and (9.62) in (9.60) now gives

$$d_n (y_n^n - y_1^n) = 0,$$

and hence

$$d_n = 0. \tag{9.63}$$

Now using (9.61), (9.62), and (9.63) in (9.58) gives $b = 0$. Hence the vectors $x^I, x^2, \dots, x^k, y^{k+1}, \dots, y^n$ are linearly independent.

In summary we have n permutation matrices $R_1, R_2, \dots, R_n$ and n linearly independent vectors $z^1, z^2, \dots, z^n$ such that $R_i A z^i = z^i$ and hence $e^T R_i A z^i = e^T z^i$ $(i = 1, 2, \dots, n)$. Therefore, since $e^T R_i = e^T$ for

$i = 1, 2, \ldots, n$,

$$e^{\mathrm{T}}(A - I)z^i = 0 \quad (i = 1, 2, \ldots, n).$$

Since $z^1, z^2, \ldots, z^n$ are linearly independent, we now conclude that $e^{\mathrm{T}}(A - I) = 0$, that is, $e^{\mathrm{T}}A = e^{\mathrm{T}}$. Thus A is column-stochastic. □

The proof of Theorem 9.14.1 provides a sufficient condition for a row-stochastic matrix to be a doubly stochastic matrix, which makes use of the eigenspaces X^P defined in its proof for each permutation matrix P. We state it for real matrices but, like Theorem 9.14.1, it applies to general fields.

Corollary 9.14.2 *Let A be a nonnegative matrix of order n. If A is row-stochastic and an eigenspace X^P of smallest dimension contains a nonzero, nonconstant vector, then A is actually a doubly stochastic matrix.* □

We now turn to magic squares. Let $n \geq 1$ and $r \geq 0$ be integers. A *magic square* A of order n is a nonnegative integral matrix all of whose row and column sums are equal.[28] If r is the common value of the row and column sums, then we call A an *r-magic square*. Let $\mathcal{H}_n(r)$ denote the *set of all r-magic squares of order n*. Clearly $\mathcal{H}_n(r)$ is a finite set and we let $H_n(r)$ denote its cardinality. We have $H_1(r) = 1$ and $H_2(r) = r + 1$ for all $r \geq 0$. Magic squares are related to doubly stochastic matrices as $A \in \mathcal{H}_n(r)$ implies that $(1/r)A \in \Omega_n$. Also letting $R = (r, r, \ldots, r)$, a vector of size n with each element equal to r, we see that $H_n(r)$ equals $\mathcal{Z}^+(R, R)$.

Stanley [134, 135] proved the following basic theorem.

Theorem 9.14.3 *Let n be a positive integer. The number $H_n(r)$ of r-magic squares of order n is a polynomial in r of degree $(n-1)^2$.*

For example, as already computed by MacMahon [86],

$$H_3(r) = \binom{r+4}{4} + \binom{r+3}{4} + \binom{r+2}{4} \quad (r = 0, 1, 2, \ldots).$$

Stanley also showed that

$$H_n(-1) = H_n(-2) = \cdots = H_n(-n+1) = 0,$$

and

$$H_n(-n-r) = H_n(r) \quad (r = 0, 1, 2, \ldots).$$

Let $\mathcal{S}_n(r)$ denote the *set of symmetric r-magic squares of order n*, and let $S_n(r)$ denote its cardinality. We have the following theorem due to Stanley [134, 135] and Jia [74].

---

[28]Note that the two diagonal sums are not included, nor do we insist that the elements of the matrix be $1, 2, \ldots, n^2$.

Theorem 9.14.4 *Let n be a positive integer. The number $S_n(r)$ of symmetric r-magic squares of order n satisfies:*

(i) $S_n(r) = P_n(r) + (-1)^r Q_n(r)$ *where $P_n(r)$ and $Q_n(r)$ are polynomials in r.*

(ii) *The degree of $P_n(r)$ equals $\binom{n}{2}$.*

(iii) *The degree of $Q_n(r)$ equals $\binom{n-1}{2} - 1$ if n is odd, and equals $\binom{n-2}{2} - 1$ if n is even.*

Stanley proved (i) and (ii) of Theorem 9.14.4 and that the numbers given in (iii) were upper bounds for the degree of $Q_n(r)$. Jia [74] showed that equality holds.

The following generalization of transportation polytopes is considered in [46]. Let R and S be positive vectors of sizes m and n, respectively, and let $\mathcal{N}^*(R, S)$ denote the set of all m by n real matrices $A = [a_{ij}]$ such that the matrix $|A| = [|a_{ij}|]$ obtained by replacing each element of A by its absolute value is in $\mathcal{N}(R, S)$. Then $\mathcal{N}^*(R, S)$ is a convex polytope of dimension mn whose extreme points are precisely those matrices $A \in \mathcal{N}^*(R, S)$ for which $|A|$ is an extreme point of $\mathcal{N}(R, S)$. In particular, by taking R and S to be the vector $(1, 1, \ldots, 1)$ of size n, the set of matrices A of order n such that $|A| \in \Omega_n$ is a convex polytope of dimension n^2 whose extreme points are the $2^n n!$ matrices obtained from the permutation matrices of order n by replacing each 1 with ± 1.

The convex hull of the m by n subpermutation matrices is studied in [35]. In [30] polyhedra defined by relaxing the nonnegativity assumptions of doubly stochastic matrices are investigated; such polyhedra need not be bounded and so are not in general polytopes. The convex polytope of doubly stochastic matrices which are invariant under a prescribed row and column permutation is investigated in [17]. The volume of Ω_n is considered in [42] and [114]; the volume of Ω_n has been computed for $n \leq 8$. Some additional material related to the topics treated in this chapter can be found in [62]. Another related polytope is the polytope $\mathcal{D}_n$ of degree sequences of simple graphs[29] with n vertices. This polytope is the convex hull of those vectors $D = (d_1, d_2, \ldots, d_n)$ such that there is a graph with degree sequence D. In [116] (see also [87]) it is shown that the extreme points of $\mathcal{D}_n$ are the degree sequences of the class of graphs known as threshold graphs.[30] The vertex–edge graph of $\mathcal{D}_n$ is determined as well as many other interesting properties.

In [94], [95], [145] a doubly stochastic matrix $S(G)$ associated with a graph G (its adjacency matrix A) of order n is considered. Let the degrees of the vertices $v_1, v_2, \ldots, v_n$ of G be $d_1, d_2, \ldots, d_n$, respectively, and let D be the diagonal matrix of order n whose main diagonal is $(d_1, d_2, \ldots, d_n)$.

---

[29]Graphs with no loops and at most one edge joining a pair of distinct vertices.

[30]Threshold graphs are graphs which are uniquely determined by their degree sequences.

The matrix $D - A$ is the *Laplacian matrix* $L(G)$ of G. The matrix $I_n + L(G)$ is a positive definite symmetric matrix, and the matrix $S(G)$ is defined to be its inverse $(I_n + L(G))^{-1}$, which can be shown to be a doubly stochastic matrix. In [94], [95], [145] the elements of $S(G)$ are investigated in relation to properties of G.

References

[1] E. Achilles, Doubly stochastic matrices with some equal diagonal sums, *Linear Algebra Appl.*, **22** (1978), 293–296.

[2] S.N. Afriat, On sum-symmetric matrices, *Linear Algebra Appl.*, **8** (1974), 129–140.

[3] K. Balasubramanian, Maximal diagonal sums, *Linear Multilin. Alg.*, **7** (1979), 249–251.

[4] K. Balasubramanian, On equality of some elements in matrices, *Linear Algebra Appl.*, **22** (1978), 135–138.

[5] M.L. Balinski, Establishing the matching polytope, *J. Combin. Theory, Ser. B*, **13** (1972), 1–13.

[6] M.L. Balinski and A. Russakoff, On the assignment polytope, *SIAM Review*, **16** (1974), 516–525.

[7] R. Bapat, Doubly Stochastic Matrices with equal subpermanents, *Linear Algebra Appl.*, **51** (1983), 1–8.

[8] R.B. Bapat and T.E.S. Raghavan, *Nonnegative Matrices and Applications*, Cambridge U. Press, Cambridge, 1997.

[9] J. von Below, On a theorem of L. Mirsky on even doubly-stochastic matrices, *Discrete Math.*, 55 (1985), 311–312.

[10] S.K. Bhandir and S. Das Gupta, Two characterizations of doubly-superstochastic matrices, *Sankhyā Ser. A*, **47** (1985), 357–365.

[11] G. Bongiovanni, D. Bovet, and A. Cerioli, Comments on a paper of R.A. Brualdi: "Notes on the Birkhoff algorithm for doubly stochastic matrices," *Canad. Math. Bull.*, **31** (1988), 394–398.

[12] A. Borobia, $(0, \frac{1}{2}, 1)$ matrices which are extreme points of the generalized transitive tournament polytope, *Linear Algebra Appl.*, **220** (1995), 97–110.

[13] A. Borobia, Vertices of the generalized transitive tournament polytope, *Discrete Math.*, **179** (1998), 49–57.

[14] A. Borobia and V.Chumillas, *-graphs of vertices of the generalized transitive tournament polytope, *Discrete Math.*, **163** (1997), 229–234.

[15] A. Borobia, Z. Nutov, and M. Penn, Doubly stochastic matrices and dicycle coverings and packings in Eulerian digraphs, *Linear Algebra Appl.*, **223/224** (1995), 361–371.

[16] R.A. Brualdi, The *DAD* theorem for arbitrary row sums, *Proc. Amer. Math. Soc.*, **45** (1974), 189–194.

[17] R.A. Brualdi, Convex polytopes of permutation invariant doubly stochastic matrices, *J. Combin. Theory, Ser. B*, **23** (1977), 58–67.

[18] R.A. Brualdi, Notes on the Birkhoff algorithm for doubly stochastic matrices, *Canad. Math. Bull.*, **25** (1982), 191–199.

[19] R.A. Brualdi, The chromatic index of the graph of the assignment polytope, *Annals Disc. Math.*, **3** (1978), 49–53.

[20] R.A. Brualdi, An interesting face of the polytope of doubly stochastic matrices, *Linear Multilin. Alg.*, **17** (1985), 5–18.

[21] R.A. Brualdi, Some application of doubly stochastic matrices, *Linear Algebra Appl.*, **107** (1988), 77–100.

[22] R.A. Brualdi, *Introductory Combinatorics*, 3rd ed., Prentice-Hall, Upper Saddle River, 1999.

[23] R.A. Brualdi, The doubly stochastic matrices of a vector majorization, *Linear Algebra Appl.*, **61** (1984), 141–154.

[24] R.A. Brualdi and G. Dahl, Majorization-constrained doubly stochastic matrices, *Linear Algebra Appl.*, **361** (2003), 75–97.

[25] R.A. Brualdi and T. Foregger, Matrices with constant permanental minors, *Linear Multilin. Alg.*, **3** (1975–76), 227–243.

[26] R.A. Brualdi and P.M. Gibson, The assignment polytope, *Math. Programming*, **11** (1976), 97–101.

[27] R.A. Brualdi and P.M. Gibson, Convex polyhedra of doubly stochastic matrices I. Applications of the permanent function, *J. Combin. Theory, Ser. A*, **22** (1977), 194–230.

[28] R.A. Brualdi and P.M. Gibson, Convex polyhedra of doubly stochastic matrices II. Graph of Ω_n, *J. Combin. Theory, Ser. B*, **22** (1977), 175–198.

[29] R.A. Brualdi and P.M. Gibson, Convex polyhedra of doubly stochastic matrices III. Affine and combinatorial properties of Ω_n, *J. Combin. Theory, Ser. A*, **22** (1977), 338–351.

[30] R.A. Brualdi and P.M. Gibson, Convex polyhedra of doubly stochastic matrices IV, *Linear Algebra Appl.*, **15** (1976), 153–172.

[31] R.A. Brualdi, D.J. Hartfiel, and S.-G. Hwang, On assignment functions, *Linear Multilin. Alg.*, **19** (1986), 203–219.

[32] R.A. Brualdi and G.-S. Hwang, Generalized transitive tournaments and doubly stochastic matrices, *Linear Algebra Appl.*, **172** (1992), 151–168.

[33] R.A. Brualdi and S.-G. Hwang, Vector majorization via Hessenberg matrices, *J. London Math. Soc.*, **53** (1996), 28–38.

[34] R.A. Brualdi, S.-G. Hwang, and S.-S. Pyo, Vector majorization via positive definite matrices, *Linear Algebra Appl.*, **257** (1997), 105–120.

[35] R.A. Brualdi and G.M. Lee, On the truncated assignment polytope, *Linear Algebra Appl.*, **19** (1978), 33–62.

[36] R.A. Brualdi and B. Liu, The polytope of even doubly stochastic matrices, *J. Combin. Theory, Ser. A*, **57** (1991), 243–253.

[37] R.A. Brualdi, S.V. Parter, and H. Schneider, The diagonal equivalence of a nonnegative matrix to a stochasic matrix, *J. Math. Anal. Appl.*, **16** (1966), 31–50.

[38] R.A. Brualdi and H.J. Ryser, Classes of matrices associated with the optimal assignment problem, *Linear Algebra Appl.*, **81** (1986), 1–17.

[39] R.A. Brualdi and H.J. Ryser, *Combinatorial Matrix Theory*, Cambridge U. Press, Cambridge, 1991.

[40] R.A. Brualdi and H.W. Wielandt, A spectral characterization of stochastic matrices, *Linear Algebra Appl.*, **1** (1968), 65–71.

[41] R.E. Burkhard, Time-slot assignment for TDMA-systems, *Computing*, **35** (1985), 99–112.

[42] C.S. Chan and D.P. Robbins, On the volume of the polytope of doubly stochastic matrices, *Experiment. Math.*, **8** (1999), 291–300.

[43] K.-M. Chao and C.S. Wong, Applications of M-matrices to majorizations, *Linear Algebra Appl.*, **169** (1992), 31–40.

[44] G.-S. Cheon and S.-Z. Song, On the extreme points on the majorization polytope $\Omega_3(y \prec x)$, *Linear Algebra Appl.*, **269** (1998), 47–52.

[45] G.-S. Cheon and I.M. Wanless, An update on Minc's survey of open problems involving permanents, *Linear Alg. Appl.* (2005), to appear.

[46] S. Cho and Y. Nam, Convex polytopes of generalized doubly stochastic matrices, *Commun. Korean Math. Soc.*, **16** (2001), 679–690.

[47] A.B. Cruse, A note on symmetric doubly-stochastic matrices, *Discrete Math.*, **13** (1975), 109–119.

[48] A.B. Cruse, On removing a vertex from the assignment polytope, *Linear Algebra Appl.*, **26** (1979), 45–57.

[49] J. Csima and B.N. Datta, The DAD theorem for symmetric nonnegative matrices, *J. Combin. Theory, Ser. A*, **12** (1972), 147–152.

[50] W.H. Cunningham and Y. Wang, On the even permutation polytope, *Linear Algebra Appl.*, **389** (2004), 269–281.

[51] G. Dahl, Tridiagonal doubly stochastic matrices, *Linear Algebra Appl.*, **390** (2004), 197–208.

[52] W.E. Deming and F.F. Stephan, On a least squares adjustment of a sampled frequency table when the expected marginal totals are known, *Ann. Math. Statist.*, **11** (1940), 427–444.

[53] D. Djoković, Extreme points of certain convex polytopes, *Monatsh. Math.*, **69** (1965), 385–388.

[54] D. Djoković, Note on nonnegative matrices, *Proc. Amer. Math. Soc.*, **25** (1970), 80–92.

[55] T. Dridi, Sur les distributions binaires associées à des distributions ordinales, *Math. Sci. Humaines*, **18** (69) (1980), 15–31.

[56] A.L. Dulmage and N.S. Mendelsohn, The term and stochastic ranks of a matrix, *Canad. J. Math.*, **11** (1959), 269–279.

[57] J. Edmonds, Maximum matching and a polyhedron with 0, 1-vertices, *J. Res. Nat. Bur. Stands.*, **69B** (1965), 125–130.

[58] G.P. Egoryčev, A solution of van der Waerden's permanent problem (in Russian), *Dokl. Akad. Nauk SSSR*, **258** (1981), 1041–1044. Translated in *Soviet Math. Dokl.*, **23** (1981), 619–622.

[59] G.M. Engel and H. Schneider, Cyclic and diagonal products on a matrix, *Linear Algebra Appl.* **7** (1973), 301–335.

[60] S. Evdokimov, M. Karpinski, and I. Ponomarenko, Compact cellular algebras and permutation groups, Sixteenth British Combinatorial Conference (London 1997), *Discrete Math.*, **197/198** (1999), 247–267.

[61] D.I. Falikman, A proof of van der Waerden's conjecture on the permanent of a doubly stochastic matrix (in Russian), *Mat. Zametki*, **29** (1981), 931–938. Translated in *Math. Notes*, **29** (1981), 475–479.

[62] M. Fiedler, Doubly stochastic matrices and optimization, *Advances in Mathematical Optimization*, Math. Res. 45, Akademie-Verlag, Berlin, 44–51, 1988.

[63] T.H. Foregger, An upper bound for the permanent of a fully indecomposable matrix, *Proc. Amer. Math. Soc.*, **49** (1975), 319–324.

[64] S. Friedland, A proof of a generalized van der Waerden conjecture on permanents, *Linear Multilin. Alg.*, **11** (1982), 107–120.

[65] P.M. Gibson, Disproof of a conjecture on assignment functions, *Linear Multilin. Alg.*, **21** (1987), 87–89.

[66] C.D. Godsil, Compact graphs and equitable partitions, *Linear Algebra Appl.*, **255** (1997), 259–266.

[67] I.S. Gopal and C.K. Wong, Minimizing the number of switchings in an SS/TDMA system, *IEEE Trans. Comm.*, **Comm-37** (1985), 497–501.

[68] M. Grötschel, M. Jünger, and G. Reinelt, Acyclic subdigraphs and linear orderings: polytopes, facets and a cutting plane algorithm, *Graphs and Orders*, ed. I. Rival, Reidel, Dordrecht, 1985, 217–264.

[69] A.J. Hoffman, A special class of doubly stochastic matrices, *Aequationes Math.*, **2** (1969), 319–326.

[70] J. Hood and D. Perkinson, Some facets of the polytope of even permutation matrices, *Linear Algebra Appl.*, **381** (2004), 237–244.

[71] A. Horn, Doubly stochastic matrices and the diagonal of a rotation matrix, *Amer. J. Math.*, **76** (1954), 620–630.

[72] S.G. Hwang, On a conjecture of Dittert, *Linear Algebra Appl.*, **95** (1987), 161–169.

[73] W. Imrich, On the connectivity of Cayley graphs, *J. Combin. Theory Ser. B*, **26** (1979), 323–326.

[74] R.-Q. Jia, Symmetric magic squares and multivariate splines, *Linear Algebra Appl.*, **250** (1997), 69–103.

[75] L.S. Joel, The convex hull of the transposition matrices, *J. Res. Nat. Bur. Stands.*, **78B** (1974), 137–138.

[76] D.N. Johnson, A.L. Dulmage, and N.S. Mendelsohn, On an algorithm of G. Birkhoff concerning doubly stochastic matrices, *Canad. Math. Bull.*, **3** (1960), 237–242.

[77] M. Katz, On the extreme points of a certain convex polytope, *J. Combin. Theory*, **8** (1970), 417–423.

[78] M. Katz, On the extreme points of the set of substochastic and symmetric matrices, *J. Math. Anal. Appl.*, **37** (1972), 576–579.

[79] P. Knopp and R. Sinkhorn, Minimum permanents of doubly stochastic matrices with at least one zero entry, *Linear Multilin. Algebra*, **11** (1982), 351–355.

[80] W. Kolodziejczyk, New characterizations of doubly superstochastic matrices, *Found. Control Engrg.*, **13** (1989) 119–124.

[81] R.B. Levow, A problem of Mirsky concerning nonsingular doubly stochastic matrices, *Linear Algebra Appl.*, **5** (1972), 197–206.

[82] J.L. Lewandowski and C.L. Liu, SS/TDMA satellite communications with k-permutation modes, *SIAM J. Alg. Discrete Methods.*, **8** (1987), 519–534.

[83] J.L. Lewandowski, C.L. Liu, and J.W.-S. Liu, An algorithmic proof of a generalization of the Birkhoff–von Neumann theorem, *J. Algorithms*, **7** (1986), 323–330.

[84] D. London, Some notes on the van der Waerden conjecture, *Linear Algebra Appl.*, **4** (1971), 155–160.

[85] D. London, On matrices with doubly stochastic pattern, *J. Math. Anal. Appl.*, **34** (1971), 648–652.

[86] P.A. MacMahon, *Combinatory Analysis*, Cambridge U. Press, Cambridge, Vol. 1 1925; Vol. 2, 1916. Reprinted in one volume, Chelsea, New York, 1960.

[87] N.V. Mahdev and U.N. Peled, *Threshold Graphs and Related Topics, Annals of Discrete Mathematics, 56*, North-Holland, Amsterdam, 1995.

[88] M. Marcus and M. Newman, On the minimum of the permanent of a doubly stochastic matrix, *Duke Math. J.*, **26** (1959), 61–72.

[89] A.W. Marshall and I. Olkin, Scaling of matrices to achieve specified row and column sums, *Numer. Math.*, **12** (1968), 83–90.

[90] A.W. Marshall and I. Olkin, *Inequalities: Theory of Majorization and Its Applications*, Academic Press, New York, 1979.

[91] N.S. Mendelsohn and A.L. Dulmage, The convex hull of subpermutation matrices, *Proc. Amer. Math. Soc.*, **9** (1958), 253–254.

[92] M.V. Menon, Matrix links, an extremization problem, and the reduction of a nonnegative matrix to one with prescribed row and column sums, *Canad. J. Math.*, **20** (1968), 225–232.

[93] M.V. Menon and H. Schneider, The spectrum of a nonlinear operator associated with a matrix, *Linear Algebra Appl.*, **2** (1969), 321–334.

[94] R. Merris, Doubly stochastic graph matrices, *Univ. Beograd. Publ. Elektrotehn. Fak. (Ser. Mat.)*, **8** (1997), 64–71.

[95] R. Merris, Doubly stochastic graph matrices, II, *Linear Multilin. Alg.*, **45** (1998), 275–285.

[96] R. Merris and W. Watkins, Convex sets of doubly stochastic matrices, *J. Combin. Theory, Ser. A*, **16** (1974), 129–130.

[97] H. Minc, On lower bounds for permanents of (0, 1)-matrices, *Proc. Amer. Math. Soc.*, **22** (1969), 117–123.

[98] H. Minc, Doubly stochastic matrices with minimal permanents, *Pacific J. Math.*, **58** (1975), 155–157.

[99] H. Minc, *Permanents*, Addison-Wesley, Reading, 1978.

[100] H. Minc, A note on Egoryčev's proof of the van der Waerden conjecture, *Linear Multilin. Alg.*, **11** (1982), 367–371.

[101] H. Minc, Theory of permanents 1978-1981, *Linear Multilin. Alg.*, **12** (1983), 227–268.

[102] H. Minc, *Nonnegative Matrices*, Wiley-Interscience, New York, 1988.

[103] L. Mirsky, On a convex set of matrices, *Arch. Math.*, **10** (1959), 88–92.

[104] L. Mirsky, Even doubly-stochastic matrices, *Math. Annalen*, **144** (1961), 418–421.

[105] L. Mirsky, Results and problems in the theory of doubly-stochastic matrices, *Z. Wahrscheinlichkeitstheorie Verw. Gebiete*, **1** (1962/63), 319–344.

[106] D. Naddef, The Hirsch conjecture is true for (0, 1)-polytopes, *Math. Programming*, **45** (1989) (Ser. B), 109–110.

[107] D. Naddef and W.R. Pulleyblank, Hamiltonicity and combinatorial polyhedra, *J. Combin. Theory, Ser. B*, **31** (1981), 297–312.

[108] D. Naddef and W.R. Pulleyblank, Hamiltonicity in (0–1)-polyhedra, *J. Combin. Theory, Ser. B*, **37** (1984), 41–52.

[109] H. Nakazato, Set of 3×3 orthostochastic matrices, *Nihonkai Math. J.*, **7** (1996), 83–100.

[110] J. von Neumann, A certain zero-sum two-person game equivalent to the optimal assignment problem, *Contributions to the Theory of Games*, Vol. 2, Princeton U. Press, Princeton, 1953, 5–12.

[111] A. Nishi, An elementary proof of Johnson–Dulmage–Mendelsohn's refinement of Birkhoff's theorem on doubly stochastic matrices, *Canad. Math. Bull.*, **22** (1979), 81–86.

[112] Z. Nutov and M. Penn, On non-$(0, \frac{1}{2}, 1)$-extreme points of the generalized transitive tournament polytope, *Linear Algebra Appl.*, **223** (1996), 149–159.

[113] Z. Nutov and M. Penn, On the integral dicycle packings and covers and the linear ordering polytope, *Discrete Appl. Math.*, **60** (1995), 293–309.

[114] I. Pak, Four questions on Birkhoff polytope, *Ann. Combin.*, **4** (2000), 83–90.

[115] B.N. Parlett and T.L. Landis, Methods for scaling to doubly stochastic from, *Linear Algbra Appl.*, **48** (1982), 53–79.

[116] U.N. Peled and M.K. Srinivasan, The polytope of degree sequences, *Linear Algebra Appl.*, **114/115** (1989), 349–377.

[117] H. Perfect and L. Mirsky, The distribution of positive elements in doubly-stochastic matrices, *J. London Math. Soc.*, **40** (1965), 689–698.

[118] A. Rämmer, On even doubly stochastic matrices with minimal even permanent, *Tartu Riikl. Ül. Toimetised* **878** (1990), 103–114.

[119] A. Rämmer, On minimizing matrices, *Proceedings of the First Estonian Conference on Graphs and Applications (Tartu–Kääriku, 1991)*, Tartu U. Press, Tartu, 1993, 121–134.

[120] G. Reinelt, *The Linear Ordering Problem: Algorithms and Applications*, Heldermann, Berlin, 1985.

[121] H.J. Ryser, A new look at the optimal assignment problem, *Linear Algebra Appl.*, **66** (1985), 113–121.

[122] E.M. de Sá, Some subpolytopes of the Birkhoff polytope, *Electron. J. Linear Algebra*, **15** (2006), 1–7.

[123] G. Sabidussi, Graph multiplication, *Math. Z.*, **72** (1960), 446–457.

[124] H. Schrek and G. Tinhofer, A note on certain subpolytopes of the assignment polytope associated with circulant graphs, *Linear Algebra Appl.*, **111** (1988), 125–134.

[125] A. Schrijver, *Theory of Linear and Integer Programming*, Wiley-Interscience, Chichester, 1986.

[126] A. Schrijver, *Combinatorial Optimization. Polyhedra and Efficiency. Vol. A. Paths, Flows, Matchings*, Chapters 1–38, Algorithms and Combinatorics, 24, A, Springer, Berlin, 2003.

[127] R. Sinkhorn, A relationship between arbitrary positive matrices and doubly stochastic matrices, *Ann. Math. Statist.*, **35** (1964), 876–879.

[128] R. Sinkhorn, Diagonal equivalence to matrices with prescribed row and column sums, *Amer. Math. Monthly*, **74** (1967), 402–405.

[129] R. Sinkhorn, Diagonal equivalence to matrices with prescribed row and column sums. II, *Proc. Amer. Math. Soc.*, **45** (1974), 195–198.

[130] R. Sinkhorn, Doubly stochastic matrices which have certain diagonals with constant sums, *Linear Algebra Appl.*, **16** (1977), 79–82.

[131] R. Sinkhorn, A problem related to the van der Waerden permanent theorem, *Linear Multilin. Alg.*, **16** (1984), 167–173.

[132] R. Sinkhorn and P. Knopp, Concerning nonnegative matrices and doubly stochastic matrices, *Pacific J. Math.*, **21** (1967), 343–348.

[133] R. Sinkhorn and P. Knopp, Problems involving diagonal products in nonnegative matrices, *Trans. Amer. Math. Soc.*, **136** (1969), 67–75.

[134] R.P. Stanley, Linear homogeneous diophantine equations and magic labelings of graphs, *Duke Math. J.*, **40** (1973), 607–632.

[135] R.P. Stanley, *Enumerative Combinatorics, Volume I*, Cambridge U. Press, Cambridge, 1997.

[136] Z. Świtalski, Half-integrality of vertices of the generalized transitive tournament polytope ($n = 6$), *Discrete Math.*, **271** (2003), 251–260.

[137] G. Tinhofer, Graph isomorphism and theorems on Birkhoff type, *Computing*, **36** (1986), 285–300.

[138] G. Tinhofer, Strong tree-cographs are Birkhoff graphs, *Discrete Appl. Math.*, **22** (1988/89), 275–288.

[139] G. Tinhofer, A note on compact graphs, *Discrete Appl. Math.*, **30** (1991), 253–264.

[140] H. Tverberg, On the permanent of a bistochastic matrix, *Math. Scand.*, **12** (1963), 25–35.

[141] B.L. van der Waerden, Aufgabe 45, *Jber. Deutsch. Math.-Verein*, **35** (1926), 117.

[142] E.T.H. Wang, Maximum and minimum diagonal sums of doubly stochastic matrices, *Linear Algebra Appl.*, **8** (1974), 483–505.

[143] R.H. Warren, Revisiting the 0, 1 assignment problem, *Linear Algebra Appl.*, **160** (1992), 247–253.

[144] M.E. Watkins, Connectivity of transitive graphs, *J. Combin. Theory*, **8** (1970), 23–29.

[145] X.-D. Zhang, A note on doubly stochastic graph matrices, *Linear Algebra Appl.*, **407** (2005), 196–200.

[146] B. Zhou and B. Liu, Some results on compact graphs and supercompact graphs, *J. Math. Study*, **32** (1999), 133–136.

Master Bibliography

E. Achilles, Doubly stochastic matrices with some equal diagonal sums, *Linear Algebra Appl.*, **22** (1978), 293–296.

N. Achuthan, S.B. Rao, and A. Ramachandra Rao, The number of symmetric edges in a digraph with prescribed out-degrees, *Combinatorics and Applications*, Proceedings of the Seminar in Honor of Prof. S.S. Shrikhande on His 65th Birthday, ed. K.S. Vijayan and N.M. Singhi, Indian Statistical Institute, 1982, 8–20.

P. Acosta, A. Bassa, A. Chaiken, *et al.*, On a conjecture of Brualdi and Shen on block transitive tournaments, *J. Graph Theory*, **44** (2003), 213–230.

S.N. Afriat, On sum-symmetric matrices, *Linear Algebra Appl.*, **8** (1974), 129–140.

R.K. Ahujo, T.L. Magnanti, and J.B. Orlin, *Network Flows: Theory, Algorithms, and Applications*, Prentice-Hall, Englewood Cliffs, 1993.

H. Anand, V.C. Dumir, and H. Gupta, A combinatorial distribution problem, *Duke Math. J.*, **33** (1966), 757–769.

I. Anderson, Perfect matchings of a graph, *J. Combin. Theory, Ser. B*, **10** (1971), 183–186.

R.P. Anstee, Properties of a class of $(0, 1)$-matrices covering a given matrix, *Canad. J. Math*, **34** (1982), 438–453.

R.P. Anstee, Triangular $(0, 1)$-matrices with prescribed row and column sums, *Discrete Math.*, **40** (1982), 1–10.

R.P. Anstee, The network flows approach for matrices with given row and column sums, *Discrete Math.*, **44** (1983), 125–138.

R.P. Anstee, Invariant sets of arcs in network flow problems, *Discrete Appl. Math.*, **13** (1986), 1–7.

S. Ao and D. Hanson, Score vectors and tournaments with cyclic chromatic number 1 or 2, *Ars Combin.*, **49** (1998), 185–191.

511

P. Avery, Score sequences of oriented graphs, *J. Graph Theory*, **15** (1991), 251–257.

K. S. Bagga, L. W. Beineke and F. Harary, Two problems on colouring tournaments, *Vishwa Internat. J. Graph Theory*, **1** (1992), 83–94.

K. Balasubramanian, On equality of some elements in matrices, *Linear Algebra Appl.*, **22** (1978), 135–138.

K. Balasubramanian, Maximal diagonal sums, *Linear Multilin. Alg.*, **7** (1979), 249–251.

M.L. Balinski, Establishing the matching polytope, *J. Combin. Theory, Ser. B*, **13** (1972), 1–13.

M.L. Balinski, On two special classes of transportation polytopes, *Math. Programming Study*, **1** (1974), 43–58.

M.L. Balinski and A. Russakoff, On the assignment polytope, *SIAM Review*, **16** (1974), 516–525.

C.M. Bang and H. Sharp, An elementary proof of Moon's theorem on generalized tournaments, *J. Combin. Theory, Ser. B*, **22** (1977), 299–301.

C.M. Bang and H. Sharp, Score vectors of tournaments, *J. Combin. Theory, Ser. B*, **26** (1979), 81-84.

R. Bapat, Doubly stochastic matrices with equal subpermanents, *Linear Algebra Appl.*, **51** (1983), 1–8.

R.B. Bapat and T.E.S. Raghavan, *Nonnegative Matrices and Applications*, Cambridge U. Press, Cambridge, 1997.

L.W. Beineke and F. Harary. Local restrictions for various classes of directed graphs, *J. London Math. Soc.*, **40** (1965), 87–95.

L.W. Beineke and J.W. Moon, On bipartite tournaments and scores, *The Theory and Application of Graphs* (Kalamazoo, Michigan, 1980), Wiley, New York, 1981, 55–71.

J. von Below, On a theorem of L. Mirsky on even doubly-stochastic matrices, *Discrete Math.*, **55** (1985), 311–312.

I. Bengtsson, Å. Ericsson, and M. Kuś, Birkhoff's polytope and unistochastic matrices $N = 3$ and $N = 4$, preprint.

C. Berge, *Graphs and Hypergraphs*, North-Holland, Amsterdam, 1973 (translation and revision of *Graphes et hypergraphes*, Dunod, Paris, 1970).

S.K. Bhandir and S. Das Gupta, Two characterizations of doubly-superstochastic matrices, *Saṅkhyā Ser. A*, **47** (1985), 357–365.

G. Birkhoff, Tres observaciones sobre el algebra lineal, *Univ. Nac. Tucumán Rev. Ser.* A, vol. no. (1946), 147–151.

E.D. Bolker, Transportation polytopes, *J. Combin. Theory*, **13** (1972), 251–262.

B. Bollobás, *Modern Graph Theory*, Grad. Texts Math. **184**, Springer, New York, 1998.

V.I. Bol'shakov, Upper values of a permanent in Γ_n^k (in Russian), *Combinatorial analysis, No. 7*, Moscow, 92–118 and 164–165 (1986).

V.I. Bol'shakov, The spectrum of the permanent on Γ_N^k (in Russian), *Proceedings of the All-Union Seminar on Discrete Mathematics and Its Applications, Moscow (1984)*, Moskov. Gos. Univ. Mekh.-Mat. Fak., Moscow, 1986, 65–73.

G. Bongiovanni, D. Bovet, and A. Cerioli, Comments on a paper of R.A. Brualdi: "Notes on the Birkhoff algorithm for doubly stochastic matrices," *Canad. Math. Bull.*, **31** (1988), 394–398.

A. Borobia, $(0, \frac{1}{2}, 1)$ matrices which are extreme points of the generalized transitive tournament polytope, *Linear Algebra Appl.*, **220** (1995), 97–110.

A. Borobia, Vertices of the generalized transitive tournament polytope, *Discrete Math.*, **179** (1998), 49–57.

A. Borobia and V.Chumillas, *-graphs of vertices of the generalized transitive tournament polytope, *Discrete Math.*, **163** (1997), 229–234.

A. Borobia, Z. Nutov, and M. Penn, Doubly stochastic matrices and dicycle coverings and packings in Eulerian digraphs, *Linear Algebra Appl.* **223/224** (1995), 361–371.

A.V. Borovik, I.M. Gelfand, and N. White, *Coxeter Matroids*, Birkhäuser, Boston–Basle–Berlin, 2003.

J.V. Brawley and L. Carlitz, Enumeration of matrices with prescribed row and column sums, *Linear Algebra Appl.*, **6** (1973), 165–174.

L.M. Brégman, Certain properties of nonnegative matrices and their permanents (in Russian), *Dokl. Akad. Nauk SSSR*, **211** (1973), 27–30. Translated in *Soviet Math. Dokl.*, **14** (1973), 194–230.

D.M. Bressoud, Colored tournaments and Weyl's denominator formula, *European J. Combin.*, **8** (1987), 245–255.

R.A. Brualdi, Convex sets of non-negative matrices, *Canad. J. Math.*, **20** (1968), 144–157.

R.A. Brualdi, Matchings in arbitrary graphs, *Proc. Cambridge Philos. Soc.*, **69** (1971), 401–407.

R.A. Brualdi, The *DAD* theorem for arbitrary row sums, *Proc. Amer. Math. Soc.*, **45** (1974), 189–194.

R.A. Brualdi, Combinatorial properties of symmeteric non-negative matrices, *Teorie combinatorie, Atti dei Convegni Lincei, 17*, ed. B. Segre, Accademia Nazionale dei Lincei, Rome, 1976, 99–120.

R.A. Brualdi, Convex polytopes of permutation invariant doubly stochastic matrices, *J. Combin. Theory, Ser. B*, **23** (1977), 58–67.

R.A. Brualdi, The chromatic index of the graph of the assignment polytope, *Ann. Disc. Math.*, **3** (1978), 49–53.

R.A. Brualdi, Matrices of zeros and ones with fixed row and column sum vectors, *Linear Algebra Appl.*, **33** (1980), 159–231.

R.A. Brualdi, Matrices of 0's and 1's with total support, *J. Combin. Theory, Ser. A*, **28** (1980), 249–256.

R.A. Brualdi, On Haber's minimum term rank formula, *Europ. J. Combin.*, **2** (1981), 17–20.

R.A. Brualdi, Notes on the Birkhoff algorithm for doubly stochastic matrices, *Canad. Math. Bull.*, **25** (1982), 191–199.

R.A. Brualdi, The doubly stochastic matrices of a vector majorization, *Linear Algebra Appl.*, **61** (1984), 141–154.

R.A. Brualdi, An interesting face of the polytope of doubly stochastic matrices, *Linear Multilin. Alg.*, **17** (1985), 5–18.

R.A. Brualdi, Some application of doubly stochastic matrices, *Linear Algebra Appl.*, **107** (1988), 77–100.

R.A. Brualdi, Short proofs of the Gale/Ryser and Ford/Fulkerson characterization of the row and column sum vectors of $(0, 1)$-matrices, *Math. Inequalities Appl.*, **4** (2001), 157–159

R.A. Brualdi, *Introductory Combinatorics*, 4th ed., Prentice-Hall, Upper Saddle River, 2004.

R.A. Brualdi, Algorithms for constructing $(0,1)$-matrices with prescribed row and column sum vectors, *Discrete Math.*, to appear.

R.A. Brualdi and G. Dahl, Majorization-constrained doubly stochastic matrices, *Linear Algebra Appl.*, **361** (2003), 75–97.

R.A. Brualdi and G. Dahl, Matrices of zeros and ones with given line sums and a zero block, *Linear Algebra Appl.*, **371** (2003), 191–207.

R.A. Brualdi and G. Dahl, Constructing (0,1)-matrices with given line sums and certain fixed zeros, to appear.

R.A. Brualdi and L. Deaett, More on the Bruhat order for $(0, 1)$-matrices, *Linear Algebra Appl.*, to appear.

R.A. Brualdi and T. Foregger, Matrices with constant permanental minors, *Linear Multilin. Alg.*, **3** (1975–76), 227–243.

R.A. Brualdi and P.M. Gibson, The assignment polytope, *Math. Programming*, **11** (1976), 97–101.

R.A. Brualdi and P.M. Gibson, Convex polyhedra of doubly stochastic matrices I. Applications of the permanent function, *J. Combin. Theory, Ser. A*, **22** (1977), 194–230.

R.A. Brualdi and P.M. Gibson, Convex polyhedra of doubly stochastic matrices II. Graph of Ω_n, *J. Combin. Theory, Ser. B*, **22** (1977), 175–198.

R.A. Brualdi and P.M. Gibson, Convex polyhedra of doubly stochastic matrices III. Affine and combinatorial properties of Ω_n, *J. Combin. Theory, Ser. A*, **22** (1977), 338–351.

R.A. Brualdi and P.M. Gibson, Convex polyhedra of doubly stochastic matrices IV, *Linear Algebra Appl.*, **15** (1976), 153–172.

R.A. Brualdi, J.L. Goldwasser, and T.S. Michael, Maximum permanents of matrices of zeros and ones, *J. Combin. Theory, Ser. A*, **47** (1988), 207–245.

R.A. Brualdi, D.J. Hartfiel, and S.-G. Hwang, On assignment functions, *Linear Multilin. Alg.*, **19** (1986), 203–219.

R.A. Brualdi and G.-S. Hwang, Generalized transitive tournaments and doubly stochastic matrices, *Linear Algebra Appl.*, **172** (1992), 151–168.

R.A. Brualdi and S.-G. Hwang, Vector majorization via Hessenberg matrices, *J. London Math. Soc.*, **53** (1996), 28–38.

R.A. Brualdi and S.-G. Hwang, A Bruhat order for the class of (0,1)-matrices with row sum vector R and column sum vector S, *Electron. J. Linear Algebra*, **12** (2004), 6–16.

R.A. Brualdi, S.-G. Hwang, and S.-S. Pyo, Vector majorization via positive definite matrices, *Linear Algebra Appl.*, **257** (1997), 105–120.

R.A. Brualdi and G.M. Lee, On the truncated assignment polytope, *Linear Algebra Appl.*, **19** (1978), 33–62.

R.A. Brualdi and Q. Li, Small diameter interchange graphs of classes of matrices of zeros and ones, *Linear Algebra Appl.*, **46** (1982), 177–194.

R.A. Brualdi and Q. Li, Upsets in round robin tournaments, *J. Combin. Theory, Ser. B*, **35** (1983), 62–77.

R.A. Brualdi and Q. Li, The interchange graph of tournaments with the same score vector, *Progress in Graph Theory*, Academic Press, Toronto, 1984, 128–151.

R.A. Brualdi and B. Liu, The polytope of even doubly stochastic matrices, *J. Combin. Theory, Ser. A*, **57** (1991), 243–253.

R.A. Brualdi and B. Liu, A lattice generated by (0,1)-matrices, *Ars Combin.*, **31** (1991), 183–190.

R.A. Brualdi and R. Manber, Chromatic number of classes of matrices of 0's and 1's, *Discrete Math.*, **50** (1984), 143–152.

R.A. Brualdi and R. Manber, Prime interchange graphs of classes of matrices of zeros and ones, *J. Combin. Theory, Ser. B*, **35** (1983), 156–170.

R.A. Brualdi and R. Manber, Chromatic number of classes of matrices of 0's and 1's, *Discrete Math.*, **50** (1984), 143–152.

R.A. Brualdi, R. Manber, and J.A. Ross, On the minimum rank of regular classes of matrices of zeros and ones, *J. Combin. Theory, Ser. A*, **41** (1986), 32–49.

R.A. Brualdi and T.S. Michael, The class of 2-multigraphs with a prescribed degree sequence, *Linear Multilin. Alg.*, **24** (1989), 81–102.

R.A. Brualdi, S.V. Parter, and H. Schneider, The diagonal equivalence of a nonnegative matrix to a stochastic matrix, *J. Math. Anal. Appl.*, **16** (1966), 31–50.

R.A. Brualdi and J.A. Ross, Invariant sets for classes of matrices of zeros and ones, *Proc. Amer. Math. Soc.*, **80** (1980), 706–710.

R.A. Brualdi and J.A. Ross, On Ryser's maximum term rank formula, *Linear Algebra Appl.*, **29** (1980), 33–38.

R.A. Brualdi and H.J. Ryser, Classes of matrices associated with the optimal assignment problem, *Linear Algebra Appl.*, **81** (1986), 1–17.

R.A. Brualdi and H.J. Ryser, *Combinatorial Matrix Theory*, Cambridge U. Press, Cambridge, 1991.

R.A. Brualdi and J.G. Sanderson, Nested species subsets, gaps, and discrepancy, *Oecologia*, **119** (1999), 256–264.

R.A. Brualdi and J. Shen, Dicrepancy of matrices of zeros and ones, *Electron. J. Combin.*, **6** (1999), #R15.

R.A. Brualdi and J. Shen, Landau's inequalities for tournament scores and a short proof of a theorem on transitive sub-tournaments, *J. Graph Theory*, (2001), 244–254.

R.A. Brualdi and J. Shen, Disjoint cycles in Eulerian digraphs and the diameter of interchange graphs, *J. Combin. Theory, Ser. B*, **85** (2002), 189–196.

R.A. Brualdi and H.W. Wielandt, A spectral characterization of stochastic matrices, *Linear Algebra Appl.*, **1** (1968), 65–71.

R.E. Burkhard, Time-slot assignment for TDMA-systems, *Computing*, **35** (1985), 99–112.

D. de Caen, D.A. Gregory, and N.J. Pullman, The Boolean rank of zero–one matrices, *Proceedings of the Third Caribbean Conference on Combinatorics and Computing*, Barbados (1981), 169–173.

D. de Caen and D.G. Hoffman, Impossibility of decomposing the complete graph on n points into $n-1$ isomorphic complete bipartite graphs, *SIAM J. Discrete Math.*, **2** (1989), 48–50.

R.M. Calderbank and P. Hanlon, The extension to root systems of a theorem on tournaments, *J. Combin. Theory, Ser. A*, **41** (1986), 228–245.

P. Camion, Chemins et circuits hamiltoniens des graphes complets, *C.R. Acad. Sci. Paris, Sér. A*, **259** (1959), 2151–2152.

E.R. Canfield and B.D. McKay, Asymptotic enumeration of 0–1 matrices with constant row and column sums, *Electron. J. Combin.*, **12** (2005), # 29.

C.S. Chan and D.P. Robbins, On the volume of the polytope of doubly stochastic matrices, *Experiment. Math.*, **8** (1999), 291–300.

K.-M. Chao and C.S. Wong, Applications of M-matrices to majorizations, *Linear Algebra Appl.*, **169** (1992), 31–40.

I. Charon and O. Hurdy, Links between the Slater index and the Ryser index of tournaments, *Graphs Combin.*, **19** (2003), 323–334.

W.K. Chen, *Applied Graph Theory*, North-Holland, Amsterdam, 1976.

R.S. Chen, X.F. Guo, amd F.J. Zhang, The edge connectivity of interchange graphs of classes of matrices of zeros and ones, *J. Xinjiang Univ. Natur. Sci.*, **5** (1) (1988), 17–25.

W.Y.C. Chen, A counterexample to a conjecture of Brualdi and Anstee (in Chinese), *J. Math. Res. Exposition*, **6** (1986), 68.

W.Y.C. Chen, Integral matrices with given row and column sums, *J. Combin. Theory, Ser. A*, **2** (1992), 153–172.

W.Y.C. Chen and A. Shastri, On joint realization of $(0,1)$-matrices, *Linear Algebra Appl.*, **112** (1989), 75–85.

G.-S. Cheon and S.-Z. Song, On the extreme points on the majorization polytope $\Omega_3(y \prec x)$, *Linear Algebra Appl.*, **269** (1998), 47–52.

G.-S. Cheon and I.M. Wanless, An update on Minc's survey of open problems involving permanents, *Linear Alg. Appl.*, **403** (2005), 314–342.

H.H. Cho, C.Y. Hong, S.-R. Kim, C.H. Park, and Y. Nam, On $\mathcal{A}(R,S)$ all of whose members are indecomposable, *Linear Algebra Appl.*, **332/334** (2001), 119–129.

S. Cho and Y. Nam, Convex polytopes of generalized doubly stochastic matrices, *Commun. Korean Math. Soc.* **16** (2001), 679–690.

V. Chungphaisan, Conditions for sequences to be r-graphic, *Discrete Math.*, **7** (1974), 31–39.

G.W. Cobb and Y.-P. Chen, An application of Markov chain Monte Carlo to community ecology, *Amer. Math. Monthly*, **110** (2003), 265–288.

L. Comtet, *Advanced Combinatorics*, Reidel, Dordrecht, 1974.

G. Converse and M. Katz, Symmetric matrices with given row sums, *J. Combin. Theory*, **18** (1975), 171–176.

A.B. Cruse, A proof of Fulkerson's characterization of permutation matrices, *Linear Algebra Appl.*, **12** (1975), 21–28.

A.B. Cruse, A note on symmetric doubly-stochastic matrices, *Discrete Math.*, **13** (1975), 109–119.

A.B. Cruse, On removing a vertex from the assignment polytope, *Linear Algebra Appl.*, **26** (1979), 45–57.

A.B. Cruse, On linear programming duality and Landau's characterization of tournament scores, preprint.

J. Csima and B.N. Datta, The DAD theorem for symmetric nonnegative matrices, *J. Combin. Theory, Ser. A*, **12** (1972), 147–152.

W.H. Cunningham and Y. Wang, On the even permutation polytope, *Linear Algebra Appl.*, **389** (2004), 269–281.

G. Dahl, Tridiagonal doubly stochastic matrices, *Linear Algebra Appl.*, **390** (2004), 197–208.

G.B. Dantzig, *Linear Programming and Extensions*, Princeton U. Press, Princeton, 1962.

W.E. Deming and F.F. Stephan, On a least squares adjustment of a sampled frequency table when the expected marginal totals are known, *Ann. Math. Statist.*, **11** (1940), 427–444.

O. Demuth, A remark on the transportation problem (in Czech, with German and Russian summaries), *Časopis Pěst. Math.*, **86** (1961), 103–110.

D. Djoković, Extreme points of certain convex polytopes, *Monatsh. Math.*, **69** (1965), 385–388.

D. Djoković. Note on nonnegative matrices, *Proc. Amer. Math. Soc.*, **25** (1970), 80–92.

A. Doig, The minimum number of basic feasible solutions to a transport problem, *Operational Res. Quart.*, **14** (1963), 387–391.

T. Dridi, Sur les distributions binaires associées à des distributions ordinales, *Math. Sci. Humaines*, **18** (69) (1980), 15–31.

J. Dubois, Polytopes de transport symétriques, *Discrete Appl. Math.*, **4** (1973), 1–27.

A.L. Dulmage and N.S. Mendelsohn, The term and stochastic ranks of a matrix, *Canad. J. Math.*, **11** (1959), 269–279.

J. Edmonds, Maximum matching and a polyhedron with $0, 1$-vertices, *J. Res. Nat. Bur. Stands.*, **69B** (1965), 125–130.

R.B. Eggleton and D.A. Holton, The graph of type $(0, \infty, \infty)$ realizations of a graphic sequence, *Combinatorial Mathematics VI* (Proc. Sixth Austral. Conf., Univ. New England, Armidale, 1978), Lecture Notes in Math. 748, Springer, Berlin, 1978, 41–54.

E.E. Eischen, C.R. Johnson, K. Lange, and D.P. Stanford, Patterns, line-sums, and symmetry, *Linear Algebra Appl.*, **357** (2002), 273–2.

G.P. Egoryčev, A solution of van der Waerden's permanent problem (in Russian), *Dokl. Akad. Nauk SSSR*, **258** (1981), 1041–1044. Translated in *Soviet Math. Dokl.*, **23** (1981), 619–622.

V.A. Emeličev, M.K. Kravcov, and A.P. Kračkovskiĭ, On some classes of transportation polytopes, *Soviet Math. Dokl.*, **19** (1978), 889–893.

V.A. Emeličev, M.K. Kravtsov and A.P. Kračkovskiĭ, Transportation polytopes with maximal number of k-faces, *Soviet Math. Dokl.*, **31** (1985), 504–508.

G.M. Engel and H. Schneider, Cyclic and diagonal products on a matrix, *Linear Algebra Appl.* **7** (1973), 301–335.

P. Erdős and T. Gallai, Graphs with prescribed degrees of vertices (in Hungarian), *Mat. Lapok*, **11** (1960), 264–274.

R. Ermers and B. Polman, On the eigenvalues of the structure matrix of matrices of zeros and ones, *Linear Algebra Appl.*, **95** (1987), 17–41.

S. Evdokimov, M. Karpinski, and I. Ponomarenko, Compact cellular algebras and permutation groups, sixteenth British Combinatorial Conference (London 1997), *Discrete Math.*, **197/198** (1999), 247–267.

D.I. Falikman, A proof of van der Waerden's conjecture on the permanent of a doubly stochastic matrix (in Russian), *Mat. Zametki*, **29** (1981), 931–938. Translated in *Math. Notes*, **29** (1981), 475–479.

S. Fallat and P. van den Driessche, Maximum determinant of $(0, 1)$-matrices with certain constant row and column sums, *Linear Multilin. Alg.*, **42** (1997), 303–318.

M. Fiedler, Doubly stochastic matrices and optimization, *Advances in Mathematical Optimization*, Math. Res. 45, Akademie-Verlag, Berlin, 1988, 44–51.

L. R. Ford, Jr. and D. R. Fulkerson, *Flows in Networks*, Princeton U. Press, Princeton, 1962.

T.H. Foregger, An upper bound for the permanent of a fully indecomposable matrix, *Proc. Amer. Math. Soc.*, **49** (1975), 319–324.

S. Friedland, A proof of a generalized van der Waerden conjecture on permanents, *Linear Multilin. Alg.*, **11** (1982), 107–120.

D.R. Fulkerson, Hitchcock transportation problem, *RAND Corporation Report*, P-890 (1956).

D.R. Fulkerson, Zero–one matrices with zero trace, *Pacific J. Math.*, **10** (1960), 831–836.

D.R. Fulkerson, Upsets in round robin tournaments, *Canad. J. Math.*, **17** (1965), 957–969.

D.R. Fulkerson, Blocking polyhedra, *Graph Theory and Its Applications*, ed. B. Harris, Academic Press, 1970, 93–112.

D.R. Fulkerson, A.J. Hoffman, and M.H. McAndrew, Some properties of graphs with multiple edges, *Canad. J. Math.*, **17** (1965), 166–177.

D.R. Fulkerson and H.J. Ryser, Widths and heights of (0,1)-matrices, *Canad. J. Math.*, **13** (1961), 239–255.

D.R. Fulkerson and H.J. Ryser, Multiplicities and minimal widths for (0,1)-matrices, *Canad. J. Math.*, **14** (1962), 498–508.

D.R. Fulkerson and H.J. Ryser, Width sequences for special classes of (0,1)-matrices, *Canad. J. Math.*, **15** (1963), 371–396.

W. Fulton, *Young Tableaux*, London Math. Soc. Student Texts 35, Cambridge U. Press, Cambridge, 1997.

D. Gale, A theorem on flows in networks, *Pacific. J. Math.*, **7** (1957), 1073–1082.

A.O. Gel'fond, Some combinatorial properties of (0,1)-matrices, *Math. USSR – Sbornik*, **4** (1968), 1.

I. Gessel, Tournaments and Vandermonde's determinant, *J. Graph Theory*, **3** (1979), 305–307.

P.M. Gibson, Facets of faces of transportation polytopes, Congressus Numerantium No. XVII (1976), *Utilitas Math.*, Winnipeg, Man., 323–333.

P.M. Gibson, A bound for the number of tournaments with specified scores, *J. Combin. Theory, Ser. B*, **36** (1984), 240–243.

P.M. Gibson, Disproof of a conjecture on assignment functions, *Linear Multilin. Alg.*, **21** (1987), 87–89.

P.M. Gibson, Disjoint 3-cycles in strong tournaments, unpublished manuscript.

C.D. Godsil, Compact graphs and equitable partitions, *Linear Algebra Appl.*, **255** (1997), 259–266.

A.J. Goldman and R.H. Byrd, Minimum-loop realization of degree sequences, *J. Res. Nat. Bur. Stands.*, **87** (1982), 75–78.

I.S. Gopal and C.K. Wong, Minimizing the number of switchings in an SS/TDMA system, *IEEE Trans. Comm.*, **Comm-37** (1985), 497–501.

M. Grötschel, M. Jünger, and G. Reinelt, Acyclic subdigraphs and linear orderings: polytopes, facets and a cutting plane algorithm, *Graphs and Orders*, ed. I. Rival, Reidel, Dordrecht, 1985, 217–264.

M. Grady, Research problems: combinatorial matrices having minimal nonzero determinant, *Linear Multilin. Alg.*, **35** (1993), 179–183.

M. Grady and M. Newman, The geometry of an interchange: minimal matrices and circulants, *Linear Algebra Appl.*, **262** (1997), 11–25.

C. Greene, An extension of Schensted's theorem, *Advances in Math.*, **14** (1974), 254–265.

D.A. Gregory, K.F. Jones, J.R. Lundgren, and N.J. Pullman, Biclique coverings of regular bigraphs and minimum semiring ranks of regular matrices, *J. Combin. Theory, Ser. B*, **51** (1991), 73–89.

B. Guiduli, A. Gyárfás, S. Thomassé and P. Weidl, 2-partition-transitive tournaments, *J. Combin. Theory, Ser. B*, **72** (1998), 181–196.

R.M. Haber, Term rank of $(0,1)$-matrices, *Rend. Sem. Mat. Padova*, **30** (1960), 24–51.

R.M. Haber, Minimal term rank of a class of (0,1)-matrices, *Canad. J. Math.*, **15** (1963), 188–192.

S.L. Hakimi, On realizability of a set of integers as degrees of the vertices of a linear graph I, *J. Soc. Indust. Appl. Math.*, **10** (1962), 496–506.

S.L. Hakimi, On realizability of a set of integers as degrees of the vertices of a linear graph II: uniqueness, *J. Soc. Indust. Appl. Math.*, **11** (1963), 135–147.

G.H. Hardy, J.E. Littlewood, and G. Pólya, *Inequalities*, Cambridge U. Press, Cambridge, 1934 (1st ed.), 1952 (2nd ed.).

V. Havel, A remark on the existence of finite graphs (in Czech), *Časopis Pěst. nat.*, **80** (1955), 477–485.

K.A.S. Hefner and J.R. Lundgren, Minimum matrix rank of k-regular $(0,1)$-matrices, *Linear Algebra Appl.*, **133** (1990), 43–52.

J.R. Henderson, Permanents of $(0,1)$-matrices having at most two 0's per line, *Canad. Math. Bull.*, **18** (1975), 353–358.

J.A. Henderson, Jr., On the enumeration of rectangular $(0,1)$-matrices, *J. Statist. Comput. Simul.*, **51** (1995), 291–313.

J.R. Henderson and R.A. Dean, The 1-width of (0,1)-matrices having constant row sum 3, *J. Combin. Theory, Ser. A*, **16** (1974), 355–370.

J.R. Henderson and R.A. Dean, A general upper bound for 1-widths, *J. Combin. Theory, Ser. A*, **18** (1975), 236–238.

G.T. Herman and A. Kuba, Discrete tomography: a historical overview, *Discrete Tomography: Foundations, Algorithms, and Applications*, ed. G.T. Herman and A. Kuba, Birkhäuser, Boston, 1999, 3–34.

A.J. Hoffman, A special class of doubly stochastic matrices, *Aequationes Math.*, **2** (1969), 319–326.

J. Hood and D. Perkinson, Some facets of the polytope of even permutation matrices, *Linear Algebra Appl.*, **381** (2004), 237–244.

A. Horn, Doubly stochastic matrices and the diagonal of a rotation matrix, *Amer. J. Math.*, **76** (1954), 620–630.

D.J. Houck and M.E. Paul, Nonsingular 0–1 matrices with constant row and column sums, *Linear Algebra Appl.*, **50** (1978), 143–152.

D.J. Houck and A.O. Pittenger, Analytic proofs of a network feasibility theorem and a theorem of Fulkerson, *Linear Algebra Appl.*, **24** (1979), 151–156.

S.G. Hwang, On a conjecture of Dittert, *Linear Algebra Appl.*, **95** (1987), 161–169.

W. Imrich, Über das schwache kartesische Produkt von Graphen, *J. Combin. Theory*, **11** (1971), 1–16.

W. Imrich, On the connectivity of Cayley graphs, *J. Combin. Theory, Ser. B*, **26** (1979), 323–326.

M. Jerrum, Mathematical foundations of the Markov chain Monte Carlo method, *Probabilistic Methods for Algorithmic Discrete Mathematics*, Algorithms and Combinatorics, 16, ed. M. Habib, C. McDiarmid, J. Ramírez-Alfonsín, and B. Reed, Springer, Berlin, 1998, 116–165.

R.-Q. Jia, Symmetric magic squares and multivariate splines, *Linear Algebra Appl.*, **250** (1997), 69–103.

L.S. Joel, The convex hull of the transposition matrices, *J. Res. Nat. Bur. Stands.*, **78B** (1974), 137–138.

C.R. Johnson and D.P. Stanford, Patterns that allow given row and column sums, *Linear Algebra Appl.*, **327** (2001), 53–60.

D.N. Johnson, A.L. Dulmage, and N.S. Mendelsohn, On an algorithm of G. Birkhoff concerning doubly stochastic matrices, *Canad. Math. Bull.*, **3** (1960), 237–242.

L.K. Jorgensen, Rank of adjacency matrices of directed (strongly) regular graphs, *Linear Algebra Appl.*, **407** (2005), 233–241.

D. Jungnickel and M. Leclerc, A class of lattices, *Ars Combin.*, **26** (1988), 243–248.

W.B. Jurkat and H.J. Ryser, Term ranks and permanents of nonnegative matrices, *J. Algebra*, **5** (1967), 342–357.

R. Kannan, P. Tetali, and S. Vempala, Simple Markov chain algorithms for generating bipartite graphs and tournaments, *Random Struc. Algorithms*, **14** (1999), 293–308.

M. Katz, On the extreme points of a certain convex polytope, *J. Combin. Theory*, **8** (1970), 417–423.

M. Katz, On the extreme points of the set of substochastic and symmetric matrices, *J. Math. Anal. Appl.*, **37** (1972), 576–579.

V. Klee and C. Witzgall, Facets and vertices of transportation polytopes, *Mathematics of Decision Sciences, Part I*, ed. G.B. Dantzig and A. Veinott, American Mathematical Society, Providence, 1967, 257–282.

D.J. Kleitman, Minimal number of multiple edges in realization of an incidence sequence without loops, *SIAM J. Appl. Math.*, **18** (1970), 25–28.

D.J. Kleitman and D.L. Wang, Algorithms for constructing graphs and digraphs with given valencies and factors, *Discrete Math.*, **6** (1973), 79–88.

A. Khintchine, Über eine Ungleichung, *Mat. Sb.*, **39** (1932), 180–189.

P. Knopp and R. Sinkhorn, Minimum permanents of doubly stochastic matrices with at least one zero entry, *Linear Multilin. Algebra*, **11** (1982), 351–355.

D.E. Knuth, Permutation matrices and generalized Young tableaux, *Pacific J. Math*, **34** (1970), 709–727.

W. Kolodziejczyk, New characterizations of doubly superstochastic matrices, *Found. Control Engrg.*, **13** (1989), 119–124.

M.K. Kravtsov, A proof of the Dubois hypothesis on the maximum number of vertices of a symmetric transportation polytope (in Russian), *Dokl. Akad. Nauk BSSR*, **33** (1989), 9–12.

M.K. Kravtsov, Transportation polytopes with a minimal number of k-faces, *Discrete Math. Appl.*, **3** (1993), 115–125.

S. Kundu, The k-factor conjecture is true, *Discrete Math.*, **6** (1973), 367–376.

C.W.H. Lam, The distribution of 1-widths of $(0, 1)$-matrices, *Discrete Math.*, **20** (1970), 109–122.

H. G. Landau, On dominance relations and the structure of animal societies. III. The condition for a score structure, *Bull. Math. Biophys.* **15** (1953), 143–148.

V.K. Leont'ev, An upper bound for the α-height of $(0, 1)$-matrices, *Mat. Zametki*, **15** (1974), 421–429.

R.B. Levow, A problem of Mirsky concerning nonsingular doubly stochastic matrices, *Linear Algebra Appl.*, **5** (1972), 197–206.

J.L. Lewandowski and C.L. Liu, SS/TDMA satellite communications with k-permutation modes, *SIAM J. Alg. Discrete Methods.*, **8** (1987), 519–534.

J.L. Lewandowski, C.L. Liu, and J.W.-S. Liu, An algorithmic proof of a generalization of the Birkhoff-von Neumann theorem, *J. Algorithms*, **7** (1986), 323–330.

M. Lewin, On the extreme points of the polytope of symmetric matrices with given row sums, *J. Combin. Theory, Ser. A*, **23** (1977), 223–232.

C.-K. Li, On certain convex matrix sets, *Discrete Math.*, **79** (1990), 323–326.

C.K. Li, J.S.-J. Lin and L. Rodman, Determinants of certain classes of zero–one matrices with equal line sums, *Rocky Mountain J. Math.*, **29** (1999), 1363–1385.

C.K. Li, D.D. Olesky, D.P. Stanford, and P. Van den Driessche, Minimum positive determinant of integer matrices with constant row and column sums, *Linear Multilin. Alg.*, **40** (1995), 163–170.

Q. Li and H.H. Wan, Generating function for cardinalities of several (0,1)-matrix classes, preprint.

X.L. Li and F.J. Zhang, Hamiltonicity of a type of interchange graphs, Second Twente Workshop on Graphs and Combinatorial Optimization (Enschede, 1991), *Discrete Appl. Math.*, **51** (1994), 107–111.

R. Loewy, D.R. Shier, and C.R. Johnson, Perron eigenvectors and the symmetric transportation polytope, *Linear Algebra Appl.*, **150** (1991), 139–155.

D. London, Some notes on the van der Waerden conjecture, *Linear Algebra Appl.*, **4** (1971), 155–160.

D. London, On matrices with doubly stochastic pattern, *J. Math. Anal. Appl.*, **34** (1971), 648–652.

W.E. Longstaff, Combinatorial solution of certain systems of linear equations involving (0, 1)-matrices, *J. Austral. Math. Soc. Ser. A*, **23** (1977), 266–274.

L. Lovász, Valencies of graphs with 1-factors, *Period. Math. Hung.*, **5** (1974), 149–151.

P.A. MacMahon, *Combinatory Analysis*, Cambridge U. Press, Cambridge: Vol. 1 1925; Vol. 2, 1916. Reprinted in one volume, Chelsea, New York, 1960.

P. Magyar, Bruhat order for two flags and a line, *J. Algebraic Combin.*, **21** (2005), 71–101.

N.V. Mahdev and U.N. Peled, *Threshold Graphs and Related Topics*, Annals of Discrete Mathematics, 56, North-Holland, Amsterdam, 1995.

E.S. Mahmoodian, A critical case method of proof in combinatorial mathematics, *Bull. Iranian Math. Soc.*, **8** (1978), 1L–26L.

M. Marcus and M. Newman, On the minimum of the permanent of a doubly stochastic matrix, *Duke Math. J.*, **26** (1959), 61–72.

A.W. Marshall and I. Olkin, Scaling of matrices to achieve specified row and column sums, *Numer. Math.*, **12** (1968), 83–90.

A.W. Marshall and I. Olkin, *Inequalities: Theory of Majorization and Its Applications*, Academic Press, New York, 1979.

F. Matúš and A. Tuzar, Short proofs of Khintchine-type inequalities for zero–one matrics, *J. Combin. Theory, Ser. A*, **59** (1992), 155–159.

K. McDougal, The joint realization of classes of $(0, 1)$ matrices with given row and column sum vectors, Proceedings of the Twenty-Third Southeastern International Conference on Combinatorics, Graph Theory, and Computing (Boca Raton, FL, 1992), *Cong. Numer.*, **91** (1992), 201–216.

K. McDougal, On asymmetric (0,1)-matrices with given row and column vectors, *Discrete Math.*, **137** (1995), 377–381.

K. McDougal, Locally invariant positions of $(0, 1)$ matrices, *Ars Combin.*, **44** (1996), 219–224.

K. McDougal, A generalization of Ryser's theorem on term rank, *Discrete Math.*, **170** (1997), 283–288.

B.D. McKay, Asymptotics for 0–1 matrices with prescribed line sums, *Enumeration and Design*, ed. D.M. Jackson and S.A. Vanstone, Academic Press, New York, 1984, 225–238.

B.D. McKay and X. Wang, Asymptotic enumeration of 0–1 matrices with equal row sums and equal column sums, *Linear Algebra Appl.*, **373** (2003), 273–287.

B.D. McKay and I.M. Wanless, Maximising the permanent of $(0, 1)$-matrices and the number of extensions of Latin rectangles, *Electron. J. Combin.*, **5** (1998), #11.

L. McShine, Random sampling of labeled tournaments, *Electron. J. Combin.*, **7** (2000), # R8.

N.S. Mendelsohn and A.L. Dulmage, The convex hull of sub-permutation matrices, *Proc. Amer. Math. Soc.*, **9** (1958), 253–254.

J. Meng and Q. Huang, Cayley graphs and interchange graphs, *J. Xinjiang Univ.*, **9** (1) (1992), 5–10.

M.V. Menon, Matrix links, an extremization problem, and the reduction of a nonnegative matrix to one with prescribed row and column sums, *Canad. J. Math.*, **20** (1968), 225–232.

M.V. Menon and H. Schneider, The spectrum of a nonlinear operator associated with a matrix, *Linear Algebra Appl.*, **2** (1969), 321–334.

R. Merris, Doubly stochastic graph matrices, *Univ. Beograd. Publ. Elektrotehn. Fak. (Ser. Mat.)*, **8** (1997), 64–71.

R. Merris, Doubly stochastic graph matrices, II, *Linear Multilin. Alg.*, **45** (1998), 275–285.

R. Merris and W. Watkins, Convex sets of doubly stochastic matrices, *J. Combin. Theory Ser. A*, **16** (1974), 129–130.

D. Merriell, The maximum permanent in Γ_n^k, *Linear Multilin. Alg.*, **9** (1980), 81–91.

T.S. Michael, The structure matrix and a generalization of Ryser's maximum term rank formula, *Linear Algebra Appl.*, **145** (1991), 21–31.

T.S. Michael, The structure matrix of the class of r-multigraphs with a prescribed degree sequence, *Linear Algebra Appl.*, **183** (1993), 155–177.

I. Miklós and J. Podani, Randomization of presence–absence matrices: comments and new algorithms, *Ecology*, **85** (2004), 86–92.

D.J. Miller, Weak Cartesian product of graphs, *Colloq. Math.*, **21** (1970), 55–74.

H. Minc, On lower bounds for permanents of $(0,1)$-matrices, *Proc. Amer. Math. Soc.*, **22** (1969), 117–123.

H. Minc, Doubly stochastic matrices with minimal permanents, *Pacific J. Math.*, **58** (1975), 155–157.

H. Minc, *Permanents*, Addison-Wesley, Reading, 1978.

H. Minc, A note on Egoryčev's proof of the van der Waerden conjecture, *Linear Multilin. Alg.*, **11** (1982), 367–371.

H. Minc, Theory of permanents 1978–1981, *Linear Multilin. Alg.*, **12** (1983), 227–268.

H. Minc, *Nonnegative Matrices*, Wiley-Interscience, New York, 1988.

L. Mirsky, On a convex set of matrices, *Arch. Math.*, **10** (1959), 88–92.

L. Mirsky, Even doubly-stochastic matrices, *Math. Annalen*, **144** (1961), 418–421.

L. Mirsky, Results and problems in the theory of doubly-stochastic matrices, *Z. Wahrscheinlichkeitstheorie Verw. Gebiete*, **1** (1962/63), 319–344.

L. Mirsky, Combinatorial theorems and integral matrices, *J. Combin. Theory*, **5** (1968), 30–44.

L. Mirsky, *Transversal Theory*, Academic Press, New York, 1971.

S.D. Monson, N.J. Pullman, and R. Rees, A survey of clique and biclique coverings and factorizations of $(0,1)$-matrices, *Bull. Inst. Combin. Appl.*, **14** (1995), 17–86.

J.W. Moon, An extension of Landau's theorem on tournaments, *Pacific. J. Math.*, **13** (1963), 1343–1345.

J.W. Moon, *Topics on Tournaments*, Holt, Rinehart, and Winston, New York, 1968.

J.W. Moon and N.J. Pullman, On generalized tournament matrices, *SIAM Review*, **12** (1970), 384–399.

R.F. Muirhead, Some methods applicable to identities and inequalities of symmetric algebric functions of n letters, *Proc. Edinburgh Math. Soc.*, **21** (1903), 144–157.

D. Naddef, The Hirsch conjecture is true for $(0,1)$-polytopes, *Math. Programming, Ser. B*, **45** (1989), 109–110.

D. Naddef and W.R. Pulleyblank, Hamiltonicity and combinatorial polyhedra, *J. Combin. Theory, Ser. B*, **31** (1981), 297–312.

D. Naddef and W.R. Pulleyblank, Hamiltonicity in (0-1)-polyhedra, *J. Combin. Theory, Ser. B*, **37** (1984), 41–52.

H. Nakazato, Set of 3×3 orthostochastic matrices, *Nihonkai Math. J.*, **7** (1996), 83–100.

J. von Neumann, A certain zero-sum two-person game equivalent to the optimal assignment problem, *Contributions to the Theory of Games*, Vol. 2, Princeton U. Press, Princeton, 1953, 5–12.

A. Nishi, An elementary proof of Johnson–Dulmage–Mendelsohn's refinement of Birkhoff's theorem on doubly stochastic matrices, *Canad. Math. Bull.*, **22** (1979), 81–86.

Y. Nam, Integral matrices with given row and column sums, *Ars Combin.*, **52** (1999), 141–151.

M. Newman, Combinatorial matrices with small determinants, *Canad. J. Math.*, **30** (1978), 756–762.

Z. Nutov and M. Penn, On the integral dicycle packings and covers and the linear ordering polytope, *Discrete Appl. Math.*, **60** (1995), 293–309.

Z. Nutov and M. Penn, On non-$(0, \frac{1}{2}, 1)$-extreme points of the generalized transitive tournament polytope, *Linear Algebra Appl.*, **223** (1996), 149–159.

Y. Odama and G. Musiker, Enumeration of $(0, 1)$ and integer doubly stochastic matrices, preprint.

A.M. Odlyzko, On the ranks of some (0,1)-matrices with constant row sums, *J. Austral. Math. Soc., Ser. A*, **31** (1981), 193–201.

J.A. Oviedo, Adjacent extreme points of a transportation polytope, *Optimization*, **37** (1996), 279–287.

A.B. Owens, On determining the minimum number of multiple edges for an incidence sequence, *SIAM J. Appl. Math.*, **18** (1970), 238–240.

A.B. Owens and H.M. Trent, On determining minimal singularities for the realization of an incidence sequence, *SIAM J. Appl. Math.*, **15** (1967), 406–418.

I. Pak, On the number of faces of certain transportation polytopes, *Europ. J. Combin.*, **21** (2000), 689–694.

I. Pak, Four questions on Birkhoff polytope, *Ann. Combin.*, **4** (2000), 83–90.

B.N. Parlett and T.L. Landis, Methods for scaling to doubly stochastic from, *Linear Algebra Appl.*, **48** (1982), 53–79.

U.N. Peled and M.K. Srinivasan, The polytope of degree sequences, *Linear Algebra Appl.*, **114/115** (1989), 349–377.

H. Perfect and L. Mirsky, The distribution of positive elements in doubly-stochastic matrices, *J. London Math. Soc.*, **40** (1965), 689–698.

H. Perfect and L. Mirsky, Extreme points of certain convex polytopes, *Monatsh. Math.*, **68** (1964), 143–149.

N.J. Pullman and M. Stanford, Singular $(0, 1)$-matrices with constant row and column sums, *Linear Algebra Appl.*, **106** (1988), 195–208.

N. Punnim, Degree sequences and chromatic numbers of graphs, *Graphs Combin.*, **18** (2002), 597–603.

N. Punnim, The clique numbers of regular graphs, *Graphs Combin.*, **18** (2002), 781–785.

J.G. Qian, On the upper bound of the diameter of interchange graphs, *Discrete Math.*, **195** (1999), 277–285.

J.G. Qian, Some properties of a class of interchange graphs, *Appl. Math. J. Chinese Univ., Ser. B*, **13** (1998), 455–462.

M. Raghavachari, Some properties of the widths of the indicence matrices of balanced incomplete block designs through partially balanced ternary designs, *Saṅkhyā Ser. B*, **37** (1975), 211–219.

A. Rämmer, On even doubly stochastic matrices with minimal even permanent, *Tartu Riikl. Ül. Toimetised* **878** (1990), 103–114.

A. Rämmer, On minimizing matrices, *Proceedings of the First Estonian Conference on Graphs and Applications (Tartu–Kääriku, 1991)*, Tartu U. Press, Tartu, 1993, 121–134.

G. Reinelt, *The Linear Ordering Problem: Algorithms and Applications*, Heldermann, Berlin, 1985.

K.B. Reid and L.W. Beineke, Tournaments, *Selected Topics in Graph Theory*, ed. L.W. Beineke and R.J. Wilson, Academic Press, New York, 1978, 385–415.

G. deB. Robinson, On the representations of the symmetric group, *Amer. J. Math.*, **60** (1938), 745–760.

H.J. Ryser, Maximal determinants in combinatorial investigations, *Canad. J. Math.*, **8** (1956), 245–249.

H.J. Ryser, Combinatorial properties of matrices of zeros and ones, *Canad. J. Math.*, **9** (1957), 371–377.

H.J. Ryser, The term rank of a matrix, *Canad. J. Math.*, **10** (1957), 57–65.

H.J. Ryser, Traces of matrices of zeros and ones, *Canad. J. Math.*, **12** (1960), 463–476.

H. J. Ryser, *Combinatorial Mathematics*, Carus Math. Monograph #14, Math. Assoc. of America, Washington, 1963.

H.J. Ryser, Matrices of zeros and ones in combinatorial mathematics, *Recent Advances in Matrix Theory*, ed. H. Schneider, U. Wisconsin Press, Madison, 1964, 103–124.

H.J. Ryser, A new look at the optimal assignment problem, *Linear Algebra Appl.*, **66** (1985), 113–121.

E.M. Sá, Some subpolytopes of the Birkhoff polytope, *Electron. J. Linear Algebra*, **15** (2006), 1–7.

G. Sabidussi, Graph multiplication, *Math. Z.*, **72** (1950), 446–457.

C. Schensted, Longest increasing and decreasing subsequences, *Canad. J. Math.*, **13** (1961), 179–191.

H. Schrek and G. Tinhofer, A note on certain subpolytopes of the assignment polytope associated with circulant graphs, *Linear Algebra Appl.*, **111** (1988), 125–134.

A. Schrijver, A short proof of Minc's conjecture, *J. Combin. Theory, Ser. A*, **25** (1978), 80–83.

A. Schrijver, Bounds on permanent and the number of 1-factors and 1-factorizations of bipartite graphs, *Surveys in Combinatorics* (Southampton 1983), London Math. Soc. Lecture Note Ser. 82, Cambridge U. Press, Cambridge, 1983, 107–134.

A. Schrijver, *Theory of Linear and Integer Programming*, Wiley-Interscience, Chichester, 1986.

A. Schrijver, Counting 1-factors in regular bipartite graphs, *J. Combin. Theory, Ser. B*, **72** (1998), 122–135.

A. Schrijver, *Combinatorial Optimization. Polyhedra and efficiency. Vol. A. Paths, Flows, Matchings*, Chapters 1–38, Algorithms and Combinatorics, 24, A, Springer, Berlin, 2003.

A. Schrijver and W.G. Valiant, On lower bounds for permanents, *Indag. Math.*, **42** (1980), 425–427.

H.A. Shah Ali, On the representation of doubly stochastic matrices as sums of permutation matrices, *Bull. London Math. Soc.*, **33** (2001), 11–15.

S.M. Shah and C.C. Gujarathi, The α-width of the incidence matrices of incomplete block designs, *Saṅkhyā, Ser. B*, **46** (1984), 118–121.

J.Y. Shao, The connectivity of interchange graph of class $\mathcal{A}(R, S)$ of (0,1)-matrices, *Acta Math. Appl. Sinica*, **2** (4) (1985), 304–308.

J.Y. Shao, The connectivity of the interchange graph of the class of tournaments with a fixed score vector (in Chinese), *Tonji Daxue Xuebao*, **15** (1987), 239–242.

J. Shen and R. Yuster, A note on the number of edges guaranteeing a C_4 in Eulerian bipartite graphs, *Electron. J. Combin.*, **9** (2001), #N6.

G. Sierksma and H. Hoogeven, Seven criteria for integer sequences being graphic, *J. Graph Theory*, **15** (1991), 223–231.

G. Sierksma and E. Sterken, The structure matrix of (0, 1)-matrices: its rank, trace and eigenvalues. An application to econometric models, *Linear Algebra Appl.*, **83** (1986), 151–166.

A.J. Sinclair, *Algorithms for Random Generation and Counting: A Markov Chain Approach*, Birkhäuser, Boston, 1993.

R. Sinkhorn, A relationship between arbitrary positive matrices and doubly stochastic matrices, *Ann. Math. Statist.*, **35** (1964), 876–879.

R. Sinkhorn, Diagonal equivalence to matrices with prescribed row and column sums, *Amer. Math. Monthly*, **74** (1967), 402–405.

R. Sinkhorn, Diagonal equivalence to matrices with prescribed row and column sums. II, *Proc. Amer. Math. Soc.*, **45** (1974), 195–198.

R. Sinkhorn, Doubly stochastic matrices which have certain diagonals with constant sums, *Linear Algebra Appl.*, **16** (1977), 79–82.

R. Sinkhorn, A problem related to the van der Waerden permanent theorem, *Linear Multilin. Alg.*, **16** (1984), 167–173.

R. Sinkhorn and P. Knopp, Concerning nonnegative matrices and doubly stochastic matrices, *Pacific J. Math.*, **21** (1967), 343–348.

R. Sinkhorn and P. Knopp, Problems involving diagonal products in nonnegative matrices, *Trans. Amer. Math. Soc.*, **136** (1969), 67–75.

E. Snapper, Group characters and nonnegative integral matrices, *J. Algebra*, **19** (1971), 520–535.

R.P. Stanley, Linear homogeneous diophantine equations and magic labelings of graphs, *Duke Math. J.*, **40** (1973), 607–632.

R.P. Stanley, *Enumerative Combinatorics, Volume I*, Cambridge U. Press, Cambridge, 1997.

R.P. Stanley, *Enumerative Combinatorics Volume II*, Cambridge Studies in Adv. Math., Cambridge U. Press, Cambridge, 1998.

M.L. Stein and P. Stein, Enumeration of stochastic matrices with integer elements, Los Alamos Scientific Laboratory Report, LA-4434, 1970.

S.K. Stein, Two combinatorial covering problems, *J. Combin. Theory, Ser. A*, **16** (1974), 391–397.

Z. Świtalski, Half-integrality of vertices of the generalized transitive tournament polytope ($n = 6$), *Discrete Math.*, **271** (2003), 251–260.

V.E. Tarakanov, The maximal depth of a class of (0,1)-matrices, *Math. Sb. (N.S.)*, **75** (117) (1968), 4–14.

V.E. Tarakanov, Correlations of maximal depths of classes of square (0, 1)-matrices for different parameters, *Mat. Sb. (N.S.)*, **77** (119) (1968), 59–70.

V.E. Tarakanov, The maximum depth of arbitary classes of (0,1)-matrices and some of its applications (in Russian), *Math. Sb. (N.S.)*, **92** (134) (1973), 472–490. Translated in *Math. USSR – Sbornik*, **21** (1973), 467–484.

V.E. Tarakanov, The depth of (0,1)-matrices with identical row and identical column sums (in Russian), *Mat. Zametki*, **34** (1983), 463–476. Translated in *Math Notes.*, **34** (1983), 718–725.

R. Taylor, Constrained switching in graphs, *Combinatorial Mathematics, VII (Geelong, 1980)*, Lecture Notes in Math., 884, Springer, Berlin–New York, 1981, 314–336.

R. Taylor, Switchings constrained to 2-connectivity in simple graphs, *SIAM J. Algebraic Discrete Methods*, **3** (1982), 114–121.

C. Thomassen, Landau's characterization of tournament score sequences, *The Theory and Application of Graphs (Kalamazoo, Michigan, 1980)*, Wiley, New York, 1963, 589–591.

G. Tinhofer, Graph isomorphism and theorems on Birkhoff type, *Computing*, **36** (1986), 285–300.

G. Tinhofer, Strong tree-cographs are Birkhoff graphs, *Discrete Appl. Math.*, **22** (1988/89), 275–288.

G. Tinhofer, A note on compact graphs, *Discrete Appl. Math.*, **30** (1991), 253–264.

H. Tverberg, On the permanent of a bistochastic matrix, *Math. Scand.*, **12** (1963), 25–35.

J.S. Verducci, Minimum majorization decomposition, *Contributions to Probability and Statistics*, Springer, New York, 1989, 160–173.

V.G. Vizing, The Cartesian product of graphs (in Russian), *Vyčisl. Sistemy*, **9** (1963), 30–43.

B.L. van der Waerden, Aufgabe 45, *Jber. Deutsch. Math.-Verein*, **35** (1926), 117.

M. Voorhoeve, On lower bounds for permanents of certain (0,1)-matrices, *Indag. Math.*, **41** (1979), 83–86.

D.W. Walkup, Minimal interchanges of (0,1)-matrices and disjoint circuits in a graph, *Canad. J. Math.*, **17** (1965), 831–838.

H.H. Wan, Structure and cardinality of the class $\mathcal{A}(R, S)$ of $(0, 1)$-matrices, *J. Math. Res. Exposition*, **4** (1984), 87–93.

H.H. Wan, Cardinal function $f(R, S)$ of the class $\mathcal{A}(R, S)$ and its non-zero point set, *J. Math. Res. Exposition*, **5** (1985), 113–116.

H.H. Wan, Cardinality of a class of $(0, 1)$-matrices covering a given matrix, *J. Math. Res. Exposition*, **6** (1984), 33–36.

H.H. Wan, Generating functions and recursion formula for $|\mathcal{A}(R, S)|$ (in Chinese), *Numerical Mathematics: A Journal of Chinese Universities*, **6** (4), 319–326.

H.H. Wan and Q. Li, On the number of tournaments with prescribed score vector, *Discrete Math.*, **61** (1986), 213–219.

B.Y. Wang, Precise number of $(0,1)$-matrices in $\mathcal{A}(R, S)$, *Scientia Sinica*, Ser. A, **31** (1988), 1–6.

Y.R. Wang, Characterization of binary patterns and their projections, *IEEE Trans. Computers*, **C-24** (1975), 1032–1035.

B.Y. Wang and F. Zhang, On the precise number of $(0,1)$-matrices in $\mathcal{A}(R, S)$, *Discrete Math.*, **187** (1998), 211–220.

B.Y. Wang and F. Zhang, On normal matrices of zeros and ones with fixed row sum, *Linear Algebra Appl.*, **275–276** (1998), 517–526.

E.T.H. Wang, Maximum and minimum diagonal sums of doubly stochastic matrices, *Linear Algebra Appl.*, **8** (1974), 483–505.

I.M. Wanless, Maximising the permanent and complementary permanent of $(0, 1)$-matrices with constant line sums, *Discrete Math.*, **205** (1999), 191–205.

I.M. Wanless, Extremal properties for binary matrices, preprint.

R.H. Warren, Revisiting the $0, 1$ assignment problem, *Linear Algebra Appl.*, **160** (1992), 247–253.

M.E. Watkins, Connectivity of transitive graphs, *J. Combin. Theory*, **8** (1970), 23–29.

W. Watkins and R. Merris, Convex sets of doubly-stochastic matrices, *J. Combin. Theory, Ser. A.*, **16** (1974), 129–120.

W.D. Wei, The class $\mathcal{A}(R, S)$ of $(0,1)$-matrices, *Discrete Math.*, **39** (1982), 301–305.

D.B. West, *Introduction to Graph Theory*, 2nd ed., Prentice-Hall, Upper Saddle River, 2001.

R.J. Wilson, *Introduction to Graph Theory*, Longman, London, 1985.

V.A. Yemelichev, M.M. Kovalev, and M.K. Kravtsov, *Polytopes, Graphs and Optimisation*, Cambridge U. Press, Cambridge, 1984.

A.-A.A. Yutsis, Tournaments and generalized Young tableau, *Mat. Zametki*, **27** (1980), 353–359.

D. Zeilberger and D.M. Bressoud, A proof of Andrew's q-Dyson conjecture, *Discrete Math.*, **54** (1985), 201–224.

F. Zhang and Y. Zhang, A type of $(0,1)$-polyhedra, *J. Xinjiang Univ.*, **7** (4) (1990), 1–4.

H. Zhang, Hamiltonicity of interchange graphs of a type of matrices of zeros and ones, *J. Xinjiang Univ.*, **9** (3) (1992), 5–10.

X.-D. Zhang, A note on doubly stochastic graph matrices, *Linear Algebra Appl.*, **407** (2005), 196–200.

B. Zhou and B. Liu, Some results on compact graphs and supercompact graphs, *J. Math. Study*, **32** (1999), 133–136.

Index